5th International Conference on Green Energy and Sustainable Development (GESD 2022)

E3S Web of Conferences Volume 358 (2022)

Shenyang City, China
24-25 September 2022

Editors:

Zhihui Tan
Lin Liu

ISBN: 978-1-7138-6270-3

Printed from e-media with permission by:

Curran Associates, Inc.
57 Morehouse Lane
Red Hook, NY 12571

Some format issues inherent in the e-media version may also appear in this print version.

For additional information, please contact EDP Sciences – Web of Conferences at the address below.

EDP Sciences – Web of Conferences
17, Avenue du Hoggar
Parc d'Activité de Courtabœuf
BP 112
F-91944 Les Ulis Cedex A
France

Phone: +33 (0) 1 69 18 75 75
Fax: +33 (0) 1 69 28 84 91

contact-edps@webofconferences.org

Additional copies of this publication are available from:

Curran Associates, Inc.
57 Morehouse Lane
Red Hook, NY 12571 USA
Phone: 845-758-0400
Fax: 845-758-2633
Email: curran@proceedings.com
Web: www.proceedings.com

TABLE OF CONTENTS

INVITED CONTRIBUTIONS

REGULAR CONTRIBUTIONS

https://doi.org/10.1051/e3sconf/202235800001

Preface: 5th International Conference on Green Energy and Sustainable Development (GESD 2022)

This proceeding includes original and peer-reviewed research papers from the 5th International Conference on Green Energy and Sustainable Development (GESD 2022), which held on September 24-25, 2022 in Shenyang City, China.

GESD 2022 is organized by Research Center of Engineering and Science, co-organized by Chongqing University, Northeast Petroleum University, Huazhong University of Science and Technology, Asian Union of Information Technology, sponsored by Wanxiang Group Technology Center, CCCC Second Harbour Engineering Company Ltd., Chishun Chemical.

The aim of GESD 2022 was to provide a platform for researchers, engineers, academicians, as well as industrial professionals, to present their research results and development activities in Green Energy and Sustainable Development. It provides opportunities for the delegates to exchange new ideas and application experiences, to establish business or research relations and to find global partners for future collaboration. This two-day conference was being organized in invited keynote talks and three Parallel Sessions - Green Energy & Sustainable Energy and Power; Resource Extraction, Utilization & Sustainable Development; Environmental Sustainability and Green Technologies.

All full paper submissions to this GESD would be sent to at least two reviewers and evaluated based on originality, technical or research content, correctness, relevance to conference, contributions, and readability. The full paper submissions were chosen based on technical merit, interest, applicability, and how well they fit a coherent and balanced technical program. GESD 2021 received 198 manuscripts. And 107 submissions had been accepted by our reviewers. The Committees of GESD 2022 invested great efforts in reviewing the papers submitted to the conference and organizing the sessions to enable the participants to gain maximum benefit.

Zhihui Tan
Lin Liu

Conference Organizing Committee

Statement of Peer review

In submitting conference proceedings to *Web of Conferences*, the editors of the proceedings certify to the Publisher that

1. They adhere to its **Policy on Publishing Integrity** in order to safeguard good scientific practice in publishing.
2. All articles have been subjected to peer review administered by the proceedings editors.
3. Reviews have been conducted by expert referees, who have been requested to provide unbiased and constructive comments aimed, whenever possible, at improving the work.
4. Proceedings editors have taken all reasonable steps to ensure the quality of the materials they publish and their decision to accept or reject a paper for publication has been based only on the merits of the work and the relevance to the journal.

Title, date and place of the conference

GESD 2022
September 24-25, 2022
Shenyang City, China

Proceedings editor(s):

Zhihui Tan
Lin Liu

Date and editor's signature October 19, 2022

17, avenue du Hoggar - PA de Courtabœuf – BP 112 - 91944 Les Ulis Cedex A (France)
Tél. : 33 (0)1 69 18 75 75 - Fax : 33(0)1 69 07 45 17 – www.edpsciences.org

 https://doi.org/10.1051/e3sconf/202235801001

Secondary focusable and uniform convergence microprism in concentrating photovoltaic module

Limin Guo [1, 2], Zhiyong Wang [2], Youqiang Liu [2], Songxuan Chen [1], Ying Zhang [1], June Liu [1, *]

[1] China ENFI Engineering Corporation, Beijing, China
[2] Institute of Laser Engineering, Beijing University of Technology, Beijing, China

Abstract. Low tracking accuracy of trackers, structure vibration by winds, and lens deformation by temperature lead to non-vertical incident irradiation of Fresnel lens, which necessitates a secondary concentrator in actual engineering application of concentrating photovoltaic module. In this study, a secondary focusable and uniform convergence microprism is added between Fresnel lens and solar cell to improve concentrating efficiency and focal spot energy uniformity. The 3D model of microprism is established by SolidWorks, and important parameters are optimized using Zemax. Results showed that combination of Fresnel lens and focusable and uniform convergence microprism achieves the highest power when upper spherical diameter of secondary microprism measures 18 mm, included angle in opposite side facets equals 116°, spherical height removed from the top is 0.1 mm, and side length of bottom reaches 2.15 mm. The highest power of solar cell surface can reach 2.4932 W, representing a 32.9% improvement; focal spot energy uniformity is 0.71; and module height with secondary microprism measures 88 mm, which reduces by 5.5 mm without secondary microprism. Experimental results show that concentrating static test generation efficiency of a 400-times concentrating module system reaches 34.9%, acceptance angle measures ±1.18°, efficiency loss of module is less than 1.23% when temperature changes from −20°C to 20°C and from 20°C to 50°C, and 400-times module maximum output power totals 142.3 W.

1. Introduction

High concentrating photovoltaic (HCPV) systems rely on large Fresnel lens, which must be precisely oriented in the direction of the sun to maintain high concentration ratio [1]. When tracking systems exhibit certain accuracy, then HCPV system features high acceptance angle and low power loss. Many photovoltaic (PV) plants are distributed in subtropic and temperate climate zones, such as Hami, Golmud, and Nyemo in China. These light intensity regions usually show large temperature differences between day and night and during summer and winter. When temperature changes, deformations of Fresnel lens lead to changes in refractive index; such changes offset focus on battery-receiving surface and result in higher energy loss. Fresnel lens exhibit poor acceptance angle and temperature adaptability; therefore, a suitable secondary optical component is designed to improve concentrating performance.

Several institutes worked on new types of secondary concentrators and their application in concentrating photovoltaic (CPV) modules [2-3]. João Mendes-Lopes et al. improved secondary concentrators, which are divided into nine centrosymmetric areas, and their simulation results showed high irradiance uniformity of focal spots [4]. Tian Gu et al. introduced the microsystem-enabled PV using non-imaging micro-concentrators and III-V solar cells, which combined toroidal primary lens and reflective cone secondary concentrator to enhance light collection and illumination onto microscale solar cells and to achieve an acceptance angle product of more than 1° [5-6]. Zeng Fei et al. designed a total internal reflection-refraction solar concentrator combined with a homogenizer. In this design, the highest irradiance intensity is less than twice of average irradiance intensity on cell surface, and acceptance angle measures 1.5°[7].

Although previously discussed secondary concentrator designs show high optical performances, difficulty arises from their machining and assembly. This problem increases cost of CPV systems and restrict further development of this technology. This study presents a focusable and energy-uniform secondary microprism, which features an inverted truncated rectangular pyramid at the bottom and spherical and flat surface on top. With its simple structure, good work ability, and stable performance, this design shows potential in practical HCPV application.

2. Concentrating results without secondary microprism

Area of solar cell totals 2.5×2.5 mm^2, area of Fresnel lens measures 50×50 mm^2, concentration ratio is set to 400, and focus of Fresnel lens is set to 94 mm. Fresnel lens substrate is an ultra-white glass with thickness measuring

* Corresponding author: 277040689@qq.com

3.2 mm. Lens material is silicon with thickness of 1 mm. Room temperature measures 20°C.

The 3D model of Fresnel lens is established by the SolidWorks software, and Zemax optical simulation software is introduced to perform computer simulation. Custom light source in Zemax is used, source parameters are set based on AM1.5D standard spectral correlation parameters, and lens material parameters are set based on silicon parameters. Solar radiation is set to 1,000 W/m^2, which is the standard intensity. When light is normally incident on the surface of Fresnel lens, power of focusing spot is the highest, i.e., 1.8764 W, as shown in Figure 1 and 2. Simulation results show that concentrating efficiency of Fresnel lens totals 75%, focal spot energy uniformity is 0.59, and acceptance angle measures $\pm$ 0.64°. At room temperature of 20°C, decrease in concentrating efficiency of Fresnel lens exceeds 15.7% when temperature changes from -20°C to 20°C and from 20°C to 50°C. Maximum output power of the 400-times module with Fresnel lens measures 111.8 W. Thus, a suitable secondary optical component should be designed to improve concentrating performance.

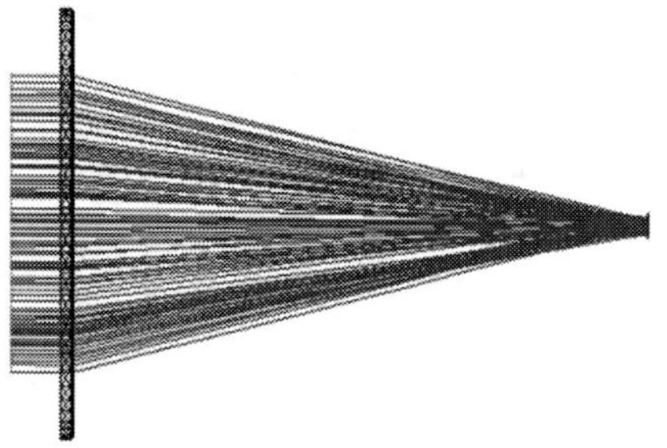

Figure 1. Concentrating effect with Fresnel lens

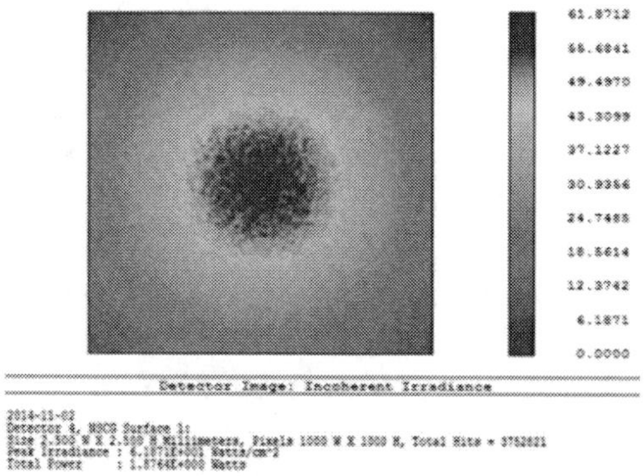

Figure 2. Energy distribution of focal spot on solar cell surface

3. Principle of secondary focusable and uniform convergence microprism

Secondary focusable and uniform convergence microprism is inserted between Fresnel lens and solar cell to improve concentrating performance. The truncated rectangular pyramid below the secondary microprism concentrates light, with the bottom rectangular facet linking the solar cell and its dimensions matching that of cell. The other four side facets work as total reflection plane, and included angle of opposite side facets is set as α. Top spherical and flat surface of secondary microprism is used as incident plane, in which the spherical part focuses solar rays, and the flat part ensures energy uniformity of solar cell surface. Fig. 3 depicts the optical path of secondary microprism, and Fig. 4 displays the 3D model of secondary microprism. Incident light is allowed to pass no more than one total reflection in the secondary microprism before it is absorbed by cell to reduce incident energy loss. In this study, air refractive index is set as n_1, and refractive index of secondary microprism is set as n_2.

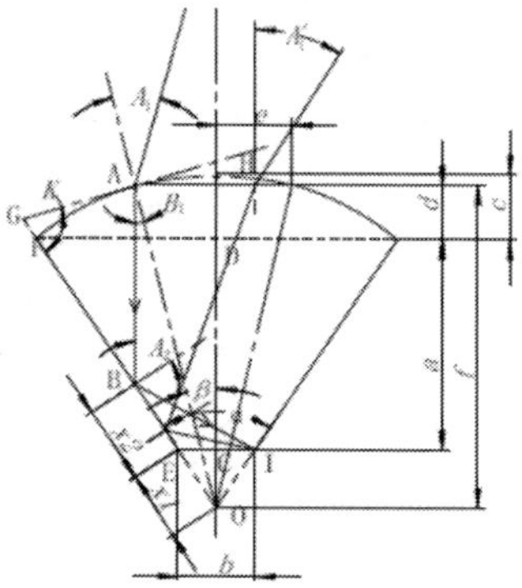

Figure 3. Optical path of secondary microprism

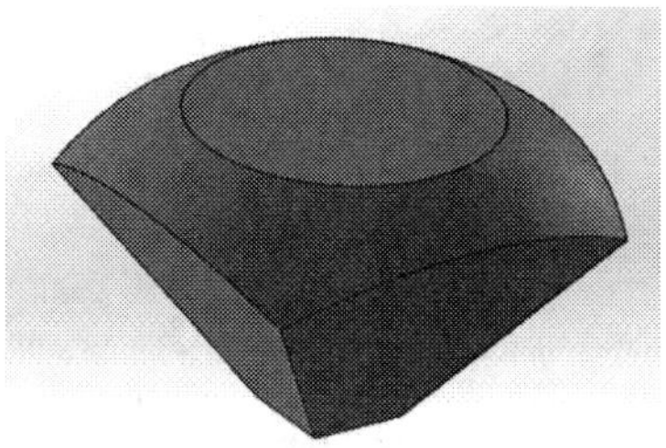

Figure 4. 3D model of secondary microprism

When light is incident on spherical surface and based on refraction and total reflection laws, this study deduces the following:

$$n_1 \sin A_1 = n_2 \sin B_1 \tag{1}$$

$$A_2 \geq \arcsin \frac{n_1}{n_2} \tag{2}$$

With Δ ABO, Δ EIO, and Δ BEI and based on geometrical relationship and law of sines, this study deduces that:

$$\frac{R}{\sin(\frac{\pi}{2} + A_2)} = \frac{x_1 + x_2}{\sin B_1} \tag{3}$$

$$\frac{b}{\sin \alpha} = \frac{x_1}{\sin(\frac{\pi - \alpha}{2})} \tag{4}$$

$$\frac{b}{\sin(\frac{\pi}{2}-A_2)}=\frac{x_2}{\sin(A_2-\frac{\alpha}{2})} \tag{5}$$

From Formulas (1) to (5), the following expression can be derived:

$$A_1 \le \arcsin\frac{n_2 b\cos(\arcsin\frac{n_1}{n_2}-\alpha)}{2n_1 R\sin\frac{\alpha}{2}} \tag{6}$$

When light is incident on flat surface and based on reference [8-11], this study deduces that:

$$\pi-\alpha \ge \arcsin\frac{n_1}{n_2}\,\pi-\alpha \ge \arcsin\frac{n_1}{n_2}+\arcsin(\frac{n_1}{n_2}\sin A_1{}') \tag{7}$$

Then, incident angle should meet the following condition:

$$A_1{}' \le \arcsin(\frac{n_2}{n_1}\sin\alpha\sqrt{1-\frac{n_1^2}{n_2^2}}+\cos\alpha) \tag{8}$$

Formulas (6) and (8) show required conditions of secondary focusable and uniform convergence microprism.

Fresnel lens can be simplified into a thin lens, and the top spherical surface of secondary microprism focusing the lights can be considered a planoconvex lens. Thus, the distance H between Fresnel lens and secondary microprism should be designed. Based on the formula of combined focal length, the following expression can be derived:

$$f'=\frac{f_1'f_2'}{f_1'+f_2'-H}=\frac{1.8L\times\frac{R}{n_{K9}(\lambda)-1}}{1.8L+\frac{R}{n_{K9}(\lambda)-1}-H} \tag{9}$$

$$\begin{cases} l_H{}'=-f'\frac{H}{f_1'}=-\dfrac{1.8L\times\frac{R}{n_{K9}(\lambda)-1}\times H}{(1.8L+\frac{R}{n_{K9}(\lambda)-1}-H)\times 1.8L} \\ l_H=f\frac{H}{f_2}=\dfrac{1.8L\times\frac{R}{n_{K9}(\lambda)-1}\times H}{(1.8L+\frac{R}{n_{K9}(\lambda)-1}-H)\times\frac{R}{n_{K9}(\lambda)-1}} \end{cases} \tag{10}$$

According to Formulas (9) and (10), the distance H between Fresnel lens and secondary microprism can be optimized.

4. Optimization design of secondary focusable and uniform convergence microprism in the 400-times module

Given the machining precision and cell dimension, in this study, parameter b is set to 2.15 mm. Zemax is used to optimized included angle α in opposite side facets, spherical diameter 2R, and spherical height g removed from the top. The material of this secondary mIcroprism is K9 or Bk7.

Assuming that the initial value of secondary microprism spherical diameter $2R$ measures 16 mm, and spherical height g removed from the top totals 0 mm, then included angle α in opposite side facets varies from 100° to 140°. Considering every 1° in this interval range and simulating them in Zemax software combined with other determined parameters, power curve received by solar cell is constructed and shown in Fig. 5. According to simulation results, maximum power received by the solar cell measures 2.4876 W when α equals 116°.

Supposing that the initial value of included angle α in opposite side facets totals 116°, and spherical height g removed from the top measures 0 mm, then spherical diameter $2R$ varies from 15 mm to 27 mm. Obtaining every 0.5 mm interval in this range and simulating them in Zemax software combined with other determined parameters, energy curve received by the solar cells is constructed and shown in Fig. 6. Based on simulation results, maximum power received by the solar cell reaches 2.4907 W when $2R$ is set to 18 mm.

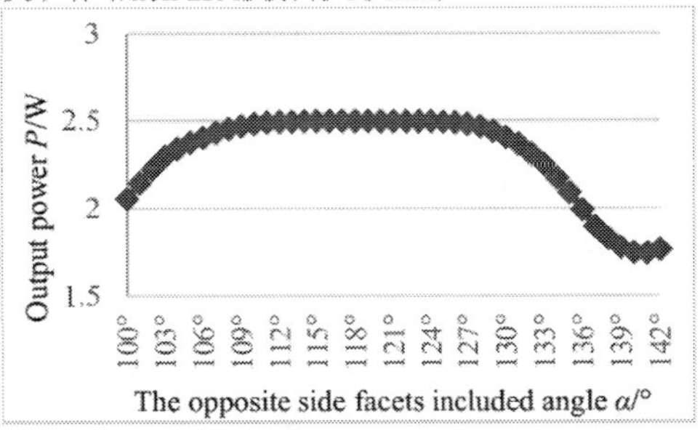

Figure 5. The α-optimized result with output power

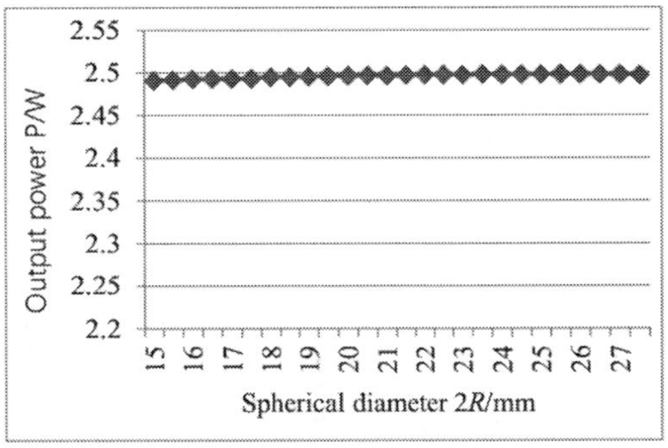

Figure 6. 2R-optimized result with output power

Supposing that initial value of included angle α in opposite side facets totals 116°, and spherical diameter $2R$ measures 18 mm, then spherical height g removed from the top ranges from 0 mm to 1 mm. Considering every 0.1 mm interval in this range and simulating them in Zemax software combined with other determined parameters, energy curve received by the solar cell is constructed, as shown in Fig. 7, and focal spot energy uniformity curve is determined, as indicated in Fig. 8. According to simulation results, maximum power received by the solar cell totals 2.4911 W, and focal spot energy uniformity measures 0.71 when g is set to 0.2 mm.

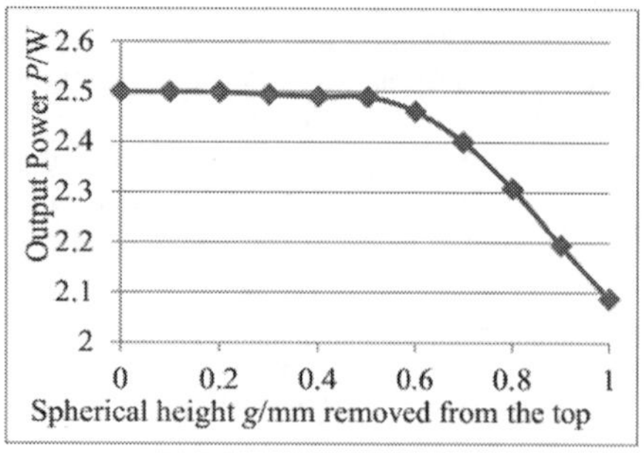

Figure 7. The g-optimized result with output power

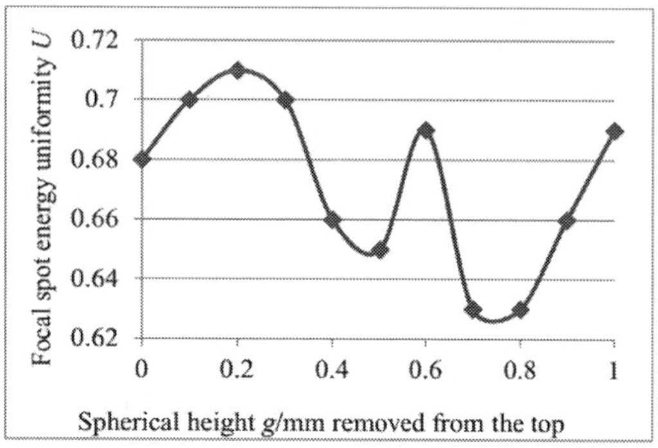

Figure 8. The g-optimized Result optimized with focal spot energy uniformity

5. Concentrating performance analysis of the 400-times module with secondary focusable and uniform convergence microprism

When parameters of secondary microprism are determined, the distance *H* between Fresnel lens and secondary microprism can be optimized using Zemax. Height *H* is changed from 80 mm to 96 mm. Considering every 2 mm interval in this range and simulating them in Zemax software combined with other determined parameters, power values received by the solar cell are determined and listed in Table 1. Based on simulation results, maximum power received by the solar cell totals 2.4932 W when *H* is set to 88 mm, which is reduced by 6 mm without the secondary microprism. Figure 9 and 10 display results of concentrating performance.

Figure 9 and 10 show that concentrating efficiency reaches 99.72%, indicating a 24.72% increase, and focal spot energy uniformity totals 0.71, presenting a 20.3% increase, without the secondary microprism. Notably, the secondary microprism hardly affects concentrating performance of the module.

Table 1. Energy received by solar cell combined with secondary microprisms of different heights

H (mm)	80	82	84	86	88
Power (W)	2.4866	2.4907	2.4913	2.4923	2.4932
H (mm)	90	92	94	96	98
Power (W)	2.4924	2.4916	2.4911	2.4874	2.4821

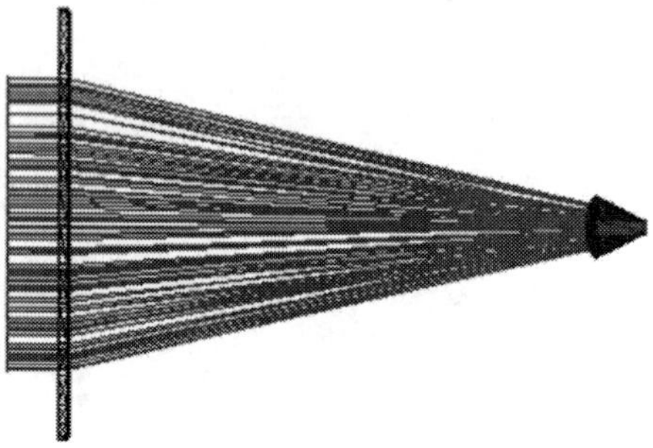

Figure 9. Concentrating effect with Fresnel lens and secondary microprism

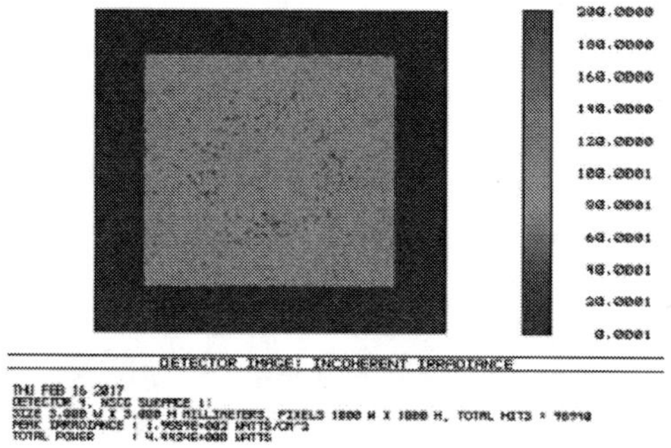

Figure 10. Energy distribution of focal spot on solar cell surface with Fresnel lens and secondary microprism

When light is normally incident on the surface of Fresnel lens, energy of focusing spot is the highest. With increasing incident light angle, some lights cannot focus on battery-receiving surface. Thus, beam energy decreases gradually. Fig. 11 shows the relationship between incident light angle and focal spot energy. When tilt angle of incident light in one direction approximates 1.18°, energy of receiving surface of battery decreases to 2.2501 W. Approximately 90% of the battery receives energy when light is normally incident. Thus, acceptance angle measures ±1.18° during simulation. Fig. 12 shows energy distribution of the focal spot on solar cell surface with incident light angle of 1.18°.

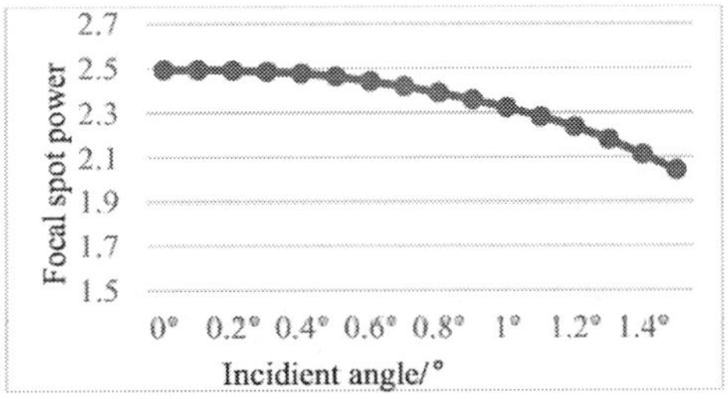

Figure 11. Dependence of focal spot power on incident angle

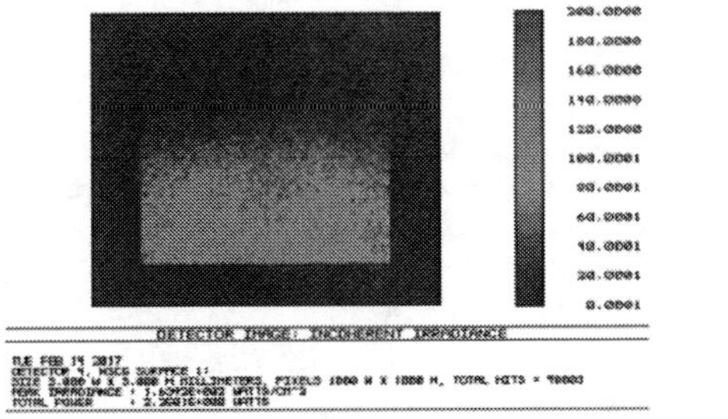

Figure 12. Energy distribution with incident angle of 1.18°

Temperature change leads to changes in teeth parameters of Fresnel lens and refractive index of silicon. Parameter changes in Fresnel lens are simulated using Zemax. Figure 13 and 14 show energy distribution of focal spots on the solar cell surface at temperatures of -20 °C and 50 °C, respectively.

Figure 9 and 10 show that energy of receiving surface of battery measures 2.462 W when temperature is -20 °C and 2.464 W when temperature is 50 °C, and efficiency loss of module measures less than 2.5% when temperature changes from -20 °C to 20 °C and from 20 °C to 50 °C.

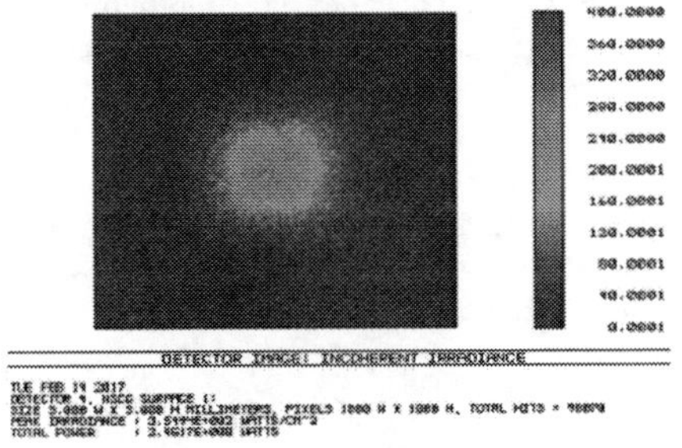

Figure 13. Energy distribution at −20°C

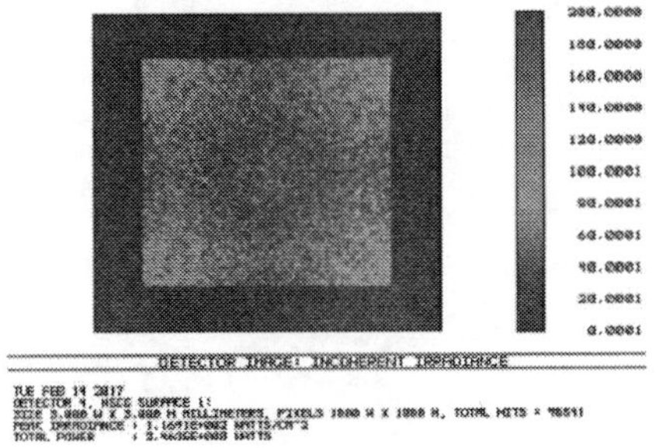

Figure 14. Energy distribution at 50°C

6. Experiment on the 400-times module with secondary focusable and uniform convergence microprism

Focusable and uniform convergence secondary microprism is constructed according to the following simulation parameters: α = 116°, $2R$ = 18 mm, b = 2.15 mm, g = 0.2 mm, as illustrated in Fig. 15. Fig. 16 shows experimental 400-times module, with a distance H of 88 mm. Fresnel lens features an optical efficiency of 75%. The solar cell is a 40% efficient GaAs-based triple-junction solar cell.

I–V curves of experimental module are tested using a sun simulator with 850 W/m^2 and parallel light intensity under two kinds of conditions, namely, vertical incidence and oblique incidence, at room temperature of 20 °C. Fig. 17 displays *I–V* curves of vertical incidence at room temperature of 20 °C and with maximum output power of 142.3W. The sun simulator can test *I–V* curves depending on different incident angles. Simulation results show that when temperature increases or decreases, actual focal length of concentrating module lengthens or shortens and changes by 1 mm for every 10 °C increment. Therefore, experiments on temperature effect are conducted by changing module height. Output power curve depends on different heights of the module, as shown in Fig. 18. When module height increases by 4 mm, output power totals 139.1 W, representing a 2.3% decrease. When module height decreases by 3 mm, output power reaches 138.7W, indicating a 2.6% decrease. These findings coincide with simulation results.

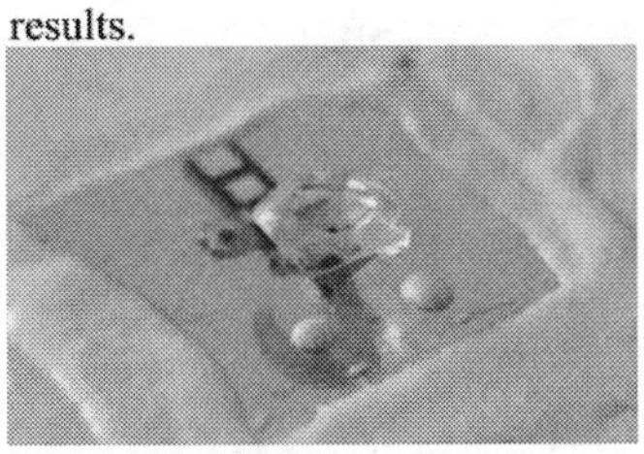

Figure 15. Energy distribution at -20°C

Figure 16. Energy distribution at 50°C

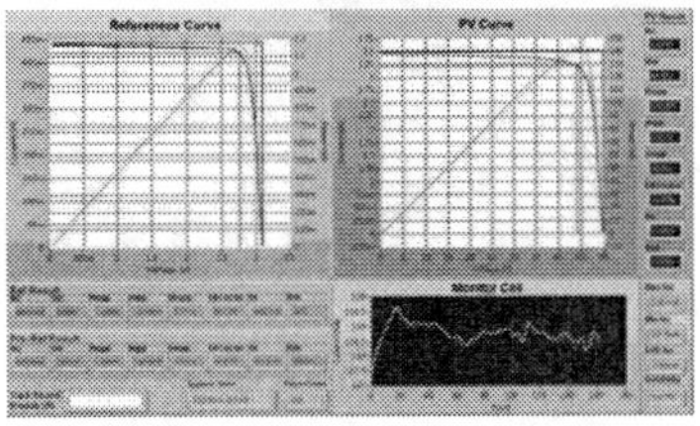

Figure. 17. I–V test curve with vertical incidence

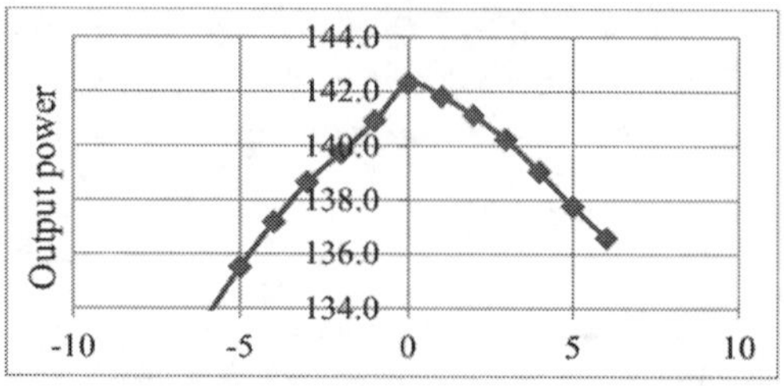

Figure 18. Relationship between module height change and output power

7. Conclusion

In this study, a focusable and uniform convergence secondary microprism is designed to improve performance of CPV module. Geometric parameters are optimized through 3D modeling and Zemax simulation. Focusable and uniform convergence secondary microprism is manufactured according to simulation parameters, and experimental 400-times module is tested to verify simulation results. Experimental observations agree with simulation results.

Results show that combination of Fresnel lens and focusable and uniform convergence microprism features the highest energy when upper spherical diameter of secondary microprism measures 18 mm, included angle of opposite side facets totals 116°, side length of bottom reaches 2.15 mm, and spherical height g removed from the top is 0.2 mm. Concentrating efficiency of the 400× concentrating module system reaches 88.67%, acceptance angle totals ±1.2°, maximum output power of 400× module measures 142.3 W, and efficiency loss of the module is less than 2.6% when temperature changes from −20 °C to 20 °C and from 20 °C to 50 °C. Compared with the case without secondary microprism, output power increases by 27.3%, acceptance angle increases by 87.5%, and efficiency loss of module decreases by 83.4%, indicating a significant improvement in generation performance of CPV module. This focusable and uniform convergence secondary microprism has been widely applied in Delingha 50MW power generation project.

References

1. W.T. Xie, Y.J. Dai, R.Z. Wang, and K. Sumathy, Concentrated solar energy applications using Fresnel lenses: A review, Renew, Sust. Energ., 15 (2011) 2588–2606.
2. B. Pablo, C.M. Juan, Z. Pablo, M. Rubén, C.Aleksandra, B. Marina, C. Julio, and H. Maikel, High performance Fresnel-based photovoltaic concentrator, Opt. Express, 18, S1 (2010) 25-40.
3. J.C. Miñano, P. Benítez, P. Zamora, M. Buljan, R. Mohedano, and A.Santamaría, Free-form optics for Fresnel-lens-based photovoltaic concentrators, Opt. Express, 21, S3 (2013) 494-502.
4. M.L. João, P. Benítez, P. Zamora, J.C. Miñano, and P.Adam, 9-fold Fresnel–Köhler concentrator with Fresnel lens of variable focal point, Opt. Express, 22, S4 (2014) 1153-1163.
5. B. Jared, M. Saavedra, B. Anderson, R. Goeke, W. Sweatt, G. Nielson, M. Okandan, B. Elisberg, D. Snively, J. Duncan, T. Gu, G. Agrawal, and M. Haney, Micro-concentrators for a microsystems-enabled photovoltaic system , Opt. Express, 22 (2014) 521-527 .
6. T. Gu, W.C. Sweatt, G. Agrawal, B.H. Jared, B.J. Anderson, R.S. Goeke, B. Elisberg, S.M. Paap, J.L. Cruz-Campa, V. Gupta, M. Okandan, G.N. Nielson, and M.W. Haney, Decentralized nonimaging micro-optical concentrator, Efficient Design for Illumination and Solar Concentration XI, Proc. of SPIE, 2014, 91910G.
7. F. Zeng, H.T. Li, and Z.B. Hao, High-concentration solar concentrator with uniform irradiation, Acta Energiae Solaris Sinica, 33, S2 (2012) 338-342 .
8. G.H. Yang, B.Z. Chen, Y.Q. Liu, L.M. Guo, Y. Shun, Z.Y. Wang, Optimized Design and Research of Secondary Microprism for Dense array Type Concentrating Photovoltaic Module, Applied Optics and Photonics, 2015, 967603.
9. R. Huang, L.M. Guo, Y.Q. Liu, C.C. Wang, M. Zhao, X.N. Zhang, H.M. Zhang, Z.Y. Wang, and Y.H. Cao, Research and Design of High Power Concentrator Module System, Chinese Journal of Luminescence, 39 S7 (2018) 974-982 .
10. L.M. Guo, Y.Q. Liu, G.M. Zhao, and Z.Y. Wang, Design and research of focusable secondary microprism in concentrating photovoltaic module, Proceeding of CPV-13 conference, Ottawa, AIP conference Proceeding, 2017, 1881.
11. L.M. Guo, Y.Q. Liu, Y.T. Qiu, and Z.Y. Wang, Design and research on Fresnel lens of combination of equal-width and equal-height of grooves. IOP Conference Series: Earth and Envionmental Science, 2017, 69:012177.

Three-dimensional fine model construction of instability in dangerous rock masses

Jingxuan Cao

College of engineering, Tibet university, Lhasa 850000, China

Abstract. Rock collapse is a common geological hazard in mountainous areas, and the prediction of the stability of dangerous rocks has always been a difficult problem in engineering practice. In this paper, a dangerous rock body in Ritu, Tibet, was scanned and measured using UAV tilt photography and 3D laser scanning technology, and the high-precision data collected by the UAV were processed and constructed into a 3D fine model, and the movement state of the dangerous rock body when it is destabilised was numerically simulated using Unity3D. In order to verify the simulation effect, a field test was conducted on a slope around Lhasa and the field test results were fitted to the numerical simulation results. By comparing the data obtained, the 3D modelling data of the high and steep slope collected by the UAV was more refined and the model was more complete than that collected by the 3D laser. Comparing the field test results with the numerical simulation results, the fit between the two is good, indicating the feasibility of using the model data obtained by UAV tilt photography for hazardous rock investigation, which provides an important reference value for future investigations of hazardous rock on high and steep slopes.

Keywords: Critical rock mass; 3D fine model; Unity3D; Model testing

1. Introduction

A dangerous rock mass is an unstable rock mass that is cut by multiple groups of rock structural faces and is located on a steep slope or cliff. The sudden, dramatic and rapid nature of the rock formations makes them one of the most difficult geological hazards to prevent, often causing traffic accidents and casualties, damaging traffic routes and other hazards, and already posing a serious threat to human life and property. The construction of three-dimensional models is a common means of studying geological hazards such as rock hazards, and this method has been used to study the destabilisation, fracture characteristics and stability of rock hazards, with good results. Therefore, three-dimensional modelling will be more effective in the analysis and characterisation of geological hazards such as rock hazards.

In the early geological modelling process, instruments such as total stations, compasses and tape measures were often used to obtain 3D geological information data, but due to the large errors in the data obtained by such instruments, they cannot yet meet the requirements of high-precision modelling data [1]. To build a more detailed 3D model, it is important to obtain effective and accurate topographic data. With the development and improvement of laser ranging technology, infrared technology and wireless remote control equipment, 3D laser scanning technology and UAV photography technology have been widely used in the geological field. The 3D laser scanning technology has been widely used in building deformation monitoring [2-3], slope monitoring [4-5], tunnel measurement [6-7], etc. because it effectively overcomes the shortcomings of traditional measurement methods such as large errors, high costs and personal safety of surveyors. UAVs have been widely used in urban construction [8-9], disaster identification and analysis [10-11] and other fields due to their higher accuracy, smaller size and flexibility, and easier and more complete access to 3D data on the surface of objects. Therefore, when modelling high and steep slopes and complex terrain areas that are beyond human reach or lack of terrain information, the data collected by the two methods should be compared and analysed first in order to obtain more effective and accurate terrain data.

Born in 1953, numerical simulation technology is widely used in the field of emergency response to emergencies such as meteorology, engineering geology, social public safety and water resources due to its outstanding decision aid function. At present, there are also a variety of numerical simulation software for studying dangerous rock masses, such as Rockfall, STONE and Ebloul. However, such software often requires a lot of time for preliminary modelling works, resulting in low efficiency, and the simulation of geological conditions differs greatly from the actual conditions, and the simulation effect is extremely unsatisfactory. In order to solve such problems, previous authors have either sought to achieve the corresponding interface between various types of numerical simulation software pre-processing [12]; or devoted themselves to the coupling of 3D geological software and numerical simulation [13]; or started to try to combine existing geological software for modelling [14]. However, these methods can only compensate to a certain extent for the shortcomings of numerical simulation in 3D modelling. After exploration, it was found that Unity3D software, which has been widely used in system simulation and virtual reality, is more suitable for numerical simulation in the geological field because it has PhysX physics engine, which can simulate collision and free fall motions more realistically.

When modelling the movement mechanism of dangerous rock masses, the previous work has not taken into account the influence of the terrain on the mapping technology and has used 3D laser scanning technology to collect data, resulting in large errors in the data; and when carrying out numerical simulations, the existing methods and software have a large gap between the simulated conditions and the field conditions, and the simulation effect is extremely unsatisfactory, resulting in some biased conclusions. It has been found that the slope data collected by UAV photogrammetry is more accurate than 3D laser scanning technology; and Unity3D can simulate collision and free fall movements more realistically, which is more suitable for numerical simulation in the geological field. In this paper, we will use UAV photogrammetry to collect data from the slopes where the rock hazards are located, and use Unity3D software to numerically simulate the movement of the rock hazards, and finally fit the field test results to the numerical simulation results to verify the fineness of the model built by this method.

2. High precision modelling

This paper presents an example of a dangerous rock along the G318 national highway in Niduo Township. The study site is located in the southeastern part of Mozhugongka County, Tibet Autonomous Region, with geographical coordinates 92°14'00"E, 29°41'30"N. The overall slope of the studied slope is about 35°, and the slope of the leading edge is 40°~45°. The surface is a loose accumulation of the Quaternary system, mostly sloping gravel block soil, and the underlying bedrock is mainly mica-quartz schist. In order to achieve high accuracy modelling of the dangerous rock masses, an unmanned aerial vehicle and a 3D laser scanner were selected to collect morphological data on the slope surface.

2.1 Data acquisition and processing

UAV tilt photogrammetry is the use of a UAV as a flying platform with a camera tilted at an angle to collect data from the study area in one vertical direction and four tilted directions. A DJI M210 industry-grade quadcopter equipped with a ZENMUSEX5S camera lens was used for the data acquisition, with an overlap of 80% in the heading and 50% in the side direction, and a course height of 150 m from the take-off point. 250 images were acquired, and 239 images were available after blurring and colour enhancement. Figure.1 shows the images of the Rideau Rock Danger obtained by the UAV.

In addition, the ILRIS-3D ground-based 3D laser scanner was used to collect data from the slope again, in order to verify the feasibility and accuracy of the data collected by the UAV. The ILRIS-3D ground-based 3D laser scanner was used to scan the observation body to obtain the 3D point cloud coordinates of the observation object. The data acquisition should be carried out in such a way that each station set up by the scanner can scan the entire area of the modelling target, while the overlap between stations should be 20%-30% to avoid point cloud loopholes. The internal data processing is to align the collected 3D coordinates with the data, data noise reduction and other processing, and finally use PolyWorks software to build a point cloud model of the region's dangerous rock body, as shown in Figure.2.

By comparing Figure. 1 and Figure. 2, it can be seen that although the 3D laser scanner has carried out a series of data processing when scanning the steep cliff and high mountain, the point cloud model established by the measured data still has more point cloud loopholes because it is easily blocked by the raised rock body and there are blind spots in the field of view, so the model established will not be fine enough, resulting in the dangerous rock body falling from the loopholes of the slope model, which cannot reflect the real movement scene of the dangerous rock body. Therefore, it has certain limitations when surveying and mapping terrain such as steep hills and complex slopes. The UAV has the advantages of small size, high mobility, high flexibility and good field of view, which makes the data collected more comprehensive and accurate, and the measured data can be more accurately modelled and better reflect the real scene. Therefore, it is more appropriate to use UAVs for data collection in the 3D modelling of slopes where dangerous rock masses are located.

2.2 Build high precision models

Due to the large size of the aerial survey data, the light effects and distortion of the image during the shooting process can make it difficult for the traditional feature point extraction algorithm based on geometric texture features to work effectively. Therefore, the data were first processed for feature point extraction and multi-view point matching to reduce the negative impact on the 3D modelling work [15]. Scale-invariant feature transform (SIFT) can quickly find the extreme points in the Difference of Gaussians (DOG) scale space and extract their position, scale and rotation invariants. feature point extraction. The formula for the difference of Gaussians numerator is:

$$COG=\frac{1}{\sqrt{2\pi}}=\left(\frac{1}{\sigma_1}\exp\left(\frac{x^2+y^2}{2\sigma_1^2}\right)-\frac{1}{\sigma_2}\exp\left(\frac{x^2+y^2}{2\sigma_2^2}\right)\right) \quad (1)$$

where x and y are the locations of the points in the image; σ_1 and σ_1 are the scale factors of the upper and lower Gaussian spaces。

After the feature points are extracted, the Euclidean distance of the feature vector is used as the criterion for determining the similarity of the feature points in multiple images, and then feature matching is performed to find a pair of matching points with high reliability.

$$d\left(R_i, S_i\right)=\sqrt{\sum_{j=1}^{n}\left(r_{ij}-s_{ij}\right)^2} \tag{2}$$

Where$R_i = (r_{i1}, r_{i2} \ldots, r_{in})$ is the feature point descriptor in the base image;$S_i = (s_{i1}, s_{i2}, \ldots, s_{in})$; $d(R_i, S_i)$is the Euclidean distance between feature points.

To identify paired feature point descriptors for the feature point descriptors in the observed image, the ratio of the Euclidean distance between the nearest and second nearest feature point feature vectors needs to be less than a set threshold. The formula is defined as follows:

$$\frac{d(R_i, S_u)}{d(R_i, S_v)}<T \tag{3}$$

Where S_u is the closest point to R_i in the observation map; S_v is the next closest point to R_i in the observation map; and T is the threshold value, usually set to 0.8.

Finally the processed UAV image data is built into a high precision model using Pix4Dmapper software. The modelling process consists of the following steps: selecting and adding images, adjusting image attributes, selecting output coordinate attributes, selecting processing option templates, adjusting output parameters and building the model. The generated 3D fine model is shown in Figure.3.

Figure. 1 Image model of the Nido Dangerous Rocks

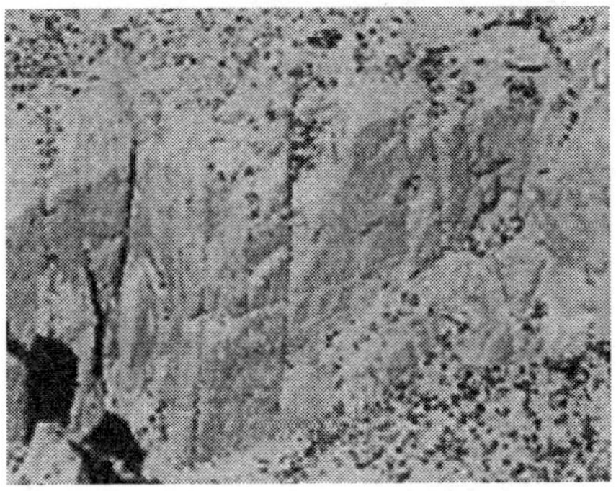

Figure. 2 Point cloud model of the Nido Dangerous Rocks

Figure. 3 3D fine model of the Nido slope

3. Numerical simulation of the movement mechanism of a dangerous rock mass

3.1 Numerical simulation during movement of dangerous rock masses

The numerical simulation of the movement mechanism of the rock hazard is mainly to simulate the interaction between the rock hazard and the slope during the movement of the rock hazard, which may involve the rotation of the rock hazard itself, the collision, rebound and rolling of the rock hazard and the slope. In the process of numerical simulation of rock movement, firstly, the rock body and slope should be set as rigid body properties, and the physical parameters such as friction, mass and volume of the rock body and slope should be set to restore the real scene to the greatest extent; secondly, the slope and rock body should be set as collision properties such as mesh collision body, box collision body and spherical collision body, so as to restore the movement of the rock body on the slope to the greatest extent. Unity3D has the advantage of being multi-lingual and can be used to write C# scripts to export the motion parameters of the simulated rock hazard. Therefore, in this paper, we will simulate the instability characteristics of five different working conditions and analyse them through Unity3D。

In the analysis of the effect of the size of the rock mass on its movement mechanism, square-shaped rock masses with diameters of 0.2m, 0.4m, 0.6, 0.8 and 1.0m were used to roll down the slope of the same material in the same unstable manner.

3.2 Movement characteristics of dangerous rock masses

Due to the different forms of rock hazards, different slope material, destabilisation and other factors, resulting in the form of movement of rock hazards: or collision with the slope in order to do flying motion, or along the slope gently rolling, or smooth and gentle sliding along the smooth slope, etc.. The formulas for solving the motion of rock hazards differ from each other. The acceleration of dangerous rock body are used to find Newton's law, not to discuss in detail, only to analyse its motion in different states [16].

The numerical simulation of the rock movement mechanism should firstly ensure that it is the only variable, the same height of release, the same slope indicating the same friction, the same density of the same rock body, etc. Secondly, the velocity, acceleration and cumulative displacement data of the rock under two working conditions with different rock body sizes and different rock body forms are extracted by means of C# scripts respectively. The extracted data were then processed into comparison curves. Because of the problem of conflict and redundancy, the data were first processed for noise reduction, and the following is the movement of the rock mass in Unity3D under different conditions.

4. Experimental Study

4.1 Field model tests

In order to verify the fineness of the 3D model built by this method, this paper uses the field rockfall test to test it. The field model tests were divided into two groups: one group of 0.2m, 0.6m and 1.0m diameter square-shaped rock hazard models, and the other group of 0.5m diameter rectangular, spherical and square-shaped rock hazard models, which were released at the test site with coded markers attached to each face of the rock hazard models. The movement data of the model at the test site was obtained by a monitoring system consisting of binocular camera stereo monitoring, a spatial 3D coordinate system for the slope, the Rocke TECH calculation system, a slope grid, spatial bounce markers and a power supply network [17]. The filming system consists of two Ac in the test site for releasing eutEye long time high speed camera system and CoaXPress high speed camera, using synchronous controller with 2 high speed cameras with high precision and synchronous filming.

Follow the tumbling orthogonal test protocol for rock movement at risk and synchronise with the binocular camera stereo vision monitoring system and save the video. Use the synchronous controller to synchronise the filming, taking care not to make any adjustments or movements to the calibrated footage and to ensure that the previous calibration results are correct before filming. To be on the safe side, take a video to verify that the whole system is working before starting the formal test process and start the field test after commissioning the equipment.

4.2 Goodness-of-fit analysis

In order to quantitatively describe the degree of fit between the field test and the numerical simulation, the test site is modelled in this paper and the numerical simulation is carried out with Unitu3D below the field conditions. The degree of agreement between the numerical simulation results and the field test results is also measured by the goodness of fit. The coefficient of determination is [18]:

$$R^2 = 1 - \frac{SSE}{SST} \tag{4}$$

SSE in equation (11) is the sum of squared residuals:

$$SSE = \sum_{i=1}^{n} \left(\hat{y}_i - y_i\right)^2 \tag{5}$$

SST is the sum of squared total deviations:

$$SST = \sum_{i=1}^{n} \left(\overline{y}_i - y_i\right)^2 \tag{6}$$

Where $\hat{y}_i$ is the fitted value; y_i is the test value; $\overline{y}_i$ is the average of y_i; The closer the value of R2 is to 1, the better the numerical simulation fit is. The results of the numerical simulation and the field test were substituted into the equation to obtain. It can be seen that the 3D model built by this method has a high degree of refinement.

5. Conclusions

This paper uses UAV and 3D laser scanner to collect slope and dangerous rock data, uses Pix4D software to generate high-precision 3D simulation model, uses Unity3D software in the analysis of various working conditions falling rock movement characteristics and accumulation position, and through the field test to verify the simulation analysis results, get the following conclusions.

(1) Comparing the data collected by UAV and 3D laser, it is found that the data collected by 3D laser has more point cloud loopholes, so the data collected by UAV is more suitable for this 3D modelling and the collected data is more accurate.

(2) By comparing the image observation of the calculation process of the high-precision 3D simulation model, the falling rock sphericity value changes from large to small, and the falling rock movement mode changes from jumping and rolling to sliding mode, so the high-precision 3D simulation calculation is more realistic and accurate than the ordinary numerical simulation software analysis.

(3) The numerical simulation fits well with the velocity, acceleration and displacement curves from the field test, thus verifying that the 3D model built by the method is more refined and beneficial to the design of highway slope rockfall hazard control projects.

Acknowledgments

Innovation experiment project of college students in Tibet University under Grant No. 202010694008.

References

1. Dong X J. Research of comprehensive application of three-dimensional image technology in geologic engineering [D], Chengdu: Chengdu University of Technology,2015. (in Chinese with English abstract).
2. Dang X B. The research and Application of 3D Laser Scanning techniques in building deformation Monitoring [D]. Xi'an: Chang'an University,2011. (in Chinese with English abstract).
3. Luo K, Wang Y C, Duan P, et al. Research on deformation monitoring of tower building based on ground 3D laser scanner[J]. Geomatics and Spatial

Information Technology,2021,44(6):76-78+83. (in Chinese with English abstract).

4. Zhao X P, Yan L L, Liu W L. Research on monitoring the shape of slope based on the three-dimensional laser scanning technology[J]. Science of Surveying and Mapping,2010,35(4):25-27. (in Chinese with English abstract).
5. E. Manda-Mvula, R. B. Kaunda. Structural Data Collection for Slope Stability Analysis Using Digital Technology—A Case Study of Melbur Pit, UK[J]. Journal of Mining Science,2019,55(1):
6. Qingquan. Li, Qingzhou Mao, Qin Zou,et al. Measurement and analysis of long-tunnel cross-sectional deformation using multi-sensor integration[J]. Annals of GIS,2010,16(2):
7. Gan L B. Tunnel engineering surveying and modeling based on 3D laser scanning technology[J]. Geotechnical Investigation and Surveying,2021,49 (6):58-61. (in Chinese with English abstract).
8. Yv Z D,Li H, Ba F, et al. 3D city model construction based on a consumer-grade UAV[J]. Remote Sensing for Land and Resources, 2018,30(2): 67-72.doi: 10.6046/gtzyyg.2018.02.09. (in Chinese with English abstract).
9. Sun S M, Huang T J, Sun Y. Construction and application of high-precision three-dimensional monomerization model for urban scenes[J] Bulletin of Surveying and Mapping,2021,(1):108-111. (in Chinese with English abstract).
10. Fang L Y, Zhao X, Wu X N, et al. Research on Oblique Photography 3D Modeling and Analysis Method of Typical Highway Geological Disasters[J].Highway, 2018,63(12):170-176. (in Chinese with English abstract).
11. Jin A B, Chen S J, Zhao A Y, et al. Numerical simulation of open-pit mine slope based on unmanned aerial vehicle photogrammetry[J]. Rock and Soil Mechanics,2021,42(1):255-264. (in Chinese with English abstract).
12. Hu B, Zhang Z Y, Huang R Q, et al. Development of pre-processing package for FLAC3D and verification of its simulating effects[J]. Chinese Journal of Rock Mechanics and Engineering,2002(9):1387-1391.
13. Wang M H, Bai Y. Study on integration of three-dimensional modeling and numerical simulation for stratified rock mass[J]. Rock and Soil Mechanics, 2005(7)1123-1126. (in Chinese with English abstract).
14. Cui F P, Hu R L, Liu Z L, Surfer software platform based complex three-dimensional geological digital models for pre-processing of FLAC3D[J]. Journal of Engineering Geology,2008(5):699-702. (in Chinese with English abstract).
15. Zhao P H, Li J J, Ye T J, et al. Application of Micro-UAV Remote Sensing in Emergency Mountain Hazards Monitoring of Alpine Valley Regio[J]. Journal of Disaster Prevention and Mitigation Engineering,2021,41(3):448-454. (in Chinese with English abstract).
16. Zhou X Y, Chen A R, Ma R J. Numerical simulation of energy dissipation mechanism on fall rocks protection nets[J]. Journal of Chang'an University (Natural Science Edition),2012,32(6):59-66. (in Chinese with English abstract).
17. Wang Z Z, Liu G Y, Li J J, et al. Experimental study on rockfall movement based on the stereo vision monitoring of binocular high-speed camera[J]. Plateau Science Research,2020,4(1):104-111. (in Chinese with English abstract).
18. Hu X L, Lin X Li M, et al. Selection strategies of hyperelastic constitutive models for carbon black filled rubber[J]. Engineering Mechanics, 2014, 31(5):34-42+48. (in Chinese with English abstract).

Influence of Goaf on Stability of Transmission Line Tower Foundation

Siyuan Zhang [1], Zhichao Fan [1], Liang Sun [1], Jiaohui Qiu [2]

[1] Ordos Power Supply Branch of Inner Mongolia Power (Group) Co., Ltd., Ordos 017010, China
[2] Transportation Institute Inner Mongolia University, Hohhot 010070, China

Abstract: There are various rock mass structures in the overlying strata of the goaf, such as cavities, cracks and separation layers. The phenomenon of under compaction with these structures still exists. Under the action of various natural forces, loads and other external forces, the broken rock mass in the relatively balanced goaf will produce structural deformation and compaction, resulting in surface subsidence. This paper expounds the surface subsidence of goaf from the aspects of surface subsidence mechanism, evaluation of surface stability of goaf and influencing factors of tower foundation stability, which provides a theoretical basis for the treatment of goaf.

Keyword: Goaf; Surface subsidence; Stability; Deformation.

1. Introduction

In the past two decades, China's economic construction has made amazing achievements. The rapid development of national economy is inseparable from the development and utilization of coal resources in China. At the same time, it also brings energy problems and sustainable development problems. It is understood that in the old energy mining bases in the central and western regions of China, there are a large range of mined out collapse areas and a large area of land subsidence. All kinds of damages cause the foundation bearing capacity of buildings and other building facilities above the goaf to decline, and even the foundation sinks seriously. The collapse of the goaf occurs from time to time. For example, in the central and western regions of China, these regions have a large population and have a greater demand for land use. With a high density of population, it is necessary to put the development of land into the use of mined out areas. In addition, in response to China's energy policy of "power transmission from west to East", many transmission line tower foundations are built on goaf. When the load of buildings, such as the foundation load of power transmission tower, is added to the goaf, the original stable goaf may be "activated", and the rock and soil mass in the caving zone above the goaf may be compacted and lack of support, resulting in the collapse of the underground residual cavity, resulting in additional displacement and deformation of the surface, thus causing the overall settlement of the surface buildings, and the buildings may have local cracking or even collapse at the same time. Such accidents have become an important problem on the road of urbanization in China, which seriously restricts the construction of railways, highways and bridges.

2. Research background

To study the mechanism of surface subsidence, it is necessary to understand the law of surface subsidence. Wang et al. obtained rock mechanics parameters by studying the geological conditions of Xingtai, and then obtained three stages of vertical deformation distribution and surface subsidence by using finite element simulation analysis, and the subsidence laws of each stage are different[1]. Lu et al. analyzed the actual data of Taiping Coal Mine and showed that in the process of mining subsidence, the strata movement followed the law of "equilibrium non-equilibrium new equilibrium", that is, the original equilibrium state was broken into non-equilibrium state, and finally reached a new equilibrium state through various processes of movement and deformation[2]. Pan et al. Found that with the continuous advancement of the working face, the relative positions of the working face and the surface points are different at different times, so the impact of mining activities on the surface is also different. The movement of the earth's surface experienced a process from the beginning of the movement, to the violent movement, and finally stopped[3]. Zhang et al. found that the form of surface deformation is not completely the same with the different mining depth, mining thickness, mining method, geological structure, seam occurrence and other factors. If the ratio of mining depth to mining thickness is large, the change of surface deformation with time and space is continuous and regular; If the ratio of mining depth to mining thickness is small or there is a large geological structure, the result is just the opposite, and large cracks or collapse pits may appear on the surface[4]. Peng found that when the length of the working face is unchanged, the coal wall compression value increases with the increase of mining depth, while the deflection value and total settlement value gradually decrease, and the total settlement value decreases faster; When the mining depth is constant, the working face length increases, and the coal wall compression value, deflection value and total settlement value increase[5]. Li et al. found through the actual measurement of three coal mines in Huaibei mining area that the overall subsidence coefficient of the mining area is large, which belongs to coal mining under the thick loose layer. The existence of the loose layer will expand the range of surface movement and increase the amount of surface subsidence. Therefore, the subsidence coefficient is larger where the loose layer is thicker[6].

From the above research results, it can be seen that there are indeed laws to follow for surface subsidence. These laws show that subsidence is a complex process that develops continuously with time and space, and has a close relationship with geological conditions, rock stratum properties and mining conditions, which provides a research direction for the further study of the causes of surface subsidence.

3. Influencing factors of tower foundation stability

With the gradual advance of underground coal mining face, generally when the distance from the opening of the cutting hole to the advancing of the mining face is equivalent to 1/4~1/2 of the mining depth, the mining impact will spread to the surface, causing surface subsidence, and a subsidence area much larger than the goaf will gradually form above the goaf, which is called surface movement basin or subsidence basin. The five indicators that often describe the movement and deformation in the surface movement basin are: subsidence, tilt, curvature, horizontal movement and horizontal deformation. Among them, subsidence, tilt and curvature belong to the movement and deformation in the vertical direction, while horizontal deformation (including tensile deformation and compression deformation) and horizontal movement belong to the movement and deformation in the horizontal direction.

These surface movements and deformations will directly cause the destruction of the original stress balance state inside the high-voltage line tower and the change of the operating conditions of the high-voltage transmission line. The parameters of the high-voltage line tower, such as the foundation spacing, the distance near the ground, the line span, the conductor sag, the deflection of the suspension insulator string, etc., may be affected and exceed the required operating parameters, thereby inducing potential safety hazards.

Different types of surface movement and deformation have different effects on the high-voltage line tower, and the high-voltage line tower structure itself has certain stiffness, which does not completely deform with the surface deformation, but also restricts the surface deformation. Therefore, the high-voltage line tower is not consistent with the surface deformation. Therefore, it is very important to study the influence and damage of various types of ground movement and deformation on the high-voltage line tower, the law of its influence and damage, and the extent and size of its influence and damage, for the study of the cooperative deformation law of the high-voltage line tower structure, foundation and foundation under the influence of mining. It is worth noting that the movement, deformation and even destruction of the high-voltage line tower due to the influence of surface movement and deformation is not a single process, but the result of the joint action and interaction of various surface movements and deformation. The movement and deformation of the surface are also affected by the high-voltage line tower, that is to say, they are synergetic deformation.

3.1 Impact of ground subsidence

The surface subsidence caused by underground mining is transmitted to the upper high-voltage line tower structure through the foundation, resulting in the subsidence of the high-voltage line tower. Subsidence includes uniform subsidence and uneven subsidence, which will be discussed separately below.

3.1.1 Influence of uniform settlement

The suspension point of the empty line of the high-voltage line tower dropped from E1 to E2 by W. after sinking, the tension of the suspension point, the sag and the distance to the ground of the conductor, the position of the maximum sag and the inclination angle changed. However, the specification clearly stipulates that the distance between the conductor and the ground shall not be less than the value L listed in table 2-1 under the maximum calculated sag. If the mining thickness is 2M and the subsidence coefficient is 0.8, assuming uniform subsidence, the ground surface will sink by 1.6m, which will have a direct impact on the ground distance of the high-voltage line tower conductor, and it is very likely to exceed the regulations, resulting in potential safety hazards for the high-voltage line tower.

Fig. 1 Influence of uniform settlement on HV Line Tower

In addition, the large, gentle and uniform subsidence of the ground surface has little impact on the high-voltage line tower, and there is almost no additional force in the high-voltage line tower structure, so it generally will not cause damage to the high-voltage line tower. However, when the subsidence value is very large, sometimes it will bring serious consequences, especially in the mining area with high diving level, when the surface subsidence causes the high-voltage line tower foundation to be close to or below the groundwater level, the foundation soil will be softened by water immersion or frost heave will occur under low air temperature, which will cause serious corrosion of the high-voltage line tower foundation and foundation, The underground metal parts of the high-voltage line tower (metal foundation, stay wire device, grounding device) are also prone to corrosion, which greatly reduces its strength and seriously threatens the safety and stability of the high-voltage line tower.

Most of the high-voltage line towers are in lattice configuration and have a split structure with four legs. The common foundation is split type. The high-voltage line tower should bear the combined action of up pulling, down pressing and horizontal shear force. The consolidation, movement and settlement of the ground surface, as well as the inelastic characteristics between the earth and the foundation, make the ground surface sink to produce pressure. In addition, the tower and various loads cause uneven settlement between adjacent foundations, resulting in a variety of combined loads, resulting in greater stress in various components of the tower.

3.1.2 Influence of uneven subsidence

At present, most foundations of high-voltage line towers are independent foundations. Compared with the overall foundation, independent foundations are more vulnerable to uneven subsidence. It can even be said that as long as the surface deformation caused by underground mining affects the foundation of the high-voltage line tower, uneven settlement will certainly occur, because the independent foundation of the high-voltage line tower has a certain root distance. If the four independent foundations are simply divided into two sides, the settlement of the independent foundations on both sides at the same time cannot be completely consistent, so uneven settlement will certainly occur in the dynamic deformation, However, the "even sinking" usually means that the high-voltage line tower sinks evenly from the beginning to the end. Relevant research shows that the settlement of high-voltage line tower foundation is basically consistent with the corresponding surface settlement. However, due to the fact that the high-voltage line tower structure has a certain stiffness and does not completely sink with the foundation settlement, the settlement of high-voltage line tower foundation is slightly smaller than that of the corresponding foundation. According to the cooperative deformation theory, if the foundation area is large, there will be partial foundation unloading and partial loading on the foundation ground, which is also one of the reasons for the difference between the settlement of foundation and foundation. However, through consulting a large number of literatures, no specific quantitative relationship between the high-voltage line tower and the corresponding surface movement and deformation has been found. The uneven settlement of the foundation will also cause redistribution of the foundation reaction of the building, and generate additional stress in the high-voltage tower structure. If the additional stress is greater than the allowable stress of the material, the high-voltage tower structure is very likely to be damaged or even collapsed. If there is a subsidence difference between adjacent towers, it will affect the near ground distance of the conductor, which mainly depends on the topographic characteristics along the line and the subsidence difference between adjacent towers. The high-voltage line tower is located in the middle of the collapse area, which is easy to cause the foundation of the high-voltage line tower to sink, retract and deform; When the high-voltage line tower is located at the edge of the collapse area, it is easy to cause inconsistent foundation subsidence on both sides of the high-voltage line tower. One side of the tower has a large subsidence value and slides down the basin, resulting in tilt deformation of the tower body.

When the foundation has uneven settlement, the high-voltage line tower may have two reactions: the first is that the high-voltage line tower structure moves and rotates as a rigid body with the foundation under its own rigid body performance, and the high-voltage line tower structure is not damaged; The second is due to the elastic properties of the materials constituting the high-voltage line tower. When the additional stress exceeds the elastic limit of the tower material, the geometry of the high-voltage line tower will change to adapt to the uneven settlement of the foundation. At this time, the high-voltage line tower structure will be damaged, and the stress of each node will also change.

3.2 Impact of surface inclination

Uneven surface subsidence is the direct cause of surface inclination. For buildings (structures) with small bottom area and large height (such as water towers, chimneys, high-voltage line towers, etc.), even if there is a slight inclination, high-rise buildings (structures) will cause a large horizontal component and overturning force distance under their own huge gravity. The overturning force distance will increase the inclination of high-rise buildings (structures), resulting in a vicious cycle, Therefore, high-rise buildings (structures) are very sensitive to tilt deformation.

The inclination of high-voltage line tower refers to the deviation of each part of the high-voltage line tower from its original position due to the inclination of the ground surface. The inclination degree of the high-voltage line tower is expressed by inclination. The inclination degree Q of the high-voltage line tower refers to the ratio of the horizontal distance (s) of the top center offset of the high-voltage line tower to the original height (H) of the high-voltage line tower.

The impact of ground tilt on high-voltage line tower is complex. The objects of impact include high-voltage line tower and its foundation, suspension string, line span, etc., which are mainly reflected in the following three aspects:

1) Impact on high voltage line tower and foundation. The ground surface inclination caused by mining will cause the corresponding inclination of the high-voltage line tower foundation. The tower inclination caused by the Tower Foundation Inclination increases the overturning moment of the foundation and the tower, thus affecting the overturning stability of the tower and the foundation.

For the stepped foundation, the inclination of the tower also increases the shear stress between the steps.

2) Influence on span, sag and suspension string. After the high-voltage line tower is tilted, the span is the distance between the projections of the ends of adjacent towers on the horizontal plane. Due to the inconsistency of the tilt value of the high-voltage line tower, the span will inevitably be expanded or reduced. The change of span will correspondingly cause the change of line sag and ground proximity. The inclination of the high-voltage line tower will also cause the stress redistribution in the conductor. In order to maintain the stress balance between the gears, the suspension string will shift to balance the unbalanced tension between the gears. The suspension string will deflect to the side with high tension, which may also cause the discharge accident of the conductor to the tower head.

3) Impact on ground proximity. The conductor relaxation caused by the inclination of the high-voltage line tower and the reduction of the actual height of the tower after the inclination may cause the insufficient distance between the conductor and the ground.

3.3 Influence of surface curvature

Uneven ground tilt causes curvature deformation of the ground surface. There are positive and negative curvature deformations, and the positive and negative curvature make the surface convex and concave respectively. In general, surface tension deformation and positive curvature deformation, compression deformation and negative curvature deformation occur at the same time. If the contact area is large enough, the ground surface will change from the original plane to a curved surface. The convex or concave surface will change the contact state between the foundation and the ground surface, break the original stress balance state of the upper building, change the foundation reaction, and generate additional stress in the building. When the curvature deformation increases to a certain extent, the building will be damaged.

Curvature deformation is one of the main factors of building damage. According to the analysis of domestic observation data of buildings in mining areas, the curvature of buildings is 21.3%~78.4% of the surface curvature. For buildings in positive curvature area, the curvature transferred to buildings (structures) is larger, while for buildings in negative curvature area, the curvature transferred to buildings is smaller. The greater the stiffness of the building, the smaller the curvature transferred from the ground surface to the building.

In particular, curvature deformation has little effect on buildings with small bottom area, and has a greater effect on buildings with large length. For the high-voltage line tower, the size of its independent foundation is generally small, so the contact area between the high-voltage line tower and the ground surface is small, and the range of redistribution by the foundation reaction is small. Therefore, the effect of surface curvature deformation is less. In addition, the high-voltage line tower structure itself has a certain stiffness and is not easy to bend with the curvature deformation of the ground surface, so the effect of curvature deformation on the high-voltage line tower can be ignored.

3.4 Impact of surface horizontal movement

The horizontal movement of the ground surface will cause the corresponding translation of the foundation of the high-voltage line tower. If the high-voltage line tower of the whole line moves horizontally at a uniform speed along the line, the operation of the line will not be affected. If the horizontal movement of adjacent high-voltage line towers along the line is inconsistent, the conductor tension and span of adjacent high-voltage line towers will increase relative to one side and decrease relative to the other side. The unbalanced tension generated by span change in adjacent spans will also cause the suspension string to deflect to the side of conductor tension. If the horizontal movement of the foundation on both sides of the high-voltage line tower is inconsistent, not only the above problems will occur, but also the root opening of the high-voltage line tower foundation will increase or decrease. The change of the root opening will cause additional stress in the structure of the high-voltage line tower, especially near the transverse plane near the foundation at the lower part of the high-voltage line tower, the stress is the largest, which is very easy to cause bending or even damage to the transverse bar.

The superstructure load acts on the foundation vertically without considering the effect of horizontal load. This is because the horizontal stress on the foundation caused by the horizontal deformation of the ground surface is certain and will not exceed the friction between the foundation and the foundation. When the friction is greater than the friction between them, the foundation and foundation slide, and the stress is no longer transmitted upward. In other words, for buildings with combined foundation, the superstructure will not be subject to additional stress caused by horizontal deformation of the ground surface. However, for the high-voltage line tower structure with independent foundation, if the horizontal movement of the independent foundation is inconsistent, the substructure will bear tension or compression and generate additional stress in the structure. Therefore, the horizontal movement and deformation of the independent foundation can be transferred to the superstructure through the foundation. Therefore, the independent foundation of truss structure such as high-voltage line tower not only transfers the load in the vertical direction, but also transfers the load in the horizontal direction. For the independent foundation, it is necessary to study the influence of its horizontal load.

4. Summary

According to the analysis, the relationship between the maximum compressive stress, the maximum tensile stress, the inclination of the tower and the surface inclination deformation is obtained.

At a certain inclination value, the maximum axial compressive stress of the high-voltage line tower is located at the position of the main material of the leg at the relative fixed side of the tower, and the maximum tensile stress is located at the position of the inclined material of the leg of the high-voltage line tower. With the increase of the surface inclination, the axial compressive stress of the compression bar and the axial tensile stress of the pull rod also increase.

There is a quadratic nonlinear relationship between the maximum compressive stress of the member and the surface tilt deformation. With the increase of the surface tilt, the maximum compressive stress of the member increases monotonically.

The maximum axial tensile stress of the member shows a quadratic nonlinear monotonic increasing trend with the increase of the surface inclination.

Acknowledgements

The authors would like to acknowledge financial support from Science and technology project of Inner Mongolia Power Group (Co., Ltd.) (2021-9)

References

1. Wang Lijuan, Wang Xiliang, Xiang min. Study on the law of surface subsidence in goaf of Xingtai Coal Mine[J]. Subgrade Engineering, 2009 (2): 2.
2. Lv Wenyu, Pan Renfei, Li Yang, cuiguoliang Study on the law of surface subsidence in goaf [J]. Open Pit Mining Technology, 2007(4):3-5.
3. Pan Chibo The Influence of mined out area on surface subsidence of waste dump site in harwusu open pit coal mine[J]. Pit Mining Technology, 2020, 35 (2): 4.
4. Zhang Yonghai Exploration on the law of surface subsidence in mines Xinjiang nonferrous metals 2011:31-33
5. Peng Linjun, Zhao Xiaodong, Li Shucai Simulation study on Surface Subsidence Law of deep mining[J]. Geotechnical mechanics, 2011,32(6):1910-1914.
6. Li Xiaoqing Study on the law of mining surface subsidence in Huaibei mining area[D]. Master's thesis of Anhui Architecture University 2015:60

Study on overwintering growth characteristics and purification effect of floating bed cultivated plants

Hongbing Chen [1], Wenna Zhang[1] and Weihong Wan [2,*]

[1] Faculty of resources and environmental science Hubei University Wuhan, China
[2] Laboratory and equipment management division Hubei University Wuhan, China

Abstract. Artificial floating bed technology has been used to repair eutrophic water. However, there are relatively few studies on the overwintering of economic plants on the floating bed. In order to realize the resource utilization of wastewater, celery, Artemisia selengensis, Garlic, Chinese yam, lettuce and other crops were selected as the test plants for the floating bed. Quantitative experimental study was conducted on the aquaculture wastewater in winter. The results showed that after 90 days of acclimation, the tested plants showed different growth characteristics, with the survival rates above 84.48%. The distribution of nitrogen and phosphorus in water and subsurface plants of 5 species were different. The contents of nitrogen and phosphorus in plants The accumulative amounts were 1.15 ~ 5.18 g·m-2 and 0.04 ~ 0.21 g·m-2, respectively, with the order of garlic > celery > seabuckthorn > sedum chrysanthemum> lettuce and celery> garlic> sesame oil> sesame oil> lettuce.

key word: Winter; artificial floating bed; economic plant; growth characteristics; absorptive capacity.

1. Introduction

Water resource is the foundation of human life and social progress. It is the central link of economic and social development, and also the prerequisite for achieving sustainable development and healthy ecosystem. At present, there are many phenomena of direct discharge of aquaculture wastewater in China, which makes the receiving water eutrophic and reduce the biodiversity. In terms of intensive aquaculture water bodies, total phosphorus and ammonia nitrogen pollution have become the main stress factors restricting the aquaculture environment [6-7]. The research shows that aquaculture organisms only absorb 23% ~ 31% of N and 10% ~ 13% of P, while 14% ~ 53% of N and 39% ~ 67% of P accumulate in the sediment. 57% of n is discharged into the environment. It can be seen that the nutrients released into the environment by breeding ponds are considerable.

In the mid-1990s, artificial floating islands were widely recognized by Japanese environmental and lake scientists. In the 1980s, the United States began to use a variety of fish culture wastewater to produce lettuce, tomato, strawberry, cucumber and other vegetables and hyacinth and other flowers. Nathalie et al. Used the commercial deep flow (NTF) hydroponic system to float and cultivate D.innoxia to purify and repair domestic sewage, and achieved good results. Biological floating island technology has also been rapidly promoted and applied. However, in the selection of biological floating island plants, the research mostly focuses on hygrophilic vegetables and flowers. Most of these plants are suitable for planting biological floating islands in spring, summer and autumn. For example, Canna has a good removal effect on pollutants in water, but it is not cold resistant. After frost beating, the stems and leaves will wither, and it is easy to freeze to death below 5℃. It is difficult to survive the winter under floating bed planting conditions, and it needs to be replanted in the second year. Therefore, the selection of overwintering plants on biological floating islands has become an important issue in current research.

2. Materials and methods

This experiment was conducted in a fish pond of a pig farm, which mainly raised four large fish. The fishpond area selected for the test and research is about 5661.5 square meters, with water quality of COD of 8 ~ 32mg/l, NH3-N of 0.06 ~ 0.48mg/l, TN of 2.2 ~ 4.7mg/l and TP of 0.17 ~ 0.23mg/l.

The research site has a subtropical monsoon climate with four distinct seasons, sufficient rainfall and mild climate. The annual average temperature is 16.9 ℃, the extreme maximum temperature is 41.4 ℃(July) and the extreme minimum temperature is -9.0 ℃ (January). The average annual sunshine duration in the area is 1669.2 hours, and the average daily sunshine is 4.57 hours.

* Corresponding author: 350765810@QQ.com

 https://doi.org/10.1051/e3sconf/202235801004

2.1 Experimental devices and materials

The extruded plate floating island carrier (1.8m long, 0.6m wide and 3cm thick) used in this study is shown in Fig. 1. Each plate is drilled with 27 holes with a diameter of 20mm, and the hole penetrations are reserved for fixing plants. Floating bed units are connected and fixed with iron wire to form 0.6m ×72m rectangular floating bed group is bound to the edge of the fish pond with nylon ropes.

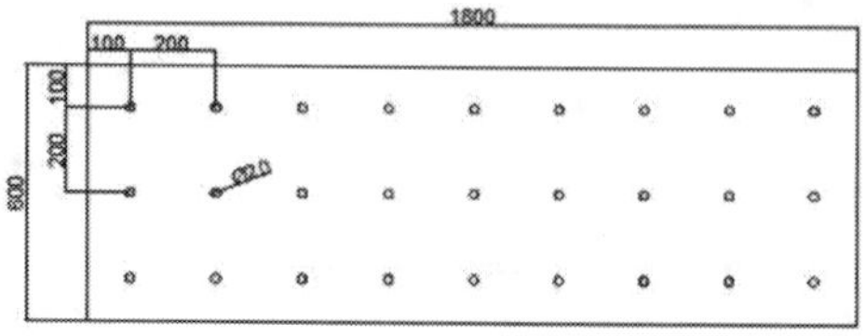

Figure 1. Top view of floating body material (unit: mm)

In the process design of plant floating bed, the selection of plants should be able to adapt to the local species growing under hydroponic conditions, with high survival rate, long evergreen period, fast sprouting and rapid growth; The plant root system is developed and the biomass is large; It has good adaptability to polluted water bodies, is resistant to pollution and has good pollution control potential. According to the screening principle of floating bed plants, celery, chrysanthemum, garlic, potherb mustard, oil lettuce and other crops were selected as floating bed test plants, and a quantitative experimental study was carried out on aquaculture wastewater in winter.

Table 1. Physiological indexes of tested plants

plant	Number of stems and leaves/Tablet	Root length/cm	Plant height/cm	Fresh weight/g·strain^{-1}
Celery	3	3.1	25.9	5.62
chrysanthemum coronarium	7	3.0	6.3	1.80
garlic	5	5.1	39.9	6.04
Potherb mustard	6	6.5	17.8	3.58
Lettuce	10	3.5	18.8	12.38
Spinach	9	1.5	19.7	6.53

2.2 Experimental method

2.2.1 Plant transplanting

Transplant five economic plants, such as celery, garlic, potherb mustard, garland chrysanthemum, and oatmeal. Before transplantation, according to the difference of samples, sampling survey was carried out and numbered respectively to investigate and count the root length, plant height, fresh weight, growth and root system of plants. The weather conditions of the experimental site were recorded on the spot. At the same time, two water samples from the fish pond were collected to measure their pH, water temperature, COD, TN, TP and other indicators.

Table 2. List of experimental water quality.

index	t °C	pH	DO mg/L^{-1}	COD mg/L^{-1}	TN mg/L^{-1}	TP mg/L^{-1}	NH_3-N mg/L^{-1}	chl-a ug/L^{-1}
Range	-2 ~ 19.8	7.07 ~ 7.73	8.80 ~ 9.53	20 ~ 82	2.4 ~ 21.6	1.14 ~ 1.71	0.72 ~ 1.46	24.46 ~ 39.79
Mean	8.7	7.54	9.05	57	9.6	1.39	1.23	33.15

2.2.2 Experimental analysis and data processing

Table 3. List of plant composition analysis

Testing object	Basic items	Analytical method	Testing equipment	Method source
plant Biomass	plant Biomass	Weighing method	1% balance	
	TN	Kelvin method	Kjeldahl Nitrogen Determiner	NY/T2017-2011
	TP	ICP	Optima 8000	

2.2.3 Data analysis and processing

Use Excel and SPSS software to process the experimental data, eliminate outliers and then calculate the average value, calculate the removal effect of pollutants in the water, the biological growth of plants and the concentration of nutrients, and compare the removal effects of various plants; Calculate the nitrogen and phosphorus content of plants, and speculate the purification effect of plants on water.

3. Results and analysis

3.1 Growth characteristics of floating bed cultivated plants

3.1.1 Growth status

Five kinds of vegetables show different growth characteristics after a period of growth:

After 90 days of domestication and adaptation, celery shows strong ecological adaptability and can grow normally from beginning to end. Each plant has new leaves, leaves grow, plants grow tall, biomass increases, and plants constantly tiller out new plants. The root system can develop normally. During the test period, new root systems have been formed continuously, and the root system has gradually increased, forming a compact and developed root system. After 90 days of domestication and adaptation, chrysanthemum garland, lettuce, garlic and Potherb Mustard all showed strong ecological adaptability and could grow normally from beginning to end. However, in the early stage of the experiment, due to the direct weather reasons and the adaptation to water temperature, some of the original stems and leaves of the plant were withered and frostbitten, but most of the leaves were bright green and grew well, with new leaves emerging. The leaves grew up, the plant grew tall and the biomass increased. Among them, Chrysanthemum chrysanthemum sprouts new plants. The root system can develop normally, and there are new root systems, and the root system also grows gradually. However, in the later stage of the experiment, aphids appeared in the leaves of potherb mustard, which was easy to cause insect damage.

Table 4. Statistics of monthly survival rate of floating bed cultivated plants

Time (d)	Statistical factors	Plant name and data					
		Celery (459)	chrysanthemum coronarium (270)	garlic (459)	Potherb mustard (567)	Lettuce (297)	Spinach (486)
30d	Number of surviving plants (plant)	440	257	459	549	286	432
	Survival rate (%)	95.59	95.18	100	96.82	96.29	88.89
60d	Number of surviving plants (plant)	421	245	448	497	243	364
	Survival rate (%)	91.72	90.74	97.60	87.65	81.81	74.89
90d	Number of surviving plants (plant)	410	235	408	479	120	0
	Survival rate (%)	89.32	87.04	88.89	84.48	41.95	0

3.1.2 Changes of plant roots

Plant roots can adsorb suspended solids and colloids in water. When growing in heavily polluted and suspended solids rich water, with the continuous growth and expansion of roots, not only the adsorption of suspended solids, particles and colloids in water is increasing, but also provide a certain place for microorganisms, leading to the proliferation of foreign microorganisms, the ability to decompose organic suspended solids, particles and colloids is strengthened, and finally the transparency of water body is improved, Water quality has been improved. When the organic nitrogen and phosphorus in sewage are transformed into inorganic ions, they can be directly absorbed and utilized by plant roots to promote plant growth. Most of the nitrogen absorbed by plant roots is in the form of nitrate ions. Part of the nitrogen is directly transported to the leaves in the form of nitrate ions and then reduced for utilization. The other part is reduced in the roots. Various amino acids and nitrogen-containing organic compounds are synthesized and transported to the upper part of the ground in the form of amino acids or other organic compounds through plant transpiration. The phosphorus absorbed by roots is mainly transported to the shoot in the form of phosphate ions, and a small amount of phosphorus is transported to the stems and leaves after the roots are synthesized into organic compounds.

The change of root length of the five plants is more obvious than that of height. Except that the net root growth of garlic is less than that of its plant height, the net root growth of the other four plants is higher than that of its plant height (Fig. 2). The roots of celery and Potherb Mustard grew the most, 14.0 cm and 11.3cml respectively; The second was the lettuce, whose root length increased by 5cm; The root growth of garland chrysanthemum and garlic were 3.9cm and 3.6cm respectively. The change of root length is mainly determined by the ecological adaptability of various plants and the environmental conditions during the experiment. The roots of celery and garland chrysanthemum belong to the radial structure of stem and root, and a large number of adventitious roots are formed in the rhizome; The root system of garlic is white and tender; Potherb mustard leaves its lateral roots from one main root; The root system of oilseed rape is relatively sparse, and there are adventitious roots in the rhizome. Celery and Potherb Mustard have developed roots. In the water environment, plants mainly rely on adventitious roots and their branch roots to complete the absorption of nutrients [7]. Therefore, the absorbed surface area is increased through the root system to meet the needs of nutrient absorption. In this way, the root system can absorb a large amount of nutrients in the water to meet its growth, and its biomass is also large. The adaptability of oilseed lettuce is poor, and the upper leaves wither and fall off, which affects the growth and development of the plant.

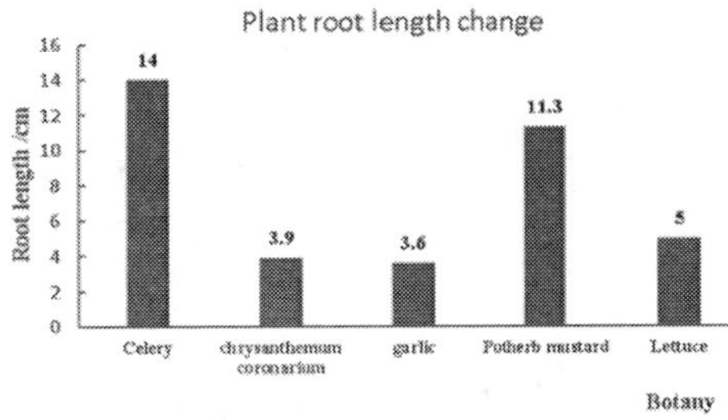

Figure 2. Change of plant root length

3.2 Absorption of nitrogen and phosphorus by floating bed plants

3.2.1 Distribution of nitrogen and phosphorus content in plants

In terms of the distribution of five plant nutrients, the nitrogen content in the aquatic part ranged from 31.31 to 56.13 $g \cdot kg^{-1}$, of which potherb mustard had the highest nitrogen content, 56.13 $g \cdot kg^{-1}$, followed by garlic (42.45 $g \cdot kg^{-1}$), celery (40.66 $g \cdot kg^{-1}$) and oilwheat (34.37 $g \cdot kg^{-1}$), and chrysanthemum garland had only 31.31 $g \cdot kg^{-1}$; The contents of nitrogen in the five plants under water ranged from 23.69 to 55.89 $g \cdot kg^{-1}$, among which garlic was the highest, followed by potherb mustard (46.45 $g \cdot kg^{-1}$) and celery (38.85 $g \cdot kg^{-1}$). There was little difference between chrysanthemum and oilwheat, which were 24.88 $g \cdot kg^{-1}$ and 23.69 $g \cdot kg^{-1}$ respectively. Celery (1.83 $g \cdot kg^{-1}$), potherb mustard (1.59 $g \cdot kg^{-1}$) and garland chrysanthemum (1.54 $g \cdot kg^{-1}$), garlic (1.26 $g \cdot kg^{-1}$) had the highest phosphorus content in the water part, followed by oilwheat (1.11 $g \cdot kg^{-1}$); The contents of phosphorus in the submerged parts of the five plants ranged from 1.19 to 1.73 $g \cdot kg^{-1}$, among which potherb mustard had the highest phosphorus content, followed by celery (1.66 $g \cdot kg^{-1}$), garlic (1.44 $g \cdot kg^{-1}$) and chrysanthemum (1.33 $g \cdot kg^{-1}$), and the smallest was lettuce. The contents of total nitrogen and total phosphorus in celery and chrysanthemum are distributed in the water part > underwater part. Nitrogen and phosphorus are transported to the water part, and phosphorus absorbed by roots is transported upward in the form of ions [68]; The distribution of total nitrogen and total phosphorus in garlic was underwater part > underwater part; The content of total nitrogen in potherb mustard and oilseed lettuce showed the distribution of above water part > under water part. Nitrogen was transported to the above ground part, while the content of total phosphorus was under water part > above water part.

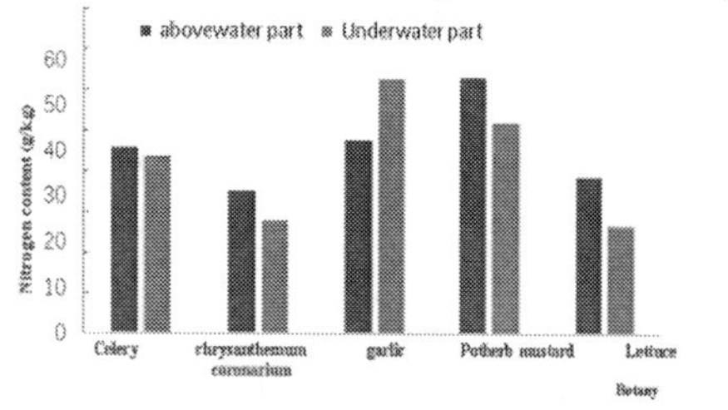

Figure 3. Distribution of nitrogen content in plants

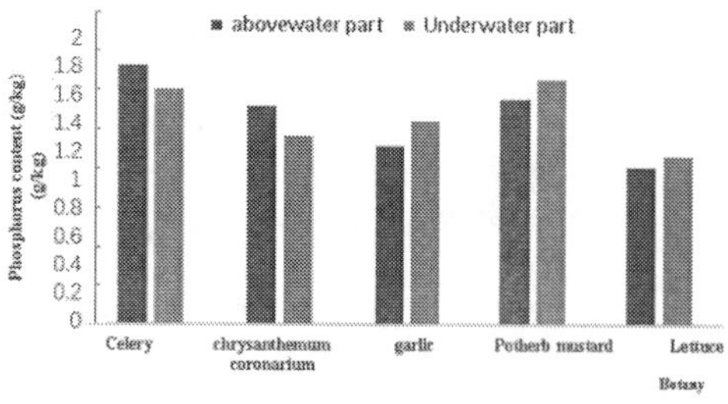

Figure 4. Distribution of phosphorus content in plants

3.2.2 Nitrogen and phosphorus accumulation in plants

Combined with the relevant data of biomass and nitrogen and phosphorus content in the body of five plants, the nitrogen and phosphorus accumulation in the above water and under water parts of plants can be obtained. The accumulated amount of nitrogen and phosphorus in the aquatic part of the planting object is 0.68 ~ 2.44 $g \cdot m^{-2}$ and 0.022 ~ 0.110 $g \cdot m^{-2}$ respectively. The nitrogen content is celery > potherb mustard > garlic > garland chrysanthemum > oil wheat, and the phosphorus content is celery > garland chrysanthemum > garlic > potherb mustard > oil wheat. The accumulated amount of nitrogen and phosphorus in celery is the largest, indicating that it has the strongest ability to absorb nitrogen and phosphorus. The accumulated amounts of nitrogen and phosphorus in the underwater part were 0.47 ~ 2.94 $g \cdot m^{-2}$ and 0.024 ~ 0.100 $g \cdot m^{-2}$, respectively. The nitrogen content was garlic > celery > potherb mustard > garland chrysanthemum > oilwheat, and the phosphorus content was celery > garlic > potherb mustard > garland chrysanthemum > oilwheat, in which garlic had the largest accumulated amount of nitrogen and had the strongest nitrogen absorption capacity on the surface; Celery has the largest phosphorus accumulation, indicating that it has the strongest ability to absorb phosphorus. Because the difference of nitrogen and phosphorus content between garland chrysanthemum and oilseed lettuce is not too large, but the nitrogen and phosphorus accumulation in plants is large, the main reason for the difference is the difference of biomass.

It can be seen that the distribution of nitrogen accumulation in each part of water celery shows a trend of 2.44 $g \cdot m^{-2}$> 2.33 $g \cdot m^{-2}$ in the water part, and the distribution of phosphorus accumulation shows a trend of 0.11 $g \cdot m^{-2}$> 0.10 $g \cdot m^{-2}$ in the water part. The nitrogen and phosphorus accumulation in the water part accounts for 51.2% and 52.4% of the total accumulation respectively; In the nitrogen accumulation of each part of chrysanthemum, the distribution of the above water part is 1.38 $g \cdot m^{-2}$> the under water part is 1.10 $g \cdot m^{-2}$, and the distribution of the phosphorus accumulation is 0.068 $g \cdot m^{-2}$> 0.059 $g \cdot m^{-2}$. The nitrogen accumulation of-the above water part accounts for 55.6% of the total accumulation, and the phosphorus accumulation of the above water part accounts for 53.5% of the total accumulation; The nitrogen accumulation of garlic above and under water was 2.94 $g \cdot m^{-2}$> 2.24 $g \cdot m^{-2}$ in the water part, and the phosphorus accumulation was 0.076 $g \cdot m^{-2}$> 0.066 $g \cdot m^{-2}$ in the water part. The nitrogen accumulation in the water part accounted for 43.2% of the total accumulation, and the phosphorus accumulation in the water part accounted for 46.5% of the total accumulation; The nitrogen accumulation of potherb mustard was 2.29 $g \cdot m^{-2}$> 1.90 $g \cdot m^{-2}$ in the water part, and the phosphorus accumulation was 0.071 $g \cdot m^{-2}$> 0.065 $g \cdot m^{-2}$ in the water part. The nitrogen accumulation in the water part accounted for 54.7% of the total accumulation, and the phosphorus accumulation in the water part accounted for 47.8% of the total accumulation; The accumulation of nitrogen in the water and underwater parts of oilseed lettuce was 0.68

$g \cdot m^{-2}$> 0.47 $g \cdot m^{-2}$ in the water part, and the accumulation of phosphorus was 0.024 $g \cdot m^{-2}$> 0.022 $g \cdot m^{-2}$ in the water part. The accumulation of nitrogen in the water part accounted for 59.1% of the total accumulation, and the accumulation of phosphorus in the water part accounted for 47.8% of the total accumulation; Although the total accumulation of nitrogen and phosphorus in the whole plant was the smallest among the five plants, the share of nitrogen accumulation in the water part was the highest, accounting for 59.1% of the total accumulation; The content of nitrogen and phosphorus in the underwater part of garlic was higher than that in the water part, and the proportion of nitrogen and phosphorus accumulation in the water part was the lowest.

It can be seen from the above analysis that the proportion of nitrogen and phosphorus accumulated in the water part of celery and chrysanthemum is more than 50%, indicating that the nitrogen and phosphorus absorbed by the two economic plants from the water through their roots migrate and accumulate more to the stems and leaves of the water part. This distribution is conducive to the removal of nitrogen and phosphorus by picking plants. There was a significant linear relationship between nitrogen and phosphorus accumulation and biomass ($p<0.05$). The correlation coefficients between nitrogen and phosphorus accumulation and biomass were 0.869 and 0.980 respectively, indicating that the correlation between nitrogen and phosphorus accumulation and biomass was good.

4. Conclusion and discussion

In this study, the growth status and the ability to absorb nitrogen and phosphorus of celery, chrysanthemum, garlic, potherb mustard and oilwheat were compared and analyzed. The main conclusions are as follows:

(1) After about 90 days of acclimation, the five tested plants showed different growth characteristics, all of which were more than 84.48%.

(2) After 90 days' growth, the fresh weight of the five plants ranged from 232.74 $g \cdot m^{-2}$ to 418.26 $g \cdot m^{-2}$, and the order was: Celery > garlic > garland chrysanthemum > potherb mustard > lettuce; In terms of the net increment of dry matter biomass, the biomass of the five plants ranged from -3.1 to 49.32 $g \cdot m^{-2}$, celery > chrysanthemum > potherb mustard > garlic > lettuce. The moisture content of the five plants ranged from 82.49% to 88.60%, and the order was as follows: lettuce > chrysanthemum > celery > garlic > potherb mustard; The water content of each vegetative organ of the five plants increased from bottom to top, and the general rule was: stem and leaf > root.

(3) 5. The content of nitrogen in the above water part of the planting object was 31.31 ~ 56.13 $g \cdot kg^{-1}$, potherb mustard > garlic > celery > oats > garland chrysanthemum, the content of nitrogen in the plant under water was 23.69 ~ 55.89 $g \cdot kg^{-1}$, garlic > potherb mustard > celery > garland chrysanthemum > oats; The phosphorus content of the above water part of the five plants was 1.11 ~ 1.83 $g \cdot kg^{-1}$, celery > potherb mustard > garland chrysanthemum > garlic > oatmeal, the phosphorus content of the under water part of the plants was 1.19 ~ 1.73 $g \cdot kg^{-1}$, potherb mustard > celery > garlic > garland chrysanthemum > oatmeal. The distribution law of nitrogen and phosphorus content in the parts of plants above and under water is different,. The contents of nitrogen and phosphorus in celery and chrysanthemum showed the distribution of above water part > under water part, and nitrogen and phosphorus were transported to the above water part; The distribution of total nitrogen and total phosphorus in garlic was underwater part > underwater part; The content of total nitrogen in potherb mustard and Brassica oleracea showed the distribution of above water part > under water part. Nitrogen was transported to the aboveground part, while the content of total phosphorus was under water part > above water part, mainly concentrated in the root.

(4) 5 the accumulation of nitrogen and phosphorus in the planting objects ranged from 1.15 to 5.18 $g \cdot m^{-2}$ and 0.04 to 0.21$g \cdot m^{-2}$, respectively. The order of magnitude was garlic > celery > potherb mustard > garland chrysanthemum > lettuce and celery > garlic > potherb mustard = garland chrysanthemum > lettuce. The accumulation of nitrogen and phosphorus of the five plants has a good correlation with the biomass. The biomass can reflect the accumulation of nitrogen and phosphorus, which is conducive to calculating the content of nitrogen and phosphorus absorbed by plants through harvesting and weighing.

References

1. Liqiufen, xinfuyan, zouyuxia, et al Study on factors affecting the growth of bacteria in shrimp ponds [J] Journal of fisheries, 2001,25 (5): 438-442
2. Zhou Jinfeng, Wen yanmao Impact of aquiculture in the foundation pond of the Pearl River Delta on the water environment [J] Journal of Sun Yat sen University (Natural science edition), 2004,43 (5): 103-106
3. Dhirendra P T.Water quality and nutrient budget in closed shrimp culture systems[J]. Aquacultural Engineering,2003,27:159-176.
4. Abdul Q S.Nutrient budgets in tanks with different stocking densities of hybrid tilapia[J]. Aquaculture,1999,170:245-252.
5. Christopher J.Nitrogen budget and effluent nitrogen components at an intensive shrimp farm[J]. Aquaculture,2003,218:397-411.
6. France S J,Nowak B F,Allan G L.The effects of nitrite on the short term growth of silver perch[J]. Aquaculture,2000,183:9-103.
7. Haoguiyu, huangminsheng, et al Department of environmental science and technology, East China Normal University [J] Shanghai: water purification technology, 2004.2
8. Network information: http://www.lbl.gov/NABIR/generalinfo/glossary.html

9. Huqingwu, Zhu Liang Bioremediation technology applied to the treatment of polluted environment [J] Shaanxi environment, 2002,9 (12): 32 ~ 35
10. Xu Yatong, Zhang Ming Microbial engineering for pollution control [M] Chemical Industry Press, Beijing 2001:146 ~ 158.
11. Loumin, liaobohan, Liuhongyu, et al Study on treatment of eutrophic water body by three aquatic floating plants [J] Chinese Journal of ecological agriculture, 2005,13 (3): 194-195
12. Wangliqing, Li Yan, zhangruilei Study on purification effect of six submerged plant systems on water quality of Dianshan Lake [J] Journal of Agricultural Environmental Sciences, 2008,27 (3): 1134-1139

Drilling Fluid Technology for the Third Interval of Sunan 23-113H2 Well

Hao Wang, Yan Wei

Drilling & Production Engineering Technology Institute, CNPC Chuanqing Drilling Engineering Company Ltd., Xi'an, Shaanxi, 710018, China

Abstract. In the construction process of Sunan 23-113H2 well, there wasYanchang well, Liujiagou well and Shihezi well, when passing through the well, the mudstone collapse of the large slope section caused the difficulty of reaming off, after repeated reaming off, the wellbore still cannot be unblocked. Finally, the well structure was changed from the conventional Third Interval to the optimized Third Interval, and the technical casing is lowered to the depth of the inclined 45° well in the inclined shaft section. After the Third Interval, the mudstone of the large inclined section and the horizontal section collapsed, and the well leakage of the inclined well section and the horizontal section was still prominent. According to the construction situation of the Third Interval of the well, this paper optimizes the drilling fluid and refines the leak prevention and plugging measures through indoor experiments to provide technical support for safe drilling.

Key words: Sunan horizontal well; Third Interval; anti-collapse; sealing up; drilling liquid

1. Introduction

In First Interval and Second Interval, there were well leakage phenomenon in Sunan 23-113H2 well, and the risk of mud and rock collapse and leakage is high, which cannot solve the stability problem of the well wall, and the collapse contradiction is prominent[1-3]. After the Third Interval, it still faces the collapse of mudstone and horizontal section, the collapse of the inclined well section and the horizontal section, and the contradiction between collapse and leakage is still prominent. According to the construction of Sunan 23-113H2 well, the best treatment agents were determined by screening the field treatment agents and classifying them according to their efficacy[4-6]. Different drilling fluid formulations were designed to test the different formula density, API water loss, HTHP-FL(Filter loss under high temperature and high pressure), apparent viscosity, plastic viscosity, dynamic shear force, etc., to determine the best drilling fluid formulation, the system formula and plugging measures were optimized, and to conduct field experiments to provide technical support for safe drilling[7-10].

2. Materials and methods

2.1 The treatment agents were classified and screened according to their different functions

Materials used for the study such as: KCl(Sinopharm Co., Ltd.); XCD; water loss agent: NAT-20, PAC-LV; wall blocking: NFA-25, FT-401, ZDS, prehydrated clay; aggravating agent: barite.

2.2 Performance control requirements of each well section

Slant well section: ρ: 1.30-1.35 g/cm^3; Fv 50-60 s; FL≤3 mL; HTHP≤12 mL; Φ6: 5-7.
Horizontal segment: ρ: 1.35-1.36 g/cm^3; Fv: 65-75 s; FL≤3 mL; HTHP≤12 mL; Φ6: 6-8.

3. Results and discussion

3.1 Indoor formula experiments, to determine the amount of treatment agent added

After rolling all drilling fluids at 120℃ for 4 h, the performance of the drilling fluids were tested, the results were shown in Table 1.

Table.1 Formulation experiment

	Formula	Density/g.cm^{-3}	API water loss/mL	HTHP-FL/mL	apparent viscosity	plastic viscosity	dynamic shear force	Φ6
1#	water+0.15%+0.1%XCD+1%NAT-20+3%NFA-25+2%ZDS+8%KCl+barite	1.30	4.2	16.8	21	15	6	3
2#	water+0.15%+0.1%XCD+1%NAT-20+3%NFA-25+2%ZDS+10%KCl+barite	1.30	4.6	17.2	14	11	3	2
3#	water+0.15%+0.1%XCD+1%NAT-20+1%PAC-LV+3%NFA-25+2%ZDS+10%KCl+barite	1.30	4.0	16.2	17	13	4	2
4#	water+0.15%+0.1%XCD+1%NAT-20+1%PAC-LV+2%NFA-25+2%FT-401+2%ZDS+10%KCl+barite	1.30	3.6	14.4	24	16	8	4
5#	water+0.15%+0.1%XCD+1%NAT-20+1%PAC-LV+2%NFA-25+3%FT-401+2%ZDS+15%KCl+barite	1.30	3.2	12.6	30	22	8	5
6#	water+0.15%+0.1%XCD+1%NAT-20+1%PAC-LV+2%NFA-25+3%FT-401+2%ZDS+15%KCl+2%Pre-hydrated drilling fluid+barite	1.30	3.0	11.8	32	24	8	6
7#	water+0.15%+0.1%XCD+1%NAT-20+1%PAC-LV+2%SMP-2+3%FT-401+2%ZDS+15%KCl+2%Pre-hydrated drilling fluid +barite	1.30	3.0	12.0	34	24	10	6

Through the above experiments, the performance of 6# and 7# drilling fluids meet the construction requirements, Therefore, the final drilling fluid formula was determined as: water+0.15%+0.1%XCD+1%NAT-20 + 1%PAC-LV +2%NFA-25 （2%SMP-2） +3%FT-401 +2%ZDS +15%KCl +2%Pre-hydrated drilling fluid.

3.2 Water loss performance test

When the drilling fluid circulates in the wellbore, with the increase of density, the liquid injection pressure will increase, if the quality of the mud cake is poor, the water loss is large, the filtrate pressed into the formation is gradually increasing, exacerbating the collapse of the debris formation prone to collapse, so while increasing the density of the drilling fluid, it is necessary to reduce the water loss, the medium pressure water loss control is less than 3.0 mL, and the high temperature and high pressure water loss is less than 12.0 mL[11,12].

Table.2 Medium-pressure water loss value at different pressures

Pressure(MPa)	Compared to 0.7 MPa pressure added value(MPa)	Medium pressure water loss(mL)
0.7		3.0
0.8	0.1	3.1
0.9	0.2	3.3
1.0	0.3	3.5

Table.3 Liquid injection pressure at the same vertical depth and different density (vertical depth 3780m)

Density (g.cm^{-3})	Compared with the added value of the density of 1.30 g/cm^3 (MPa)	Fluid Injection Pressure(MPa)
1.30		48.16
1.32	0.74	48.90
1.34	1.48	49.64
1.36	2.22	50.38

3.3 Optimize the displacement and reduce the annular air pressure consumption

Ring-air pressure consumption was calculated at a well depth of 4000 m, a density of 1.35 g/cm^3, and a plastic viscosity of 40 mPa·s.

Table.4 Different displacement and annular return speed

Cylinder size(mm)	Pump rushed	Displacement(L/S)	Ring empty return speed(m/s)	Ring air pressure consumption(MPa)
100	108	12.2	1.20	3.574
110	108	15	1.48	5.184
	95	13	1.28	4.334
120	108	17	1.68	6.494

Through the above calculation, the displacement plays a decisive role in the size of the annular air pressure consumption, so when drilling normally, the annular air return speed can reach more than 1.2 m/s, and the displacement is 12-13 L/s.

3.4 Anti-spill and plugging measures

Circulating tank is always prepared with 35 m^3 sealing drilling fluid, formula: drilling fluid+ 10% fine plugging, immediately after the leakage is lost, immediately pumped into the plugging drilling fluid, all replace the water eyelet, drilled into the casing for squeezing or static plugging. When increasing the density, each 1 ton of barite mixed with 200 kg of leak plugging agent with drilling, 1 ton of plugging agent with drilling is added to each tank of supplementary pulp, and single seal and KSD-1 are selected for plugging agent with drilling[13-15].

Table.5 Different effects of drilling plugging agent on the performance of drilling fluid

Formula	Density (g/cm³)	Medium pressure water loss (mL)	High temperature and high pressure water loss (mL)	Apparent viscosity (mPa·s)	Plastic viscosity (mPa·s)	dynamic shear force (Pa)
Drilling fluid	1.36	2.6	10.8	40	30	10
Drilling fluid+0.5%DF-A	1.36	2.6	11.2	43.5	32	11.5
Drilling fluid+0.5%KSD-1	1.36	2.6	11.2	48	35	13
Drilling fluid+0.5%GT-MF	1.36	2.6	11.4	51	36	15

3.5 Solid phase control requirements

All drilling fluids were sieved, and the vibrating sieve screen cloth is selected at more than 180 meshes. Check the sediment volume of the conical tank per shift, reach 1/2 and immediately clear the tank. After 3-4 hours of use of the high-speed centrifuge, it is necessary to remove harmful solid phases such as inferior soil in the system. When preparing the supplemental drilling fluid, glue + KCL is mainly used for system maintenance, less barite is consumed, and the supplementary drilling fluid density is adjusted according to the system density[16-18].

4. Field application

4.1 Prepare drilling fluid before Third Interval: 1.31/60/4.0/9, HTHP-FL: 12.2 mL, Then a small experiment was carried out to further adjust the performance. After adjusting the performance: 1.32/90/3.2/9, HTHP-FL: 11.2 mL, then start the Third Interval[19].

Table.6 Performance of Third Interval drilling fluid in Sunan 23-113H2 well

Well deflection	Density (g/cm³)	Medium pressure water loss(mL)	High temperature and high pressure water loss(mL)	Apparent viscosity (mPa·s)	Plastic viscosity (mPa·s)	dynamic shear force(Pa)	Φ6
45	1.32	3.2	11.2	67	47	20	6
55	1.33	3.0	11.0	71	51	20	7
65	1.34	3.0	10.0	65	47	18	6
75	1.35	2.8	10.0	64	46	18	7
Entrance window	1.35	2.8	9.8	59	42	17	7
Horizontal segment	1.35-1.36	2.6-2.8	8.0-9.0	58	43	15	6-8

4.2 After tuning the performance in the Third Interval, the viscosity of the ground funnel is high (82-98s), heated to 50 °C, 95 °C, rheology was measured, there was no obvious change, so as to analyze the performance of the drilling fluid in the well after circulating heating, and insist on monitoring the rheological parameters at the same temperature of heating on a daily basis during the later construction, and adjust it in time after the abnormality occurs[10].

Table.7 The rheological energy of third drilling fluid in the Sunan 23-113H2 well at the same temperature

Well total depth (m)	Heat the raw drilling fluid to 50℃				Heat the raw drilling fluid to 95℃			
	Funnel viscosity(s)	Apparent viscosity (mPa·s)	Plastic viscosity (mPa·s)	dynamic shear force (Pa)	Funnel viscosity (s)	Apparent viscosity (mPa·s)	Plastic viscosity (mPa·s)	dynamic shear force (Pa)
3700	86	55	40	15	63	41	30	11
3800	88	59	44	15	62	40.5	30	10.5
3900	82	64	40	14	64	40.5	30	10.5
4000	85	56	41	15	62	40	29	11
4200	87	58	44	14	61	39	29	10
4400	85	54	39	15	62	40	29	11
4600	78	53	37	16	57	36	27	9
4800	79	49	36	13	57	38	29	9
4900	78	51	36	15	59	37	28	9
5000	76	54	38	16	56	33	24	9
5100	76	56	41	15	58	36	27	9
5200	77	59	42	17	59	38	28	10
5300	78	59	43	16	56	37	28	9

4.3 The third drilling process occurred two leakage in the drilling process, one squeeze and seal, one time with the drilling, only the drill pipe is drilled to plug the leak without drilling..

Table.8 Statistics on the Third Interval leakage of Sunan 23-113H2 Well

Well total depth (m)	Leakage speed (square / hour)	Properties of drilling fluid	Treatment measure	Lost time (h)
3965	10-12	1.35/86/2.8/9	With 30 m^3 plugging drilling fluid, 1 ton KSD-1 + 1 ton single seal + 1 ton of sawdust, plugging drilling fluid will fully cover the naked eye section into the casing extrusion, the maximum sleeve pressure 9.0MPa, hold pressure 30min sleeve pressure 4.8MPa, 1 hour sleeve pressure does not drop, well cycle, normal consumption, resume drilling.	9
4727	10-20	1.36/88/2.6/9	Add 1 ton of DF-A with drill and drive high speed centrifuge for 3h, consumption is normal and resume drilling.	3

4.4 After the Third Interval session, monitored the solid phase content every day, opened the centrifuge to clear the solid phase in time according to the solid phase content, and cleaned up the conical tank for each drill.

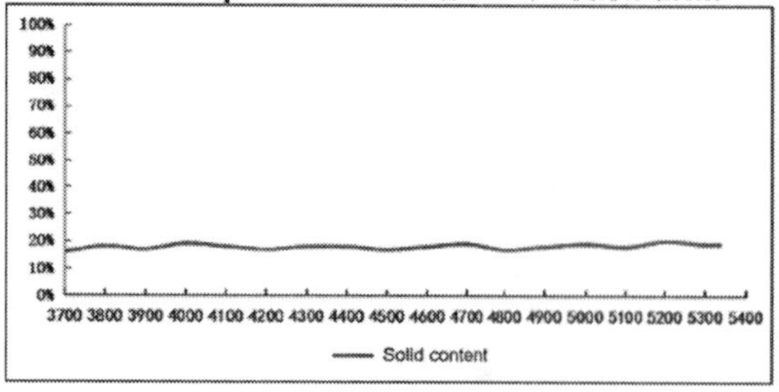

Fig.1 Solid phase content of Sunan 23-113H2 Well

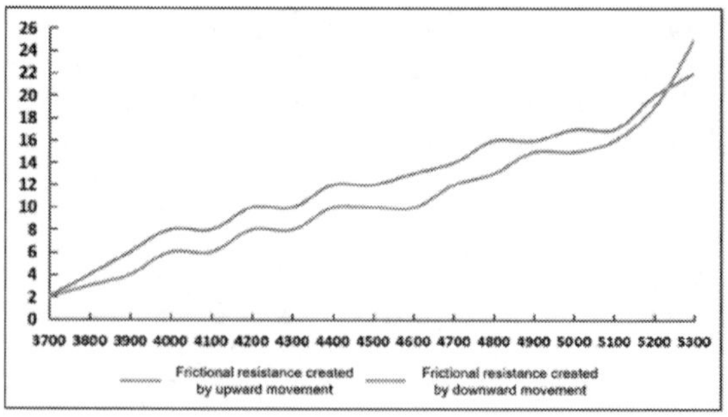

Fig.2 Frictional resistance created by upward movement and downward movement in the third drill of Sunan 23-113H2 Well

4.5 After the well is drilled into the window for 20m, it encountered dark gray mudstone, and the drilling fluid treatment ensured the KCL content of the system at 15% at first, while enhancing the blockage of the system, the medium pressure water loss < 3.0mL, the high temperature and high pressure water loss < 12mL, the mud cake was dense and smooth, and there was no obstacle in the drilling mud rock well section.[20].

Table.9 Horizontal section mudstone condition of Sunan 23-113H2 Well

Lithology	Well section(m)	Segment length (m)	Gamma	Function	HTHP-FL/mL	Up and down the drill
Dark gray mudstone	4047-4062	15	170-190	1.36/90/2.6/9	9.6	Normal
Dark gray mudstone	4206-4269	63	160-195	1.36/92/2.6/9	9.2-9.4	Normal
Dark gray mudstone	4280-4297	17	170-190	1.36/89/2.6/9	9.0-9.2	Normal
Gray mudstone	4297-4306	9	150-170	1.36/86/2.6/9	9.0-9.2	Normal
Dark gray mudstone	4306-4318	12	170-193	1.36/88/2.6/9	9.0-9.2	Normal
Gray mudstone	4332-4364	32	160-170	1.36/85/2.6/9	8.8-9.0	Normal
Dark gray mudstone	4364-4470	106	170-190	1.36/88/2.6/9	8.8-9.0	Normal
Dark gray mudstone	4730-4734	4	160-180	1.36/86/2.6/9	8.7-8.8	Normal

https://doi.org/10.1051/e3sconf/202235801005

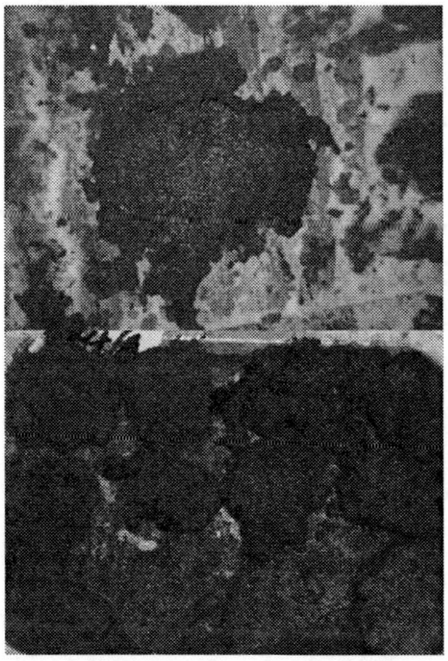

Fig.3 Dark grey mudstone of the horizontal section of Sunan 23-113H2 Well

Fig.4 High temperature and high pressure for mud cake and water loss

Fig.5 Medium-pressure water loss and mud cake

5. Conclusions and suggestions

The drilling fluid formula was determined as: water +0.15% +0.1%XCD + 1%NAT-20+1% PAC-LV +2%NFA-25(2%SMP-2) +3%FT-401 +2%ZDS +15%KCl + 2%Pre-hydrated drilling fluid. The KCl content of large slope well section reacheed 15%, and the system had strong inhibition and can effectively prevent mudstone collapse. The drilling fluid improved the system pluggability, reduced water loss, and can effectively reduce the collapse pressure of the formation, which can reduce the density of drilling fluid and reduce the risk of well leakage. It is very important to reasonably select the displacement of small wellbore. On the premise of meeting the requirements of sand carrying, the displacement should be reduced as far as possible, which can greatly reduce the annular circulation pressure consumption and prevent the leakage of horizontal section Wells. When adjusting the performance of the system, it is necessary to carry out small experiments first to avoid blind operation and causing performance adjustments to be in place for a long time, causing downhole complexity.

References

1. Chi Z, Dong FX, Xue XG, et, al. Typical geological characteristics of volcanic mechanism of Yingcheng Formation in southern Songliao Basin [J]. Journal of Jilin University (Earth Science edition). 2019(06).
2. Liu YX. Songyang gas field low density low damage with drilling plugging drilling fluid technology [J]. Drilling fluid and completion fluid. 2019(04).
3. Wang Y, Deng SW, Fan J, Zou Xp, Yang J. Geological conditions, resource potential and exploration direction of key fractured natural gas in southern Songliao Basin [J]. Natural Gas Earth Sciences. 2018(10).
4. Cao Y, Gao Sl, Qiao Xy, et, al. Natural gas formation and accumulation of Changling Broken yingcheng Formation in southern Songliao Basin [J]. Journal of Xi'an Shiyou University (Natural Science edition). 2018(04).
5. Wang LL, Wang WZ, Zhang K, et, al. Synthesis and performance evaluation of high-temperature filter loss lowering agent for water-based drilling fluid [J]. Drilling fluid and completion fluid. 2016(02).
6. Wang PJ, Chen CY, Zhang Y, et, al. Characteristics and effective reservoir distribution rules of Changling Broken Volcanic Rock Reservoir in Songliao Basin [J]. Gas industries. 2015(08).
7. Mao H, Qiu ZS, Fu JG, et, al. reparation of polymer-based nano-SiO2 and drilling fluid properties [J]. Journal of Central South University (Natural Science Edition). 2015(07).
8. Xia P, Cai JH, Fan ZJ, Wang CH. Effect of nanosilica on the properties of saline drilling fluid [J]. Drilling fluid and completion fluid. 2015(03).
9. Shi BZ. Underbalanced drilling fluid technology of igneous rock formation of YS-2 in [J]. Drilling fluid and completion fluid. 2014(06).
10. Liu X, Luo YF, Wang J, Li L. Test method and treatment technology for CO2 contamination of drilling fluid [J]. Drilling process. 2009(06).
11. Zhu K, Liu XF. Effect of inorganic salts on the performance of non-dispersed drilling fluids [J]. Journal of Lanzhou Institute of Technology. 2018(06).

12. Qiu CY, Ye HC, Wang XS, et, al. Technology and complex treatment of No 1 [J]. Drilling fluid and completion fluid. 2017(06).
13. Yao LX, Gan ZG. Research and Application of Composite Salt Drilling Fluid System [J]. China Science and Technology Information. 2017(23).
14. Li K, Zhang LW, Qi SB, Peng SB. Application of Sodium chloride in Bohai Drilling and Completion Fluid System [J]. Chemical industry management. 2017(03).
15. Liu P, Hu Y, Li WP. Research status and development trend of low-permeability reservoir injury and release technology [J]. Inner Mongolia Petrochemical Industry. 2016(08).
16. Yang YD, Xue LY, Sun XR, et, al. Multiple cooperative inhibition mechanism of water-based drilling fluid on the southern edge of Junggar [J]. Drilling fluid and completion fluid. 2015(05).
17. Huang QW, Xu JW, Wang XH, et, al. organic salt suppresses clay hydration expansion application [J]. oilfield chemistry. 2013(04).
18. Li GH, Shen JW, Jiang P. Optimization preparation of emulsified paraffin and its filtration loss reduction [J]. Drilling fluid and completion fluid. 2012(02).
19. Zhang QZ, Li CC, Zhang Y, Gu XF, Xiong XY. Progress of humic acid [J]. Journal of Xi'an Petroleum University (Natural Science edition). 2010(04).
20. Zhang JT, Zuo FJ, Zhang YX, et, al. Development and evaluation of emulsified paraffin [J]. Field, Drilling fluid and completion fluid. 2008(04).

Effects of road slope on emission characteristics for a heavy-duty diesel vehicle based on engine-in-the-loop methodology

Su Li [1], Xiaowei Wang[1], Tao Gao[1], Yongzhan Huo[1], Yang Lu[2], Mingda Wang[3,*]

[1] CATARC Automotive Test Center (Tianjin) Co., Ltd., Tianjin 300300, China.
[2] Beijing Municipal Ecological and Environmental Monitoring Center, Beijing, 100012, China 3- Chinese Research Academy of Environmental Sciences, Beijing, 100012, China

Abstract: A dump truck with a maximum designed total mass of 24500 kg was selected to measure the emission characteristics of pollutants and carbon dioxide (CO2) under fully-loaded and unloaded conditions at the same simulated road with different road slope, same driver and vehicle model by using the engine-in-the-loop methodology. The results show that driving at gradient road will result in an increase in power and total emissions. The brake specific emissions of gaseous pollutant and CO2 are lower at gradient road, while the brake emission of particle number are higher at gradient road. Road slope affects exhaust temperature so as to affect nitrogen oxide (NOx) emissions. Under unloaded conditions, long-slope conditions are more likely to cause an increase in NOx emissions.

Key words: Heavy-duty diesel vehicle, Pollutant emission, CO2 emission, Engine-in-the-loop, Road slope

1. Introduction

In 2020, carbon monoxide (CO), hydrocarbons (HC), nitrogen oxides (NOx), particulate matter (PM) emissions from commercial vehicles accounts for 29.8%, 26.6%, 84.3% and 90.9% respectively of the total vehicle emissions in China[1]. Meanwhile, the carbon dioxide (CO2) emissions of commercial vehicles accounted for 61.5% of the total vehicle CO2 emissions in 2019 [2]. Therefore, It is urgent to control the pollutant emissions and carbon emissions from commercial vehicles. The Chinese government issued the "Limits and measurement methods for emissions from diesel fueled heavy-duty vehicle (Chinese VI)" in 2018 which significantly tightened various pollutant emission limits [3]. The future fuel consumption standard of heavy-duty commercial vehicles will also significantly tighten the fuel consumption limits [4].

Many factors will affect pollutant emissions and CO2 emissions [5-6], including driving behavior, traffic flow conditions, road conditions, environmental conditions, vehicle loads, etc. But among them, the influence of road gradient is easier to ignore. This is mainly because the current fuel consumption and emission standards in China do not take into account the influence of road gradients. However, in the calculation of carbon emissions based on Vehicle Energy Consumption Calculation Tool (VECTO) in Europe, the road slope is taken into account [7]. Moreover, the current traditional vehicle energy consumption model is mainly based on the vehicle specific power (VSP) method, which also considers the influence of the road slope. Changes in road slope will affect the road resistance, resulting in changes in power and CO2 emissions. However, the impact of road slope on emissions is more complex. ZHANG et al. [8] studied the relationship between road slope less than 4% and heavy vehicle emissions and found that CO emission is the most sensitive to road slope. Prati et al. [9] conducted an actual road test on a light-duty diesel vehicle and found that uphill slopes lead to an 85% and 33% increase in CO2 and NOx emissions, while downhill slopes lead to a 45% and 60% reduction in CO2 and NOx emissions respectively. Xu et al. [10] studied the actual road emissions of a light-duty vehicle and a heavy-duty vehicle and found that the CO2 and NOx emissions of both vehicles increased with the increase of the road slope. However, these studies are based on actual road tests which are greatly affected by factors such as traffic conditions, environmental conditions, and driving behaviors. It is difficult to ensure the repeatability and consistency of the tests, and it is not suitable to conduct research on the impact of a single variable such as road slope on emissions. Moreover, these studies only considered the emission of gaseous pollutants, but did not consider the emission of particulate matter. Because the slope resistance is closely related to the vehicle loading, the effect of vehicle loading on emissions under different road slope still need to be studied.

In this context, this paper selected a dump truck with a maximum designed total mass of 24500 kg to study the emission characteristics of pollutants and CO2 under different loading and road slope based on an engine-in-the-loop (EIL) methodology.

* Corresponding author: wangmingda@vecc.org.cn

https://doi.org/10.1051/e3sconf/202235801006

2. Experimental setup

2.1 Methodology and test equipment

EIL is a special hardware-in-the-loop method which regards the engine as the actual physical hardware, while the vehicle, the driver and the road is defined by virtual model [11-13]. The EIL can then finish the vehicle tests on the engine test bed with good repeatability and consistency due to the accurate control on the laboratory environment. EIL is the most suitable method for studying the impact of single variable such as road slope on emissions. The engine-in-the-loop test platform can be seen in Figure 1.

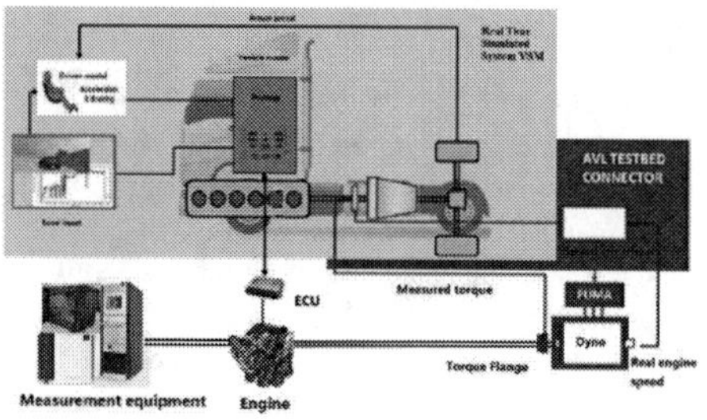

Figure 1 Engine-In-the-Loop test platform

The vehicle and driver models were constructed through the AVL VSM™ real-time system. And the road model considered the road slope. Boundary conditions such as environmental temperature and humidity were well controlled to ensure a consistent environment for each test. Pollutant emissions and CO2 emissions are obtained through the AVL AMA i60 and AVL 489 particle counter. Details of the test equipment and software are listed in Table 1.

Table 1 Test equipment and software

Equipment name	Equipment Type and Manufacturer	Accuracy
AC Dynamometer	AVL INDY P44	Torque: ± 0.3%F.S. Speed: ±1rpm
Intake air temperature conditioning	AVL Air Conditioning System 2400	Pressure: ± 1mbar Temperature: ± 0.5℃ Humidity: ± 3[%RH]
Gaseous emission measurement	AVL Emission Bench AMA i60	±2%
Particle number (PN) measurement	AVL 489	±10%
Fuel consumption measurement	AVL 753C/735S	±0.12%
Vehicle model system	AVL VSM™	
Real time system	AVL Testbed CONNECT™ (RT)	
Coastdown evaluation software	AVL Coastdown manager	

2.2 Vehicle and engine specifications

The vehicle is an N3 non-city dump truck, with a curb weight of 9000 kg, a maximum total mass of 24500 kg and a manual transmission with 10 forward gears. The rated power and the maximum torque of the equipped engine is 180kW and 1140Nm, respectively. Exhaust gas re-circulation (EGR), diesel oxidation catalyst (DOC), diesel particulate filter (DPF), selective catalytic reducer (SCR) and ammonia catalyst (ASC) make up the emission control system to meet the China VI emission standard. The road resistance coefficients under fully-loaded and unloaded are measured in the test field following the procedures of GB27840-2011 [12]. Table 2 listed the parameters of vehicle and engine used in this paper.

Table 2 Main parameters of tested vehicle and engine

Engine capacity	6.234L
Cylinder number	6
Compression ratio	17.2
Rated power/speed	180 kW/2300 rpm
Maximum torque/speed	1140 Nm/1100 - 1800rpm
Idle speed	650 rpm
Emission Control Technology Route	EGR+DOC+DPF+SCR+ASC
Emission Standards	China 6
Curb mass	9000 kg
Drive type	6×2
Tire type/number	11.00R22.5/8
Maximum total mass	24500 kg
Transmission system	10-speed manual
Final drive ratio	5.286

2.3 Test scenarios

The selected road is a PEMS test route of the corresponding N3 non-urban vehicle, located in Jianshui County, Yunnan Province, as shown in Figure 2. The total distance of the PEMS route is about 120 kilometers, which ran back and forth four times along a section of about 30 kilometers trip. The maximum altitude change for this section of the trip was more than 200 meters which brings a large slope change.

Figure 2 Schematic of the actual road at Jianshui County, Yunnan Province

The variation of vehicle speed, altitude and road slope with running distance is shown in Figure 3. The total running distance is 31.58 km, the average speed is 33.36 km/h, and the maximum speed is 75.53 km/h.

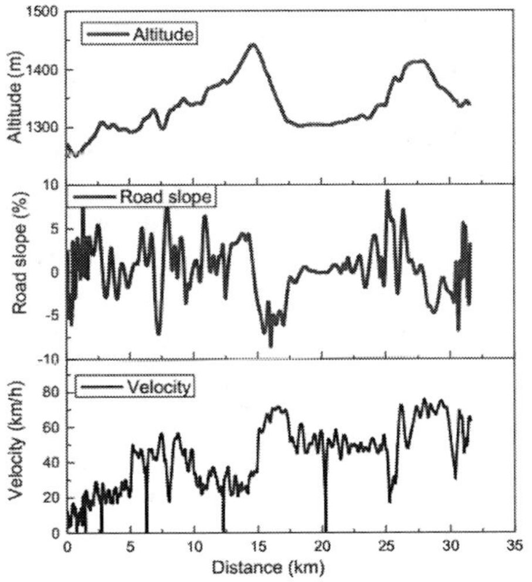

Figure 3 The variation of vehicle speed, altitude and road slope with running distance

The road slope is calculated using the road modeling software of AVL VSM^{TM}, which constructed a terrain surface model based on GPS data to correct altitude data. The model construction schematic diagram is shown in Figure 4. The parameter selection for road modeling is based on the AVL recommendation without modification.

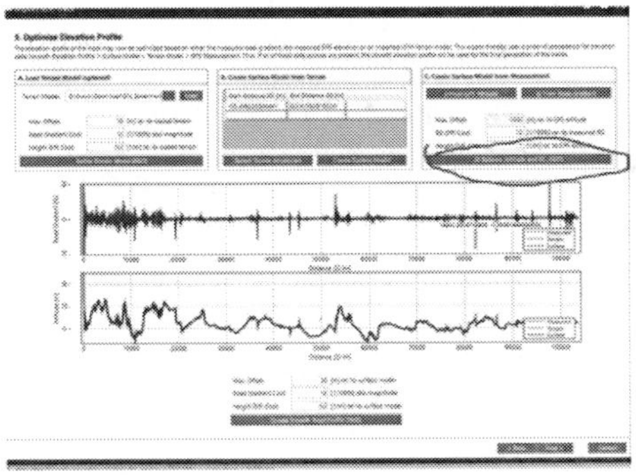

Figure 4 Schematic diagram of slope calculation and correction based on AVL VSMTM

Four test scenarios were selected, namely, fully-loaded with slope (Test 1), fully-loaded without slope (Test 2), unloaded with slope (Test 3), and unloaded without slope (Test 4). Where no slope means directly replacing the slope value with 0 in the road model. For each test scenario, the same vehicle model (but different vehicle weight and road resistance coefficients) and the same driving strategy were used for testing. Calculation methods of pollutant emissions and CO2 emissions followed the requirements of GB17691-2018 [3].

3. Results and Discussions

3.1 Pollutant and CO2 emissions

Figure 5 shows the power, brake specific emissions of CO2 and pollutant of the whole trip for the four test scenarios. Under fully-loaded conditions (corresponding to Test 1 and Test 2), the power produced by running the same trip with and without slope is 47.85 and 29.82 kW, respectively. Running at gradient road increases power by 60% compared to running without road slope, while total CO2 emissions only increase by 50%. Therefore, the CO2 brake specific emission of running with slope is 6% lower than that of no gradient. The CO, THC and NOx brake specific emissions of running with gradient are 80%, 63% and 73% lower than those of running without gradient, respectively. While the PN ratio is 63% higher for running with slope than running without slope.

Under unloaded conditions (corresponding to Test 3 and Test 4), the tendency is similar to fully-loaded conditions. Running at gradient road increases power by 39% compared to running at road without slope, while total CO2 emissions increased by only 21%. Therefore, the CO2 brake specific emission of running with gradient is 13% lower than that of no gradient. The CO, THC and NOx brake specific emissions of running with slope are 88%, 67% and 79% lower than those of running without slope, respectively. While the PN brake specific emission is 65% higher for running with slope than running without slope. The above results show that the increase in gaseous pollutant emissions caused by the slope is not as great as the increase in power, so that the gaseous pollutant brake specific emissions of running with a slope are lower. The effect of slope on particulate emissions is greater than the effect on power.

In addition, under the condition with slope (corresponding to Test 1 and Test 3), the power, brake specific emissions of CO and PN at full load are 107%, 41% and 68% higher than those at unloaded conditions, respectively, and the specific emissions of CO2, THC and NOx at full load are 16%, 82% and 65% lower than at unloaded conditions, respectively. Similar tendency exist for the conditions without slope (corresponding to Test2 and Test 4). The power, brake specific emissions of CO and PN at full load are 80%, 25% and 71% higher than those at no load, respectively. And the brake specific emissions of CO2, THC and NOx at full load are 22%, 89% and 72% lower than those at no load, respectively.

https://doi.org/10.1051/e3sconf/202235801006

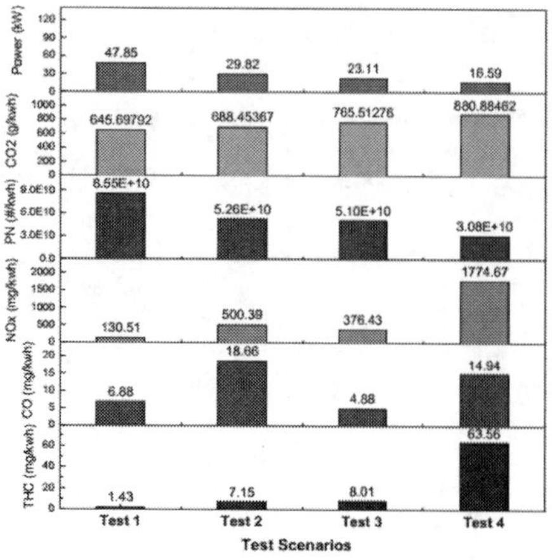

Figure 5 Power and brake specific emissions in four scenarios

3.2 Pollutant and CO_2 transient emissions

Due to the low emissions of THC and CO from diesel engines, these two emissions are no longer analyzed, and only transient emissions of NOx, PN and CO2 are analyzed. The cumulative emissions of NOx, PN and CO2 under full load conditions are shown in Figure 6. Compared with Test 2 without slope, the PN and CO2 emissions have a consistent tendency. For example, in the circled areas A, B and C in the figure, PN and CO2 emissions have increased significantly. The common feature is that these three positions are located in the acceleration phase which increase the fuel injection to meet the requirements of increased load. The corresponding circle areas D, E, and F increase significantly in NOx. The common features of D and E areas are lower vehicle speed and lower exhaust temperature. The F zone is for vehicle acceleration and hill climbing. From the characteristics of transient emissions, the effect of road slope is not as great as that of vehicle acceleration. The main reason may be that the vehicle is fully loaded at this time, and more power is required to accelerate the vehicle, so more fuel needs to be injected to achieve vehicle acceleration, resulting in an increase in PN and CO2. In addition, the influence of slope on the change of exhaust temperature is greater. In the case of no slope, the exhaust temperature is relatively stable. Only when the vehicle decelerates and idling, the exhaust temperature decreases, and when the vehicle accelerates, the exhaust temperature increases. Under the condition with slope, the change of slope will cause the change of exhaust temperature, especially under the condition of long down slope. The decrease of exhaust temperature may affect the conversion efficiency of SCR. Under full load conditions, due to the high exhaust temperature itself, the effect of this decrease in exhaust temperature on the conversion efficiency is not very significant.

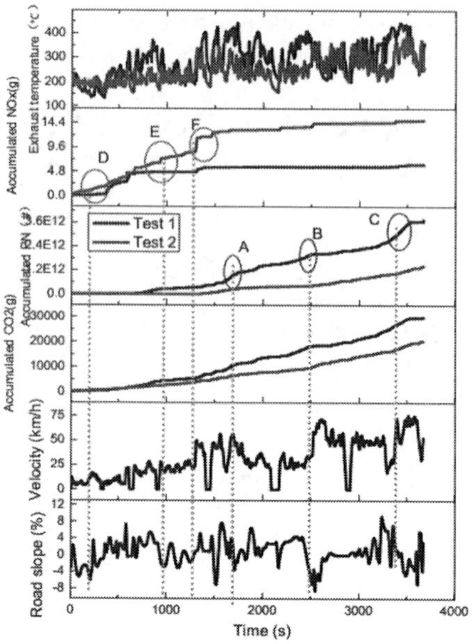

Figure 6 Cumulative emissions and exhaust temperature at full load

The cumulative emissions of NOx, PN and CO2 under no-load conditions are shown in Figure 7. Compared to the no-slope scenario, it can be seen that the areas with more obvious PN and CO2 emission growth rates are still consistent with the full load conditions, such as the circle areas A, B and C. In addition to the increase in NOx emissions in areas with lower exhaust temperatures, there is also a more significant feature that when the slope decreases but the vehicle speed increases. The exhaust temperature decreases and NOx emissions increase, as shown in the circled area F in the figure.

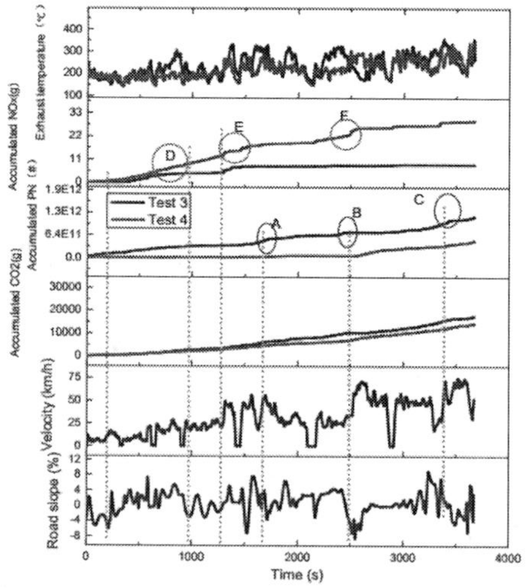

Figure 7 Cumulative emissions and exhaust temperature at no-load

4. Conclusion

The impact of the road slope on the real driving pollutant emissions and CO2 emissions of heavy-duty vehicles is studied based on EIL methodology. At the same trip, running with a slope will cause an increase in power and total emissions, but the increase in gaseous pollutant emissions is not as great as the increase in power, so the gaseous pollutants in running with slopes are lower than the emissions. PN emission of slope scenario is are higher than the scenario without slope. Slope changes affect exhaust temperature, which in turn affects NOx emissions. Under no-load conditions, long-slope driving conditions will result in increased NOx emissions.

Acknowledgments

This work was supported by Beijing Municipal Commission of Science and Technology under Grant Z191100009119014.

References

1. China mobile source environmental management annual report in 2021-Part 1: the situation of vehicle emission [J]. Ministry of Ecology and Environment of the People's Republic of China, 2021.
2. HUANG Z H, JI L, YIN J, et al. Peak Pathway of China's Road Traffic Carbon Emissions[J]. Research of Environmental Sciences, 2022, 35(02):385-393.
3. GB17691-2018 Limits and measurement methods for emissions from diesel fueled heavy-duty vehicle (Chinese Ⅵ) [S]. Ministry of Ecology and Environment of the People's Republic of China, 2018.
4. GB/T 27840-2011, Fuel consumption test methods for heavy-duty commercial vehicles[S]. China Standardization Administration, 2011.
5. WANG X W, YAN Y, JING X J, et al. Effect of Driving Pattern on RDE of Vehicle Equipped with a Port Fuel Injection Gasoline Engine[J]. Automotive Engineering, 2018 ,40(10):1146-1150.
6. SUNIL K, VINEET S, YOGRAJ S, et al. Real world vehicle emissions: Their correlation with driving parameters [J]. Transportation Research Part D 2016 (44): 157–176.
7. REGULATION (EU) 2017/2400 of 12 December 2017 implementing Regulation (EC) No 595/2009 of the European Parliament and of the Council as regards the determination of the CO2 emissions and fuel consumption of heavy-duty vehicles and amending Directive 2007/46/EC of the European Parliament and of the Council and Commission Regulation (EU) No 582/2011[S], European Commission,2017.
8. ZHANG W, LU J, XU P, et al. Moving towards Sustainability: Road Grades and On-Road Emissions of Heavy-Duty Vehicles — A Case Study. Sustainability. 2015; 7(9):12644-12671.
9. PRATI M, MECCARIELLO G, DELLA R et al. Real Driving Emissions of a Light-Duty Vehicle in Naples. Influence of Road Grade, SAE Technical Paper 2015-24-2509, 2015.
10. XU B, XING Y Y, LU J. Study on influence of road slope on exhaust emission of light and heavy vehicles [J]. Journal of Wuhan University of Technology (Transportation Science & Engineering), 2021 ,5(45):834-839.
11. WANG X W FU T Q, WANG C Q, et al. Fuel consumption and emissions at china automotive test cycle for a heavy duty vehicle based on engine-in-the-loop methodology [C]. 2020 J. Phys.: Conf. Ser. 1549 022119.
12. WANG X W, JING X J, GAO T, et al. Study on real driving fine particles emission characteristics for a heavy-duty diesel vehicle based on engine-in-the-loop methodology[J]. Automotive Engineering, 2022,44(1):58-63.
13. WANG X W, LING J, YAN F. Review on engine-in-the-loop simulation: application and development [J]. Small Internal Combustion Engine and Vehicle Technique.2019,48 （5） :79-84.

Risk evaluation of tailings dam based on game theory and matter-extension theory

Huaqiang Song, Guan Wang, Tian Luo, Zongda Yang, Jingyuan Xu, Wei Jiang

Norin Mining Limited, Beijing 100053, China

Abstract: The safety and stability of tailings dam are directly related to the safety of life and property of downstream residents and the ecological and environmental safety of downstream rivers and farmland. During the operation stage of the tailings pond, safety assessment shall be carried out from seepage damage of the tailings dam, instability of the dam foundation, structural damage of the dam body, flooded roof, and safety management, etc. Aiming at a tailings pond in the running stage of Guangxi Province in China, by establishing the evaluation index system and its grading standard of the tailings dam-break, the subjectivity and objectivity of the evaluation index are taken into consideration by the combination of matter-element extension theory and game theory. Meanwhile, in order to solve the problem of the defect of the correlation function when the eigenvalue exceeds the controlled field and of the incommensurability for different units, a linear dimensionless method is adopted. The results show that the grade of tailings dam-break is I, and the eigenvalue of grade variable is 1.3, that is, the tailings pond is a "normal tailing reservoir", but there is a risk of evolution to the "sickness tailing reservoir". Thus, corresponding measures must be taken to prevent. In addition, the result of the two methods is consistent by comparing with the unascertained measure method, which not only validates the accuracy of the evaluation method, but also embodies the unique advantage of possessing the eigenvalue of grade variable to alert its evolution direction.

Keywords: Tailings dam; game theory; dam-break; risk evaluation; matter-element extension.

1. Introduction

Tailing pond is an important facility of metal or nonmetal mining mineral concentrator used to store "waste" and is the significant major hazard in mine production. The tailing ponds in China have the characteristics of large quantity, high dam body, and large potential energy. And the number of "Trilateral Tailing Pond" and "Head of Tailing Pond" is a lot. Once an accident occurs, it will directly threaten the lives and property of the surrounding residents and cause social instability. According to statistics[1], by the end of 2012 China had a total of 12,273 tailing ponds, of which 1223 were dangerous and sick, and 2213 were inactivated, accounting for about 28%. So, the security situation is quite severe for tailing pond. Document [2] counted accidents of tailing ponds for nearly 12 years, which the dam-break accident accounted for about 56%, and other types of accidents can induce a dam-break accident. Compared with other types of accident, damage caused by the dam-break accident is the largest. For example, September 8, 2008, Xinta mine in Xiangfen County, Linfen City, Shanxi Province occurred the dam-break accident, resulting in that 271 people was killed, and 35 others were injured, and the direct economic loss was as high as several hundred million [3]. Therefore, it is necessary to study the stability of tailings dam.

Safety and stability of tailings dam are always the research focus of scholars around the world. LI Xibing et al. [4] by using fault tree analysis of tailing dam-break during flood season, found the way caused by dam-break accident and taken effective measures to prevent the flooding accident; Chen Jianhong et al. [5] through use of SLIDE software, combining with Fellenius method and Bishop method, checked and verified SLIDE software which can efficiently, accurately, intuitively reflect the slope stability of tailings dam. Wang YingBo et al. [6] through the effective identification of the safety risk of the tailings dam and the necessary analysis, the risk assessment index system of the tailings dam is established, and then the fuzzy analytic hierarchy process is used to evaluate the risk. Cai Sijing et al. [7] studied the risk of a dam break under the action of the earthquake, and used the fuzzy evaluation method.Zheng Xin et al. [8] combined with the set pair analysis theory and analytic hierarchy process to evaluate the tailings dam, which is evaluated from both qualitative and quantitative aspects, and get its security level. On the basis of the theory of vulnerability, Mei Guodong et al. [9] established the evaluation system of dam break vulnerability and the evaluation method of disaster risk loss assessment.

The safety stability and risk evaluation of tailings dam are discussed in the previous paragraph. The fault tree evaluation method can only give the risk factor which has a great impact on the dam, but does not draw substantial conclusions that the tailings dam is safe. Fuzzy comprehensive evaluation method can quantify fuzzy information, but it is more subjective. The set-pair analysis theory, although evaluated from the qualitative and quantitative aspects, was over-reliance on subjective experience. The unascertained measure evaluation method not only objectively reflects the influence degree of each evaluation index on the dam break, but can get the safety grade of the tailings dam. However, it completely ignores the influence of subjective factors such as the characteristics of each tailings dam on the evaluation according to objective data. At present, these methods have a certain effect on the prevention of the dam break, and also has some guiding significance to the actual project. However, in the process of evaluation, they exist a variety of ambiguity, complexity, and uncertainty, especially in the determination of the weight, either by subjective experience or completely objective, and risk assessment should consider both objective facts and subjective factors. Therefore, the evaluation results obtained by using these methods will inevitably be lack fidelity to some extent.

Due to its ambiguity and complexity, that is, the contradictory problem, the problem not only from the qualitative point of view that the tailings dam break caused by the impact of indicators, but also from the quantitative point of view given the value of the indicators, and ultimately the safety grade was determined. The matter-element extension method is a kind of method to solve the contradiction between the qualitative and quantitative point of view. It cannot only reflect the influence degree of each evaluation index on the dam break, but also can more accurately determine the safety grade of risk assessment. The method has been widely used in mining engineering, geotechnical engineering and other fields[10][11][12]. In addition, in order to solve the problem of complete objectivity or complete subjectivity in the evaluation, the game theory is introduced, which in the process of determining each evaluation index, not only considers the objective facts of the evaluation indexes, but also takes the subjective factors of the evaluation indexes into account.

2. Methods

The theory of matter-element extension [12,13] was put forward by Cai Wen in 1983 for how to use formal tools to solve contradictory issues and was from two aspects of qualitative and quantitative to solve contradictory problems. It mainly consists of three pillars: the basic-element theory, extension set theory and extension logic, and matter-element extension model solved contradictory problems is based on basic-element as logic cells to represent the process of contradictions. Matter-element extension method was that, firstly by determining the evaluation index system of tailings dam, a risk evaluation criteria of tailings pond dam-break was established; and classical domain, section domain, and matter-element to be evaluated were ensured; secondly, the elementary correlation function value and the weight of each evaluation index was determined by matter-element to be evaluated; finally, the risk grade of tailings pond dam-break was confirmed.

2.1 Classical domain, section domain, and matter-element to be evaluated

1) Classical domain

The theory of matter element extension is based on names of things, its attributes and values as three basic orderly elements to describe things, that is, matter element R= (names of things, characteristic of things, values) = (N, C, V).

If the risk grade of tailings dam-break is m, the classical domains can be expressed as:

$$R_j = (P_j, C_k, v_{jk}) = \begin{bmatrix} P_j & C_1 & v_{j1} \\ & C_2 & v_{j2} \\ & \vdots & \vdots \\ & C_n & v_{jn} \end{bmatrix} = \begin{bmatrix} P_j & C_1 & (a_{j1}, b_{j1}) \\ & C_2 & (a_{j2}, b_{j2}) \\ & \vdots & \vdots \\ & C_n & (a_{jn}, b_{jn}) \end{bmatrix} \quad (1)$$

Where P_j is the j-th level of risk evaluation grade of tailings dam-break (j=1,2,…,m) ; C_k is the k-th evaluation indicator of risk evaluation indicators of tailings dam-break (k = 1,2, ..., n); $v_{jk} = (a_{jk}, b_{jk})$ is the value domain of risk evaluation grade Pj of tailings dam-break for Ck.

2) Section domain

For the whole evaluation index system, section Rp is determined by evaluation indicator value of risk evaluation indicators of tailings dam-break.

$$R_p = (P, C_k, v_{pk}) = \begin{bmatrix} P & C_1 & v_{p1} \\ & C_2 & v_{p2} \\ & \vdots & \vdots \\ & C_n & v_{pn} \end{bmatrix} = \begin{bmatrix} P & C_1 & (a_{p1}, b_{p1}) \\ & C_2 & (a_{p2}, b_{p2}) \\ & \vdots & \vdots \\ & C_n & (a_{pn}, b_{pn}) \end{bmatrix} \quad (2)$$

Where P is the whole evaluation levels of risk evaluation of tailings pond dam-break; the v_{pk} is the value range of evaluation indicator C_k.

3) Matter-element to be evaluated

For risk grade P of certain tailings to be evaluated, the collection and analysis of the evaluation index data can be expressed as matter-element R_w.

$$R_w = (P_w, C_k, v_{wk}) = \begin{bmatrix} P_w & C_1 & v_{w1} \\ & C_2 & v_{w2} \\ & \vdots & \vdots \\ & C_n & v_{wk} \end{bmatrix} \quad (3)$$

Where P_w is risk grade of certain tailings to be evaluated; v_{wk} is the specific value of evaluation index C_k.

2.2 Index specifications based on membership functions

From the past research, application of matter-element extension method often encountered two problems [11]: Firstly, the eigenvalue of the matter-element to be evaluated may exceed the range of section domain, resulting in correlation functions failure; Secondly, the evaluation index of different units caused incommensurability. In order to solve these two problems, this paper adopted linear dimensionless method [14], the membership functions was expressed as:

1) Economic benefits index (the bigger, the better)

$$v_{wk}' = \begin{cases} 1 & v_{wk} \geq \max(v_{pk}) \\ \dfrac{v_{wk} - \min(v_{pk})}{\max(v_{pk}) - \min(v_{pk})} & \min(v_{pk}) < v_{wk} < \max(v_{pk}) \\ 0 & v_{wk} \leq \min(v_{pk}) \end{cases} \quad (4)$$

2) Cost index (the litter, the better)

$$v_{wk}' = \begin{cases} 1 & v_{wk} \leq \min(v_{pk}) \\ \dfrac{\max(v_{pk}) - v_{wk}}{\max(v_{pk}) - \min(v_{pk})} & \min(v_{pk}) < v_{wk} < \max(v_{pk}) \\ 0 & v_{wk} \geq \max(v_{pk}) \end{cases} \quad (5)$$

Where v'_{wk} is the k-th evaluation index value after dimensionless; v_{wk} is the k-th evaluation index value before dimensionless; max (v_{pk}) and min (v_{pk}) are respectively the maximum value and minimum value of section domain of the k-th evaluation index.

2.3 Elementary correlation functions

Elementary correlation functions values of risk evaluation of tailings pond dam-break index and the corresponding weight value were used to determine extension correlation degree, elementary correlation functions at the j-th level of the matter-element to be evaluated (v_{wk}) can be expressed as:

$$K_j(v_{wk}) = \begin{cases} \dfrac{\rho(v_{wk}, v_{jk})}{\rho(v_{wk}, v_{pk}) - \rho(v_{wk}, v_{jk})} & v_{wk} \notin v_{jk} \\ \dfrac{-\rho(v_{wk}, v_{jk})}{|v_{jk}|} & v_{wk} \in v_{jk} \end{cases} \quad (6)$$

$$\rho(v_{wk}, v_{dk}) = \left| v_{wk} - \frac{1}{2}(a_{dk} + b_{dk}) \right| - \frac{1}{2}(b_{dk} - a_{dk}) \quad (7)$$

Where $d=j$ or p, and $\rho(v_{wk}, v_{dk})$ is used to represent the distance of a point to the interval in the extension theory, that is, "extension distance".

2.4 The comprehensive weighting method based on game theory

Matter-element extension evaluation method to determine the weight values commonly used subjective weighting method (AHP, expert surveys, etc.) or objective weighting method (simple correlation function, entropy, etc.). Whether the subject weighting method or objective weighting method, only consider effects of unilateral factors. However, in the actual evaluation, the weight of evaluation index is not only affected by objective factors, but is influenced by subjective factors. In order to integrate the two influencing factors, the paper drew on the comprehensive weighting method based upon the game theory from the literature [15]. It was to optimize and integrate the objective and subjective weighting method, and finally got the comprehensive weight of matter-element extension method.

In order to improve science and rationality of evaluation index weight, and avoid subjective and objective one-sidedness, y different methods can respectively be used empowerment of evaluation index weights. So y evaluation indicators' weight vectors ($u_i = (u_{i1}, u_{i2}, \cdots, u_{im}), (i = 1, 2, \cdots, k)$) were obtained. The weight vector of arbitrary linear combination can be expressed as:

$$u = \sum_{i=1}^{k} \alpha_i \cdot u_i^T, (\alpha_i > 0) \quad (8)$$

In order to obtain a satisfactory index weight vector, formula (8) can be optimized so that deviation of u and each u_i is minimization. Thus, the game model can be expressed as:

$$Min \left\| \sum_{j=1}^{k} \alpha_j u_j - u_i^T \right\|_2 \ (i = 1, 2, \cdots, k) \quad (9)$$

And then based on the differential nature of the matrix, formula (9) can be converted into linear equations of first derivative.

$$\begin{bmatrix} u_1u_1^T & u_1u_2^T & \cdots & u_1u_k^T \\ u_2u_1^T & u_2u_2^T & \cdots & u_2u_k^T \\ \vdots & \vdots & \vdots & \vdots \\ u_ku_1^T & u_ku_2^T & \cdots & u_ku_k^T \end{bmatrix} \begin{bmatrix} \alpha_1 \\ \alpha_2 \\ \vdots \\ \alpha_k \end{bmatrix} = \begin{bmatrix} u_1u_1^T \\ u_2u_2^T \\ \vdots \\ u_ku_k^T \end{bmatrix} \quad (10)$$

Using Matlab software calculated formula (10), the solution was $(\alpha_1, \alpha_2, \cdots, \alpha_k)$. u can be expressed as:

$$u = \sum_{i=1}^{k} \alpha_i \cdot u_i^T \quad (11)$$

Then formula (11) was normalized. So, u^* can be expressed as:

$$u^* = u_i \Big/ \sum_{i=1}^{k} u_i \quad (12)$$

2.5 Extension correlation of the j-th level for tailings dam-break risk assessment

By formula (6) and (12), extension correlation with the j-th level for tailings dam-break risk assessment can be calculated by:

$$K_j(p) = \sum_{k=1}^{n} u_k^* K_j(v_{wk}) \quad (13)$$

Where u_k^* is the weight value of the k-th risk evaluation index of a tailings dam in the whole evaluation index system; K_j (v_{wk}) is elementary correlation functions value of the k-th risk evaluation index for matter-element (v_{wk}) in the j-th level.

2.6 Risk grade of tailings dam-break

According to the principle of maximum correlation degree, matter-element to be evaluated depending on the maximum value of extension correlation degree. So, the risk grade of tailings dam-break is:

$$K_{j_0}(p) = \max_{j=1}^{m} K_j(p) \quad (14)$$

By formula (14), the risk assessment grade of tailings dam to be evaluated is j_0. To more accurately determine the risk grade of the tailings dam, it can be measured by the level variable eigenvalue j^*.

$$j^* = \frac{\sum_{j=1}^{m}(m-j+1)\times \bar{K}_j(p)}{\sum_{j=1}^{m}\bar{K}_j(p)} \tag{15}$$

Where $\bar{K}_j(p) = \dfrac{K_j(p) - \min K_j(p)}{\max K_j(p) - \min K_j(p)}$ (16)

3. Results and Analyses

3.1 Geological setting

A tailing pond was built in 1998 and is a valley-type and upstream, and its annual tailings output of the concentrator is 50,000 t. Bulk density of the tailings is $2.0t/m^3$, and catchment area of the tailing pond is $1.26km^2$, and overall design height of the dam is 33.5m, and total storage capacity is 1.5 million m^3, and freeboard is 1.02 m, and dry sands length is 51 m, and the current tailings pile approximately is 320,000 m^3. The initial dam is the impervious masonry dam, and crest elevation is 458.5m, and height of the dam is 22m, and the downstream slope ratio is 1:2. Late the tailings adopted tailings dam constructed with the upstream method, and the average accumulation slope is 1: 3. The final tailings' accumulation elevation is 470m and the total height of dam about 33.5 m. At present, the tailings' accumulation elevation is 460.5m, and the total height of the dam is 24m.

3.2 Evaluation index system and its grading standard

The selection of risk evaluation indicators for the tailings dam-break is based on the previous research results and combined with the dam-break accident of tailings in the past. It could be discovered that the cause of tailings dam-break resulted from the interaction of multiple factors. However, risk evaluation indicator composed of various factors often tended to have some relevance. Thus, the relevance of risk evaluation indicators for tailings dam-break was studied, and excluded greater relevance of evaluation indicators. Besides, due to the emphasis of the actual situation and influence factors of each tailing pond was different, and combined with the characteristics of tailings, the tailings eventually formed a relatively independent risk assessment indexes system. The risk evaluation indicators of tailings dam-break were divided into 5 categories and 14 items [16,17,18].

1) Seepage failure

Seepage field has a great impact on tailings dam, and can directly induce dam-break caused by instability of the dam. Seepage failure is usually caused by seepage field occurring damage, and height of saturation line and bulk density of the tailings are the main reason for seepage failure. Therefore, the height of saturation line (denoted as C1, similarly hereinafter) and bulk density (C2) were used as the evaluation indicators to measure the seepage failure.

2) Instability of dam foundation

Instability of the dam foundation is usually a very important reason for tailings dam-break. The size of anti-sliding ability for tailings dam can directly affect the stability of the dam, while the size of anti-sliding ability is generally determined by the anti-sliding ability of dam and current situation of tailings' accumulation. However, the anti-sliding ability depended on average particle size (C3), and the existing situation of tailings' accumulation was determined by downstream slope ratio (C4), current dam height (C5) and seismic intensity (C6).

3) Damage to dam structure

The extent of damage to dam structure is an important factor influencing tailings dam-break and is generally used to measure the size of cracks for the dam. Therefore, the transverse crack coefficient (C7), the longitudinal crack coefficient (C8) and horizontal crack coefficient (C9) were used as the evaluation indicators to measure damage dam structure.

4) Flood overtopping

Due to poor permeability of tailings dam, and if the problems of flood-control and drainage emerged, when floods, it was likely to cause flood overtopping, resulting in tailings dam-break. Thus, the flood control capacity coefficient (C10), the height of beach top and the water level (C11), and drainage facility's intact coefficient (C12) were as the measuring indexes.

5) Safety management

Safety management of tailings was through the establishment of a sound safety management system, and early investigation to find various hazards of the tailings dam, then take effective preventive measures to prevent dam -break accident. It can be divided into daily safety management and emergency safety management, respectively corresponding to routine management coefficient (C13) and emergency management coefficient (C14).

Based on relevant regulations and parameters of China's "Technical Specification for Tailings pond", "Design Guidelines for Tailings Dams", "Safety Supervision Regulations for Tailings Dams", and "Technical Specifications for Safety Monitoring of Tailings Dams", and so on , and reference to the results of research on dam break risk classification of tailings reservoirs [5][7][9][17][18][20], and according to the stability, flood-control capacity and safety management of the tailings and other safety factors, the corresponding risk levels of tailings dam can be divided into four levels: Ⅰ (normal tailing reservoir), Ⅱ (sickness tailing reservoir), Ⅲ (risk tailing reservoir), Ⅳ (danger tailing reservoir). Besides, according to the membership functions, the height of saturation line, bulk density, average particle size and downstream slope ratio, etc. were economic benefits index, while the cost index contained only current dam height. So, risk evaluation criteria and membership range of the dam-break for tailings pond were shown in table 1.

 https://doi.org/10.1051/e3sconf/202235801007

Table 1 Risk evaluation criteria of dam-break for tailings reservoir

Evaluation indicators	Risk levels			
	I	II	III	IV
Bulk density (t/m³)	2~2.5(0.8~1)	1.7~2 (0.68~0.8)	1.4~1.7(0.56~0.68)	0~1.4(0~0.56)
Height of saturation line (m)	8~20(0.4~1)	6~8 (0.3~0.4)	5~6(0.25~0.3)	0~5(0~0.25)
Average particle size(mm)	0.5~1(0.5~1)	0.2~0.5 (0.2~0.5)	0.05~0.2(0.05~0.2)	0~0.05(0~0.05)
Downstream slope ratio (1: n)	5.0~10(0.5~1)	3.0~5.0 (0.3~0.5)	1~3(0.1~0.3)	0~1.4(0~0.1)
Current dam height(m)	0~20(0.8~1)	20~50 (0.5~0.8)	50~80(0.2~0.5)	80~100(0~0.2)
Seismic intensity(°)	8~12(0.75~1)	6.5~8 (0.54~0.75)	5~6.5(0.42~0.54)	0~5(0~0.42)
Transverse crack coefficient	0.75~1(0.75~1)	0.5~0.75 (0.5~0.75)	0.25~0.5(0.25~0.5)	0~0.25(0~0.25)
Longitudinal crack coefficient	0.75~1(0.75~1)	0.5~0.75 (0.5~0.75)	0.25~0.5(0.25~0.5)	0~0.25(0~0.25)
Horizontal crack coefficient	0.75~1(0.75~1)	0.5~0.75 (0.5~0.75)	0.25~0.5(0.25~0.5)	0~0.25(0~0.25)
Flood control capacity coefficient	0.75~1(0.75~1)	0.5~0.75 (0.5~0.75)	0.25~0.5(0.25~0.5)	0~0.25(0~0.25)
Height of beach top and water level(m)	1.5~2(0.75~1)	1~1.5 (0.5~0.75)	0.5~1(0.25~0.5)	0~0.5(0~0.25)
Drainage facilities intact coefficient	0.75~1(0.75~1)	0.5~0.75 (0.5~0.75)	0.25~0.5(0.25~0.5)	0~0.25(0~0.25)
Daily management coefficient	0.75~1(0.75~1)	0.5~0.75 (0.5~0.75)	0.25~0.5(0.25~0.5)	0~0.25(0~0.25)
Emergency management coefficient	0.75~1(0.75~1)	0.5~0.75 (0.5~0.75)	0.25~0.5(0.25~0.5)	0~0.25(0~0.25)

Note: the membership range of each evaluation indicator value in the brackets

3.3 Calculations

1) Classical domain

According to formula (1) and table 1, the classical domain of tailings dam-break based on matter-element extension evaluation method is

$$R_0 = \begin{vmatrix} & P_1 & P_2 & P_3 & P_4 \\ C_1 & (0.8, 1.0) & (0.68, 0.8) & (0.56, 0.68) & (0, 0.56) \\ C_2 & (0.4, 1.0) & (0.3, 0.4) & (0.25, 0.3) & (0, 0.25) \\ C_3 & (0.5, 1.0) & (0.2, 0.5) & (0.05, 0.2) & (0, 0.05) \\ C_4 & (0.5, 1.0) & (0.3, 0.5) & (0.1, 0.3) & (0, 0.1) \\ C_5 & (0.8, 1.0) & (0.5, 0.8) & (0.2, 0.5) & (0, 0.2) \\ C_6 & (0.75, 1.0) & (0.5, 0.75) & (0.25, 0.5) & (0, 0.25) \\ C_7 & (0.75, 1.0) & (0.5, 0.75) & (0.25, 0.5) & (0, 0.25) \\ C_8 & (0.75, 1.0) & (0.5, 0.75) & (0.25, 0.5) & (0, 0.25) \\ C_9 & (0.75, 1.0) & (0.5, 0.75) & (0.25, 0.5) & (0, 0.25) \\ C_{10} & (0.75, 1.0) & (0.5, 0.75) & (0.25, 0.5) & (0, 0.25) \\ C_{11} & (0.75, 1.0) & (0.5, 0.75) & (0.25, 0.5) & (0, 0.25) \\ C_{12} & (0.75, 1.0) & (0.5, 0.75) & (0.25, 0.5) & (0, 0.25) \\ C_{13} & (0.75, 1.0) & (0.5, 0.75) & (0.25, 0.5) & (0, 0.25) \\ C_{14} & (0.75, 1.0) & (0.5, 0.75) & (0.25, 0.5) & (0, 0.25) \end{vmatrix}$$

2) Section domain

According to the membership range in table 1, its section domain is

$$R_p = \begin{vmatrix} P & C_1 & (0, 1.0) \\ & C_2 & (0, 1.0) \\ & C_3 & (0, 1.0) \\ & C_4 & (0, 1.0) \\ & C_5 & (0, 1.0) \\ & C_6 & (0, 1.0) \\ & C_7 & (0, 1.0) \\ & C_8 & (0, 1.0) \\ & C_9 & (0, 1.0) \\ & C_{10} & (0, 1.0) \\ & C_{11} & (0, 1.0) \\ & C_{12} & (0, 1.0) \\ & C_{13} & (0, 1.0) \\ & C_{14} & (0, 1.0) \end{vmatrix}$$

3) Matter-element to be evaluated

According to the current actual situation with the tailings pond, and basing on its investigation and study, its parameter values and membership degree of 14 evaluation indexes for the tailings pond were shown in table 2.

Table 2 Parameter values and its membership degree of risk evaluation index

Evaluation indicators	parameter values (membership degree)
Bulk density （t/m^3）	2 （0.8）
Height of saturation line (m)	10 （0.5）
Average particle size(mm)	0.08 （0.08）
Downstream slope ratio （1: n）	2 （0.2）
Current dam height(m)	24 （0.76）
Seismic intensity(º)	7 （0.58）
Transverse crack coefficient	0.8 （0.8）
Longitudinal crack coefficient	0.8 （0.8）
Horizontal crack coefficient	0.9 （0.9）
Flood-control capacity coefficient	0.81 （0.81）
Height of beach top and water level(m)	1.6 （0.8）
Drainage facilities intact coefficient	0.8 （0.8）
Daily management coefficient	0.87 （0.87）
Emergency management coefficient	0.7 （0.7）

The membership degree of risk evaluation index for tailings dam-break in table 2, the matter-element to be evaluated for the membership degree is

$$R_w = \begin{vmatrix} P & C_1 & 0.8 \\ & C_2 & 0.5 \\ & C_3 & 0.08 \\ & C_4 & 0.2 \\ & C_5 & 0.76 \\ & C_6 & 0.58 \\ & C_7 & 0.8 \\ & C_8 & 0.8 \\ & C_9 & 0.9 \\ & C_{10} & 0.81 \\ & C_{11} & 0.8 \\ & C_{12} & 0.8 \\ & C_{13} & 0.87 \\ & C_{14} & 0.7 \end{vmatrix}$$

4) Calculation of elementary correlation function values

According to formula (6) and formula (7), elementary correlation function values of risk evaluation index for tailings dam-break at all levels can be calculated. For example, if $v_{w1} = v_{11} = 0.8 \notin (0.8, 1]$,

$$K_j(v_{wk}) = K_1(v_{11}) = \frac{\rho(v_{wk}, v_{jk})}{\rho(v_{wk}, v_{pk}) - \rho(v_{wk}, v_{jk})}$$

was selected to calculate. When $v_{11} \in (0.68, 0.8]$, $K_2(v_{11}) = \frac{-\rho(v_{wk}, v_{jk})}{|v_{jk}|}$ was chosen to calculate.

Elementary correlation function values of risk evaluation index for tailings dam-break in all levels were shown in table 3.

Table 3 Elementary correlation function values of risk evaluation index for tailings dam-break

Elementary correlation function values	Risk grades			
	Ⅰ	Ⅱ	Ⅲ	Ⅳ
K_j （v_{w1}）	0.00	0.00	-0.38	-0.55
K_j （v_{w2}）	0.17	-0.17	-0.29	-0.33
K_j （v_{w3}）	-0.84	-0.60	0.20	-0.27
K_j （v_{w4}）	-0.60	-0.50	1.00	-0.33
K_j （v_{w5}）	-0.14	0.13	-0.52	-0.70
K_j （v_{w6}）	-0.29	0.19	-0.09	-0.28
K_j （v_{w7}）	0.20	-0.20	-0.60	-0.73
K_j （v_{w8}）	0.20	-0.20	-0.60	-0.73
K_j （v_{w9}）	0.40	-0.60	-0.80	-0.87
K_j （v_{w10}）	0.24	-0.24	-0.62	-0.75
K_j （v_{w11}）	0.20	-0.20	-0.60	-0.73
Kj （vw12）	0.20	-0.20	-0.60	-0.73
Kj （vw13）	0.48	-0.48	-0.74	-0.83
Kj （vw14）	-0.14	0.20	-0.40	-0.60

5) Determination of weight

In the engineering application, AHP was used to determine the subjective weights of evaluation indexes, and the objective weights of evaluation indexes were determined by the elementary correlation function. The specific steps respectively referenced document [19] and [10]. According to the formula (10), $\alpha_1 = 0.109, \alpha_2 = 0.9075$ .Thus, the comprehensive weight value of each evaluation index after normalization was as shown in Table 4.

Table 4 Weight values of risk evaluation index for tailings dam-break

Evaluation indexes	Subjective weight values	Objective weight values	Comprehensive weight values
C1	0.0226	0.0425	0.0247
C2	0.0986	0.0755	0.0961
C3	0.0643	0.0396	0.0616
C4	0.0395	0.0566	0.0414
C5	0.0609	0.0538	0.0601
C6	0.0395	0.0587	0.0416
C7	0.0609	0.0793	0.0629
C8	0.0395	0.0793	0.0438
C9	0.0609	0.1019	0.0653
C10	0.0643	0.0838	0.0664
C11	0.0643	0.0793	0.0659
C12	0.0986	0.0793	0.0965
C13	0.1430	0.1110	0.1396
C14	0.1430	0.0595	0.1341

3.4 Results

The extension correlation of risk level for tailings dam-break can be obtained by inputting the elementary correlation function values and the comprehensive weight values in table 3 and table 4 to the formula (13). Besides, according to the principle of maximum correlation degree(formula (14)), the risk grade of tailings dam-break can be calculated. Finally, by inputting calculated values of formula (13) and (14) to formula (15) and (16), the level variable eigenvalue can be acquired (See Table 5).

Table 5 Assessment results of dam-break for tailings reservoir

Extension correlation degree	K_1 (p)	K_2 (p)	K_3 (p)	K_4 (p)	max	0	*
Values	0.0626	-0.2069	-0.4297	-0.6290	0.0626		.3

As Table 5 shown, the risk grade of the tailings dam-break was Ⅰ (normal tailing reservoir), but the level variable eigenvalue was 1.3. That is to say, the actual risk grade of the tailings dam-break was 1.3 of between I and II. By the level variable eigenvalue, the tailings pond may evolve into sickness tailing pond if decision makers did not take preventive measures. According to the comprehensive weight values of Table 4, daily management coefficient, emergency management coefficient, drainage facility's intact coefficient, the height of saturation line and flood control capacity coefficient of proportion was quite large. So, the five aspects should be strengthened to manage, preventing its evolution to the low grade.

3.5 Comparison and Analyses

In the literature [5], the concrete theory and detailed step of using the unascertained measure evaluation method to carry out the tailings dam risk assessment are introduced in detail. First, the method provides a quantitative analysis of the tailings dam by establishing an unascertained measure function of each evaluation factor; Secondly, the weight of each evaluation index is calculated by using the information entropy, and the grade is determined according to the confidence criterion. And its evaluation indicators, evaluation criteria, and weight values were shown in Table 1, Table 2 and Table 4. So, the single index evaluation matrix of the tailings is

$$(u_{ijk})_{14\times4} = \begin{bmatrix} 1 & 0 & 0 & 0 \\ 1 & 0 & 0 & 0 \\ 0 & 0 & 0.4 & 0.6 \\ 0 & 0 & 1 & 0 \\ 0.73 & 0.27 & 0 & 0 \\ 0 & 0.83 & 0.17 & 0 \\ 1 & 0 & 0 & 0 \\ 1 & 0 & 0 & 0 \\ 1 & 0 & 0 & 0 \\ 1 & 0 & 0 & 0 \\ 1 & 0 & 0 & 0 \\ 1 & 0 & 0 & 0 \\ 1 & 0 & 0 & 0 \\ 0.6 & 0.4 & 0 & 0 \end{bmatrix}$$

By single index evaluation matrix and the comprehensive weight values in Table 4, the comprehensive evaluation vector of multi-index was {0.7855, 0.1044, 0.0731, 0.037}. If confidence level $\lambda = 0.7$ and $k_0 = 07855 > 0.7$, the risk grade of the tailings dam-break was Ⅰ. Besides, $k_0 = 1 > \lambda$, the risk grade of the tailings dam-break was Ⅰ. The two discriminant results of the unascertained degree evaluation method are consistent, and its the risk grade is I.

Thus, the results of matter-element extension method and the unascertained measure method are consistent, which verifies the correctness of matter-element extension method applied to risk assessment of tailings dam-break, and is in accordance with the actual situation of the project. The difference of matter-element extension method and other methods is that it has the eigenvalue of grade variable, and can be called "the evolution of value". It not only can accurately determine the risk grade of tailings dam-break, but can determine the evolution direction of risk grade for tailings dam-break, so as to take effective measures to prevent the evolution of the risk grade from high to low.

4. Conclusions

In this paper, on the basis of China's safety standards, specifications, and regulations for tailings ponds, on-site investigation and research were conducted on a tailings pond in Guangxi Province, China. The risk of tailing dam-break is mainly embodied in five aspects, such as seepage failure, dam foundation failure, dam structure destruction, flood overtopping and safety management. By establishing evaluation index system and evaluation criteria, the theory of matter element extension and game theory are combined to evaluate its safety a. are applied to evaluate its safety level.The results show that the grade of tailings dam-break is grade I, which is normal tailings reservoir, and its the eigenvalue of grade variable is 1.3. Therefore, there is a risk of evolution to dam-break level II (ie, sickness tailing reservoir). From the evaluation results, we need to strengthen safety management, flood control facilities and flood control capacity, and take necessary preventive measures to prevent their evolution to the sickness tailing reservoir.

The method based on game theory and matter-element extension theory has advantages over other evaluation methods, and its superiority is embodied in two aspects. On one hand, the method based on game theory and matter-element extension theory takes into account the influence of both subjective factors and objective facts on evaluation; On the other hand, it has the level variable eigenvalue, namely "the evolution of value",which cannot only accurately determine the tailings dam risk grade, but also determine the evolution direction of risk level for tailings dam-break, so as to take effective measures to improve the management of technology and risk level of tailings dam, and prevent evolution of the risk grade from high to low.

The safety and stability evaluation of tailings dam is a complicated problem. It is necessary to combine monitoring data and experimental analysis data of the tailings pond, and choose an excellent evaluation method to accurately assess it. The purpose is to find out in advance the problems, so as to take necessary measures to prevent dam-break accidents.

References

1. State Administration of Work Safety and the other six departments issued work summaries about the special rectification action of the tailings in China and a notice that scheduled key work of next action about comprehensive management of the tailings [EB] [2013.12.26] http://www.chinasafety.gov.cn/
2. Zhou Keping, Liu Fuping, Hu jianhua, et al. Research of Tailings Dam-break Disaster Chain and Chain-cutting Disaster Mitigation Control Technology[J]. JOURNAL OF CATASTROPHOLOGY, 2013, 3 (28) : 24-29.
3. WANG Fei-yue. Based on the uncertainty theory of tailings dam stability analysis and comprehensive evaluation[D]. Changsha: Central South University, 2009.
4. LI Xi-bing, JIANG Wei-dong, et al. FAULT TREE ANALYSIS OF TAILINGS DAM BREAK DURING FLOOD SEASON[J]. Journal of Safety and Environment, 2001, 1 (5) : 45-48.
5. CHEN Jian-hong, ZHANG Tao, ZENG Xiang-nong, et al. Slope stability simulation modeling and safety analysis of tailings dam[J]. Journal of Central South University (Science and Technology), 2008, 39 (4) : 635-640.
6. Li Gen, Wang Laigui, Chen Lei, et al. Numerical Simulation of the Seepage Stability of A Tailing Reservoir[J]. METAL MINE, 2008 (9) : 123-125.
7. PENG Kang, LI Xi-bing, WANG Shi-ming, et al. Optimization model of unascertained measurement for dam-break risk evaluation in tailings dams[J]. Journal of Central South University (Science and Technology), 2012 (4) : 1447-1452.
8. LI Qiang, ZHANG Li-ting, QI Qing-lan, et al. Instability characteristics and stability analysis of a tailings dam based on fluid-solid coupling theory[J]. Rock and Soil Mechanics, 2012, 33 (Supp.2) :243-250.
9. JIA Chao, XIAO Shufang, LIU Ning. Application of extenics theory to evaluation of tunnel rock quality[J].Chinese Journal of
10. Rock Mechanics and Engineering, 2003, 22(5): 751 756.
11. AN Yong-lin, PENG Li-min, WU Bo, et al. Comprehensive extension assessment on tunnel collapse risk[J]. Journal of Central South University: Natural Science Edition, 2011, 42(2): 514-520.
12. SHANG Jun-long, HU Jian-hua, MO Rong-shi, et al. Predication model of game theory-matter-element extension for blast ability classification and its application[J]. Journal of Mining & Safety Engineering, 2013 (1) : 86-92.
13. CAI Wen, YANG Chun-yan, LIN Wei-chu. Extension Engineering Method [M] . Beijing: Science Press, 2000: 207 – 209.
14. YANG Chun-yan, CAI Wen. Study on extension engineering [J] . Engineering Science, 2000, 2(12): 90 – 96.
15. HE Yong-xiu, DAI Ai-ying, ZHU Jiang, et al. Risk assessment of urban network planning in China based on the matter-element model and extension analysis[J]. Electrical Power and Energy Systems, 2011, 33(3): 775-782.
16. CHEN Jia-liang. Research on evaluation of combined weight based on game theory[J]. Fujian Computer, 2003(9): 15-16.
17. LI Quan-ming, ZHANG Xing-kai, WANG Yun-hai, et al. Risk index system and evaluation model for failure of tailings dams[J]. Shuili Xuebao, 2009 (08) : 989-994.
18. WEI Yong, XU Kai-li. Application of set pair analysis on safety evaluation of tailing dam stability[J]. Journal of Safety Science and Technology, 2012, 8 (9) : 101-105.
19. ZHANG Li-ting. Summary on the dam-break of tailing pond[J]. Shuili Xuebao, 2013 (5) : 594-600.
20. WANG Tao, HOU Ke-peng, GUO Zhen-shi, et al. Application of analytic hierarchy process to tailings pond safety operation analysis[J]. Rock and Soil Mechanics, 2008, 29 (z1) : 680-686.

https://doi.org/10.1051/e3sconf/202235801008

Well Trajectory Measurement Error Based on Gyro Survey

Binbin Diao*, Yan Zhang, Deli Gao, Zhe Liu

China University of Petroleum, Beijing, China

Abstract. Gyroscopic survey technology is one of the key technologies to promote the progress of directional drilling technology. It is of great practical significance to study the error of well trajectory measurement based on gyro surveys and identify the main error sources. This paper introduces the error sources which lead to the uncertainty of borehole position in gyroscopic survey, six gyroscopic survey error models, and the calculation model of the contribution rate of the error sources to the error ellipse. The differences between gyro and MWD survey errors and the main error sources of gyro surveys are investigated with an example. The results show that gyroscopic surveys cannot reduce the uncertainty of borehole position compared to MWD surveys in the environment without magnetic interference. The static measurement by a three-axis gyro inclinometer is more beneficial for the improvement of the measurement accuracy of the borehole position. The error sources that contribute to the measurement error of the well trajectory are also different in the vertical section, build-up section, and horizontal section. Unlike the main error source of the MWD survey, the g-dependent error sources of the gyro sensor significantly contribute to the measurement error of the borehole trajectory. Therefore, the measurement error of the well trajectory can be significantly reduced by improving the gyroscope survey tool and reducing the influence of spin mass unbalance.

Key words: Directional drilling, gyro, measurement error, contribution rate

1. Introduction

Advanced measurement instruments and downhole control tools have played an important driving role in the development of oil and gas well types from shallow and medium-deep wells to deep and ultra-deep wells, and also from straight wells to directional wells, horizontal wells, large displacement wells, complex structural wells, and cluster wells. Although the magnetic follow-up drilling (MWD) instrument is the most commonly used instrument for borehole trajectory measurement, the research and application of gyroscopic inclination measurement technology in the oil and gas field has never stopped. In particular, the gyroscopic inclinometer for drilling invented in recent years has developed the gyroscope inclination measurement technology in drilling engineering to a new level [1]. To break the monopoly of foreign technical service companies on the market of inclinometer products, relevant research institutions and universities in China have carried out a lot of research and developed some inclinometers with independent intellectual property rights, including dynamic tuning gyroscopes, fiber optic gyroscopes, and other inclinometers [2-7].

The existence of errors in the measurement of borehole trajectory geometric parameters and trajectory calculation, etc., determines that the measurement error of borehole trajectory is difficult to avoid. Since the 1970s, foreign borehole trajectory error analysis theories have undergone development from random measurement error theory to systematic measurement error theory, mainly forming the WdW model [8] and the ISCWSA (Industry Steering Committee on Wellbore Survey Accuracy) model [9,10]. In particular, the ISCWSA model has gradually become an industry standard, and the ISCWSA model is a collective term for the survey error analysis models applicable to MWD and GMWD (Gyroscopic MWD). To date, the model has been revised four times and more than 80 independent error sources have been identified [11].

The gyroscopic inclinometer is resistant to magnetic interference, and GWD is more often used in drilling engineering projects such as cluster well anti-collision operations or rescue wells. In these projects, it is necessary to control the relative position of the orthogonal well and the drilled well, so it is of great practical importance to scientifically evaluate and describe the error of borehole trajectory measurement based on gyroscopic inclination.

2. Gyroscopic inclination error source

Unlike MWD tools, gyro inclinometers are available in several different types and modes of downhole operation. The gyro inclinometer also has internal acceleration sensors to determine the inclination angle and high side tool face angle, while internal gyro sensors are used to

* Corresponding author: diaobinbin@126.com

measure the azimuth of the borehole relative to the positive ratio. To be able to measure inclination in all cases, the gyro inclinometer requires a full set of gyro and acceleration sensors to provide tri-axial component measurements of angular velocity and acceleration. If a gyro inclinometer uses a smaller number of sensors, its operating range will be significantly limited.

The working mode of a gyro inclinometer can be roughly divided into two cases: one is that the inclinometer is in a static state, fixed at the measurement point to record data, which is the static measurement mode; the other is that the tool keeps moving in the borehole while recording data, which is the continuous measurement mode. When the gyro inclinometer is in static measurement mode, the tool measures the geometric parameters of the trajectory at designed intervals along the borehole. While the gyro inclinometer is in continuous measurement mode, the initial orientation of the gyro inclinometer needs to be known, and then the gyro sensor tracks the change in the attitude of the inclinometer as it moves through the borehole.

Since there can be different combinations of sensors and measurement modes inside the gyro inclinometer, Torkildsen et al [10] proposed six models: ① Biaxial Gyro Compassing Model (GYRO1), which is suitable for tools containing XY-axis acceleration sensors and XY-axis gyro sensors for static ② Level Rotor Gyro Model (GYRO2), which applies to the tool containing XY-axis accelerometer and Z-axis gyro sensor for continuous measurement; ③ Biaxial Semi-Continuous Gyro Model (GYRO3), which applies to the tool containing XYZ-axis accelerometer and Z-axis gyro sensor for continuous measurement. Triaxial Continuous Gyro Model (GYRO5), which applies to the tool containing XYZ triaxial accelerometer and XYZ triaxial gyro sensor for continuous measurement; ⑥ Triaxial Gyro Compassing Model (GYRO6), which applies to the tool containing XYZ triaxial accelerometer and XYZ triaxial gyro sensor for continuous measurement. XYZ triaxial accelerometer and XYZ triaxial gyro sensor for static measurements. In the gyrocompass error model, not only the errors generated by the tool but also the external reference errors, tool misalignment errors, and well-depth errors are considered, totaling 10 error groups and 48 error sources. Among them, only 9 error sources of tool misalignment errors and measurement depth errors are the same as those of the MWD error model, and the remaining 39 error sources are different from the MWD error model. The error sources considered in each of the six models are different. For example, 25 error sources are considered in the three-axis compass gyro model, as shown in Table 1.

Table 1 Error sources in the three-axis compass gyro model

Error Code	Description
DREF	Depth random error
DSR	Depth systematic reference
DSF	Depth scale
DSTR	Depth stretch type
MXY1	xy misalignment 1
MXY2	xy misalignment 2
MXY3	xy misalignment 3
MXY4	xy misalignment 4
SAG	BHA sag
XYACB	xy accelerometer bias
ZACB	z accelerometer bias
ACSFE	Accelerometer scale factor error
ACMIS	Accelerometer misalignment
XYGB1	xy gyro bias 1
XYGB2	xy gyro bias 2
XYGRN	xy gyro random noise
XYGGD1	xy gyro g-dependent error 1
XYGGD2	xy gyro g-dependent error 2
XYGGD3	xy gyro g-dependent error 3
XYGGD4	xy gyro g-dependent error 4
ZGBIAS	z gyro bias
ZGRN	z gyro random noise
ZGGD1	z gyro g-dependent error 1
ZGGD2	z gyro g-dependent error 2
GSFE	Gyro scale factor

3. Calculation of borehole trajectory measurement error

Although the error sources considered in the gyroscopic error model are very different compared to the MWD error model, the error propagation, summation equation, and conversion framework for calculating the borehole trajectory measurement error are identical. The core equation for borehole trajectory measurement error calculation is the error propagation equation [11,12].

$$\boldsymbol{e}_j = \sigma_j \frac{\boldsymbol{dr}}{\boldsymbol{dp}} \frac{\partial \boldsymbol{p}}{\partial \varepsilon_j} \tag{1}$$

where $\boldsymbol{e}_j$ denotes the magnitude of the error caused by the error source j; σ_j denotes the standard deviation of the error source j; $\boldsymbol{dr}/\boldsymbol{dp}$ denotes the effect of the measurement errors of bathymetry, borehole slope, and azimuth on the borehole position in the geodetic coordinate system; and $\partial p/\partial \varepsilon_j$ denotes the weight function of the error source j.

In the ISCWSA model, four error propagation modes are considered not only for systematic and random errors but also for inter-well and global errors to represent different cumulative effects of errors. The 48 error sources considered in the model can also be classified into four categories based on the error propagation patterns. Therefore, the covariance matrix C_i at the location of the measurement point i can be expressed as:

$$C_i = C_{i,rand} + C_{i,syst} + C_{i,well} \tag{2}$$

where $C_{i,rand}$ denotes the covariance matrix considering all random errors at the ith measurement point location; $C_{i,syst}$

denotes the covariance matrix considering all systematic errors at the measurement point i location; and $C_{i,well}$ denotes the covariance matrix considering all inter-well and global errors at the measurement point i location. Also, the calculation of these covariance matrices is given in the ISCWSA model. Then, the eigenvalues and eigenvectors of the covariance matrix C_i are calculated and combined with the predetermined confidence factors to obtain the ellipsoidal size and spatial attitude of the errors at the location of the metered point i.

4. Contribution of the error source to the error ellipse

A total of 48 error sources are considered in the six gyroscopic error models, and to identify the main error sources among them, the contribution of each error source to the uncertainty of the borehole position needs to be analyzed. The measurement error of a measurement point on the borehole trajectory can be described by an error ellipsoid whose three semi-axis lengths are determined by the three eigenvalues of the covariance matrix, respectively. Since error ellipsoids of different sizes necessarily have different eigenvalues of the covariance matrix as well as semi-axis lengths of the error ellipsoid, volumes, and error ellipsoids, the contribution of each error source to the measurement error of the borehole trajectory can be analyzed by calculating the contribution of each error source to them. The contribution of each error source to the eigenvalues of the covariance matrix, the semi-axis length of error ellipsoid, and the error ellipsoid volume or surface area can be obtained by calculating their rates of change after defaulting that error source [13]. However, it is that the calculation of the borehole trajectory measurement error is more applied to the clump well drilling project, and the directional drilling engineers are generally more concerned with the errors in the direction of the high side of the borehole and on the lateral direction. Therefore, it is more relevant to study the contribution of each error source to the error ellipse.

The influence of the error source on the error ellipse area can be expressed as:

$$A_{i,n} = \frac{S_i - S_{i,n}}{S_i} \times 100\% \qquad (n = 1, 2, 3, \cdots, N) \tag{3}$$

where, $A_{i,n}$ is the contribution of the nth error source to the error ellipse of the measurement point i; S_i is the area of the error ellipse of the measurement point i; $S_{i,n}$ is the area of the error ellipse of the measurement point i after the error source n has defaulted.

According to the formula of the error ellipse area, the contribution of the error source to the error ellipse area can be obtained as follows:

$$R_{i,n} \doteq \frac{r_{i,h} r_{i,l} - r_{i,n,h} r_{i,n,l}}{\sum_{n=1}^{N} \left(r_{i,h} r_{i,l} - r_{i,n,h} r_{i,n,l} \right)} \times 100\% \qquad (j = 1, 2, 3, \cdots, N) \tag{4}$$

where, $R_{i,n}$ is the contribution of the normalized error source n to the area of the error ellipse at the measurement point i; $r_{i,h}$ and $r_{i,l}$ are the high-side direction and lateral semi-axis length of the error ellipse at the measurement point i, respectively; $r_{i,n,h}$ and $r_{i,n,l}$ are the high-side direction and lateral semi-axis length of the error ellipse at the measurement point i after the default of the error source n, respectively.

5. Example calculation and analysis

The design data of a horizontal well JY29HF track in an oil field are shown in Table 2. The geomagnetic field strength at the location of this well is 49054.5 nT, the geomagnetic inclination is 46.53°, and the geomagnetic declination is -5.45°. Take the confidence factor equal to 1, the measurement error of JY29HF is calculated based on the ISCWSA Gyro Model [10] and the Standard ISCWSA MWD Model [9].

Table 2 Design data of JY29HF track

Measured Depth (m)	Inclination (°)	Azimuth (°)	Vertical Depth (m)	+N/-S (m)	+E/-W (m)
0.00	0.00	0.00	0.00	0.00	0.00
1,000.00	0.00	0.00	1,000.00	0.00	0.00
1,246.71	39.47	347.52	1,227.65	79.75	-17.64
2,560.69	39.47	347.52	2,241.93	895.36	-198.09
2,892.86	92.62	347.66	2,372.00	1,181.34	-260.93
3,353.95	92.45	3.04	2,351.49	1,639.07	-298.14
3,991.22	92.45	3.04	2,324.30	2,274.87	-264.34
4,099.30	95.30	0.84	2,317.00	2,382.62	-260.68

The magnitudes of the high-side errors in the borehole trajectory are shown in Figure 1. The calculation results show that the Standard ISCWSA MWD Model has the smallest calculation results, and the difference between the calculation results of each model in the straight and inclined sections is little, and the calculation results of the Biaxial Gyro Compassing Model and Level Rotor Gyro Model in the horizontal section are significantly larger than other models. The magnitudes of the lateral errors in the borehole trajectory are shown in Figure 2. The calculation results show that the calculation results of the Standard ISCWSA MWD Model are also the smallest, the difference between the calculation results of each model in the straight section is small, and the difference between the calculation results of each model increases gradually with the increase of measurement depth after entering the inclined section. Moreover, among the six gyroscopic error models, the calculation results of the Level Rotor Gyro Model are significantly larger than other models, and the calculation results of the Triaxial Gyro Compassing Model are the smallest. As shown in Figure 3, the vertical errors and the high-side errors have similar characteristics.

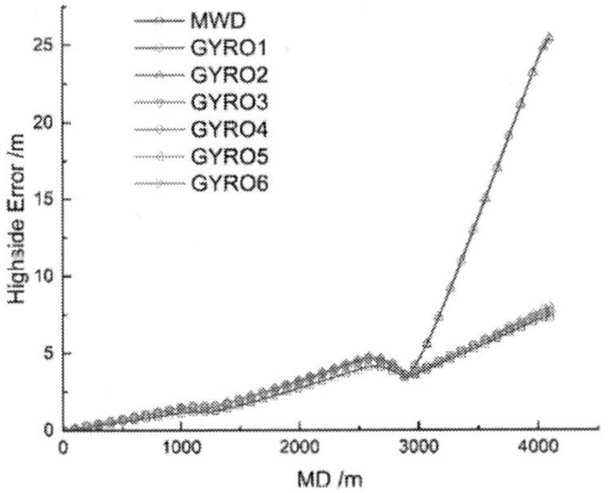

Figure 1 Highside Errors

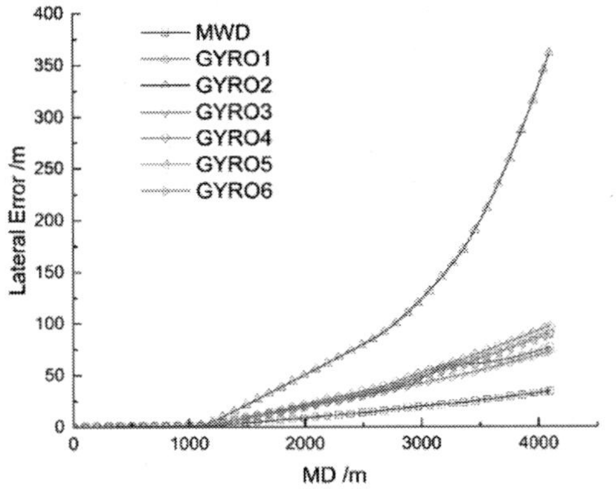

Figure 2 Lateral Errors

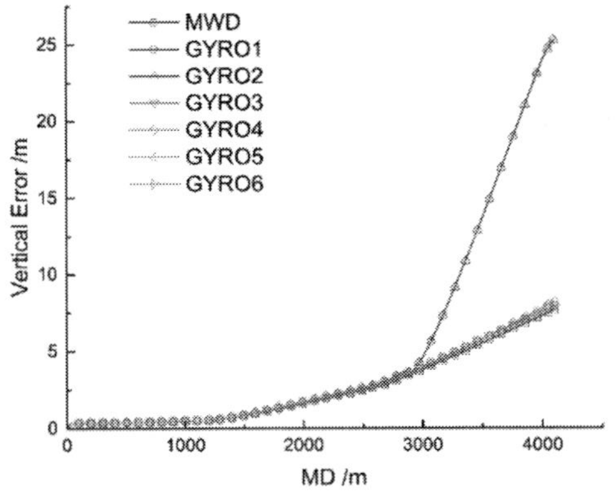

Figure 3 Vertical Errors

The calculation results from the error ellipsoid show that the uncertainty of the borehole position is greater when applying the gyro suitable for ISCWSA Gyro Model to measure the borehole trajectory compared with MWD; the angle of borehole inclination has a significant influence on the results of gyro inclination measurement, especially those suitable for Biaxial Gyro Compassing Model and Level Rotor Gyro Model The gyro is not suitable for horizontal section; among the gyroscopes suitable for six gyroscopic error models, the application of three-axis gyroscope for static measurement is more beneficial to reduce the measurement error of well borehole trajectory.

The contribution rates of each error source to the error ellipse are shown in Figs. 4 and 5, respectively. The calculation results show that in the straight section, the contribution of each error source to the error ellipse varies less with the measurement depth, and the main error sources are MXY3, MXY4, ACMIS, and XYACB; entering the inclined section, the contribution of these four error sources gradually decreases with the increase of the measurement depth, but MXY3, ACMIS, and XYACB are still the main error sources, and SAG, XYGGD1, and XYGGD2 also quickly become the main error sources; after entering the horizontal section, the contribution of SAG and XYGGD1 continues to increase with the increase of measurement depth, while the contribution of XYGGD2, MXY3, ACMIS, and XYACB decreases with the increase of measurement depth.

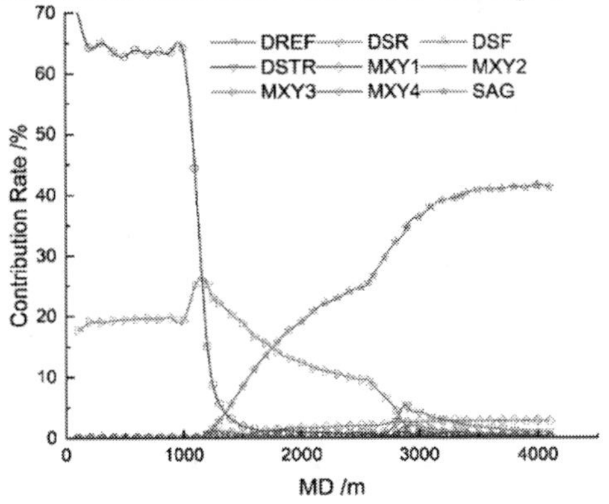

Figure 4 Contribution rates of error sources common to MWD survey

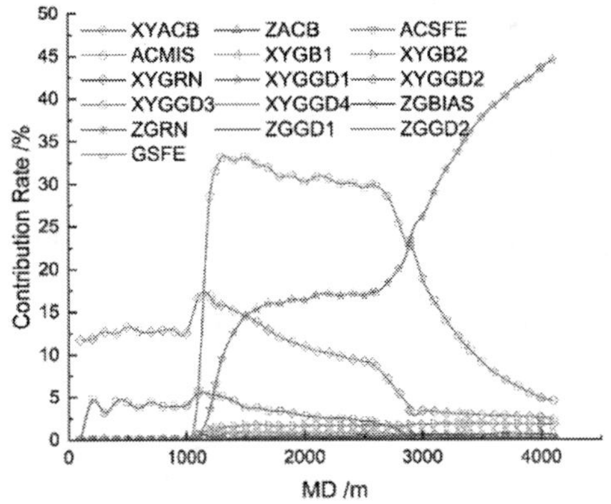

Figure 5 Contribution rates of error sources associated with sensors

From the calculation results of the contribution rate, the error sources contributing more than 5% in the horizontal section include SAG, XYGGD1, and XYGGD2. Therefore, the uncertainty of borehole location can be significantly reduced by improving the gyroscopic logging tool and reducing the standard deviation of XYGGD1 and XYGGD2.

6. Conclusion

(1) Since there is no magnetic sensor, gyroscopic inclinometers can measure inclination normally in a magnetically disturbed environment. However, in an environment without magnetic interference, the uncertainty of borehole position is not reduced using conventional gyroscopic inclination compared to MWD inclination.
(2) Among the inclinometers suitable for six gyroscopic error models, the application of three-axis gyroscopic inclinometers for static measurements is more conducive to reducing the measurement error of borehole trajectories.
(3) The main error sources in the straight section include MXY3, MXY4, ACMIS, and XYACB; the main error sources in the build-up section are MXY3, ACMIS, XYACB, SAG, XYGGD1, and XYGGD2; and the error sources contributing more than 5% in the horizontal section only include SAG, XYGGD1 and XYGGD2.
(4) The uncertainty of borehole location can be significantly reduced by improving the gyroscopic survey tool and reducing the standard deviation of XYGGD1 and XYGGD2.

Acknowledgements

The authors gratefully acknowledge the financial support of the Natural Science Foundation of China (NSFC, 51974336).

References

1. Tang Guoqi and Xiao Shengsi. Borehole inclinometer technology status and development prospect. Exploration Engineering, 1999 (Supl): 235-238.
2. Xu Liuwan, Wang Yanli and Wu Xiaolong. New pattern of development, introduction and application of drilling technology and equipment in China. Equipment Geotechnical Engineering, 2013, 14(05): 11-14.
3. Zheng WenCheng, Yan Li, Ren Qinyong, et al. Design and implementation of navigation system of dtg north seeker based on DSP. Chinese Ship Research, 2010, 5(1): 76-80.
4. Hu Xunbo, Chen Guangcong, Bu Jijun, et al. Optimum design of high-precision dynamically tuned gyroscope north finding system. Piezoelectrics & Acoustooptics, 2014, 36(04): 580-583.
5. Xie Mujun, Tan Xuguang and Wang Zhiqian. Multi-position north seeker and error analysis of position transformation based on dynamically tuned gyro. Opto-Electronic Engineering, 2008(11): 4-7+12.
6. Hu Xudong, Ji Shitao, Hong Wei, et al. Relative intensity noise reduction technology of high precision fiber optic Gyroscope. Scientific and Technological Innovation, 2022(01): 25-28.
7. Shang Kejun, Lei Ming, Li Haowei, et al. Design, manufacturing and future development of the integrated fiber optic gyroscope. Journal of Chinese Inertial Technlogy, 2021, 29(04): 502-509.
8. Wolff C J M, De Wardt J P. Borehole position uncertainty - analysis of measuring methods and derivation of systematic error model. SPE Journal of Petroleum Technology, 1981, 33(12): 2339-2350.
9. Williamson H S. Accuracy prediction for directional measurement while drilling. SPE 67616, 1999.
10. Torgeir T, Havardstein T S, Weston L J, et al. Prediction of wellbore position accuracy when surveyed with gyroscopic tools. SPE 90408, 2004.
11. Jamieson A. Introduction to wellbore positioning. Scotland: University of the Highlands & Islands, 2017.
12. Liu Xiushan. Borehole trajectory uncertainty and its characterization. Petroleum Exploration and Development, 2019, 46(02):201-206.
13. Diao Binbin, Gao Deli, Hu Degao, et al. Identification of main error sources from well trajectory measurement based on contribution rate analysis. Drilling & Production Technology, 2021, 44(1): 1-6.

Economic Analysis of Two Heat Pumps to Recover Heat from Circulating Water

Hu Zhang *, Jinxu Lao

State grid Shandong Electric Power Research Institute, Jinan, China

Abstract: The use of heat pump to recover the condensing heat of the unit can realize the cascade utilization of energy, improve the utilization efficiency of energy, not only can effectively increase the heating capacity of the unit, but also can improve the use efficiency of heating extraction steam, so as to improve the economy of the thermal power plant. Heat pumps suitable for use in power plants can be divided into absorption heat pumps and compression heat pumps. Which type of heat pump is more economical needs specific judgment for the specific application environment. In this paper, a specific example is used to compare the economy of the two heat pump systems. The calculation results show that the economy of the compressed heat pump is better than that of the absorption heat pump only when the COP of the compressed heat pump is greater than the critical value, and the waste heat recovery ability and water saving performance of the compressed heat pump are better.

Key word: absorption heat pump; compression heat pump; heating capacity; evaporation loss; economic analysis.

1. Introduction

Central heating is the steam and hot water produced by the central heat source,which supplies the heat needed by production,heating and life of a city or part of the area through the pipe network.Combined heat and power generation is a production method that thermal power plants use the steam that has done work in the steam turbine to heat users while producing electric energy. Combined heat and power plants are important heat sources for central heating.Regulating steam extraction heating unit is the main unit of cogeneration.Part of the steam that has done work in the steam turbine is extracted and sent to the first station of the heat network. It is used to heat the circulating water of the heat network in the heater of the heat network. The heat carried by the steam is transferred to the heat user through the medium of the circulating water of the heat network.A large amount of waste heat will still be generated in the process of regulating the extraction type heating unit. The usual treatment is to release this waste heat to the natural environment through the condenser and cooling tower, which is the cold source loss of the steam turbine.The amount of waste heat in this part is very large, but the grade is very low. It is usually impossible to obtain mechanical work from this part of waste heat, but this part of waste heat can be used for heating and heating, so as to realize the cascade utilization of energy and improve the utilization efficiency of energy.Heat pump technology plays an important role in the utilization of waste heat.

The working principle of heat pump is based on the inverse "Carnot cycle", which consumes a small amount of circulating net work, extracts heat from the circulating cooling water of the self-condenser, and sends it to the circulating water of the heat network to realize the purpose of waste heat utilization.Using heat pump heat recovery of condenser circulating water is used for heating of the heat on the one hand, can be improved by recycling waste heat heating system of heating capacity, on the other hand can improve the heat supply network water temperature of the heater, heater heat transfer temperature difference, reduce the network heater irreversible loss reduce network so as to improve the efficiency of heating extraction.

The heat recovered from the circulating water of the condenser can be used for heat supply by either lithium bromide absorption heat pump or compression heat pump [1-2].Absorption characteristics of the lithium bromide absorption heat pump units using the lithium bromide solution, based on the latent heat of vaporization of steam drive energy drive low temperature low pressure of water vapor in the evaporator condenser cooling water to extract the heat, the two parts of heat combined into absorption heat pump system to heat, and the water vapor carry heat to heat supply network in the condenser circulating water.The working medium of the compression heat pump absorbs the heat from the condenser circulating cooling water in the evaporator, flows through the heat pump compressor to absorb the compression work, and is cooled by the circulating water of the heat network in the

* Corresponding author: 13075336627@163.com

condenser, condenses into liquid, releases the latent heat of vaporization, heats the circulating water of the heat network, and then enters the evaporator for the next cycle. The most important performance index of heat pump is the heating coefficient, represented by the symbol COP, which is defined as the ratio of the energy obtained by the user to the energy consumed by the driving heat pump.According to the working principle of heat pump, the COP of compression heat pump and absorption heat pump is always greater than 1.Under current technical conditions, the COP of compression heat pump is generally around 3 ~ 6, while that of absorption heat pump is generally around 1.65 ~ 1.85 [3].In reference [4-6], the equivalent heat drop method is used to analyze the economy of electrically driven compression heat pump system, steam driven compression heat pump system and absorption heat pump system respectively. Reference [7-9] illustrates the contribution of absorption heat pump system to energy saving and emission reduction based on actual cases.It can be seen that the energy saving benefit of both absorption heat pump system and compression heat pump system is far greater than that of direct extraction heat supply production mode.

Absorption heat pump system and compression heat pump system have their own applicable conditions and advantages and disadvantages.Single heat pump heating coefficient, the heating coefficient of compression heat pump is far higher than that of absorption heat pump, but electric compression heat pump in the process of production needs to consume large amounts of high grade of electricity, reference [10, 11], respectively, compared the COP is 1.7 absorption heat pump system and the COP is 4.5 and the COP for 6 electric compression heat pump system efficiency difference, It is concluded that the economy of absorption heat pump is better than that of compression heat pump.Combined with the analysis and comparison of actual operation parameters of power plants, lreference [12] also draws the conclusion that the economy of absorption heat pump is better than that of electrically driven compression heat pump.reference [3] compares the economic difference between the absorption heat pump and the electrically driven compression heat pump by introducing the electric energy conversion coefficient, and obtains the critical condition that the economy of the electrically driven compression heat pump system is higher than that of the absorption heat pump system.However, these analyses only consider the change of the heat pump system to the power generation of the system or the calculation is complex and not intuitive, and do not consider the contribution of the heat pump system in water saving, so it is not comprehensive enough.This paper comprehensively considers the differences of various heat pump systems in power generation, heating and water saving, comprehensively and intuitively compares the economy of electrically driven compression heat pump, steam driven compression heat pump and absorption heat pump, and discusses the economic difference between compression heat pump and absorption heat pump.

2. Economic analysis model

In the original system without the use of a heat pump, the heat required to heat the circulating water in the heat grid is provided entirely by the heating extractor in the heat grid heater.Usually, the hot water outlet temperature produced by the heat pump cannot reach the required water supply temperature of the circulating water in the heat network. In the heat pump production mode listed in this paper, the heating system adopts the mode of heat pump and heat network heater running in series.The heat extraction steam from the medium pressure cylinder of the steam turbine further heats the hot water from the heat pump in the heat network heater to the required water supply temperature of the heat network and then supplies it to the municipal pipe network.

In order to compare the economy of compression heat pump and absorption heat pump more easily and intuitively, the following assumptions are made: the parameters of the turbo-generator set are consistent; The supply and return temperatures of the circulating water in the heat network are consistent; The flow of circulating water in the heat network is consistent; The heat of the heat pump is equal; After the condenser circulating cooling water flows through the evaporator of the heat pump, the temperature drop is 5°C. Generator efficiency ηd is 0.99.

Because the heat production of heat pumps is the same, the operation parameters of heat network heaters in all kinds of heat pump production methods are the same, so it is not necessary to consider the heat network heaters and the corresponding heat recovery system in the economic comparative analysis, thus simplifying the calculation.

2.1 Absorption heat pump

In the production mode of absorption heat pump heating, the steam source driving the heat pump is taken from the exhaust steam of the medium pressure cylinder, and the condensate water at the shaft plus outlet is used as the desuperheated water to reduce the temperature and decompress the exhaust steam from the medium pressure cylinder to saturation steam, and then it is sent to the generator of the absorption heat pump as the driving steam.The condensate generated after heat release in the generator flows through the water-water heater and is partially recovered by the condensate at the outlet of the shaft and returned to the condenser.

The temperature reduction water required for cooling 1 kg/s extraction steam is:

$$f = \frac{h_{zp} - h_{bh}}{h_{bh} - h_{zj.out}} \qquad (1)$$

Where: hzp, hbh and hzj.out are the enthalpy of middle exhaust, saturated steam and axial added water, respectively.

Exhaust steam from 1 kg/s medium pressure cylinder will reduce the power generation:

$$w_1 = (h_{zp} - h_c) \bullet \eta_d \qquad (2)$$

Where: hc is the exhaust enthalpy of steam turbine.

The heat release of driving steam in the generator is:

$$Q_G = (1+f)\times(h_{bh} - h_{bh.s}) \quad (3)$$

Where, hbh.s is saturated enthalpy of water.
The heat absorption of the working medium in the evaporator is:

$$Q_E = Q_G \times (\mathrm{COP}_x - 1) \quad (4)$$

Where: COPx is the heating coefficient of the absorption heat pump.
The heat production of the absorption heat pump is:

$$Q_C = Q_G + Q_E \quad (5)$$

The extraction efficiency of No.8 low pressure heater is α_8, and the heat released by saturated water in water-water heater is used for No.8 low pressure heater. Due to the exclusion of part of No.8 pumping steam, the added value of turbine energy generation is:

$$w_2 = (1+f)\bullet(h_{bh.s} - h_{gl})\bullet\alpha_8\bullet\eta_d \quad (6)$$

Where: hgl is the enthalpy of supercooled water.
After the exhaust steam of the medium pressure cylinder of 1 kg/s is extracted, the power generation reduction value of the unit is:

$$w_x = w_1 - w_2 \quad (7)$$

Decrease in electricity revenue

$$E_{x2} = e_d \bullet w_x \quad (8)$$

Where: ed is the feed-in price.
The condenser circulating water flow through the evaporator is

$$f_n = \frac{Q_E}{cp\times\Delta t} \quad (9)$$

Where: cp is the specific heat of water at constant pressure 4.1868kJ/kg℃, Δt is the temperature drop of circulating water in the evaporator.
The reduction value of evaporation loss of circulating water per hour is:

$$f_z = 3.6\times k_{zf}\times\Delta t\times f \quad (10)$$

Where: kzf is the evaporation loss coefficient, and its value is 0.001
Water cut

$$E_{x3} = e_s \bullet f_z \quad (11)$$

Where: es is the industrial water price.

2.2 Compression heat pump

The production mode of compressed heat pump can be divided into two kinds: mode 1, extract part of the exhaust steam from the medium pressure cylinder to expand and do work in the condensing small steam turbine, the exhaust steam of the small steam turbine is recovered to the condenser, and the small steam turbine drives the compressor of the heat pump through the coupling; Method 2, the use of plant electricity to drive the compressor of the heat pump.

2.2.1 Steam driven compression heat pump

The steam source of the condensing steam turbine driving the compression heat pump comes from the exhaust steam of the medium pressure cylinder. The exhaust steam volume of the medium pressure cylinder that is equal to that of the absorption heat pump is:

$$f_{y1} = \frac{Q_C}{\mathrm{COP}_{y1}\times(h_{zp} - h_{xj.c})} \quad (12)$$

Where: hxj.c and COPy1 are the exhaust enthalpy of small machine and the heating coefficient of compression heat pump respectively.
The reduction value of turbine power generation is:

$$w_{y1} - f_{y1}\times w_1 \quad (13)$$

Decrease in electricity revenue

$$E_{y2} = e_d \bullet w_{y1} \quad (14)$$

The hourly evaporation loss of circulating water is reduced

$$f_{zy1} = \frac{Q_C\times 3.6\times\Delta t\times k_{zf}\times(\mathrm{COP}_{y1}-1)}{\mathrm{COP}_{y1}\times\Delta t\times cp} \quad (15)$$

Water cut

$$E_{y3} = e_s \bullet f_{zy1} \quad (16)$$

2.2.2 Electric energy driven compression heat pump

The electrical energy required to drive the heat pump is

$$w_{y2} = \frac{Q_C}{\mathrm{COP}_{y2}} \quad (17)$$

Where: COPy2 is the heating coefficient of the compression heat pump.
Decrease in electricity revenue

$$E'_{y2} = e_d \bullet w_{y2} \quad (18)$$

The hourly evaporation loss of circulating water is reduced

$$f_{zy2} = \frac{Q_C\times 3.6\times\Delta t\times k_{zf}\times(\mathrm{COP}_{y2}-1)}{\mathrm{COP}_{y2}\times\Delta t\times cp} \quad (19)$$

Water cut

$$E'_{y3} = e_s \bullet f_{zy2} \quad (20)$$

2.3 Revenue from heat pump systems

Under the premise of equal heat supply, the use of heat pump can reduce the flow of heating extraction steam. The contribution of heat pump to power plant revenue can be analyzed from the following three aspects:
Earnings 1: Power generation increased by crowding out heating extraction steam;
Earnings 2:Generation loss generated by driving heat pump;
Earnings 3:The evaporation loss of condenser circulating cooling water is reduced.

https://doi.org/10.1051/e3sconf/202235801009

Table 1 Revenue from each heat pump system

Items	absorption heat pump	compression heat pump	
		Small steam turbine drive	Electricity drive
Earnings 1	E_{x1}	E_{y1}	E'_{y1}
Earnings 2	$E_{x2}=-e_d \cdot w_x$	$E_{y2}=-e_d \cdot w_{y1}$	$E'_{y2}=-e_d \cdot w_{y2}$
Earnings 3	$E_{x3}=e_s \cdot f_z$	$E_{y3}=e_s \cdot f_{zy1}$	$E'_{y3}=e_s \cdot f_{zy2}$
Total earnings	$E_x = \sum_1^3 E_{xi}$	$E_y = \sum_1^3 E_{yi}$	$E'_y = \sum_1^3 E'_{yi}$

2.4 Economic comparison

Because the heat load of the system is equal and the heat produced by the heat pump is equal, the quantity of heating extraction steam is equal in the production mode of various heat pumps, and this part of steam returns to the turbine to do work, so the power generation increased by the system is equal.

$$E_{x1} = E_{y1} = E'_{y1} \tag{21}$$

Order

$$E_y \geq E_x \tag{22}$$

and

$$E'_y \geq E_x \tag{23}$$

Solving equations (21), (22) and (23) can obtain the corresponding critical COP value when the system revenue obtained by the compression heat pump production mode is greater than that obtained by the absorption heat pump production mode.

3. The calculation example

Taking an ultra-supercritical 600 MW unit as an example, the exhaust pressure of the medium pressure cylinder at rated load is 1.171Mpa, the temperature is 394.4 ℃, the enthalpy value h_{zp} is 3249.7kJ /kg, the exhaust enthalpy of the steam turbine h_c is 2316.4kJ /kg, and the benchmark electricity price e_d is 0.4 yuan /kWh. The industrial water price e_s is 2 yuan/T. The relevant parameters required for calculation are as follows:

Table 2 Calculate the required parameters

Items	Symbol	Unit	Value
Exhaust steam enthalpy of medium pressure cylinder	h_{zp}	kJ/kg	3249.7
Saturated vapor enthalpy	h_{bh}	kJ/kg	2738.06
Saturated water enthalpy	$h_{bh.s}$	kJ/kg	604.723
Exhaust enthalpy of steam turbine	h_c	kJ/kg	2316.4
Outlet enthalpy of shaft seal heater	$h_{zj.out}$	kJ/kg	146.1
Supercooled water enthalpy	h_{gl}	kJ/kg	166.218
Exhaust steam enthalpy of Small steam turbine	$h_{xj.c}$	kJ/kg	2438.83
Extraction efficiency of No. 8 low pressure heater	---	---	0.1145
The benchmark price of electricity	e_d	yuan/kWh	0.4
The price of industrial water	e_s	yuan/t	2

According to the relevant standards, the COP of the absorption heat pump used in the power plant shall not be lower than 1.7, and the COP of the absorption heat pump shall be 1.7. Put the value of COP and the value in the above table into equations (1) to (11) to obtain the relevant data of the absorption heat pump

Table 3 Calculation data of absorption heat pump

Items	Symbol	Unit	Value
Medium exhaust steam extraction	f_{zp}	kg/s	1
Desuperheating water flow	f	kg/s	0.1974
The working fluid absorbs heat	Q_E	kW	1788.122
Heat production by heat pump	Q_C	kW	4342.582
Driving steam reduces power eneration	W_1	kW	923.97
Squeeze out steam to increase power generation	W_2	kW	59.52
Quantity of electricity reduction	W_x	kW	864.45
Amount of water	f_z	t/h	1.538

By substituting the values of Qc and W1 into equations (12) to (20), and further solving equations (21), (22) and (23) simultaneously, COPy1 ≥ 5.67 and COPy2 ≥ 4.98 are obtained.

By comparing the calculation, the compression heat pump for electricity, electricity only when the COP is greater than 4.98 gains the compression heat pump system is greater than the absorption heat pump system, and for small condensing steam turbine drive of compression heat pump system only when the COP is greater than 5.67 compression heat pump system of revenue is greater than the absorption heat pump system.

The benefit of compressed heat pump driven by condensing steam turbine is not as good as that driven by electric energy because the former still has cold source loss.

The above is just a case for illustration. The economic difference between the absorption heat pump and the compression heat pump is affected by the electricity price, water price, system connection mode and the heating coefficient of the heat pump.

4. Conclusion

Because of the heat pump COP is greater than 1, so from the perspective of energy use, no matter what type of driven heat pump recovery condenser cooling water waste heat can be implemented without increasing thermal power plant capacity expansion of capacity reduction of central heating energy consumption of heating purpose, relative to the direct extraction steam heating has good energy saving effect, At the same time, it can reduce the evaporation loss of condenser circulating cooling water, reduce the thermal pollution caused by thermal power plant to the environment, save water resources and protect the ecological environment. It has significant economic, social and environmental benefits, as well as significant energy saving and emission reduction effects, in line with the national energy saving and emission reduction policy requirements.

The COP of the compression heat pump is larger than that of the absorption heat pump. Under the same heat production condition, the compression heat pump can recover more waste heat of the condenser circulating cooling water, and accordingly, the evaporation loss of the condenser circulating cooling water will be less.Therefore, the ability of compression heat pump to recover waste heat is stronger than that of absorption heat pump, and the amount of water saving is greater than that of absorption heat pump, which is of great significance for some areas lacking water resources.

The operation cost of absorption heat pump is lower than that of compression heat pump. The structure of compression heat pump is simple and more convenient to control.In actual operation, with the change of the return water temperature of the heat network and the temperature of the condenser circulating cooling water, the performance of the heat pump will change, and the performance of the compression heat pump is more sensitive to the change of water temperature than that of the absorption heat pump.Therefore, the use of heat pump system in the field of district heating should be based on the actual production situation of the site to choose the most appropriate heat pump technology application scheme.

Acknowledgements

This work was supported by the science and technology project of State grid shandong electric power research institute (ZY-2022-16), which studied the peak shaving capacity of different types of cogeneration units in Shandong power grid

References

1. Huizhong ZHao, Xingang ZHao, Overview of Thermal Power Plant Waste Heat Utilization Technology and Engineering Example[J]. GAS & HEAT, 2018,(07):0A01-0A05.
2. Xiaolan Ge, Cong Yu,CHang Wu,et al,Analysis of steam turbine waste heat recovery technology[J]. Internal Combustion Engine & Parts, 2018,(09):57-58.
3. Jianglong Guo,SHuping CHang,Aihua Feng,et al,Economic Analysis for Compression-type and Absorption-type Heat Pump Recycling the Waste Heat of Circulating Water in Power Plant[J].TURBINE TECHNOLOGY ,2012,(10):379-380.
4. CHuanling Liu, Ang Song,Economic Research of Heating Supply Using Circulating Water-source Heat Pumps in Thermal Power Plant[J].SHANDONG DIANLI JISHU, 2016,(3):30-33.
5. Sike Hu,Ruzhou Liu,De CHen,et al,Analysis of Recycling Afterheat of Circulating Cooling Water from Thermal Power Plant by Utilizing Compression Heat Pump[J].Journal of Northeast Dianli University ,2015,(8):14-19.
6. Qunli Long, Study on the Thermoeconomics of a Cooling Water Waste Heat Recycled System in Power Plant[J]. SHANDONG DIANLI JISHU, 2013, (2):68-72.
7. Xiaoguang Guan,Application Research on Waste Heat Recovery Technology of Absorption Heat Pump of HeatSupply Unit[J].Mining Engineering, 2017, (2):68-69.
8. ZHigang ZHang, SHuguo Wang,Rongpeng Zeng, Recycling waste heat of circulating water by absorption heat pump in power plant[J].Huadian Technology, 2016, (4):45-47.
9. Yang ZHao,Lei Liang,Qingchuan Ma,et al,Application of Lithium Bromide Absorption Heat Pump to Recycle Waste Heat of Cooling Water in Power Plant[J].Journal of SHanghai University of Electric Power, 2016, (10):454-458.
10. ZHenqi ZHou, CHunhui Cui, Meng Yuan,et al, Economic analysis of two kinds of heat pump heating system with waste heat of circulating water[J].ENE R GY CONSE R VATION,2015,(12):56-59.
11. Yun ZHang, Binshuai ZHu,An Economic Comparison of Absorption Heat Pump and Compression Heat Pump in Central Heating Area of Power Plant[J].ENE R GY CONSE R VATION TECHNOLOGY, 2016,(9):440-443.
12. Biaojun ZHong, Guanhua Deng,Xueqing CHen,et al,Comparision of Economy between two Heat

Pumps for Recycling Waste Heat in Power Plant[J].REFRIGERATION,2015,(3):80-84.

Research on the area of mechanized construction of transmission lines

Tiejun Zhou [1], Jingshuang Gan [2], Lingyun Li [1], Zhongyuan Xia [2]

[1] State Grid Henan Electric Power Company Economic and Technological Research Institute, Zhengzhou 450007, China
[2] PowerChina Henan Electric Power Survey and Design Institute Co., Ltd., Zhengzhou 450007, China

Abstract. In order to continuously improve the construction capacity of power grid projects, improve the level of construction technology, and promote the transformation and upgrading of construction enterprises, power companies actively promote the mechanized construction of transmission lines. On the basis of summarizing the existing calculation rules of mechanized construction area, through on-site investigation and extensive collection of funds, research and excavate indicators that adapt to the new situation and new regulations. According to the actual area composition of the mechanized construction site, scientific calculation rules are formulated by modeling analysis and other methods. Compare and verify the calculated area of the new rule with the actual compensation area and the calculated area of the original rule, and propose a complete set of calculation rules for mechanized construction area, which has the conditions and value for popularization and application.

Keywords: Transmission line; mechanized construction; compensation area; scientific model.

1. Introduction

The mechanized construction of the transmission line is to use the mechanized construction equipment in the whole process of the construction of the transmission line to complete the construction tasks of each construction process. The area occupied by the mechanized construction of transmission lines refers to the area occupied during the construction of transmission lines due to the use of machinery for transportation, stacking of materials, equipment, excavation of earth and stone, construction of foundations, erection of towers, and deployment of guide and ground wires. This area corresponds to the area for compensation for young crops, which is related to how to calculate compensation for young crops.

At present, the construction area of the transmission line is calculated by referring to the "Detailed Rules for the Compilation of Estimated Estimates for Power Transmission Projects". According to different voltage levels and line length, after deducting the proportion of the area without young crops to the entire length of the line, it is multiplied by the "comprehensive width of the temporary construction site"[1]. This construction area calculation method is suitable for the case where the line conductor cross-section, towers and foundation works are small, and the construction is carried out manually, and it fails to reflect important technical parameters such as the line loop, foundation form, tower type, average span and other important technical parameters. Differences, without considering the area added by the line crossing and spanning construction. With the increase of social demand for electricity consumption, the transmission capacity of lines of various voltage levels has increased significantly, the cross-section of conductors, towers and foundation works have increased significantly, and the number of objects along the line has also increased significantly. Coupled with the extensive promotion of mechanized construction, and manpower The calculation method of the construction site area of the transmission line that matches the construction has not matched the on-site situation of the mechanized construction, and cannot meet the requirements of the accurate calculation of the budget site area for the mechanized construction of the transmission line project. It is urgent to conduct an in-depth analysis of the area of the mechanized construction of the transmission line. and calculations.

On the basis of the current mechanized construction area calculation rules, through on-site research and extensive collection of funds, according to the actual area composition of the mechanized construction site, using modeling analysis and other methods to formulate scientific calculation rules, and calculate the new rules. The area is compared with the actual compensation area and the original calculation area, and a set of perfect calculation rules is proposed for the calculation of the area of the mechanized construction site, which has the conditions and value of popularization and application.

https://doi.org/10.1051/e3sconf/202235801010

2. Analysis on the composition of mechanized construction area

According to the on-site investigation and actual measurement, the analysis shows that the area occupied by the mechanized construction site is closely related to the technical indicators such as the number of tower bases, the single-base engineering volume and the average span of the line.The composition of the site area is mainly divided into six parts: construction road, foundation site, tower site, spanning frame site, wire channel and accessory installation site and traction tension field, as shown in Table 1.

Tabel 1 The composition of mechanized construction area

serial number	The composition of mechanized construction area
1	construction road area
2	foundation site area
3	tower site area
4	spanning frame site area
5	wire channel and accessory installation site area
6	traction tension field area

The analysis of the area of each part is as follows:
(1) The construction road area refers to the area of the road that needs to be kept for the passage of the machinery itself when large and medium-sized machinery is used for construction. This area is related to factors such as the number of towers, the width of the road occupied by machinery, and the total length of the line.
(2) The area of the foundation site refers to the area occupied by the foundation materials, construction machinery, and excavation earthwork of the stacking site. The area is related to the foundation type, the amount of earthwork excavation and other factors.
(3) The tower site area refers to the area occupied by the stacking tower materials and construction machinery. The area is related to factors such as the opening of the tower and the amount of material of the tower. Generally, after the foundation construction is completed, the tower construction is carried out, and the area of the tower site basically coincides with the area of the foundation site.
(4) The area of the spanning rack site refers to the area occupied by the spanning rack that needs to be erected due to the line crossing over the high-speed rail, high-speed rail, etc. The area is related to the length, width, etc. of the span.
(5) The area of the installation site for the wiring channel and accessories refers to the area occupied by the work that needs to be coordinated during the installation of the wiring and accessories. The area is related to factors such as cross arm length and path length.
(6) The area of the traction tension field refers to the area occupied by the construction of the stretcher. The area is related to factors such as the number of stretch sites and the area of stretch sites.

3. Calculation of mechanized construction area

According to the composition of the area of mechanized construction, the calculation rules of the area of each part are put forward.

3.1 Construction road area

The formula for calculating the floor area of the construction road is:
Construction road area = number of towers × average road length × average road width × seasons
According to the on-site measurement, the road width is between 3-5.5m. The road width of the high-voltage grade will increase slightly due to the large tonnage of the transport vehicle and the wide wheelbase. The average width is calculated as 4m. The 35kV line project Generally, the transportation volume is small, and some projects can use three motorized wheels for the secondary transportation of materials, so the width of the road can be calculated as 3m [2].
The road length of each project and each tower location is quite different. According to the on-site compensation data and the mechanized construction plan of each project, the average road length is calculated as 80m for flat land, 120m for hills, 160m for river network and mountain.
Since each construction stage requires access roads, the occupied area of the road is basically the same as the number of quarters corresponding to the construction period. The construction period of individual 35kV and 110kV line projects is within one season, and most of the construction period is two quarters. Projects with a duration of more than one year are calculated in three quarters.

3.2 Foundation site area

According to the type of foundation, the foundation construction site is divided into slab foundation, cast-in-place pile foundation and excavation pile foundation. Through the analysis of the factors affecting the area of the foundation construction site of different foundation types and different voltage levels and the comparison of the corresponding site area, it can be seen that the amount of concrete and the floor area of the tower (including the two variables of foundation root opening and column width) are the most important factors affecting the area of the foundation construction site. Therefore, in order to measure the area of the foundation construction site more accurately, it can be considered to use modeling to predict the area of the construction site based on the amount of concrete of different foundation types and the area of the tower, so as to provide a more scientific basis for the calculation of the area of the foundation construction site.

3.2.1 Computational Model Building of the foundation site area

The multiple linear regression model is one of the methods often used to explore the law that the dependent variable follows the changes of multiple independent variables. [5] In this paper, the amount of concrete and the floor area of the tower are used as independent variables, and the site area is used as the dependent variable, and the multiple linear regression method is used to establish a model and predict the area. Taking the area of the foundation construction site as the dependent variable y, and the single foundation concrete volume and the floor area of the tower as the independent variables x1 and x2, respectively, the following binary linear regression model is constructed.

$$y=\beta_0+\beta_1x_1+\beta_2x_2+\varepsilon \tag{1}$$

In the formula, y is the area of the foundation construction site, x1 is the amount of single-base foundation concrete, x2 is the floor area of the tower, β0、β1and β2 are the regression coefficients, and ε is the random model error. For the multiple linear regression model, the least squares method is usually used first—minimizing the sum of squares of the residuals to solve the estimation of the regression coefficients, and then the established regression equations and regression coefficients are calculated by means of the F test and the t test. The significance is tested. Given the test significance level α (usually 0.05), the model verification shows that the influence of the independent variable on the dependent variable y is significant.

Set $(y_i,\ x_{i1},\ x_{i2}), i=1,2,\ldots,n$ as the n-time sample observations of the variable, and use the least squares estimation method to give the estimated values of the unknown parameters β0、β1and β2 in the model, so that the sum of squares of errors is minimized, and the sum of squares of errors is:

$$Q(\beta_0,\beta_1,\beta_2)=\sum_{i=1}^{n}(y_i-\beta_0-\beta_1x_{i1}-\beta_2x_{i2})^2 \tag{2}$$

The following conditions are met:

$$Q(\hat\beta_0,\hat\beta_1,\hat\beta_2)=\sum_{i=1}^{n}(y_i-\hat\beta_0-\hat\beta_1x_{i1}-\hat\beta_2x_{i2})^2 = \min_{\beta_0,\beta_1,\beta_2}\sum_{i=1}^{n}(y_i-\beta_0-\beta_1x_{i1}-\beta_2x_{i2})^2 \tag{3}$$

The calculation from the above formula is an extreme value problem. Since Q is a non-negative quadratic function aboutβ0、β1、β2, its minimum value always exists. According to the principle of finding extreme values in calculus, only the partial derivative of Q with respect to the parameter to be estimated $\hat\beta_0$、$\hat\beta_1$、$\hat\beta_2$ is required, and set it equal to 0, the normal equation system for the parameter to be estimated can be obtained:

$$\begin{cases} \left.\dfrac{\partial Q}{\partial\beta}\right|_{\beta_0=\hat\beta_0}=-2\sum_{i=1}^{n}(y_i-\hat\beta_0-\hat\beta_1x_{i1}-\hat\beta_2x_{i2})^2=0 \\ \left.\dfrac{\partial Q}{\partial\beta}\right|_{\beta_1=\hat\beta_1}=-2\sum_{i=1}^{n}(y_i-\hat\beta_0-\hat\beta_1x_{i1}-\hat\beta_2x_{i2})^2=0 \\ \left.\dfrac{\partial Q}{\partial\beta}\right|_{\beta_2=\hat\beta_2}=-2\sum_{i=1}^{n}(y_i-\hat\beta_0-\hat\beta_1x_{i1}-\hat\beta_2x_{i2})^2=0 \end{cases} \tag{4}$$

$\hat\beta_0$、$\hat\beta_1$、$\hat\beta_2$ are obtained as follows:

$$\hat\beta_1=\frac{\sum Y_iX_{1i}\sum X_{2i}^2-\sum Y_iX_{2i}\sum X_{1i}X_{2i}}{\sum X_{1i}^2\sum X_{2i}^2-\left(\sum X_{1i}X_{2i}\right)^2} \tag{5}$$

$$\hat\beta_2=\frac{\sum Y_iX_{2i}\sum X_{1i}^2-\sum Y_iX_{1i}\sum X_{1i}X_{2i}}{\sum X_{1i}^2\sum X_{2i}^2-\left(\sum X_{1i}X_{2i}\right)^2} \tag{6}$$

$$\hat\beta_0=\bar y-\hat\beta_1\bar x_1-\hat\beta_2\bar x_2 \tag{7}$$

where , $Y_i=y_i-\bar y, X_{1i}=x_{1i}-\bar x_1, X_{2i}=x_{2i}-\bar x_2$, $\bar y$ is the sample mean of the dependent variable, $\bar x_1$ and $\bar x_2$ are the sample mean of the two independent variables, respectively.

Therefore, the calculation formula can be obtained:

$$y=\hat\beta_0+\hat\beta_1\bar x_1+\hat\beta_2\bar x_2 \tag{8}$$

3.2.2 Calculation results

Among the projects completed and put into operation in the past two years of a company, 70 representative transmission line projects are selected, including 5 500kV projects, 18 220kV projects, 29 110kV projects, and 18 35kV projects. For different voltage levels and different foundation types, the constructed binary linear regression model was used to measure and calculate, and the calculation formula of the foundation site area was obtained as follows:

1 、 500kV transmission line engineering foundation construction site area

（1）Slab foundation construction site area calculation formula:

$y=507.26+10.46x_1+2.32x_2$

Among them: x1 is the concrete volume of the new construction, x2 is the floor space of the new construction towers (the same below)

（2）Calculation formula of construction site area of cast-in-place pile foundation:

$y=380.81+2.36x_1+2.31x_2$

2 、 220kV transmission line engineering foundation construction site area

（1）Slab foundation construction site area calculation formula:

$y=149.91+12.91x_1+4.04x_2$

（2）Calculation formula of construction site area of cast-in-place pile foundation:

$y=330.51+4.18x_1+2.12x_2$

（3）Calculation formula of construction site area of steel pipe cast-in-place pile foundation:

$y=498.75+1.57x_1$

3 、 110kV transmission line engineering foundation construction site area

（1）Slab foundation construction site area calculation formula:

$y=40.98+14.37x_1+5.25x_2$

(4)

(2) Calculation formula of construction site area of cast-in-place pile foundation:

$y=383.65+3.97x_1+1.88x_2$

(3) Calculation formula of construction site area of steel pipe cast-in-place pile foundation:

$y=313.27+3.65x_1$

4、35kV transmission line engineering foundation construction site area

(1) Slab foundation construction site area calculation formula:

$y=3.85+9.07x_1+8.34x_2$

(2) Calculation formula of construction site area of cast-in-place pile foundation:

$y=287.08+6.52x_1+1.60x_2$

(3) Calculation formula of construction site area of steel pipe cast-in-place pile foundation:

$y=291.63+4.68x_1$

3.3 Pole tower assembly construction site

According to the on-site investigation, after the mechanized construction is adopted, the on-site hoisting mainly relies on cranes, and the traditional construction method of "external pull wire and internal suspension holding pole" is cancelled. For the site, it is important to consider the increase in the number of compensation seasons caused by different construction periods[3]. When the construction site of the excavated piles or some cast-in-place pile foundations is small, the construction site may be appropriately expanded during the construction of the towers. It is also possible to adjust the construction organization, adopt the construction method of partial grouping and hoisting in batches to reduce the construction site occupied, but it will affect the efficiency of mechanized construction and significantly increase the number of crane shifts used. In order to take into account the site area requirements in the construction stage of the tower group, the excavation foundation construction site needs to be appropriately enlarged on the basis of formula calculation.

3.4 Spanning frame site area

spanning frame site area=The length of the spanning frame × the width of the spanning frame × the adjustment factor of the average intersection angle × the increase coefficient of transportation, stacking, and cable。

According to the design data, the width of the spanning frame is determined as the total width of the wire cross-arm. Considering the wind deviation allowance, the two sides are expanded by 1 meter. The form of spanning frame is determined according to the site conditions. The average spacing of each row is considered to be 1.5 meters, and the front and rear of the construction site are expanded by 1 meter. The inclined span is considered at an average angle of 70°, and the area is increased by 6.4%. The area of transportation, stacking and pulling wire during the erection of the spanning frame shall be calculated as 40% of the net area.

According to the above calculation principles, the spanning area of each voltage level is determined as shown in Table 2.

Table 2. The spanning area of each voltage level

No.	Crossed line	Cross frame form	Cross frame height	35kV	110kV	220kV	500kV
				7.2m	10m	16.8m	27.2m
1	Low-voltage line	Two rows on one side	3.5m	38	52	88	142
2	10kV	Double side two rows	7m	75	104	175	284
3	35kV	Double side three rows	10m	107	149	250	405
4	110kV	Double-side three-row, four-row each 50%	11.5m	\	171	288	466
5	220kV	Double-side four-row	13m	\	\	325	527
6	500kV	Double Side Five Row	16m	\	\	\	648

3.5 Wire channel and accessory installation site area

Wire channel and accessory installation site area=Transmission line path length × comprehensive channel width.

It can be seen from the on-site investigation and young crop compensation data that the occupied area of the on-site pay-off channel is small. When the width of the cross arm is small, the single-circuit or double-circuit line can be combined into one construction along the center line of the line during the process of unwinding the traction rope. Therefore, the width of the wiring construction channel of 35-220kV voltage level can be calculated as 1m wide.

The installation of inter-phase spacers in the stalls needs to occupy the construction area of the pressure channel. Due to the small number of inter-phase spacers, the increase of the site area is not obvious, so they can be incorporated into the wire construction channel.

A set of conductor spacer bars is installed at an average of 40-60m per phase conductor, and the number of installations is large. The 500kV conductor generally adopts four splits, and the weight of the spacer bar is obviously doubled compared to 110kV and 220kV (see the table below for the weight of the spacer bar for each voltage level). It is installed on the tower by the climbing operator and needs to be installed by the ground operators at the installation point, so the increase in the construction area is obvious. The length of the 500kV wire cross-arm is long, and it is difficult to merge the traction rope into the same channel during the deployment process. Therefore, the width of the construction channel width for single and double circuits of 500kV voltage level is calculated as 2m wide.

The attachment installation construction site at the base of the straight tower is obviously smaller than the construction site of the group tower, and the installation of the hanging line of the corner tower and the installation of the insulator string basically overlaps the construction site of the pole tower. After the construction of the tower group is completed, in order to reduce the difficulty of on-site coordination and avoid secondary compensation, the construction site is generally reserved until the attachment is installed and completed. Due to the different construction periods, the number of construction site compensation seasons may be increased for each project.

3.6 Traction tension field area

Traction tension field area=The total number of venues across the board × venue area

The traction field is mainly used to pull the wire, and also store the ground wire or optical cable; the tension field needs to store the wire, and carry out the work of tension spreading and wire crimping, and two consecutive pay-off segments can share the same tension field, so The site area is significantly larger than the traction field.

According to the past billing habits, combined with the actual compensation data, the traction and tension field site area of each voltage level is set to 2000 ㎡/place for 35kV line projects, 2500 ㎡/place for 110kV line projects, 3000 ㎡/place for 220kV line projects, and 500kV line projects. Line engineering 4000 ㎡/place. According to the third chapter of "110kV ~ 750kV Overhead Transmission Line Tension Line Construction Technology Guidelines" (DL T5343-2018), the tension line section should generally be controlled within 20 base towers, and high-speed railways, railways, and high-voltage power lines should be comprehensively considered. Discontinuous construction caused by special spanning and separate lay-out segments are considered on average as 15 bases/place, and are calculated by rounding up to an integer. [4]

The normal split length of the optical cable is about 4km. It is considered that half of the optical cable stretch site shares the same site with the wire stretch site, and the other half needs to be set up separately. Therefore, the optical cable stretch site is added at 8km/place (or according to the designed single site. Divide the number of root sub-discs by 2 and round up to an integer), and the area of a single place is calculated as 900 square meters according to the on-site investigation.

4. Empirical Analysis

4.1 Sample selection

Among the projects completed and put into operation in the past two years of a company, 70 representative transmission line projects are selected, including 5 500kV projects, 18 220kV projects, 29 110kV projects, and 18 35kV projects. The sample receipts mainly cover the following:

(1) List of towers in the as-built drawings, foundation construction drawings, grounding construction drawings, and list of towers and towers.

(2) Commencement, completion and transition reports, approved construction organization design and special construction plans related to mechanized construction.

(3) Statistical table of completed engineering quantities (focusing on foundation opening, concrete, earthwork quantities, tower type, tower height and spanning objects).

(4) Scanned copies of compensation data for young crops and tower bases, trees, nurseries, green belts, road occupations and other compensation data within the range of the access road, and fill in the "statistical table of actual value of mechanized construction site area" according to the actual compensation data.

(5) At each construction stage, there are about 10 electronic photos that can reflect the occupancy of the construction site on site.

4.2 Mechanized construction area calculation

According to the income data of each voltage level project and the calculation rules of mechanized construction area proposed in this paper, the construction site area of each voltage level is calculated and compared with the actual compensation area, as shown in Table 3.

Table 3 Comparison of the area calculated under the new rules for a single kilometer, the actual compensation area, and the area calculated under the original rules

No.	Voltage level	Compensation area calculated by the original rule (per acre/km)	Compensation area calculated by this rule (per acre/km)	Actual compensation area (per acre/km)	Deviation rate (the original rule is more realistic)	Deviation rate (the rule in this article is more realistic)
1	35kV	10	9.56	10.44	-4.21%	-8.43%
2	110kV	13.41	14.27	16.68	-19.60%	-14.45%
3	220kV	11.90	15.57	15.08	-21.09%	3.25%
4	500kV	22.79	15.72	16.38	39.13%	-4.03%

It can be seen from Table 3 that the deviation of the compensation area calculated by the rules in this paper compared with the actual compensation area is within 15%, and except for the 110kV voltage level, the other deviations are within 10%. The compensation area calculated by the original rules has a larger deviation rate than the actual compensation area. Except for the 35kV voltage level deviation within 10%, the other deviations are greater than 15%, and the 500kV voltage level deviation rate is as high as 39.13%. The calculation rules of mechanized construction area proposed in this paper are effective.

5. Conclusion

Mechanized construction is an important measure to improve the construction efficiency and safety quality of transmission line projects. Based on the current calculation method of the construction site area of transmission lines, combined with specific projects and typical design schemes, and fully considering the difference between mechanized construction and traditional construction modes, this paper proposes the composition and calculation rules of the mechanized construction area, and draws the following conclusions. :

1) The occupied area of mechanized construction can be divided into six parts: construction road, foundation site, tower site, spanning frame site, wire channel and accessory installation site and traction tension field.

2) Clarify the calculation rules for the occupied area of each part of the mechanized construction, for example, the construction road area is equal to the product of the number of towers, the average road length (or base-by-base measured calculation) and the average road width; The product of the integrated channel width, etc.

Compared with the original calculation rule, the calculation rule of mechanized construction area proposed in this paper is more suitable for the situation that the transmission capacity of the current line is significantly increased, and the cross-section of the conductor, the tower, and the amount of foundation works are greatly increased, which makes up the blank of the calculation of the mechanized construction area.

References

1. General Electric Power Planning and Design Institute of the State Power Corporation. Detailed rules for the preparation of estimates for power transmission projects [M]. Beijing: China Electric Power Press, 1998.
2. Zhang Aihu, Su Xiaoqing, Li Feng. Deepening application of mechanized construction scheme of 110kV transmission line [J]. Jiangxi Electric Power, 2018,42(02):29-32.
3. Liu Xingzhong, Jia Lei. Differentiation Analysis of Mechanized Construction Measures for Transmission Line Engineering [J]. Science and Technology Innovation and Application, 2020(35):87-88.
4. Shao Dongliang, Wang Long. Application of mechanized construction technology in the whole process of power grid engineering [J]. Journal of State Grid Technology College, 2016,19(03):25-29.
5. Feng Xiaojuan. Uncertainty Evaluation Method of Binary Linear Regression [J]. Journal of Xi'an University of Architecture and Technology (Natural Science Edition), 2000(03):304-306.

Path selection for carbon peaking and carbon neutralization based on uncertainty theory

Hai Chi, Xincheng Wang

Jinan Preschool Education College, Jinan 250307, China

Abstract: Formulating reasonable policies to help achieve the goal of carbon peaking and carbon neutrality while ensuring economic growth is a key issue for the government since the goal was proposed. Combining with the relevant realistic background, we consider carbon emission reduction policies including green bond issuance and carbon emission allowances, analyze how the government should choose a method of strategic optimization to maximize corporate profits under the specified carbon emission reduction rate, and construct a related model of uncertain programming called optimal model of carbon emission reduction strategy, in order to obtain the optimal carbon emission reduction policies that the government should formulate under different circumstances while ensuring sustainable economic development.

Keywords: Uncertainty theory; The goal of carbon peaking and carbon neutrality; Green bonds; Carbon emission allowances

1. Introduction

"Carbon Neutrality" is an important strategic decision to promote China's energy and economic transformation and to build a community of human destiny, which is also called the "Carbon Neutrality" goal. China is a large energy-consuming country, and the whole society is facing a huge challenge under the goal of carbon peaking and carbon neutrality. On October 24, 2021, the "Opinions of the Central Committee of the Communist Party of China (CPC) and the State Council on the Complete and Accurate Implementation of the New Development Concept to Achieve Carbon Neutrality" stated that China will reach the peak of carbon emissions by 2030 and achieve carbon neutrality by 2060. Therefore, how to achieve the goal of carbon peaking and carbon neutrality has become a hot issue for scholars at home and abroad to study.

A large number of researchers focused on the process of carbon emission reduction, Ding et al. [1] studied China's carbon emission trading policy and considered the classification of emission reduction policies, Rebredo and Ugolin [2] focused on green bonds and financial market interaction through empirical analysis, Rannou et al.[3] studied how green bonds reduced levels of carbon emission and investigated the green bond market and the European carbon market correlation, they stated that green bonds can be used for future carbon reduction but do not consider it as a carbon neutral instrument in carbon tax, Hussain et al. [4] considered the important role of emission allowances in carbon emissions, and Li et al. [5] studied the allocation of emission allowances at regional and provincial levels in China and determined the feasibility of national emission allowance allocation. In addition, Lee et al. [6] considered the role played by green financing in the process of carbon emission reduction.

Integrating the research of scholars and the requirements of the new development concept, we consider carbon emission reduction policies including green bond issuance and carbon emission allowances, and analyze the problem of how the government should choose a strategic optimization method to maximize corporate profits under a prescribed carbon emission reduction rate in this paper. Moreover, uncertainty theory is applied to solve the problem.

2. Preparatory knowledge

2.1 Uncertainty theory

Definition 2.1.1 [7] Uncertainty measure

Let Γ be a non-empty set and the set L consisting of subsets of Γ be called a σ algebra if it satisfies

(1) $\Gamma \in L$;

(2) If $\Lambda \in L$, then $\Lambda^c \in L$;

(3) If $\Lambda_1, \Lambda_2, \cdots, \Lambda_n \in L$, then the elements in $\Lambda_1 \cup \Lambda_2 \cup \cdots \cup \Lambda_n \in L$, L are called events.

The uncertainty measure is a function from L to [0,1]. To obtain an axiomatic definition of the uncertainty measure, a value $M\{\Lambda\}$ is assigned to each event Λ to indicate the confidence that the event Λ occurs. To ensure that $M\{\Lambda\}$ has certain mathematical properties, Liu[7] proposed the following three axioms.

Axiom 1: (Normative axiom) for the set of argument domains $\Gamma, M\{\Gamma\}=1$;

Axiom 2: (Duality axiom) For any $\Lambda \in L$, $M\{\Lambda\}+M\{\Lambda^c\}=1$;

Axiom 3: (Subadditivity axiom) For any sequence of countable events $\{\Lambda_i\}$, there is

$$M\left\{\bigcup_{i=1}^{\infty}\Lambda_i\right\} \le \sum_{i=1}^{\infty} M\{\Lambda_i\} \tag{1}$$

in this case, (Γ, L, M) is called the uncertain space.

Definition 2.1.2[7] Uncertain variables

Uncertain variables are measurable functions from the uncertainty space (Γ, L, M) to the set of real numbers. For any Borel set B in the set $\{\xi \in \mathrm{B}\}=\{\gamma \in \Gamma \mid \xi(\gamma) \in \mathrm{B}\}$ is an event.

Definition 2.1.3[7]Uncertainty distribution

The uncertainty distribution of the uncertainty variable ξ is defined as

$$\Phi(x)=M\{\xi \le x\} \tag{2}$$

where x is any real number.

Definition 2.1.4[7] :The uncertain variable ξ is called a normal uncertain variable when its uncertainty distribution is

$$\Phi(x)=\left(1+\exp\left(\frac{\pi(e-x)}{\sqrt{3}\sigma}\right)\right)^{-1}, x \in R \tag{3}$$

is denoted as $\xi \sim N(e,\sigma)$, where e and σ are both real numbers and $\sigma > 0$.

By the nature of uncertain variables, if $\xi_1, \xi_2, \cdots, \xi_n$ is all normal uncertain variables, i.e.

$$\xi_i \sim \mathrm{N}(e_i, \sigma_i) \qquad i=1,2,\cdots,n \tag{4}$$

then the variable $\eta=\sum_{i=1}^{n}\lambda_i\xi_i$ after their weighted summation remains a normal uncertain variable for any $\lambda_i \ge 0, i=1,2,\cdots,n$

$$\eta \sim \mathrm{N}\left(\sum_{i=1}^{n}\lambda_i e_i, \sum_{i=1}^{n}\lambda_i\sigma_i\right) \tag{5}$$

Theorem 2.1.1[7]: Let ξ be an uncertain variable with an uncertain inverse distribution $\Phi^{-1}(\alpha)$, then

$$M\{\xi \le c\} \ge \alpha \tag{6}$$

when and only when

$$\Phi^{-1}(\alpha) \le c \tag{7}$$

where α and c are both constants and $0 < \alpha < 1$.

Theorem 2.1.2[7]: Let $\xi_1, \xi_2, \cdots, \xi_n$ be mutually independent uncertain variables, each with a regular uncertain distribution $\Phi_1, \Phi_2, \cdots, \Phi_n$, if $f(x_1, x_2, \cdots, x_n)$ is continuous and strictly increasing for $x_1, x_2, \cdots, x_m$ and continuous and strictly decreasing for $x_{m+1}, x_{m+2}, \cdots, x_n$, then

$$\xi=f(\xi_1, \xi_2, \cdots, \xi_n) \tag{8}$$

is an uncertain variable with the following uncertain inverse distribution function

$$\Psi^{-1}(\alpha)=f\left(\Phi_1^{-1}(\alpha), \Phi_2^{-1}(\alpha), \cdots, \Phi_m^{-1}(\alpha), \Phi_{m+1}^{-1}(1-\alpha), \cdots, \Phi_n^{-1}(1-\alpha)\right) \tag{9}$$

Theorem 2.1.3[7]: Let ξ be an uncertain variable that obeys an uncertain distribution function of Φ , if there exists a mathematical expectation of this uncertain variable, then

$$E[\xi]=\int_0^1 \Phi^{-1}(\alpha)d\alpha \tag{10}$$

where the uncertainty inverse distribution function of the normal uncertainty variable $\xi \sim \mathrm{N}(e,\sigma)$ is

$$\Phi_1^{-1}(\alpha)=e+\frac{\sigma\sqrt{3}}{\pi}\ln\frac{\alpha}{1-\alpha} \tag{11}$$

Theorem 2.1.4[7]: If ξ_1, ξ_2 are two mutually independent uncertain variables with definite expected values, for any real numbers a_1 and a_2 , then we have

$$E[a_1\xi_1+a_2\xi_2]=a_1E[\xi_1]+a_2E[\xi_2] \tag{12}$$

2.2 Uncertain programming

2.2.1 Model

Suppose x is a decision vector and ξ is an uncertain vector, since the uncertain objective function $f(x,\xi)$ cannot be directly maximized, we can maximize its expected value

$$\max \mathrm{E}[f(x,\xi)] \tag{13}$$

In addition, since the uncertainty constraint $g_j(x,\xi) \le 0, j=1,2,\cdots,p$ does not define an explicit feasible set, it is natural to use the confidence level $\alpha_1, \alpha_2, \cdots, \alpha_p$ to describe the uncertainty constraint

$$M\{g_j(x,\xi) \le 0\} \ge \alpha_j, j=1,2,\cdots,p \tag{14}$$

To maximize the expected value, Liu [8] proposed the following model.

$$\begin{cases} \max \mathrm{E}\left[f(x,\xi) \right] \\ s.t. \\ M\left\{ g_j(x,\xi) \le 0 \right\} \ge \alpha_j, j=1,2,\cdots,p \end{cases} \quad (15)$$

Theorem 2.2.1[8] :Assume that the objective function $f(x,\xi_1,\xi_2,\cdots,\xi_n)$ is strictly increasing for $\xi_1,\xi_2,\cdots,\xi_m$ and strictly decreasing for $\xi_{m+1},\xi_{m+2},\cdots,\xi_n$, and if $\xi_1,\xi_2,\cdots,\xi_n$ is an independent uncertain variable with uncertain distribution $\Phi_1,\Phi_2,\cdots,\Phi_n$, then the expectation $\mathrm{E}\left[f(x,\xi_1,\xi_2,\cdots,\xi_n) \right]$ is equivalent to

$$\int_0^1 f\left(x,\Phi_1^{-1}(\alpha),\Phi_2^{-1}(\alpha),\cdots,\Phi_m^{-1}(\alpha),\Phi_{m+1}^{-1}(1-\alpha),\cdots,\Phi_n^{-1}(1-\alpha)\right)d\alpha \quad (16)$$

Theorem 2.2.2[8] : Assuming that the constraint function $g(x,\xi_1,\xi_2,\cdots,\xi_n)$ is strictly increasing for $\xi_1,\xi_2,\cdots,\xi_k$ and strictly decreasing for $\xi_{k+1},\xi_{k+2},\cdots,\xi_n$, and assuming that the uncertain variable $\xi_1,\xi_2,\cdots,\xi_n$ has an independent uncertainty distribution $\Phi_1,\Phi_2,\cdots,\Phi_n$, the chance constraint

$$M\left\{ g(x,\xi_1,\xi_2,\cdots,\xi_n) \le 0 \right\} \ge \alpha, j=1,2,\cdots,p \quad (17)$$

holds when and only when

$$g\left(x,\Phi_1^{-1}(\alpha),\cdots,\Phi_k^{-1}(\alpha),\Phi_{k+1}^{-1}(1-\alpha),\cdots,\Phi_n^{-1}(1-\alpha)\right) \le 0 \quad (18)$$

2.2.2 Equivalent form of the model

From theorem 2.2.1, for the model (15), the expected value of the objective function $\mathrm{E}\left[f(x,\xi_1,\xi_2,\cdots,\xi_n) \right]$ is equivalent to

$$\int_0^1 f\left(x,\Phi_1^{-1}(\alpha),\Phi_2^{-1}(\alpha),\cdots,\Phi_m^{-1}(\alpha),\Phi_{m+1}^{-1}(1-\alpha),\cdots,\Phi_n^{-1}(1-\alpha)\right)d\alpha \quad (19)$$

By theorem 2.2.2, the chance constraint $\mathrm{M}\left\{ g(x,\xi_1,\xi_2,\cdots,\xi_n) \le 0 \right\} \ge \alpha, j=1,2,\cdots,p$ holds when and only when

$$g_j\left(x,\Phi_1^{-1}(\alpha_j),\cdots,\Phi_k^{-1}(\alpha_j),\Phi_{k+1}^{-1}(1-\alpha_j),\cdots,\Phi_n^{-1}(1-\alpha_j)\right) \le 0 \quad (20)$$

Thus, the uncertainty planning model (2-15) is equivalent to the following clear model.

$$\begin{cases} \max \int_0^1 f\left(x,\Phi_1^{-1}(\alpha),\Phi_2^{-1}(\alpha),\cdots,\Phi_m^{-1}(\alpha),\Phi_{m+1}^{-1}(1-\alpha),\cdots,\Phi_n^{-1}(1-\alpha)\right)d\alpha \\ s.t. \\ g_j\left(x,\Phi_1^{-1}(\alpha_j),\cdots,\Phi_k^{-1}(\alpha_j),\Phi_{k+1}^{-1}(1-\alpha_j),\cdots,\Phi_n^{-1}(1-\alpha_j)\right) \le 0, j=1,2,\cdots,p \end{cases} \quad (21)$$

3. Model construction

3.1 Problem description and assumptions

The achievement of the goal of carbon peaking and carbon neutrality requires the participation of both business and government. In order to ensure that carbon emission reduction does not hinder economic development, the government must consider the profits of enterprises when making decisions. The goal of enterprises is to maximize their profits at a specified rate of carbon reduction, and the issue considered by the government is how to develop reasonable carbon reduction policies while ensuring that maximize enterprises' profits.

In this paper, we consider the optimization problem of carbon emission reduction strategies involving n enterprises and build a relevant model to find the optimal solution of the relevant carbon emission reduction policy, that is, the amount of green bonds and carbon emission allowances issued by the government to each enterprise, by first making the following assumptions:

Hypothesis 1: Decisions regarding carbon reduction are made at the beginning of the production process.

Hypothesis 2: The price of carbon is a standard determined by the whole market.

Hypothesis 3: The price of the product produced by an enterprise is determined by the demand of market.

3.2 Parameter introduction and model construction

The profit of an enterprise consists of sales revenue, cost of production, green bonds and carbon emission allowances[9]. The expression of sales revenue is $p_i D_i$, which is the product of the price and demand of products; the expression of cost is $v_i D_i$, which is the product of cost of unit product's production and products' demand. The number of green bonds is represented by G_i and the expression of carbon emission allowances is $p_c(D_i e_i(1-rgG_i m)-mQ_i)$. In summary, the expression of the profit function of for the i th enterprise can be obtained as follows

$$z_i = (p_i - v_i)D_i + gG_i - p_c(D_i e_i(1-rgG_i m)-mQ_i) \quad (21)$$

Then the expression for the sum of the profits of n enterprises is

$$\sum_{i=1}^{n}\left((p_i - v_i)D_i + gG_i - p_c(D_i e_i(1-rgG_i m)-mQ_i)\right) \quad (23)$$

which is the uncertain objective function. Since this function cannot be minimized directly, we can minimize its expected value with the expression

$$\min \ \mathrm{E}\left(\sum_{i=1}^{n}\left((p_i - v_i)D_i + gG_i - p_c(D_i e_i(1-rgG_i m)-mQ_i)\right)\right) \quad (24)$$

In an uncertain economic and social environment, the profits of enterprises are always changing. To ensure the development of the economy, the government expects the chances that the profits of enterprises are higher than the minimum threshold to be higher than a certain probability, so this section uses the opportunity constraint to manage the profits of enterprises, then the expression of the enterprise profit constraint is

$$M\left\{\sum_{i=1}^{n}\left((p_i - v_i)D_i + gG_i - p_c(D_i e_i(1-rgG_i m)-mQ_i)\right) \ge nz_0\right\} \ge \alpha \quad (25)$$

where M is the uncertainty measure, z_0 is the minimum threshold for each enterprise's profit, and α is the confidence level.

Therefore, the optimal model of carbon reduction strategy is constructed as follows.

$$\begin{cases} \min \quad \mathrm{E}\left(\sum_{i=1}^{n}\left((p_i - v_i)D_i + gG_i - p_c(D_i e_i(1-rgG_i m) - mQ_i)\right)\right) \\ s.t \\ M\left\{\sum_{i=1}^{n}\left((p_i - v_i)D_i + gG_i - p_c(D_i e_i(1-rgG_i m) - mQ_i)\right) \geq nz_0\right\} \geq \alpha \\ G_i, Q_i \geq 0, i = 1,2,\cdots,n \end{cases} \tag{26}$$

The main parameters used in this paper are shown in Table 1, where v_i is the uncertain variable. Because the economic environment is more complex and in constant change, historical data often do not reflect the future trend of products' production costs well, in this case, the government must artificially estimate rather than use historical data to describe the production cost of per unit of products, but there will be a certain error between the artificial estimate and the true value, and if the probability theory is used will magnify this error, while using uncertain variables to describe the artificial estimate will not produce error amplification, so v_i is considered as an uncertain variable.

Table 1 Introduction of main parameters

Name	Meaning
p_i	The products' price of the i th enterprise
v_i	The i th enterprise's production cost of per unit of product
D_i	The market demand for the products of the i th enterprise
g	Whether or not to set green bond terms (takes the value of 0 or 1)
G_i	Number of i th enterprise allowed to issue green bonds
p_c	Price of Carbon
e_i	Carbon emissions of the i th enterprise's unit product
r	Carbon reduction rate
m	Whether to implement carbon emission allowances(takes the value of 0 or 1)
Q_i	Carbon allowances for the i th enterprise

3.3 Equivalent form of the model

Theorem 3.3.1: Assuming $\Phi_i, i = 1,2,\cdots,n$ is a continuous monotonically increasing uncertain distribution function of v_i, the model can be rewritten in the following form.

$$\begin{cases} \min \quad \sum_{i=1}^{n}\left(\left(p_i - \int_0^1 \Phi_i^{-1}(\alpha)d\alpha\right)D_i + gG_i - p_c(D_i e_i(1-rgG_i m) - mQ_i)\right) \\ s.t \\ \sum_{i=1}^{n}\left(\left(p_i - \Phi_i^{-1}(\alpha)\right)D_i + gG_i - p_c(D_i e_i(1-rgG_i m) - mQ_i)\right) \geq nz_0 \\ G_i, Q_i \geq 0, i = 1,2,\cdots,n \end{cases} \tag{27}$$

Proof: First, by Theorem 2.1.4, we know that the expectation of the firm's profit is

$$\sum_{i=1}^{n}\left(\left(p_i - \mathrm{E}(v_i)\right)D_i + gG_i - p_c(D_i e_i(1-rgG_i m) - mQ_i)\right) \tag{28}$$

then, because of Theorem 2.1.3, we have.

$$\mathrm{E}(v_i) = \int_0^1 \Phi_i^{-1}(\alpha)d\alpha \tag{29}$$

Therefore, the objective function is proven.
Second, according to Theorems 2.1.1 and 2.1.2 it follows that

$$\sum_{i=1}^{n}\left((p_i - v_i)D_i + gG_i - p_c(D_i e_i(1-rgG_i m) - mQ_i)\right) \tag{30}$$

the inverse function at α is

$$\sum_{i=1}^{n}\left(\left(p_i - \Phi_i^{-1}(1-\alpha)\right)D_i + gG_i - p_c(D_i e_i(1-rgG_i m) - mQ_i)\right) \tag{31}$$

then, based on the monotonicity of the uncertain variables, we can obtain the constrained expression for the firm's profit:

$$\sum_{i=1}^{n}\left(\left(p_i - \Phi_i^{-1}(\alpha)\right)D_i + gG_i - p_c(D_i e_i(1-rgG_i m) - mQ_i)\right) \geq nz_0 \tag{32}$$

Therefore, Theorem 3.3.1 is proved.
Theorem 3.3.2: Assuming that the production cost per unit of product is a normal uncertain variable with a distribution of $v_i \sim \mathrm{N}(\varepsilon_i, \omega_i)$, the equivalent form of the model is

$$\begin{cases} \min \quad \sum_{i=1}^{n}\left((p_i - \varepsilon_i)D_i + gG_i - p_c(D_i e_i(1-rgG_i m) - mQ_i)\right) \\ s.t \\ \sum_{i=1}^{n}\left(\left(p_i - \left(\varepsilon_i + \frac{\omega_i\sqrt{3}}{\pi}\ln\frac{\alpha}{1-\alpha}\right)\right)D_i + gG_i - p_c(D_i e_i(1-rgG_i m) - mQ_i)\right) \geq nz_0 \\ G_i, Q_i \geq 0, i = 1,2,\cdots,n \end{cases} \tag{33}$$

Proof: Since v_i is a normal uncertain variable, the expression of its inverse distribution function is

$$\Phi_i^{-1}(\alpha) = \varepsilon_i + \frac{\omega_i\sqrt{3}}{\pi}\ln\frac{\alpha}{1-\alpha} \tag{34}$$

according to Theorem 2.1.3, we can calculate the expectation of v_i as

$$\mathrm{E}(\xi_i) = \varepsilon_i \tag{35}$$

Therefore, Theorem 3.3.2 is proved.

4. Numerical calculation example

4.1 Problem description

In this paper, two companies producing batteries in Shenzhen are selected as numerical examples, Li Kexing Battery (Co., Ltd.) and Shenzhen Gree Battery (Grep). Since the two products are similar and both are established in Shenzhen, these two companies are selected for numerical experiments.

The production costs per unit of product for the two companies were estimated by experts based on experience, and the relevant distributions are shown in Table 2, and the values of the remaining parameters are shown in Table 3.

Table 2 Uncertainty distribution of v_i

Company Name	The uncertainty distribution of v_i
Li Kexing Battery	$v_1 \sim \mathrm{N}(2.2385, 0.1)$
Shenzhen GREE Battery	$v_2 \sim \mathrm{N}(2.5232, 0.1)$

Table 3 Values of remaining parameter

Name	Take value
p_1	3
p_2	4
v_1	2.2385
v_2	2.5232
D_1	2.8
D_2	2.5
p_c	25
e_1	0.000086787
e_2	0.00007598
r	0.0941

4.2 Results

The optimal values of the government's strategies for achieving the goal of carbon peaking and carbon neutrality were solved using the function called fmincon in MATLAB software. The optimal issuance of green bonds and optimal allowances for carbon emissions are shown in Table 4 and Table 5.

Table 4 Optimal issue size of green bonds

Company Name	Green Bond Issuance
Li Kexing Battery	59997
Shenzhen GREE Battery	59996

Table 5 Optimal carbon emission allowances

Company Name	Carbon Emission Allowance
Li Kexing Battery	2399.9
Shenzhen GREE Battery	2399.8

5. Conclusion

In order to achieve the goal of carbon peaking and carbon neutrality while ensuring economic development, we innovatively apply uncertainty theory, consider the maximization of enterprises' profit under the premise of the specified carbon emission reduction rate, construct a optimal model of carbon emission reduction strategy by combining the relevant realistic background, and solve the model in this paper. Secondly, two battery manufacturing companies in Shenzhen with similar products are selected for numerical experiments, and the optimal quantities of green bonds and carbon emission allowances are solved for the two companies under the corresponding constraints. We provide a rational strategy for the optimization of policies related to the realization of the goal of carbon peaking and carbon neutrality.

References

1. Ding G, Deng Y, Lin S. A study on the classification of China's provincial carbon emissions trading policy instruments: Taking Fujian Fujian province as an example [J]. Energy Reports, 2019, 5: 1543-1550.
2. Reboredo J C, Ugolini A. Price connectedness between green bond and financial markets [J]. Economic Modelling, 2020, 88: 25-38.
3. Rannou Y, Boutabba M , Barneto P . Are Green Bond and Carbon Markets in Europe complements or substitutes? Insights from the activity of power firms[J]. Post-Print, 2021.
4. Hussain J, Pan Y, Ali G, et al. Pricing behavior of monopoly market with the implementation of green technology decision under emission reduction subsidy policy[J]. Science of the Total Environment, 2020, 709: 136110.
5. Li Z, Zhao T, Wang J, et al. Two-step allocation of CO2 emission quotas in China based on multi-principles: Going regional to provincial[J]. Journal of Cleaner Production, 2021, 305: 127173.
6. Lee C C, Lee C C. How does green finance affect green total factor productivity? Evidence from China[J]. Energy Economics, 2022, 107: 105863.
7. B Liu. Uncertainty Theory, 2nd edn [M]. Springer-Verlag, Berlin, 2007.
8. B Liu. Theory and Practice of Uncertain Programming,2nd edn [M]. Springer-Verlag, Berlin, 2009.
9. Hussain J, Lee C C. A green path towards sustainable development: Optimal behavior of the duopoly game model with carbon neutrality instruments [J]. Sustainable Development, 2022.

Decoupling Energy Consumption from Economic Growth in Beijing-Tianjin-Hebei Region from 2005 to 2019

Jingxin An

School of Economics and Management, Tianjin University of Science and Technology, 29 13th Street, Binhai New Area Economic Development Zone, Tianjin, 300457, PR. China

Abstract: As the capital economic circle in China, Beijing-Tianjin-Hebei region has experienced rapid economic development depending on a large amount of fossil fuel energy consumption in decades. Based on Tapio decoupling index, this paper logically analyzes the nexus between economic growth and energy consumption in Beijing-Tianjin-Hebei region from 2005 to 2019. An interesting result is obtained: except for the strong decoupling state in 2012-2013, the relationship between economy and energy consumption showed weak decoupling state for the rest of the research period. In addition, the decoupling elasticity coefficient of weak decoupling varies greatly in different years. The empirical results suggest that adjust economic development mode and transform energy consumption structure are the most necessary means for the future coordinated development of energy-economy-environment in Beijing-Tianjin-Hebei region.

Keywords: Beijing-Tianjin-Hebei region; economic growth; energy consumption; decoupling.

1. Introduction

Energy is the power of economic development and the material basis for human survival and development. Since the implementation of the reform and opening-up policy, China's economy has experienced rapid development. During the past 20 years from 2001 to 2020, China's GDP rose from 11.08631 trillion yuan in 2001 to 101.3567 trillion yuan in 2020, with an average annual growth rate of 8.3% [1]. However, China's obvious economic achievements are mainly supported by a large amount of fossil fuel energy consumption. During the same period, China's energy consumption increased from 1555.47 million tons of standard coal in 2001 to 4980 million tons of standard coal in 2020, with an average annual growth rate of 6.0 percent [2]. Although China is actively developing hydropower, nuclear power, solar energy, Marine energy and other clean energy according to local conditions [3], fossil fuel energy consumption still plays a major role in China's energy consumption with the proportion of 84.1% [4]. Massive consumption of fossil energy will inevitably lead to environmental pollution, affect energy security and harm China's future sustainable development. Therefore, China must change from an extensive economic development model that simply pursues economic development speed and GDP to a new normal economic development model.

Beijing-Tianjin-Hebei region is the "capital economic circle" in China, located in the heart of The Bohai Rim in Northeast China. In 2021, the GDP of Beijing-Tianjin-Hebei region reached 9.6 trillion yuan, accounting for 8.3% of China's total GDP. Among them, Beijing's GDP is 4,026.96 billion yuan, Hebei's 4,039.13 billion yuan, and Tianjin's 1,569.51 billion yuan, with an average annual increase of 6.3%, 6.3% and 5.0% respectively compared with 2013 [5]. Despite the rapid economic development in this region, energy consumption is also supported by fossil energy consumption. In 2019, energy consumption in The Beijing-Tianjin-Hebei region totaled 481.46 million tons of standard coal, accounting for 9.9% of national total energy consumption [6]. Li Guoping, dean of Peking University Capital Development Institute, mentioned that the Beijing-Tianjin-Hebei region has accounted for 11%-12% of China's total carbon emissions in recent years [7], both higher than the proportion of population and GDP in the country at the 2021 Beijing-Tianjin-Hebei Collaborative Development Forum under the Carbon Neutral Vision (The Beijing-Tianjin-Hebei region has 7.81% of population and 8.5% of GDP) [8-9]. A large amount of energy consumption has caused enormous environmental pressure to Beijing-Tianjin-Hebei region and posed a great challenge to the sustainable development and low-carbon green development. It is of great significance to study relationship between economic growth and energy consumption in Beijing-Tianjin-Hebei region. The purpose of the research is to provide targeted guidance and suggestions in order to change the economic development mode, improve the energy consumption structure, reduce carbon emissions, build world-class urban agglomeration, and promote the high-quality development of China's economy in the future.

https://doi.org/10.1051/e3sconf/202235801012

Up to now, many scholars have used a variety of methods to study the relationship between China's regional energy consumption and economic growth. The methods commonly involves regression equation, correlation analysis, grey correlation analysis, causality, cointegration test, VAR model, decoupling model, *etc.*

When it comes to regression equation, Zhao Jianhui used principal component analysis to extract the main factors affecting China's energy consumption, and then conducted regression analysis, and found that economic development, industrial structure and energy prices are the main factors affecting China's energy consumption [10]. Han Jun and Zhang Huinan used spatial autocorrelation analysis to analyze the spatial correlation between regions, and then used geographically weighted regression model to analyze the spatial difference between regions. They found that under the rapid economic development, the per capita energy consumption level and total energy consumption of different regions were significantly different [11].

As for correlation analysis, Zhu Jian and Li Weihua used correlation analysis to study the relationship between economic growth and total energy consumption, energy consumption elasticity coefficient and energy consumption per unit GDP of Jiangsu province and Xinjiang Province respectively, and found that the economic development of Jiangsu province and Xinjiang Province has a strong dependence on energy consumption. According to the coordinated development of energy and economy, different cities in the province are divided into 4 grades [12-13].

As for grey correlation analysis, Shao Minlan analyzed the correlation between economic growth and energy consumption from two aspects of economic growth and energy consumption variety structure, economic growth and energy consumption industrial structure by constructing grey correlation model, and found that the grey correlation between energy consumption and economic growth is different in different industries [14]. Du Junhui et al. used grey correlation analysis to study the relationship between energy consumption and economic growth in Hebei Province, and found that there was a significant grey correlation between economic growth and energy consumption in Hebei Province [15].

In the aspect of causality and co-integration test, Jian Xin, Zhang Lihua and Xu Weilan et al. used co-integration test and Granger causality to study the relationship between energy consumption and economic growth in Zhejiang province, Shandong Province and Jiangsu Province respectively, and concluded that energy consumption and economic growth are mutually two-way causality [16-18]. Guo Ke and Wang Liqun used co-integration analysis and state space model to study the relationship between energy consumption and economic growth in the Beijing-Tianjin-Hebei region. In addition to the mutual causal relationship between the two, they also found that the economic development in this region has a direct demand for higher energy consumption [19].

In the aspect of VAR model, Zhang Zirong used VAR model to study the relationship between Energy consumption and GDP in China and found that energy consumption has always been positively correlated to economic development [20]. Fan Yaxuan used VAR model to study the energy consumption in Hebei Province and found that the current energy consumption in Hebei province far exceeds the national average level, with coal as the main energy consumption and clean energy consumption accounting for a small proportion [21].

In terms of decoupling model, Liu Huimin used Tapio decoupling model to study the decoupling relationship between China's economic growth and energy consumption, and found that there was a repeated process of decoupling [22]. Ling Liwen et al. used Tapio decoupling model to analyze the decoupling state between industrial waste and economic growth in Guangdong Province and found that there were significant temporal and spatial differences in the decoupling state [23]. Hu Huaimin et al. used Tapio decoupling model to study the decoupling between economic development of transportation industry and carbon emission of transportation energy in the Yangtze River Economic Belt, and found that the decoupling of carbon emission of transportation in the Yangtze River Economic Belt is mainly "weak decoupling" [24]. Sun Han et al. used Tapio decoupling model and LEAP system to forecast and analyze the decoupling between output value of various sectors and energy consumption in China under baseline and prospect scenarios. The research shows that the decoupling state of output value of China's three major industries from 2020 to 2030 is different from that of energy consumption [25].

Given the extent of literature, there are few studies on the decoupling relationship between energy consumption and economic growth in Beijing-Tianjin-Hebei region. Therefore, the main contribution of this paper is that the Tapio decoupling theory is adopted to study the relationship between economic growth and energy consumption in Beijing-Tianjin-Hebei region from 2005 to 2019. Moreover, the research work can provide the targeted solution of improving regional low-carbon development in economy and energy consumption.

2. Decoupling theory

Decoupling has become a new research hotspot in recent years as a measure of the coupling relationship between human economic activities and energy consumption or environmental pollution. Decoupling theory is the basic theory proposed by the Organization for Economic Cooperation and Development (OECD) to "destroy the link between economic growth and resource consumption or environmental degradation" [26]. The economic growth of a country or region usually leads to resource consumption and environmental pollution. However, after taking some measures, with the same amount of economic growth, the resource consumption and environmental pollution are reduced, which is called decoupling [27].

At present, decoupling models are mainly divided into two types, one is the OECD decoupling index model based on initial value and final value, and the other is the Tapio decoupling model based on elastic change of growth [28]. Tapio decoupling model combines relative quantity change and total quantity change to study the asynchronous change of growth elasticity of economic development and resources and environment in the same period. The decoupling state is subdivided into eight types: weak decoupling, expansion negative decoupling, strong negative decoupling, weak negative decoupling, recessive decoupling, strong decoupling, expansion connection and recessive connection. Tapio decoupling model established a relatively perfect decoupling index evaluation system, which effectively corrects the possible errors caused by the OECD decoupling index model due to the extreme selection of only the initial value and the final value [22]. Therefore, more scholars prefer to use Tapio decoupling model to study the relationship between economic development and energy consumption.

In this paper, the Tapio decoupling model is used to construct the decoupling analysis model of economic growth and energy consumption in the Beijing-Tianjin-Hebei region. The elastic decoupling analysis is expressed as follows:

$$\varepsilon_{E,GDP} = \frac{\triangle E/E_{t-1}}{\triangle GDP/GDP_{t-1}} = \frac{E_t - E_{t-1}/E_{t-1}}{GDP_t - GDP_{t-1}/GDP_{t-1}} \quad (1)$$

Where $\varepsilon_{E,GDP}$ denotes the decoupling elasticity coefficient, E is the total energy consumption, GDP is the gross product, E_t is the total energy consumption in year t, GDP_t is the gross product in year t, E_{t-1} is the total energy consumption in year $t-1$, GDP_{t-1} is the gross product in year $t-1$, $\triangle E$ is E_t minus E_{t-1}, $\triangle GDP$ is GDP_t minus GDP_{t-1}, $\triangle E/E_{t-1}$ is the rate of change of E, and $\triangle GDP/GDP_{t-1}$ is the rate of change of GDP. Based on the difference of decoupling elastic coefficients, the decoupling elastic coefficients are divided into three states: decoupling, negative decoupling and connection, and these three states can be subdivided into eight different states [22]. Specific indicators are shown in Table 1.

Table 1. Tapio decoupling elasticity comparison table

		$\triangle E$	$\triangle GDP$	Elastic ε
Decoupling	Weak Decoupling	>0	>0	(0, 0.8)
	Strong Decoupling	<0	>0	(−∞, 0)
	Recessive Decoupling	<0	<0	(1.2, +∞)
Negative Decoupling	Weak Negative Decoupling	<0	<0	(0, 0.8)
	Strong Negative Decoupling	>0	<0	(−∞, 0)
	Expansion Negative Decoupling	>0	>0	(1.2, +∞)
Connection	Expansion Connection	>0	>0	(0.8, 1.2)
	Recessive Connection	<0	<0	(0.8, 1.2)

From Table 1, the relationship between economy and energy consumption in Beijing-Tianjin-Hebei shows weak decoupling, strong decoupling, and recessive decoupling. Strong decoupling indicates that energy consumption remains unchanged or decreases with the economic growth; weak decoupling means that economic growth is not constrained by energy consumption, but the growth rate of energy consumption is slower than that of economic growth; recessive decoupling means that energy consumption decreases faster than that of economic recession. Negative decoupling includes weak negative decoupling, strong negative decoupling and expansion negative decoupling. Among them, strong negative decoupling means that economic recession and energy consumption increases, weak negative decoupling means that energy consumption decreases slower than economic recession, and expansion negative decoupling means that economic growth and energy consumption increases substantially. Connection states include expansion connection and recessive connection. The expansion connection means that economic growth and energy consumption increase simultaneously, and the recessive connection means that economic growth and energy consumption decrease simultaneously. It can be seen that in different states, strong decoupling is the most ideal decoupling state between economic growth and energy consumption, while strong negative decoupling is the least ideal decoupling state between economic growth and energy consumption.

3. Data

In this paper, GDP data of Beijing-Tianjin-Hebei region from 2005 to 2019 are collected from China National Bureau of Statistics (2005-2019) [29], and the energy consumption data are extracted from China Energy Statistical Yearbook (2006-2020) [30]. The original data are shown in Table 2.

Table 2. Original data of economic growth and energy consumption in Beijing-Tianjin-Hebei region (2005-2019)

Year	Total Energy Consumption (10^4 tons of standard coal)	Total GDP (100 million RMB)
2005	29443	19081.8
2006	32198	21968.2
2007	34813	26736.8
2008	36013	31195.6
2009	37863	33917.4
2010	41303	39798.4
2011	44091	46686
2012	45636	51145.2
2013	44270	55339.6
2014	44296	58775.5
2015	46159	62057
2016	46453	66992.5
2017	47003	72974.5
2018	47428	78963.5
2019	48146	84479.2

4. Decoupling results and analysis

Taking 2005 as the base year, the decoupling index between economic growth and energy consumption in Beijing-Tianjin-Hebei region can be calculated according to Equation (1), which are shown in Table 3.

https://doi.org/10.1051/e3sconf/202235801012

Table 3. Decoupling analysis results of economic growth and energy consumption in Beijing-Tianjin-Hebei region (2005-2019)

Year	$\Delta E/E_{t-1}$	$\Delta GDP/GDP_{t-1}$	Decoupling Elastic	Decoupling State
2005-2006	0.094	0.151	0.623	Weak Decoupling
2006-2007	0.081	0.217	0.373	Weak Decoupling
2007-2008	0.034	0.167	0.204	Weak Decoupling
2008-2009	0.051	0.087	0.586	Weak Decoupling
2009-2010	0.091	0.173	0.526	Weak Decoupling
2010-2011	0.068	0.173	0.393	Weak Decoupling
2011-2012	0.035	0.096	0.365	Weak Decoupling
2012-2013	−0.03	0.082	−0.366	Strong Decoupling
2013-2014	0.001	0.062	0.016	Weak Decoupling
2014-2015	0.042	0.056	0.75	Weak Decoupling
2015-2016	0.006	0.08	0.075	Weak Decoupling
2016-2017	0.012	0.089	0.135	Weak Decoupling
2017-2018	0.009	0.082	0.11	Weak Decoupling
2018-2019	0.015	0.07	0.214	Weak Decoupling

4.1 Analysis of variation of decoupling index

As can be seen from Table 3, the relationship between economy and energy consumption in Beijing-Tianjin-Hebei showed weak decoupling state for the research period except for the strong decoupling state in 2012-2013. In addition, the decoupling elasticity coefficients in 2013-2014 and 2015-2016 were less than 0.1, which was very close to strong decoupling. The decoupling elasticity coefficients from 2005 to 2012 and 2013 to 2019 were all in the interval (0, 0.8), and both ΔE and ΔGDP were positive. Therefore, economic growth and energy consumption in The Beijing-Tianjin-Hebei region showed weak decoupling in these two periods. The decoupling elasticity coefficient from 2012 to 2013 was within the interval $(-\infty, 0)$, with negative ΔE and positive ΔGDP. Therefore, economic growth and energy consumption in the Beijing-Tianjin-Hebei region showed a strong decoupling state during this period. In Table 3, the change trend of energy consumption change rate, GDP change rate and decoupling elastic are shown in Figure 1.

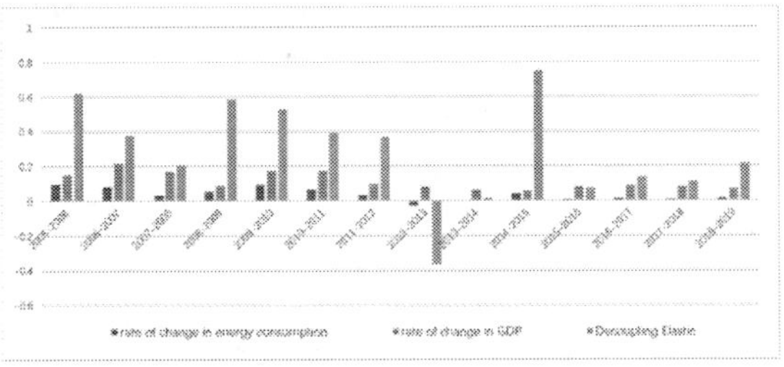

Figure 1. Change rates of energy consumption and GDP and decoupling elasticity coefficient in Beijing-Tianjin-Hebei Region (2005-2019)

As shown in Figure 1, the decoupling elasticity coefficients from 2005 to 2008 showed a decreasing trend year by year, which were 0.623, 0.373 and 0.204, respectively. The decoupling elasticity coefficients increased from 2008 to 2009, but showed a decreasing trend in the four years after the increase, which were 0.586, 0.526, 0.393 and 0.365, respectively. The decoupling value during 2012-2013 was −0.366 showing strong decoupling state, during which the energy consumption decreased by 3%, while GDP increased by 8.2% compared with the previous year. Energy consumption in 2013-2014 increased only 260,000 tons of standard coal compared to the previous year with the growth rate of 0.1%, so it is very close to strong decoupling. From 2014 to 2015, the decoupling elasticity coefficient increased sharply and reached 0.75, the highest value in the research period. The decoupling state improved from 2015 to 2019. Although there was no strong decoupling, the elasticity coefficients of decoupling remained at a low level, which were 0.075, 0.135, 0.11 and 0.214, respectively.

4.2 Analysis of the reasons for the great variation of decoupling elasticity coefficient in several special periods

Although the decoupling state of economic growth and energy consumption in Beijing-Tianjin-Hebei region was relatively stable from 2005 to 2019 except for strong decoupling in 2012-2013, the decoupling elasticity coefficient varied greatly in different periods. In particular, the three periods of 2008-2009, 2012-2013 and 2014-2015 showed the most dramatic changes. The following reasons for the great variation of decoupling elasticity coefficients in these three periods are described in detail.

4.2.1 Analysis of reasons for the increase of decoupling elasticity coefficient in 2008-2009

In 2008, after the outbreak of the global financial crisis, to cope with the crisis, Chinese government launched "four trillion plan", to invest four trillion yuan to expand domestic demand and stimulate economic growth by the end of 2010. Beijing-Tianjin-Hebei region has already stopped production of energy-intensive, highly polluting enterprises and back into the production, leading to the consumption of energy increase demand. This is also the main reason why the elasticity coefficient of decoupling between energy consumption and economic growth in the Beijing-Tianjin-Hebei region rebounded in 2008-2009. With the resolution of the financial crisis, the decoupling elasticity coefficient from 2010 to 2012 decreased to a relatively low level of 0.393 and 0.365 respectively, and the GDP growth rate in both years reached 17.3%.

4.2.2 Analysis of the causes of strong decoupling in 2012-2013

In 2013, the energy consumption in Beijing, Tianjin and Hebei was 67.24, 78.82 and 296.64 million tons of standard coal, decreasing 4.54, 3.26 and 5.86 million tons of standard coal, respectively, compared with 2012 [30].

The main reason is that serious environmental pollution forced the government to take measures to interfere with the production of high energy consumption and high pollution enterprises. In 2013, smog pollution in the Beijing-Tianjin-Hebei region reached an unprecedented level, with the average annual PM2.5 and PM10 concentrations exceeding the standard in all cities at prefectural level and above, with the average annual PM2.5 concentration at 106 micrograms per cubic meter and the average PM10 concentration at 181 micrograms per cubic meter. The proportion of days with air quality meeting the standard was 37.5% on average, far lower than 64.2% in the Yangtze River Delta and 76.3% in the Pearl River Delta [31]. In order to reduce the concentration of smog, the governments of Beijing, Tianjin and Hebei, under the leadership of the central government, have carried out coordinated efforts to control smog, increased the promotion of clean and renewable energy, such as liquefied petroleum gas and solar energy, and controlled the total coal consumption. Beijing is required to cut 13 million tons of raw coal, Tianjin 10 million tons, and Hebei province 40 million tons by the end of 2017 [32], which also keeps the growth rate of energy consumption in the Beijing-Tianjin-Hebei region at a relatively low level after 2012.

4.2.3 Analysis of the reasons for the sharp increase of decoupling elasticity coefficient during 2014-2015

The decoupling elasticity coefficient from 2014 to 2015 was 0.75, which was close to the expansion connection state. This was not because of the significant increase in energy consumption, but because of the GDP growth rate during this period declined and reached the lowest value from 2005 to 2019. As a result, energy consumption growth rate is close to GDP growth rate. In August 2015, a major explosion occurred in Ruihai Company's dangerous goods warehouse in Tianjin Binhai New Area, which was the main reason for the slowdown of GDP growth. Tianjin Binhai New Area is the international shipping center and international logistics center in northern China, as well as the largest port in northern China, which occupies an important position in foreign trade. In 2015, the total GDP of Tianjin was 1,08795 trillion yuan, with a growth rate of only 2.2% compared with 2014, which was the lowest GDP growth rate of Tianjin during the study period from 2005 to 2019 [29]. In 2014, the total import and export of goods in Tianjin was 160846.57 million US dollars. In 2015, the total import and export volume of Tianjin goods was 11,4282.8 million US dollars, down 28.9% compared with 2014 [33]. Therefore, the explosion accident caused huge economic losses to Tianjin, and then had a certain impact on the overall development of the Beijing-Tianjin-Hebei region.

4.3 Analysis on the causes of weak decoupling in Beijing-Tianjin-Hebei region for many years

4.3.1 Economic growth plays a more significant role in decoupling

As can be seen from Table 3, although the total energy consumption in the Beijing-Tianjin-Hebei region showed an upward trend from 2005 to 2019, the growth rate of total energy consumption was much lower than that of GDP. It can be calculated from Table 3 that the average annual growth rate of energy consumption in Beijing-Tianjin-Hebei region was 3.6% from 2005 to 2019, while the average annual growth rate of GDP in Beijing-Tianjin-Hebei region was 11.3% during the same period. This difference is more obvious during the period of 2012-2019. The average annual growth rate of energy consumption and GPD is 0.8% and 7.4% respectively. Therefore, the economic growth plays a more significant contribution in driving decoupling on the whole, which is one of the reasons why economic growth and energy consumption in the Beijing-Tianjin-Hebei region have remained weak decoupling for many years.

4.3.2 Industrial restructuring in the Beijing-Tianjin-Hebei region

In China's "11th Five-Year Plan", development concepts, such as promoting the optimization and upgrading of industrial structure and building a resource-saving and environment-friendly society, were proposed to promote the adjustment of industrial structure. Meanwhile, the National Development and Reform Commission formulated the "Regional Planning of Beijing-Tianjin-Hebei Metropolitan Area", which is also one of the important plans of the 11th Five-Year Plan. According to the plan, Beijing-Tianjin-Hebei region will develop "8+2" model, *i.e.* the two municipalities directly under the Central government of Beijing and Tianjin and the eight prefecture-level cities of Shijiazhuang, Baoding, Qinhuangdao, Langfang, Tangshan, Zhangjiakou, Chengde and Cangzhou in Hebei province. Under this model, Beijing, Tianjin and Hebei have a clear industrial development orientation. Beijing focuses on the development of the tertiary industry with service industry and high-tech industry, gives full play to its talent advantages, and gradually transfers low-end manufacturing and heavy pollution enterprises to other cities. On the basis of traditional processing and manufacturing, Tianjin vigorously develops advanced manufacturing industries such as new energy vehicles and modern medicine, vigorously supports the development and construction of Binhai New Area, and builds Tianjin Port into a modern international port. Hebei serves as a supporting base for the high-tech industries of Beijing and Tianjin, and at the same time takes advantage of its geographical advantages to develop agriculture to ensure the grain supply of Beijing and Tianjin. This collaborative development mode has made the tertiary industry in Beijing and Tianjin develop rapidly in recent years. In

2019, the added value of China's tertiary industry was 53.5371 trillion yuan, accounting for 54.3% of the total GDP [34]. Among them, the added value of Beijing's tertiary industry accounted for 83.5% of the total GDP [35], 29.2 percentage points higher than the national total. The added value of Tianjin's tertiary industry accounted for 63.5% of its GDP [36], 9.2 percentage points higher than the national total. The added value of Hebei's tertiary industry accounted for 51.3 % of its GDP [37], 3 percentage points lower than the national total. Therefore, Beijing and Tianjin have an obvious driving effect on the development of the tertiary industry in the Beijing-Tianjin-Hebei region, and the developed tertiary industry in these two places has provided a driving force for the sustained and rapid economic growth of the Beijing-Tianjin-Hebei region.

4.3.3 Binding by international treaties

With the global warming, more and more people realized the harm caused by industrial activities to the earth's natural environment. The global carbon emission control mechanism represented by the Kyoto Protocol officially came into effect on February 16, 2005, which is the first time in human history to limit global greenhouse gas emissions in the form of regulations. In 2005, China was the largest emitter of carbon dioxide in the world. If China does not undertake the responsibility of reducing greenhouse gas emissions for a long time, it will come under increasing pressure in the United Nations Framework Convention on Climate Change, which will affect its international reputation and status. Beijing-Tianjin-Hebei region, as one of China's major industrial areas, must control carbon emissions, so the growth rate of energy consumption is bound to slow. This was evident in the first few years of the agreement, when the total energy consumption in the Beijing-Tianjin-Hebei region increased by 9.4%, 8.1% and 3.4 % from 2006 to 2008, showing a downward trend for three consecutive years.

5. Policy Suggestions and Conclusions

5.1 Policy Suggestions

Based on the decoupling analysis of economic growth and energy consumption in Beijing-Tianjin-Hebei region from 2005 to 2019, this paper provides the following targeted suggestions for the region's future development:

5.1.1 Create a green and low-carbon energy consumption model

The governments of Beijing, Tianjin and Hebei should strengthen the publicity of energy conservation and emission reduction, and promote green and low-carbon consumption to citizens. In the aspect of transportation, improve the urban transportation system, increase the number of bus and subway stops in non-central urban areas of Beijing and Tianjin, plan more routes, and lower the charging standard for different numbers of passengers. As for the management of private cars, in addition to limiting the number of vehicles, cross-departmental cooperation should be strengthened to strictly control the emissions of motor vehicles from the production and circulation links. Improve factory standards, strengthen the supervision of environmental protection departments, and eliminate motor vehicles with high energy consumption and high emissions. At the same time, encourage people to buy new energy vehicles, reduce taxes on new energy vehicles, and give policy support to enterprises producing new energy vehicles.

5.1.2 Adjust the energy consumption structure

First, accelerate the development and utilization of renewable energy, reduce the consumption of fossil energy, especially coal, close down outdated production facilities, and increase the share of renewable energy in energy consumption. Beijing, Tianjin and Hebei should develop different kinds of renewable energy according to their geographical advantages. For example, Zhangjiakou city in Hebei province is located in the windward zone between the North China Plain and the Inner Mongolia Plateau. The government should develop wind energy and improve the construction of wind energy infrastructure in Zhangjiakou city. Second, the government should promote the clean revolution of traditional energy, actively promote clean coal technology, and realize the clean development, production and use of coal. Speed up the production sector to replace coal, oil and other traditional fossil fuels with clean energy such as liquefied petroleum gas and coal-to-natural gas. The quality of coal should also be strictly controlled, and the sale of coal with high ash and inorganic sulfur content should be prohibited to prevent the production of large quantities of gases harmful to human health.

5.1.3 Make use of Beijing and Tianjin's scientific and technological advantages to promote the energy technology revolution

Beijing and Tianjin should fully rely on the advantages of universities and research institutes to promote the transformation of traditional energy use technology. The government should increase investment in scientific research institutions, coordinate the application and promotion of energy technologies, demonstration and experiment, and focus on tackling key problems. Optimize the coal processing process, improve the overall performance of coal combustion equipment, and reduce the coal consumption by reducing the proportion of direct combustion of raw coal. Actively promote the research and development of CCUS technology, reduce the carbon emitted by fossil energy combustion, and achieve large-scale low-carbon utilization of fossil energy, in line with the 2030 carbon peak target proposed by the State Council. Improve the matching mechanism between innovation chain and industrial chain to fully utilize the innovation achievements actual industries. At the same time, Beijing, Tianjin and Hebei should also pay attention to exchanges with other countries on clean energy technologies, actively participate in various forums, and establish multilateral cooperation mechanisms to achieve mutual benefit and win-win results.

5.1.4 Change the economic development model of the Beijing-Tianjin-Hebei region

The Beijing-Tianjin-Hebei region should continue to promote the adjustment of industrial structure. The government should evacuate Beijing's non-capital functions, orderly transformation of high-energy-consuming industries, increase the proportion of tertiary industry, maintain a reasonable scale of secondary industry, and deeply integrate advanced manufacturing with modern service industry. Beijing, Tianjin and Hebei will jointly explore the coordinated layout of the industrial chain in the region and build industrial clusters. Adhere to the development of Beijing and Tianjin's emerging industries and Hebei's traditional energy industry transformation and upgrading. The government should promote the construction of supporting bases for Beijing and Tianjin advanced manufacturing industry in Hebei, strengthen the incubation of scientific and technological achievements and the sustainable development of advanced manufacturing industry, give full play to the radiating and driving role of Beijing and Tianjin excellent scientific and technological achievements, and build an intelligent city group.

5.2 Conclusions

Based on the Tapio decoupling theory, this paper analyzed the decoupling state of economic growth and energy consumption in Beijing-Tianjin-Hebei region from 2005 to 2019, and the following conclusions can be drawn:

(1) According to the results of the decoupling elasticity coefficient, it can be clearly found that the decoupling state of economic growth and energy consumption in Beijing-Tianjin-Hebei region from 2005 to 2019 showed overall weak decoupling. The reason for the sustained weak decoupling for many years mainly rely on the strong driving effect of rapid economic growth on decoupling. Through research, it is found that the high-speed economic development of the Beijing-Tianjin-Hebei region mainly benefits from the significant proportion of the tertiary industry in Beijing and Tianjin. In addition, international constraints such as the Kyoto Protocol have also contributed to the weak decoupling of the Beijing-Tianjin-Hebei region over the years.

(2) According to the elasticity coefficient of decoupling during the weak decoupling period, it can be clearly found that although the state of weak decoupling has lasted for many years, the elasticity coefficient of weak decoupling varies greatly in different periods. The main reason for the change is the impact of unstable factors such as the international financial crisis and major events in the Beijing-Tianjin-Hebei region.

(3) According to the results of the decoupling elasticity coefficient, it can be clearly found that there is a strong decoupling between economic growth and energy consumption in Beijing-Tianjin-Hebei region from 2012 to 2013. The main reason is that severe air pollution in Beijing-Tianjin-Hebei region has prompted the government to introduce policies to cut energy consumption.

Based on the above conclusions, the following suggestions are put forward, such as: create a green and low-carbon consumption model, adjust the energy consumption structure, give full play to the scientific and technological advantages of Beijing and Tianjin to improve energy technology, and change the economic development model.

Finally, although this paper has made contributions to the study on the decoupling relationship between economic growth and energy consumption in Beijing-Tianjin-Hebei region, there are still some limitations in this study. For example, this paper does not examine the decoupling between economic growth and energy consumption in Beijing, Tianjin and Hebei, respectively. In addition, different energy consumption structures may have different impacts on the decoupling state. In this regard, more complex and subtle methods can be used to study the decoupling relationship between economic growth and energy consumption in the Beijing-Tianjin-Hebei region, which is a possible future research direction.

References

1. National Bureau of Statistics, 2001 – 2020, gross domestic product, http://data.stats.gov.cn.
2. National Bureau of Statistics, 2001 – 2020, total energy consumption, https://data.stats.gov.cn.
3. Huang, B.B., Zhang, Y.Z., Wang, C.X., New Energy Development and Issues in China during the 14th Five-Year Plan [J]. Electric Power, 2020, 53 (01): 1-9.
4. National Bureau of Statistics, 2020, total energy consumption, http://data.stats.gov.cn.
5. China News, 2022, Over the past eight years, the GDP of the Beijing-Tianjin-Hebei region has increased by 70%, http://www.chinanews.com.cn/gn/2022/02-25/9685901.shtml.
6. China Energy Statistical Yearbook. China Statistics Press. 2020.
7. Mu, A.Y., LU, Y. The "Dual Carbon" Goal Leads the Coordinated Development of Beijing-Tianjin-Hebei Industrial Chain Cluster [J]. Journal of Hebei University of Economics and Trade (Comprehensive Edition), 201, 21(04):90-95.
8. Central People's Government of the People's Republic of China, 2021, Bulletin of the seventh National Census, http://www.gov.cn/guoqing/2021-05/13/content_5606149.htm.
9. National Bureau of Statistics, 2020, gross regional product, http://data.stats.gov.cn.
10. Zhao, J.H., Analysis of influential factors of energy consumption in China based on principal component

regression model [J]. China Mining Magazine, 2014, 23(01):44-49.

11. Han, J., Zhang, H.N., Measurement of Regional Energy Consumption under the Background of Economic High-quality Development in China [J]. Journal of quantitative and technical economics, 2019, 36(07):42-61.
12. Zhu, J., Zhan Y.J., Correlation Analysis of Energy Consumption and Economic Development in Jiangsu Province [J]. Energy of China, 2009, 31(06):42-45.
13. Zhang, Y.H., Li, W.H., Lin, M., Feng, G., Correlation analysis on energy consumption and economic growth in Xinjiang [J]. Renewable Energy Resources, 2010, 28(06):137-139.
14. Shao, M.L., Grey correlation analysis of energy consumption, energy structure and economic growth in China [J]. China Collective Economy, 2021(18):5-6.
15. Du, J.H., Wang, H.X., Liu, W., Research on the relationship between energy consumption and economic growth in Hebei Province [J]. Journal of North China Institute of Science and Technology, 2021, 18(05):92-99.
16. Jian, X., An empirical study on relationship between energy consumption and economic growth of Zhejiang province [J]. China Market, 2021 (17): 22-24.
17. Zhang, L.H., Wu, L.Y., Wang, S.Y., Empirical Analysis of Energy Consumption and Economic Growth in Shandong Province [J]. Energy and Energy Conservation, 2022(01):45-48.
18. Xu, W.L., Guo, C.L., Analysis on Two-way Long-term Regression Relationship between Energy Consumption and Economic Growth in Jiangsu Province [J]. Energy of China, 2021, 43(04):75-79.
19. Guo, K., Wang, L.Q., Summarization on the interactive relation between energy consumption and economic growth in Beijing-Tianjin-Hebei Area [J]. Urban Problems, 2015 (05): 52-59.
20. Zhang, Z.R., Empirical analysis of the relationship between Economic growth and energy consumption in China [J]. Journal of Commercial Economics, 2018(17):36-39.
21. Fan, Y.X., Research on the relationship between energy consumption and economic development in Hebei province [J]. Business & Economy, 2020 (01): 33-34.
22. Liu, H.M., Can economic growth decouple with energy consumption: a temporal and spatial study of Eastern [J]. China Population, Resources and Environment, 2016, 26(12):157-163.
23. Ling, L.W., Cai, C.M., Yu, P.X., Zhang, D.B., Research on the Relationship between Economic Growth and Industrial Waste of Guangdong Province——based on the Quantity Decoupling and Speed Decoupling Perspectives [J]. Chinese Journal of Management Science, 2016, 24(S1):948-954.
24. Hu, H.M., Zuo, W., Xu, S.Y., Decoupling Effect and Driving Factors of Transportation Energy Carbon Emission in Yangtze River Economic Belt [J]. Resources and Environment in the Yangtze Basin, 2022, 31(04):862-877.
25. Sun, H., Zhang, H.Y., Fu, X.L., Forecast of Multiple-Sectoral Decoupling of China's Economic Development with Energy Consumption [J]. Resources and Industries, 2021, 23(05):11-20.
26. OECD. Environmental Indicators-Development, Measurement and Use[R]. Paris: OECD, 2003.
27. Song, W., Chen, B.M., Chen, X.W., Decoupling Evaluation between Cultivated Land Occupation and Economic Growth in Changshu City [J]. Journal of Natural Resources, 2009,24(09):1532-1540.
28. Peng, J.W., Huang, X.J., Zhong, T.Y., Zhao, Y.T., Decoupling Analysis of Economic Growth and Energy Carbon Emissions in China [J]. Resources Science, 2011, 33(04):626-633.
29. National Bureau of Statistics, 2005-2019, GDP of The Beijing-Tianjin-Hebei region, http://data.stats.gov.cn.
30. China Energy Statistical Yearbook. China Statistics Press. 2006-2020.
31. Chinese government website, 2014, In 2013, all cities in the Beijing-Tianjin-Hebei region exceeded the annual average concentration of PM2.5 and PM10, www.gov.cn/xinwen/2014-03/25/content_2645501.htm.
32. Chinese government website, 2013, Rules for implementing the Action Plan on Air Pollution Prevention and Control in the Beijing-Tianjin-Hebei region and surrounding areas, www.gov.cn/gongbao/content/2013/content_2541901.htm.
33. National Bureau of Statistics, 2014-2015, total import and export of goods, http://data.stats.gov.cn.
34. National Bureau of Statistics, 2019, value-added of tertiary industry, http://data.stats.gov.cn.
35. Beijing Statistical Yearbook. China Statistics Press. 2020.
36. Tianjin Statistical Yearbook. China Statistics Press. 2020.
37. Hebei Statistical Yearbook. China Statistics Press. 2020.

Experimental study on aerosol dry deposition characteristics in wind tunnel simulation

Sibo Lyu[1,2], Wusheng Liu[1,2], Ruojie Li[1,2] and Zhijie Tian[1,2]

[1] China Institute for Radiation Protection, No. 102, Xuefu Street, Taiyuan City, Shanxi Province, China
[2] CNNC Key Laboratory for Radiation Protection Technology, China

Abstract. The radioactive aerosols are produced during the operation of nuclear power plants will cause certain harm to human body. Dry deposition is one of the important factors affecting the aerosols deposition to the ground, and radioactive aerosols and common aerosols have similar properties in terms of dry deposition. Based on the similarity theory, with the help of the newly built environmental wind tunnel of the China Institute for Radiation Protection, with a part of a nuclear power plant site as a research object, and the wind tunnel simulation experiment of aerosol dry deposition characteristics was carried out in groups. The particle dynamic analyzer (PDA) was used to measure and calculate the dry deposition velocity of aerosol particles, and the results are grouped, analyzed and discussed. The research results show that when the variables are controlled, when the underlying surfaces are the same, compared with aerosol particles with particle sizes of 0-2 μm and 2-5 μm, the dry deposition velocity of aerosol particles greater than 5 μm is more likely to change with the increase of wind speed.

1. Introduction

Following the Fukushima nuclear accident in Japan, concerns about the health problems caused by radiation from nuclear power plants have grown rapidly. During the operation of nuclear power plants, radioactive aerosols are generated, which are an important source of inhalation hazards for people. The median particle size of radioactive aerosols in nuclear power plants, nuclear fuel reprocessing plants and some nuclear plants is similar, about 4 μm [1]. Radioactive aerosols are harmful to the public and the environment due to their high toxicity and wide range of migration in the air. Therefore, it is of great significance to study the migration behavior of radioactive aerosols in the atmospheric environment. During the transport of radioactive pollutant plumes, dry deposition is one of the important factors affecting the deposition of aerosols to the ground, and the amount of pollutants transferred from the atmosphere to the surface caused by dry deposition is usually quantitatively described by the dry deposition velocity. At present, monitoring and sampling research on the deposition of radioactive aerosols in the atmosphere relying on their own gravity and swaying force [2-3] is an extremely important part of radioactive aerosol research[4-6]. At present, the construction of radioactive aerosol database in my country is still in its infancy, lacking real experimental data as technical support. Therefore, the dry deposition velocity is the key research object of this paper. Radioactive aerosols and ordinary aerosols have a certain common mechanism of action, and there is little difference between them during the dry deposition process. Considering that simulating the dry deposition of radioactive aerosols requires specially suitable sites and devices for operating radioactivity, it is possible to first carry out the deposition experiments of ordinary aerosols to accumulate the basic data of dry deposition of radioactive aerosols. Dry deposition of aerosols depends on many factors: aerosol properties (diameter, density, etc.), deposition surface variables (roughness, temperature, electrostatic properties, etc.) and meteorological variables (wind speed, turbulence, relative humidity, etc.) The deposition methods on different underlying surfaces are also different[7]. In order to facilitate the evaluation of the total effect of various physical mechanisms on dry deposition, the dry deposition velocity is defined, which is comprehensively affected by the underlying surfaces, aerosol particle characteristics, and meteorological conditions[8]. It has been confirmed by many field experiments and laboratory experiments that the effect of aerosol particle size on the dry deposition velocity is particularly strong[9]. Radioactive aerosols inhaled into the human body are difficult to discharge, and the harm to human health is also closely related to their particle size[10]. Air kinematic viscosity is a strong function of temperature, and temperature will affect the degree of atmospheric mixing and cause changes in air turbulence, which has a great impact on the dry deposition velocity[11]. The greater the wind speed, the greater the friction speed of the transported aerosol particles, and the dry deposition speed and the wind speed are positively correlated to a certain extent[12]. The increase of relative humidity will increase the particle size and also affect the dry deposition velocity of aerosol particles to a certain extent[13]. Different types of underlying surfaces also have different effects on the dry deposition velocity. Surface roughness of water and sea surfaces increases due to corrugations, and wind tunnel studies have shown that the presence of wave surfaces increases deposition velocitys[14]. The focus of this experiment is to study and discuss the dry deposition velocity in the wind tunnel simulation experiment.

Compared with field experiments, physical simulation experiments are not limited by weather conditions, have short experimental periods, cost less and can be repeated many times. Therefore, a wind tunnel simulation experiment was chosen to reproduce the flow field of the actual prototype in the wind tunnel according to the similarity criterion, and on this basis, the diffusion and deposition of aerosol particles were studied.

2. Instruments and Equipment

The simulation experiment was completed in the Atmospheric Boundary Layer Wind Tunnel No. 2 of the China Institute of Radiation Protection. This wind tunnel is currently the largest DC blowing type environmental wind tunnel in the Asian test section, which can simulate the flow and diffusion of the atmospheric boundary layer. The test section of the cave is 28 m long, 5 m wide and 3.5 m high, and the simulated wind speed range is 0.2 m/s~20 m/s.

The flow field measurement instrument uses the StreamLine hot-wire anemometer (ConstantTemperature Anemometer, CTA) produced by DANTEC Company in Denmark. After the wind speed and angle of the probe are calibrated, the instantaneous flow velocity at any position in the wind tunnel can be measured and the average wind speed, turbulent flow can be measured. Strength and other statistics.

The aerosol diffusion and deposition measurement instrument uses the Particle Dynamics Analyzer (PDA) produced by DANTEC Company in Denmark, which according to the Doppler effect generated when the particles pass through the laser beam, it measures the speed of the particles passing through. Based on the principle of phase Doppler, it can realize the synchronization of particle size, one-dimensional to three-dimensional flow velocity and the number of particles in a non-contact real-time manner measurement.

Considering the scale of the wind tunnel and the measurement conditions and range of the PDA, the 10E03 large-capacity liquid particle generator produced by DANTEC, Denmark was selected to generate and discharge aerosol particles that meet the experimental requirements.

The raw material for generating aerosol particles is a colorless, odorless, harmless to human body and a tracer with good followability, Liquid Diisooctyl Sebacate (DEHS).

3. Experimental method

In this experiment, the atmospheric boundary layer wind tunnel was used for the simulation study of dry deposition. The purpose was to measure the dry deposition velocity of aerosol particles under different conditions, and to analyze and summarize the aerosol dry deposition behavior and characteristics.

A certain number of wedges and roughness elements are placed at the entrance of the wind tunnel test section to generate the velocity profile and turbulence intensity distribution required in the experiment. In the middle of the test section, the site model is made into a model. The ratio of the site model to the prototype selected for this experiment is 1:1000, and the radius of the model simulation area is 6.7 km. A three-dimensional mobile measuring mechanism is installed above it, which can be equipped with various measuring instruments. The mobile measuring mechanism can move in three directions to measure data. The flow and diffusion characteristics of aerosols are studied according to the similarity theory.

Adjust the working pressure of the liquid particle generator so that the particle size of the aerosol particles produced is approximately Gaussian distribution, and the two measuring lenses (laser transmitter and receiver) of the PDA are fixed on the three-dimensional mobile measuring vehicle with profiles. 3D PDA collaborative measurement. This measurement mode can simultaneously obtain the three-dimensional velocity data of the aerosol particles. After coordinate transformation, the velocity of the aerosol particles in all directions in the wind tunnel coordinate system can be obtained, and the aerosol dry deposition velocity V_d can be calculated iteratively through the formula. However, during the experiment, it was found that the effective measurement height of the PDA is 100 mm from the surface of the model, because the 3D PDA is used for collaborative measurement. In order to ensure sufficient data volume, the laser energy must be adjusted to at least 1.6W, and the laser brightness is very large under such power, the diffuse reflection light irradiated on the surface of the model will be received by the PDA, and a large amount of noise signals will be generated to interfere with the data acquisition. Therefore, the aerosol dry deposition data obtained at lower measurement heights are of no practical significance. Therefore, in this experiment, the minimum measurement height was set to 100 mm.

In this experiment, the topography of the underlying surfaces of the model is divided into five types: flat terrain, windward slope, mountain peak, leeward slope and valley. The dry deposition velocity at the lowest measurement height of each terrain under each wind speed is measured and calculated. Different particle sizes of aerosol particles have different effects on human health, and the particle sizes are classified accordingly. schematic diagram of the wind tunnel simulation experiment of aerosol dry deposition is shown in Figure 1. The specific experimental scheme is shown in Table 1.

https://doi.org/10.1051/e3sconf/202235801013

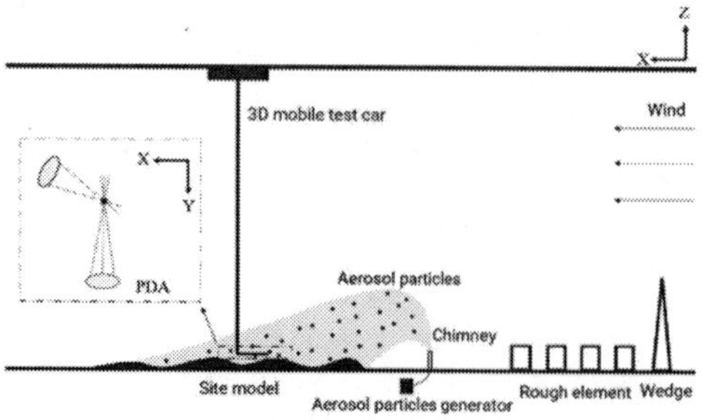

Fig. 1. Schematic diagram of wind tunnel simulation experiment.

Table 1. The specific experimental scheme

experimental method	instrument	Purpose	project	content
Measuring data from aerosol particles with a laser	PDA	Measuring the size and velocity of aerosol particles	wind direction	NE
			wind speed	1 m/s, 2 m/s
			Measurement content	NE wind direction model, on the axis passing through the chimney point, there are representative position points at different positions in front, top and rear of the mountain, 100 mm (flat terrain), 200 mm (flat terrain), 300 mm (flat terrain), 500 mm mm (windward slope), 600 mm (peak), 700 mm (leeward slope), 1000 mm (flat terrain), 2000 mm (flat terrain), 2900 mm (valley), 3000 mm (peak), 3200 mm (valley) , 3400 mm (windward slope), 3500 mm (peak), 3600 mm (valley), 3800 mm (windward slope), 4050 mm (peak), 4300 mm (leeward slope), a total of 18 selected points were measured.

4. Result

Data comparison of aerosol particles with particle sizes of 0~2 μm and larger than 5 μm at 1 m/s and 2 m/s wind speeds. The five types of terrain are shown in Figure 2, Figure 3, Figure 4, Figure 5 and Figure 6.[WS(Wind Speed);PS(Particle Size)].

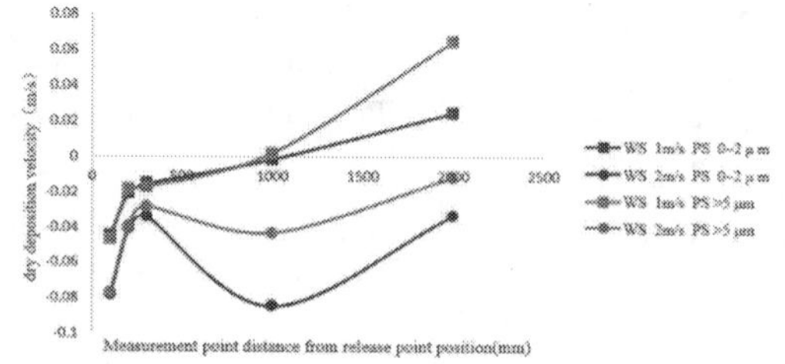

Fig. 2. Dry deposition velocities of aerosol particles of 0~2μm and larger than 5μm in different wind speeds under flat terrain.

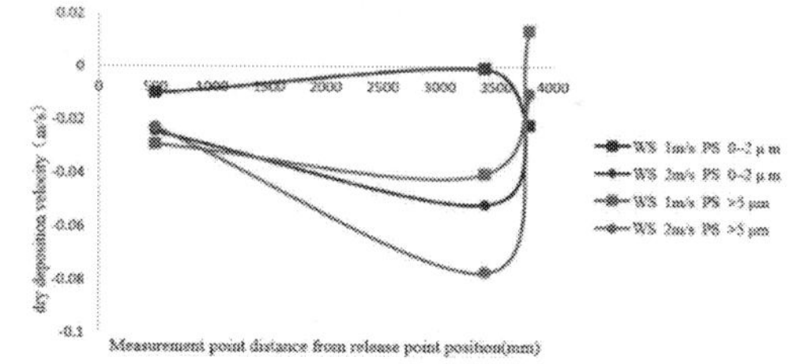

Fig. 3. Dry deposition velocities of aerosol particles of 0~2μm and larger than 5μm in different wind speeds under windward slope.

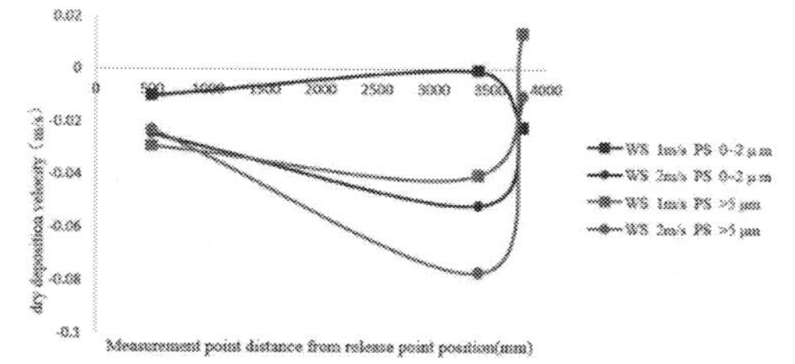

Fig. 4. Dry deposition velocities of aerosol particles of 0~2μm and larger than 5μm in different wind speeds under mountain peak.

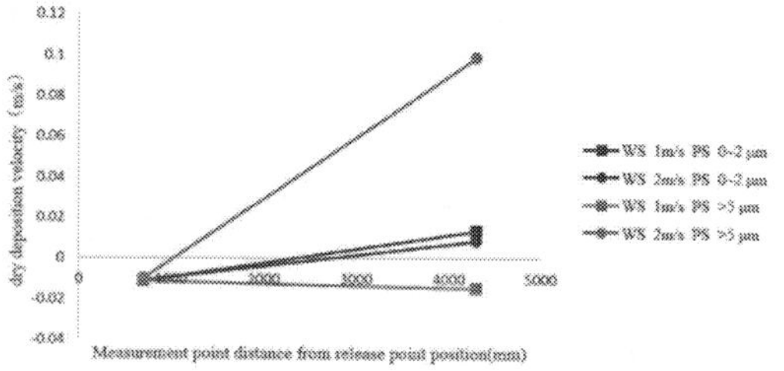

Fig. 5. Dry deposition velocities of aerosol particles of 0~2μm and larger than 5μm in different wind speeds under leeward slope.

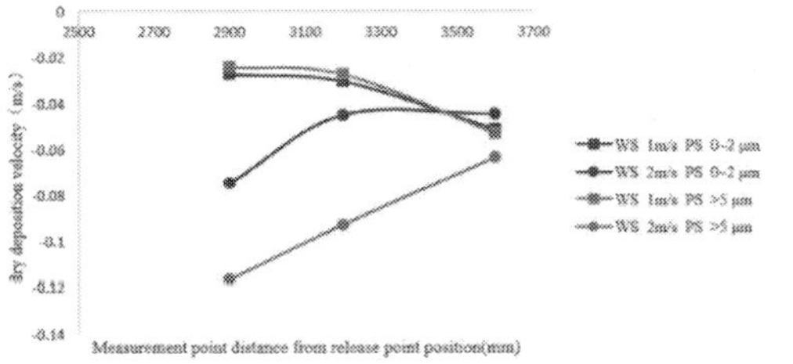

Fig. 6. Dry deposition velocities of aerosol particles of 0~2μm and larger than 5μm in different wind speeds under valley.

It can be seen from the figure that different wind speeds have different effects on the dry deposition speed. The results measured in this experiment show that, except for the leeward slope terrain, when the wind speed increases, the aerosol particles will obtain a larger negative deposition speed. , that is, the higher the wind speed, the less likely the aerosol particles will be deposited. For particles with different particle sizes, the dry deposition velocity also changes with the increase of wind speed. For different underlying topography, the trend of change is different, and the direction and size change.

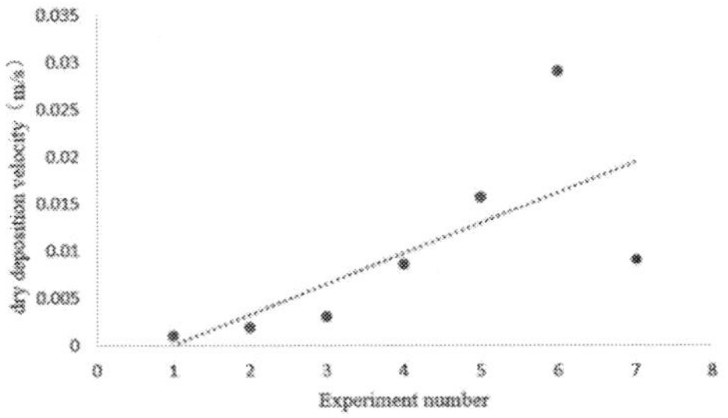

Fig. 7 The red point is the measurement result of the dry deposition velocity this time, and the blue point is the dry deposition velocity value obtained from previous experiments

Under the conditions of similar underlying surfaces and wind speed, the dry deposition speed of the obtained aerosol particles is in the same order of magnitude compared with the previous experimental results[15-17], As shown in Figure 7. The red point is the measurement result of the dry deposition velocity this time, and the blue point is the dry deposition velocity value obtained from previous experiments That is the data obtained in this experiment are basically correct and reliable.

5. Conclusions

In this paper, wind tunnel experiments are carried out in the No. 2 atmospheric boundary layer wind tunnel of the China Institute of Radiation Protection, taking a site with some complex terrain as a representative. A flow field experiment was carried out, and on this basis, the PDA laser measurement method was used to measure and study the dry deposition velocity of aerosol particles simulated in a wind tunnel. By comparing with previous experimental results, it is confirmed that the measurement results of this experiment are true and reasonable. When the underlying surfaces are the same, in general, when the wind speed increases, the aerosol particles are less likely to be deposited, and individual terrains may appear different situations. In the case of different underlying surfaces, the dry deposition velocity of aerosol particles with different particle sizes is affected by the wind speed differently.

References

1. Yu Shui. Determination of radioactive aerosol particle size and representativeness of assumed ICRP values [J]. Radiation Protection Communications, 1997, 017(005):18-23.
2. GRUNDEL M, PORSTEND J. Differences between the activity sizedistributions of the different natural radionuclide aerosols in outdoorair [J]. Atmospheric Environment, 2004, 38: 3 723- 3 728.
3. SACHCHIDA N, TRIPATHI R, GILES HARRISON. Poster sessionI: radioactive aerosols atmospheric removal of radioactive aerosols[J]. J Aerosol Sci, 2000, 31(1): S472- S473.
4. He Huasang. Measurement of long-lived alpha radioactive aerosol concentration in ambient air. Radiation Protection, 1993, 13(4): 303-307.
5. Fu Cuiming, Li Jingtao, Zhang Zhilong, et al. Testing of a new type of filter paper for continuous monitoring of radioactive aerosols [J]. Radiation Protection Communications, 2003, 23(5):32-34.
6. Wu Xiaofei, Liu Hongshi. Aerosol sampling method for radioactive environmental investigation in nuclear power plants [J]. Nuclear Electronics and Detection Technology, 2004(05):112-114.
7. ROUPSARD, Pierre, et al. Measurement in a wind tunnel of dry deposition velocities of submicron aerosol with associated turbulence onto rough and smooth urban surfaces. Journal of Aerosol Science, 2013, 55: 12-24.
8. CHAMBERLAIN, A. C. at CHADWICK RC (1953): Deposition of airborne radio-iodine vapour. Nucleonics, 11: 22.
9. Lin Guanming, Cai Xuhui, Hu Min, et al. Research progress of atmospheric aerosol dry deposition [J]. China Environmental Science, 2018, 38(9):10.
10. Pope CA,Muhlestein JB, May HT,et al.Ischemic heart disease events triggered by short-term exposure to fine particulate air pollution[J]. Circulation, 2006,114(13):2443-2448.
11. Pryor S C. Size-resolved particle deposition velocities of sub-100 nm diameter particles over a forest[J]. Atmospheric Environment, 2006, 40(32):6192-6200.
12. ZHANG, L., et al. Dry deposition fluxes and deposition velocities of seven trace metal species at five sites in Central Taiwan--a summary of surrogate surface measurements and a comparison with model

estimation. Atmospheric Chemistry & Physics Discussions, 2011, 11.12.

13. OKUBO, Ayako; TAKEDA, Shigenobu; OBATA, Hajime. Atmospheric deposition of trace metals to the western North Pacific Ocean observed at coastal station in Japan. Atmospheric Research, 2013, 129: 20-32.
14. Zufall M J, Dai W, Davidson C I. Dry deposition of particles to wave surfaces: II. Wind tunnel experiments [J]. Atmospheric Environment, 1999, 33(26):4283-4290.
15. Slinn S A , Slinn W G N . Predictions for particle deposition on natural waters[J]. Atmospheric Environment, 1980, 14(9):1013-1016.
16. Zhang L , Gong S , Padro J , et al. A size-segregated particle dry deposition scheme for an atmospheric aerosol module[J]. Atmospheric Environment, 2001, 35(3):549-560.
17. Zhang J, Shao Y. A new parameterization of particle dry deposition over rough surfaces[J]. Atmospheric Chemistry and Physics, 2014, 14(22): 12429-12440.

Design scheme of metal wastewater purification device in industrial park

Tianqi Wang [1], Fangchen Dong [1], Peiling Xiang [2], Shanshan He [3], Tielin Wang [4]

[1] School of Software Engineering, Chongqing University of Posts and Telecommunications 400065, China
[2] School of Media Arts, Chongqing University of Posts and Telecommunications
[3] School of Economics and Management, Chongqing University of Posts and Telecommunications 400065, China
[4] School of Modern Post Institute, Chongqing University of Posts and Telecommunications 400065, China

Abstract: Due to the limitation of chemical process technology and management system, the traditional treatment mode of heavy metal waste liquid pollution has many problems, such as high power consumption in the process, high wear of consuming steel, high management cost, and secondary pollution caused by complex solid waste residue in data detection. Team based on the front end of deep learning module and back-end soft measurement model of collaborative structures, design and improvement of the learning algorithm combined with nested integration depth control series new front end treatment module, the successful implementation of heavy metal ions in industrial wastewater concentration of accurate prediction, and according to the predicted results and the analysis of liquid ingredients, targeted treatment after intelligent classification.

Keywords: Deep learning; Heavy metal concentration measurement; Ion exchange technology; Intelligent water;

1. The introduction

1.1 Current situation of existing products

The team conducted field investigations in panzhihua, Kunshan and other iron and steel forging bases and found the following problems in existing processes and assembly lines.

[Limitations of chemical method technology]

The traditional chemical process treatment has technical limitations. With the launch of discharge standards or about to reach their limits, there is an urgent need for innovation in the treatment technology of heavy metal wastewater, as well as the development of efficient and cost-effective devices and new schemes.

[Secondary processing produces dangerous waste, which cannot be reused]

Traditional chemical processes turn metal ions in industrial heavy metal wastewater into precipitated sludge and other hazardous wastes, requiring retreatment and landfill of secondary hazardous wastes, and cannot be recycled.

[High comprehensive cost of chemical method]

Chemical method has low initial cost, dangerous waste deposition after long-term use, and there is no matching intelligent monitoring system, which requires manual maintenance, high maintenance and repair cost, and short service life.

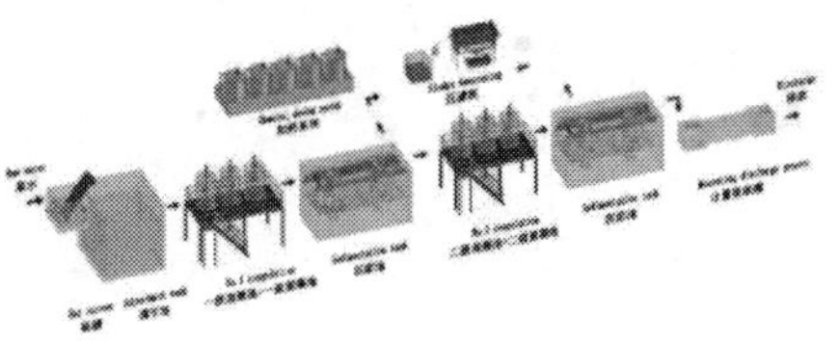

Figure 1. Traditional wastewater treatment process

simple operation and convenient maintenance, but it has some shortcomings such as low sensitivity, easy to be interfered by coexisting ions, and can not be used to determine multiple ions at the same time.[1]

[Lack of "wisdom" management, high management cost]

Most of the existing smart water concepts are only for regional monitoring, inspection, data management, etc. The traditional industrial management mode is single, the effluent management mode is chaotic, the phenomenon of illegal discharge, leakage and discharge is not rare, and the relevant enterprises need to spend a lot of manpower and time to manually manage and detect the wastewater treatment situation.

Front-line technical personnel urgently need a safe intelligent management and detection system to assist water management and purification, and collect relevant data for visual management.

https://doi.org/10.1051/e3sconf/202235801014

[Lack of management innovation, single mode]

There are many existing wastewater treatment processes. Whether the chemical method which has been mature for many years or the biological method which is in the early stage of development, they are all aimed at the technical innovation of chemical process, but ignore the management system of electroplating waste liquid treatment, and the mode innovation and design of control system.

At present, most one-stop water treatment suppliers only provide simple automation devices and control consoles with single functions, such as remote control, which only meet simple functions such as operation of control devices. This is in contradiction with the ratio of the industrial Internet of Things in the future development.

Figure 2 Manual guided operating equipment

Figure 3 Large acid-base neutralization

1.2 Product design objectives

In the disposal process, industrial ion exchange technology is combined with the integration of deep learning algorithm and distributed control of exchange bin module to achieve a technical solution of high efficiency, low pollution and low comprehensive cost from the perspective of process control.

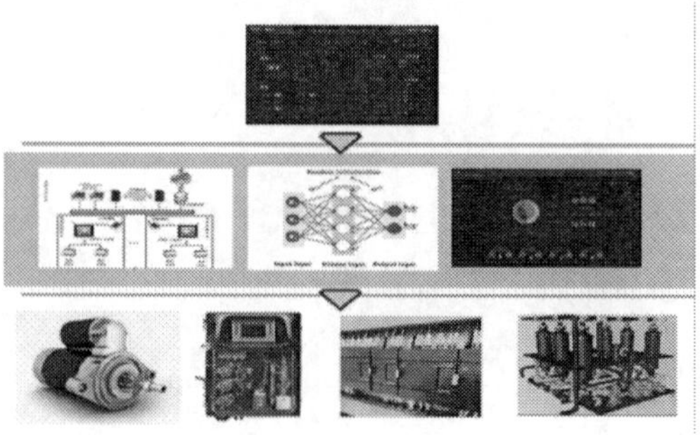

Figure 4 Front and rear Cloud Trinity mode tray

1.3 Process Route

The device is deeply treated by ion exchange adsorption system, which makes the effluent reach the standard stably, reduces waste solids and realizes resource reuse. At the same time, the heavy metal wastewater without organic matter and cyanide can be directly into the ion exchange adsorption system for comprehensive treatment, discharge standards. The ion exchange adsorption system concentrates heavy metal ions through regeneration process. According to actual requirements, the reclaimed liquid can be processed by returning to the collection tank or recycling, and the real-time interaction with the back-end data module can be maintained.

2. Targeted ion exchange bin

2.1 Function Design

[Structural design]

Heavy metal pollution has caused more and more harm to the environment and human beings, with the further implementation of the global sustainable development strategy, the heavy metal wastewater treatment requirements will be increasingly strict. Heavy metal wastewater is a very complicated mixed system, and it is difficult to meet the treatment requirements with single treatment technology.[2,3]

The prototype should be small in size, light in weight, convenient in transportation, and have strong emergency handling ability. The device has continuous treatment capacity and high treatment efficiency for heavy metal (electroplating) wastewater. It can treat wastewater with concentration ranging from 0.1mg/L to 500mg/L. The whole device is standardized design, each column is designed as a standardized column, the column between the general, easy to replace the material in the tank; The main parts are standardized parts, convenient for maintenance and disassembly repair work.

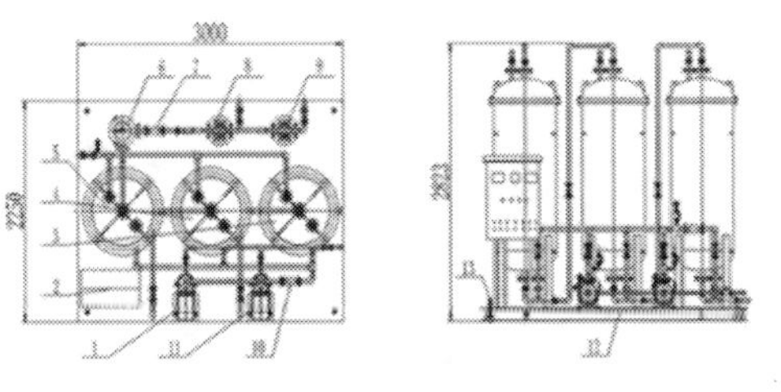

Figure 5 Structure and layout of the mobile device

The device uses standardized cylinder as filling container of ion exchange material. The cylinder is composed of cylinder and upper and lower end cover with the same design parameters. The material is low alloy steel, and the inner wall of the cylinder is lined with rubber, which has the performance of acid and alkali resistance and corrosion resistance.

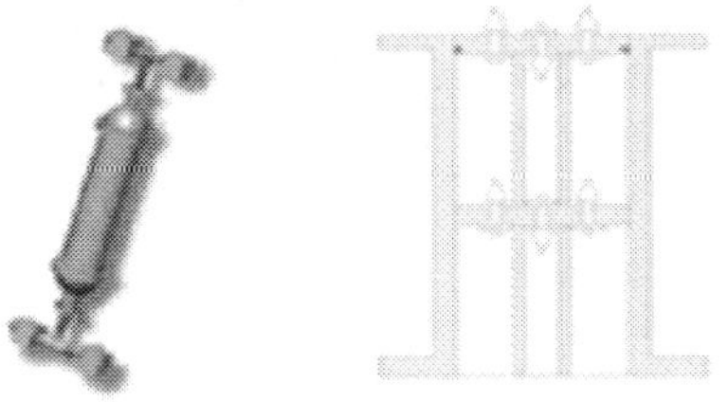

Figure 6 Standardized single cylinder

[Automation design]

The system adopts force control configuration, Siemens S7-200 PLC and touch screen as operation interface and logic control unit. According to the different ions and concentration, water quantity, standard column adsorption saturation capacity and other parameters in the field wastewater, the corresponding column passing time is calculated, and based on this, the automatic switching function of the system is realized, and the online storage, long-distance transmission and fault alarm functions of key data are completed.

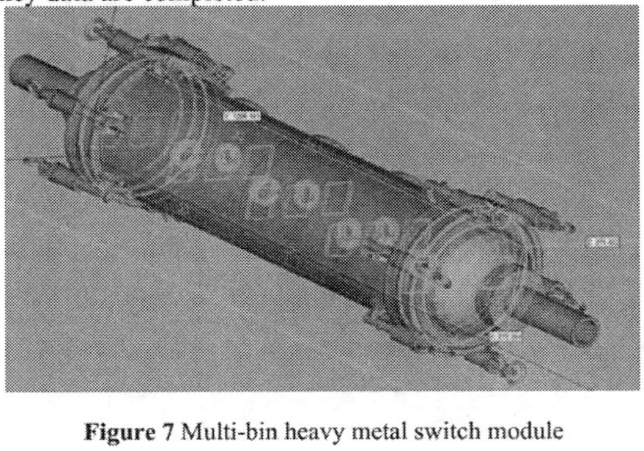

Figure 7 Multi-bin heavy metal switch module

[Standard modularization]

The device adopts standard modular design, with three single columns and their supporting pipes as a module group, and the columns in the module group are connected in series through pipes. The whole device is currently provided with two module groups.The computing model of soft-sensing neural technology has been verified based on neural network, and the modeling principle of BP neural network has been elaborated in detail in Literature [5].Simple contains three layers BP neural network input layer, hidden layer and output layer, each node in the diagram represents a neurons; The neurons of the preceding layer and the following layer are connected by corresponding weights. BP neural network training is mainly divided into two processes: positive transfer and error signal of the reverse[5]

In order to cooperate with the module group, a series of common accessories are configured for the two groups of modules. The module group is connected in parallel through pipes, and the connecting pipes between the module group and the module group are provided with solenoid valves. Various working modes of the module group can be realized by controlling the solenoid valve

2.2 Material adsorption experimental verification

To check the strength of the column, the design pressure of the device is P = 0.5mpa, and the yield strength of low alloy steel [σ]t=345MPa. Then the diameter ratio coefficient K of the tank is:

$$K = \frac{2[\sigma]^t + p}{2[\sigma]^t - p} = 1.001 \leq 1.2$$

Therefore, the internal pressure cylinder strength can be checked according to the maximum tensile stress criterion. Refer to the national standard GB150-2011 "Fixed pressure vessel" can be obtained, the cylinder cylinder welding joint coefficient ϕ=0.85, cylinder diameter Di=250mm, the actual wall thickness of the cylinder δ I =15mm, and the cylinder calculated wall thickness, namely the minimum wall thickness of the cylinder is:He Dan et al. from South China Normal University established a soft sensor model (PSO-BP model) by optimizing BP neural network through particle swarm optimization algorithm (PSO), and used the nonlinear approximation ability of BP neural network and the global search ability of PSO to accurately predict CODEFF and SSEFF in the process of sewage treatment[4]

$$\sigma = \frac{p_c D_i}{2[\sigma]^t \phi - p_c} = 0.2mm \leq \delta_i$$

In the formula, PC is the calculated pressure of the cylinder, and its value is PC =1.1× P = 0.5mpa. It is known from above that the cylinder design of the tank meets the requirements of working strength. Similarly, with reference to the national standard GB150-2011 "Solid 6 Type Pressure Vessel", the tank end cover can be checked for strength, and the check result is: the tank end cover design meets the working strength requirements. [6]Because the external pressure is small when the tank works, there is no instability of the tank, so the tank can not be stability check. At this point, the overall structure of the device tank design check completed, in line with GB150-2011 "Fixed pressure vessel" standard requirements.

The saturated adsorption capacity of the fiber for a specific metal ion is set as D mg/g, the fiber column is

https://doi.org/10.1051/e3sconf/202235801014

filled with X Kg, the flow rate of the equipment is Y m3/h, the concentration of specific metal ions in the original liquid is Z mg/L, and the saturation adsorption time of metal ions is T h.

The following equation can be obtained:

$$8D \bullet X \bullet 1000 = Z \bullet Y \bullet 1000 \bullet T$$

It can be concluded that:

$$T = (D - X) / (Z \bullet Y)$$

$$(2 \bullet D \bullet X \bullet 1000) / (B(\mathrm{Cr}) \bullet 1000) = C(\mathrm{OH}_{-}) \bullet K \bullet t \bullet 1000$$

Available:

$$t = (D \bullet X) / (B(\mathrm{Cr}) \bullet C(\mathrm{OH}_{-}) \bullet K \bullet 500)$$

Copper was adsorbed by cation exchange fiber and regenerated with 5%HCl solution. The saturated adsorption capacity of the fiber to a specific metal ion is D mg/g, C(H+)=1.44mol/L, the fiber column is filled with X Kg, the regeneration flow rate is K m3/ H, and the regeneration time is T H. B represents the relative atomic mass of the metal element, and the relative submass of copper is 64. According to the regeneration principle, one molecule of copper needs two molecules of hydrochloric acid solution to regenerate,

The equation obtained is:

$$(2 \bullet D \bullet X \bullet 1000) / (B(\mathrm{Cu}) \bullet 1000) = C(\mathrm{H}+) \bullet K \bullet t \bullet 1000$$

Available:

$$t = (D \bullet X) / (B(\mathrm{Cu}) \bullet C(\mathrm{H}_{+}) \bullet K \bullet 500)$$

Nickel was adsorbed by cation exchange fiber and regenerated with 5%HCl solution. The saturated adsorption capacity of the fiber to a specific metal ion is D mg/g, C(H+)=1.44mol/L, the fiber column is filled with X Kg, the regeneration flow rate is K m3/ H, and the regeneration time is T H. B represents the relative atomic mass of the metal element, and the relative submass of nickel is 59. According to the regeneration principle, 1 molecule of nickel needs 2 molecules of hydrochloric acid solution to regenerate,

The equation obtained is:

$$(2 \bullet D \bullet X \bullet 1000) / (B(\mathrm{N}_{i}) \bullet 1000) = C(\mathrm{H}+) \bullet K \bullet t \bullet 1000$$

Available:

$$t = (D \bullet X) / (B(\mathrm{Ni}) \bullet C(\mathrm{H}_{+}) \bullet K \bullet 500)$$

According to the saturation regeneration formula above, the prediction time of ion concentration of each module is calculated, and the results are input into the automatic control program, and the device can complete the intelligent allocation of targeted modules at the corresponding time.

3. Index verification of front-end device processing process

3.1 Training samples based on concentration prediction results

Experimental process: Dayu front-end switch module was used for adsorption regeneration process. Sodium hydroxide solution with a mass fraction of 5% was configured with caustic soda. After soaking for 30min, the regeneration process was completed by washing and discharge with water. The water sample taken is the outlet water sample of the front-end disposal module.

After repeated process verification for other switch modules, the content of hexavalent chromium is 4.87mg/ L, the PH value is 12.36, and the total content of copper and zinc is 0.36mg/ L and 0.475mg/ L, respectively, far better than the national emission threshold. Therefore, it is verified again that the wastewater entering the switch module can be purified with higher efficiency after concentration prediction and assisted intelligent distribution.

Due to the low concentration of chromium-containing wastewater, the material cannot be saturated in a short time, so the saturation time of each adsorption tank is set as 2 hours. The experimental raw water is chromium-containing wastewater, the concentration of raw water is 11.97mg/ L, the content of other heavy metal ions is < 0.05mg/ L, the PH value of raw water is 8.66, and the ambient temperature is 30°C. The experimental raw water was treated by connecting the exchange bin in series. The experimental flow was set at $1m^3/h$, and the actual flow was 0.9566-1.005m3 /h. The adsorption process lasted for 180min, and the chromium containing wastewater was treated from $2.868m^3$ to $3.015m^3$. All the tested water samples after treatment are the total outlet water samples of the device.

▼	单位	mg/L	mg/L	mg/L	mg/L	mg/	——	mg/L	s/cm	[illegible]
1	实验原水	11.97	<0.05	<0.05	<0.05	<0.05	8.66	16	1654	
2	ZA-1-1	0.04	<0.05	<0.05	<0.05	<0.05	9.63			3
3	ZA-1-2	0.033	<0.05	<0.05	<0.05	<0.05	9.64			8
4	ZA-1-3	0.036	<0.05	<0.05	<0.05	<0.05	9.67			10
5	ZA-1-4	0.04	<0.05	<0.05	<0.05	<0.05	9.65			20
6	ZA-1-5	0.04	<0.05	<0.05	<0.05	<0.05	9.56			30
7	ZA-1-6	0.036	<0.05	<0.05	<0.05	<0.05	9.47			63
8	ZA-1-7	0.047	<0.05	<0.05	<0.05	<0.05	9.74	11	1257	68
9	ZA-1-8	0.035	<0.05	<0.05	<0.05	<0.05	10	16	1272	103
10	ZA-1-9	0.072	<0.05	<0.05	<0.05	<0.05	9.4	11	1260	138

Figure 8 Experimental data of adsorption function verification

3.2 Intelligent scheduling platform construction

Basic platform architecture:

The dispatching system is mainly equipped with concentration prediction and intelligent distribution technology, which plays an intelligent control of the front-end water control module on the whole. The system is divided into four parts, including perception layer, transmission layer, application layer and terminal layer.

(1) Perception layer. Mainly for the acquisition of enterprise data set, these data through the acquisition system of each module. The module consists of water pressure sensor, detection sensor, hydraulic detection, dynamic fan, lifting pump, and fault detector and other hardware equipment. It is mainly used to detect safety information, water quantity and other information data sets used for research.

(2) Transport layer. It is composed of various communication modules, including gateways, Purple Bee wireless connection equipment, and various wireless and wired transmission networks, so as to achieve remote control.

(3) Application layer. Here, the switching device is assembled in a distributed architecture mainly by outsourcing customized parts through selective selection.
(4) Terminal layer. IPC and PC can obtain the real-time operation data of the equipment in the system in time, and the concentration prediction report and intelligent distribution information can be visually presented here.

4. Conclusion

This chapter makes a more detailed analysis of the front-end water treatment module from two perspectives, and shows the heavy metal wastewater treatment process based on back-end intelligent analysis through pilot test verification and comparison of effluent data indicators:
1. Briefly explain the drawbacks and pain points of the general assembly line operation found by the team through field research. And detailed analysis of many pain points under the need to solve the core problems.
2. Through a series of experiments and data verification design optimization of the traditional process, determine the front-end processing device design concept and indicators.
3. Under the intelligent control of the back-end platform, a complete cycle is completed, and the results are compared with the set indicators, which are in line with the design standards.

References

1. Zheng Xiaohong. The application status and development prospect of heavy metal on-line monitoring system in wastewater pollution source monitoring [J]. Instrumentation & Analysis and Monitoring,2019,(3):44-46.
2. Zou Zhaohua, He Sufang, Han Caiyun, et al. Research progress of heavy metal wastewater treatment technology [J]. INDUSTRIAL WATER TREATMENT, 2010, 30(5): 9-12. (in Chinese)
3. Ma Qian, Zhang Xiaolong. Research progress in new technologies of heavy metal wastewater treatment at home and overseas. Chinese Journal of Environmental Engineering, 2007, 1(7): 10-14. (in Chinese).
4. Jan He, Laipeng Lin, Guoqiang Niu, Xiaohui Yi, Mingzhi Huang. A soft sensor model for wastewater treatment based on improved particle swarm optimization BP neural network [J]. Journal of South China Normal University (Natural Science Edition), 2021, 53(2): 114-120.
5. Lu Jiandong. Research on the application of genetic algorithm and neural network in wastewater treatment control and monitoring system [D]. Chongqing: Chongqing University, 2007 .
6. Feng Yuzhao, Ma Pingping. A BP algorithm for soft sensing of heavy metal ion concentration in wastewater [D]. University of Logistics Engineering, 2011,27(3):53-57. (in Chinese)

Landslide susceptibility prediction using C5.0 decision tree model

Qiangqiang Shua [1], Xiaogang Chen [2], Zhipeng Lian [3,*], Gengzhe Liu [4], Siyu Tao [2]

[1] School of Architecture Engineering, Sichuan Universit of Arts and Science, Dazhou City, China
[2] School of Infrastructure Engineering, Nanchang University, Nanchang, China
[3] Wuhan Center, China Geological Survey, Wuhan, China
[4] University of Pennsylvania school of design, Philadelphia, PA,19104, United States

Abstract: Regional landslide susceptibility prediction (LSP) research is of great significance to the prevention and control of landslides. This study focuses on the LSP modelling based on the decision tree model. Taking the northern part of An'yuan County of Jiangxi Province as an example, 14 environmental factors including elevation, gully density and lithology are obtained based on geographical information system (GIS) and remote sensing satellite. Frequency Ratio method and C5.0 decision tree (DT) model are coupled to build DT model for LSP modelling. Then the predicted results are graded into five attribute intervals. Finally, LSP performance of DT model is evaluated by comparing the area value under the receiver operating characteristic curve (ROC) and classification of landslide susceptibility. The results show that the AUC accuracy of the C5.0 DT model is 0.805, and the LSP results of the C5.0 DT model are consistent with the actual distribution pattern of landslides in this County.

Keywords: Landslide susceptibility prediction; Frequency ratio; C5.0 decision tree; Geographical Information System

1. Introduction

The characteristic of landslides occurrence in patches makes it impossible to fully explain and predict the spatial distribution of landslides only by the study of single landslides. Therefore, many scholars have begun to pay attention to the research on the regional landslide susceptibility prediction (LSP)[1-3]. LSP can be considered as the probability of landslides occurrence at a particular location under the action of various environmental factors[4, 5]. It can be seen from the pattern of geographical similarity that the LSP model can be established by the known environmental factors of landslides, and the spatial location of potential landslides can be accurately predicted[6, 7]. The relationship between environmental factors and LSP is the key to establish LSP model, and the establishment of LSP model will directly affect the accuracy of LSP[8, 9].

The steps of LSP mainly include the acquisition of landslide inventory, selection of environmental factors, selection of models and setting of parameters, landslide susceptibility mapping (LSM), uncertainty analysis and accuracy evaluation of LSP results[10-12]. At present, mathematical statistics, machine learning and heuristic models are mainly used in research[13, 14]. Compared with the other two models, machine learning models have more excellent generalization ability and LSP effect, so machine learning models are now widely used in LSP modelling. Machine learning models usually refer to support vector machine[15, 16], decision tree (DT)[17], logistic regression[18], artificial neural network[19, 20], grey correlation degree[21] and random forest[22, 23]. At present, DT model has been widely concerned in various fields because of its excellent nonlinear fitting performance. The widely used C5.0 DT model is used to analyze the LSP in the study area in this study.

In the establishment of C5.0 DT model, it is necessary to analyze the connection between landslides and environmental factors. Frequency ratio (FR) method is adopted to establish the connection between landslides and environmental factors in this study. FR values are taken as input variables of C5.0 DT and RF models, and the accurate LSP results are predicted by nonlinear fitting of input and output variables of the models[22, 24, 25]. Taking An'yuan County of China as area, combines with 221 landslides in the northern part of An'yuan County, 14 original environmental factors are selected. Finally, LSP results and the accuracies are carried out.

* Corresponding author: lianzhipeng1985@163.com

2. LSP modelling analysis

2.1 Modeling study

(1) basic data are analyzed and 14 environmental factors are selected for the construction of LSP modelling.
(2) input variables of the model are obtained by processing environmental factors based on FR method.
(3) all grids of landslide areas are assigned a value of 1, and grids with the same number as those of landslide grids are selected from all grids in the study area and assigned a value of 0, which are randomly divided into training sets and testing sets in a ratio of 7 to 3.
(4) the FR values of 14 environmental factors and the original basic data are used as input variables of the C5.0 DT model to conduct modelling analysis and obtain the LSM of the study area.
(5) ROC is used to evaluate the accuracy of C5.0 DT model.

2.2 C5.0 DT

Decision tree is a classification or decision set expressed by tree structure according to the characteristics of sample sets[9, 26]. It can find a nonlinear partition method, and use a top-down recursive method to organize a large number of datasets from classification to generate visual classification rules[17, 27]. The C5.0 DT model can take the maximum information gain rate as a classification, and elicit it through information entropy according to the specific calculation method[28]. Its average uncertainty is determined before the information source is sent, and is the ratio of node information gain to node splitting information measure. Take a sample set D, the total number of categories of set D is K, and C is a group of target variables, then entropy is calculated as Eq (1).

$$Ent(D) = -\sum_{K=1}^{K} P_i(C_i / D)\log_2 P_i(C_i / D) \qquad (1)$$

Where Pi (Ci/D) is the relative probability of target variable Ci to sample set D. For information gain, it is assumed that there are N possible values in a discrete attribute variable a, (a1,a2,a3...,..aN), if data set D is used for partitioning, N branch nodes will be generated. an means that data set D contains all the n nodes in feature a, Dn means the total number of samples with the value of an, and DN is the same. Therefore, the information gain calculation formula of sample set D after the introduction of conditions is shown in Eq (2):

$$Gain(D,a) = Ent(D) - \sum_{n=1}^{N} \frac{|D^N|}{|D|} Ent(D^n) \qquad (2)$$

In general, $Ent(D) > \sum_{n=1}^{N} \frac{|D^N|}{|D|} Ent(D^n)$, information gain can be represented as the degree of information complexity reduction under a certain condition, so information gain can be used to select the attributes of decision tree partition. When the information gain is larger, the purity of dataset partition is higher by using features[28]. The growth of decision tree is achieved by selecting the optimal grouping variable with the maximum information rate, the calculation of information gain rate is shown in Eq (3):

$$GainRatio(D,a) = \frac{Gain(D,a)}{IV(a)} \qquad (3)$$

IV (a) is a specific value of the specific attribute a. Generally, the larger N is (that is the greater the possibility of a possible value), the larger IV (a) is, IV (a) can be calculated as Eq (4) :

$$IV(a) = -\sum_{n=1}^{N} \frac{|D^n|}{|D|} \log_2 \frac{|D^n|}{|D|} \qquad (4)$$

2.3 Accuracy evaluation of LSP results

Area under ROC (AUC) is widely used to evaluate the LSP accuracy. In this method, Sensitivity is used as the vertical coordinate and 1-specificity is used as the horizontal coordinate, and AUC is used as the specific value of the LSP modelling accuracy evaluation[24, 29, 30]. The larger AUC value is, the higher the model accuracy is.

3. Study area introduction and landslide environmental factors analysis

3.1 An'yuan County introduction and landslide inventory

An'yuan County is in the south of China, Located in the subtropical monsoon humid climate, with an annual rainfall of 1600~1800 mm. An'yuan County is a typical hilly county with Ganjiang River system and Pearl River system. The northern part of An'yuan County, which covers an area of 970 km2, is selected as the study area. According to the distribution of landslides in the north of An'yuan County (Fig 1) and the lithology, it can be seen that the distribution of landslides is relatively scattered. The possible reason is the north of An'yuan County is basically low mountains, and the distribution of water system is relatively uniform. Landslides are mainly distributed along the ravines in the middle and upper elevation area. In addition, influenced by local lithology, landslides are mainly distributed in siltstone, shale and mudstone areas.

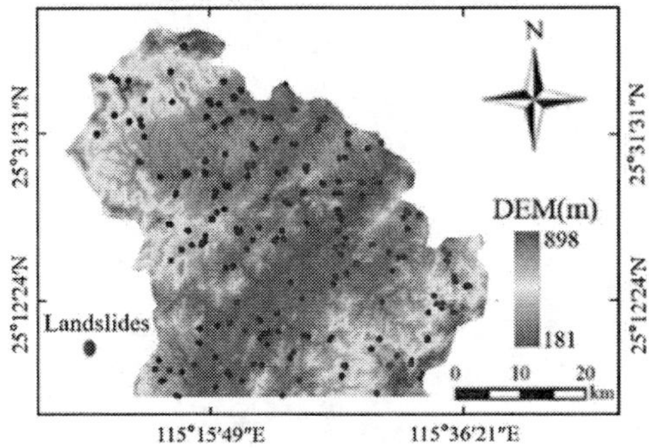

Figure 1. The evolution and landslide inventory in north of An'yuan country.

https://doi.org/10.1051/e3sconf/202235801015

3.2 Landslide data sources and environmental factors analysis

The relevant basic data are as follows: 1) historical landslide inventory map and field survey data are obtained; 2) digital elevation model (DEM) with 30m resolution is downloaded from geographic data cloud platform; 3) the geological conditions of the study area are obtained from the 1:100,000 geological map of Hydrogeology Atlas of China; 4) normalized difference vegetation index (NDVI) and other land cover factors are obtained from landsat-8 remote sensing satellite images.

14 environmental factors are selected as input variables, including 13 continuous environmental factors such as elevation, slope, profile curvature, gully density, and the discrete environmental factors, lithology. The continuous environmental factors are divided into 8 attribute intervals by the natural break point method and lithology is classified according to the original natural state. The FR values are used as the input variable to the model after the environmental factors obtained, as shown in Table 1(only partial factors are shown); The larger the FR values are, the more favorable the attribute intervals are to the occurrence of landslide[31, 32].

3.2.1 Topographic and hydrologic environmental factors.

Topographic factors such as elevation, slope, aspect and profile curvature are obtained (Fig 2), as shown in Table 1. Elevation is negatively correlated with FR values, when elevation is 181.92-361.40 m, FR value is greater than 1, indicating that elevation has a favorable influence on landslide development. When elevation values increase, FR values are less than 1, indicating that elevation has little relation with landslides. When the slope range is 7.65°-22.51°, FR value is greater than 1, indicating that the slope range has a favorable influence on landslide development.

MNDVI mainly reflects the distribution characteristics of surface water, which can scour and infiltrate the slope, making the landslides more prone to dislocation[33-35]. When MNDVI is in the range of 0-0.50, FR value is greater than 1, indicating that MNDVI plays a role in promoting landslides development in this range. When the gully density is between 0.46-0.60 and 0.72-1.46, the FR values are greater than 1, indicating that the intervals have a favorable influence on the development of landslides.

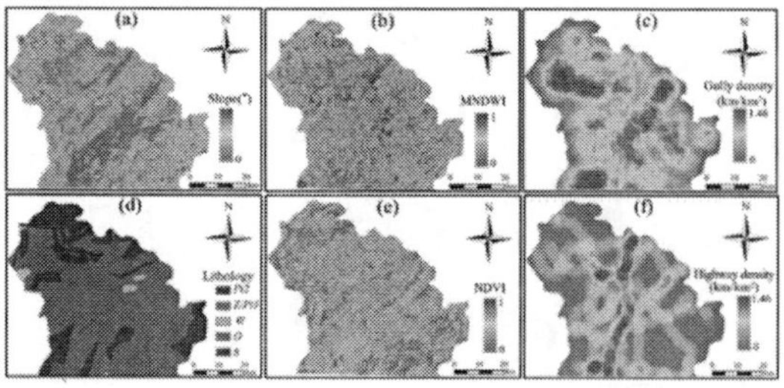

Figure 2. Environmental factors: (a) Slope, (b) MNDWI, (c) Gully density, (d) Lithology, (e) NDVI, (f) Highway density

3.2.2 Basic geology and land cover factors.

The strata in the northern part of An'yuan County are mainly composed of Middle Proterozoic (Pt2), Late Paleoproterozoic (Z、Pt3), Cambrian (∈), Ordovician (O) and Silurian (S), among which the FR value of Z、Pt3 is greater than 1, indicating that the strata in Z、Pt3 is conducive to the development of landslides. For the clay and sand content in the study area, when the clay content attribute range is 22.93-29.22, FR value is greater than 1, and when the clay content is in this range, landslides are more likely to occur. When the content of sand is in the range of 31.22-35.17, 54.43-61.68 and 31.22-35.17, FR values are greater than 1, indicating that the probability of landslide is greater when the content of sand in the range than other ranges.

Land cover factors include NBVI, NDBI and road density. Landslides are more likely to occur when NDVI is between 0.31-0.68. NDBI can better reflect the situation of building land, which can reflect the concentrated area of local human activities[36]. NDBI is more likely to occur landslide in the range of 0.25-1.00, the FR values are obviously greater than 1 when the road density ranges from 0.59 to 1.53 and from 1.89 to 2.52. Artificial slope cutting is often carried out in highway excavation, which will change the geological conditions beside the road and promote the landslides evolution [37, 38].

Table 1. Frequency ratio of each evaluation factor

Environmental factors	Value	Total number of grids	Proportion of grids (%)	Grids number in the landslide	proportion of grids in the landslide (%)	FRs
Elevation(m)	181.92-250.55	119026	11.04	790	15.58	1.41
	235.55-260.71	129221	11.99	854	16.85	1.41
	260.71-361.30	407714	37.83	2040	40.24	1.06
	361.30-454.31	185880	17.25	530	10.46	0.61
	454.31-510.13	72976	6.77	250	4.93	0.73
	510.13-591.57	79202	7.35	318	6.27	0.85
	591.57-679.86	48372	4.49	184	3.63	0.81
	679.86-898.01	35308	3.28	103	2.03	0.62
NDVI	—0.01-0.31	2576	0.24	1	0.02	0.08
	0.31-0.47	16441	1.53	108	2.13	1.40
	0.47-0.57	49141	4.56	323	6.37	1.40
	0.57-0.63	110779	10.28	633	12.49	1.21
	0.63-0.68	200967	18.65	1063	20.97	1.12
	0.68-0.73	276952	25.70	1276	25.17	0.98
	0.73-0.79	291987	27.09	1318	26.00	0.96
	0.79-1.01	128856	11.96	347	6.85	0.57

4. LSP modelling analysis in northern An'yuan County

4.1 Spatial dataset analysis

The study area is divided into a total of 1077699 grid units based on 30m resolution, and the internal area of known landslides are divided into 5069 grid units. Before taking the FRs of 14 environmental factors as the input variables of the model, the 5069 landslide grid units are assigned a value of 1, and 5069 grid units are randomly selected from the entire grid units in the northern part of this County and assigned a value of 0 to form the dataset as the output variables. The dataset is randomly divided into model training set and testing set in a ratio of 7/3. Finally, 14 kinds of environmental factors treated by FR method are substituted into the trained C5.0 DT model to obtain the LSM of the northern Part of An'yuan County. LSIs are divided into five attribute intervals of very low, low, medium, high and very high by the natural break point method [3, 39].

4.2 LSP results and AUC accuracy evaluation

The LSP results are shown in Figure 3. The levels of LSP results are mainly very low and low, accounting for about 70% of the study area. In the C5.0 DT model, the areas of very low, low, medium, high and very high prone areas are 58.08%, 14.84%, 7.40%, 14.02% and 5.67% respectively, and the areas within the landslide are 15.21%, 14.84%, 7.40%, 14.02% and 5.67%. If the range of LSIs is very low, it indicates that the probability of landslides in this region is very small [40]. It shows that the C5.0 DT model can reflect more landslides with less susceptibility value and LSP performance is excellent.
According to Figure 3, it can be seen that the levels of LSP results of most areas in the study area predicted by C5.0 DT model are low and very low. The results show that the high-occurrence area of landslides is mainly concentrated in the middle of the study area, and the other areas are scattered. The low-occurrence area of landslides is mainly distributed in northwest and southeast regions. The main reason is that the northern part of An'yuan county is a typical hilly area, the northwest and southeast are mainly mountainous areas with higher elevations, and the central part is the most densely hilly area, which is more conducive to the development of landslides. The lithologic of the central region is late Paleoproterozoic, mainly metamorphic rocks. The high density of gullies, large degree of rock fissures and dense distribution of river network in this region further promote the development of landslides [41, 42]. AUC value is used to evaluate the accuracy of C5.0 DT model. As shown in Figure 3, AUC of C5.0 DT model is 0.805. The C5.0 DT model uses nonlinear partition method and recursive method to classify and organize dataset, and generates visual classification rules, so that the classification results are consistent with the actual situation as much as possible [26].

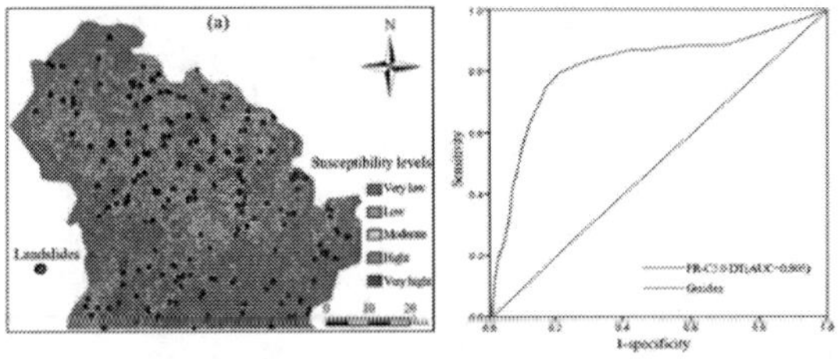

Figure 3. Landslide susceptibility maps of DT model and ROC curves of C5.0 DT model

5. Conclusion

The C5.0DT model is established to perform LSP modelling in the northern part of An'yuan County of China. The C5.0 DT model can accurately predict the distribution pattern of LSIs in the study area based on the accuracy analysis. The LSP results show that the high and very high LSP regions are mainly concentrated in the middle of the study area, and the low and very low LSP regions are mainly concentrated in the northwest and southeast regions. The evolution of landslides is mainly concentrated in the area with moderate elevation and slope with dense gullies, and lithology is mainly siltstone, shale and mudstone.

References

1. Jente Broeckx, Matthias Vanmaercke, Rica Duchateau, et al. A data-based landslide susceptibility map of Africa[J]. Earth-Science Reviews, 2018, 185: 102-121.
2. Langping Li, Hengxing Lan, Changbao Guo, et al. A modified frequency ratio method for landslide susceptibility assessment[J]. Landslides, 2017, 14(2): 727-741.
3. Faming Huang, Jianbo Yang, Biao Zhang, et al. Regional Terrain Complexity Assessment Based on Principal Component Analysis and Geographic Information System: A Case of Jiangxi Province, China[J]. ISPRS International Journal of Geo-Information, 2020, 9(9): 539.
4. Bernhard Rabus, Manuele Pichierri. A new InSAR phase demodulation technique developed for a typical example of a complex, multi-lobed landslide displacement field, Fels Glacier Slide, Alaska[J]. Remote Sensing, 2018, 10(7): 995.
5. Tao Hu, Xin Fan, Shuo Wang, et al. Landslide susceptibility evaluation of Sinan County using logistics regression model and 3S technology[J]. Geological Science and Technology Bulletin, 2020, 39(2): 113-121.
6. Wu Huayong, Dong Yue, Gao Lei, et al. Identifying nitrate sources in surface water, regolith and groundwater in a subtropical red soil Critical Zone by using dual nitrate isotopes[J]. CATENA, 2020, (prepublish)

7. Lei Liu, KL Yin, JJ Wang, et al. Dynamic evaluation of regional landslide hazard due to rainfall: a case study in Wanzhou central district, Three Gorges Reservoir[J]. Chinese Journal of Rock Mechanics and Engineering, 2016, 35(3): 558-569.
8. Guruh Samodra, Guangqi Chen, Junun Sartohadi, et al. Comparing data-driven landslide susceptibility models based on participatory landslide inventory mapping in Purwosari area, Yogyakarta, Java[J]. Environmental Earth Sciences, 2017, 76(4): 1-19.
9. ZZ Guo, KL Yin, FM Huang, et al. Landslide displacement prediction based on surface monitoring data and nonlinear time series combination model[J]. Chinese Journal of Rock Mechanics and Engineering, 2018, 37(S1): 3392-3399.
10. Badal Pokharel, Massimiliano Alvioli, Samsung Lim. Assessment of earthquake-induced landslide inventories and susceptibility maps using slope unit-based logistic regression and geospatial statistics[J]. Scientific reports, 2021, 11(1): 1-15.
11. Xinzhi Zhou, Haijia Wen, Yalan Zhang, et al. Landslide susceptibility mapping using hybrid random forest with GeoDetector and RFE for factor optimization[J]. Geoscience Frontiers, 2021, 12(5): 101211.
12. Faming Huang, Siyu Tao, Zhilu Chang, et al. Efficient and automatic extraction of slope units based on multi-scale segmentation method for landslide assessments[J]. Landslides, 2021, 18(11): 3715-3731.
13. Recep Can, Sultan Kocaman, Candan Gokceoglu. A comprehensive assessment of XGBoost algorithm for landslide susceptibility mapping in the upper basin of Ataturk dam, Turkey[J]. Applied Sciences, 2021, 11(11): 4993.
14. Faming Huang, Jun Yan, Xuanmei Fan, et al. Uncertainty pattern in landslide susceptibility prediction modelling: Effects of different landslide boundaries and spatial shape expressions[J]. Geoscience Frontiers, 2022, 13(2): 101317.
15. Gwo-Fong Lin, Ming-Jui Chang, Ya-Chiao Huang, et al. Assessment of susceptibility to rainfall-induced landslides using improved self-organizing linear output map, support vector machine, and logistic regression[J]. Engineering Geology, 2017, 224: 62-74.
16. Faming Huang, Jinfeng Li, Junyu Wang, et al. Modelling rules of landslide susceptibility prediction considering the suitability of linear environmental factors and different machine learning models[J]. Geological Science and Technology Bulletin, 2022, 41(2): 44-59.
17. Zizheng Guo, Yu Shi, Faming Huang, et al. Landslide susceptibility zonation method based on C5. 0 decision tree and K-means cluster algorithms to improve the efficiency of risk management[J]. Geoscience Frontiers, 2021, 12(6): 101249.
18. Rui Liu, Luyao Li, Saied Pirasteh, et al. The performance quality of LR, SVM, and RF for earthquake-induced landslides susceptibility mapping incorporating remote sensing imagery[J]. Arabian Journal of Geosciences, 2021, 14(4): 1-15.
19. Paraskevas Tsangaratos, Andreas Benardos. Estimating landslide susceptibility through a artificial neural network classifier[J]. Natural hazards, 2014, 74(3): 1489-1516.
20. Xiaojing Wang, Faming Huang, Xuanmei Fan, et al. Landslide susceptibility modeling based on remote sensing data and data mining techniques[J]. Environmental Earth Sciences, 2022, 81(2): 1-19.
21. Qiong Wu, Huiming Tang, Xiaohan Ma, et al. Identification of movement characteristics and causal factors of the Shuping landslide based on monitored displacements[J]. Bulletin of Engineering Geology and the Environment, 2019, 78(3): 2093-2106.
22. Alireza Arabameri, Biswajeet Pradhan, Khalil Rezaei, et al. Assessment of landslide susceptibility using statistical-and artificial intelligence-based FR – RF integrated model and multiresolution DEMs[J]. Remote Sensing, 2019, 11(9): 999.
23. Zengzhen Qian, Mingqiang Sheng, Faming Huang, et al. Uplift Performance of Plate Anchors in Cement-stabilised Aeolian Sand[J]. Frontiers in Earth Science, 2021, 9: 1100.
24. Yijing Li, Ping Liu, Huokun Li, et al. A comparison method for 3D laser point clouds in displacement change detection for arch dams[J]. ISPRS International Journal of Geo-Information, 2021, 10(3): 184.
25. Faming Huang, Lihan Pan, Xuanmei Fan, et al. The uncertainty of landslide susceptibility prediction modeling: suitability of linear conditioning factors[J]. Bulletin of Engineering Geology and the Environment, 2022, 81(5): 1-19.
26. Burak F Tanyu, Aiyoub Abbaspour, Yashar Alimohammadlou, et al. Landslide susceptibility analyses using Random Forest, C4. 5, and C5. 0 with balanced and unbalanced datasets[J]. Catena, 2021, 203: 105355.
27. Faming Huang, Kunlong Yin, Jinsong Huang, et al. Landslide susceptibility mapping based on self-organizing-map network and extreme learning machine[J]. Engineering Geology, 2017, 223: 11-22.
28. Junwei Ma, Huiming Tang, Xiao Liu, et al. Establishment of a deformation forecasting model for a step-like landslide based on decision tree C5. 0 and two-step cluster algorithms: a case study in the Three Gorges Reservoir area, China[J]. Landslides, 2017, 14(3): 1275-1281.
29. Roger Sayre, Charlie Frye, Deniz Karagulle, et al. A new high-resolution map of world mountains and an online tool for visualizing and comparing characterizations of global mountain distributions[J]. Mountain Research and Development, 2018, 38(3): 240-249.
30. Ionut Cristi Nicu, Andrei Asăndulesei. GIS-based evaluation of diagnostic areas in landslide

susceptibility analysis of Bahluieţ River Basin (Moldavian Plateau, NE Romania). Are Neolithic sites in danger?[J]. Geomorphology, 2018, 314: 27-41.

31. Taskin Kavzoglu, Ismail Colkesen, Emrehan Kutlug Sahin. Machine learning techniques in landslide susceptibility mapping: a survey and a case study[J]. Landslides: Theory, practice and modelling, 2019: 283-301.
32. Faming Huang, Jiawu Chen, Weiping Liu, et al. Regional rainfall-induced landslide hazard warning based on landslide susceptibility mapping and a critical rainfall threshold[J]. Geomorphology, 2022, 408: 108236.
33. Amit Kumar Batar, Teiji Watanabe. Landslide susceptibility mapping and assessment using geospatial platforms and weights of evidence (WoE) method in the Indian Himalayan Region: recent developments, gaps, and future directions[J]. ISPRS International Journal of Geo-Information, 2021, 10(3): 114.
34. Chenglong Yu, Jianping Chen. Application of a GIS-based slope unit method for landslide susceptibility mapping in Helong city: comparative assessment of ICM, AHP, and RF model[J]. Symmetry, 2020, 12(11): 1848.
35. Faming Huang, Songyan Hu, Xueya Yan, et al. Landslide susceptibility prediction and identification of its main environmental factors based on machine learning models[J]. Geological Science and Technology Bulletin, 2022, 41(2): 79-90.
36. Wenbin Li, Yu Shi, Faming Huang, et al. Uncertainties of collapse susceptibility prediction based on remote sensing and GIS: effects of different machine learning models[J]. Frontiers in Earth Science, 2021, 9: 635.
37. Viet-Ha Nhu, Danesh Zandi, Himan Shahabi, et al. Comparison of support vector machine, Bayesian logistic regression, and alternating decision tree algorithms for shallow landslide susceptibility mapping along a mountainous road in the west of Iran[J]. Applied Sciences, 2020, 10(15): 5047.
38. Faming Huang, Jing Zhang, Chuangbing Zhou, et al. A deep learning algorithm using a fully connected sparse autoencoder neural network for landslide susceptibility prediction[J]. Landslides, 2020, 17(1): 217-229.
39. Abdelaziz Merghadi, Ali P Yunus, Jie Dou, et al. Machine learning methods for landslide susceptibility studies: A comparative overview of algorithm performance[J]. Earth-Science Reviews, 2020, 207: 103225.
40. Tao Chen, Li Zhu, Rui-qing Niu, et al. Mapping landslide susceptibility at the Three Gorges Reservoir, China, using gradient boosting decision tree, random forest and information value models[J]. Journal of Mountain Science, 2020, 17(3): 670-685.
41. A-Xing Zhu, Yamin Miao, Junzhi Liu, et al. A similarity-based approach to sampling absence data for landslide susceptibility mapping using data-driven methods[J]. Catena, 2019, 183: 104188.
42. Faming Huang, Zhongshan Cao, Jianfei Guo, et al. Comparisons of heuristic, general statistical and machine learning models for landslide susceptibility prediction and mapping[J]. CATENA, 2020, 191: 104580.

Research progress of waste retaining dam expansion project technology for large domestic waste landfills

Shuxiang Song, Chao Zheng, Qiufeng Huang

Guangzhou Grantop Environment Service co., ltd, Guangzhou, Guangdong province, 510000, China

Abstract: In view of the problem of insufficient storage capacity of the original landfill site, it is a very economical and effective way to use the horizontal, longitudinal and composite extension of garbage dam with geotextile reinforced earth dam. In this paper, the types of garbage dam, dam extension technology, dam reinforcement method and stability of garbage dam are introduced and discusses the types and advantages and disadvantages of dam retaining, the construction technology of dam retaining, the stability analysis and influencing factors of garbage dam retaining and the safety monitoring of dam body.

Key words: garbage dam; extension technology of dam retaining; stability analysis; leachate.

1. Introduction

With the rapid development of urban construction, garbage disposal technology has gradually improved. From 1989 up to the present, it has been revised four times, and now it is used "Technical Specification for Sanitary Landfill Treatment of Domestic Waste" (GB 50869-2013)[1]. The increasing improvement of people's quality of life has led to a sharp increase in urban garbage. From 2013 to 2019, the annual output of domestic garbage increased from 16.979 million tons to 236 million tons, as shown in Fig. 1[2, 3]. The main disposal methods of domestic waste in developed countries are landfill and incineration. For domestic waste in China, sanitary landfill is the preferred way. Therefore, landfills have entered our lives. According to incomplete statistics, China has built about 1000 landfills by 2017[4]. With the increasing number of domestic waste, the early constructed landfills have entered into a state of near saturation from the operation stage. According to the original plan, the landfills that will reach a fully saturated state will be closed. However, there are many problems in the site selection of landfills, such as difficulties in state approval of land use, large area, terrain conditions, and landfill leachate invading underground contaminated groundwater, which make the site selection of landfills difficult. Therefore, it is a feasible scheme to maximize the utilization rate of early landfills, and can effectively increase the garbage storage capacity and solve the problem of urban land use.

There are several ways to expand the capacity of landfills. The first is the excavation and sorting method, which excavates the originally landfill garbage and re-filters and uses it; the second method is compression expansion method, the original garbage compression landfill; these two methods can increase the storage capacity of landfills, but the disadvantage is time-consuming and labor-intensive with high economic cost[5]. The third is the construction of the original landfill dam, usually in the original dam horizontal extension, vertical extension and composite extension, and they are based on the early landfill garbage dam[6-8]. Many scholars have studied and applied it, Zheng et al[9] conducted numerical and experimental studies on the upgrading and expansion project of a large landfill in the south, and found that the upgrading and expansion could increase the storage capacity of the landfill. Under the action of geotextile, when the geotextile strength was 50%, the safety factor of the dam body could meet the safety specification. Zhou et al[7]conducted a comparative analysis of non-reinforced soil and reinforced soil through centrifugal test and numerical simulation, and found that compared with non-reinforced soil, reinforced soil can reduce horizontal displacement, and geotextile can significantly improve the stability of dam body. Chen et al[10]studied the expansion of Xingfeng Municipal Solid Waste Sanitary Landfill Site in Guangzhou, and found that when the original dam with horizontal displacement was expanded, special attention should be paid to the reinforcement method of the original dam, the material selection of the new dam and the connection between the new and old dam. After the construction of the existing landfill, it is also necessary to analyze the stability of the garbage dam and the garbage pile. The stability is verified by laboratory test and numerical simulation analysis, that is, the stability of the garbage dam under normal conditions is analyzed by the indoor centrifuge model test, and the simulation analysis under various sudden conditions is carried out by numerical simulation.

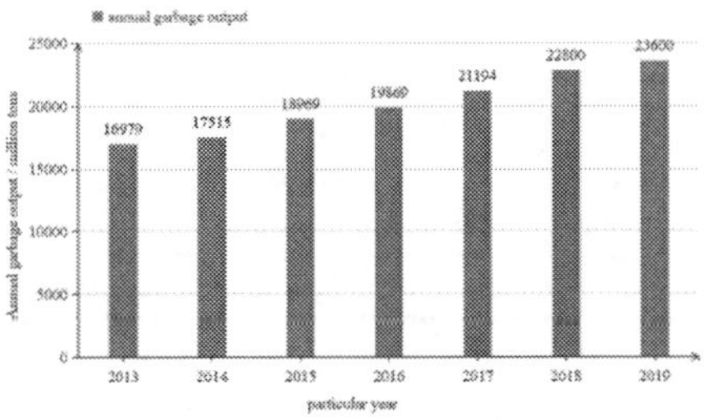

Fig. 1 Domestic waste production in large and medium-sized cities in China from 2013 to 2019

2. Study on Characteristics of Garbage Dam and Connection between New and Old Dams

2.1 Types and Type Selection of Garbage Dam

The main function of the garbage dam is to maintain the stability of the garbage pile, prevent the sliding of the garbage pile caused by rainwater scouring and prevent the pollution of groundwater caused by the discharge of landfill leachate. It can also improve the height of the garbage pile and increase the storage capacity[11-12]. Although landfills have been implemented on a large scale throughout the country, there is no specific industry specification for the design of landfill dams, which is considered according to the hydraulic structure specification, civil building specification and other industry specifications.

Earth-rock dams began to appear in the 1930s[13]. With the rapid development of dam engineering, from the continuous engineering practice of concrete gravity dams and arch dams, the structural design and construction of the dam body are gradually improved, and the form of dam retaining is gradually increasing[14]. In order to apply various needs, gravity dams, flexible dams, homogeneous earth-fill dams, steel dams, rubber dams, masonry gravity dams, concrete face rockfill dams, wide-wall earth-rock dams, etc. Among them, the earth-rock dam has strong foundation adaptability, simple dam construction materials and convenient construction, which makes the earth-rock dam have certain advantages in the construction of dam engineering. The common types of garbage dam are gravity dam and flexible dam. Gravity dam is represented by masonry dam and concrete dam, and flexible dam is represented by rolling earth-rock dam.

Through practical application and economic benefits : masonry dam and concrete dam are gravity dam, mainly rely on self-weight to provide anti-sliding force. Compared with earth-rock dams, masonry dams have smaller volume and will not affect the storage capacity of landfills; when the dam foundation condition is good or the stone transportation distance is close, the masonry dam with small volume is the most suitable for the valley landfill site limited by the site. Concrete dams are also available in areas lacking stone; compared with concrete dam, masonry dam consumes less cement and steel, and has good economy; concrete dam has the advantages of high strength and good waterproof, and the disadvantages of poor economic applicability and high cost. Compared with concrete dam, gravel earth dam has the advantages of good economic applicability, local materials and low cost. However, the slope slope of earth-rock dam should reach 1:2~1:3. The slow slope will cause the volume of garbage dam to increase. When the original dam is upgraded and expanded, it will affect the storage capacity of landfill.Compared with concrete dam and gravel dam, reinforced earth dam has strong economic applicability, low comprehensive cost, steep slope about 1:1~1:1.5, which can reduce the volume of the dam and increase the storage capacity. Roller compacted earth-rock dam is compaction by rolling machine to improve the shear strength of filler, so as to ensure the stability of the dam. The unit price of roller compacted earth-rock dam is relatively low, but its volume is relatively large, so the overall cost is high. Compared with the gravity garbage dam, the roller compacted earth-rock dam has the advantages of low process requirements, simple construction and short construction period. The roller compacted earth-rock dam has the disadvantages of relatively complex construction technology and long construction period. In comparison, the advantages and disadvantages of refuse dam determine that gravity dam is commonly used in refuse dam. According to the construction requirements of landfill, comprehensive comparison of geotextile reinforced earth dam is the best way to build old landfill dam.

2.2 Research on construction technology of garbage dam

(1) Transverse extension

Transverse extension of the original landfill usually requires sufficient transverse space[15]. Transverse extension has two ways : the first one extends outward from the original landfill body, as shown in Fig. 2 (a); the second is to build a new landfill next to the original landfill, which is equivalent to the design of a new landfill, as shown in Fig. 2 (b) and Fig. 2 (c). Both methods increase the working area of the garbage pile to achieve the purpose of increasing the storage capacity. According to the terrain conditions, it can also be divided into plain horizontal expansion and valley horizontal expansion. As shown in Fig. 3, the expansion projects of Hangzhou Tianziling Second Landfill and Guangzhou Xingfeng Landfill in China are both valley horizontal expansion landfills.

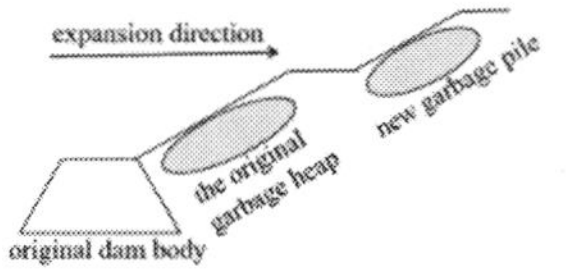

(a)

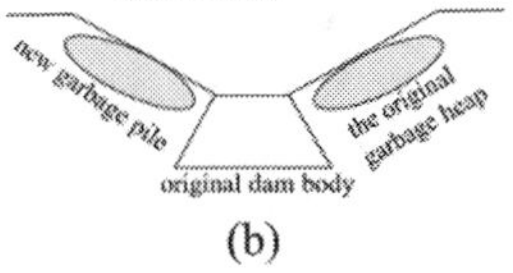

(b)

https://doi.org/10.1051/e3sconf/202235801016

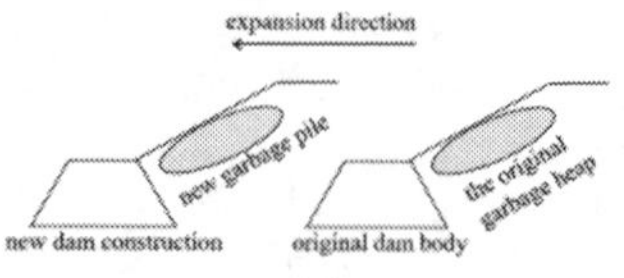

(c)

Fig. 2 Horizontal expansion schematic

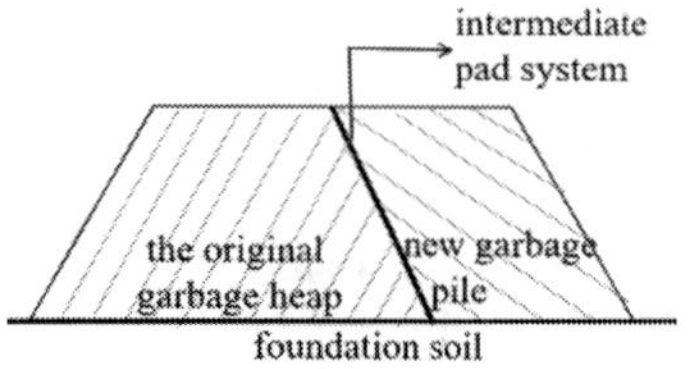

(a) plain type

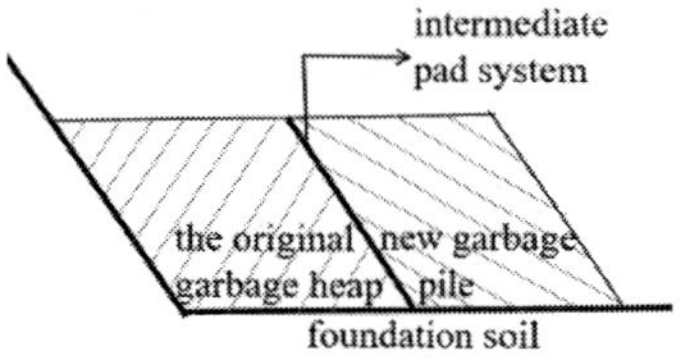

(b) valley type

Fig. 3 Horizontal expansion schematic

Vertical expansion

For the insufficient horizontal space and insufficient storage capacity of the original landfill site, two ways of vertical expansion and vertical expansion can be adopted, the first is to increase the slope of the pile on the external surface of the original landfill site when the stability requirements are met, as shown in Fig. 4 (a); the second is the heightening design on the original garbage retaining dam, as shown in Fig. 4 (b). Both methods are expanded by increasing the height of the garbage pile, but the vertical expansion by changing the slope of the garbage pile will have a certain impact on the stability of the slope. According to the terrain conditions, it can be divided into plain vertical expansion and valley vertical expansion. As shown in Fig. 5, the expansion project of Pudong Dawn landfill belongs to the plain vertical expansion landfill[18]. Vertical expansion on the original refuse dam is the most commonly used method. In the design of landfill expansion, attention should be paid to the settlement caused by the load of new refuse piles.

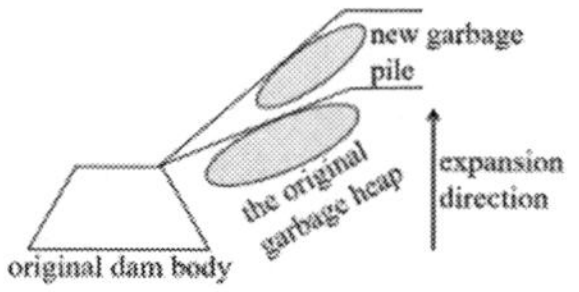

(a)

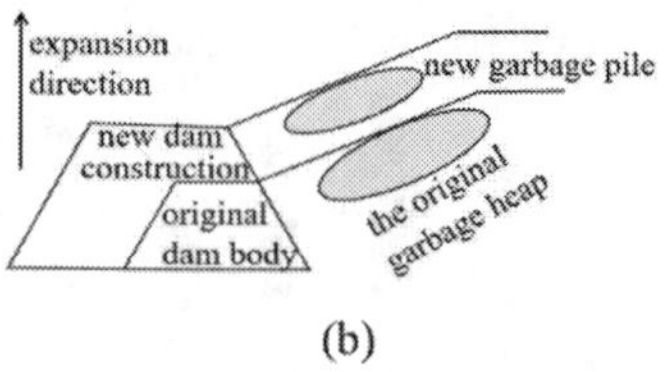

(b)

Fig. 4 Vertical expansion schematic

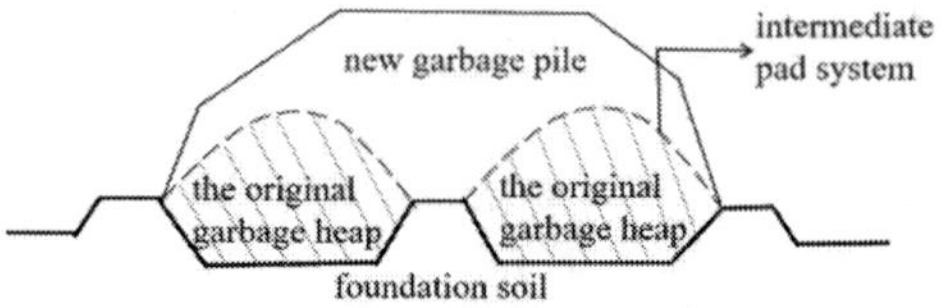

(a) plain type

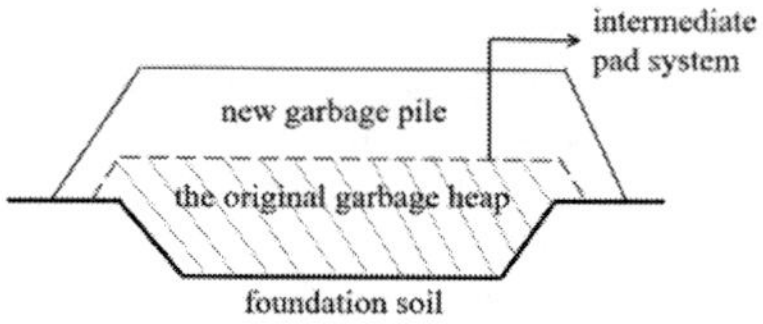

(b) valley type

Fig. 5 Vertical expansion schematic

(3) Composite expansion

When the landfill space is sufficient, the composite expansion technology extends both horizontally and vertically, as shown in Fig. 6. According to the terrain conditions, it can also be divided into plain-type composite expansion and valley-type composite expansion. As shown in Fig. 7, the expansion project of Suzhou Qizishan landfill belongs to valley-type composite expansion landfill mainly with vertical expansion[16].

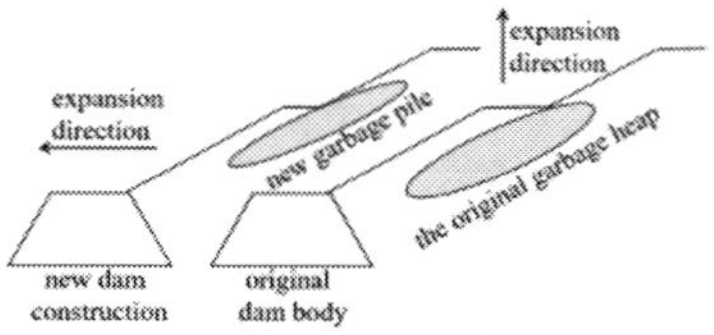

Fig. 6 Composite expansion schematic

(a) plain type

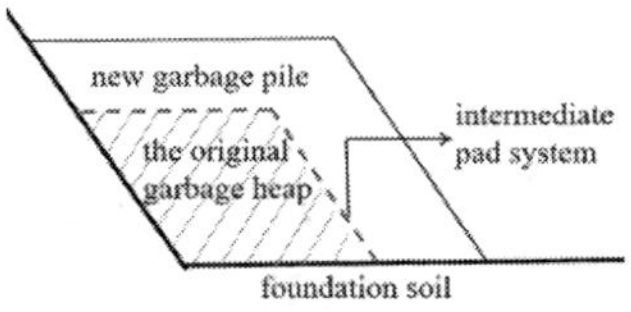

(b) valley type

Fig. 7 Composite expansion schematic

2.3 Connection between original dam body and new dam body of refuse dam

The original dam reinforcement methods are:
(1) The dam is heightened and thickened. For the dam project with better quality, the original dam is heightened and thickened. (2) The anti-seepage layer and protective layer are added on the dam body, and the original dam body cannot be reinforced by large-scale excavation. The anti-seepage wall can be constructed in the dam body by cement mixing pile, high-pressure jet grouting, and vertical membrane laying. The anti-seepage geomembrane can also be laid or the anti-seepage panel can be constructed by concrete. (3) To enhance the anti-sliding performance of the dam, a row of anti-sliding piles can be set at the foot of the dam. (4) Consolidation grouting. If holes appear in the original dam, cement slurry grouting reinforcement should be carried out for the original dam.

If the original dam body is earth-rock dam and the new dam body is concrete dam, due to the rigid structure of the concrete structure, the deformation of the new and old dam body will not be coordinated and the gap will be generated. The excessive weight of the concrete structure will also lead to the collapse of the original dam body due to the lack of bearing capacity. The connection between the original dam body and the new dam body is as follows: (1) When the sliding surface is located at the connection between the original dam body and the new dam body, the safety factor here will be low and do not meet the requirements of the specification. Therefore, it is necessary to insert the anchor into the original dam body and establish a connection with the new dam body through the steel pipe, so as to increase the tension and anti-sliding force of the original dam body and the new dam body. (2) Since the original dam body and the new dam body are not constructed at the same time, the original dam body will be deformed due to the influence of rainfall and weathering and the load of the new dam body, thereby reducing the strength. Therefore, in the extension project of the garbage retaining dam, the construction cushion of the new dam body can improve the bearing capacity of the original dam body and increase the stability of the dam body[17]. (3) The appropriate dam form is selected. Although the concrete dam has the advantages of high strength and good waterproof, it is not suitable for all the new and old dam reinforcement. Geotextile reinforced earth dam is the best form for the construction of the old landfill.

3. Factors Affecting Stability of Landfill Waste Dam

3.1 Leachate

Leachate is a kind of high concentration wastewater produced by rainwater erosion and groundwater immersion in the process of landfill and landfill. It has the characteristics of complex substances, high ammonia nitrogen concentration and organic pollutants. For landfill dams, the water level of landfill leachate directly affects the anti-sliding stability of landfill dams. When the water level of landfill leachate is low, the anti-sliding stability of landfill dams is better. With the gradual increase of leachate water level, the overall stability of the dam body will also be affected, resulting in the deepening of the potential sliding surface position. The anti-seepage measures of landfill leachate are as follows: (1) Do a good job of bottom liner system, adopt single layer, double layer or composite liner system, choose high strength and grand tensile properties of geomembrane or other geosynthetics. (2) The leachate can be collected by setting up a leachate drainage well to prevent the rise of water level. Therefore, it is necessary to monitor the height of leachate water level for the overall anti-sliding stability of landfill dams.

3.2 Waste dump landslide

The rise of leachate will not only affect the overall stability of garbage dam, but also cause slope instability, resulting in garbage dump landslide. Generally, when the leachate level is low, the refuse pile has a certain shear strength, and the load generated by the liquid on the pile slope is also relatively small, which has little harm to the slope stability and dam stability. When the leachate level is high, the heap slope will have a large load due to the effect of leachate. When the leachate overflows, the garbage heap will produce a large area of tensile stress, and there will be a potential instability area. This area will reduce the safety factor of the heap slope, leading to the landslide of the heap and affecting the overall stability of the garbage dam. Therefore, it is necessary to ensure that there will be no sliding and tilting problems under the influence of the external load of the garbage and its own gravity, so as to effectively improve the overall stability of the garbage dam and ensure the safety of the landfill.

3.3 Other factors

Combined with practical engineering, it is found that hydrogeological conditions, appropriate dam form, uneven settlement of garbage pile, slope ratio of outer slope of dam body, compactness of dam body, local collapse of flood intercepting ditch and dam abutment all affect the safety and stability of dam body.

4. Study on the stability of landfill waste dam

4.1 Analysis method of landfill stability

The factors affecting the stability of landfills should be considered from the slope stability of landfills, the stability of landfills, the overall anti-slip and anti-dip stability of landfill dams, and the stability under earthquake [17-19]. Through centrifuge model test and numerical simulation analysis. Centrifuge model test simulates the deformation and failure process of prototype by increasing centrifugal force to simulate gravity. At the same time, the rationality and reliability of theoretical analysis and numerical calculation can be verified, and the effect of field test can be reflected intuitively[20].

(1) Slope stability of landfills: Arc sliding method is often used in slope stability analysis of landfills[21], Wedge limit equilibrium method[22-23] and finite element analysis method[24]; the strength of the slope model is calculated and analyzed according to the Mohr-Coulomb strength formula and the total stress formula; stability Analysis of Slope Engineering by Finite Element Method[25]; study on failure mechanism of reinforced soil slope by centrifuge model test[26-27].

(2) The stability of garbage pile: Geostu-dio-Slope finite element calculation software or Lizheng geotechnical software is used to analyze the overall stability of the dam and garbage pile. The safety factor of non-circular sliding surface is calculated by Morganstein-Price method. The Mohr-Coulomb strength formula is used as the calculation criterion, and the total stress method is used to calculate[28].

(3) The overall anti-sliding and anti-overturning stability of refuse retaining dam: the anti-sliding and anti-overturning stability of refuse retaining dam are calculated by Lizheng software and Geo studio-Slope finite element software[29].

(4) Stability under seismic action: pseudo-static method was used to simplify the seismic load, and a five-wedge analysis model was established to solve the safety factor under seismic action, and the stability and failure mechanism of landfill under seismic load were analyzed[22].

(5) Specification for earth-rock dams in China[30] usually adopts rigid body limit equilibrium analysis method, and simplified Bishop method is recommended for homogeneous earth dams; for the stability calculation of complex conditions, numerical algorithms such as slice method and finite element method are needed. For the stability analysis of reinforced earth dams, the arc sliding surface limit equilibrium method is usually adopted[7].

4.2 Study on the stability of refuse dam

The constructed landfill site should not only meet the requirements of no landslide, no sliding and tilting of the dam body, normal settlement and good slope stability[31-32], but also meet the stability requirements of numerical simulation under various working conditions, such as considering the influence of seismic load, considering hydrostatic pressure in saturated state or considering hydrodynamic pressure. The minimum safety factor obtained by simulation calculation should meet the requirements of the specification. With the increase of landfill, there will be landfill leachate in the landfill, which needs to be guided and discharged in time to ensure that the leachate level does not cause the head difference before and after the dam body, resulting in seepage failure of the dam body. Low leachate level can also ensure the anti-sliding stability of the garbage dam.

The stability of the garbage dam is very important, which affects the operation and use of the garbage landfill. In order to avoid the instability of the garbage dam, we can observe the dam deformation, settlement and safety monitoring of the garbage dam and garbage pile, and the slope monitoring method can be used for monitoring[33]. Slope deformation monitoring methods mainly include slope surface measurement (range finder, total station, leveling, theodolite, etc.), multi-point displacement meter, GPS measurement, SAR interferometry, microseismic monitoring, distributed optical fiber displacement measurement (OTDR and BOTDR) and time domain reflection test (TDR). Centrifuge test, theoretical analysis and numerical analysis can be used to analyze the slope and dam stability, and long-term monitoring of landfill dam deformation, slope displacement, leachate water level and settlement can be carried out. Artificial intelligence algorithms such as neural network, grey algorithm and support vector machine can be used to predict.

5. Conclusions and outlook

Due to the rapid growth of urban garbage, the storage capacity of landfill can no longer meet people' s needs. It is necessary to carry out the construction on the basis of the existing landfill. This paper refers to the current situation of the storage capacity of the landfill, how to select the form of the garbage dam, introduces three ways of extension on the original dam, and summarizes the analysis method of the stability of the garbage dam.

(1) The garbage dam is designed with reference to water conservancy projects. The common garbage dam is gravity dam and flexible dam, which are represented by masonry dam or concrete dam and rolling earth-rock dam respectively. Geotextile reinforced earth dam is often used in dam extension projects. Concrete, gravel soil, reinforced soil, stone chips or gravel are generally used as the filling materials of waste dam. Geomembranes, geotextiles, HDPE geogrids and high toughness polyester geotextiles with high strength and good tensile properties are used as the reinforcement materials.

(2) The construction methods of garbage retaining dam include horizontal expansion, vertical expansion and composite expansion. According to the terrain conditions, there are plain expansion and valley expansion. The connection between the original dam and the new dam should be well protected.

(3) The factors affecting the stability of waste dam include leachate level, displacement and settlement of waste pile, hydrogeological conditions, dam form and slope ratio,

dam compactness, local collapse of flood intercepting ditch and dam abutment.

(4) The stability of refuse retaining dam is analyzed by centrifuge test and numerical simulation. The displacement of refuse pile, slope stability, sliding and tilting of dam body, leachate level and settlement displacement should be analyzed.

Through the above research combined with Hangzhou Tianziling Second Landfill, Guangzhou Xingfeng Landfill, Pudong Liming Landfill Extension Project, Suzhou Qizishan Landfill Extension Project and other practical projects, it is proved that the dam retaining construction project can increase the storage capacity of the landfill and save the cost without relocating the site. Through the study found and put forward the following points: (1) Leachate has a great influence on the stability of the dam, usually through the laying of leachate drainage wells to collect and discharge the filtrate, when the drainage well clogging occurs, what way can be quickly solved without affecting the garbage heap; (2) The refuse heap has a certain shear strength, and what kind of material can be added to enhance the strength of the heap and reduce the harm of landslide; (3) Whether there is a more convenient and higher connection strength method between the original dam and the new dam. In short, it is hoped that this study can provide certain theoretical basis and guidance for the future construction of waste dam.

References

1. GB 50869-2013, Technical code for municipal solid waste sanitary landfill[S].
2. Zhang Jinghong, Chen Dongyang. Flexible governance, a new trend of NIMBY governance in China: A case study of the Xiantao unrest[J]. Journal of Jiangsu Ocean University (Humanities& Social Edition), 2021, 19(6): 41–53.
3. Lin Jingjing , Mao Yajing , Lin Zongping . A tripartite evolutionary game study on the promotion mechanism of the "Internet+Recycling Resources" model[J]. Ecological Economy, 2022, 38(3): 202-209.
4. Wang Qinglong , Yang Xing . Seepage analysis and anti-seepage repair measures for refuse landfill[J]. Design of Hydroelectric Power Station, 2019, 35(4): 45-49.
5. Ju Alian , Zhao Lijie . Lifecycle management of waste landfill sites in Japan[J]. Recyclable Resources and Circular Economy, 2022, 15(2): 41-46.
6. Chen Liang . Discussion on expansion patterns of old municipal solid waste landfill site[J]. Environmental Engineering, 2013, 31(5): 117-120.
7. Zhou Xiaowen , Jiang Haoran , Zhou Mi , et al. Centrifuge test and numerical simulation on stability of vertically extended earth dam in landfill project[J]. Journal of Harbin Institute of Technology, 2020, 52(11): 157-166.
8. Zhu Bin , Chen Yunmin , Ke Han . Seismic stability analysis of extended municipal solid waste landfills[J]. Rock and Soil Mechanics, 2008, 150(6): 1483-1488.
9. Zheng Chao , Zhou Weidong , Zhou Xiaowen , Xue Xieluan . Numerical simulation and experimental study of the behavior of steep reinforced dam for a municipal waste landfill[J/OL]. Journal of Engineering Geology: 1-10[2022-06-13].DOI : 10.13544 / j.cnki.jeg.2020-303.
10. Chen Weihong , Xiong Bo , Zhu Jiangying . Design of rubbish dam heightening in existing confined landfill site[J]. Special structures, 2018, 35(06): 53-59.
11. Zize Peng . Discussion on structural design of stone waste dam[J]. Urban Roads Bridges & Flood Control, 2012, 162(10): 16, 174-176.
12. Ding Yun . Design and analysis of waste dam[J]. Dam & Safety, 2013, 78(4): 59-62.
13. Wang Xiaosong , Li Baozhu . Developments in the practices of high earth dam and rockfill dam[J]. Journal of Engineering of Heilongjiang University, 1995(3): 16-19.
14. Zhu Hancheng , Zhang Jie . Discussion on engineering geological survey of small dams of small and medium rivers[J]. Yangtze River, 2021, 52(S1): 101-103.
15. Zhang Ganfei , Wang Yanming , Xu Yongfu , et al. Key geotechnical problems in the reconstruction and expansion of existing MSW landfills[J]. China Civil Engineering Journal, 2007(4): 73-81.
16. Ren Hongjuan , Shan Hualun , Zhang Jianping . Application of vertical anti-seepage technology in extension project of Qizi Hill waste landfill site in Suzhou city[J]. Environmental Sanitation Engineering, 2009, 17(4): 34–36, 39.
17. Yang Yue , He Yin Shi , Chen Feng , et al. Stability of old landfill dams under the action of leachate and load[J]. Journal of Liaoning Technical University (Natural Science), 2022, 41(1): 7-15.
18. He Jianqing , Shi Shuangyan , Gao Wenhua . Quasi-static analysis for the earthquake stability assessment of the landfills with a retaining dam along the underlying liner system[J]. Journal of Safety and Environment, 2020, 20(5): 1736-1742.
19. Liu Jianguo , Nie Yongfeng , Bai Qingzhong . Analysis of structural stability in sanitary landfill[J]. Chongqing Environmental Science, 2001(1): 62-66.
20. Wu Aiqing , Yu Wuhua , Li Bo . Comprehensive analysis method with centrifuge tests and numerical simulation on high and steep reinforced slopes[J]. Port & Waterway Engineering, 2013, 482(8): 33-38.
21. Zhu Xiangrong , Wang Zhaohui , Fang Pengfei . Study on feasibility of enlarging capacity in Tianziling waste landfill[J]. Chinese Jounal of Geotechnical Engineering, 2002(3): 281-285.
22. Feng Shijin , Chen Yunmin , Gao Guangyun . Analysis on translational failure of landfill along the underlying liner system[J]. Chinese Journal of Geotechnical Engineering, 2007(1): 20-25.

23. Tu Fan , Cui Guangqiang , Lin Congmou , et al. Sensibility analysis of factors aeffecting stability of waste landfill based on neural network[J]. Rock and Soil Mechanics, 2010, 31(4): 1168-1172.
24. Filz George-M , Esterhuizen Jacob-J-B , Duncan J.-Michael . Progressive Failure of Lined Waste Impoundments[J]. Journal of Geotechnical and Geoenvironmental Engineering, 2001, 127(10): 841-848.
25. Griffiths D V , Lane P A . Slope stability analysis by finite elements[J]. Géotechnique, 1999, 49(3): 387-403.
26. Yan Zhigang , Wang Dejun , Zhang Kexu , et al. Deformation analysis for geotextile reinforced earth structure[J]. Journal of Harbin Institute of Technology, 2003(5): 576-580.
27. Wang Liping , Zhang Ga , Zhang Jian-Min. Centrifuge model tests of geotextile-reinforced soil embankments during an earthquake[J]. Geotextiles & Geomembranes, 2011, 29(3): 222-232.
28. Chen Weihong , Zhu Jiangying . Application of high-tenacity geotextile in in-situ elevation of high and seep garbage dams[J].Special structures, 2020, 37(3): 22-28.
29. Wang Rui . Stability analysis and treatment measures of a garbage dam[J]. Huabei Natural Resources, 2020, 98(5): 104-105,108.
30. DL/T 5395-2007, Design specification for rolled earth-rock fill dams[S].
31. Wang Yanming , Zhang Ganfei . Settlement Calculation and Analysis of aging waste in the vertical expansion of existing Landfill[J]. Environmental Engineering, 2007, 148(4): 4, 59-62.
32. He Haijie, Lan Jiwu, Chen Yunmin , et al. Monitoring and analysis of slope slip process at a landfill in Northwest China[J]. Chinese Journal of Geotechnical Engineering, 2015, 37(9): 1721-1726.
33. Yang Jie , Ma Chunhui , Cheng Lin , et al. Research advances in the deformation of high-steep slopes and its influence on dam safety[J]. Rock and Soil Mechanics, 2019, 40(6): 2341-2353, 2368.

E3S Web of Conferences **358**, 01017 (2022) https://doi.org/10.1051/e3sconf/202235801017
GESD 2022

Study on The Test Methods of Heavy Metal in Exhaust Gas for a Copper Melting Factory

Jing Wang, Jilu Wang and Zhenyu Liu *

Tianjin Eco-Environmental Monitoring Center, Tianjin, China

Abstract. The raw materials, auxiliary materials and production process of a typical copper smelting factory were investigated, the emission characteristics of exhaust gas pollutants produced by the production process were studied. To test the samples by graphite furnace, firstly in order to unify the digestion system in the experiment, the applicability of three different method was verified by experiments. According to the results, the recovery of nitric acid-perchloric acid digestion system is a relatively good, so it can be recommended. The testing results show linear range, detection limit and detectability of the method could meet the requirements of environmental monitoring.

1. Introduction

At present, the issue of environmental pollution in China has become a public focus, while heavy metal pollution among them has attracted more attention due to the characteristics of being difficult to degrade and hazardous to human health. Heavy metals refer to metal elements whose density is greater than 4.5g/cm3. In terms of environmental pollution, the heavy metals mainly refer to the biological toxic elements such as copper, lead, chromium and metal-like selenium and arsenic, which can have strong interaction with protein and enzyme in human body, can cause chronic poisoning. Copper in exhaust gas mainly comes from smelting, metal processing, iron and steel manufacturing, among which smelting is the main source of copper pollution in the atmosphere [1].

At present, with the continuous strengthening of air pollution control in China, there are a number of pollutant emission standards in China that stipulate the emission limits of heavy metals. Relevant standards related to the emission of copper pollutants, such as:" emission standards of pollutants for inorganic chemical industry" (GB31573-2015) [2]. "Standard for pollution control on the municipal solid waste incineration" (GB18485-2014) [3]. "Standard for pollution control on hazardous waste incineration" (GB 18484-2020) [4]. However, compared with the emission control elements in developed countries, the required monitoring methods lack uniformity and systematization. In addition to the uniform sample collection method in the monitoring process, the pre-treatment, analysis and testing, quality control and other aspects need to be further improved and standardized. The monitoring ability of environmental monitoring departments varies greatly among regions, with western regions obviously lagging behind eastern regions. Even among monitoring stations at the same level, there are also great differences in personnel and technical level, instruments and equipment [5]. Therefore, in the future, it is necessary to focus on the study and introduction of operational and high accuracy monitoring methods for heavy metals in exhaust gas to meet the increasing requirements of environmental management.

2. Typical production process

The main processes of this project are melting, casting, rolling and wire drawing, annual operation 8760h, annual production of 6000 tons of high frequency resistance welding wire. The production of the enterprise is equipped with a leading furnace, which puts the electrolytic copper raw materials into the furnace body in batches and electrically heats the furnace body to 1100~1200°C for continuous heating and melting. During the melting process of electrolytic copper, a certain amount of copper-containing exhaust gas will be produced. At the same time, a small amount of flake graphite should be added as a covering agent to prevent copper oxidation. After melting into copper water, the mold is used to pull out 17 mm copper rod, the 17mm copper rod is rolled into 8mm copper rod through the mill, the 8mm copper rod is made into 3 mm copper wire through mechanical wire drawing, and then the 3 mm copper wire is made into 1.3 mm, 1.8mm and 1.5mm copper wire respectively. Raw and auxiliary materials used in the production process are shown in Table 1.

* Corresponding author: 305472630@qq.com

https://doi.org/10.1051/e3sconf/202235801017

Table 1. List of raw and auxiliary materials used in the factory

No. number	Name	Specification /Model	Annual Consumption	Maximum Storage
1	Electrolytic Copper	/	6100t	300t
2	Emulsion	200kg/barrel	2t	0.5t
3	Engine oil	200kg/barrel . 32#	0.5t	0.2t
4	Crystal line Flake Graphite	scaly	8t	1t
5	Water	/	1330 m³	/
6	Electricity	/	300million KW.h	/

Note: for electrolytic copper, the thick plate made of crude copper (including 99% copper) is used as anode, the pure copper is made into thin sheet as cathode, and the mixed solution of sulfuric acid and is used as electrolyte. After power on, copper dissolves from the anode into copper ions and moves to the cathode.

3. Pollutant emission analysis

The pollutant of the exhaust gas in the factory is copper. that produced in the process of copper melting. According to the"The First National Survey of Pollution Sources Manual on The Coefficient of Production and Discharge of Industrial Pollution Sources" (Part II). Using electrolytic copper raw material to produce copper wire products, the process is "casting + tying process" soot emission coefficient is 1.62 kg/t – product. The annual output of high frequency resistance welding wire of the enterprise is 6000t. According to the calculation, the soot production amount of melting process is 9.72t/a, and the annual working hours of melting process is 8760h, so the particle production rate is 1.11kg/h. An air collecting hood is set above the upper leading furnace, and the dust generated is collected by an induced draft fan into the pulse and bag dust collector for treatment, and the exhaust gas after treatment is discharged by a 15m high exhaust cylinder. The air volume of the air collecting hood fan is 20000m3/h, the air volume of the air collecting hood is close to the exhasut gas generation part, and the air volume can meet the collection efficiency requirements, so the exhaust gas collection efficiency of this project can reach 90%. The calculation process is only for particulate matter, and the emission of copper containing exhaust gas cannot be known. Therefore, on-site sampling and graphite furnace method should be used to monitor the concentration of copper emission in exhaust gas.

4. Analysis and test

4.1 Method and Principle

Acquisition stationary sources with the quartz fiber filter cylinder exhaust particulate matter, after nitric acid - perchloric acid digestion preparation into the solution, the solution of copper in graphite furnace atomization device, via high temperature atomization, the ground state copper atoms of hollow cathode lamp emission characteristic spectral line (324.7 nm) absorption, within a certain range, the absorbance value has a linear relation with the mass concentration.

4.2 Main Equipment and Reagent Materials

4.2.1 Main Equipment.

The main equipment includes: graphite furnace atomic absorption spectrophotometer, graphite tube, copper hollow cathode lamp, smoke sampler, quartz fiber filter cartridge, temperature regulating electric heating plate, suction filtration device, etc.

4.2.2 The reagent.

(1) Nitric acid: ρ (HNO3) = 1.42 g/ml, superior pure.
(2) Perchloric acid: ρ (HClO4) = 1.68g/ml, superior pure.
(3) Hydrochloric acid: ρ (HCL) = 1.19 g/ml, superior pure.
(4) Nitric acid solutions: 1+99(v/v), preparing by nitric acid(ρ= 1.42 g/ml) .
(5) Mixed solution of nitric acid and hydrochloric acid: add 55.5 ml nitric acid(ρ= 1.42 g/ml) and 167.5 ml hydrochloric acid(ρ= 1.19 g/ml) to 500 ml pure water, and then diluted to 1 L with pure water.
(6) Copper standard stock solution: ρ (Cu)=1000mg/L.
(7) Copper standard intermediate solution: ρ (Cu)=10mg/L.
(8) Copper standard solution: ρ (Cu)=100μg/L.

4.3 Sample Collection

The distribution and sampling of heavy metals in exhaust gas shall comply with the relevant provisions of GB/T 16157 and HJ/T 397. The filter cartridge shall collect at least 0.6m3 of exhaust gas sample (in standard dry state). If the concentration of heavy metal copper is low, the sampling volume shall be increased appropriately. When the temperature of flue gas in the flue is higher than the melting point of copper, cooling measures should be taken. The temperature of flue gas before entering the filter cylinder is lower than the melting point of metal elements. During sampling, each batch of samples should carry at least two filters of the same batch, which are not connected with the sampler after being brought to the sampling site. After sampling, they should be brought back to the laboratory for testing. After sampling, take out the filter cartridge carefully, fold the seal inward, put it back into the filter cartridge box vertically, and take it back to the laboratory for analysis.

4.4 Sample Digestion

In order to unify the digestion system, the applicability of different digestion systems of nitric acid-perchloric acid, nitric acid- hydrofluoric acid-perchloric acid and nitric acid- hydrochloric acid was verified by experiments. Electrothermal plate digestion was used to verify the digestion system for the determination of copper digestion effect.

4.4.1 Nitric acid-perchloric acid.

Cut the filter cartridge into pieces with ceramic scissors, place in a 150 mL conical flask and moisten with a little water. add 30ml nitric acid and 5ml perchloric acid (it is appropriate to dip the sample in acid, not enough nitric acid), Insert a short-necked glass funnel into the mouth of the bottle, heat it on the electric heating plate until boiling, steam it until nearly dry, remove it and cool it. Add another 10mL nitric acid and continue to heat until nearly dry and slightly cool. Add a small amount of water for extraction and filtration, transfer to a 50mL volumetric flask, and then dilute with water to scale.

4.4.2 Nitric acid- hydrofluoric acid-perchloric acid.

Cut the filter cartridge into pieces with ceramic scissors, put it into the digestion tank, add 20.0 mL nitric acid, shake the digestion tank, immerse the filter cartridge fragments, let stand for 2 min, heat to 140 °C and hold for 60 min, cool for 5 min, add 5.0 mL hydrofluoric acid, shake for 2 min, heat again to 140 °C, hold for 60 min, cool for 5 min and add 2.0 mL perchloric acid, shake for 2 min, heat to 160 °C to maintain 150 min to catch the acid to nearly dry. Cooled to a fixed capacity of 50 mL.

4.4.3 Nitric acid- hydrochloric acid.

Cut the filter cartridge into pieces with ceramic scissors, put it into the digestion tank, add 25.0 mL of nitric acid-hydrochloric acid mixed solution (5.55% $HNO3$/16.75% HCl), immerse the filter cartridge fragments in it, cover the digestion lid, heat the reflux at 100 °C for 120 min, remove the digestion cap after cooling, add 10 mL of water, and let stand for 30 min for maceration. The extract was filtered into a 50 mL volumetric flask, volumeted with water, to be measured.

4.4.4 The data of contrast experiment for three methods.

The comparative experiments of the above three digestion methods were carried out on two blank filters and three blank standard addition filters. The data are shown in table 2:

Table 2. Test value in the different digestion system

Nitric Acid-Hydrochloric Acid			Nitric Acid-Perchloric Acid			Nitric Acid-Hydrofluoric Acid-Perchloric Acid		
Labeling (μg/L)	Measured Value (μg/L)	The Recovery (%)	Labeling (μg/L)	Measured Value (μg/L)	The Recovery (%)	Labeling (μg/L)	Measured Value (μg/L)	The Recovery (%)
0 (Blank)	1.23	/	0 (Blank)	1.28	/	0 (Blank)	12.4	/
	1.40	/		1.36	/		12.2	/
Average	1.32	/	/	1.32	/	/	12.3	/
0.5 (Low)	1.61	58.0	0.5 (Low)	1.62	60.0	0.5 (Low)	13.4	222
	1.73	82.0		1.87	110		13.3	194
	1.71	78.0		1.82	100		13.3	206
Average	/	72.7	/	/	90.0	/	/	207
10 (Middle)	11.1	97.8	10 (Middle)	11.3	99.4	10 (Middle)	24.8	125
	10.2	88.8		11.9	106		25.4	131
	10.4	90.8		11.8	105		24.9	126
Average		92.5	/	/	103	/	/	127
20 (High)	17.08	78.8	20 (High)	23.2	110	20 (High)	32.9	103
	16.54	76.1		23.4	110		33.4	106
	16.48	75.8		23.5	111		33.2	105
Average	/	76.9	/	/	110	/	/	104

According to the results, the recovery of nitric acid-perchloric acid digestion system is a relatively good. Therefore, nitric acid-perchloric acid system is recommended.

4.5 Analysis and Measurement

4.5.1 Measuring instrument condition

(1) Drying stage.The purpose of drying is to remove the water and acid in the solution. If the temperature is too low or the holding time is not enough, the water and acid in the graphite tube will damage the structure of the graphite tube in the high temperature ashing and atomization stage. If the temperature is too high or the temperature rises too fast, the solution bursts into boiling and a double absorption peak may occur during atomization. The effect of low temperature and long time drying is better.

(2) Ashing stage.The purpose of ashing is to remove low boiling compounds before atomization, reduce matrix interference, reduce background absorption, and increase the ashing temperature as much as possible on the premise of ensuring the non-volatile loss of the elements to be

measured. If the matrix is complex and the background absorption is high, the ashing time can be appropriately prolonged.

(3) Atomization stage.The selection of atomization time and temperature is to maximize the atomization degree of the element to be measured, which depends on the properties of the element to be measured. Too low temperature, insufficient atomization time and incomplete atomization will reduce the signal, widen the peak shape and produce memory effect; If it is too high, the atomization is too fast, the response of the detection system can not keep up with the atomization signal, the reproducibility and sensitivity are reduced, and the service life of the graphite tube is shortened.

(4) Disability removal stage.The residue removal stage is to remove the matrix and nonatomically elements to be tested and remove the memory effect. The atomization temperature of chromium and nickel is high. On the premise that the instrument allows and ensures the service life of graphite tube, try to increase the residue removal temperature to ensure that all residual substances volatilize.

According to the curve of copper absorbance value, the absorbance value is the largest when the ashing temperature is 1100 °C and the ashing time is 20 s; The absorbance is stable when the atomization temperature is 2000 - 2100 °C, 2000 °C is recommended, and the atomization time is recommended to be 5s.

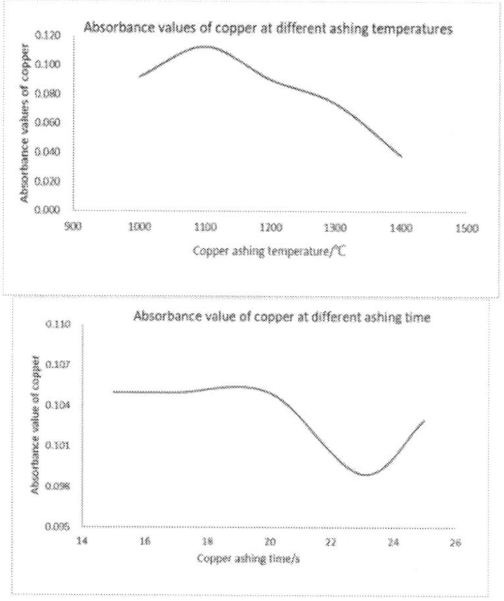

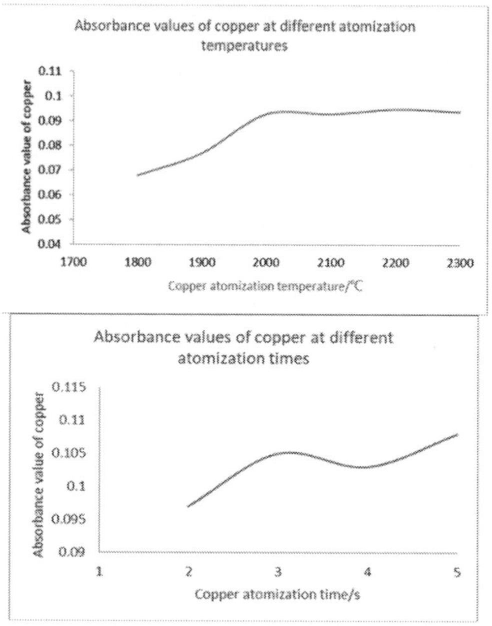

Figure 1. Temperature and time curve of the ashing and atomization

Table 3. Optimum instrument determination conditions

Test Conditions	Value
Measuring Wavelength(nm)	324.7
Lamp Current(mA)	4.0
Passband Width(nm)	0.7
Drying temperature(°C)/time(s)	110/30
pyrolysis temperature(°C)/time(s)	1100/20
Atomization temperature(°C)/time(s)	2000/5
Residue removal temperature(°C)/time(s)	2450/3
Whether the gas is stopped in atomization	yes
Argon flow rate(ml/min)	250
Injection volume(μL)	20
Integral mode	Peak
Instrument model	PE900T

4.5.2 Drawing the calibration curve.

Absorb 100 μg/L copper solution 0,2.50, 5.00, 7.50, 10.00, 15.00, 20.00 ml in a 50 ml volumetric flask, shake well with 1% nitric acid solution at constant volume. This standard series contains 0,5.0, 10.0, 15.0, 20.0, 25.0, 30.0, 40.0 μg/L copper. Then the absorbance of the standard solution series was measured in sequence from low to high according to the instrument reference measurement conditions.

Table 4. The of data calibration curve

Standard Solution Concentration (μg/L)	0	5.0	10.0	15.0	20.0	25.0	30.0	40.0
Response Signal value	-0.0001	0.0248	0.0467	0.0764	0.1035	0.1319	0.1572	0.215
Calibration Curve Regression equation	y =0.00536 x − 0.00243							
Linear correlation coefficient	R^2=0.999476							

4.6 Detection limit and detectability of the method

The method detection limit of each element in this standard revision is determined in accordance with the requirements of technical guidelines for the preparation and revision of environmental monitoring and analysis methods (HJ 168-2010). Since the used filter cartridge contains various target elements, according to the requirements of (HJ 168-2010), the blank filter cartridge can be detected. Directly use the blank filter cartridge to determine the detection limit, determine it according to the pretreatment method of fixed source exhaust gas filter cartridge, measure it in parallel for 7 times, and calculate the detection limit (MDL) according to formula (1).

$$\mathrm{MDL} = t_{(n-1.0.99)} \times S \qquad (1)$$

n-Determination times of parallel samples;
$t_{(n-1.0.99)}$ --T-distribution with N-1 degree of freedom and 99% confidence (one side)
S—Standard deviation of n parallel measurements
Technical Guidelines for The Formulation and Revision of Environmental Monitoring and Analysis Method Standards (HJ 168) stipulates that the detection limit of 4 times is the detectability. When the sampling volume is 0.6m3 and the volume is constant to 50 mL, the method detection limit and determination limit of copper are shown in the table below.

Table 5. Detection limit and Detectability by nitric acid-perchloric acid electrothermal plate digestion method

Serial Number	Value(μg/L)
1	1.48
2	1.57
3	1.43
4	1.53
5	1.49
6	1.61
7	1.63
The Average (μg/L)	1.53
The Standard Deviation	0.073
Detection Limit(μg/L)	0.3 μg/L/0.025 μg/m^3
Detectability(μg/L)	1.2 μg/L/0.1 μg/m^3

4.7 Test Results

By sampling and testing the exhaust gas containing copper collected at the melting process, the monitoring point is at the outlet of the purification facility. A total of 7 samples and 2 blank filter cartridges were collected on site. The emission concentration of copper in the exhaust gas of 7 samples was respectively 0.047. 0.042. 0.050. 0.041. 0.039. 0.045. 0.037 mg/m^3,RSD value is 10.6%, the test method have good stability.

5. Conclusions

(1) In order to unify the digestion system, the applicability of three different method was verified by experiments. According to the results, the recovery of nitric acid-perchloric acid digestion system is a relatively good. Therefore, nitric acid-perchloric acid system is recommended.
(2) To test the samples by graphite furnace, the results show the linear range, detection limit and detectability of the method could meet the requirements of environmental monitoring. The testing results of samples that collected in the factory show that the method has a good stability.
(3) The exhaust gas containing copper produced by the melting process is monitored, and the emission concentration of copper meets the requirements of relevant standard limits.

Acknowledgments

This work was financially supported by environmental protection standards project-"Stationary source emission - Determination of copper, nickel, cadmium- Graphite furnace atomic absorption spectrometry".

References

1. Zhang Peili, Hong Bo, Cui Huizhen. Study on pollutant emission characteristics of waste gas from reclaimed copper smelting [J]. Green technology, 2019 (10): 145-149.
2. Discharge standard of pollutants for inorganic chemical industry (GB 31573-2015) [S].
3. Department of science and technology standards, Ministry of environmental protection. Pollution control standard for domestic waste incineration: gb18485-2014 [S]. Beijing: China Environmental Science Press, 2014.
4. Department of science and technology standards, Ministry of environmental protection. Pollution control standard for hazardous waste incineration: gb14848-2020 [S]. Beijing: China Environmental Science Press, 2014.
5. Liu Jie. Chen Minmin . Construction of Heavy Metal Monitoring Technical System during Twelfth Five-Year Plan in China[J] . Environmental Science and Management. 2014. 39(6): 125-128.

Study on intelligent analysis algorithm for achieving standard of polymer flooding well group

Hanqiang Wang

No.1 Oil Production Plant of Daqing Oilfield Co., LTD, Daqing 163111, Heilongjiang, China

Abstract: In order to master the development effect of polymer flooding well group, it is necessary to accurately analyze the influence of different factors in the whole polymer flooding process on the development index. Combined with the principle of big data analysis, based on the neighborhood rough set theory and Kmeans clustering algorithm, an intelligent analysis algorithm is proposed to determine the achievement of development indicators of polymer flooding well group. Firstly, the neighborhood rough set was used to reduce the attributes of the influencing factors of the Wells with and without the standards. Secondly, Kmeans algorithm was used to cluster the reduced influencing factors to delete the data inconsistent with the actual compliance. Finally, the clustering model is used to judge the standard status of other well groups, and the practical application effect is very good.

Key words: Polyflooding well group; Rough set; Kmeans; Intelligent analysis.

1. Introduction

The geological static factors, production dynamic factors and actual development factors of oilfield development blocks play an important role in the process of oilfield polymer development [1,2]. In order to carry on the long-term reasonable exploitation of the oilfield, it is very necessary to study and analyze the law and development effect of various influencing factors in the development process, and keep abreast of the development effect standard of the polymer flooding well group. Therefore, the prediction method of polymer flooding development index has been paid more and more attention by the oilfield enterprises. Shi Chengfang et al. [3] established a prediction model for fluid production of produced Wells in polymer blocks and production dynamic changes in the initial stage and process of polymer injection by analyzing the corresponding relationship between fluid production of produced Wells and water absorption index of produced reservoirs. Zhao Guozhong et al. [4] conducted index prediction based on the three-layer CBP neural network model by studying the change of water content and its influencing factors in the polymer flooding stage. Qiu Haiyan et al. [5] combined the advantages and disadvantages of HCZ, Weibull and Weng's forecasting models and optimized them, and proposed a weighted combination forecasting model to guide the actual production. Hou Jian et al. [6-7] used numerical simulation technology to analyze the factors affecting the change of polyflooding oil augmenting effect, studied the relationship between characteristic parameters and influencing factors through regression statistical model, and then obtained the prediction model of characteristic parameters to predict the change trend of polyflooding oil augmenting. In this paper, according to the actual dynamic and static data of the development process of polymer flooding in oil field, combined with the principle of big data analysis, based on the neighborhood rough set theory and Kmeans clustering algorithm, an intelligent analysis algorithm is proposed to determine the development index compliance of polymer flooding well group, which has achieved good practical results.

2. Neighborhood rough set theory

2.1 Basic rough set

Rough set [8-10], proposed by Professor Pawlak in 1982, can effectively analyze and process all kinds of incomplete data information, such as data with some characteristics of imprecision, inconsistency and incompleteness. Through rough set theory, the hidden knowledge in data information can be mined, and the hidden law inside data information can be mined. The main principle of rough set theory is to use the method of knowledge reduction to get the classification rules of the problem to be solved without changing the reasoning ability of knowledge classification. The basic idea of rough set theory is to classify things by equivalence relation to recognize knowledge.

1.1.1 Construct knowledge systems. The target of rough set theory is information knowledge system, which is represented by quadruple: $S=<U,A,V,f>$.Among them, U represents a finite set of objects, $U=\{x_1,x_2,...x_n\}$; A represents A finite set of attributes, V represents the set of property values, f Is a function of properties and objects, Attribute set A of information system S is divided into conditional attribute set C and decision attribute set D.

1.1.2 The approximate set. Given a knowledge base $K=(U,R)$, And there are arbitrary subsets $X\subseteq U$, Equivalence relation $R\in IND(K)$, There are:

$\underline{R}(x)=\cup\{Y\in U/R:Y\subseteq X\}$,

$\bar{R}(x)=\cup\{Y\in U/R:Y\subseteq X\}$, Among them $\underline{R}(x)$ is X Lower approximation, $\bar{R}(x)$ is X The approximate. $Pos_R(X)=\underline{R}(X)$ Referred to as a subset X R is the domain, $Neg_R(X)=U-\bar{R}(X)$ this is called the R negative domain of the subset X, We can find the boundary of the subset X: $BN_R(X)=\bar{R}(X)-\underline{R}(X)$.

Calculate the information attribute dependence. That is, relative to the set of conditional attributes B, Calculate the dependence degree of decision attribute set D on it.To determine how important set D is to set B. The calculation method of dependence degree is shown in Formula (1):

$$\gamma_B(D)=\frac{Pos_B(D)}{|U|} \quad (1)$$

As can be seen from formula (1), the so-called dependence degree of D on B subset is actually the proportion of the positive domain set determined by B subset in the domain U.

1.1.4 Calculate the importance of the attribute. In information knowledge decision system, attribute importance is defined as the influence degree of conditional attribute on decision attribute. Let the information system be denoted by $S=(U,C\cup D,V,f)$, $\forall B\subseteq C$, If the attributes $\partial\in B$, Then the importance degree formula of conditional attribute ∂ to decision attribute D:

$$Sig(a,B,D)=\gamma_B(D)-\gamma_{B-\{a\}}(D) \quad (2)$$

1.1.5 I'm going to reduce the attribute.Take any subset B of attributes of an information system S, The ambiguous relation $IND(B)$ corresponding to A is defined as::

$$IND(B)=\{(x,y)\in U^2|\forall a\in B[a(x)=a(y)]\} \quad (3)$$

Let the set of attributes $B\in A$, For any property $a\subseteq B$, If you have $IND(B)=IND(B-\{a\})$, So call A unnecessary in B, Otherwise, A is said to be necessary in B.If a collection of properties $B\subseteq A$ meet $IND(B)=IND(A)$, So B is A reduction of A.

2.2 Neighborhood rough set

Basic rough set theory is aimed at discrete data processing, and so on the continuous data processing, need to first discretization operation, it will introduce error lead to change the original data attributes, which can express the information of the original set of properties, cause the loss of information of information system, and thus the information system of classification performance. Therefore, the neighborhood rough set model [11-13] is adopted in this paper to directly process continuous data to avoid information loss caused by data discretization.

1.2.1 neighborhood. $<U,\Delta>$ A nonempty metric space, $x\in U,\delta\geq 0$, According to point set $\delta(x)=\{y|\Delta(x,y)\leq\delta,y\in U\}$ is the δ neighborhood of x.

1.2.2 Neighborhood decision system.Assuming that $U=\{x_1,x_2,...x_n\}$ it's the set of all the samples, D is a set of conditional attributes that describe the sample, $D=\{l_1,l_2,...l_n\}$ is a set of classification decision attributes, $<U,A,D>$ is a neighborhood decision system.

1.2.3 The dependency and importance of the attribute.The dependence degree of decision attribute D on conditional attribute B is:

$$k_D=r_B(D)=\frac{|Pos_B(D)|}{|U|} \quad (4)$$

The importance degree of decision attribute D to condition attribute B is:

$$Sig(a,B,D)=\gamma_B(D)-\gamma_{B-\{a\}}(D) \quad (5)$$

1.2.4 Attribute reduction. When the obtained attribute importance degree is greater than the set lower limit of importance degree, Output the set of reductions red, red is the set that holds the reduction result.Neighborhood rough set algorithm flow:

Step1: Input decision system $<U,A,D>$, Set neighborhood radius calculation parameters and lower importance limit efc;

Step2: Preprocessing, normalizing the original data, and calculating the neighborhood radius δ;

Step3: Initializes the reduction set $red=\phi$;

Step4: For each property a_i in $A-red$, Calculate its positive domain: $Pos_{red}(D)=\underline{N}_{red}D$;

Step5: Calculate the dependency and importance of each attribute;

Step6: IF is greater than efc; otherwise, output the reduction result red, else, return to Step4.

https://doi.org/10.1051/e3sconf/202235801018

3. Attribute data screening based on K-means clustering

After the rough set algorithm is reduced, it is necessary to remove the unqualified data in the classified data. The basic idea is to classify the data based on the reduced data attributes, and delete the data inconsistent with the original standard classification. K-means clustering algorithm is an unsupervised clustering algorithm [14-15], which has the advantages of simple principle, easy implementation and fast clustering speed and is widely used in various fields. However, the algorithm itself is also sensitive to the initial cluster centroid and the comparison of noise and outliers. The clustering principle of K-means algorithm is to continuously divide the data set into different categories according to the centroid through iteration, and verify the clustering effect by evaluating the criterion function, so as to obtain independent inter-class and compact intra-class clustering results.

3.1 Algorithm principle

2.1.1 Select the similarity measurement method between samples. K-means clustering algorithm is not easy to deal with discrete data, but it is very suitable for continuous data. A hypothetical set of data samples $X=\{x_m|m=1,2,...,total\}$, The sample in X has d properties, Let's call them A1,A2…Ad, And these attributes are continuous data.Then the sample data I and j can be expressed as Xi=(Xi1,Xi2,…Xid) , Xj=(Xj1,Xj2,…,Xjd).d(Xi,Xj) is used to represent the similarity between samples Xi and Xj. The smaller the value of d(Xi,Xj), the smaller the distance between samples, and the more similar the two samples are. On the contrary, if the sample distance is larger, it indicates that the two samples are more dissimilar. In the aspect of sample similarity calculation, Euclidean distance and Manhattan distance can be selected according to the specific situation to measure the similarity between data samples. The more commonly used measurement method is Euclidean distance, and the specific calculation method is shown in Formula (6):

$$d(x_i,x_j)=\sqrt{\sum_{k=1}^{d}(x_{ik},x_{jk})^2} \tag{6}$$

2.1.2 Set the criterion function of clustering effect evaluation. The classical K-means clustering algorithm uses the sum of squared error as the criterion function to evaluate the clustering effect. Suppose the data set X is partitioned into k subsets X1,X2... Xk, the number of samples contained in the cluster subset is denoted by N1, N2... , NK, and the cluster centroids are denoted by M1,m2... Mk is denoted by. Then the formula of error sum of squares criterion function is:

$$E=\sum_{k=1}^{k}\sum_{p\in X_i}\|p-m_i\|^2 \tag{7}$$

2.1.3 Calculate the centroid of each cluster subset

1) In the initial state, k centroids are randomly generated, and the sample data are assigned to K clusters according to formula (6);

2) Calculate the average value of the sample data in each cluster and replace the centroid of the cluster with this value;

3) Redistribution was carried out according to the distance between each sample and the centroid of each cluster;

4) Judge whether the evaluation criterion is met, and stop clustering if it is. Otherwise, go to 2) and recalculate k cluster centroids.

3.2 K-means clustering algorithm flow

Input: number of sample clusters K and data set. Output: K clustering results for the dataset. Specific implementation process:

1) Randomly generate K cluster centers;

2) According to the principle of minimum distance, the samples in the data set to be clustered are divided into the nearest clustering set;

3) Calculate the average value of sample data in each cluster set, and replace the center of the original cluster as the cluster center of the next iteration;

4) Repeat Step 2 and Step 3 continuously until the stopping rule is met or the cluster center does not change, then K clusters of the data sample set will be returned.3 Real Data Simulation

3.3 Attribute reduction of influencing factors based on neighborhood rough sets

Was the first to classify a block of polymer flooding well group, and then according to the monthly statistics a standard well group of the well group and not done well group production days, polymer concentration, oil production, to produce liquid, flowing pressure, effective thickness, water cut, oil of nissan, nissan fluid, formation pressure, oil intensity, fluid producing intensity, oil change, water change, flowing pressure changes, and polymer concentration changes, Oil production index, fluid production index, geological reserves, pore volume, injection rate, polymer usage, cumulative injection-production ratio, compliance identifier (1 for compliance, 0 for noncompliance). The statistics of the first well group in January 2018 are shown in Table 1. The theory in Section 2 was used to reduce rough set attributes, delete redundant attributes, and finally determine the factors related to achieving the standard of polymer flooding well group, including water cut change, liquid yield change, flow pressure change, concentration change, polymer dosage change, injection-production ratio, injection rate and liquid extraction index.

Table 1 Influencing factors of Class I well group

NO	Production days (day)	Mining polyethylene concentration (mg/L)	In oil production (t)	Flowing pressure (MPa)	Oil change (%)	Water cut change (%)	Flowing pressure change (%)	Change of concentration of harvesting poly (%)	Fluid productivity index, (t/d·m·MPa)	Injection speed (PV/a)	Amount of polymer (mg/L.PV)	Injection-production ratio	Standard logo
1	25	331	21	5.25	0.95	1.03	0.92	0.98	5.84	0.449	509	1.16	1
2	31	463	365	5.21	0.85	0.99	0.94	0.96	7.18	0.381	584	0.95	1
3	31	71	182	6.79	0.99	1.02	1.16	1.15	6.11	0.304	372	0.9	1
4	31	277	87	2.2	0.95	1.02	1.03	1.07	3.04	0.651	818	1.22	0
5	31	529	241	6.99	0.99	1.00	1.0	1.11	13.64	0.383	549	1.48	0
6	31	42	182	6.19	1.02	1.00	0.91	0.79	8.17	0.486	609	1.1	1

3.4 Actual data screening and well group standard determination based on Kmeans clustering

Kmeans clustering was performed on the 8 attributes after rough set attribute reduction according to the theory in Section 3. Some results are shown in Table 2. It can be seen that the clustering result of well 4 is inconsistent with the standard identification. According to the cluster analysis, there are 142 data of the well group reaching the standard in the first class, among which 15 data of the well group have inconsistent clustering results. There are 120 data in the non-standard well group of the first class, and the clustering results are consistent with the standard identification. Through data analysis and comparison, the correctness and effectiveness of the algorithm are verified. Through the influence factors of 3.1 and 3.2 reduction and data filtering, the intelligent analysis model was established based on clustering algorithm, select the block type of well group in recent 3 months up to standard data to verify the validity of the intelligent analysis algorithm, the block type of well group, 45, which the standard well group 32, not done well group 13, through the method of intelligent analysis, The recognition rate of the well group reaching the standard is 87.5%, and the recognition rate of the well group not reaching the standard is 84.6%.

Table 2 Comparison of actual compliance and clustering results

No	Water cut change (%)	Fluid productivity index, (t/d·m·MPa)	Flowing pressure change (%)	Standard logo	Kmeans Clustering results
1	1.03	5.84	0.92	1	1
2	0.99	7.18	0.94	1	1
3	1.02	6.11	1.16	1	1
4	1.02	3.04	1.03	0	1

4. Conclusion and Understanding

This paper proposes a judging standard of a kind of intelligent polymer flooding well group development analysis algorithm, this algorithm can be combined with data of oil field, has the advantages of adaptive with the increase of oilfield development data, the proposed intelligent analysis method can get a better mark identification model, and improve the accuracy of the identification model, has good popularization value.

References

1. LI Xingrong, TONG Le, Wang Lu, et al. Review of polymer flooding technology [J]. Contemporary Chemical Industry, 2017, 46(6):1228-1230.
2. XU Hao. Optimization of alternate polymer injection parameters in polymer flooding oilfield [J]. Chemical Management, 2017(11).
3. Shi Chengfang, Xiao Wei, Wang Fenglan. Prediction model of polymer flooding development index [J]. Acta petrolei sinica,2005,26(5):78~84.
4. Zhao Guozhong, Meng Shuguang, Jiang Xiangcheng. A neural network method for predicting water content in polymer flooding [J]. Acta petrolei sinica,2004,25(1):70~73.
5. Qiu Haiyan, Ding Xianfeng, Hu Xiaoyun, et al. Prediction of oil production by polymer flooding using combinatorial model [J]. Xinjiang petroleum geology,2010,31(2):194~196.
6. Hou Jian, Zhao Hui, Du Qingjun, et al. Study on sensitive parameters affecting the performance of polymer flooding [J]. Journal of oil and gas technology,2007,29 (1):118~121.
7. Hou Jian, Guo Lanlei, Yuan Fuqing, et al. Quantitative characterization of polymer flooding production performance of different types of reservoirs in shengli oilfield [J]. Acta petrolei sinica,2008,29 (4):577~581.
8. GUO Qing. Research on Uncertain Information System and Decision Making Based on Rough Set Theory [D]. Hefei University of Technology, 2017.
9. Zhang Qinghua, Xue Yubin, Wang Guoyin. Optimal Approximation Sets for Rough Sets [J]. Journal of Software, 2016, 27(2):295-308.
10. Li Jia, Liang Jiye, Pang Tianjie. A multi-attribute decision ranking method based on Dominant Rough Set [J]. Journal of Nanjing University (Natural Sciences), 2016, 52(5):844-852.
11. An Ruoming, So bright. Application of Neighborhood Rough Set in Attribute Reduction and Weight Calculation [J]. Computer Engineering and Applications, 2016, 52(7):160-165.
12. XU Feng, Yao Sheng, Ji Xia, et al. Uncertainty measurement method of information system based on Fuzzy Neighborhood Rough Set [J]. Journal of Nanjing University (Natural Science),2017(5):926-936.

13. Shen Lin, Chen Jianhui. Variable Accuracy Neighborhood Rough Set attribute Reduction Algorithm based on lower approximate distribution [J]. Journal of Guizhou University (Natural Science Edition), 2017, 34(4):53-58.
14. Zhao Wei, Lin Nan, Han Ying, et al. An improved collaborative filtering algorithm based on K-means clustering [J]. Journal of Anhui University (Self-Science Edition), 2016, 40(2):32-36.
15. Ou Hui, Xia Zhuoqun, Wu Zhiwei. Rough Set K-means Clustering Algorithm based on Improved manifold Distance [J]. Computer Engineering and Applications, 2016, 52(14):84-89

Research on Comprehensive Evaluation Method of Competitiveness of Energy-saving Low-carbon Value-added Service

Qibo Yan*, Meiqi Sheng, Ailin Zhao, Kailin Zhang, Xiaochun Zhang

School of North China Electric Power University, Beijing, China

Abstract. Traditional electricity sales services present a development trend of diversification and integration, and will gradually evolve into comprehensive energy services in the future. Energy-saving, low-carbon value-added services are an important part of it, which is important for grid companies to cultivate core competitiveness in the electricity sales market. effect. In this regard, this article conducts a comprehensive evaluation of the competitiveness of energy-saving low-carbon value-added services based on the perspective of power grid companies. First, this article constructs an energy-saving low-carbon value-added service competitiveness evaluation system from the aspects of energy-saving, intelligent, personalized, and comprehensive levels; secondly, constructs a comprehensive evaluation model for the competitiveness of energy-saving and low-carbon value-added services, using AHP weighting method Empower the supervisor, and then conduct a comprehensive evaluation based on the fuzzy comprehensive evaluation; finally, select the Y city power company to conduct an empirical analysis of the comprehensive evaluation model, and provide a favorable reference for other companies to carry out energy-saving low-carbon value-added services in the future.

1. Introduction

In March 2015, the policy of "Several Opinions on Further Deepening the Reform of the Electric Power System (Zhongfa [2015] No. 9)" (hereinafter referred to as "Circular 9") was issued, and a new round of power reform in my country was officially launched. As the key content of the new electricity reform, the reform of the electricity sales side is bound to bring about profound changes to the electricity market in my country, especially the electricity sales market environment. As social capital begins to allow participation in the electricity sales business, a diversified competition pattern of electricity sales entities will surely be formed, which will force the innovation of the terminal sales model of the electricity market and improve the value-added services of electricity sales. Literature [1] analyzes the competitiveness of power sales services of China's power grid companies under the background of the new power reform, and analyzes the factors affecting the competitiveness of power sales services from the four dimensions of value-added services, user stickiness, corporate image and market environment, indicating that it will be open in the future In the electricity sales market environment, electricity sales companies are designated as service companies and need to develop differentiated value-added services according to user needs; literature [2] emphasizes power generation companies, high-tech industrial parks, energy-saving service companies, public service companies, and distributed power sources Market entities such as users have become competitors of traditional power grid companies in the electricity sales market; due to different perspectives on the connotation of low-carbon competitiveness of enterprises, academic research on the factors affecting low-carbon competitiveness of enterprises also has their own focus, literature [3] China believes that low-carbon competitiveness includes many factors such as operational competitiveness, management competitiveness, resource competitiveness, and environmental competitiveness; literature [4] analyzes the influencing factors of its low-carbon competitiveness from the perspective of the manufacturing industry, specifically including customer low-carbon demand , Develop low-carbon products/services, establish low-carbon value creation systems, low-carbon management models, and maintain innovation; literature [5] from the perspective of carbon intangible assets, build enterprises from the five dimensions of human resources, technology, organization, market, and social relations Low-carbon competitiveness evaluation index system.

After establishing different index systems for different industries or regions, there are also differences in the comprehensive evaluation methods adopted by academia

* Corresponding author: 2174443547@qq.com

for research objects. Literature [6] uses the TOPSIS method to comprehensively analyze and evaluate the low-carbon competitiveness of some steel companies in China; Literature [7] uses the DPSIR model to establish a low-carbon competitiveness evaluation index system based on the low-carbon economic theory, which is determined by the entropy weight method Index weights, combined with the TOPSIS method to evaluate China's inter-provincial low-carbon competitiveness, comparative analysis of its spatial differences and regional model structure, and in the literature [8] through the entropy method and panel data model research and analysis of China's carbon emission reduction The impact of regional low-carbon competitiveness. However, there are relatively few researches on the comprehensive evaluation of power grid enterprises' competitiveness in energy-saving and low-carbon value-added services in academia, and there is much room for improvement. Therefore, based on the background of the new round of power reform, this article builds an energy-saving and low-carbon value-added service competitiveness evaluation system. It has certain practical significance for promoting power grid enterprises to develop low-carbon value-added services.

2. Energy-saving low-carbon value-added service competitiveness evaluation model

2.1 Energy-saving low-carbon value-added service competitiveness evaluation index system

New Electricity Reform Document No. 9 clearly states that it encourages the establishment of electricity sellers, encourages electricity sellers to innovate services, expand business models, and provide users with value-added services. Driven by the new power reform, we will provide users with a wider range of services and a better service attitude. We will tailor power packages according to the needs of different users; provide users with clean and environmentally friendly power as a priority, and vigorously promote energy-saving technologies for users Save electricity; use the Internet to create a power information network platform to allow users to experience the convenience brought by innovative technology; comprehensively use heat, cold, air, water and other energy sources to provide "one-stop" integrated energy management.

Based on the above analysis, the entire indicator system will evaluate the business competitiveness of energy-saving, low-carbon value-added services through four primary indicators of energy conservation, intelligence, personalization, and integration of the service, and expand the corresponding secondary indicators. As shown in Table 1.

Table 1 Energy-saving low-carbon value-added service competitiveness evaluation index system

First level indicator	Secondary indicators
Energy saving level	Renewable energy quota ratio
	Electric energy substitution completion rate
	Annual carbon dioxide emissions
Intelligent level	Smart meter coverage
	Convenience level of service process
	Construction level of network information platform
	Functional experience level
	Usability experience level
Personalization level	Differentiated package
	Customized experience level
	Guide users to cut peaks and fill valleys
	Auxiliary services such as frequency modulation peak shaving, backup, etc.
Comprehensive level	Completeness of technical support
	Quality of metering, collection and settlement
	Quality of guaranteed power supply business
	Timeliness of emergency repair
	Brand

The energy-saving level mainly considers the environmental benefits brought by power grid companies after promoting energy-saving low-carbon value-added services, and mainly considers the proportion of renewable energy quotas, the completion rate of electric energy substitution, and the annual carbon dioxide emissions to reflect the three secondary indicators. Among them: renewable energy Percentage of quota = annual electricity generation from renewable energy/total annual electricity generation, electric energy replacement completion rate = annual replacement electricity/annual electricity consumption of the whole society;

Smart power consumption is supported by smart devices, Internet information and innovative technologies. Therefore, the level of intelligence will be indexed from smart meter coverage, service process facilitation, network information platform construction, functional experience, and sustainable experience. These five three-level indicators are considered, among which, the coverage of smart meters = the annual number of smart meters installed/total number of meters;

The level of personalization is mainly based on the comprehensive experience of energy-saving and low-carbon value-added services perceived by users to select indicators, from four aspects: differentiated packages, customized experience levels, guiding users to cut peaks and fill valleys, frequency modulation peak shaving, backup and other auxiliary services. consider;

Comprehensive level is embodied in contract management for major customers, operation and maintenance services of related technical equipment, and integrated energy management services. Therefore, integrated power service indicators will include the integrity of technical support, the quality of measurement, collection and settlement, and the guaranteed power supply business. The five three-level indicators of quality, timeliness of troubleshooting and brand image are considered.

2.2 Analytic hierarchy process to determine the index weight

Based on the relationship between the indicators, the comprehensive benefit evaluation indicators of multi-site integration are divided into four levels. After stratifying the indicators, compare the indicators of the same level with each other to examine their importance. Use 1 to 9 to indicate the degree to which index A is more important than B. If the importance of index A is equal to that of index B, it is represented by 1; if the importance of index A is very strong than that of index B, it is represented by 9. The meaning of the other numbers is the middle level of 1-9.

According to the result of the comparison, the judgment matrix A can be constructed.

$$A = (a_{ij}) n \times n \tag{1}$$

In the next step, the relative importance V' is obtained according to the square root method, and the normalization standard processing is performed on the relative importance, and the weight W_i is calculated. It is the eigenvector of the judgment matrix A.

$$W = (W_1, W_2, \cdots, W_n) \tag{2}$$

We use $\lambda \max - n$ to represent the degree to which matrix A deviates from compatibility, here is a $C.I$ representing the deviation from compatibility.

$$C.I = \frac{\lambda \max - n}{n-1} \tag{3}$$

It is also necessary to introduce an $R.I$ value to relax the consistency requirements of the high-latitude matrix, as shown in Table 2.

Table 2 $R.I$ value table

Dimension	1	2	3	4	5	…
$R.I$	0	0	0.52	0.89	1.12	…

At this time, the calculation formula for the consistency of the judgment matrix is as follows:

$$C.R = \frac{C.I}{R.I} \tag{4}$$

If the calculated result is less than 0.10, it means there is good agreement.

3. Fuzzy comprehensive evaluation method for comprehensive benefit evaluation

There are multiple complex indicators involved in the evaluation of the competitiveness of energy-saving and low-carbon value-added services, and the fuzzy comprehensive evaluation is suitable for studying problems containing multiple complex factors, adding the idea of fuzzy mathematics, and making fuzzy linear transformations and subordination levels of things. Divide, transform the evaluation of qualitative problems into evaluation of quantitative problems, and obtain more accurate and credible evaluation results. The model looks like this:

3.1 Determining factor collection and comment collection

First, the evaluation factors are formed into a set of factors U, $U = \{u_1, u_2, u_3, \cdots, u_n\}$, where $u_i (i = 1, 2, 3, \cdots, n)$ is the evaluation index factor, and n is the total number of evaluation factors in the same level. The evaluation results obtained by the evaluation system constitute the comment set V, $V = \{v_1, v_2, v_3, \cdots, v_n\}$, where $v_j (j = 1, 2, 3, \cdots, m)$ refers to the j evaluation result, and m is the number of evaluation grades.

3.2 Construct membership matrix

The meaning is the degree of membership of the i evaluation factor u_i to the evaluation result v_j, and the degree of membership $R_1 = \{r_{i1}, r_{i2}, r_{i3}, \cdots, r_{im}\}$, $i = 1, 2, 3, \cdots, n$ comprehensive evaluation of n elements results in a membership matrix with n rows and m columns:

$$R = \begin{bmatrix} r_{11} & r_{12} & \cdots & r_{1m} \\ r_{21} & r_{22} & \cdots & r_{2m} \\ \vdots & \vdots & \ddots & \vdots \\ r_{n1} & r_{n2} & \cdots & r_{nm} \end{bmatrix} \tag{5}$$

3.3 Multi-level fuzzy comprehensive evaluation

First, perform the first-level fuzzy comprehensive evaluation, that is, comprehensively evaluate the lowest-level factor $u_i (i = 1, 2, 3, \cdots, k)$, and obtain the evaluation index membership degree matrix as:

$$R_i = \begin{bmatrix} r_{i11} & r_{i12} & \cdots & r_{i1m} \\ r_{i21} & r_{i22} & \cdots & r_{i2m} \\ \vdots & \vdots & \ddots & \vdots \\ r_{in1} & r_{in2} & \cdots & r_{inm} \end{bmatrix} \tag{6}$$

Through fuzzy mathematical operation, the weight coefficient vector and the membership degree matrix can be calculated to obtain the evaluation result vector of each index factor under the index factor i:

$$B_i = W_i * R_i = (W_{i1}, W_{i2}, \cdots, W_{in}) \begin{bmatrix} r_{i11} & r_{i12} & \cdots & r_{i1m} \\ r_{i21} & r_{i22} & \cdots & r_{i2m} \\ \vdots & \vdots & \ddots & \vdots \\ r_{in1} & r_{in2} & \cdots & r_{inm} \end{bmatrix} = (b_{i1}, b_{i2}, \cdots, b_{im}) \tag{7}$$

Finally, a multi-level fuzzy comprehensive evaluation is carried out. The comprehensive evaluation of the upper layer is based on the calculation results of the evaluation of the lower layer. The final evaluation result vector is obtained through fuzzy operation:

$$B = W * (B_1, B_2, \cdots, B_k)^T = \{w_1, w_2, \cdots, w_k\} * (B_1, B_2, \cdots, B_k)^T \tag{8}$$

3.4 Evaluation result analysis

The evaluation result of fuzzy comprehensive evaluation obtains a fuzzy vector. Multiply the fuzzy comprehensive evaluation result B and the transposed matrix of the evaluation set V to obtain the score value S.

$$S = BV^T \tag{9}$$

4. Case analysis

Select Y city power company for empirical analysis. The area where the power company of Y city is located has a large power load and extensive power business. In addition, there are many power sales companies in city Y. The power market is very competitive. The leaders of the Y city electric power company attach great importance to the cultivation of the core competitiveness of the power grid company, and adopt various measures to hope that the electric power company can gain an advantage in the stimulated power market competition. Therefore, it is necessary to evaluate the core competitiveness of the Y city power company, find the company's weaknesses in the core competitiveness and propose improvement measures.

4.1 Determine the weight

According to the comprehensive evaluation steps of the core competitiveness in the previous section, the 1-9 scale method is used to construct the judgment matrix of the first-level indicators, as shown in Table 3. And meet the consistency test, CR=0.056<0.10. The calculation method for the weights of the secondary indicators of the primary indicators is the same, and the results of the weights of the secondary indicators are shown in Table 4. After judgment, the weights of the secondary indicators meet the consistency test.

Table 3 First-level index judgment matrix and weight

A	B_1	B_2	B_3	B_4	W
B_1	1	0.33	0.20	0.17	0.062
B_2	3	1	0.33	0.25	0.121
B_3	5	3	1	0.33	0.257
B_4	6	4	3	1	0.560

Table 4 Primary weight and secondary weight

First level indicator	First-level indicator weight W1	Secondary indicators	Secondary index weight W2
Energy saving level	0.062	Renewable energy quota ratio	0.333
		Electric energy substitution completion rate	0.570
		Annual carbon dioxide emissions	0.097
Intelligent level	0.121	Smart meter coverage	0.200
		Convenience level of service process	0.113
		Construction level of network information platform	0.287
		Functional experience level	0.244
		Usability experience level	0.156
Personalization level	0.258	Differentiated package	0.352
		Customized experience level	0.284
		Guide users to cut peaks and fill valleys	0.148
		Auxiliary services such as frequency modulation peak shaving, backup, etc.	0.216
Comprehensive level	0.560	Completeness of technical support	0.113
		Quality of metering, collection and settlement	0.244
		Quality of guaranteed power supply business	0.287
		Timeliness of emergency repair	0.200
		Brand	0.156

4.2 Establish membership matrix

By inviting a total of 5 experts from power grid companies, government departments and research institutions, according to the situation of the target company, combined with expert experience to score the core competitiveness of the power company in city Y. After statistical analysis and processing of the expert scoring situation, the fuzzy membership degree of each secondary index is obtained, and the fuzzy evaluation matrix $R_1 - R_4$ is established.

$$R_1 = \begin{bmatrix} 0 & 0.667 & 0.167 & 0.100 & 0.067 \\ 0.233 & 0.400 & 0.200 & 0.167 & 0 \\ 0 & 0.233 & 0 & 0.267 & 0.500 \end{bmatrix} \tag{10}$$

$$R_2 = \begin{bmatrix} 0.067 & 0.067 & 0.100 & 0.767 & 0 \\ 0.100 & 0.467 & 0.300 & 0.133 & 0 \\ 0.067 & 0.133 & 0.067 & 0.733 & 0 \\ 0.400 & 0.400 & 0.133 & 0.067 & 0 \\ 0.067 & 0.333 & 0.467 & 0.133 & 0 \end{bmatrix} \tag{11}$$

$$R_3 = \begin{bmatrix} 0.500 & 0.220 & 0.150 & 0.090 & 0.040 \\ 0.300 & 0.330 & 0.230 & 0.090 & 0.050 \\ 0.290 & 0.400 & 0.110 & 0.100 & 0.100 \\ 0.430 & 0.400 & 0.100 & 0.070 & 0 \end{bmatrix} \tag{12}$$

$$R_4 = \begin{bmatrix} 0.467 & 0.333 & 0.133 & 0.067 & 0 \\ 0.533 & 0.433 & 0.033 & 0 & 0 \\ 0.467 & 0.267 & 0.200 & 0.067 & 0 \\ 0.433 & 0.367 & 0.200 & 0 & 0 \\ 0.700 & 0.233 & 0.067 & 0 & 0 \end{bmatrix} \quad (13)$$

4.3 Comprehensive evaluation analysis

The calculated comprehensive evaluation vector A_i of the secondary index is shown below.

$$\begin{aligned} A1 &= (0.025, 0.25, 0.575, 0.13, 0.02) \\ A2 &= (0.176,0.318,0.200,0.307,0) \\ A3 &= (0.450, 0.166, 0.150, 0.119, 0.115) \\ A4 &= (0.445, 0.275,0.182, 0.098, 0) \end{aligned} \quad (14)$$

From the above formula, the comprehensive evaluation vector A of the power company's core competitiveness can be obtained:

$$A = (0.062, 0.121, 0.258, 0.560) \begin{bmatrix} A_1 \\ A_2 \\ A_3 \\ A_4 \end{bmatrix} = (0.388, 0.251, 0.200, 0.130, 0.031) \quad (15)$$

According to the principle of maximum membership degree, for the evaluation result of the core competitiveness of the power company in city S, among the five levels of membership, the value of 0.388 is the largest, and the comment value is "strong core competitiveness", which is similar to that of the city power company in Y The actual operating conditions are consistent. It can be seen from this that the power company in Y city has a strong core competitiveness in the local power market and is in a dominant position in the power market. Therefore, the measures taken by the Y city power company to enhance the company's core competitiveness are effective, and it is worth learning from the experience of power companies in other regions.

5. Conclusion

First, build a comprehensive evaluation index system for the competitiveness of energy-saving and low-carbon value-added services. Considering the four aspects of energy-saving level, intelligence level, individualization level and comprehensive level, a comprehensive evaluation index system for energy-saving low-carbon value-added service competitiveness including four first-level indicators and 17 second-level indicators has been constructed. Then a comprehensive evaluation model for the competitiveness of energy-saving value-added services is constructed. Use AHP weighting method to carry out subjective weighting, and then carry out comprehensive benefit evaluation based on fuzzy comprehensive evaluation. Finally, choose Y city power company for empirical analysis. Through the AHP-fuzzy comprehensive evaluation method, it is found that the comprehensive benefit evaluation result of the power company's energy-saving and low-carbon service competitiveness in Y City is consistent with its actual operating conditions, indicating that the energy-saving low-carbon value-added service competitiveness evaluation index system and model have Realistic feasibility can help other companies to further enhance the comprehensive competitiveness of energy-saving and low-carbon services in the future.

Acknowledgments

This work was financially supported by The National Social Science Fund of China(19ZDA081)and The Fundamental Research Funds for the Central Universities(2020MS067).

References

1. Xu Jieyan, Lu Yue, Qu Hong, et al. Analysis of my country's power grid companies' power sales service competitiveness under the background of new power reform[J]. Power Demand Side Management, 2020, 22(4): 89-93.
2. Xiao Xianyong, Ma Yuanqian, Mo Wenxiong, et al. High-quality power value-added service mode of power grid companies under the background of power sales side liberalization[J]. Journal of Electric Power Science and Technology,2016,31(04):4-10.
3. Zhu Liming. Evaluation system and training path for low-carbon competitiveness of large power generation companies[D]. Wuhan: China University of Geosciences, 2013.
4. Huang Shan,Wu Xiaojie,Zong Qijun.The source of China's manufacturing low-carbon competitiveness and ways to improve[J].East China Economic Management,2013(27).
5. Jiang Yuguo,Fan Lili.Research on the Evaluation of Low-carbon Competitiveness of Enterprises from the Perspective of Carbon Intangible Assets[J].Business Economic Management,2014(09).
6. Jiang Yuguo, Chun Weide. Application of a dynamic evaluation method in the evaluation of low-carbon competitiveness of steel enterprises[J].Operations Research and Management,2018,27(12):165-171.
7. Li Jian,Piao Shengren,Wang Zheng.Evaluation and spatial difference analysis of inter-provincial low-carbon competitiveness based on DPSIR-ENTRORY-TOPSIS model[J].Arid Land Resources and Environment,2016,30(12):40-46.
8. Cao Qingren, Zhou Siyu. Analysis of the Impact of China's Carbon Emission Reduction Policy on Regional Low-Carbon Competitiveness——Based on the Analysis of Provincial Panel Data[J]. Ecological Economy, 2020,36(11):13-17.

https://doi.org/10.1051/e3sconf/202235801020

In-situ SEM characteristics of dispersed organic matter in continental shale with its implication for dessert evaluation--A case study of Paleogene shale in the Cangdong Sag, Bohai Bay Basin, China

Delu Xie*, Xianzheng Zhao, Xiugang Pu, Fengming Jin, Changwei Chen, Wenya Jiang, Wenzhong Han, Shunyao Song, Wei Zhang, Zhannan Shi, Xiongying Dong, Pengtao Wang, Chen Zhao, Ling Ma, Cuiping Li, Haodong Li, Chuan He, Na Wang, Jing Zhou

PetroChina Dagang Oilfield Company, Tianjin 3002802, China

Abstract. Organic matter (OM) in continental shale serves as both the source of oil and gas and the storage space in unconventional petroleum systems. However, directly identifying the types of organic matter under SEM is challenging when simultaneously observing minerals and pores. Kong2 Member(E2k2) of Paleogene in Cangdong sag of Bohai Bay basin is a typical continental shale oil layer in China. Based on the positioning observation technology combining field emission scanning electron microscope (FE-SEM) and fluorescence microscope, the in-situ SEM identification and observation of macerals were carried out, and the identification methods and characteristics of organic macerals were summarized. The results show that: (1) Organic macerals in E2k2 shale are divided into vitrinite, inertinite, liptinite and solid bitumen by external morphology, hardness, brightness, color, protrusion, pore and fracture development of organic matter, and further subdivided into multiple subcategories. Based on the SEM charging effect of the remaining oil, it is further confirmed that the shale movable oil and oil generation potential developed by lipoid group is the largest, while the shale movable oil and oil generation potential developed by vitrinite group and inertinite group is the worst; (2) The organic pores include primary pores and secondary pores. The pores of primary organic matter are derived from the biological structure of primary organic matter, and the secondary organic pores are developed during the thermal maturation of oily organic matter. Clay mineral catalysis, difference of hydrocarbon generation potential and residual pores of primary organic matter control the development of organic pores; (3) Calcareous-dolomitic shale and felsic shale are typical lithology formed in relatively dry and humid climate respectively, and the types of organic macerals are significantly different. Although the former has weak total hydrocarbon generation, it has stronger oil generation potential and is worthy of attention in dessert prediction and exploration.

Keywords. In-situ SEM characteristics; Dispersed organic matter; Maceral group; Shale oil; Bohai Bay Basin

1. Introduction

Dispersed organic matter (DOM) plays an important role in determining hydrocarbon generation potential of continental shale(Ko et al., 2018; Vandenbroucke and Largeau, 2006). Different from marine shale, continental shale oil reservoirs in China are mixed with extensive terrigenous organic debris, which is new in age and low in maturity, resulting in a variety of macerals of organic matter coexists in continental shale. Therefore, the strong heterogeneity and complexity of DOM in continental shale hinder the correct understanding and evaluation on shale reservoir (Curtis et al., 2012; Loucks et al., 2012). Since that the continental shale has the characteristics of mixed fine particle deposition at the micro-scale, high-resolution field emission scanning electron microscopy (FE-SEM) observations are an indispensable tool for identifying shale minerals and pores(Guo et al., 2018; Milliken et al., 2013). However, the DOM displays in black because of low density, directly identifying the types of organic matter under SEM is challenging when simultaneously observing minerals and pores. The SEM backscatter cross section secondary imaging method based on charging effect of electron beam is used to identify the residual oil (i. e. crude oil not lost in the experimental vacuum) in the micro-nano pore of the reservoir, owing to the charging effect of non-conductive

* Corresponding author: dg1629016@smail.nju.edu.cn (D. Xie)

samples in the SEM imaging process (Jin et al., 2021; Wang et al., 2015).

Bohai Bay Basin is an important petroliferous rift basin in eastern China, taking the lead in realizing the industry discovery of shale oil (Fig. 1). The shale of the second member of the Paleogene Kongdian Formation (E2k2) in Cangdong sag develops a semi-deep lake deep lake facies shale oil reservoir system formed in a semi-arid humid climate, with felsic shale (FS), calcareous dolomitic shale (CDS) and mixed shale (MS) alternately deposited spatially. A lot of researches have been done in the aspects of mineral composition, geochemical characteristics of organic matter, automatic mineral identification by AMICS and high-resolution SEM imaging at an early stage. High TOC, felsic, high-frequency lamina and low clay minerals are important component characteristics of E2k2 shale, and the laminar FS was regarded as the main exploration lithology of Cangdong Sag (Zhao et al., 2017, 2019). What are the similarities and differences in the in-situ characteristics of the DOM between the CDS formed in a relatively arid climate and the FS formed in a humid climate? How to identify the types and characteristics of DOM in-situ by high-resolution SEM? What impact will it have on the evaluation of shale oil desserts? The above issues, on the one hand, are seriously restricting the evaluation of shale oil resource potential and the dessert prediction in the Bohai Bay Basin; on the other hand, are also the weak field of basic research of continental shale. In view of this, a combination of optical microscopy and FE-SEM were used to systematically analyze the types and characteristics of DOM and the distribution of movable oil in E2k2 shale, particularly the CDS and FS formed in arid and humid climate respectively, with discussion on implication of macerals on the hydrocarbon distribution, migration and and dessert evaluation. This study is expected to provide a theoretical reference for shale oil resource evaluation and exploration and development in sedimentary basins with similar geological background.

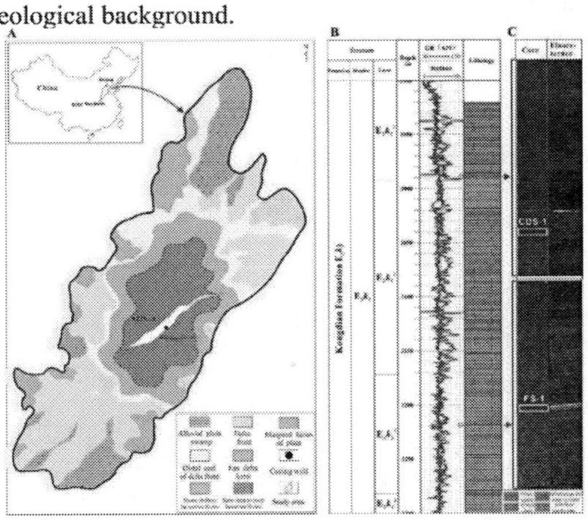

Fig. 1. (A) Sedimentary facies distribution of semi-deep to deep lacustrine shale of the Paleogene E2k2 in Cangdong Sag, Bohai Bay Basin (Zhao et al., 2017); (B) Generalized stratigraphy of Paleogene E2k2 in Cangdong Sag, showing that three main fine-grained sedimentary shales developed in frequent interbedding way including felsic shale (FS), calcareous-dolomitic shale (CDS) and mixed shale(MS).

2. Geological background

Bohai Bay basin is a Cenozoic inland lake basin developed under tensile tectonic background. Cangdong sag in the study area is sandwiched between Cangxian uplift, Xuhei uplift and Kongdian uplift (Fig.1). During the sedimentary period of Paleogene E2k2, Cangdong sag was an inland closed lake basin, which was in a subtropical semi-arid to humid climate. A set of sedimentary formation with a thickness of 400 ~ 600m, mainly dark shale, mixed with thin to medium siltstone, medium-fine sandstone and argillaceous dolomite. The sedimentary facies change regularly from the edge of the lake basin to the cente, showing a delta front subfacies in the edge of the lake basin with mainly developed conventional medium-fine sandstones dominated, a pre-delta subfacies and semi-deep lake subfacies in the middle of the lake basin, which is the main source rock composed of shale and argillaceous dolomite, and a far end of the delta front and gravity flow sedimentation in the transitional zone of both above.

Preliminary research results reveal that (Han et al., 2021): (1) X-ray diffraction (XRD) shows that E2k2 shales are mainly composed of quartz, feldspar, dolomite, calcite, analcite, pyrite, clay and other minerals and lithology mainly includes FS, CDS and mixed shale (MS) (Fig.1B); (2) CDS, formed in a relatively dry climate, has relatively lower TOC (average = 2.23%), S1 (average = 3.01 mg/g) and S2 (average = 10.25 mg/g), but higher OSI(TOC/S1*100) (average = 125.38 mg/g TOC). However, FS formed in a relatively humid climate, has relatively higher TOC (average = 5.61%), S1 (average = 3.89 mg/g) and S2 (average = 31.87 mg/g), but lower OSI (average = 68.80 mg/g TOC); (3) The analysis of organic matter types showed that Tmax and hydrogen index (average = 442.95°C and average = 418.61 mg/g respectively) of CDS were slightly lower than those of FS (average = 442.95 °C and average = 418.61 mg/g respectively). The main type of organic matter in both shales were kerogen type I and II1, but the organic matter type of FS was better; (4) In-situ imaging under SEM showed that the laminar CDS and FS had a high abundance of organic matter along the laminar distribution, with complex and indistinguishable organic macerals.

3. Materials and methods

The main workflow involved in this study as followed:

Firstly, macerals are identified under an optical microscope with reflected white light and oil immersion. The surfaces of petrographic pellets are subsequently cleaned to remove immersion oil, and then the pellets are cut for SEM sample preparation. The same surfaces as observed under reflected-light microscopy are Ar ion-milled for SEM observations, and the same fields of view need to be located under the SEM. Microfractures, pyrite framboids, or fossils can be used as markers for easier identification. Alternatively, OM can be first examined on ion-milled surfaces under SEM, and then samples can be moved to the reflected-light microscope for maceral

 https://doi.org/10.1051/e3sconf/202235801020

identification. In this case, additional sample preparation is unnecessary.

For the identification method of shale movable oil based on charging effect, seen in Jin et al., 2021; Wang et al., 2015. ImageJ is mainly used for the BSD image processing, and the proportion of shale movable oil is obtained through the statistics of 100 field of view.

Microscope observation, Ar-ion polishing and FE-SEM experiment were carried out at theKey Laboratory of Deep Oil and Gas, China University of Petroleum. In addition, the basic research data of 23 CDS and 35 FS were provided by PetroChina Dagang Oilfield Research Institute.

4. Results

The petrographic classification of DOM in continental shale is mainly based on that of coal rock. In general, there are four macerals developed in E2k2 lacustrine shale, among which vitrinite, inertinite and liptinite are primary organic matter, while the secondary organic matter is mainly solid bitumen(SB) and movable oil, which is the result of the transformation of oil prone macerals during thermal maturation. Different from the coal rocks, there is almost no zooclasts in E2k2 shale.

4.1 Petrographic classification of DOM

4.1.1 Vitrinite

Vitrinite is formed by the gelation of roots, stems and bark of plants. The different degree of gel leads to the different preservation of primitive plant cell structure. It has low oil generation potential and high methane generation potential. OM rich in vitrinite belongs to a typical type-III kerogen. It is usually dispersed in the mineral matrix of shale in the form of small particles (Fig. 2A-A').

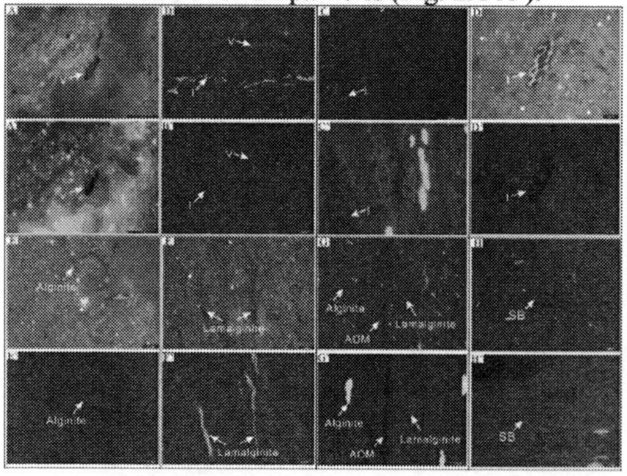

Fig. 2. Photomicrographs of DOM in Paleogene E2k2 shales in reflected white light and oil immersion(A-H) and in fluorescence mode (A'-H'), including primary OM of vitrinite(A-A'), inertinite (B-B', C-C'), zooclasts(D-D') and liptinite(E-G, E'-G'), and secondary OM of SB(H-H'). V=vitrinite; I=inertinite; Z=zooclasts; AOM=amorphous organic matter; SB=solid bitumen; OM= organic matter.

4.1.2 Inertinite

Inertinite is a kind of macerals formed from wood microstructure through filament carbonization produced by fire or oxidation before deposition. It has little hydrocarbon generation potential and is usually classified as type IV kerogen. Even in the early mature stage, inertinite has high reflectivity. Under the optical microscope, it is best to distinguish between inertinite and vitrinite by the color of reflected light (Fig. 2B-3D').

4.1.3 Liptinite

Lipinite is one of maceral that is generally easy to produce oil, with high hydrocarbon generation potential and usually classified as type I/II kerogen. The lipinite macerals in E2k2 shale are mainly including alginite (such as telalginite and lamalginite) and amorphous organic matter(AOM). Alginite appears as slender rods. Under the blue light irradiation, the Alginite at the early mature stage showed strong yellow-green fluorescence. AOM refers to unstructured OM in shale, which is derived from microbial degradation of phytoplankton, zooplankton and bacteria (Fig. 2E-G').

4.1.4 Solid bitumen

SB is a kind of secondary organic matter formed during kerogen transformation and hydrocarbon generation, and its shape is limited by the pore shape it fills (Fig. 2H-H').

4.2 In-situ characteristics and petrography of DOM under SEM

Based on the identification and verification by optical microscopy, the microscopic features of in organic macerals of the argon ion-polished shale were obtained under FE-SEM observation. Specifically, the types of microcomponents can be identified according to comprehensive factors such as the external morphology of organic matter, contact relationship with surrounding minerals, organic matter color, brightness, protrusions, organic pore development characteristics and fracture development characteristics.

4.2.1 Vitrinite

Vitrinite can be further divided into provitrinite, collinite and vitrodetrinite (Fig.3 A1-A4).

(1) Provitrinite, a kind of vitrinite with relatively well-preserved microstructure of primitive plants, shows various styles of crisscross network, with obvious cell cavities and walls (Fig.3 A1). The cell cavities of provitrinite are usually round, oval and regularly arranged. Under the formation pressure, the cell cavity is often deformed and irregular. The typical in-situ SEM characteristics of provitrinite are high protuberance, dense and uniform texture, dark color, dark brightness, clean and smooth surface, and generally undeveloped pores.

(2) Collinite is one of vitrinite whose original plant cell structure can no longer be seen through strong gel. The color shade of unstructured vitrinite is similar to that of

https://doi.org/10.1051/e3sconf/202235801020

structural vitrinite, but the protuberance is slightly lower than that of provitrinite. Transverse cracks are often seen inside the collinite, and the protrusion is significantly higher than that of SB, which is an effective sign to distinguish the collinite from the SB (Fig.3 A2).

(3) Vitrodetrinite, was clastic particle of vitrinite that cannot identify the primitive cell structure, formed by mechanical crushing of structural vitrinite and unstructured vitrinite. It is irregular in shape, similar to poorly sorted sedimentary clastic particles, with dark brightness, flat and smooth contact line with minerals, high protuberance, and generally undeveloped pores and fractures (Fig.3 A3-A4).

4.2.2 Inertinite

Inertinite can be further divided into fusinite and semifusinite, sclerotinite, and inertodetrinite (Fig.3 B1-B4).

(1)Fusinite and semifusinite, similar to provitrinite and also derived from plant microstructures, such as roots, stems and bark of plants, are a type of microscopic component with the best-preserved cell structure among inertinite. On the whole, the protrusions of fusinite are higher than that of the structural vitrinite, and the color is slightly lighter than that of the vitrinite, with undeveloped pores and fine stomata distribution on cell walls(Fig.3 B1).

(2)Sclerotinite, darker in color, 70% black, are usually in a regular round shape, with high protrusions, dense edges, no pores, and regular circles of different sizes often developed in the middle (Fig.3 B2).

(3)Inertodetrinite, similar to vitrodetrinite, the clastic particles of inertinite, whose cell structure can no longer be identified under the scanning electron microscope, are all irregular in shape, uniform and dense in texture, smooth and glossy, dark gray in color, slightly lighter than the vitrodetrinite, and slightly higher in brightness than the vitrodetrinite. The contact lines between minerals and inertodetrinite are straight and smooth, with high protrusions, and generally no pores and no cracks are found (Fig.3 B3-B4).

4.2.3 Liptinite

Under the scanning electron microscope, algae can be identified according to the different shapes of the alginite (Fig.3 C1-C3). Mixed with clay-sized mineral particles and others, AOM has an uneven surface (Fig.3 C4). At the early maturity stage, it may be difficult to distinguish between AOM and SB, but SB usually has a uniform surface filling pores.

4.2.4 Secondary organic matter

According to the order of appearance of SB in the evolution stage of organic matter, it can be divided into pre-SB, post-SB and pyrobitumen.

The E2k2 shale SB of Well GX-A is dominated by the former two, and the pyrobitumen increases with the increase of burial(Fig.3 D1-D4). The most notable feature of SB is that it has no specific particle shape, and is often filled in the space composed of inorganic minerals in the form of cement, which is greatly affected by the distribution of surrounding materials. Under SEM, the SB is 40% to 50% black, lighter in color than vitrinite and intertinite, and has lower protrusions, with some even lower than the protrusions of surrounding minerals. The pores of pre- SB and post- SB are not developed (Fig.3 D1-D2). Pyrobitumen, as a type of insoluble organic matter, is the solid residual bitumen after the pyrolysis of crude oil. Thus, a large number of sponge-like or honeycomb-like micro-pores often developed in the pyrobitumen, and the pores are fine and evenly distributed, irregular in shape, developed in network, with the phenomenon of small pores appearing in large pores, which is the main difference between pyrobitumen and other SB.

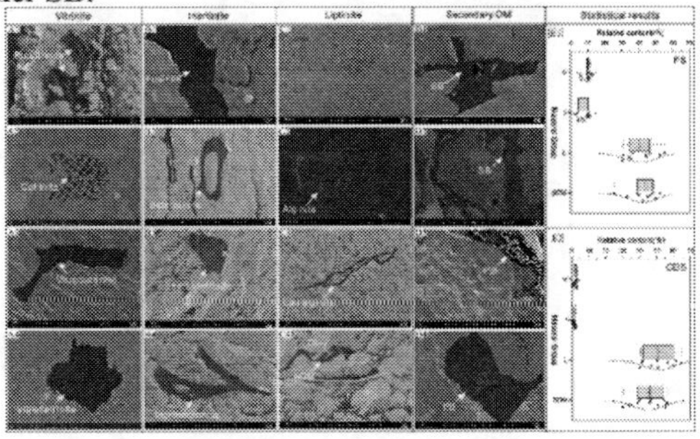

Fig. 3. Photomicrographs of DOM in Paleogene Kong2 Member shales under SEM, including vitrinite (A1-A4), inertinite (B1-B4), liptinite (C1-C4) and secondary OM (D1-D4). AOM= amorphous organic matter; SB= Solid bitumen; PB=pyrobitumen; OM=organic matter.

4.3 Quantitative results of DOM under SEM

Quantitative analysis of E2k2 shale maceral group based under SEM, shows that FS has higher vitrinite (average =9.56%), inertinite (average = 7.41%), and lower liptinite (average =40.91%) and secondary organic matter (average = 42.12%)(Fig.3 E1). However, in calcareous-dolomitic shale (CDS), there are lower vitrinite (average =2.11%), inertinite (average =1.46%), and higher liptinite(average =51.36%) and secondary organic matter (average = 45.07%)(Fig.3 E2).

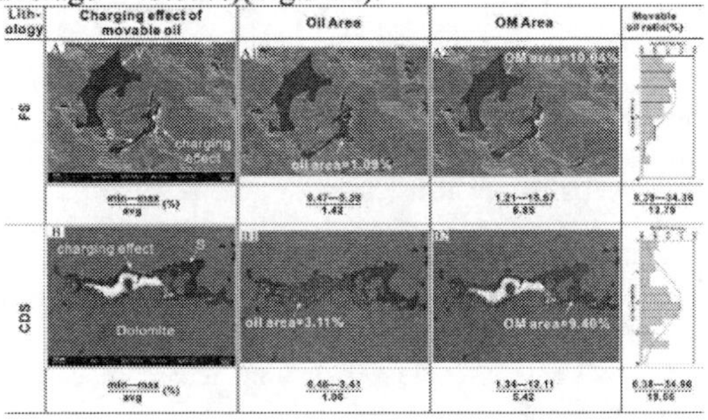

Fig. 4. Distribution of movable oil in FS (A-A2) and CDS (B-B2) under SEM based on charging effect. Image processing was completed in the software of ImageJ. V=vitrinite; S=sporinite; FS= felsic shale; CDS= calc areous- dolomitic shale; OM= organic matter. Movable oil ratio=oil area/OM area(%).

4.4 SEM petrography of Movable oil

The charging effect of SEM can be used to identify the residual movable oil in shale (Fig.4). BSD imaging shows that charging effect on the surface of vitrinite and inertinite was hardly difficult to observe under SEM (Fig.4A), while that of alginite and AOM is the most significant (Fig.4B), which is consistent with the hydrocarbon generation tendency of the macerals. Through ImageJ processing of SEM images with charging effect, it is found that, CDS has a higher distribution of movable oil (average =19.56%) than that of FS (average =12.79%).

5. Discussion

5.1 Controls on organic pore development

The controls of shale organic macerals on organic pores are manifested in three aspects:

(1) Residual pores of primary organic matter: the organic matter in shale includes primary organic matter and secondary organic matter, and organic matter pores can also be divided into primary organic matter pores and secondary organic matter pores (Gao et al., 2021, 2019; Hu et al., 2021; Teng et al., 2022). The pores of primary organic matter are mainly distributed in the provitrinite, fusinite and semifusinite. The common feature of these microscopic components is that the lignocellulosic structure of the original higher plants is well preserved, with visible cell walls and cell cavities. Under the action of vertical compaction or lateral extrusion, the pores of primary organic matter are easily deformed.

(2) Differences in hydrocarbon generation potential. The organic maceral, with sapropel and chitin as the main parent materials, has the greatest potential of hydrocarbon generation, while the vitrinite is mainly the parent materials for gas-generating maceral. The intertinite can generate neither gas nor oil, but it is different from “dead carbon”. As secondary organic matter, SB has the potential of further cracking to produce light oil and natural gas, therefore it has better hydrocarbon generation potential than vitrinite and intertinite.

(3) Catalysis of clay minerals. Organoclay complexes are widespread in shale, and almost all SB therein developed abundant honeycomb pores, especially the SB distributed around illite has significantly more pores than the SB that coexists with other clay minerals. The catalysis of clay minerals can be resolved into two processes (Xie et al., 2021, 2020)- when kerogen is converted into hydrocarbons, organic acids are generated, which promote the conversion of montmorillonite to illite; and the H+ precipitated during the illiteization of clay minerals and H2O can create conditions for the hydrocracking of crude oil, having a certain catalytic effect on the generation of hydrocarbons and developing pores.

5.2 Implications on movable oil and dessert evaluation

Different from marine shale, the organic matter of continental shale is controlled by lake basin area, water properties, climate and structure, especially the influence of climate is significant. In the CDS formed by a relatively arid climate, the input intensity of terrigenous detrital rocks is low, and the content of terrigenous organic matter, such as vitrinite and inertinite, is significantly lower; instead, under strong evaporation conditions, liptinite especially algite, such as telalginite and lamalginite, develop and thus form favorable organic matter types with stronger oil-generating potential. On the contrary, in the FS formed by a relatively humid climate, the input intensity of terrigenous detrital rocks is high, the content of terrigenous organic matter such as vitrinite and inertinite is high, and the total organic matter abundance of shale is significantly higher than that of CDS, but the relatively lower liptinite means that the oil generation potential of FS is not higher than that of CDS(Fig.5) .

Due to the development of high TOC, the FS in Cangdong Sag has always been regarded as the exploration priority zone (Zhao et al., 2019, 2017). However, the above understandings indicate that CDS, despite its weak total hydrocarbon generation, has stronger oil generation potential and a higher proportion of movable oil, which is worthy of attention in sweet spot prediction and exploration.

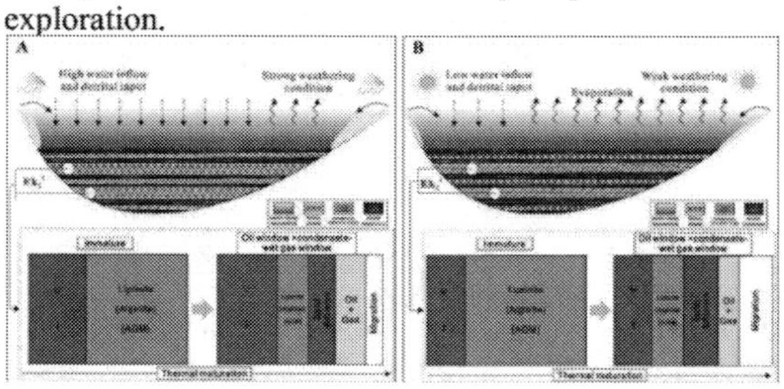

Fig. 5. Evolution model of DOM in typical continental shale oil reservoirs, including FS (A), calcareous-dolomitic shale(B).

6. Conclusions

Taking the E2k2 shale in Cangdong sag of Bohai Bay Basin as an example, the following conclusions are obtained through microscope, fluorescence observation and FE-SEM:

Organic macerals in E2k2 shale are divided into vitrinite, inertinite, liptinite and SB by external morphology, hardness, brightness, color, protrusion, pore and fracture developmentof organic matter, and further subdivided into multiple subcategories. Based on the SEM charging effect of the remaining oil, it is further confirmed that the shale movable oil and oil generation potential developed by lipoid group is the largest, while the shale movable oil and oil generation potential developed by vitrinite group and inertinite group is the worst;

The organic pores include primary pores and secondary pores. The pores of primary organic matter are derived from the biological structure of primary organic matter, and the secondary organic pores are developed during the thermal maturation of oily organic matter. Clay mineral catalysis, difference of hydrocarbon generation potential and residual pores of primary organic matter control the development of organic pores;

Calcareous-dolomitic shale and FS are typical lithology formed in relatively dry and humid climate respectively, and the types of organic macerals are significantly different. Although the former has weak total hydrocarbon generation, it has stronger oil generation potential and is worthy of attention in dessert prediction and exploration.

Acknowledgments

This study was funded by the PetroChina Science and Technology Major Project (no. 2019E-26), the China Postdoctoral Science Foundation (no. 2022M710567) and the Postdoctoral Foundation, PetroChina Dagang Oil Field (no. 2022BO54).

References

1. Curtis, M.E., Cardott, B.J., Sondergeld, C.H., Rai, C.S., 2012. Development of organic porosity in the Woodford Shale with increasing thermal maturity. Int. J. Coal Geol. 103, 26–31. https://doi.org/10.1016/j.coal.2012.08.004
2. Gao, F., Song, Y., Zhikai, L., Li, Z., Yuan, Y., Yinghan, Z., Chen, L., Guo, W., 2019. Development characteristics of organic pore in the continental shale and its geneticmechanism a case study of
3. Shahezi Formation shale in the Changlingfault depression of Songliao Basin. Acta Pet. Sin. 40, 1030–1044.
4. Gao, F., Wang, C., Song, Y., Chen, Z., Liu, Q., Li, Z., Jiang, Z., Zhang, X., 2021. Ar-ion polishing FE-SEM analysis of organic maceral identification. Pet. Geol. Exp. 43, 360–367.
5. Guo, H., He, R., Jia, W., Peng, P., Lei, Y., Luo, X., Wang, X., Zhang, L., Jiang, C., 2018. Pore characteristics of lacustrine shale within the oil window in the Upper Triassic Yanchang Formation, southeastern Ordos Basin, China. Mar. Pet. Geol. 91, 279–296. https://doi.org/https://doi.org/10.1016/j.marpetgeo.2018.01.013
6. Han, W., Zhao, X., Jin, F., Pu, X., Chen, S., Mu, L., Zhang, W., Shi, Z., Wang, H., 2021. Sweet spot evaluation and exploration practice of lacustrine shale oil of the second member of Kongdian Formation in Cangdong sag, Bohai Bay Basin. Pet. Explor. Dev. 48, 900–910. https://doi.org/https://doi.org/10.1016/S1876-3804(21)60075-7
7. Hu, T., Pang, X., Jiang, F., Wang, Q., Liu, X., Wang, Z., Jiang, S., Wu, G., Li, C., Xu, T., Li, M., Yu, J., Zhang, C., 2021. Movable oil content evaluation of lacustrine organic-rich shales: Methods and a novel quantitative evaluation model. Earth-Science Rev. https://doi.org/10.1016/j.earscirev.2021.103545
8. Jin, X., Li, G., Meng, S., Wang, X., Liu, C., Tao, J., Liu, H., 2021. Microscale comprehensive evaluation of continental shale oil recoverability. Pet. Explor. Dev. 48. https://doi.org/10.1016/S1876-3804(21)60021-6
9. Ko, L.T., Ruppel, S., Loucks, R., Hackley, P., Zhang, T., Shao, D., 2018. Pore-types and pore-network evolution in Upper Devonian-Lower Mississippian Woodford and Mississippian Barnett mudstones: Insights from laboratory thermal maturation and organic petrology. Int. J. Coal Geol. 190, 3–28. https://doi.org/10.1016/j.coal.2017.10.001
10. Loucks, R., Reed, R., Ruppel, S., Hammes, U., 2012. Spectrum of pore types and networks in mudrocks and a descriptive classification for matrix related pores. Am. Assoc. Pet. Geol. Bull. 96, 1071–1098. https://doi.org/10.1306/08171111061
11. Milliken, K.L., Rudnicki, M., Awwiller, D.N., Zhang, T., 2013. Organic matter-hosted pore system, Marcellus Formation (Devonian), Pennsylvania. Am. Assoc. Pet. Geol. Bull. 97. https://doi.org/10.1306/07231212O48
12. Teng, J., Liu, B., Mastalerz, M., Schieber, J., 2022. Origin of organic matter and organic pores in the overmature Ordovician-Silurian Wufeng-Longmaxi Shale of the Sichuan Basin, China. Int. J. Coal Geol. 253. https://doi.org/10.1016/j.coal.2022.103970
13. Vandenbroucke, M., Largeau, C., 2006. Kerogen origin, evolution and structure. Org. Geochem. 38, 719–833. https://doi.org/10.1016/j.orggeochem.2007.01.001
14. Wang, X., Sun, L., Zhu, R., Jin, X., Li, J., Wu, S., Bi, L., Liu, X., 2015. Application of charging effects in evaluating storage space of tight reservoirs: A case study from Permian Lucaogou Formation in Jimusar sag, Junggar Basin, NW China. Pet. Explor. Dev. 42. https://doi.org/10.1016/S1876-3804(15)30044-6
15. Xie, D., Yao, S., Cao, J., Hu, W., Qin, Y., 2020. Origin of calcite cements and their impact on reservoir heterogeneity in the Triassic Yanchang Formation, Ordos Basin, China: A combined petrological and geochemical study. Mar. Pet. Geol. 117. https://doi.org/10.1016/j.marpetgeo.2020.104376
16. Xie, D., Yao, S., Cao, J., Hu, W., Wang, X., Zhu, N., 2021. Diagenetic alteration and geochemical evolution during sandstones bleaching of deep red-bed induced by methane migration in petroliferous basins. Mar. Pet. Geol. 127. https://doi.org/10.1016/j.marpetgeo.2021.104940
17. Zhao, X., Pu, X., Han, W., Zhou, L., Shi, Z., Chen, S., Xiao, D., 2017. A new method for lithology identification of fine grained deposits and reservoir sweet spot analysis: A case study of Kong 2 Member in Cangdong sag, Bohai Bay Basin, China. Pet. Explor. Dev. 44. https://doi.org/10.1016/S1876-3804(17)30061-7
18. Zhao, X., Pu, X., Jin, F., Han, W., Shi, Z., Cai, A., Wang, A., Guan, Q., Jiang, W., Zhang, W., 2019.

 https://doi.org/10.1051/e3sconf/202235801020

Geological characteristics and key exploration technologies of continental shale oil sweet spots: A case study of Member 2 of Kongdian Formation in the Cangdong sag in the Huanghua depression, Bohai Bay Basin. Pet. Res. 4, 97–112. https://doi.org/10.1016/j.ptlrs.2019.01.006

E3S Web of Conferences **358**, 01021 (2022) https://doi.org/10.1051/e3sconf/202235801021
GESD 2022

Improvement and application of drilling fluid technology for gas exploration wells in Yanchi block

Hao Wang, Yan Wei *

Drilling & Production Engineering Technology Institute, CNPC Chuanqing Drilling Engineering Company Ltd., Xi'an, Shaanxi, 710018, China

Abtract. Currently, the exploration and development scope of Changqing oilfield has gradually developed to edge blocks and deep systems, typically of Yanchi block. However, in the construction process, there are some problems such as poor well wall stability, frequent well leakage and prominent collapse contradiction, which seriously restricts the increase of drilling speed. In this paper, the measures such as well structure optimization, system formulation optimization, separate management of collapse and leakage, refined performance control, and pressure-bearing leakage plugging were taken by a combination with the construction situation in 2021 so as to improve drilling speed and reduce the complexity of downhole failures.

Key words: Yanchi block; exploratory well; collapse prevention; leakage prevention; drilling fluid

1. Introduction

With the gradual development of exploration and development scope in Changqing oilfield to edge blocks and deep systems, the exploration and development intensity of Yanchi block has increased year by year in the past two years[1-4]. In 2020, the Fourth Project Department completed 7 gas exploration wells in the Yanchi block with poor production water quality and difficult drilling fluid performance maintenance. It is found that the stability of the strata near the thrust belt on the west margin is extremely poor[5-6]. Among them, reverse reaming is required for Li 56H well after three times. The double-rock layer of Li 47 well collapses seriously, and the pump is held back during the reaming process, so that the lifting and lowering are blocked. The collapse of the Anding, Zhiluo and double-rock layers is serious, and the drill resistance is serious. In addition, there are many leakage layers and frequent well leakage below the Yanchang stratum. The average single well plugging time exceeds 200 h, and the downhole failures are complex and time-efficient, showing a failure complexity rate as high as 13.56%. Moreover, the drilling fluids of Li 47 well, Li 69 well and Li 79 well constructed in 2020 have increased viscous shear force and increased filter loss, making it difficult to maintain and adjust performance on site[7-12].

Table. 1 Statistics of complicated conditions of construction wells in 2020

Well	Well depth (m)	Finish drilling layer position	Complex time (h)	Complex limitation (%)
Li 56H	5040	Shihezi	637	22.04
Li 47	4345	Krimori	986	26.27
Li 57-4	3936	Lashzhong	114	5.7

Table. 2 Exploration well leakage in the Yanchi block in 2020

Well	Well leakage		Measures			Leakage time (h)
	Leakage stratum	Leakage speed (m^3/h)	Static plugging	Squeeze plugging	Squeeze cement	
Li 57-4	Shihezi	30-35	3	0	1	104
Li 69	Krimori	20	1	2	1	136
Li 63	Liujiagou/Shiqianfeng/Shanxi	20	3		2	200
Li 64	Shanxi	30	2	2	1	210
Li 47	Yanchang/Liujiagou/Shiqianfeng/Shihezi	15-60	3	8	7	417
Li 56H	Heshanggou/Liujiagou/Shiqianfeng/Shihezi	5-25	16	4	2	323

* Corresponding author: zjwyan@cnpc.com.cn

https://doi.org/10.1051/e3sconf/202235801021

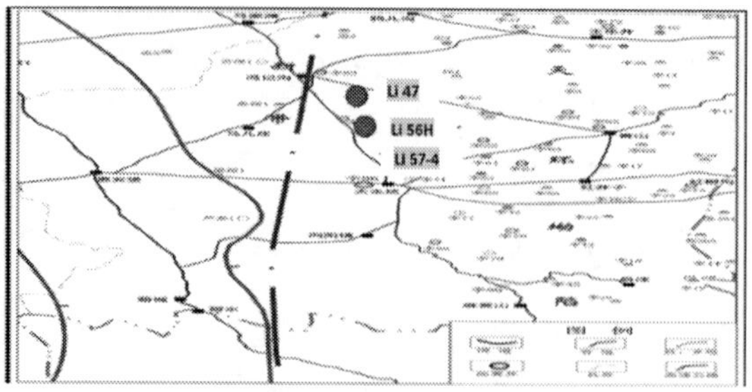

Fig. 1 Location of Li 47, Li 57-4 and Li 56H

2. Drilling fluid technology

2.1 Formula experiment with anti-salt treatment agent

Water-based drilling fluid was prepared from polymer protoslurry, 0.2% sodium hydroxide, 0.1% sodium carbonate, 0.15% PAC, 3% ZDS and 2% prehydrated clay. As shown in Table 3, it is found that the formulas of 5# and 6# meet the construction requirements, and the addition of potassium chloride improves the system inhibition, so that formula is determined as polymer protoslurry+0.2% sodium hydroxide+0.1% sodium carbonate+0.15% PAC+3% ZDS+2% prehydrated clay+3% SMP-2+3% SFT+2% SMC+5-7% KCl+0.05% K-PAM.

Table. 3 Formulation experiment

Number	Formula	FL (mL, 7.5min)	AV (mPa·s)	PV (mPa·s)	YP (Pa)	YP/PV (Pa/mPa·s)
1#	Black	10.4	25	20	5	0.25
2#	Black+2%SMP-2+1%SFT-1+1%SMC	7.6	22	18	4	0.22
3#	Black+3%SMP-2+2%SFT+2%SMC	6.2	23	18	5	0.28
4#	Black+3%SMP-2+3%SFT+2%SMC	5.2	23	17	6	0.35
5#	Black+3%SMP-2+3%SFT+2%SMC+5%KCl+0.05%K-PAM	5.2	22	16	6	0.38
6#	Black+3%SMP-2+3%SFT+2%SMC+7%KCl+0.05%K-PAM	5.0	25	18	7	0.38

2.2 Optimizing system transformation points

The well transformation point completed in 2020 was Zhifang-Heshanggou group. It is found that the lower part of the Yanchang collapses seriously as displayed by electrical measurement and the drilling encounters resistance. Therefore, the transformation point is optimized to enter 400 m into the Yanchang so as to control filter loss, enhance inhibition, and prevent the collapse of the Yanchang in advance.

Table. 4 Statistics of exploration well transfer system of Yanchi block in 2021

Well	Transformation point	Post-transformation performance
Li 81	2960 m (Yanchang)	1.04/36/5.8/9
Li 76	1970 m (Yan'an)	1.04/40/5.4/9
Li 77	2670 m (Yanchang)	1.06/41/5.0/9
Li 88	2750 m (Yanchang)	1.04/39/4.4/9
Li 96	2760 m (Yanchang)	1.05/40/5.2/9
Li 97	2650 m (Yanchang)	1.08/40/5.4/9
Li 87	2600 m (Yanchang)	1.06/41/4.8/9

2.3 Standardizing the performance of each stratum and strengthening the daily performance monitoring.

During the system transformation, it is the first time that KCl was added by 5%, and the content was gradually increased according to the field drilling fluid in the Yanchang group and Zhifang group. Subsequently, the drilling fluid performance of each stratum was standardized (seeing in Table 5). After entering Shiqianfeng group, the system enhances the plugging, the content of prehydrated clay reaches more than 3%, the content of SFT reaches 5%, filter loss (FL) is less than or equal to 5 mL, filter loss under high temperature and high pressure (HTHP-FL) is less than 16 mL. The strong system plugging is used to reduce the stratum collapse pressure.

Table. 5 Performance requirements of exploration Wells in the Yanchi block

Performance	Technical requirement
Density	After transformation: 1.10 g/cm³; Yanchang: 1.15g/cm³; Liujiagou: 1.17-1.18 g/cm³; Shiqianfeng: ≥1.18g/cm³; Before taking the core: ≥1.22 g/cm³; Finishing drilling: 1.23-1.26 g/cm³
FL	Yanchang-Liujiagou: ≤6 mL; Shiqianfeng-Shihezi: ≤5 mL; Shanxi-finish drill: ≤4mL
HTHP-FL	Yanchang-Shihezi: ≤20 mL; Shanxi-finish drill: ≤15 mL
Bentonite content	Yanchang-Shihezi: 20-30 g/L; Shanxi-finish drill: 30-40 g/L

2.4 Transforming joint management of collapse and leakage into separate management of collapse and leakage

After the system transformation, the plugging agent was blocked with the drill, select particles, sheets and fibers, and different types to gradually improve the pressure capacity of the stratum, and add the drill plugging agent by 1-2%. Preequipped 30 m^3 plugging slurry (8-10%), it was timely pumped into the well for plugging leakage. When plugging effect is not good, cement is used to plug the leakage. For the leakage layer that appears during the drilling period, it must be completely blocked and then continue to drill, to avoid the recurrence of the leakage layer. After drilling through Liujiagou group, the pressure plugging was carried on, and the drilling fluid density should be greater than 1.20 g/cm^3. If the requirements are not met, no drilling will be conducted, and the light drill rod is started for plugging.

Table. 6 Blocking situation of exploration well in the Yanchi block

Well	Leakage depth	Stratum	Lease speed	Prformance	Treatment measure
Li 81	4551	Shihezi	10-15	1.20/52/7.8/9	Plugging slurry
Li 76	3169	Yanchang	Lost back	1.09/49/5.6/9	Plugging slurry while drilling
	3731	Liujiagou	20	1.17/52/5.4/9	Plugging slurry while drilling Cemented once in Yanchang and Liujiagou
Li 77	3144	Yanchang	40-50	1.15/46/5.2/9	Cemented once under light drill rod
	4319	Shanxi	20	1.24/70/4.8/9	Cemented once in Shanxi and Liujiagou
Li 88	3478	Zhifang	80	1.16/43/5.6/9	Cemented once
Li 97	3370	Zhifang	Lost back	1.11/44/5.6/9	Cemented once under Light drill rod
	3541	Heshanggou	10	1.10/44/5.8/9	Cemented once under light drill rod

Table. 7 Pressure-bearing and leakage plugging of exploration wells in the Yanchi block

Well	Pressure-bearing situation
Li 76	At Shiqianfeng with plugging slurry for pressure bearing, failed to bear the pressure, cemented in Liujiagou, and the density is increased to 1.23 g/cm^3
Li 77	Leakage plugging under pressure at Liujiagou, and the density is increased to 1.22-1.23 g/cm^3 during drilling, and the consumption is 1.6-2.4 m^3/h
Li 81	Leakage plugging under pressure at Liujiagou, the density is 1.22 g/cm^3, and the density is 1.21 g/cm^3 before coring
Li 88	Leakage plugging under pressure at Liujiagou, and the density is 1.24 g/cm^3, the density is 1.23 g/cm^3 before coring, plugging while drilling
Li 87	Leakage plugging under pressure at Liujiagou, and the density is 1.23 g/cm^3, the density is 1.23 g/cm^3 before coring
Li 96	Leakage plugging under pressure at Liujiagou, and the density is 1.25 g/cm^3, the density is 1.23 g/cm^3 before coring
Li 97	Leakage plugging under pressure at Liujiagou, and the density is 1.24 g/cm^3, the density is 1.24 g/cm^3 before coring

2.5 Making full use of small experiments to understand the efficacy of each treatment agent, fine to adjust the performance

The need to adjust the performance of the original slurry was to first carry out indoor experiments, according to the targeted formula adjustment of the experimental results, to achieve fine maintenance. Li 76 well was drilled to 3300 m, and the block appeared in the Yanchang, and the performance of drilling fluid was 1.16/45/6.2/9. The adjustment idea is to reduce the filter loss first, then increase the density, carry out small experiments to optimize the formula, and select the treatment agent, and finally get better results.

Table. 8 Performance adjustment experiment of Li 76 well

Order number	Formula	FL (mL)	HTHP-FL (mL)
1	Black	6.2	21.6
2	Black+1%SMP-2+1%SFT	6.0	20.2
3	Black+1.5%SMP-2+1%SMC	5.8	19.6
4	Black+1%SMP-2+1%SFT+0.5%SMC+1% ZDS+1% prehydrated clay	4.8	15.6

3. Field application

By optimizing the system, the potassium polysulfonic acid system has relatively stable performance, good salt and calcium resistance, strong inhibition and good sealing performance, and meets the construction requirements of gas exploration wells in Yanchi block, and there is no sudden change in performance.

Table. 9 Statistics of core extraction and finished drilling performance in Yanchi block in 2021

Well	Core taking performance	Before taking the core HTHP-FL (mL)	Finish drilling performance	Finishing drilling HTHP-FL (mL)
Li 81	1.20/52/6.0/9	18.8	1.23/62/5.8/9	17.6
Li 76	1.22/62/4.8/9	14.2	1.23/81/4.4/9	12.8
Li 77	1.23/65/4.8/9	14.8	1.24/85/4.2/9	13.2
Li 88	1.24/65/5.0/9	15.6	1.25/83/4.6/9	14.2
Li 96	1.24/63/4.8/9	16.2	1.25/86/4.4/9	14.6
Li 97	1.23/68/4.8/9	15.8	1.25/88/4.6/9	14.4
Li 87	1.24/66/4.4/9	15.2	1.25/85/4.2/9	13.4

Through the transfer system in the middle of the Yanchang, the downhole collapse is greatly reduced, no serious jamming phenomenon occurs, and the phenomenon of drilling and tripping jamming is less.

From the completion of the well coring to the completion of the trip, the drilling is normal. Among them, 5 wells are opened for the second time to the completion of the trip without encountering obstacles, and the average well diameter after the electrical measurement does not exceed 10%.

Table. 10 Statistics of average well diameter in Yanchi block in 2021

Well	Finishing drilling depth (m)	Finishing drilling stratum	Average well diameter expansion rate (%)
Li 96	4435	Krimori	5.23
Li 81	4830	Krimori	3.63
Li 88	4690	Krimori	6.36
Li 76	4510	Krimori	5.22
Li 77	4560	Krimori	6.11
Li 97	4470	Krimori	4.98
Li 87	4260	Krimori	4.59

Through the measures of plugging while drilling in advance and dividing and controlling the collapse and leakage, there is no well collapse caused by well leakage, and the success rate of the plugging has been greatly improved. The average plugging time for a single well was shortened by 106.3 h compared with 2020, a decrease of 44.29%, and the success rate of one-time plugging was 100%.

Table. 11 Statistics of plugging of construction Wells in Yanchi block in 2021

Time	Well number	Number of leakage wells	Plugging time time (h)	Average plugging time in a single well (h)
2021	7	5	668.5	133.7
2020	6	6	1440	240
Comparison			-771.5	-106.3（-44.29%）

The overall speed-up effect is obvious. Compared with 2020, the average drilling cycle is shortened by 19.69 days, the well construction cycle is shortened by 21.83 days, the monthly speed of drilling rigs is increased by 50.15%, and the complexity rate is reduced by 54.8%.

Table. 12 Comparison of conventional gas exploration Wells in Yanchi block in 2021

Time	Finishing drilling numbe	Average well depth (m)	Average drilling cycle (d)	Average well building cycle (d)	Average monthly speed of drilling rig (m/ty)	Complex rate (%)
2021	7	4536	41.34	65.19	2494	5.79
2020	6	4299	61.03	87.02	1661	12.81
Comparison	+1	+237	-19.69	-21.83	+833	-7.02
%			-32.26	-25.09	50.15	-54.8

4. Conclusions

In this paper, the poly-sulfur system has good inhibition, blocking nature, strong salt and calcium resistance ability and stable performance, which meets the safe construction of gas exploration wells in Yanchi block. The advance transformation system can effectively prevent the collapse of the stratum above the Yanchang group and greatly reduce the complex downhole aging. Precise maintenance of drilling fluid performance and adjustment of performance according to the timely characteristics of the stratum can greatly reduce the complexity of downhole failures. By plugging leakage while drilling in advance, gradually increasing stratum pressure, dividing and controlling collapse and leakage, and plugging leakage under pressure in Liujiagou, the contradiction between collapse and leakage can be alleviated in essence, and the drilling speed can be accelerated. When adjusting the performance of the system, it is very necessary to carry out small-scale experiments to avoid blind operation, which will cause the performance to be unadjusted for a long time and cause downhole complexity.

References

1. Shi CD, Wang WQ, Shi ZM, et al. Research on key drilling technologies of deep shale gas horizontal wells in Yanchi block[J]. Petroleum Drilling Techniques, 2021,49 (6): 23-28.
2. Jing WF, Yan RH, Liu T, et al. Identification method of complex oil-water relationship reservoir logging in Yanchi block of northern Shaanxi[J]. Mud logging engineering, 2020, 31(3): 99-104.
3. Wang H, Wei Y, Zhao Y, et al. Research and application of poly drilling fluid system[J]. Liaoning Chemical Industry, 2017, 46(10): 994-996.
4. Han ZB, Liu HB, Zhang JT, et al. Study on the mechanical properties and well wall stability of deep brittle shale[J]. Special oil and gas reservoir, 2020, 27(5): 167-174.
5. Liu YZ, Gang WZ, Chen G, et al. Geochemical characteristics of aromatic hydrocarbons in the 7th section in Ordos Basin[J]. Acta Sedimentologica Sinica, 2018, 36(4): 818-828.
6. Liu W, He L, Hu DL, et al. Key technology of Marine deep shale gas drilling in southern Sichuan[J]. Petroleum Drilling Techniques, 2019,47(6): 9-14.
7. Jiang YQ, Fu YH, Xie J, et al. Development trend and comprehensive evaluation system of Marine shale gas reservoir evaluation[J]. Natural Gas Industry, 2019, 39(10): 1-9.
8. Zhao SJ, You YW, Liu HB, et al. High-performance water-based drilling fluid technology in the horizontal section of Fuling jiaoye well 18-10HF well[J]. Drilling fluid & completion fluid, 2019, 36(5): 564-569.
9. Li MZ. Stainding technology of horizontal well pipe in Hangjin banner region of Ordos Basin[J]. Fault-Block Oil & Gas Field, 2018, 25(5): 661-664.

10. Lin YX, Zhen JW. Horizontal well water-based drilling fluid technology of deep shale gas in Weiyuan block[J]. Petroleum Drilling Techniques, 2019, 47(2): 21-27.
11. Ning YP, Cui YC, Xue B, et al. Improvement and application of drilling fluid of natural gas exploration well in Yanchang block of Ordos Basin[J]. Drilling fluid and completion fluid, 2009, 26(2): 126-128+140.
12. Fan HF, Zang YB, Zhang JC, et al. Technical difficulties and countermeasures of deep shale gas drilling[J]. Drilling & Production Technology, 2019,42(3): 20-23+7.

The effect of injection speed on the law of heavy oil initiation and seepage flow research

Jiajia Zhu, Jing Liu, Zhengbin Li, Yimin Jiang, Zhihao Jiang, Xiujie Liu, Zhiqiang Zhao

College of Petroleum Engineering, China University of Petroleum (East China), Qingdao, Shandong 266580, China

Abstract. In order to make the heavy oil flow, the repulsion pressure gradient should be larger than the start-up pressure gradient. The results show that the smaller the injection speed is, the smaller the start pressure gradient is, and the subsequent seepage process has a smaller and smoother replacement pressure gradient, and there is an effective limit of the start pressure gradient. Using the non-stationary method to calculate the relative oil-water permeability and water saturation, the bound water saturation does not differ much with increasing injection speed, and the residual oil saturation increases. Therefore, in order to start the heavy oil more easily in the mine production can be started at a lower production rate, the study of injection speed is important for the heavy oil start-up and development effect.

Keywords: Heavy oil; injection speed; initiation pressure gradient; seepage characteristics.

1. Introduction

At present, heavy oil resources account for about two-thirds of the global oil and gas resources reserves[1]. The development of heavy oil reservoirs occupies a very important position in the development of China's oil fields and has a high development potential.[2] The development of heavy oil reservoirs occupies a very important position in the development of China's oil fields and has high development potential. Although China's heavy oil reservoir resources are abundant, the reservoir environment is complex, with high viscosity, poor mobility, low permeability, and low reservoir permeability.[3][4] It is characterized by low permeability and complex pore structure of the reservoir. Heavy oil is rich in gum, asphaltene and other macromolecular compounds, and the molecular structure of heavy oil is an inhomogeneous colloidal dispersion system, which leads to the characteristics of heavy oil fluids with high viscosity and non-Newtonian fluid characteristics in the flow, and the seepage characteristics in porous media are different from those of conventional crude oil, and the seepage law changes, generally showing non-linear seepage[5] . There may be a start-up pressure gradient, and the heavy oil can start to flow only when the displacement pressure gradient exceeds the start-up pressure gradient.[6][7] The start-up pressure gradient may exist. The domestic research on heavy oil start-up pressure gradient is limited but not deep enough, although the relationship between start-up pressure gradient and flow in heavy oil reservoirs of low permeability carbonate rocks has been explained[8] , but there is a lack of research on the law of the effect of injection speed on heavy oil start-up and and seepage flow, while start-up pressure gradient is an important link to study nonlinear seepage flow. At present, the methods to measure the start-up pressure gradient of heavy oil include indoor physical simulation experiments, numerical experiments and well test interpretation. The main methods to measure the start-up pressure by indoor physical simulation experiments are: steady-state method, unsteady-state method, capillary equilibrium method, bubble method, etc.[9][10][11] Compared with other methods, the bubble method is faster and more accurate. In this experiment, the bubble method is used to determine the pressure gradient of heavy oil start-up pressure. Then the relative permeability curve is calculated by the non-stationary method.[12][13] The study of oil-water phase permeability law is the key to reservoir development. The study of the effect of injection speed on the start-up and seepage of heavy oil is important for the efficient development of heavy oil reservoirs, which can improve the production and recovery of heavy oil[14] , and also provide the theoretical basis for the development of reasonable heavy oil reservoir exploitation plan in the later stage.

2. Bubble method experiment

2.1 Experimental apparatus and materials

Experimental instruments: artificial core making device, Jiangsu Hongbo Machinery Manufacturing Co., Ltd; advection pump, Shanghai Sanwei Scientific Instruments Co., Ltd; sand filling tube, Nantong Xinhua Cheng Scientific Research Instruments Co., Ltd; constant temperature box (HX- Ⅱ type), Nantong Huaxing Petroleum Instruments Co., Ltd; electronic balance, intermediate vessel, etc.

Experimental materials: simulated formation water, formation dewatered crude oil (a domestic oil field, basic properties as shown in Table 1), petroleum ether, kerosene, etc.

Table 1. Basic data of heavy oil test samples

number	Sampling wells	Saturation fraction/%	Aromatic fraction/%	Colloid /%	Asphaltene /%	Viscosity (50℃)/ mPa·s
1#	EZwell	47.65	15.49	19.09	0.48	328

The physical parameters of the artificial cores are shown in Table 2.

Table 2. Table of core physical parameters

Artificial rock core	Length /cm	Diameter /cm	Porosity /%	Permeability /×10^{-3}μm^2
Meso-osmosis	40	2.5	39.8	500

2.2 Experimental methods

(1) Injection speed determination

According to the actual production of the reservoir, the seepage velocity at the casing section of the well was taken as the basis for the injection velocity of the sand-filled pipe experiment, and the variation of fluid seepage velocity with radius at the section around the well was calculated as shown in Figure 1. By comparing the experimental settings of injection speed and seepage velocity (Figure 2), the injection speed was set to 0.25 mL-min^{-1} , 0.5 mL-min^{-1} , and 1 mL-min^{-1}.

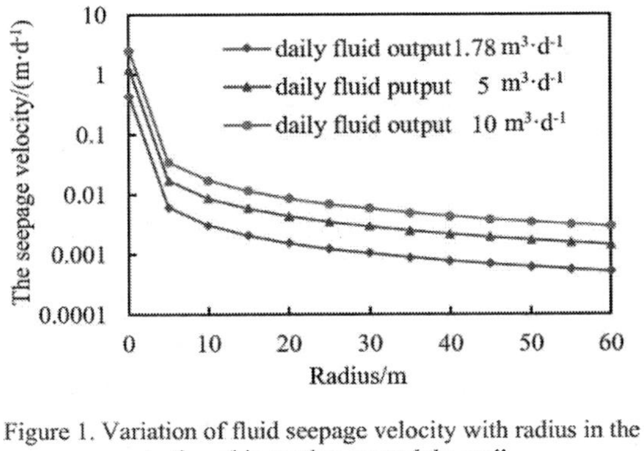

Figure 1. Variation of fluid seepage velocity with radius in the stratigraphic section around the well

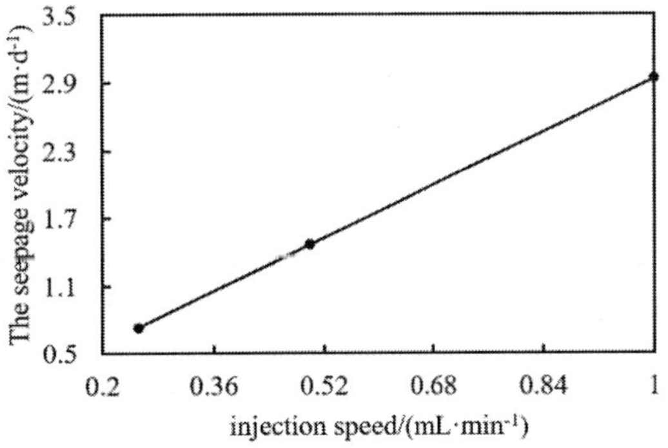

Figure 2. Experimental setup of injection velocity versus seepage velocity

(2) Start-up stress test method

In this experiment, the bubble method (Figure 3) is used. By setting a certain injection speed to repel the core, a small amount of liquid is injected into the line at the outlet end to form droplets (using methyl blue staining for easy observation), and when the droplets in the line start moving, the pressure at the injection end is read as the initiation pressure and the time used is the initiation time.

Figure 3. Air bubble method to measure the starting pressure

(3) Experimental protocol

Three sets of injection speeds were set: 0.25 mL-min^{-1}, 0.5 mL-min^{-1}, and 1 mL-min^{-1} to investigate the effect of injection speed on the start-up and seepage characteristics of heavy oil.

2.3 Experimental steps

The experimental flow chart is shown in Figure 4, and the experimental steps are.

(1) Record the basic data of the selected sand-filled pipe core: length, cross-sectional area, etc.

(2) The fabricated sand-filled tube model was placed on an electronic scale, weighed, and the dry weight was recorded.

(3) Place the sand-filled tube in a thermostat, constant temperature at 30℃, add back pressure at the outlet end, inject saturated water at the injection speed of 0.25 mL-min^{-1} until the pressure gradient is stable, then record the pressure difference and put it on an electronic scale and weigh the wet weight.

(4) Put the sand-filled tube core in the thermostat again for saturated oil operation until no water comes out at the outlet, at which time the water in the sand-filled tube model is the bound water, and the volume of the water driven out is the volume of saturated in oil.

 https://doi.org/10.1051/e3sconf/202235801022

(5) In the thermostat, the injection speed was set to 0.25 mL-min^{-1} , 0.5 mL-min^{-1} , and 1 mL-min^{-1} in order to repel the cores, and the starting pressure gradient and starting time were measured by the bubble method and recorded.
(6) The temperature of the thermostat was kept at 30℃, the permeability of the core was 500×10^{-3} μm^2 , and the saturated core of 1# oil sample (viscosity 936 mPa-s at 30℃) was selected, and the injection speeds were set to 0.25 mL-min^{-1} , 0.5 mL-min^{-1} , and 1 mL-min^{-1} , respectively. The oil and fluid production were recorded by a stopwatch, a pressure transducer, and a measuring cylinder.

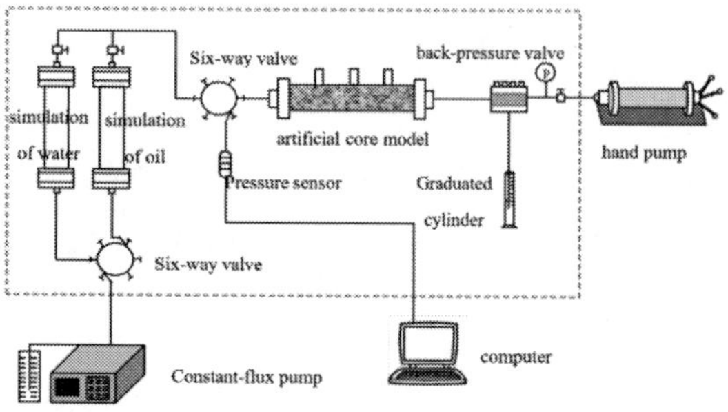

Figure 4. Experimental flow chart

3. Experimental results and discussion

3.1 Influence of different injection speeds on the start-up of heavy oil

The experimental results are shown in Figure 5. It can be seen that, in the core of the same permeability, with the increase of injection speed, the start-up time of heavy oil decreases, and the decrease increases gradually; while the start-up pressure gradient increases with the increase of injection speed, and the increase of magnitude also increases accordingly; when the injection speed increases from 0.25 mL-min^{-1} to 0.5 mL-min^{-1} , the change of start-up pressure gradient and start-up time is small, but from 0.5 mL- min^{-1} to 1 mL-min^{-1} , the starting pressure gradient increased by 9.8 times and the starting time decreased by only 1.3 times, indicating the existence of an effective starting pressure gradient limit.

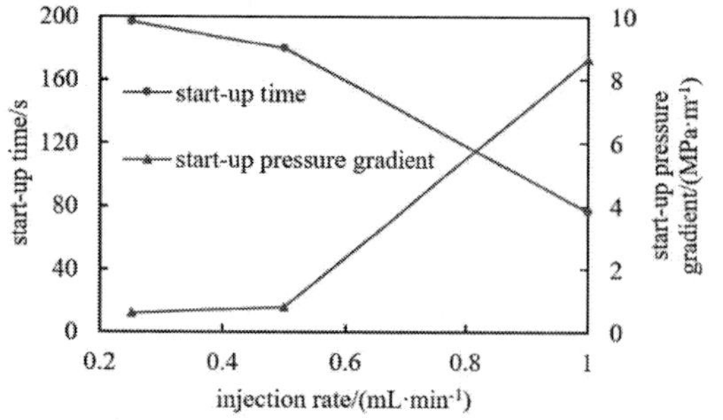

Figure 5 Start-up of heavy oil at different injection speeds

Figure 6 shows the changes of the repulsion pressure gradient with the number of PV injected at different injection speeds when the permeability is certain, it can be seen that as the number of PV injected increases the pressure gradient gradually rises to the maximum value and then begins to decline, and after reaching a certain magnitude, it gradually stabilizes. This indicates that the heavy oil has not yet started at the beginning, when the repulsion pressure gradient is larger than the heavy oil start pressure gradient, the heavy oil starts to start, and after starting the repulsion pressure gradient starts to decline and gradually reaches a stable level; and the larger the injection speed, the larger the pressure gradient at stabilization.

As in the heavy oil flow, heavy oil flow resistance mainly comes from internal friction (between oil molecules, between gum molecules, between asphaltene dispersion phase, between oil molecules and gum molecules, between oil molecules and asphaltene dispersion phase, between gum molecules and asphaltene dispersion phase, etc.) and interfacial tension[15] . In the heavy oil from stationary to flowing state, the frictional resistance changes from static friction to kinetic friction, and the molecular structure in the heavy oil is damaged to a certain extent, so that the kinetic friction is smaller than the static friction, so after the heavy oil is started, the repulsion pressure gradient needs to drop by a certain amount.

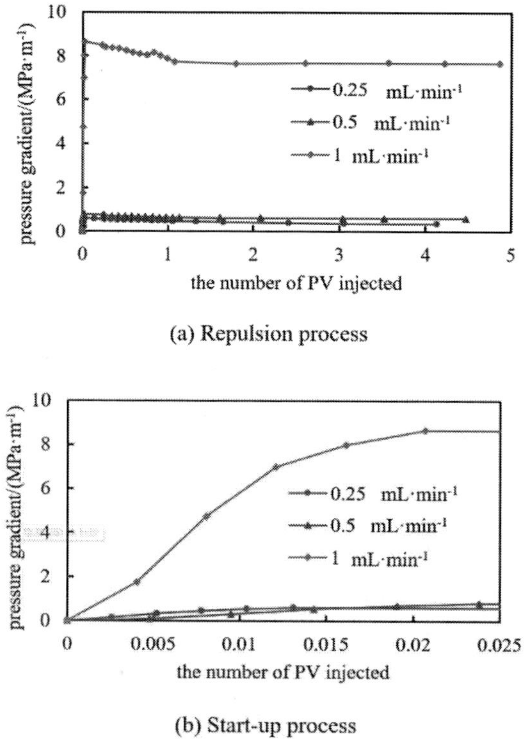

Figure 6. Variation of repulsion pressure gradient at different injection speeds

Figure 7. shows the relationship between the ramp-up rate and pressure gradient in the ramp-up stage. It can be seen that there is a good linear relationship between the pressure gradient and the ramp-up rate, indicating that when the ramp-up rate is too fast, the heavy oil in the core does not have time to flow, resulting in a higher start-up pressure. In order to be able to start the heavy oil more easily in the mine production can be started at a lower production rate.

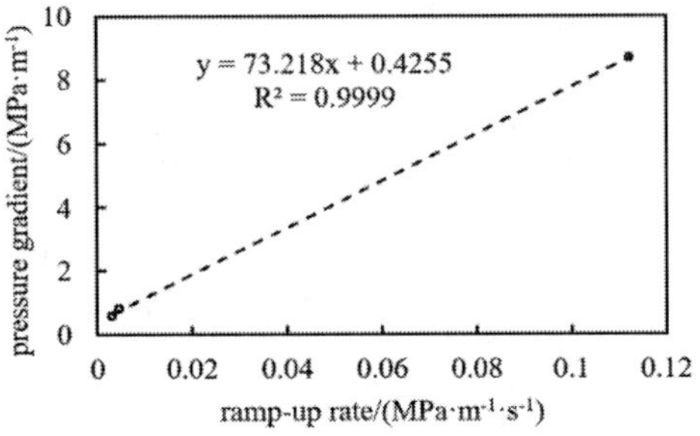

Figure 7. boost rate versus pressure gradient

3.2 Influence of injection speed on the seepage characteristics of heavy oil

The relative permeability curves calculated using the non-stationary method are shown in Figure 8. It can be seen that: with the increase of injection speed, the iso-permeability point of oil-water relative permeability curve moves to the left and the two-phase co-permeability zone decreases. Figure 9 shows the bound water saturation and residual oil saturation at different injection speeds. It can be seen that the bound water saturation does not differ much with the increase of injection speed, and the residual oil saturation increases.

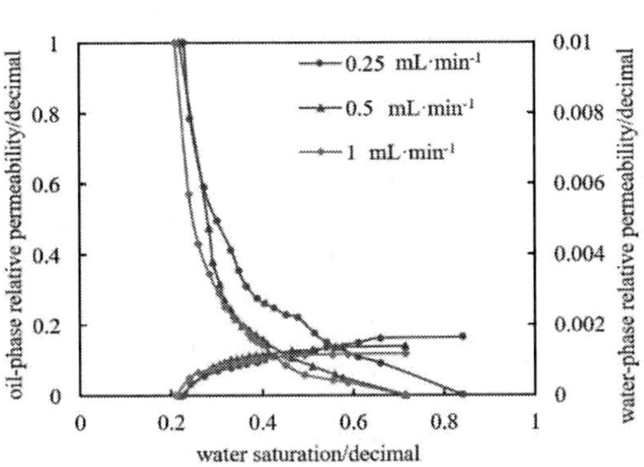

Figure 8. Oil-water relative permeability curves at different injection speeds

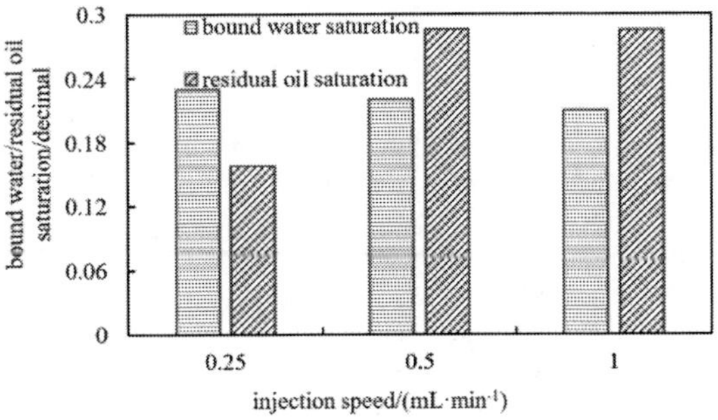

Figure 9. Bound water saturation and residual oil saturation at different injection speeds

Figure 10 shows the time to see water and the degree of water-free recovery at different injection speeds, and Figure 11 shows the recovery rate at different injection speeds. It can be seen that as the injection speed increases, the time to see water becomes shorter, the water-free recovery rate first becomes faster and then stabilizes, and the water drive recovery rate decreases. Since the larger the injection speed, the larger the corresponding replacement pressure difference, the faster the advancement speed of the leading edge of the water drive, and the faster the time to water.

Therefore, the injection speed (replacement pressure difference) should be properly controlled to slow down the advancing speed of the leading edge of the water drive, so as to increase the wave area and achieve late water or low water content in the recovery fluid as much as possible, thus indicating that a reasonable injection speed is conducive to improving the recovery rate of heavy oil reservoirs.

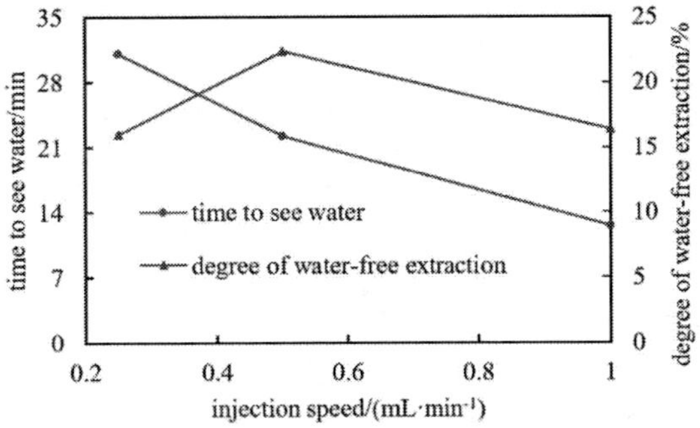

Figure 10. Time to see water and degree of water-free extraction at different injection speeds

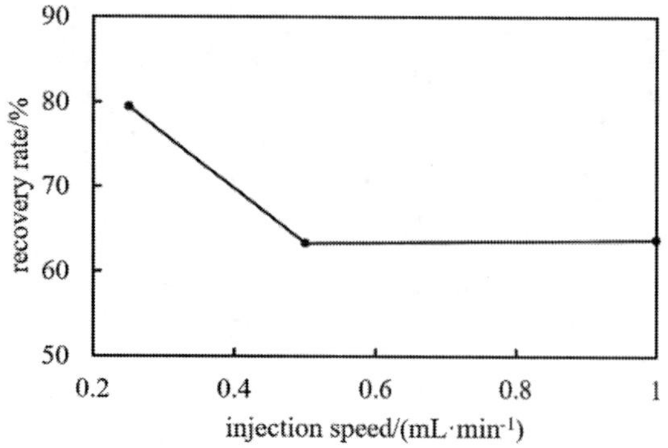

Figure 11. Recovery rate at different injection speeds

4. Conclusion

The indoor sand-filled pipe replacement experiment simulates the start-up process and seepage development effect of heavy oil under certain environment, and the experimental results show that.

(1) The smaller the injection speed, the smaller the start-up pressure gradient, and the smaller and smoother the repulsion pressure gradient in the subsequent seepage process. There is a good linear relationship between the pressure gradient and the ramp rate at different injection speeds. When the ramp rate is too fast, the heavy oil in the core does not have time to flow, resulting in a higher start-up pressure. In order to start the heavy oil more easily in the mine production, it can be started at a lower production rate.

(2) Using the non-stationary method to calculate the oil-water relative permeability curve, the heavy oil percolation law deviates from Darcy's law, and the oil drive efficiency becomes better as the injection speed decreases, which is expressed in the phase percolation curve as the area of oil-water two-phase co-percolation zone increases, the bound water saturation does not differ much, and the residual oil saturation decreases. Try to achieve late water or low water content in the recovery fluid, while a reasonable injection speed is beneficial to improve the recovery rate of heavy oil reservoir.

References

1. Zhang Wenrui, Dai Yujie, Liu Guiman. Numerical simulation of heavy oil seepage pressure field in annular well network[J]. Contemporary Chemical Industry, 2022, 51(01): 196-201+205.
2. Jiang RZ, Ni QD, Zhang CHG, et al. Stress-sensitive seepage model and numerical simulation of variable initiation pressure gradient in heavy oil reservoirs[J]. Oil and Gas Geology and Recovery, 2021, 28(06): 54-62.
3. Jiao Huan, Dai Yujie, Wang Xuehui, et al. Current status and development trend of heavy oil seepage research[J]. Petrochemicals, 2019, 48(12) : 1283-1288.
4. Jia Yan. Research on viscosity reduction and efficiency enhancement process technology for heavy oil extraction[J]. Petrochemical Technology, 2019, 26(11): 295-296.
5. Wang Bu'e, Zheng Huiguang. Experiments on improving the rheological properties of heavy oil seepage[J]. Petroleum and Natural Gas Geology, 2005(03): 370-373.
6. Jeffery Kamov, Wang Chuan-Yu. Anomalous oil [M]. Petroleum Industry Press, 1983.
7. Sun Jianfang. Experimental study on single-phase seepage characteristics of super-heavy oil in Shengli Zheng411 block[J]. Petroleum Drilling Technology, 2011, 39(06): 86-90.
8. Cao LY, Liu CH, Sun JF, et al. Study on the initiation pressure gradient of low-permeability carbonate heavy oil reservoirs[J]. Science, Technology and Engineering, 2015, 15(01): 215-219.
9. Zhang Lei, Liu Chao, Chen Jianbo, et al.. Calculation of start-up pressure gradient in heavy oil reservoirs using dynamic data[J]. China Offshore Oil and Gas, 2016, 28(03): 101-105.
10. Shem, C.X., Wang, D.C.. Experimental study on fluid start-up pressure gradient[J]. Oil and Gas Field Surface Engineering, 2010, 29(04): 30-32.
11. Ke W-L, Yu G-M, Zhou W-S, et al. Experimental study of nonlinear seepage initiation pressure gradient in heavy oil[J]. Petroleum Drilling Process, 2016, 38(03): 341-346.
12. Liu Dongqing, Wang Shantang, Bai Yanli, et al. Research and application of seepage mechanism of Shengli heavy oil[J]. Inner Mongolia Petrochemical, 2012, 38(02): 3-6.
13. Chen Z, Yin YP, Chen H. Exploration of the non-stationary method for calculating oil-water relative permeability[J]. Fractured Block Oil and Gas Field, 2005(01): 41-43+91.
14. Fu Yunchuan, Wang Chengsheng, Zhou Winchao, et al. Study on oil-water seepage law between multilayer combined recovery layers in offshore heavy oil fields[J]. Contemporary Chemical Industry,2020,49(05):798-801.
15. Chen Zhaojun. Design and Application of Flow Improver for Heavy Oil in Tahe, Xinjiang[D]. China University of Petroleum (East China), 2014.

Quantitative analysis of the driving factors of carbon emissions and the evolution of carbon neutrality

Ben Niu[1], Qidi Cao[2], Zijian Xu[3], Yujie Zhang[4,*]

[1] School of mechanical engineering and automation, Dalian Polytechnic University, Dalian, 116034, Liaoning Province, China.
[2] School of Information Science and Engineering, Dalian Polytechnic University, Dalian, 116034, Liaoning Province, China.
[3] School of Information Science and Engineering, Dalian Polytechnic University, Dalian, 116034, Liaoning Province, China.
[4] Dalian Polytechnic University, Dalian, 116034, Liaoning Province, China.

Abstract. Based on the selection of quantitative indicators and divisions, this paper conducts an overall quantitative analysis. And some hard-to-measure energy structure and industrial structure carbon emission driving factors, combined with the laws of carbon emission and absorption are analyzed. According to the definition and measurement standards of low-carbon economic development, we select 18 energy structure indicators and 47 industrial structure indicators for quantitative analysis. The driving factors of the energy structure are mainly carbon emission intensity and energy conversion efficiency. The driving factors of industrial structure are mainly divided into industrial proportion and economic scale. The rapid growth of GDP level is asynchronous with the total amount of carbon emissions, and gradually shows a state of "relative decoupling"

1. Brief introduction

1.1 Background

In order to save energy, promote the transformation of scientific and technological achievements, and enhance the scientific and technological content of our economy. At the same time, the state also attaches great importance to it and will adopt more powerful policies and measures to promote energy conservation and emission reduction. While energy conservation is important, emission reduction cannot be ignored. Carbon neutrality is a "hot topic" corresponding to the direction of emission reduction. It refers to the total amount of carbon dioxide or greenhouse gas emissions directly or indirectly produced by a country, enterprise, or individual in a certain period. By using low-carbon energy to replace fossil fuels, afforestation, energy conservation and emission reduction, etc., it can offset its own carbon dioxide or greenhouse gas emissions. In the end, positive and negative offsets are achieved to achieve relatively "zero emissions" measures.[1]

In the current era of advocating innovative green development, each country has also made great efforts to reduce emissions in its own direction and policy. As of December 2020, many countries have announced their intentions and targets for net-zero emissions. Six of these countries have completed legislation regulating net-zero emissions by 2050, including the UK, France, Denmark, Sweden, New Zealand, and Hungary. Six countries or regions have proposed draft legislation, namely the European Union, Canada, South Korea, Spain, Chile and Fiji. An additional 14 countries are on the policy agenda. Of course, our country's position is also very clear. my country has also written the development goal of "promoting green development and promoting the harmonious coexistence of man and nature" into the 14th Five-Year Plan for my country's national economic and social development and the outline of the 2035 long-term goals. It is enough to see the importance of "carbon neutral" promotion.

2. Driver and Outcome Relationships

2.1 Data analysis method

The data in this paper comes from the carbon emission statistics from 2000 to 2018 provided in the China Carbon Accounting Database (CEADs). The data includes energy structure parameters such as raw coal, clean coal, and other washed coal. The comprehensive statistics in this paper are divided into coal, oil and natural gas for data analysis.

According to the carbon emission calculation guidelines provided by IPCC, combined with my country's characteristics, the calculation formula is:

$$C = \sum_{N=1}^{3} E_N \times \eta_N$$

Among them, C is the carbon emission, E_N is the N th energy consumption, and η_N is the N th energy emission coefficient.

The unit is million tons.

* Corresponding author: zhangyj@dlpu.edu.cn

https://doi.org/10.1051/e3sconf/202235801023

Table 1. Carbon emission coefficients of various energy sources

	coal	Petro	natural gas
carbon emission factor	0.75	0.57	0.5

Carbon emission intensity is the ratio of carbon emission per unit of GDP. The China Carbon Accounting Database (CEADs) contains energy consumption data for various industries. Therefore, the carbon emission of each industry is the product of the energy consumption of each industry and the number of carbon emission intensity.

2.2 Descriptive statistics

Since the 21st century, our country has developed rapidly. The following is an analysis of our country's energy structure and economic development trends by filtering some data through the data panel.

Table 2. Panel data statistics

item	2000	2002	2004	2006	2008	2010	2012	2014	2016	2018
coal (million tons)	2332.7	2714.7	3702.7	4858.8	5513.5	6463.5	7427.1	7586.5	7540.9	7540.9
Petro	590.7	660.6	845.1	962	1002.8	1129.9	1259.8	1382.9	1529.8	1529.8
natural gas	79.9	97	132.6	188.1	244.5	311.3	393.8	468.9	550.5	550.5
GDP/100 million	94730.61	114808	165350.7	228008.1	323016.4	431604.5	569046.5	675076	770420.3	902508.4
population	12.63	12.8	12.96	13.11	13.25	13.38	13.51	13.64	13.79	13.93
GDP per capita	243906.	291467	422540	563507	779301	1011208	1311193	1532378	1719189	1973385

Since the reform and opening up, our country's economy has developed rapidly. The rapid economic development is inseparable from the support of our country's energy structure and industrial structure. It can be seen from the data that carbon emissions increased rapidly from 2000 to 2013. Since the 2013-2018 reform and opening up, our country's economy has developed rapidly, and the rapid economic development cannot be separated from the support of my country's energy structure and industrial structure. It can be seen from the data that carbon emissions increased rapidly from 2000 to 2013, and the growth gradually flattened from 2013 to 2018. The development curve is shown in the figure below.

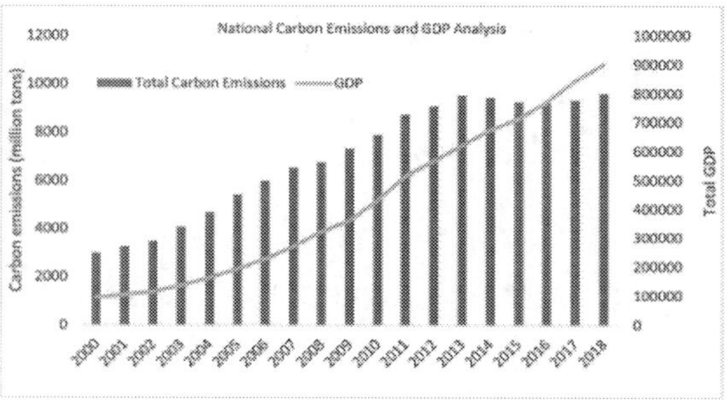

Fig. 1. National carbon emissions and GDP analysis

2.3 Analysis of the relationship between industrial structure and carbon emissions

2.3.1 Analysis of the relationship between economic scale and carbon emissions

Statistics on the relationship between my country's GDP and carbon emissions are shown in the figure below

Table 3. Relationship between GDP and carbon emissions

years	2000	2002	2004	2006	2008	2010	2012	2014	2016	2018
carbon consumption	3003	3472	4680.4	6008.7	6761	7904.5	9080.5	9438.4	9621.1	9621.1
GDP	94730	114807	165350	228008	323016	431604	569046	675076	770420	902508
carbon emission growth rate	8.21%	6.83%	17.67%	14.56%	15.40%	11.25%	8.95%	3.28%	8.47%	7.78%
GDP growth rate	9.70%	10.5%	18.06%	21.99%	18.04%	16.82%	19.34%	18.71%	11.77%	19.55%

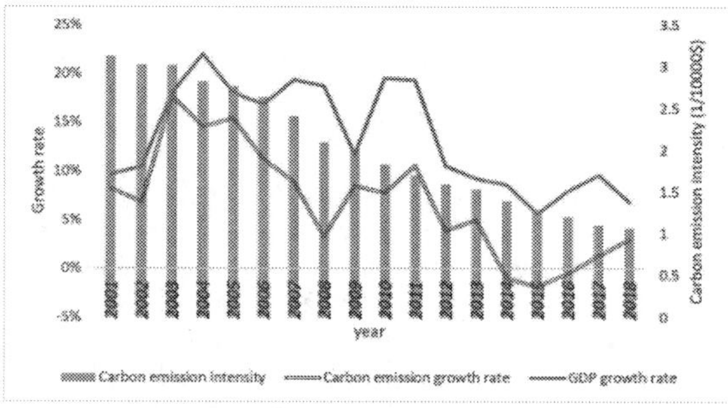

Fig. 2. Changes in carbon emissions, GDP, and carbon emission intensity

https://doi.org/10.1051/e3sconf/202235801023

In recent years, my country's economy has developed rapidly. From 2001 to 2008, the annual growth rate of national GDP reached 13.46%. The national GDP growth from 2001 to 2018 can be roughly divided into two stages: the first stage is the rapid growth stage from 2001 to 2011; the latter stage is the slow growth stage from 2012 to 2018. From 2001 to 2018, the average annual growth rate of carbon emissions was 6.83%. It has roughly gone through four stages. The first stage is 2001-2011, which is the stage of rapid growth. The second stage is 2011-2014, which is a slow growth stage. The third stage is 2015-2016, which is a negative growth stage. The last stage is the slow growth stage in 2017-2018.

Combining GDP growth rate and carbon emission growth rate, changes in carbon emission intensity can be divided into the following two stages. The first phase was a slow decline in 2001-2003. The second stage is the rapid decline stage from 2004 to 2018, with an average annual decline of 6.65%.

Among them, GDP and carbon emission values show high correlation characteristics. The correlation coefficient between the total GDP and the total carbon emission is R^2 =0.985, and the fluctuation is obviously asynchronous. However, it can be seen from the analysis that GDP growth has accelerated, while the growth rate of total carbon emissions has slowed down, and there is a negative growth. This shows a state of "relative decoupling" between GDP and carbon emissions.[2]

2.3.2 Industry proportion and carbon emission structure analysis

The data of each industry's share of the total economic output are as follows:

Table 4. The proportion of each industry

years	2000	2002	2004	2006	2008	2010	2012	2014	2016	2018
primary industry/%	17.74	16.10	14.18	12.42	11.79	10.35	9.93	9.40	9.22	8.21
Secondary industry/%	42.49	42.71	42.71	45.50	47.12	47.05	46.70	44.30	40.20	38.60
Tertiary Industry/%	36.54	37.96	39.88	38.85	37.86	39.38	40.14	43.08	47.36	49.96

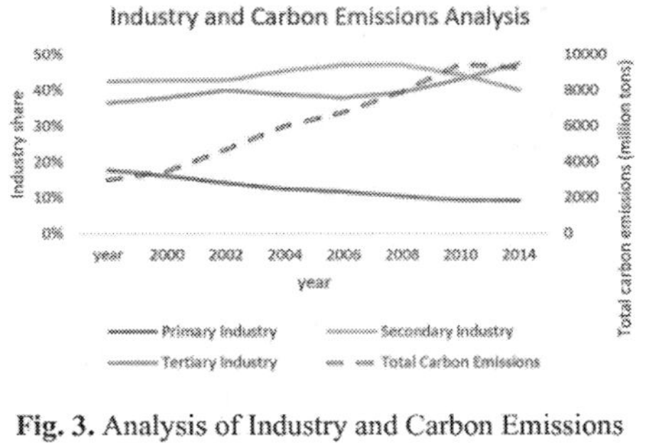

Fig. 3. Analysis of Industry and Carbon Emissions

After the third industry classification, agriculture, forestry, animal husbandry, fishery, and water conservancy are classified as primary industries. Industry and construction are the secondary industries. The transportation industry and warehousing industry belong to the tertiary industry. From 2000 to 2018, the proportion of the national primary industry decreased by 9.53%; the secondary industry decreased by 3.89%; the tertiary industry increased by 13.42%. The three industries reflect the rising trend of the tertiary industry and the downward trend of the primary and secondary industries. The trend is shown above. The industry with the highest carbon emissions is the secondary industry, with an annual consumption of 8,000 million tons after 2013. In recent years, the proportion of the secondary industry has continued to rise, the proportion of the primary industry has declined, and the secondary industry has shown a downward trend from 2010 to 2016. The average annual decrease of the primary industry was 1.06%; the average annual decrease of the secondary industry was 0.43%; the average annual increase of the tertiary industry was 1.49%.

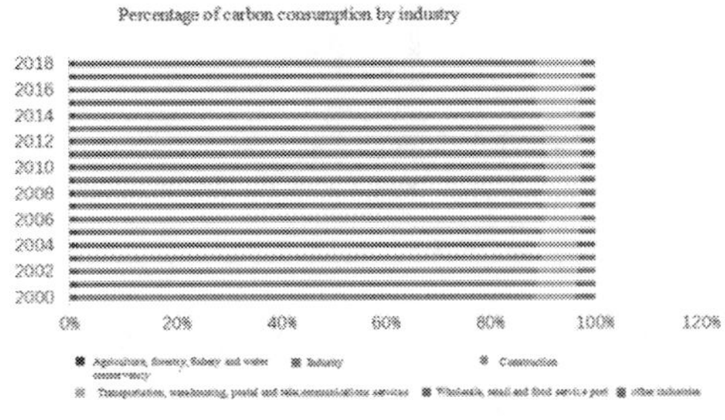

Fig. 4 Proportion of carbon consumption by industry

Among the three different industries, the secondary industry has the largest proportion of carbon consumption, accounting for about 89% of the total carbon consumption. Among them, in the secondary industry, industrial production is the main factor of carbon emissions. Among them, industry accounts for the largest proportion all year round, and the proportion is stable at close to 90%. The transportation industry, warehousing industry, and post and telecommunications service industry can be analyzed from the figure, and the proportion of 2000-2018 has increased slightly, and the proportion is close to 90%. From 2000 to 2013, industrial carbon consumption continued to grow steadily, and in 2013, carbon consumption reached 80 million tons.

2.4 Energy Structure and Carbon Emission Analysis

2.4.1 Analysis of energy types

Filter the energy structure information from the data panel, and analyze the statistical results as follows:

Table 5. Analysis of the proportion of various energy sources

years	2000	2002	2004	2006	2008	2010	2012	2014	2016	2018
coal	2332	2714	3702	4858	5513	6463	7427	7586	7240	7540
Petro	590	660	845	962	1002	1129	1259	1382	1485	1529
natural gas	79	97	132	188	244	311	393	468	491	550
Coal share	77.67%	78.18%	79.11%	80.86%	81.55%	81.77%	81.79%	80.38%	78.55%	78.38%
oil share	19.67%	19.02%	18.06%	16.01%	14.83%	14.29%	13.87%	14.65%	16.12%	15.90%
Natural gas proportion	2.66%	2.79%	2.83%	3.13%	3.62%	3.94%	4.34%	4.97%	5.33%	5.72%

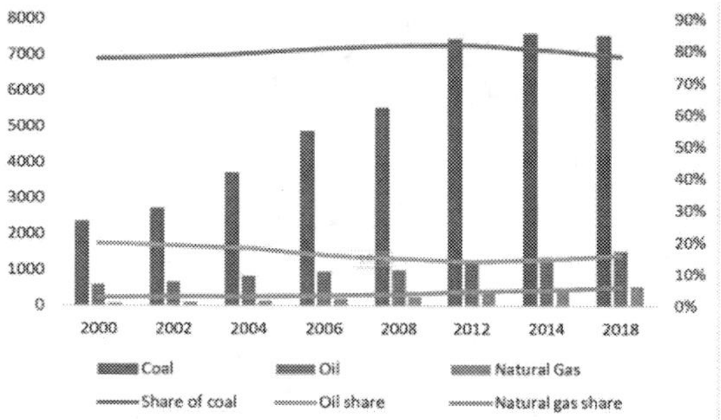

Fig. 5. Analysis of carbon emissions of various energy sources

Among them, coal carbon emissions account for about 80%. The proportion increased slowly from 2000 to 2012 and showed a downward trend from 2012 to 2018. Carbon emissions from petroleum-based energy showed a decreasing trend from 2000 to 2018. Carbon emissions from natural gas energy are on the rise.

2.4.2 *Analysis of the relationship between carbon emission intensity, energy conversion rate and carbon emission*

According to the Statistical Yearbook of Energy, the relationship between energy conversion rate, carbon intensity and total carbon emissions is shown below.

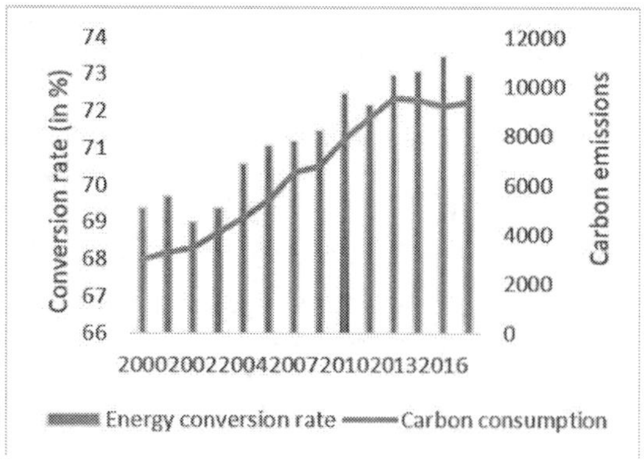

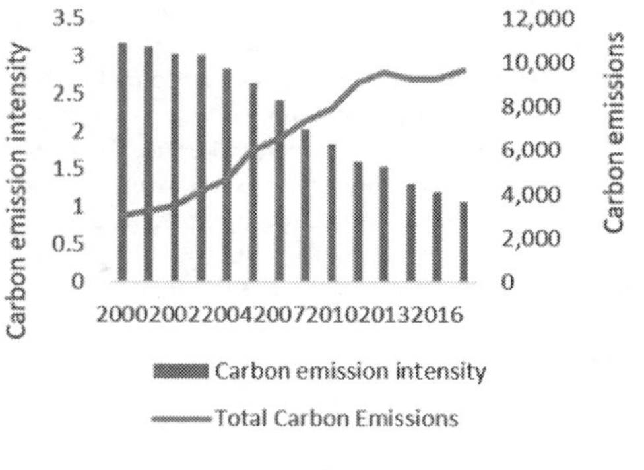

Fig.6. Relationship between energy conversion rate, carbon emission intensity and total carbon emission

Energy conversion efficiency is the ratio of final energy consumption to total energy consumption. Improving energy conversion efficiency can save energy. Carbon emission intensity is the ratio of carbon emission per unit of GDP. Carbon emission intensity is decreasing year by year. In 2008, the carbon emission intensity decreased by 13%, and the annual average carbon emission intensity decreased by 5.82%.

2.5 Related analysis

We use the Spearman correlation coefficient to analyze the correlation of each industry, GDP, energy conversion rate, carbon emission intensity and carbon consumption to verify their correlation.

Define A and B as two sets of different data, then the spearman correlation coefficient can be expressed as:

$$r_s = 1 - \frac{6\sum_{i=1}^{n} d_i^2}{n(n^2-1)}$$

Where, d_i is the grade difference between $A_i B_i$, and the grade of number is the position after ranking in ascending order.

The calculation results are as follows:

Table 6. Spearman correlation coefficient results

	Carbon emission intensity	energy conversion rate	primary industry	secondary industry
Correlation	-.970**	0.884**	-1.000**	-0.158
	Tertiary Industry	GDP	Coal	Petroleum
Correlation	0.879**	0.970**	0.982**	0.970**

According to the calculation results, carbon emission intensity, energy conversion rate, primary industry, tertiary industry, GDP, coal, oil and natural gas are significant at 95% confidence level.

3. Summarize

China's energy structure is mainly coal energy, but the proportion of natural gas carbon emissions is on the rise. The driving factors of the energy structure are mainly carbon emission intensity factors and energy conversion efficiency factors. China's carbon emission intensity has decreased year by year, and the conversion efficiency has increased year by year, which is strongly related to carbon emissions.

China's industrial structure is mainly dominated by the secondary industry and the tertiary industry. The driving factors of industrial structure are mainly divided into industrial proportion factors and economic scale factors. China's tertiary industry has begun to develop rapidly, which is strongly related to carbon emissions. The rapid growth of GDP level is asynchronous with the total amount of carbon emissions, and gradually shows a state of "relative decoupling".

3.1 Global advice:

- Resource-based industries need to vigorously innovate and optimize the industrial structure.

In essence, energy conservation and emission reduction requires continuous innovation in energy development and use technology to reduce carbon emissions. Strengthen the effectiveness of technology to ensure the optimization and adjustment of industrial structure and scientific and rational development. It is also necessary to fully adjust the internal structure of the energy. Make full use of the innovation of engineering technology, realize low-carbon economic development, reduce environmental pollution, and fully solve the adverse impact of social and economic development on the environment.[3]

- Optimize energy structure and improve energy efficiency.

At the current stage, the main supporting energy for the development of my country's industry and economy is coal, oil, and natural gas-based fossil fuel energy. Among them, most industries in resource-based provinces and developing provinces and cities mainly rely on natural resources such as coal to generate GDP effects. In the process of mining, refining, purification, and use, more greenhouse gases, mainly carbon dioxide, will be produced, which will increase carbon emissions and be detrimental to the carbon cycle. Therefore, my country's energy consumption structure should be transformed from a structure dominated by fossil fuels to a structure dominated by clean energy and renewable energy. Change the relatively single energy structure, develop and utilize new energy and alternative energy, and realize the diversification of the energy structure.

References

1. Low-carbon economy has broad prospects for development in China [J]. Transportation Enterprise Management, 2010: 9.
2. Li Yangyang. Research on the relationship between energy consumption, carbon emissions and economic growth in Gansu Province [D]. Lanzhou University of Finance and Economics, 2020.
3. Yang Yingming, Sun Jiandong, Li Quansheng. Research status and prospect of energy structure optimization in my country [J]. Coal Engineering, 2019: 159-163.

E3S Web of Conferences 358, 01024 (2022)
GESD 2022
https://doi.org/10.1051/e3sconf/202235801024

Analysis of the tectonic characteristics of the western tectonic complex of N2 area

Yingwei He

Daqing Oilfield Co., Ltd. 2nd Oil Production Plant, Daqing 163000, Heilongjiang, China

Abstract: The western part of N_2 area is located at the tectonic high point of D oil field, with faults cut into each other, staggered distribution and complex fracture structure. In this paper, based on 3D seismic interpretation, the research on fault geometry, kinematic characteristics, fracture deformation period and grade division are systematically carried out. The quantitative discriminatory criteria of fault plane segmental growth are applied to implement the reliability of the fault plane combination method.

Keywords: Faults, fault distance, developmental characteristics, segmental growth.

1. Presentation of the problem

The western part of N2 area is located at the high point of SRT back-slope structure. Its development area is 26.23km2, and the structure is very complex. There are 71 developed faults, accounting for 46.4% of the whole area, and the density of faults is as high as 2.7 faults/km2. The type of fault combination is complex, and the planes are mostly developed with intersecting faults, so it is necessary to systematically carry out the study on the characteristics of fault development, formation mechanism and genesis period, and finely delineate the way of fault planes combination, so as to provide geological basis for the adjustment and tapping of the block.

2. Analysis of fault development characteristics and genesis mechanism

(I) Fault development characteristics

The tectonic complex area in the west of N2 area is cut by north-northwest and north-east faults on the plane, among which 29 north-east faults are developed, mainly with a fault distance less than 5 meters and an average extension length of 802 meters; 42 north-northwest faults are developed, mainly with a fault distance greater than 10 meters and an average extension length of 1272 meters. The north-northwest fault has a large fault distance and a long plane extension. Fracture profiles are mostly nested V-shaped, Y-shaped and anti-Y-shaped.

Table 1 General characteristics of the top surface fault of Sa II oil formation in the western tectonic complex area of N2 area

Number	Fault alignment	Total number of articles	Fault scale			Average extension length
			>10m	5-10m	<=5m	
I	North East	29	6	2	21	802m
II	north-northwest	42	21	4	17	1272m

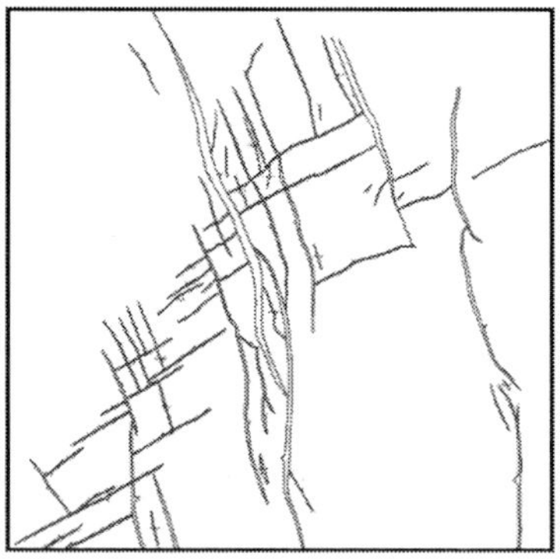

Figure 1. Polygon distribution of fractures on the top surface of S II oil formation

https://doi.org/10.1051/e3sconf/202235801024

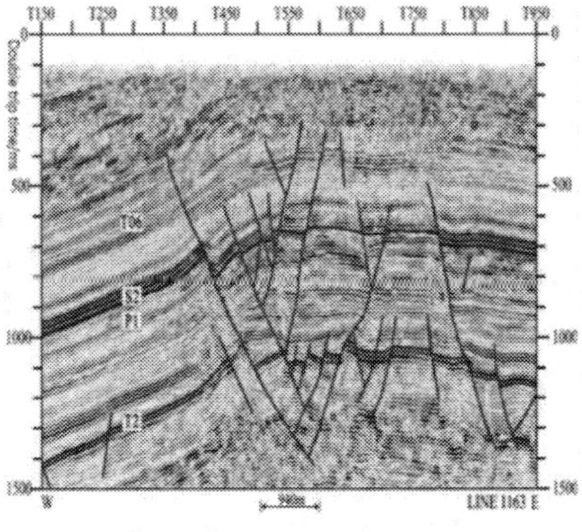

Figure 2. Fault profile in the western part of N2 area

The fault burial depth-distance curve is applied to quantitatively characterize the fault nucleation point, and the maximum distance marks the fault nucleation point; the maximum distance of the north-northwest and north-east faults is in Sa Ⅱ or Sa Ⅲ group, and is commonly nucleated in Sa Ⅱ and Sa Ⅲ group.

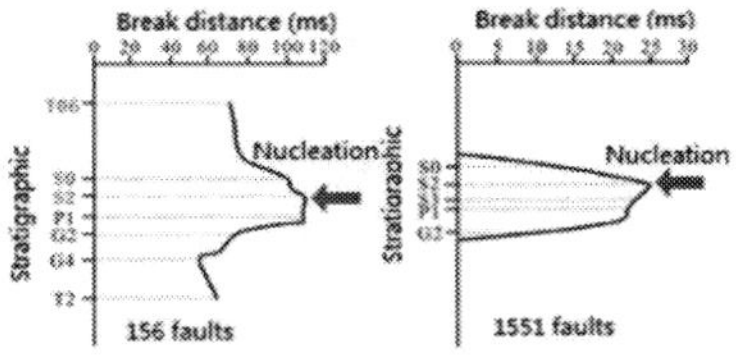

Figure 3. Fault distance-burial depth curve of typical north-northwest trending faults

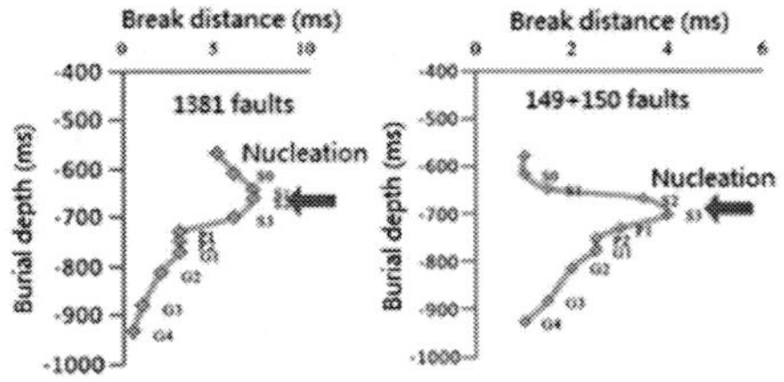

Figure 4. Fault distance-burial depth curve of typical north-east trending faults

The small layer fine comparison quantitatively characterizes the fault syngenesis. From the results of the fine comparison of small layers of the over north-northwest trending 139-01 fault connecting wells, the fault stratigraphy and fault thickening coefficients are different and show syngenetic characteristics.

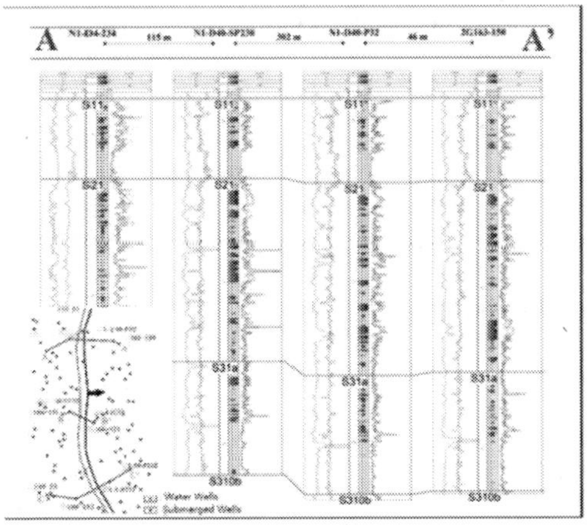

Figure 5. Fine comparison of small layers in the Saltu formation of the 139-01# fault with connected wells

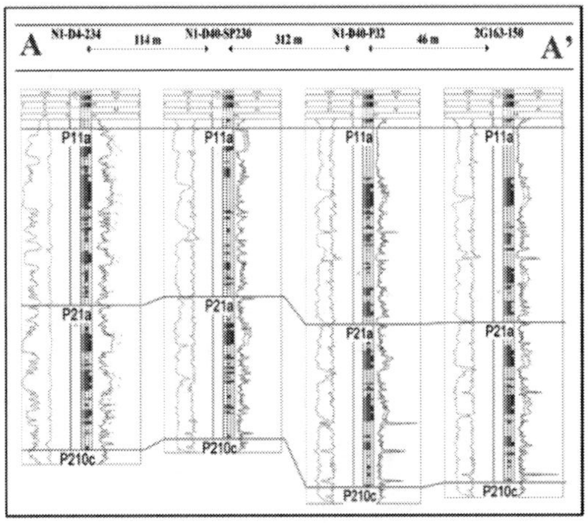

Figure 6. Fine comparison of small layers in a continuous well of the Grapevine formation of the 139-01# fault

Table 2 Comparison of the thickness of the trans-northwest trending faults

Oil layers	Stratigraphic thickness (West)	Stratigraphic thickness (East)	Thickness difference	Between East and West Distance	Fracture thickening factor	Stratigraphic thickening factor
SRT	119.6m	125.8m	6.2m	302m	20.52m/km	4.89m/km
PTH	71m	81.4m	10.4m	312m	33.33m/km	1.32m/km

From the fine comparison results of the small layer of the over north-east-oriented 1513 fault connecting wells, the fault stratigraphy and fault thickening coefficients do not differ much and show non-synclinal characteristics. The overall comparison of faults in the whole area shows that the north-northwest fault is syngenetic and the north-east fault is non-syngenetic.

https://doi.org/10.1051/e3sconf/202235801024

Table 3 Comparison of the thickness of the trans-northeast-facing 1513# fault

Oil layers	Stratigraphic thickness (North)	Stratigraphic thickness (South)	Thickness difference	Between East and West Distance	Fault thickening factor	Stratigraphic thickening factor
SRT	123m	123.6m	0.6m	130m	4.62m/km	2.46m/km
PTH	76m	76m	0m	127m	0m/km	2.18m/km

(II) Analysis of fault genesis mechanism

Combining the contemporaneous fauld, active stage and interbedding, the SII oil layer is divided into five sets of fault systems: the Nen-late-Ming-late Paleogene inversion period active fault, the Yaojia-Nen 1st and 2nd member active fault, the Yaojia-Nen 1st and 2nd member active fault, Jia-Nen 1st and 2nd member inversion period active faults, Quantou Formation-Yaojia-Nen 1st member inversion period active faults, Denglouku Formation-Yaojia-Nen 1st member 2nd member rifting period, inversion period active faults .

Table 4 General characteristics of faults of each fracture system on the top surface of Sa II oil formation

Number	active stage	Stress Direction	Fracture systems	Fault Trending towards	Contemporaneous fauld	Fault penetration level	
						Breakthrough Quan Tou Group	Broken to Nen 2nd
I	Reversal period	North-south left rotation pressure twist	Nen-late-Ming-late Paleogene inversion period active fault	North-East	Non-Syngenetic	No	Yes
II	depressional period	east-west direction	Yaojia-Nen 1st and 2nd member active fault	North-Northwest	Syngenetic	No	No
III	depressional period	east-west direction	Yaojia-Nengyi II section reversal period active fracture	North-Northwest	Syngenetic	No	Yes
IV	depressional period	east-west direction	Quantou Formation-Yaojia-Nen 1st member inversion period active faults	North-Northwest	Syngenetic	Yes	Yes
V	freezing period	North-west facing	Dengluku Formation-Yaojia-Nengyiji Section Faulting Period and Reversal Period Active Faulting	North-East	Syngenetic	Yes	Yes

Reversal period: Type V faults are revived and re-activated, Type I faults (north-east) are formed, Type III and Type IV continue to be active; depressional period (Yaojia Formation-Nen I and II section): Type II and Type III faults are formed, Type IV faults continue to be active; depressional period (Quantou Group depositional period) Type IV faults are formed, Type V faults continue to be active; depressional period: Type V faults begin to form. North-northwest trending faults are active in multiple phases and continue to develop successively. The north-east oriented faults are formed by late extrusion stress. By comparing the results of physical simulation experiments in different orientations with the results of the study area, the stress direction suffered is 45° with the north-northwest oriented fault.

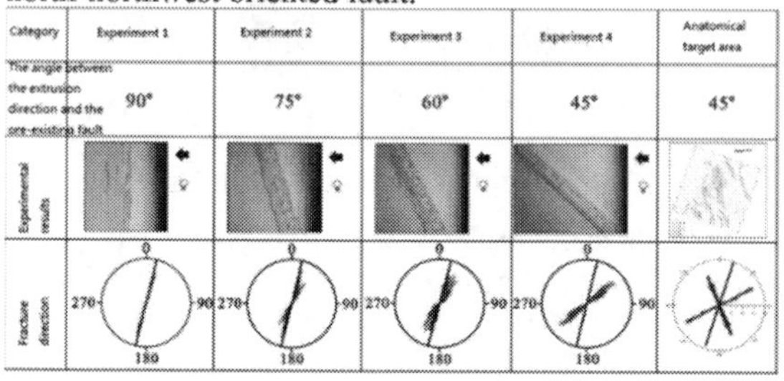

Figure 7. Comparison of the results of physical simulation experiments in different orientations with the results of the study area

3. Fault plane connection method

Fault growth is a dynamic process that undergoes a continuous process of destruction-connection-re-destruction. Segmental growth is a process in which a large number of smaller displacement positive faults gradually interact with each other to grow and connect in segments to form a small number of larger scale large faults as the remote stress increases. Fault displacement-distance curve is one of the important methods to identify the segmental growth of faults, and the presence of low value zone of displacement is a segmental growth fault, otherwise it is an isolated fault. It is possible that the segmented growth fault is interpreted as two faults due to the limitation of seismic resolution. In order to finely portray the fault plane connection mode, 33 faults with fault distance greater than 5m in the whole area were identified for development type according to the above method, and 10 segmental growth faults were identified.

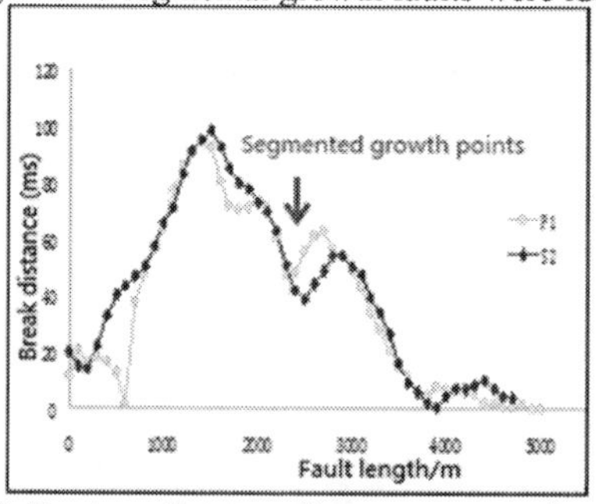

Figure 8. Fracture distance-distance curves of segmented growth faults

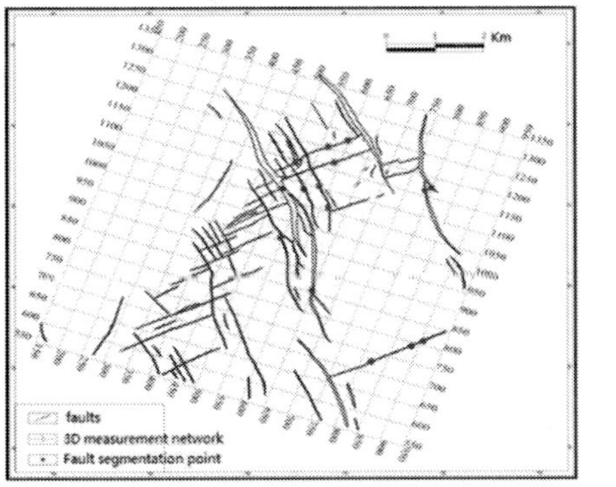

Figure 9. Distribution of fault types on the top surface of SaII Group

The method of quantitative determination of fault growth stage by applying "conversion displacement (D)/distance (S)" - quantitative criteria of fault segment growth process. For "soft connection" lateral stacked faults, the conversion displacement (D) is the sum of the displacements of the two faults at the center of the stacked fault section, and the separation distance is the vertical distance between the two faults at the center of the stacked fault section. For hard-connected faults, the conversion displacement is the sum of the two fault displacements at the center of the overlapping section of fault A and extinction fault B. The separation distance is the distance between the two faults at the center of the overlapping section of fault A and extinction fault B. If due to seismic resolution limitation, the extinction fault B may not be developed, the conversion displacement is equal to the displacement in the middle of the abruptly changed section of the fault strike, and the departure distance is equal to half of the distance between the faults in the sub-parallel section.

Based on the three-dimensional seismic data of SL basin, the data of fault transition displacement (D) and separation distance (S) are counted to improve the quantitative criteria of fault segment growth connection; when D/S is less than 0.27, the fault segment is in the "soft connection" stage of lateral stacking, when D/S is between 0.27 and 1, it is in the stage of starting rupture The "transition" stage, when D/S is greater than 1, the fracture is in the stage of complete rupture "hard connection" stage, that is, divided into three stages: lateral stacking, the beginning of damage and complete destruction.

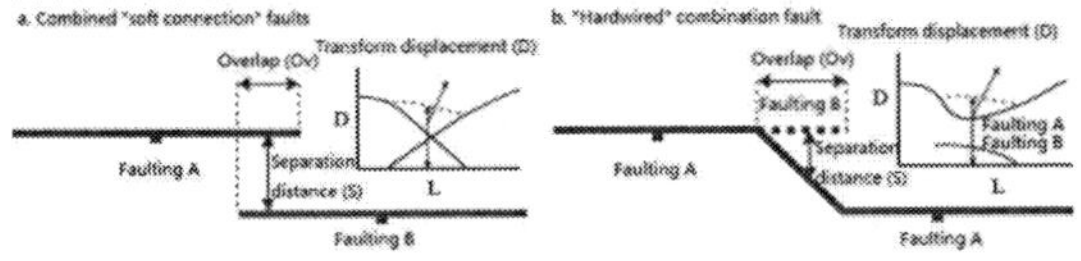

Figure 10. Terms related to "soft" and "hard" connection faults

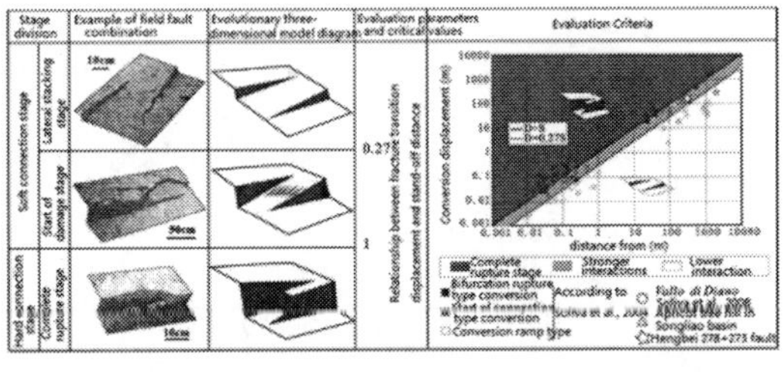

Figure 11. Quantitative discriminatory criteria for fault segment growth

Applying the quantitative fault growth discrimination criteria, 10 segmental growth faults in the block were finely delineated in terms of fault plane connection: 7 in the hard connection stage, 2 in the transition stage, and 1 in the soft connection stage.

4. Awareness

(1) The tectonically complex area is cut by north-northwest and north-east faults on the plane, and the north-northwest faults have large fault distances and long plane extensions. Fracture profiles are mostly nested V-shaped, Y-shaped and anti-Y-shaped.

(2) The north-northwest and north-east faults are commonly in the S II and S III component nuclei, and the north-northwest fault is a synclinal fault, while the north-east fault is a non-synclinal fault.

(3) The north-northwest fault is active in multiple phases and continues to develop successively. The north-east fault is formed by late extrusion stress, and the stress direction is 45° from the north-northwest fault.

(4) There are 10 segmental growth faults developed in the tectonically complex area, and the fault growth quantitative discrimination criteria are applied to finely delineate the fault plane connection mode: 7 in the hard connection stage, 2 in the transition stage and 1 in the soft connection stage.

References

1. Fu Xiaofei, Xu Peng, Wei Changzhu et al. Internal structural characteristics of tensor fracture zones and their relation to oil and gas transport and preservation[J]. Geological Foreground,2012,19(6): 57-69
2. GIBSON R G. Fault-zone seals in siliciclastic strata of the Columbus basin, Offshore Trinidad[J]. AAPG, 1994, 78(9): 1372-1385
3. Lv Yanfeng et al. eds. Fault closure study[M]. Petroleum Industry Press,2002
4. Fu Xiaofei, Fang Deqing, Lv Yanbang, Fu Guang, Sun Yonghe. A method to evaluate the vertical closure of faults from the internal structure of fracture zones[J]. Earth Science,2005(03)
5. VERMILYE J M, SCHOLZ C H. Fault propagation and segmentation:insight from the microstructural examination of a small fault [J].Journal of Structural Geology,1999,21(11):1623-1636.

6. MCGRATH A G, DAVISON I. Damage zone geometry around fault tips [J]. Journal of structural geology,1995,17(7):1011-1024.
7. CAINE J S, EVANS J P, FORSTER C B. Fault zone architecture and permeability structure [J]. Geology,1996,24:1025-1028.
8. KNIPE R J. Juxtaposition and seal diagrams to help analyze fault seals in hydrocarbon reservoirs [J]. AAPG,1997,81(2):187-195.
9. KIM Y S, PEACOCK D C P, SANDERSON D J.Fault damage zones [J].Journal of Structural Geology, 2004,26(3):503-517.
10. Fan Jichang, Liu Mingjun. A method for determining the internal structural and physical parameters of fracture zones[J]. Petroleum Geophysical Exploration,2007,42(2):164-169.
11. Wu Zhiping, Chen Wei,Xue Yan,et al. Structural characteristics of fracture zones and their transport and sealing properties for oil and gas[J]. Journal of Geology,2010,84(4):570-578.
12. Gao Songyang. Application of logging data in fracture identification: An example of sandstone reservoir in H area[J]. Journal of Petroleum and Natural Gas,2009,31(2):272-274.
13. Shen Zhongshan, He Xin, Mei Mei, Li Ping. Geological characteristics of the fracture zone in the northern Xingshugang oil field in the Songliao Basin[J]. Daqing Petroleum Geology and Development, 2016,35(3):22-25.

Analysis on the distribution characteristics of oil-water layer and potential-tapping effect of add-perforation in XX oilfield

Yilin Guo

Victory oil and gas project management department, Daqing Oilfield Co., Ltd, Daqing 163000, Heilongjiang, China

Abstract: XX oilfield has a total of 10,561 oil and water wells, 6,790 wells were drilled in the oil-water layer, but only 10.3% of the total wells were drilled in the oil-water layer, so the development degree of the same layer is relatively low. In the stage of water cut and stable production, the geological reserves of oil-water layer are considerable. Based on a large number of data statistics of oil-water layer, this paper understands the developed situation of oil-water layer and the characteristics of plane and vertical distribution, and discusses the feasibility of oil-water layer to add perforation and tap potential from the effect of added perforation well.

Key words: oil-water layer; distribution characteristics; add-perforation and potential-tapping.

1. General situation of oil and water stratum in XX oilfield

1.1 Reserves status

XX oilfield has eight oil-water transitional zones with GIII, GII, GI, PII, PI4-7, PI1-2, SIII and SII. Up to now, the recovery degree of oil-water layer is 15.84%, and there is still a large scale of remaining reserves, among which the reserves of GII oil layer group are the highest and SIII oil layer group is the lowest.

1.2 Development status

Among the 10561 wells in XX oilfield, 6,790 wells were drilled in the oil-water layer, including 3989 producers and 2801 water injectors. The average individual well drilled 6.6 small layers, the sandstone thickness is 8.6m and effective thickness is 3.4m, the oil-water layer of individual well develops small thickness.

Table 1 Statistical table of wells drilled the oil-water layer in XX oilfield

Total number of wells	Drilled well number	Development mode	Well number	Producer	Water injector
10561	6790	Water flooding	3938	2390	1548
		Follow-up water flooding	482	319	163
		Polymer flooding	2370	1280	1090
		Total	6790	3989	2801

1.3 Perforation status

By the end of 2020, 700 wells in the oil-water layer were perforated, of which 621 were put into production with perforation, and 79 were later developed with add-perforation. Only 10.3% of wells drilled in the oil-water layer, which has the potential for further development with add-perforation.

Table 2 Statistical table of the wells perforated the oil-water layer in XX oilfield

Perforated well number	development mode	Well number	producer	Water injector	Put into production with perforation		add-perforation development	
					Producer	Water injector	Producer	Water injector
700	Water flooding	672	210	462	154	441	56	21
	Follow-up water flooding	3	2	1	2	1		
	Polymer flooding	25	15	10	14	9	1	1
	Total	700	227	473	170	451	57	22

2. Characteristics of oil-water layer potential distribution in XX oilfield

In order to ensure the scale and effect of add-perforation, the developed thickness of oil-water layer is mainly analyzed when selecting potential wells of oil-water layer.

2.1 The plane potential is mainly concentrated in two transitional zone

According to the statistical results of plane transitional zones, the average individual well developed sand body thickness in the oil-water layer of the second transitional zone is the largest, the thickness of sandstone is 9.2m and the effective thickness is 6.3m. Followed by the first transitional zone and the third transitional zone, with effective thickness of 4.7m and 4.2m, respectively.

Table 3 Statistical table of developed status of oil-water layer in different transitional zone of XX oilfield

Transitional zone	Producer			Water injector			Total		
	Well number	Sandstone thickness (m)	Effective thickness (m)	Well number	Sandstone thickness (m)	Effective thickness (m)	Well number	Sandstone thickness (m)	Effective thickness (m)
First transitional zone	324	7.8	4.6	274	8.8	4.8	598	8.3	4.7
Second transitional zone	159	9.3	6.8	120	9.1	5.6	279	9.2	6.3
Third transitional zone	117	7.3	5.1	106	6.4	3.1	223	6.9	4.2
Fourth transitional zone	174	6.7	4.3	96	6.5	3.2	270	6.6	3.9
Net oil zone	3215	9.0	3.1	2205	8.5	3.1	5420	8.8	3.1
Total	3989	8.7	3.5	2801	8.4	3.3	6790	8.6	3.4

2.2 The vertical potential is mainly concentrated in groups PⅠ, SⅡ and GⅡ.

According to the statistical results of different vertical oil layer groups, the thickness of oil-water layer in group PI and group SII is large. The average effective drilled thickness is 5.1m in group PI and 3.5m in group SII. The number of wells drilled in group G is the largest, and the number of wells drilled in group GII accounts for 36.6%.

Table 4 Statistical table of oil-water layer drilled in different oil layer group of XX oil field

Oil layer group	Drilled well	Proportion (%)	The average individual well		
			Drilled sublayers	Sandstone thickness(m)	Effective thickness(m)
SI	23	0.3	1.6	2.1	0.7
SII	309	3.6	3.3	5.8	3.5
SIII	324	3.8	2.8	4.9	2.7
PI	571	6.7	2.7	6.9	5.1
PII	1014	11.9	3.1	5.3	3.1
GI	1758	20.6	5.0	5.5	2.3
GII	3122	36.6	6.9	7.5	2.0
GIII	1406	16.5	5.9	9.0	3.4
Total			6.6	8.6	3.4

2.3 The remaining oil potential is mainly concentrated in groups SⅡ and GⅡ.

At present, oil saturation is an important index to evaluate reservoir potential, and its value reflects the degree of remaining oil potential of reservoir. The higher the oil saturation, the greater the potential of reservoir. Taking the northeast block as an example, the optimized CIF Log platform interpretation module is used to screen the layer with current oil saturation ≥40%. Among the selected layers, the saturation in SII is the largest, and the GII is the second.

Table 5 Statistical table by oil layer group

Oil layer group	Current oil saturation≥40%					
	Well number	Number of sublayers	Thickness (m)	Current oil saturation(%)	Ratio of total wells(%)	Ratio of total thickness(%)
S2	46	101	69	56.4	52.9	20.4
S3	31	67	45.8	54.1	50.0	26.1
P1	57	154	174.9	54.0	47.1	25.8
P2	111	302	270.6	54.3	44.4	27.7
G1	121	271	168	53.0	31.5	15.4
G2	178	517	193.8	55.9	35.0	20.9
G3	6	14	6.7	53.9	14.6	6.3

2.4 Analysis of potential of add-perforation in scale

The development thickness of oil-water layer is generally small, and most wells do not meet the potential of stimulation tapping. This study focuses on the potential analysis of wells with effective thickness of oil-water layer greater than 4 meters. There are 2,157 such wells, accounting for 31.8% of the total number of wells, among which 1,279 are producers, which are the main potential tapping objects for the next step.

Table 6 Statistical table of effective thickness classification of oil-water layer in XX oilfield

Effective thickness classification (m)	Producer			Water Injector			Total		
	Well number	Sandstone thickness (m)	Effective thickness (m)	Well number	Sandstone thickness (m)	Effective thickness (m)	Well number	Sandstone thickness (m)	Effective thickness (m)
m≤2	1722	5.7	0.9	1249	5.4	0.9	2971	5.5	0.9
2 < m≤4	988	9.1	3.0	674	8.7	2.9	1662	8.9	2.9
4 < m≤6	568	10.8	5.0	413	10.6	4.9	981	10.7	5.0
6 < m	711	14.1	9.1	465	14.4	9.2	1176	14.2	9.2
Total	3989	8.7	3.5	2801	8.4	3.3	6790	8.6	3.4

Based on the above statistical analysis results, the average effective thickness of oil-water layer is only 3.4m, and the thickness scale is too small, which is not suitable for scale add-perforation. However, considering the distribution characteristics of plane, vertical and oil saturation, the second transitional zone and groups PI, SII and GII can be selected as the main objects to realize the potential, and the well locations in local areas can be selected to add perforation and tap the potential.

3. Analysis on the potential tapping effect of oil-water layer in XX oilfield

3.1 Early effect of add-perforation

In 2019, through a comprehensive comparative analysis of the logging curve data of the oil-water layer and the net oil layer, six wells were selected to independently perforated the oil-water layer to produce. From the add-perforation effect of three individual well in the same layer, the average daily fluid production and oil production per well were 41t, 1.0t respectively, and water cut is 97.0% in the initial stage. In November 2021, the average daily liquid production per well is 31t, daily oil production per well is 0.8t, and water cut is 97.3%.

Table 7 Effect of single-production wells after add-perforation in the oil-water layer

No	Add-perforation layers	Completion date	Perforate the same layer		Before perforation			Initial stage of perforation			November 2021		
			Sandstone thickness (m)	Effective thickness (m)	fluid production (t/d)	oil production (t/d)	Water cut (%)	fluid production (t/d)	oil production (t/d)	Water cut (%)	fluid production (t/d)	oil production (t/d)	Water cut (%)
1	GIII1-3 ~ GIII10	20.01	9.8	2.8	30	0.5	98.2	42	0.8	98.1	19	1.0	94.6
2	GII34 ~ GIII9-12	19.12	12.5	3.6	31	0.8	97.2	31	0.8	97.4	33	0.6	98.3
3	GII6-7 ~ GII14(b)	19.12	8.7	5.0	40	0.9	97.6	51	1.8	96.5	41	0.9	97.8
Average individual well			10.3	3.8	/	/	/	41	1.0	97.0	31	0.8	97.3

In terms of the add-perforation effect of the three wells in the same layer, the average daily increase of liquid production and oil production per well in the initial stage was 14t, 0.5t respectively, and the comprehensive water cut increased by 0.3%.

Table 8 Effect of commingled production wells after add-perforation in the oil-water layer

No.	Add-perforation layers	Completion date	Perforate the same layer		Before perforation			Initial stage of perforation			difference			November 2021		
			Sandstone thickness (m)	Effective thickness (m)	fluid production (t/d)	oil production (t/d)	Water Cut (%)	fluid production (t/d)	oil production (t/d)	Water Cut (%)	fluid production (t/d)	oil production (t/d)	Water Cut (%)	fluid production (t/d)	oil production (t/d)	Water Cut (%)
1	PI1-4	20.04	7.7	7.3	50	2.5	94.9	79	2.8	96.5	29	0.2	1.6	94	2.5	97.3
2	SII15-2 ~ SIII4-7(b1)	21.01	8.1	6.3	21	0.4	98.2	34	1.4	95.8	13	1.1	-2.5	31	1.2	96.2
3	SII1-4 ~ SII4+6	20.12	7.7	6.1	6	0.1	98.6	7	0.2	97.9	1	0.1	-0.8	10	0.2	97.6
Average individual well			7.8	6.6	7.8	1.0	96.1	40	1.4	96.4	14	0.5	0.3	45	1.3	97.1

The effect of add-perforation reflects the phenomenon of low liquid production capacity and high water cut, and the water cut of the layer is more than 97%.

3.2 Field test of potential tapping in oil-water layer

In 2021, in order to evaluate the oil-bearing capacity of oil-water layer and water layer, and verify the liquid production capacity and water cut status of oil-water layer in different oil layer groups, the CIF Log platform's interpretation module results of the same layer were used to select representative producers to carry out add-perforation in oil-water layer and separated layer production test.

3.2.1 Design of test well scheme

In this test, the daily oil production is less than 2t and the bottom depth of the add-perforation interval is more than 4m away from the artificial bottom. The wells with oil saturation greater than 45% should be developed with oil-water layer.

Taking well L as an example, the scheme design is as follows:

In order to verify the liquid production scale of different oil-water layers, the well was stratified with pressure switch to verify the oil-bearing characteristics of oil-water layers. Specific technological process: 1 downhole controllable electric regulator is installed in the plugging layer, production layer 1, production layer 2 and production layer 3, and there are four in total in the whole well.

Table 9 Operating process of separated layer production after add-perforation in oil-water layer test

Classification	Operation process
Single production in oil-water layer	Open production zone 1 downhole controllable electric regulator, plugging layer, production zone 2, production zone 3 closed
Single production in oil-water layer	Shut down the downhole controllable electric regulator of production layer 1, plugging layer and production layer 3, and open the downhole controllable electric regulator of production layer 2
Single production in oil-water layer	Shut down the downhole controllable electric regulator of production layer 2, plugging layer and production layer 1, and open the downhole controllable electric regulator of production layer 3
Commingled production in oil-water layer	The downhole controllable electric regulators of plugging layer, production layer 1, production layer 2 and production layer 3 were all opened

Table 10 Summary of field test wells for potential tapping in oil-water layer

NO	Before stimulation: Daily fluid production(t)	Daily oil production(t)	Water cut (%)	Original production layer	Add-perforation layer	Sandstone thickness of add-perforation(m)	Effective thickness of add-perforation(m)	Production layer	Property
1	51	0.7	98.6	SI1(a) ~ SIII3(b)	SIII8-10	5.3	4.3	Production layer 1	good oil layer
					PI1-2	9.6	8.6	Production layer 2	good oil layer
					PI2	6.7	6.7	Production layer 3	good oil layer
					Total	21.6	19.6	commingle	
2	47	0.4	99.0	SI1(a) ~ SII13-16	SIII8	1.9	1.9	Production layer 1	good oil layer
					SIII9 ~ PI1	3.6	2.8	Production layer 2	good oil layer
					PI2	2.7		Production layer 3	water layer
					Total	8.2	4.7	commingle	
3	36	0.7	98.0	GI8 ~ GII17	GII17-18	2.4	1.9	Production layer 1	good oil layer
					GII22	1.4	1.0	Production layer 2	good oil layer
					GII24-26	4.2		Production layer 3	water layer
					Total	8.0	2.9	commingle	
4	88	1.0	98.8	SI1(a) ~ SIII3	PI5-7	6.1	5.3	Production layer 1	good oil layer
					PII1-3	1.8	1.2	Production layer 2	poor oil layer
					PII1-4	3.3	2.9	Production layer 3	Medium oil layer
					Total	11.2	9.4	commingle	
5	104	1.0	99.0	SI1 ~ SII15+16(a)	PI5+6(a)	4.0	3.8	Production layer 1	Good, thick layer
					PI5+6(b)	1.6	1.6	Production layer 2	Good, thick layer
					PII1-4	3.5		Production layer 3	water layer
					Total	9.1	5.4	commingle	
6	28	0.8	97.1	SI3(a) ~ SII5-8	SII10+11(b)	2.8	2.6	Production layer 1	good oil layer
					SII13+14	1.2	1.1	Production layer 2	Medium oil layer
					SII15+16(b)	2.7		Production layer 3	water layer
					Total	6.7	3.7	commingle	
7	112	1.2	98.9	SI1(a) ~ GI11+12	GI17-18	3.3	2.8	Production layer 1	good oil layer
					GI19-20	3.7	1.5	Production layer 2	good oil layer
					GII5	1.7		Production layer 3	water layer
					Total	8.7	4.3	commingle	
8	58	1.0	97.5	SI1(a) ~ PI2	PI2(e1)	2.3	2.0	Production layer 1	Medium oil layer
					PI2(e2)	3.0	3.0	Production layer 2	Medium oil layer
					PI5+6	4.2		Production layer 3	water layer
					Total	9.5	5.0	commingle	

3.2.2 Implementation effect analysis of test wells

At present, six test wells have been implemented for add-perforation and separated layer production in the oil-water layer, with an average thickness of 10.8m and an effective thickness of 8.1m. At the initial stage of add-perforation layer, the daily production of liquid production and oil production is 45t and 2.8t, and the comprehensive water cut is 93.8%, compared with the initial stage of water flooding, the daily oil production is 0.2t less than the daily oil production increase, and the effect is better.

Table 11 Statistical table of individual well effect of add-perforation in oil-water layer

NO	add-perforation layer	Sandstone thickness of add-perforation(m)	Effective thickness of add-perforation (m)	Production data of add-perforation layer (initial stage): Daily fluid production (t)	Daily oil production (t)	Water cut (%)	Effective liquid production intensity
1	SIII8-10 ~ PI2	21.6	19.6	62	4.5	92.8	3.2
2	SIII8(b) ~ PI2(a1)	8.2	7.4	104	8.3	92.0	14.1
3	GII17-18 ~ GII24-26	8.0	2.9	32	0.7	97.8	11.0
4	PI5-7 ~ PII1-4	11.2	9.4	19	0.5	97.3	2.0
5	PI5+6 ~ PII1-4	9.1	5.4	11	0.2	98.2	2.0
6	SII10+11(b) ~ SII15+16(b)	6.7	3.7	43	2.5	94.2	11.6
average		10.8	8.1	45	2.8	93.8	5.6

Wells were classified according to the initial production and water cut conditions: three wells with better results had an initial production of 70t liquid production per day, 5.1t oil production per day and a comprehensive water cut of 92.7%; three poor wells produced 20t of fluid production and 0.5t of oil production per day at the initial stage, with water cut of 97.5%. Among the poor wells, there were two wells with low liquid production and high water cut, and one with high liquid production and high water cut. According to the field analysis of the two plugging removing wells, some switches are not open, half open, sand buried and scaling.

Table 12 Statistical table of the effect of add-perforation wells in oil-water layer

type	Well number	Average sandstone thickness of add-perforation (m)	Average effective thickness of add-perforation (m)	Average daily fluid production (t)	Average daily oil production (t)	Average water cut (%)
Good effect	3	12.2	10.2	70	5.1	92.7
Bad effect	3	10.2	9.4	20	0.5	97.5

There are six oil-water layer wells perforated 18 layers, of which 14 are oil-water layers. The average thickness of each interval is 3.7m, and the effective thickness is 3.5m. In the initial stage, the daily fluid production and oil production are 15t and 1.1t, and the comprehensive water cut is 93.2%. There are four water-bearing layers, with an average thickness of 3.3m in a single interval, and a daily production of 14t liquid and 0.5t oil at the initial stage. The comprehensive water cut is 96.4%. The water cut in oil-water layer is low, and the water cut in water-bearing layer is relatively high.

Table 13 Statistical table of the effect of add-perforation wells in oil-water layer by layer properties

Layer properties	Layer number	Average sandstone thickness of add-perforation (m)	Average effective thickness of add-perforation (m)	Average daily fluid production (t)	Average daily oil production (t)	Average water cut (%)
Oil-water layer	14	3.7	3.5	15	1.1	93.2
Water layer	4	3.3		14	0.5	96.4

There are five oil layer groups in six oil-water layer wells, among which SII group has the highest production intensity, PI group has the largest effective thickness and the lowest water cut.

Table 14 Statistical table of the effect of add-perforation wells in oil-water layer by oil layer groups

Oil layer group	Layer number	Average sandstone thickness of add-perforation (m)	Average effective thickness of add-perforation (m)	Average daily fluid production (t)	Average daily oil production (t)	Average water cut (%)	Effective liquid production intensity (t/m)
SII	2	2.0	1.8	16	1.1	93.1	8.9
SIII	2	3.6	3.1	22	1.2	94.6	7.1
PI	6	5.3	4.5	19	2.4	87.2	4.2
PII	2	2.6	2.1	9	0.3	96.8	4.4
GII	2	1.9	1.5	4	0.2	95.5	2.9
Total	14	3.3	3.5	15	1.1	93.2	4.4

There are seven types of sand bodies in the six oil-water layer add-perforation wells, among which channel sand, channel sand margin and main sheet sand have higher liquid production intensity, while banded sand body has lower liquid production intensity.

Table 15 Statistical table of the effect of add-perforation wells in oil-water layer by sand properties

Sand properties	Layer number	Average effective thickness of filling hole (m)	Average daily fluid (t)	Average daily oil (t)	The average water cut (%)	Effective liquid production strength (t/m)	Effective liquid production intensity (t/m)
Channel sand	3	6.4	6.0	18	1.4	92.3	3.1
Channel sand side	4	3.2	2.9	14	0.8	94.2	4.8
Banded channel sand	1	6.1	5.3	0			0.1
Main sheet sand	1	3.6	2.8	71	6.8	90.4	25.4
Main sheet sand side	3	2.1	1.7	9	0.3	96.5	5.0
Banded main sheet sand	1	2.4	1.9	0			0.2
Non banded main sheet sand	1	1.4	1.0	9	0.3	96.5	8.5
Total	14	3.3	3.5	15	1.1	93.2	4.4

Based on the above statistical analysis results, it is found that:

(1) The liquid production capacity of SII and SIII groups was high; Other oil layers are relatively low.

(2) The liquid production intensity of channel sand, channel sand side and main sheet sand is higher than that of banded sand body.

(3) For producers with low water cut and fast production decline after add-perforation, add-perforation should be carried out corresponding to layers of surrounding water injectors.

(4) The oil saturation of the six test wells in the same layers is above 45%, but the effect difference is large and the rule is not clear, so it is not implemented as a scale add-perforation for the time being.

4. Conclusion

(1) The oil-water layer of XX oilfield has large reserves but small thickness, and the average effective thickness is only 3.4m. Combined with the distribution characteristics of plane, vertical and oil saturation, the second transitional zone and group PI, group SII and group GII are mainly taken as the main objects to realize the potential, and well locations are selected in local areas to add perforation and tap the potential.

(2) Considering the thickness distribution characteristics of oil-water layer and the situation of add-perforation, it is considered that it is not suitable for scale add-perforation and potential tapping in oil-water layer. However, according to the characteristics of electrical logging curve, it is feasible to select oil-bearing partial oil section in local well group to implement add-perforation, so as to improve the contribution of stimulations.

(3) Oil increasing capacity of the same layer can be evaluated more directly by single production after the same layer is perforated. However, in combination with the development fact, when the same layer and the original production layer are commingled production, the impact of inter-layer contradictions will lead to changes in the production condition of each layer. Therefore, targeted tests should be carried out later for further evaluation and verification.

References

1. Lin Jingye, et al. Petroleum geology & oilfield development in Daqing, 1998,02
2. liu wei, analysis of oil-water reservoir potential [J]. Petroleum geology & oilfield development in Daqing, 2008,27 (6): 80-82
3. Zhao Guozhong, Li Jie. Multidisciplinary integrated reservoir research method and application. Petroleum Industry Press [J], 2009: 39-97

Application of a new method of reservoir interface prediction in geological steering of horizontal Wells

Chenpu Liu

Exploration Division of Daqing Oilfield, Daqing 163453, Heilongjiang, China

Abstract. New method in this paper, we study the reservoir interface prediction in the application of horizontal well geological guidance function, the article in the process of study, theoretical analysis of the horizontal well geosteering and reservoir prediction interface, new methods and combining the theory and practice, studies the reservoir prediction interface new method in the application of horizontal well geosteering.

Key words: Reservoir interface; New prediction methods; Horizontal well; geosteering.

1. Introduction

Geological steering of horizontal Wells is of great significance to the exploitation of oilfield resources. In the traditional geological steering research of horizontal Wells, the prediction method of reservoir interface is not accurate enough, and the calculation of formation dip Angle is not accurate, which leads to the deviation in the "landing" stage of horizontal Wells, and affects the subsequent oilfield exploitation. Therefore, based on technical data and practical cases, this paper proposes a new reservoir interface prediction method, which can improve the calculation accuracy of formation dip Angle, and then realize accurate reservoir calculation, laying a good foundation for the subsequent horizontal well "landing" stage.

2. Geosteering analysis of horizontal Wells

Horizontal Wells refer to oil Wells with horizontal distribution of boreholes in the reservoir. Under normal circumstances, Wells with large parallel hole extension distance and well deviation of about 85° are regarded as horizontal Wells. The biggest characteristic of horizontal well is that the empirical trajectory of horizontal well is in parallel with the distribution of the whole reservoir. In order to ensure the efficient production of oil reservoir, the geological data of oil reservoir area should be known in real time and the trajectory of the well should be adjusted. The work of obtaining geological data, adjusting boreholes, and surveying experience is also known as geosteering.

Through the theoretical analysis, it can be found that the geological steering of horizontal Wells is the key to the exploitation of horizontal Wells. Whether the geological steering of horizontal Wells is reasonable or not and the efficiency of horizontal Wells directly determines the oilfield exploitation efficiency of horizontal Wells.

3. A new method for reservoir interface prediction

Reservoir interface prediction method is an important work in horizontal well steering, the core of which is the acquisition and calculation of formation dip Angle. In the process of horizontal well work, reservoir can be predicted by seismic data. However, in general, the vertical resolution of seismic data is affected by certain factors, and there is a certain gap between it and the actual value of reservoir. In horizontal well exploitation, the distribution of horizontal well borehole and the whole reservoir is almost one, so the error of reservoir prediction will affect the later exploitation work. In addition, after the horizontal well "lands", the deviation Angle of the whole horizontal well is about 85°, even if the reservoir prediction error of 1m, it will also cause great rigidity to the horizontal well, resulting in the problem of reservoir drilling rate decline. Judging from the practical application effect of the current reservoir interface prediction method, although the reservoir interface prediction method has a certain effect. However, the calculation error of dip Angle in the whole method is large, which will directly affect the whole reservoir prediction. Therefore, this paper analyzes the calculation and optimization of formation dip Angle and proposes a new method for reservoir interface prediction [1].

3.1 Multiple methods are applied to obtain dip Angle

The dip Angle is the key factor affecting the prediction of the whole reservoir interface. Therefore, in the whole process of reservoir interface prediction and analysis, accurate acquisition of formation dip Angle is required. At present, the methods to obtain formation dip Angle mainly include imaging logging calculation of formation dip Angle, offset well logging calculation and analysis method, and formation thickness variation calculation method. The three dip calculation methods have different application ranges, so in the actual application process, it is necessary to apply different logging techniques according to the known situation of reservoir interface, so as to ensure reasonable logging application and directly improve the application effect of technology. The following is an analysis of the three dip calculation methods:

1.1.1 The dip Angle is calculated from the variation of formation thickness. The formation thickness is the main factor affecting the dip Angle, and the change of the thickness will inevitably change the dip Angle. Therefore, it is necessary to clarify various construction techniques in the whole process of analysis of formation thickness change to ensure more reasonable measurement application and improve the comprehensive application effect of thickness measurement. The stratum may have different states such as horizontal and dip in dip measurement. Therefore, the dip Angle of the whole formation is different. In the dip calculation model, the horizontal well traverses the formation including the upper and lower vertices. In the calculation process, the upper vertex is named as the top layer entry point, and the lower vertex is the bottom boundary layer. As shown in Figure 1, the position of each point can be considered as a plane, where the relationship between dip Angle and overall formation thickness becomes more obvious. In Figure 1, the dip Angle of the stratum is A, the thickness of the stratum is H, the distance between the upper entry point and the exit point is H2, and the bottom thickness, the horizontal distance between the two points and the included Angle have constructed a triangle. Therefore, in the calculation of formation dip Angle, the inverse trigonometric function can be used to calculate:

a=arctan（H2 / S2）

In the formula, the vertical distance of H2 formation.

In the publicity, S2 represents the horizontal distance between the entry point and the exit point.

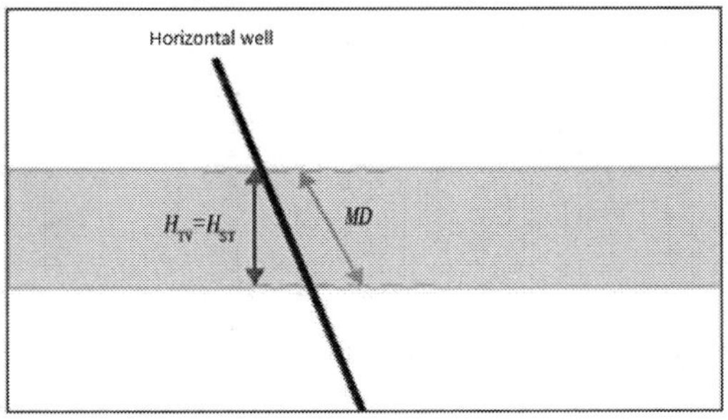

Fig. 1 Calculation of dip Angle in horizontal formation

Note: in the figure, Htv is the apparent thickness of formation, Hst is the true thickness of formation, and MD is the parallel track footage (straight-line distance between two points).

However, in practice, it can be found that the thickness of the whole reservoir may not be completely horizontal, but also may appear inclined state. Under the condition of inclined stratum, there will be a certain gap between the visual thickness of stratum plane and the actual thickness of stratum. As shown in Figure 2, there is a certain gap between the visual thickness of Htv and the real thickness Hst after the occurrence of stratum inclination, resulting in different inclination conditions. At this time, the calculation of stratum inclination Angle will also change, and a= Arctan (H2 / S2) is still used for calculation. However, in the formula, H2 represents the formation thickness - the distance from the entry point to the exit point.

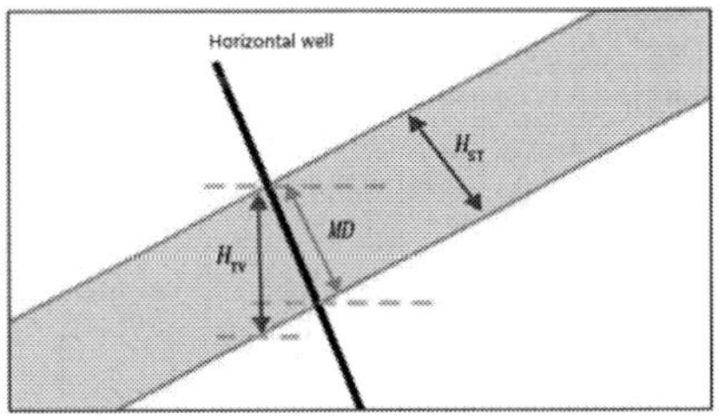

Fig. 2 Calculation of dip Angle under dip condition

Using stratum thickness to calculate, its application is simple, simple and direct, and can effectively control the error. However, if the stratum is thin, it is difficult to determine the apparent thickness and the specific thickness of the stratum, and the accuracy of the determination will be greatly reduced, which directly affects the determination effect.

1.1.2 Imaging logging calculates formation dip. Imaging logging is one of the widely used techniques in geosteering. When it is used in the calculation of formation dip Angle, the calculation principle is relatively simple. It is mainly used for logging with the azimuth of ground well. By picking up the distance between the two measurement points and wellbore dimensions, the dip Angle can be measured accurately and effectively using geometric formulas. For example, in a density imaging log, there will be an Angle between the wellbore diameter and the density imaging detection range. See Figure 3. In the figure, L represents the interval between Wells and A represents the diameter of induction area at the lower end of density imaging, and the two form A boundary, where Angle B represents the included Angle between horizontal well and formation, and the inverse trigonal function can be constructed between the three lines, whose specific calculation formula is β = Arctan [(L + 2R)/A]. At this time, at the well inclination Angle O, the formation inclination A is β+ θ -90 °. In the formula, R represents the primary radius of density imaging. Density imaging is used to predict and analyze reservoir interface, and the measurement accuracy is good. However, considering the high cost of imaging logging, imaging detection technology is seldom used in geosteering.

 https://doi.org/10.1051/e3sconf/202235801026

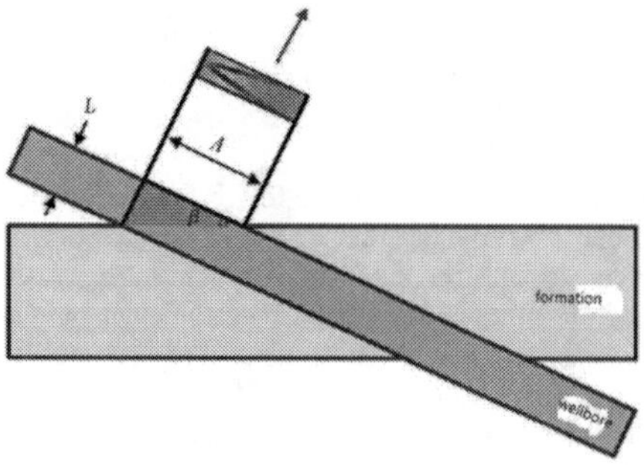

Fig. 3 Angle between density imaging logging and formation

1.1.3 Use offset Wells to calculate dip angles. The offset well calculation method is also one of the common calculation methods, which is mainly applied to the data related to the marker layer of the disabled offset well. For example, offset well zoning, thickness stability, lithologic characteristics, etc., after obtaining relevant information, the formation dip Angle can be calculated based on altitude and horizontal distance. In the actual calculation process, it is necessary to understand the main trend of the formation along the well, which mainly includes downdip mode along the well track and updip mode along the well track. In different modes, the formation dip Angle is different. FIG. 4 and FIG. 5 show different dips in different modes. The dip Angle of the formation is mainly affected by the horizontal distance between offset Wells and the vertical height difference. Therefore, the following formula is constructed: α = arctan (H1 / S2). In the whole formula, the distance between the horizontal well S1 traversing the zone and the offset well, and H1 represents the vertical elevation difference between the two Wells.

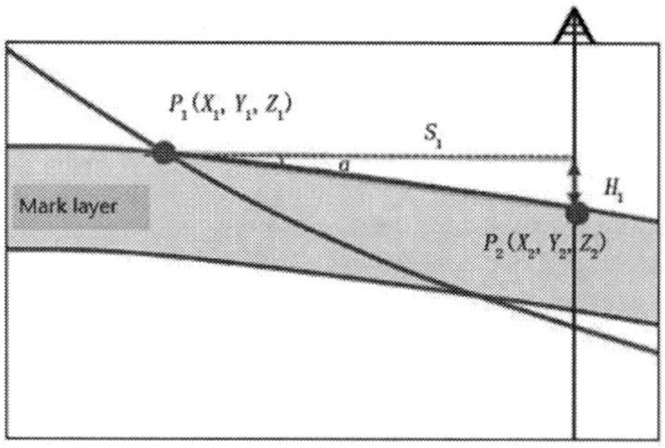

Fig. 4. Formation dip pattern along well trajectory

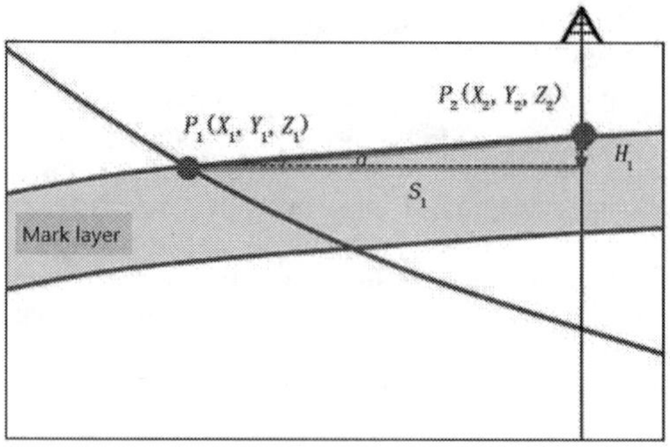

Fig. 5 Updip pattern of the formation along the well trajectory

Offset logging is a comprehensive computing technique that can be used to quickly estimate dip angles for different conditions. In the process of application of this calculation technology, it is necessary to measure the distance between Wells, but in fact, the error of the distance between Wells is relatively large, which will seriously affect the logging problems and is not conducive to the development of various logging work.

3.2 Solution of boundary distance equation

After the calculation of dip Angle, the boundary distance equation needs to be built, and the boundary distance equation can be used to continue the prediction of reservoir interface. Similarly, boundary formation prediction also includes down-hole dip and up-hole dip modes. In the whole process of boundary distance calculation, the formation is mainly calculated comprehensively along the down-hole dip mode of well trajectory to ensure the good implementation of calculation and improve the application effect of well calculation. The calculation of boundary distance is also very important. The position point of horizontal well bit, the entry point and exit point of marker layer and the thickness between target layers are the main factors affecting the boundary distance. Therefore, a variety of data application values can be mixed to construct trigonometric functions, and the boundary distance can be reverse-computed by using trigonometric functions. The actual analysis of boundary distance calculation is as follows, and the distance study of boundary distance formula is as follows:

1.2.1 In the downward dip state of the ground layer, L 1 = (Z3 − Z2) − S 3 tanα

In the formula, Z3 represents the bit position point.

In the formula, S3 represents the horizontal distance between the bit and the exit point.

1.2.2 When the ground layer is tilted upwards, L 1 = (Z3 − Z2) +S 3 tanα

In the formula, Z3 represents the bit position point.

In the formula, S3 represents the horizontal distance between the bit and the exit point. See Figure 6/7.

https://doi.org/10.1051/e3sconf/202235801026

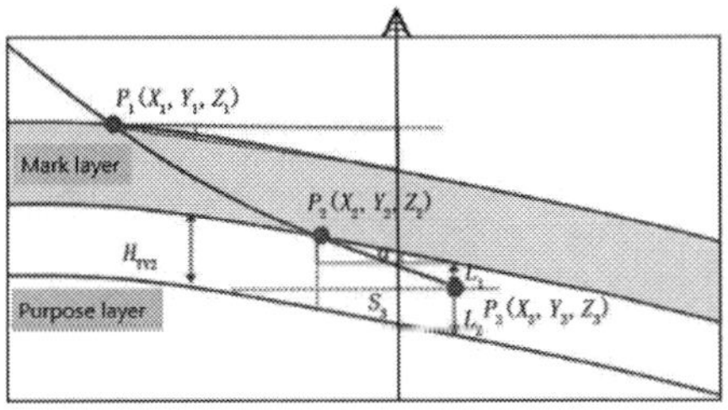

Fig. 6 Formation dip pattern along well trajectory

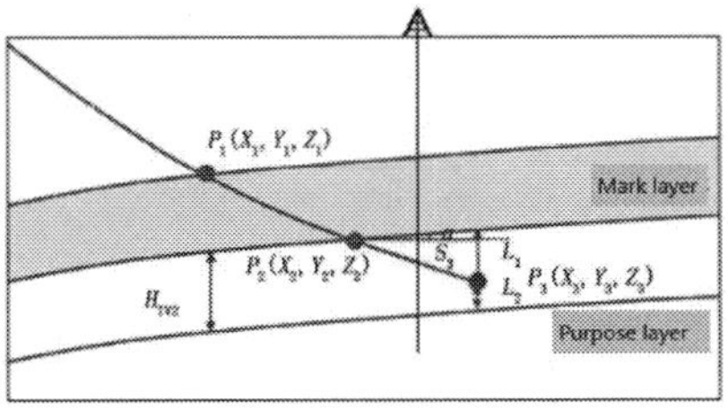

Fig. 7 Formation upwelling pattern

3.3 Application of integrated geosteering process

Integrated geosteering processes are also used in the new forecasting method. In the implementation of geological steering of horizontal Wells, it is difficult to adjust the work in the later stage if there is initial deviation. So, in the prediction of the new method is proposed in the integration process of geosteering, main is to establish a geosteering pretreatment scheme, after the deviation problem, can according to the integration of geosteering process to finish the work, the following is a new method of prediction of geosteering technology process, geosteering technology process analysis:

1.3.1 Complete data collection. Data collection is the premise of geological traverse. In the whole process of geological data collection, geological data, reservoir data, geophysical data and marker layer data in horizontal well area should be collected.

1.3.2 Geological design analysis shall be completed, and geological design analysis shall be required after data collection to ensure the good development of various geological designs and improve the subsequent geological guidance management. In the whole stage of geological design, it is required to analyze the original structural formation dip Angle, calculate the formation dip Angle by adjacent Wells, calculate the formation dip Angle by itself and pick up the formation dip Angle by imaging, mainly through the analysis and calculation of the formation dip Angle to determine the basic situation of the reservoir [2].

1.3.3 Optimizing the well trajectory and analyzing the reservoir boundary distance according to formation dip Angle, formation correlation and formation thickness plays a very important role in the application of formation dip Angle and can improve the application effect of formation.

1.3.4 Finally, analyze and understand the actual situation of the geological reservoir to confirm whether the reservoir meets relevant standards and ensure the good development of subsequent oil Wells [3].

4. The main role of the new method of reservoir interface prediction in geological steering of horizontal Wells

The new prediction method of horizontal well reservoir interface is introduced above, including integrated geosteering process, formation dip Angle calculation combination and boundary distance calculation and analysis. In the specific research, in order to verify the application of the new prediction method proposed in this paper, the logging of an oilfield as an example, summed up the specific application of 13 oil Wells in the oilfield geological guidance.

3.1 The field consists of 13 horizontal Wells with a 100% hit rate and 90% penetration rate and a drilling cycle time of 24 hours.

3.2 In the process of drilling, in order to verify the new method of reservoir interface prediction, a new type of reservoir interface prediction software is constructed by combining information technology. The prediction software is used for comprehensive drilling prediction to ensure the good implementation of various prediction work [4].

3.3 In this well, a new method of reservoir interface prediction was used to understand dip and other conditions. Through calculation and analysis, it is found that ZJ09, ZJ08 and ZJ10 are 22.3m, 5.8m and 6.5m respectively from landing. The prediction showed that the entire interval was drilled in 12.25 minutes. The dip angles were 89°, 89.5° and 0° in different cases throughout the drilling operation. The logging was a fine fluorescent sandstone, and the drilling was successfully completed. The plane position of the drilling target was less than 20m away from the design position, which was shown by relevant technical statistics to ensure the completion of the whole calculation and application, and played a good role in the application of logging calculation [5].

5. Conclusion

This paper summarizes the application of the new method of reservoir interface prediction in horizontal well geosteering based on theory and practice. It is hoped that the research of this paper can be helpful to the application of horizontal well formation drilling method and promote the good implementation of formation drilling.

References

1. HUANG Yujin. Application of a new method of reservoir interface prediction in geological steering of horizontal Wells [J]. Logging Engineering, 2022, 33(1):6-6
2. Luo Shengyuan, Chen Xiaohong, Ni Fangjie. Application of natural gamma spectrometry logging

in organic-rich shale evaluation and geological guidance [J]. Well logging Technology, 2021, 45(3):8-8

3. du. Application of geosteering technology in horizontal well development in Linxing Middle block [J]. Frontiers in Earth Science (Hans), 2020, 10(1):8-8
4. Yang Bin, LI Linyan, KONG Jian, et al. Special Oil and Gas Reservoirs, 2020, 27(2):7-7
5. XIE Fei, WANG Yun, MEI Junwei, et al. A new calculation method of formation apparent dip Angle based on formation rotation: Application to geological steering of horizontal Wells [J]. Natural Gas Industry, 2020, 40(6):8-8

Research on fast identification model of water-flooded layer in old oilfield-- Taking Xingbei area of Daqlng Oilfield as an example

Xiaoyu Shang

No.4 Oil Production Plant in Daqing Oilfield Co Ltd., Daqing 163511, China

Abstract: After long-term water flooding development in old oilfields, oil layers are generally flooded. Accurate and rapid recognize the water flooding layer is the key to later infill well layout and development plan adjustment. In this paper, taking Xingbei area of Daqing Oilfield as an example, on the basis of clarifying the characteristics of the water-flooded layer curve, through logging curve optimization, data preprocessing and algorithm model optimization processes, a rapid identification model of water-flooded layer suitable for this block is established. The results show that the HAC, CAL, RLLS and RMG curves with hidden duplicate information can be removed through the correlation screening of logging curves and the importance score of the tree model, which can reduce the amount of data calculation. When the four algorithms are used to identify the flooding level of each layer, the recognition rate of the XGboost algorithm can reach up to 95.45%; the reliability of this result has been confirmed in the model verification process (87.89%), which further shows that the model can be used to identify Xingbei area flooded.

Key words: oil field; water-flooded layer; logging curve; data processing; algorithm model.

1. Introduction

The continuous development of oil and gas resources in old oilfields makes the water cut of oil Wells up to 99%, and the current situation of flooding is serious [1]. Setting infill well is one of the main means to improve the development effect of old oil fields, which can control the water cut of oil Wells and improve the recovery rate at the same time [2-3]. Therefore, it is of great practical significance to quickly identify the distribution characteristics of water-flooded layer for the arrangement of infill well, the adjustment of development plan and the improvement of water flooding efficiency in old oil fields. At present, the water-flooded zone identification of sandstone reservoir mainly relies on logging curve identification technology [4]. Reservoir waterlogging characteristics are closely related to sedimentary microfacies, sedimentary rhythm and other factors of sand bodies. The heterogeneity within and between reservoir layers has corresponding changes in the degree of waterlogging, which is reflected in the logging curve [5]. For example, the amplitude of deep and shallow lateral resistivity decreases, while the amplitude of acoustic curve increases. Therefore, the key to identify the characteristics of flooded layer is to excavate the hidden information of logging curve quickly and accurately. Traditional methods generally identify and classify aquifer levels based on the undulatory characteristics of logging curves in flooded zones combined with intersection graph [6], deletion graph analysis [7] and other methods, which are time-consuming and have low accuracy. With the development of artificial intelligence technology, data mining has been applied in the identification of flooded layer. Yang Mingren et al. used adaptive enhancement algorithm to identify the water-flooded layer characteristics of Changqing tight sandstone reservoir [8]. Li Jianping et al. used quantum neural network technology to automatically identify the flooded layer and improve the accuracy of the flooded layer identification [9]. Of course, there are also logging technologies such as C/O ratio spectral logging [10] and C/H ratio logging that can directly identify the characteristics of the flooded layer [11]. Although the recognition accuracy is higher, the cost is relatively high. Therefore, based on the data mining technology, this paper proposes an intelligent classification method combining the optimization of the corresponding curve of the water-flooded layer, data standardization processing and XGboost algorithm to realize the rapid identification of the water-flooded layer in the old oilfield in Xingbei area of Daqing.

 https://doi.org/10.1051/e3sconf/202235801027

2. Classification of waterlogged layer in Xingbei area of Daqing

Xingbei area in Daqing is located on the secondary structure of Daqing placantide in the central depression of Songliao Basin, and the main oil and gas reservoirs (Saertu and Tuohua oil reservoirs) are typical fluvial delta facies sedimentary system [12]. After more than 40 years of development, the current comprehensive water cut of the reservoir has exceeded 90%[13]. At present, it is urgent to clarify the distribution characteristics of the water-flooded reservoir and establish a rapid identification method, so as to provide a basis for the setting and adjustment of the infill-well in the later stage. During reservoir flooding, the injected fresh water continuously dilutions the original formation water and increases the formation water saturation, resulting in the continuous decrease of the formation resistivity, density and other logging curves, while the acoustic time difference and spontaneous potential curves continue to increase. According to the degree of flooding and the corresponding characteristics of logging curve, the water-flooded layer in Xingbei area of Daqing has been divided into three levels by using movable water saturation: low water flooding, medium water flooding and high water flooding (Table 1). In this paper, according to the existing waterlogged layer classification standard, the algorithm classification calibration is carried out in the later stage.

Table. 1 The division standard of water-flooded layer in sandstone reservoir in Xingbei area of Daqing Oilfield

Flooding level	Mobile water saturation （%）	Classification of algorithms
Not flooded	＜12	0
Low water	12~24	1
Water flooded	24~36	2
High water	＞36	3

3. Rapid identification of flooded layer

3.1 Log curve optimization

The Wells in the study area contained 14 logs: Acoustic time difference (HAC), borehole compensated acoustic wave (BHC), hole diameter (CAL), density (DEN), gamma ray (GR), spontaneous potential (SP), 2.5m bottom gradient resistivity (R25), compensated neutron (CNL), eight lateral resistivity (RFOC), deep lateral resistivity (RLLD), shallow lateral resistivity (RLLS)), microgradient (RMG), micropotential (RMN) and flushing zone resistivity (RXO). In order to reduce the repeatability indication information of the logging curve in the flooded layer, the logging curve with hidden similarity information was firstly identified through the correlation screening of the logging curve, and then the similar logging curve was removed through the importance score of the tree model, so the logging curve was optimized and the amount of calculated data was reduced.

(1) Well log correlation screening

Correlation screening is an important data processing method to remove repeated computation and reduce computation time [14]. Correlation screening of well logging curves in the study area can remove one or more curves and improve the calculation efficiency. The correlation coefficient between two curves can be obtained by Pearson function calculation. The higher the correlation coefficient is, the information implied by two curves is about similar. Generally, a correlation coefficient greater than 0.8 is considered to be highly similar [14]. FIG. 1 shows the correlation analysis of 14 known logging curves, among which the strong correlation (R>0.8) includes: HAC and BHC, CAL and DEN, RLLD and RLLS, RMG and RMN

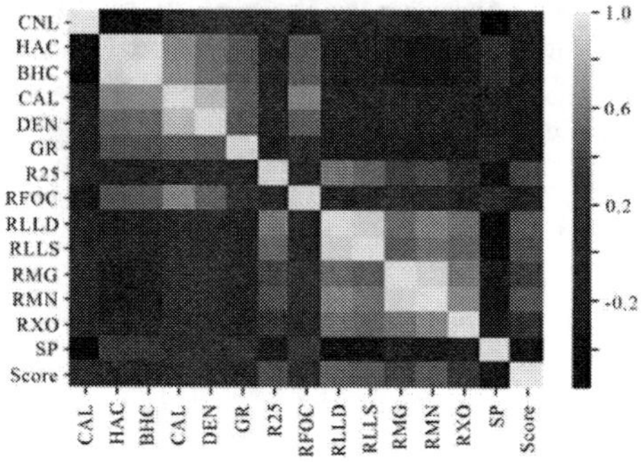

Fig. 1 Correlation screening of 14 logging curves in the study area

(2) Importance screening of logging curve tree model

In order to delete one of the similarity curves, we applied the tree model method to perform the importance score [15] to further optimize the logging curve. Figure 2 shows the scoring results, in which the corresponding degree of the flooded layer is good, and the score is often higher. We deleted the low scores from the well-correlated logs: HAC, CAL, RLLS, and RMG. Ten curves, BHC, DEN, GR, R25, RFOC, CNL, RLLD, RMN RXO and SP, were selected as the data mining research of flooded layer identification.

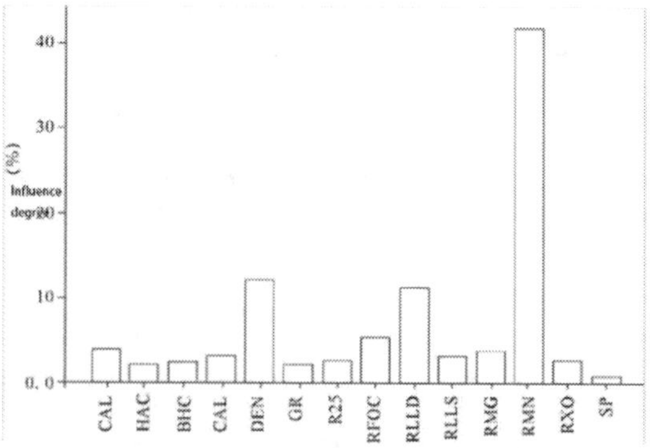

Fig. 2 Importance tanking of decision tree for 14 logging curves

3.2 Log curve preprocessing

Standard deviation normalization is to conduct center translation transformation and dimensionless compression processing on logging curve data, so that each dimension of data has zero mean and unit variance, which can be used to eliminate the influence of large amplitude curve in logging curve [16]. It is divided into three steps :(1) calculate the mean value of the data in each dimension (calculate with all the data); (2) The mean value is subtracted from each dimension; (3) Divide each dimension of the data by the standard deviation of the data on that dimension. As can be seen from the comparison figure before and after treatment (FIG. 3), the change shape of the curve does not change, but the details of the change amplitude are reduced, which reduces the influence of abnormally high or low values.

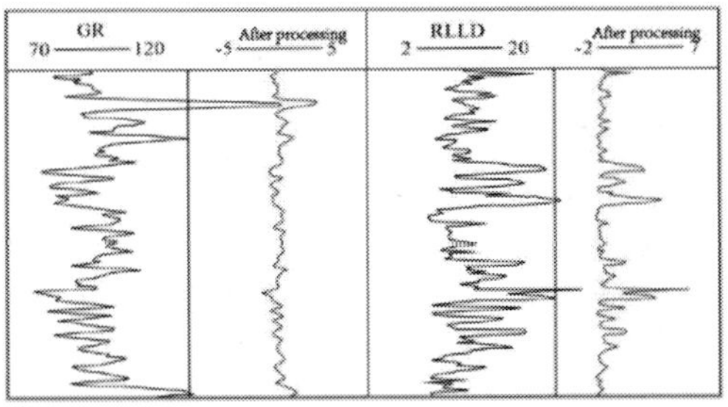

Fig. 3 The comparison diagram of logging curve of X5-1-2008 well before and after preprocessing

3.3 Classification and recognition model construction of flooded layer

A variety of classification and recognition models are applied here to achieve the purpose of algorithm optimization, which are as follows: Support Vector Machines (SVM) [17], eXtreme Gradient Boosting (XGboost) [18], Gradient Boosting Decision Tree (GBDT) [19] and Random Forest (RF) [20] algorithm. Firstly, 10 Wells were randomly selected, including 825 low-flooded Wells, 1156 medium-flooded Wells and 1086 high-flooded Wells, which had completed water-flooded zone identification and curve pretreatment. Then, logging datasets of the same flooding level are pooled and mixed. Finally, "cross validation" was applied in the actual calculation, and the flooded layer data was divided into training set (TRAIN, 90%) and test set (test, 10%). The classification and recognition calculation process was repeated for 10 times, and the average recognition rate was finally used as the evaluation standard of the model. All algorithmic procedures were completed on MATLAB 2010B (The Mathworks Inc., USA). The flow chart of the completed data processing module is shown in Figure 4.

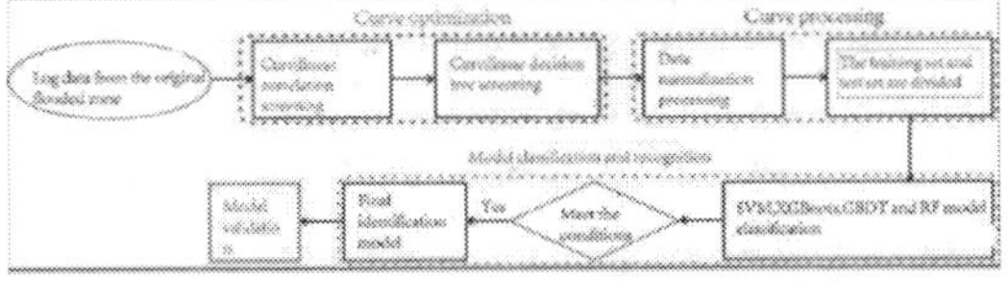

Fig. 4 The flow chart of data processing model

4. Evaluation and validation of identification result of flooded layer

4.1 Evaluation of identification result of flooded layer

We selected 14 logs for each well using correlation screening and tree model importance scoring. It can be seen from FIG. 1 that the correlation coefficient R2 greater than 0.8 is the same type of logging curves, such as HAC and BHC. Despite the differences in their implied information, their role in indicating the flood level is basically the same. In addition, it can be found from Figure 2 that similar curves, such as RLLS and RLLD, are more indicative of the flooded layer than those with shallow detection depth. Therefore, we finally deleted 4 kinds of curves, namely HAC, CAL, RLLS and RMG, and used the remaining 10 kinds of curves for flooded layer identification.

Table 2 is the comparison of the recognition results of SVM, XGboost, GBDT and RF algorithms when the selected flooded layer data of 10 Wells are used for model recognition. Through comparison, it is found that the recognition accuracy of all the classification algorithm models is high, and the recognition accuracy is more than 85%. Among them, XGboost has the highest accuracy, which can reach 95.45%. Therefore, XGboost algorithm is selected to establish a fast identification model when identifying the flooded layer level in Xingbei area based on logging curve. This is because XGboots algorithm is a novel tree learning algorithm that can deal with sparse data. It approximates the learning tree through distributed weighted histogram algorithm and reduces the contingency of sampling results combined with cross-validation [18], so it has good model prediction performance for water-flooded layer identification of such unbalanced sparse data.

Table. 2 The comparison of calculation results of different classification algorithms

Algorithm model	Ordinary precision (%)	Cross validation accuracy (%)	The recall rate (%)
SVM	87.98	86.44	87.98
XGboost	95.45	95.30	95.45
GBDT	90.03	89.37	90.03
RF	95.04	94.80	95.04

4.2 Validation of waterlogged layer identification model

In order to further verify the reliability and universality of the selected model, we randomly selected an untrained well to identify its flood level. In the actual prediction process, we put the data into SVM, XGboost, GBDT and RF algorithm models respectively, and the classification results are shown in Figure 5. The classification results were 87.57% (SVM), 87.89% (XGboots), 87.02% (GBDT) and 86.35% (RF) respectively. The XGboost model still had the highest recognition rate. This result is consistent with the recognition result after the previous model training. Therefore, we believe that the optimized XGboost model is most suitable for rapid identification of flooded layer level in Xingbei area.

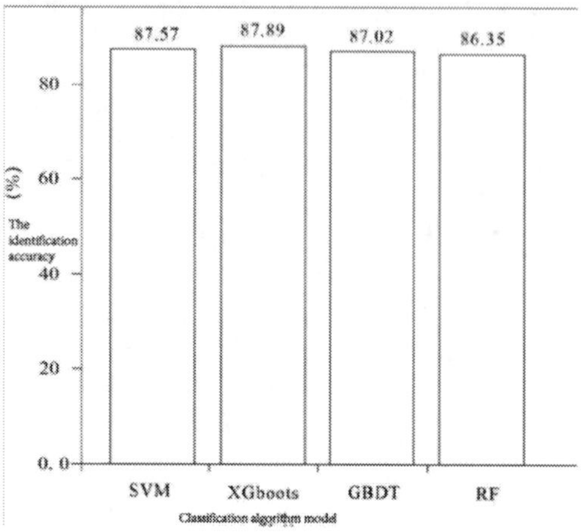

Fig. 5 The comparison of different model results during verification of water-flooded layer

hat the model building process here can be used as a reference for other blocks, but the established model may not be applicable to other blocks. Of course, if there is sufficient data training, the universality of the model will be higher. In addition, learning techniques such as data transfer [21] can also improve the universality of the model.

5. Conclusion

Based on the logging curve, this paper established the waterlogged layer identification model in Xingbei area through algorithm optimization, curve processing and classification model comparison. The conclusion is as follows:

(1) HAC, CAL, RLLS and RMG curves with implicit duplicate information were deleted by logging correlation screening and tree model importance scoring methods, and the remaining 10 curves were normalized by standard deviation to reduce the influence of abnormally high or low values.

(2) When identifying the flooding level, the recognition accuracy of SVM, XGboost, GBDT and RF is higher than 85%, and the XGboost algorithm has the highest recognition rate (95.45%).

(3) When verifying the established algorithm model, the difference between the results of the four algorithm models is small, but the highest accuracy is still the XGboost algorithm model (87.89%), which confirms the reliability of this model for identifying the flooded layer in Xingbei area.

References

1. Liu Yanhong, Luo Xingwang, Zhang Wu, Li Chen, Zhang Ruixue, Luo Qiang, Wang Bin, Zhang Yanmei. Study on logging identification method and influencing factors of waterlogged layer in tight sandstone reservoir. Special hydrocarbon reservoirs.
2. Ma Tao. Analysis of flume delta depositional system in continental large depression basin under arid climate conditions: A case study of Quan-4 Member of Fuxin Uplift belt in southern Songliao Basin [D]. Beijing: China University of Geosciences (Beijing), 2006.
3. Andrew D. Miall. Reconstructing the architecture and sequence stratigraphy of the preserved fluvial record as a tool Reservoir development: A reality check[J].AAPG Bulletin,2006,90(7): 989-1002.
4. Ju Wu, Shen Huilin, Yang Hong, et al. Research on logging evaluation method of water-flooded layer in conglomerate reservoir in junggar basin [J]. Progress in geophysics, 2009,24(3):974-980
5. Tian Zhongyuan, Mu Longxin, Sun Deming, et al. Study on logging characteristics and mechanism of conglomerate water-flooded layer [J]. Acta Petrolei Sinica, 2002, (6) : 50-55+3
6. Chen Ganghua, Liang Shasha, Wang Jun, et al. Application of machine learning AdaBoost.M2 algorithm in conglomerate fluid identification. Oil geophysical prospecting, 2019,54 (6) :1357-1362.
7. Chen Xin, Chen Kegui, Wang Zhaofeng, Dai Xiongjun, Fan Xiangyu. Logging response characteristics and waterlogging analysis of flooded zone in Aryskum oilfield. Progress in geophysics.
8. Mingren Yang, Huilin Shen, Sa Qu, Qipeng Sun, Limin Zhang, Shuming Xiao. Application of AdaBoost algorithm in water-flooded layer identification of tight sandstone. China offshore oil and gas (in Chinese), 2021,33(4):1673-1506
9. Li Jianping, Yang Du, Fan Yougui. Quantum AD hoc network model and its application in oilfield waterlogged layer identification [J]. Microcomputer applications, 2020, 36 (12): 9-11+15.
10. Wang B T, Gao Y F, Fan Y X, et al. Study on logging evaluation method of remaining oil in Lamadian Oilfield [J]. Journal of oil and gas technology, 2014, 36 (7): 84-88+6.
11. He Zhongsheng, Cui Zhigang, Chen Guanghui, et al. Evaluation of residual oil and water flooding degree using hydrocarbon ratio logging technology [J]. Journal of southwest petroleum university (science & technology edition), 2016,38 (5) : 41-49.
12. The white deer. Study on main controlling factors of Fuyang reservoir formation in Xingbei Oilfield, Daqing Placanticline, Northeast Petroleum University. 2015.
13. Zhang Haiying. Development strategy of Xingbei Oilfield in ultra-high water cut stage. Scientific management. 2019.3:256.
14. Sun Peipei, Zhang Ning, Zhao Linjing, Peng Yuanyuan, Liu Xijian, Liu Xiaohui, Zhai Yassen, Feng Meiqing. Effects of subinhibitory concentration of cinnamon essential oil on lipid homeostasis in

Candida albicans [J]. Chinese Journal of Analytical Chemistry,202,50(01):92-107.

15. Cao Taoyun. Research on variable importance based on random forest [J]. Statistics and Decision, 222,38(04):60-63.
16. Li Sihan, Wang Changquan, Wu Hua, Liu Tao, Zhao Xu. Normalization method of capillary pressure curve based on effective reservoir thickness [J]. China science and technology papers, 2020,15(05):593-598.
17. Zhang Junhua, Ren Xiongfeng, Zhao Jie, Tan Mingyou, Yu Zhengjun. Reservoir prediction method and application based on cross-validation support vector machine [J]. Science technology and engineering, 2020,20(13):5052-5057.
18. Yan Xingyu, Gu Hanming, Xiao Yifei, Ren Hao, Ni Jun. Application of XGBoost algorithm in logging interpretation of tight sandstone gas reservoir [J]. Oil geophysical prospecting,2019,54(02):447-455+241.
19. Gu Yufeng, Zhang Daoyong, Bao Zhidong, Feng Zhigang, Li Jinggong. Prediction of permeability by gradient lifting Decision Tree (GBDT) : A case study of tight sandstone reservoir in the west Chang 4+5 member of Jiyuan Oilfield [J]. Progress in geophysics (in Chinese),2021,36(02):585-594.
20. Wang Yan, Wang Ruogu, Wei Keying, Lu Haijiao, Xiong Xiaowei, Zhang Tianjie. Classification of tight reservoir based on random forest: A case study of the 8th Member of Eastern Hehe Formation in Yan 'an Gas Field [J]. Journal of xi 'an shiyou university (natural science edition),2021,36(06):1-8.
21. Yang C, Zhao H, Bruzzone L, Benediktsson JA, Liang Y, Liu B, Lunar Impact Crater Identification and Age Estimation with Chang 'e Data by Deep and Transfer learning. Nat Commun 2020; 11 (1).

Study on the ambulatory monitoring and optimization measures of the Gas wells in block A

Shujuan Yan

Geological Brigade,No.7 oil production plant, Daqing Oilfield Co., Ltd, Daqing, China

Abstract. The analysis of the Gas well is the core work and accompanying with the whole process of gas field development. The dynamic monitoring technology of gas well is an effective way to provide theoretical basis for dynamic analysis. Now there are 3 normal production gas Wells in the production test area of block A. In the process of production, pressure recovery, hydrostatic pressure and corrosion monitoring were carried out for the changes of production, pressure, fluid physical properties and underground surface engineering.We know the fluid accumulation, the corrosion of the downhole string, the reservoir characteristics, the reservoir physical parameters by these monitoring activities. So we guide the reasonable production of gas Wells timely and effectively, in order to better provide a basis for gas well production management and scientific decision-making.We should carry out rectification and strengthen management to problematic gas Wells and loopholes in dynamic monitoring, and provide better assurance for the development of gas fields.

Key words: Development of Gas field; Ambulatory Monitoring; Corrosion Monitoring.

1. Introduction

Block A of the oilfield is located in the south of the Nose structure belt of Q oilfield, namely the No.1 trap.This fault-nosed structure is formed by the no. 8 fault cutting the Pu Nan nose-shaped structural belt.The Hei Di Miao oil layer is developed in the 3rd and 4th member of Nen formation, the 4th member of Nen formation is group H Ⅰ, and the 3rd member of Nen Formation is Group H Ⅱ.According to the rock cycle and section structure, the fourth member of Nen formation can be divided into six cycles, and the third member of Nen formation can be divided into four cycles.The reverse cycle of Hei Di Miao reservoir is characterized by the lithology changing from fine to coarse and from bottom to top.The bottom of the cycle is the stable distribution of mud stone, and the upward is mud silt, silt, silt-sand, fine sand, sometimes visible the middle sand.And the dark mud stone segment gradually thinned from bottom to bottom.The effective porosity of the reservoir ranges from 28.1% to 40.1%, with an average of 33.9%.Air permeability ranges from 80×10-3μm2 ~ 9427×10-3μm2 with an average of 1736×10-3μm2. The clastic material in the oil layer has a low grain size, and is generally silt grade.The rock type belongs to feldspar lithic sandstone.The quartz sand ranges from 23.0% to 25.0% with an average of 25%.The feldspar ranges from 28.0% to 34.0% with an average of 31.0%.The detritus ranges from 27.0% to 38.0%.The average particle size (Mz) is 0.082mm,the standard deviation (σ1) is 1.711,the sorting is poor,and the mud content is 9.8%.

According to the development stage and actual situation of gas field in block A ,Production Wells are mainly monitored.The main content of dynamic monitoring includes pressure, temperature, production, underground surface engineering, fluid properties and composition, etc.The system has been optimized and improved through practical application, which basically meets the needs of dynamic tracking analysis and production management of gas field development.

2. Analysis of Dynamic Monitoring Effect in Test Area

2.1 The Test of Flow Temperature, Flow Pressure, Static Temperature and Static Pressure

In order to understand the variation law of formation pressure and well bore fluid in the test area.Pressure-temperature gradient tests were carried out for single production Wells and grasp formation pressure change of gas reservoir timely and accurately. So in the future, dynamic reserve calculation, reasonable production allocation of gas well and evaluation of gas reservoir development effect play an important role.

a well

Combined with pressure-gradient test results from the well in 2009 and 2019, the fluid density in the wellbore is roughly evenly distributed in the shut-in state, and there is no fluid accumulation in the manometer measuring point depth to the wellhead.The test data show that the flow pressure gradient is slightly greater than the static pressure gradient, indicating that the water and recombination produced in the well opening state are mixed with the flow. After shut-in, water and recombination are subsiding and there is pure gas column in the wellbore (Figure 1-Figure 2).

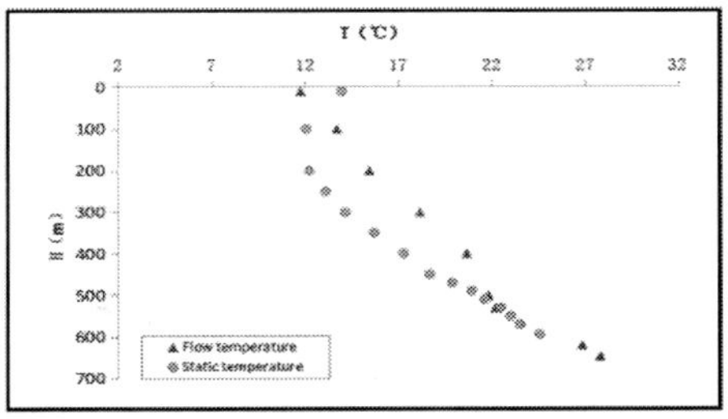

Figure 1. Comparison of flow temperature and static temperature of well a in block A

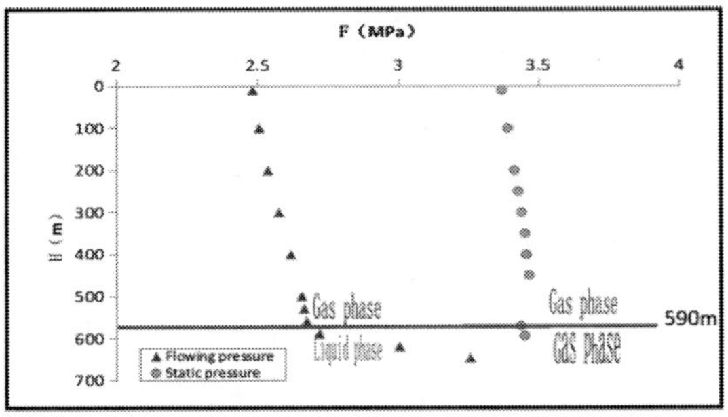

Figure 2. Comparison of flow pressure and static pressure of well a in block A

(2) b well

After the geological scheme, engineering design and all field work for pressure - temperature gradient test of well b, there is no influence of wellbore fluid accumulation in well b in both open and closed state (Figure3).

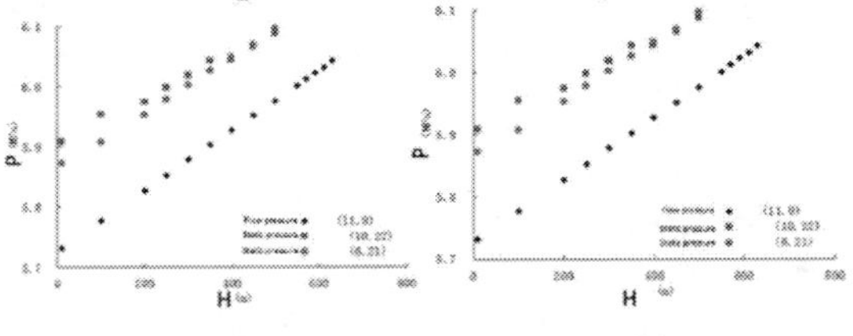

Figure 3 Distribution diagram of wellbore flow temperature, flow pressure, static temperature and static pressure in well b

c well

After the geological scheme, engineering design and all field work for pressure - temperature gradient test of well c,there is no influence of wellbore fluid accumulation in well c in both open and closed state (Figure4).

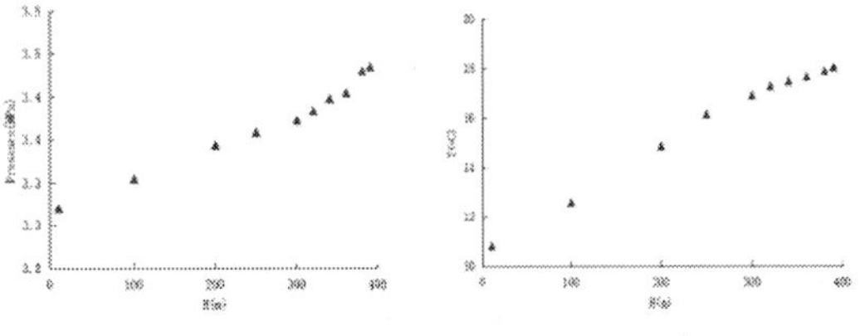

Figure 4. The distribution map of flow temperature and flow pressure in well c wellbore

Through the test of flow temperature, flow pressure and static temperature and static pressure, it can be concluded that there is no influence of wellbore fluid accumulation in the opening state of three gas Wells. The small change of pressure indicates that the gas reservoir has good gas supply capacity.

2.2 Buildup Test

In order to understand the gas well pressure recovery situation, study the reservoir characteristics of gas reservoirs, determine reservoir physical parameters and single well gas supply range and the reservoir pollution near the bottom of the gas well. So well b is selected for pressure recovery test (Figure5-Figure8).

The gas pressure will change regularly and propagate in all directions like water waves When the gas well is shut in for pressure measurement. In the range of ripples, pressure sweeps the micro and macro structure of gas layer at each point. According to the scanning information , the macroscopic characteristics and the parameters of the gas layer within the scanning range can be judged. Therefore, pressure recovery test data and other information was used to measure many characteristics of the test zone and well.

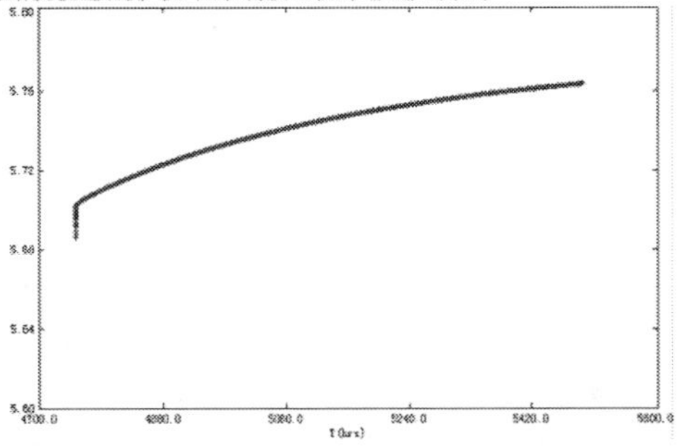

Figure 5. The pressure recovery curve of well b

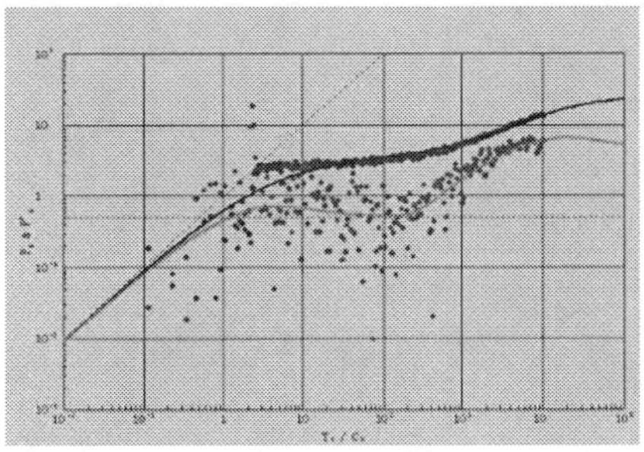

Figure 6. The double log fitting curve of well b

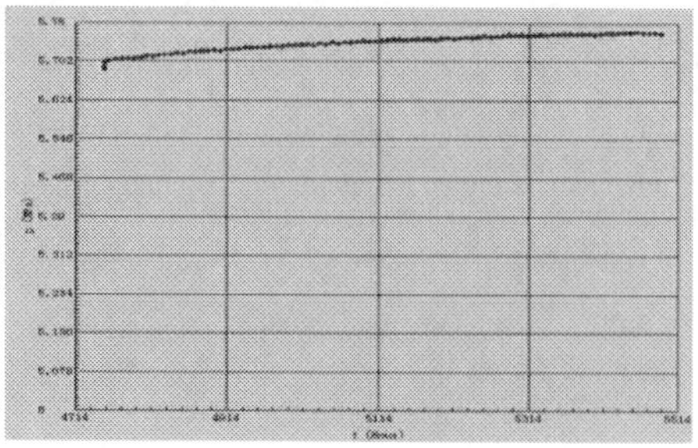

Figure 7. The history fitting curve of well b

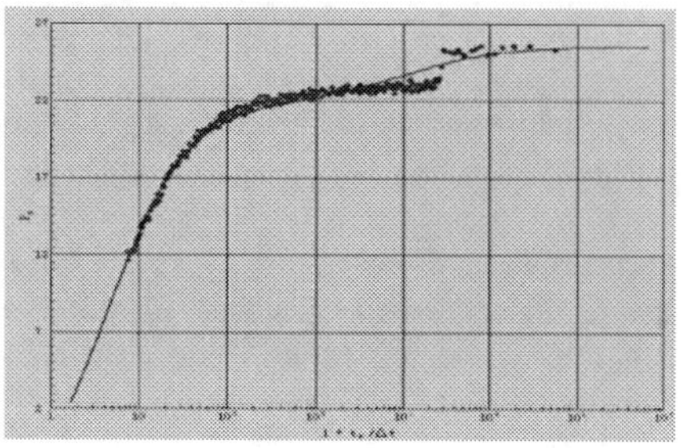

Figure 8. The Horner Dimensionless test curve of well b

By means of pressure recovery well test, combined with geological dynamic and static data, the formation parameters of well b are obtained by using variable well reservoir+radial coincidence+infinite formation. The flow coefficient is 158.23 μ m^2 .m.Effective permeability of formation was 0.20μm^2.The skin coefficient was -4.25. And the detection radius was 2224.15. The production status of the well was analyzed through the study of various parameters. The skin coefficient of the well decreased from 30.08 in 2011 to -4.52 at present. It shows that the pollution degree of the well has been effectively improved (Table1,Figure9).

Table 1. Pressure recovery well test data sheet of well b

model	Variable well reservoir + radial coincidence + infinite formation	
Average formation pressure	5.8058	MPa
Permeability k	0.1977	μm^2
Flow coefficient kh/μ	158.228	μm^2 •m/mPa•s
Formation coefficient kh	2.0165	μm^2 •m
Fluidity k/μ	15.5126	μm^2 •m/mPa•s
Well store constant C	257.417	m^3/MPa
Well bore skin factor Sw	-4.2529	
mobility ratio	8.299	
storativity ratio	10.643	
Interface radius r	315.943	m
radius of investigation ri	2224.152	m

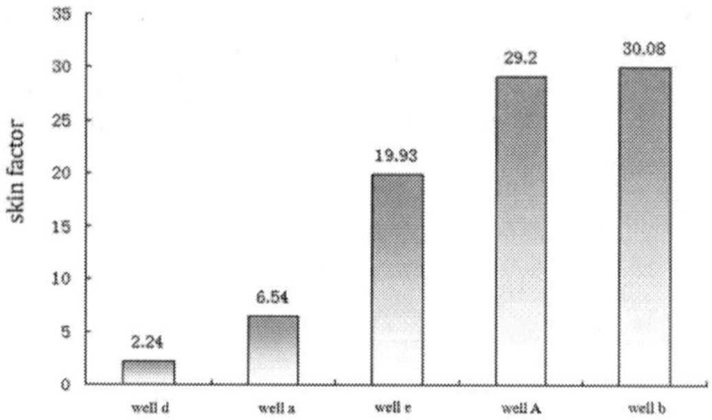

Figure 9-Skin coefficient distribution diagram of gas well test stage in block A

2.3 The Corrosion Monitoring of Gas Well

Now, the corrosion monitoring of gas Wells is mainly carried out by electrochemical method. temperature and pressure are important related factors of corrosion rate For gas well corrosion monitoring. But corrosion rate is also affected by humidity (or partial pressure of water vapor), flow rate, etc.

Corrosion monitoring showed high corrosion rates in all five Wells. The corrosion rates ranged from 0.05mm/a~0.1543mm/a.The corrosion degree is moderate and severe according to the standard of NACE RP-0775-91 stipulated by the American Society of Corrosion Engineers (NACE)(Table 2).

Table 2. The table of Asme(American Society of Corrosion Engineers) Nace RP-0775-91 Corrosion degree specification

classification	Uniform corrosion rate(mm/a)	pitting rate(mm/a)
Mild corrosion	< 0.025	< 0.127
Moderate corrosion	0.025~0.125	0.127~0.201
serious corrosion	0.125~0.254	0.201~0.308
Extremely severe	> 0.254	> 0.308

The wellbore corrosion of N80 and P110 materials is serious according to the current downhole production conditions of Wells a and c. The well life is less than 10 years based on current corrosion rate monitoring. However, partial pitting or perforation may occur in some sections. Measures should be taken to protect the parts with high corrosion rate. Periodic corrosion monitoring is used to monitor the corrosion. In the production process, corrosion inhibitor is added periodically with the corrosion monitoring method which is combined to protect the wellbore. Ensure the normal and safe production of gas Wells (Table 3).

Table 3. The table of corrosion monitoring of gas well string by electrochemical method

gas field	serial number	well number	corrosion rate(mm/a)		The depth of main corrosion well (m)
			2020Y	2021Y	
the test area of A gas production	1	a	0.0955-0.1751	0.0520-0.1462	480-630
	2	c	0.0941-0.1563	0.05-0.15	300-390
	3	b		0.053-0.148	520-650
	4	d		0.05-0.147	500-650
	5	e		0.05-0.1543	375-500
Total (Wells)			2	5	

3. Study on the Optimization of Gas Well Dynamic Monitoring Measures

At present, the gas production test area A of Pu Nan Oilfield is still in the initial stage of testing. The underground geological conditions of single production well are complex. And the well condition can not fully meet the needs of production dynamic monitoring. The main performance has the following aspects:

(1) The phenomenon of gas and water flowing out together and oil and gas flowing out together occurs in production Wells. All of these will affect gas well production. In view of the above problems, the analysis of geological characteristics and production dynamic characteristics should be strengthened. A reasonable production system is established for the production of single well through the study of geological conditions and rational production allocation method. So maximize resource utilization and ensure long-term stable production of gas Wells (Figure10).

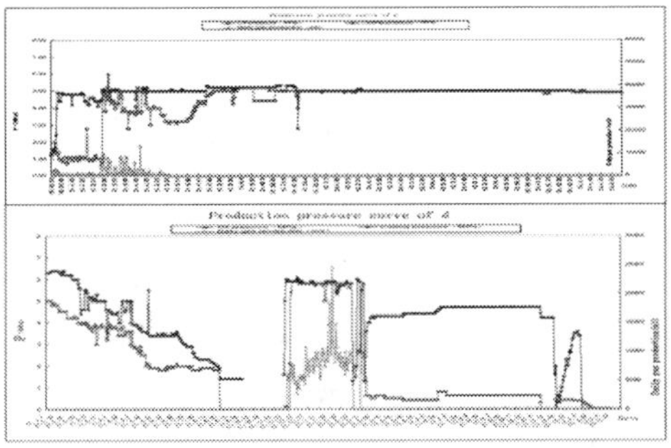

Figure10. The production curves for wells c and d

(2) The gas wells are usually low-yield industrial gas wells or middle industrial gas wells in the test area. Single-layer production capacity is difficult to meet the requirements of gas production profile testing. At present, the gas production profile test results are obvious for the single-layer daily gas production above 10,000 cubic meters. If the daily gas output is less than 10,000 cubic meters, the test results cannot be accurately accepted due to the insufficient accuracy of the test instrument. In order to provide technical support for reservoir data acquisition and analysis in the future, the study of gas production profile testing technology in middle and low pay zones was carried out for producing Wells in the experimental area.

(3)Because the perforation test technology of pipe transport is adopted in the gas well test in the test area, the perforation gun and screen tube of tubing are not lifted out after perforation. At present, the perforating gun and screen are still suspended on the wellbore tubing , and the bottom end of the tubing is sealed in some Wells. So the result of that cannot be properly tested through the tubing in some of the project's test instruments.

4. Conclusion and understanding

1.The results of dynamic monitoring show that there is no fluid accumulation at the bottom of the well and the corrosion degree is moderate or severe. And the formation parameters such as formation flow coefficient, effective permeability and skin coefficient are obtained. These results provide a more powerful theoretical basis for stratigraphic analysis. These also provide a better basis for gas well production management and scientific decision-making.

2. The analysis of geological characteristics and production dynamic characteristics should be strengthened for the problem Wells. In this way, a reasonable production system can be established for the production of individual Wells. So this method can not only ensure the long-term stable production of gas Wells but also maximize the resource utilization.

3. Downhole gun and screen should be removed by operational construction and install a bell mouth in order to test the future work smoothly. For the planned gas test Wells in the future, the perforating gun and screen pipe should be removed after the gas is discharged from the perforation. And the bell mouth should be installed. Then the system will be tested and put into production.

References

1. Jiang Fucong, Li Zhongquan et al. Journal of Chengdu university of technology (science & technology edition),2016,43(06):671-678.
2. Jin Xue-Yan, LIANG Jiang-ping et al. Sedimentary characteristics and accumulation response of Hei Di Miao oil reservoir in the southern Changyuan and Gulong area [J]. Petroleum geology & oilfield development in Daqing, 2018,37(02):1-8.
3. Wu Heyong, Liang Xiaodong, Xiang Caifu, et al. Syncline reservoir characteristics and reservoir-forming mechanism in Song Liao Basin. Science in China (series D), 2007,37 (2) : 185-191.
4. Xu Jianjun, Wang Bao'e, Yan Limei, et al. The strategy of the smart home energy optimization control of the hybrid energy coordinated control.

Transactions of China Electrotechnical Society, 2017, 32(12) 214-223.

5. Xu, J., Huang, L., Yin, S. et al. All-fiber self-mixing interferometer for displacement measurement based on the quadrature demodulation technique. Opt Rev. 2018, 25(1):40-45.
6. Heuer J K, Stubbins J F. An XPS characterization of FeCO3filmsfrom CO2corrosion[J].CorrosSci, 1999, 41(7):1231-1243.
7. Nesic S, Postlethwaite J, Olsen S. An electrochemical model for prediction of corrosion of mild steel in aqueous carbon dioxide solutions[J].Corrosion, 1996, 52 (4):280.
8. Gang Zhenbao, Hou Dujie, He Haijun et al. Prediction method of CO2 corrosion in deep gas Wells in northern songliao basin [J]. Natural gas exploration and development, 2008, 31 (2):46-52.

Study on asynchronous injection and production of horizontal Wells in ultra-low permeability thin layer low abundance oil field

Guang Yang

Institute of Geology, No.9 Oil Production Plant, PetroChina Daqing Oilfield Company Limited, Daqing 163853, China

Abstract: The ultra-low permeability, thin layer and low abundance GLN area is developed by water-injection with horizontal and straight combined well pattern. With the extension of development time, the water appears in multiple sections and directions in horizontal Wells, and the well area gradually enters the stage of medium and high water cut, which shows the characteristics of rising when water is injected, falling when liquid is stopped, and limiting the profile control pressure in multiple rounds of Wells. Therefore, developing the "note when mining, mining without note" asynchronous injection-production adjustment of horizontal well technology, make full use of the artificial fracture of high permeability channel and matrix suction effect between low permeable formation permeability, increase the crack near the matrix suction swept volume, to achieve the purpose of the horizontal well steady liquid water control, up to the level of the special low permeability thin layer low abundance wellblock improve recovery factor. Through the mechanism of injection-production, the use of pore throat boundaries, the respect such as injection-production timing and parameter optimization research, combined with the reservoir numerical simulation and theoretical formula, conduct asynchronous injection-production field practice, improved the development effect of horizontal well, and have achieved good economic benefits, for the next class of reservoir level wellblock middle and high water cut stage to explore a new way of governance.

Key words: ultra-low permeability oil field; Horizontal well; Asynchronous injection and production; Numerical simulation; Enhanced oil recovery

1. The introduction

The target reservoir in the peripheral GLN area is the Putaohua oil reservoir with an average buried depth of 1560m, which is a structure-lithologic reservoir. The reservoir is mainly composed of underwater distributary channel and mat sand deposits, with poor physical property and thin thickness. The reserve abundance is 12.8×104t/km2, the average porosity is 15.3%, and the average air permeability is 2.1mD. It is a low-porosity, ultra-low-permeability and low-abundance reservoir. In view of the fact that the vertical well development benefit is poor, the horizontal well is developed by water-injection combined with horizontal well pattern, and the horizontal well is completed by longitudinal zone-fracturing. The initial oil production per well is 10.6t, which is 3.0 times of the surrounding vertical well. However, with the prolongation of development time, the horizontal well is gradually exposed to water in multiple sections and directions, and the well area gradually enters the stage of medium and high water cut, showing the characteristics of rising when water is injected, falling when water is stopped, and limiting the profile control pressure of multiple rounds of Wells. Therefore, the asynchronous injection-production adjustment technology of "no injection at the time of injection, no injection at the time of production" was developed and perfected to make full use of the imbibition effect between the high permeability channel of artificial fractures and the low permeability layer of the matrix, increase the imbibition and volume of the matrix near the fractures, achieve the purpose of stabilizing fluid and controlling water, and improve the recovery efficiency. A new management method is explored for the middle and high water cut stage in horizontal well area of similar reservoir in the future.

2. Study on asynchronous injection and production

2.1 Study on injection and production mode and mechanism

The asynchronous injection-production adjustment technology is a special periodic water injection mode, which adopts the mode of "no injection at injection time and no injection at production time", which can be divided into three stages: water injection, stuffy well and oil

 https://doi.org/10.1051/e3sconf/202235801029

recovery. In the stage of water injection, the injection well is opened and the horizontal well is closed. In the stuffy well stage, the injection well and the horizontal well are shut in simultaneously. In the production stage, the injection well is shut in and the horizontal well is opened [1-3].

In the stage of water injection, the injected water flows into the fracture sufficiently to achieve the initial oil-water displacement. With the increase of injection pressure, the formation energy is continuously replenished, and the injected water preferentially spreads to the matrix macropore throat which is far away from the fracture. At this time, the water driving force plays a dominant role. In the stuffy stage, the fluid velocity in the pore becomes slow, and the capillary force imbibition is enhanced, and the smaller the pore throat is, the stronger the capillary force is. In this case, the injected water is imbibed into the pore throat by the crack and the macropore throat, and the remaining oil in the matrix pore throat is replaced. Capillary force plays a leading role. During the production phase, imbibition displacement of remaining oil into large pore throats and fractures is more conducive to recovery as formation pressure increases and increases the differential production pressure, while reducing inefficient circulation. Therefore, from the perspective of microscopic oil displacement mechanism, asynchronous injection-production adjustment technology can play a role in improving displacement pressure difference, enhancing imbibition displacement and reducing inefficient circulation [4-7]

2.2 Study on pore-throat boundary of asynchronous injection and production

In the tight well stage of asynchronous injection-production development mode, capillary force is the main controlling factor of oil-water interpermeability, and its size determines the strength of impibition displacement. It is particularly important to study the production performance of different capillary forces in the reservoir. Combined with the laboratory core experiment in GLN area, numerical simulation is applied to observe the development effect of different injection and production development methods by changing different capillary forces [8]. Meanwhile, according to the capillary force formula, different capillary forces were transformed into corresponding pore-throat radii, and the relationship between different pore-throat radii and recovery degree was obtained.

Capillary force formula:

$$Pc=\frac{2\sigma\cos\theta}{r} \quad (1)$$

Type in the: σ—The interfacial tension, mN/m; θ—Wetting Angle; r—The pore radius, mm; Pc—Capillary force, Pa.

According to the research results in Table 1, at the initial stage of capillary force increase, the recovery degree of asynchronous injection and production is similar to that of synchronous injection and production. At this time, the development effect of asynchronous injection and production is not different from that of synchronous injection and production when capillary force is smaller in the reservoir. However, when the capillary force is greater than 0.3MPa, that is, the pore throat radius is less than 2.35μm, the capillary force imbibition is enhanced and plays a leading role. The recovery degree of synchronous injection and production decreases sharply, and the development effect becomes worse. However, the recovery degree of asynchronous injection and production rebounded, and then decreased, but the overall development effect was better than that of synchronous injection and production. However, the pore-throat radius in ultra-low permeability GLN area is less than 2.35μm, accounting for up to 73.8%, which is suitable for asynchronous injection and production to excavate the remaining oil in fine pore-throat [9-10].

Table 1 Quantitative characteristics of capillary force and pore-throat radius in different injection and production methods

Capillary force /MPa	The pore throat radius (μm)	Asynchronous injection and production /t	Asynchronous injection production degree /%	Simultaneous injection and production of cumulative oil production /t	Degree of synchronous injection and production /%
0.01	52.0	34005.6	12.12	33513.2	12.03
0.03	17.3	32585.3	11.93	32123.2	11.87
0.05	10.4	32020.3	11.86	31456.3	11.78
0.08	6.5	31456.3	11.78	30456.3	11.64
0.1	5.2	31254.2	11.75	30216.2	11.61
0.3	2.35	34611.1	12.20	29232.2	10.72
0.5	1.04	34201.3	12.15	26554.2	10.39
0.7	0.74	32013.2	11.85	21565.3	9.76
1.0	0.52	18652.4	9.39	15658.3	9.02
2.0	0.26	15331.1	8.98	10645.3	8.39
3.0	0.17	14023.3	8.81	9321.5	8.22

2.3 Study on timing of asynchronous injection and production

In view of the fact that the pore structure in GLN area is highly heterogeneous and the water in horizontal Wells is fast, it is considered to establish well groups with different pore throat structures to optimize the timing of asynchronous injection and production. The reservoir numerical simulation model is established according to the actual geological parameters of the reservoir, and the timing of asynchronous injection and production is optimized for reservoirs with different pore structures.

Table 2 Classification of different pore-throat structures

Pore types	Throat category	Throat radius /μm
A	Large throat	2.0-4.0
B	In the throat	1.0-2.0
C	Small throat	0.5-1.0

Nine asynchronous injection and production timing schemes are designed for reservoirs with pore structures of class A, B and C. The recovery degree and water cut at the end of 10 years of asynchronous injection and production of horizontal Wells are taken as evaluation indexes to optimize the asynchronous injection and production timing of horizontal Wells.

Table 3 Numerical simulation and prediction results of extraction degree and water cut at the end of 10 years with different injection timing

plan	Injection-production plan	Class A Recovery degree (%)	Class A The moisture content (%)	Class B Recovery degree (%)	Class B The moisture content (%)	Class C Recovery degree (%)	Class C The moisture content (%)
1	Continuous injection	21.18	87.23	20.27	89.23	18.39	87.56
2	When the moisture content is 10%, it is transferred to asynchronous injection and production	20.95	89.25	19.35	90.25	17.64	89.64
3	When the moisture content is 20%, it is transferred to asynchronous injection and production	21.86	91.13	19.16	90.13	18.86	85.13
4	When the moisture content is 30%, it is transferred to asynchronous injection and production	22.19	90.75	21.52	87.75	19.39	84.62
5	When the moisture content is 40%, it is transferred to asynchronous injection and production	23.15	89.43	22.15	85.43	19.94	84.43
6	When the moisture content is 50%, it is transferred to asynchronous injection and production	24.87	91.45	22.85	84.45	20.88	82.17
7	When the moisture content is 60%, it is transferred to asynchronous injection and production	24.45	89.13	23.39	79.68	21.53	80.29
8	When the moisture content is 70%, it is transferred to asynchronous injection and production	23.18	90.12	22.97	81.27	22.78	79.15
9	When the moisture content is 80%, it is transferred to asynchronous injection and production	22.98	91.28	21.45	90.22	20.31	83.57

According to the numerical simulation results, the smaller the throat, the later the suitable time for asynchronous injection and production. The optimal asynchronous injection and production rate was 50% for large pore throat (2.0-4.0μm), 60% for medium pore throat (1.0-2.0μm), and 70% for small pore throat (0.5-1.0μm).

Table 4 Results of asynchronous injection and production time with different pore-throat structures

The pore structure	Throat radius (μm)	Capillary force (MPa)	Optimal timing of asynchronous injection and production
Large pore throat	2.0-4.0	0.12	water 50%
The pore throat	1.0-2.0	0.21	water 60%
Pore throat	0.5-1.0	0.40	water 70%

2.4 Optimization of asynchronous injection and production parameters

2.4.1 Optimization of injection and production cycle

The injection-production period is optimized according to the reservoir conductance coefficient and stage injection-production balance. In the aspect of optimizing the water injection cycle, in order to ensure the water injection sweep within the distance between injection and production Wells, the formula of reservoir conductivity coefficient is applied to optimize the water injection cycle of 2 months. In terms of optimizing oil production cycle, combined with field tests, when the optimized oil production cycle is 4.2 months, the cumulative injection-production ratio of the stage is 1.0 to maintain the injection-production balance.

Reservoir conductivity coefficient formula:

$$T=\frac{\mu\cdot Ct\cdot\Phi\cdot L^2}{2K} \tag{2}$$

Type in the: T—Half cycle of water injection, s; μ—Fluid viscosity, mPa•s; Ct—Integrated coefficient of compression, Pa-1; Φ—porosity, %; L—Injection-production well spacing, m; K—Permeability of air, mD

At the same time, according to the actual situation of the oil field, 12 different injection and production cycles are designed for numerical simulation, so as to optimize the injection and production cycle reasonably. According TO the numerical simulation results, the recovery degree of 60 injection and 120 production period is the highest, the water content is the lowest, and the development effect is the best. Therefore, based on the above two methods, in order to ensure the development effect, the optimal injection and production cycle is February injection and April production.

2.4.2 Stuffy well cycle optimization

According to the fluid flow velocity formula from fracture to matrix:

$$V=\frac{K_m}{\mu}\left(\frac{P_1-P_2}{L}-G\right) \tag{3}$$

Type in the: V—Mean seepage velocity, m/s; Km—Permeability of matrix, mD; μ—Fluid viscosity, mPa·s; P1—Fracture pressure, Pa; P2—The matrix

https://doi.org/10.1051/e3sconf/202235801029

stress, Pa; L—The distance between the fracture and the substrate, m; G—Formation initiation pressure, MPa/m.

The time of fluid seepage from the fracture to the matrix rock is:

$$t = \frac{L}{V} \tag{4}$$

You can get it by combining the two equations:

$$t = \frac{L^2\mu}{K_m(P_1 - P_2 - GL)} \tag{5}$$

Combined with oilfield development experience, L is 8m, fluid viscosity is 1mPa•s, matrix permeability is 2.1md, P1 is 14.9mpa when fracturing, P2 is 10.21mpa when developing formation pressure before measures, G is 0.6mpa /m when starting pressure, and the theoretical soaking time is 33D. At the same time, according to the actual situation of the oil field, seven different stuffy well cycles are designed for numerical simulation to optimize the stuffy well cycle reasonably. According to the numerical simulation results, the stuffy well has the highest recovery degree, the lowest water content and the best development effect in 30 days. Therefore, by integrating the above two methods, the stuffy well period is optimized to be 1 month in order to ensure the development effect.

Table 5 Prediction results of development indicators at the end of 10 years for different injection and production cycles

Soak period /d	Recovery degree /%	The moisture content /%
0	30.15	73.24
15	30.36	72.77
30	31.28	71.58
45	30.25	72.88
60	30.95	73.15
75	29.98	75.36
90	29.68	73.99

3. Field test effect

According to the idea of "theoretical support, trial first and scale promotion", 24 Wells were promoted in the GLN area from 2018 to 2019, and the daily oil production per well increased by 1.8T, with an additional 156T in the cycle. From 2020 to 2021, it was gradually promoted to 60 injection and 36 production in GLN area, and achieved good results: First, the formation pressure was gradually recovered. The formation pressure increased from 9.9MPa to 10.09MPa, and the injection-production ratio increased from 0.9 to 1.8. The second is to increase the production of individual Wells. The oil per day per well increased by 1.5t, the average oil per round increased by 126T, and the cumulative oil increased by 9293T. The third is to reduce the invalid cycle. Oil change rate per thousand tons of water increased from 15T to 73T; Fourth, better economic benefits have been achieved. Without increasing the input, the efficiency is 23.06 million yuan, and good economic benefits have been achieved. The final recovery efficiency is estimated to increase by 3.58 percentage points.

4. Conclusion

(1) Asynchronous injection and production can make full use of the full imbibition effect between high and low permeability layers through the adjustment mode of no injection at the time of injection and no injection at the time of production, achieve the purpose of stabilizing the liquid and controlling the water of the horizontal well, achieve the role of improving the recovery efficiency, and explore a new treatment method for the horizontal well in the middle and high water cut stage.

(2) In the asynchronous injection and production stage, the injected water fully flows into cracks and matrix macropore throat, and the water driving force plays a dominant role; In the stuffy stage, the fluid velocity in the pore slows down, and the injected water imbibs into the pore throat, and the capillary force plays a leading role. In the stage of oil recovery, the remaining oil of imbibition displacement will be more favorable to be recovered due to the increase of production pressure difference.

(3) At the initial stage when the pore throat radius gradually decreases, the water driving force plays a dominant role, and the development effect of asynchronous injection and production is similar to that of synchronous injection and production. However, when the pore throat radius is less than 2.35μm, capillary force imbibition is enhanced and plays a leading role. The development effect of asynchronous injection and production is better than that of synchronous injection and production.

(4) The smaller the throat, the later the time for asynchronous injection and production, the best asynchronous injection and production of large pore throat (2.0-4.0μm) is 50%, the best asynchronous injection and production of medium pore throat (1.0-2.0μm) is 60%, and the best asynchronous injection and production of small throat (0.5-1.0μm) is 70%.

(5) The optimization of asynchronous injection and production parameters is injection in February, stuffy in January, and production in April. The development effect of asynchronous injection and production is better than that of synchronous injection and production, and the final recovery degree is expected to increase by 3.58 percentage points.

References

1. Yu Haiyang, Yang Zhonglin, Ma Tian, et al. Feasibility study on different well asynchronous injection and production in multistage fractured tight reservoir [J]. Bulletin of petroleum science,2018,3(1):32-44.
2. Gong Yaojin. Li Qingchun. Feasibility of asynchronous injection and production in heterogeneous multi-layer sandstone reservoir [J]. Special oil and gas reservoirs, 2015,12(1):63-65.
3. Gao Bo, CAI Yuchuan, Zhang Xin, et al. Experimental study on asynchronous injection and production in late high water cut of rehetai reservoir in jin 1-04 block [J]. Petroleum geology and engineering, 2013,27(2):50-52.

4. Wang Meizhu, Yang Zhengming, Wang Xuewu, et al. Research on nonlinear seepage periodic water injection in extra-low permeability reservoirs in the periphery of Daqing [J]. Fault-block Oil & Gas Field, 2012, 19(3): 327-331.
5. Wang Meizhu, Yu Rongze, Cui Maolei, et al. A new method for periodic water injection in ultra-low permeability longitudinal heterogeneous reservoirs [J]. Science and Technology Review, 2012, 30(30): 34-38.
6. Xie Wei, Shi Lihua, Lv Yinghong, et al. Research on periodic water injection scheme for ultra-low permeability heterogeneous reservoirs [J]. Unconventional Oil and Gas, 2016, 3(1): 47-52.
7. Wang Rui, Yue Xiangan, You Yuan, et al. Experiment on periodic water injection and imbibition effect of fractured low permeability reservoir [J]. Journal of xi 'an shiyou university (natural science edition),2007,22(6):56-58
8. It is very green. Research on mechanism of periodic injection and production and mathematical model establishment [D]. Daqing: Applied Mathematics Major of Daqing Petroleum Institute, 2010.
9. Yin Daiyin, Wang Guofeng, Liu Guannan. Numerical model of imbibition oil recovery in low permeability reservoir [J]. Journal of daqing petroleum institute,2004,28(2):31-33
10. Song Xin. Numerical simulation study on development adjustment of Chao55 Block in Chaoyanggou Oilfield [D]. Daqing: Theory and Technology of Oil and Gas Field Development, Northeast Petroleum University, 2014.

Research on the Application of Computer Vision Technology in Power System UAV Line Inspection

Xuchang Cheng*

Shandong Polytechnic College Department of mechanical and Electrical Engineering Shandong Polytechnic College Jining, China

Abstract. The stability of transmission lines is one of the important factors to ensure stable power supply. In this paper, a set of UAV intelligent power line inspection flight control system is designed according to the special flight requirements of UAV in power line inspection operation based on the principle of photogrammetry. The key technologies of the system include UAV flight attitude control, high-precision time synchronization between multiple sensors, defect detection and intelligent diagnosis, airborne sensor calibration, wireless communication, and ground data processing. In this paper, a camera and an on-board computer are innovatively used for real-time image processing and pattern recognition for wire feature acquisition. At the same time, this paper proposes an improved Canny edge detection algorithm combined with image segmentation technology to achieve wire feature extraction and identification. The field test results show that the system can complete the remote control of the UAV and realize the intelligent navigation and inspection of the transmission line UAV.

Keywords: Computer; vision technology; power system; unmanned aerial vehicle; line patrol.

1. Introduction

The stable supply of electricity is a key factor to ensure the stable development of people's life and production. Large-scale power outages caused by line failures will cause serious economic losses. Due to the extremely harsh working environment of the equipment in the transmission line, it has been affected by exposure to sunlight, acid rain, freezing and typhoons over the years, and failures may occur within the service period. In order to prevent large-scale power outages due to failure of electrical equipment, it is necessary to regularly inspect the transmission lines and replace damaged equipment in a timely manner. The detection methods of transmission lines mainly include manual line inspection, human-helicopter inspection, robot inspection and unmanned aerial vehicle inspection. Manual line inspection is not only cost-effective and inefficient, but also threatens the personal safety of maintenance personnel. Although manned helicopter patrol line has high work efficiency, it requires high technical level of operators. Although the robot patrol line has a long battery life and no patrol blind spots, it will cause certain damage to the power line due to its own weight [1]. UAV has the advantages of low working cost, easy operation, and low operation risk. Therefore, the use of UAV line inspection can effectively reduce the operating cost of the power grid and improve the efficiency of inspection work. UAV line inspection has now become one of the important means of transmission line inspection. In this paper, the work process of UAV line inspection based on machine vision is that the UAV is equipped with camera equipment (optical, remote sensing, infrared), and machine vision technology is used to identify, locate and shoot the detection target and determine the fault. This paper adopts the existing technical products and modules, and comprehensively uses the millimeter-wave radar, ultrasonic radar, RTK high-precision fusion positioning and improved image recognition technologies to achieve a breakthrough in the key technologies of UAV intelligent tracking, navigation and inspection.

2. UAV line inspection system design

2.1 System process design

The electric line patrol flight control system is a flight control software customized for the line patrol requirements of the UAV. Through simple and reasonable route design, the purpose of making the supporting UAV realize automatic or semi-automatic operation is achieved. The technical process is as follows as shown in Figure 1 (the picture is quoted in FAILURE ANALYSIS OF A UAV FLIGHT CONTROL SYSTEM USING MARKOV ANALYSIS).

*Corresponding author: 280190430@qq.com

https://doi.org/10.1051/e3sconf/202235801030

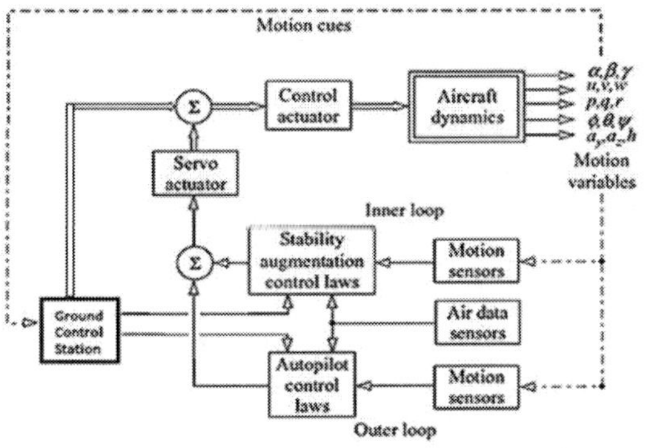

Fig. 1. The flow chart of the UAV flight control system in the detection of tree obstacles in power line inspection

Figure 1 describes the working principle and specific process of the power line patrol flight control system. First enter the coordinates of the tower, then set parameters such as flight height and photo overlap, select the head and tail towers that need to perform the line patrol task, then upload the task and wait for the UAV to load, click Start task to make the UAV start fully automatic flight, the flight status can be monitored throughout the flight, and the task is completed when the drone returns and lands.

2.2 System Architecture

The structure of the power line safety inspection system based on remote sensing technology is divided into four parts: UAV platform and data acquisition system, ground measurement and control station, data communication link system, and ground data processing system [2]. The main block diagram of the drone inspection system is shown in Figure 2 (the picture is quoted from Present state and future prospect of autonomous control technology for industrial drones).

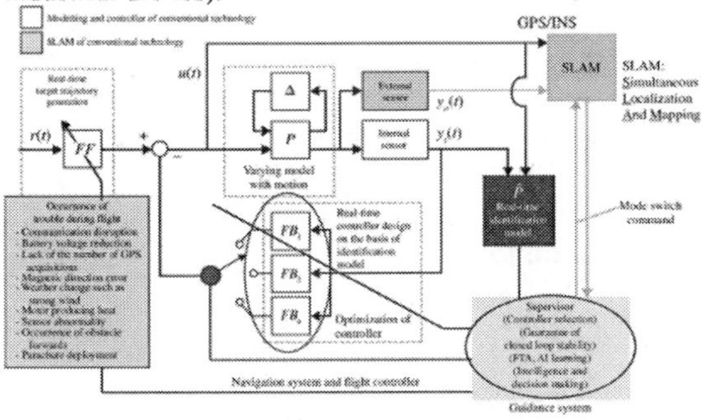

Fig. 2. Block diagram of the main components of the UAV inspection system

2.2.1 UAV platform and multi-sensor data acquisition system

This subsystem is the main platform for power line inspection data collection and acquisition, including UAV flight platform, multi-sensor data acquisition system, stable platform, positioning and attitude system (POS), autonomous early warning and obstacle avoidance system, data encoding and decoding system and other equipment. The composition and functions of the main equipment are as follows: The UAV flight platform can be divided into two categories: fixed wing and rotary wing. The UAV flight platform is responsible for carrying sensor systems, stable platforms, POS, autonomous early warning and obstacle avoidance systems and other necessary airborne equipment [3]. According to the specific requirements of inspection tasks, flight distance, flight speed, etc., different flight platforms should be selected. For example, to carry out rapid inspection of visible light shooting, a small fixed-wing UAV platform can be selected; For fine inspections including infrared, ultraviolet, visible light, etc., and at the same time need to have sufficient flight distance, a large rotary-wing UAV platform should be selected. POS is mainly used for the real-time determination of the position and attitude information of the UAV. In the power line safety inspection system based on remote sensing technology, it is also used to locate and mark the data obtained by the UAV, so that various sensor data can be unified. It is convenient to carry out multi-source data collaborative identification of defects and hidden dangers in the same spatial coordinate system.

2.2.2 Ground TT&C Station

The ground measurement and control station subsystem are mainly used for the monitoring and control of the flight status of the UAV during the flight process, the piloting of the flight, and the control of the sensor data acquisition method. The subsystem includes UAV platform ground control system, multi-sensor ground control system, data encoding and decoding system, and real-time data analysis system and other equipment.

2.2.3 Data communication link system

This subsystem is the key equipment for the communication connection of the whole system, including the signal receiving/transmitting equipment of the airborne and ground measurement and control stations and the UAV communication relay equipment. This subsystem is mainly responsible for ensuring the data communication between the ground measurement and control station and the UAV flight platform, especially in the environment with harsh communication conditions such as mountainous terrain with complex terrain to ensure reliable communication between the UAV and the ground [4]. The data link should first ensure the real-time download of UAV positioning and attitude data and the upload of control commands, so as to complete the tracking and control of the working state of the UAV. The transmission rate of the data link is customized according to the actual needs. Generally, at the rate of 4Mbps to 8Mbps, the real-time download of video data can be realized, so that the ground measurement and control personnel can intuitively understand the flight scene.

 https://doi.org/10.1051/e3sconf/202235801030

2.2.4 *Ground data processing system*

The subsystem is a post data processing, storage and application system, including a multi-sensor data preprocessing and geometric processing system, an intelligent expert system for power line safety inspection based on laser, optical, infrared, ultraviolet and other sensors, as well as a three-dimensional system of line corridors. visualization system, etc. The ground data processing system uses photogrammetry and remote sensing data processing methods and technical processes to perform high-precision geometric processing on various images, point clouds, videos, coordinates, and attitude data. On this basis, according to the characteristics of power lines, towers, corridor features, etc. and their appendages, through the artificial intelligence, pattern recognition and various visualization technologies of the expert system, the identification and confirmation of transmission line and channel defects and hidden dangers are realized. Quickly locate the accident point of the transmission line, and realize the timely diagnosis and troubleshooting of the line safety status.

2.3 Realization of line patrol function

Design a cascade PID controller to control the position, speed and acceleration of the UAV at the horizontal and vertical heights. The structure of the controller is shown in Figure 3 (picture quoted from Recent Advances in Unmanned Aerial Vehicles: A Review). Compared with the ordinary position controller, the cascade PID controller simultaneously performs closed-loop control on the position, speed and acceleration of the UAV. The position information obtained according to the accelerometer data is used as the feedback value of the position, velocity and acceleration PID control. This constitutes a three-level closed-loop control, which can greatly enhance the anti-interference ability of the UAV. The quadrotor drone must be unlocked before use. The take-off process is that after the drone is unlocked, the flight controller controls the altitude and speed of the drone by controlling the governor according to the control signal [5]. After entering the detection area, you can choose to enter the automatic inspection mode.

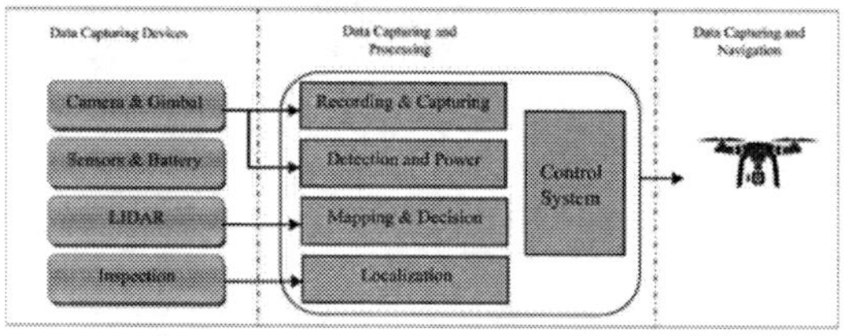

Fig. 3. The realization of the line-following function

2.3.1 *Line-following flight*

After the drone reaches the desired height that needs to be detected, it will hover at the desired height, and then enter the automatic line inspection mode. The drone will automatically conduct automatic inspection according to the direction of the power line. And we will know the location of the fault. At this time, we only need to send a special person for targeted maintenance, which can greatly save manpower and time [6]. The specific process is that when the drone enters the route flight mode, the flight controller will read the current position information, and then use the barometer to determine the current altitude information, and also obtain the current position information through GPS, and then use the Karl Mann filter fusion, the PID is controlled by the fused signal to change the flying height of the UAV, and control the change of the horizontal position of the UAV, and finally realize the inspection of the UAV.

2.3.2 *Automatic landing of UAV*

After the drone completes the line patrol action, it needs to return. In this design, a one-key return is designed. In the seventh channel on the remote control, a one-key return channel operation is designed. When the remote control is dialed to automatically land the gear is, the drone will obtain the position of the drone, and the barometer will obtain the air pressure state at the altitude and compare it with the initial air pressure state. If it is different from the initial state, the flight control will perform PID tuning. To control the altitude state of the drone, and then obtain the position information through GPS, if the desired horizontal position is not reached, the flight control will perform PID tuning to control the horizontal position of the drone, and finally perform PID altitude tuning to finally achieve landing.

3. Design and implementation of line patrol algorithm

When the drone is patrolling the line, no single sensor can get accurate control feedback information. The multi-sensor data fusion of UAV is actually a process of learning from each other's strengths and complementing each other's weaknesses. At the same time of data fusion, image recognition is used to further improve the accuracy of fusion data [7]. Therefore, the system adopts data fusion algorithm and image recognition algorithm to realize the intelligent line patrol process of UAV.

3.1 Data fusion

The attitude calculation adopts the complementary filtering algorithm, which integrates the data of the gyroscope, accelerometer and GPS/RTK to obtain the angle between the UAV and the horizontal plane. Among them, the gyroscope can obtain the rotation angle within the integration time Δt, that is, the rotation quaternion $Q_{\Delta t}$, and if the current attitude is Q, the attitude at the next moment can be predicted as $Q_{\Delta t} \bullet Q$. The acceleration data $a_{ccbody}, GPS/RTK$ of the machine system can be obtained from the accelerometer, the velocity data $\vec{v}_{enu}$ of the geographic system can be

 https://doi.org/10.1051/e3sconf/202235801030

obtained, and the acceleration data a_{ccenu} of the geographic system can be obtained after the difference. Therefore, the gravity vector under the machine system can be expressed as:

$$\vec{G}_{body} = \vec{a}_{ccbody} - Q^{*} \bullet \vec{a}_{ccenu} \bullet Q^{*} \quad (1)$$

The angle between the vectors $\vec{G}_{body}$ and $Q \bullet \overline{(0,0,1)} \bullet Q^{*}$ is the angle between the drone and the horizontal plane. The correct attitude angle can be estimated by taking the complementary correction of the included angle predicted by the gyroscope.

The position data adopts the Kalman filter algorithm, fuses the data of the position sensor (GPS / RTK , millimeter wave, ultrasonic), and estimates the real-time position and velocity information. Set the state variable $X = \begin{cases} s \\ v \end{cases}$, Observational amount$Z = \{S\text{observation}\}$, where s represents position and v represents velocity. Build the observation and prediction equations:

$$\begin{cases} Z(x) = [1 \bullet 0] X(k) + v \\ X(k) = \begin{bmatrix} 1 & t \\ 0 & t \end{bmatrix} X(k-1) + w \end{cases} \quad (2)$$

Where v and w are observation noise and process noise, respectively. The revised state estimate is then updated using the Kalman state update equation:

$$X(k \mid k-1) = A \bullet X(k-1 \mid k-1) + B \bullet U(k) \quad (3)$$

3.2 Image recognition algorithm

For the currently acquired video data of the power line, the system image recognition algorithm flow is shown in Figure 4 (the picture is quoted from A UAV-based machine vision method for bridge crack recognition and width quantification through hybrid feature learning). Among them, the steps of Gauss filtering, Canny edge detection and morphological filtering are image preprocessing, and the last two steps are feature detection and feature screening respectively, and finally realize wire detection. The image recognition algorithm uses grayscale image preprocessing, and according to the environmental characteristics of the detected object, uses Gauss filter to convolve the image to smooth the image, reduce the obvious noise effect on the edge detector, and effectively improve the next step. Edge detection effect.

The system uses a depth camera and an ultra-wide-angle lens for image recognition, and uses the improved Canny algorithm to detect the edge of the image. After processing, not only can the positioning accuracy of edge detection be improved, but also the image edge will not be blurred. The core of Canny is to transform the edge detection problem into a local maxima problem of image gradient. The Gaussian function can be used for the best approximation.

$$d(x,y) = \frac{1}{\sqrt{2\pi\sigma_1}} e^{-(\frac{x_i^2 + y_j^2}{2\sigma_1^2})} \quad (4)$$

In the formula, the standard deviation σ_1 takes 2.4; x_i and y_i to represent the fractional distance relative to the center of the filter window. If there is a difference greater than or equal to the distinguishable color difference threshold, the filter window needs to be reduced to 3 × 3, and then recalculated; if there is no difference greater than or equal to the distinguishable color difference threshold, the definition domain of $\sigma_1 = 1.2$ is used. Value Kernel Filtering

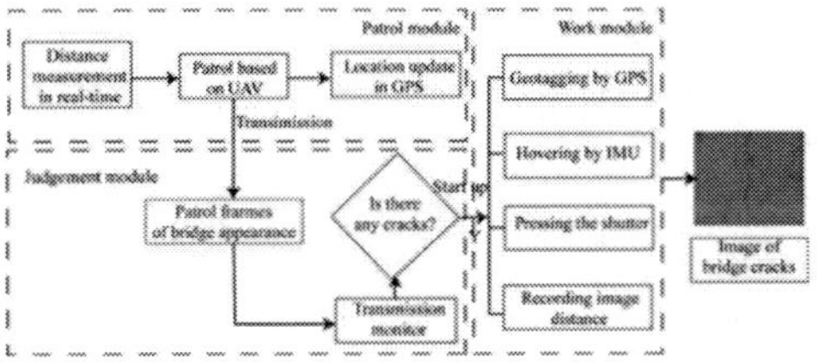

Fig. 4. Image recognition process

4. System Check

In April 2021, the author's unit used a multi-rotor drone to inspect the overhead transmission line channel of a 220kV line No. 093-134, with a total length of 20 kilometers and a line corridor width of 400 meters. The landforms in this area are dominated by mountains and hills, making it difficult to fly. In this experiment, the multi-rotor UAV was used for aerial flight operations, and the PTZ camera with an effective pixel of 20 million was used for vertical aerial photography, with a relative altitude of 115 meters. The 3-axis stabilization system mounted on the gimbal can greatly reduce the jitter during shooting, improve flight stability and ensure the accuracy of subsequent aerial triangulation of images. In the case of using the newly developed power line patrol UAV flight control system throughout the process, the UAV always flies according to the preset trajectory, image overlap, and relative altitude, and at the same time flies stably and has a good attitude. The clarity of the captured images fully meets the requirements of subsequent use of various post-processing software. The precise position and attitude of the images in the air are determined through aerial triangulation, and then the ground 3D point cloud and wire point cloud are obtained through intensive matching. The distance error between the measuring wire and the three-dimensional point cloud on the ground meets the obstacle distance detection standard of power inspection. The subsequent process of aerial triangulation through software is shown in Figure 5 below:

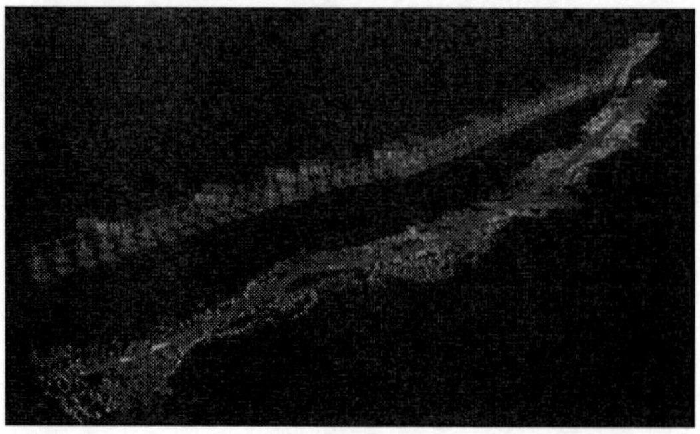

Fig. 5. Photo positions recovered from aerial triangulation

5. Conclusion

This paper proposes a set of power inspection system based on UAV. The UAV will complete the line inspection task according to the pre-established flight route, and the staff can also perform power inspection by manual operation. During the line patrol mission, the UAV and the ground server will transmit images and videos with high bandwidth and high stability through the operator's wireless network card, and support up to 1080p60i full HD video live broadcast. For the received logs, images, videos and other files, the system will intelligently archive and save them for easy retrospection. At the same time, the system will automatically perform temperature measurement, calibration and analysis on the images, and alarm for abnormally high temperature components and towers, ensuring that the power patrol It improves the real-time and accuracy of inspection tasks, and also improves the quality and efficiency of inspection.

References

1. Yang Chunfeng, Yu Ronghua, Huang Wei, et al. Research on UAV power line inspection technology based on lidar and visible light stereo measurement. Automation Technology and Application, vol. 40, pp.49-51, October 2021.
2. Zhu Tielin. Application analysis of satellite communication in UAV power line patrol. Inner Mongolia Electric Power Technology, vol. 38, pp.59-62, March 2020.
3. Huang Jing. Research and implementation of UAV inspection training and examination system based on large-view simulation technology. Southern Energy Construction, vol. 6, pp.153-158, January 2020.
4. Zhou Yi, Li Dongwu, Meng Hao, et al. Application of Genetic Algorithm Path Planning in UAV Power Line Patrol. Automation Technology and Application, vol. 40, pp.55-62, February 2021.
5. Yang Hao, Lin Nan, Yuan Chen, et al. Information security risk analysis and protection of UAVs for power line patrolling. Journal of Electric Power Science and Technology, vol. 36, pp.99-104, April 2021.
6. Peng Hong, Li Hongli, Du Yu, et al. Aerial image stitching algorithm of power line patrol UAV. Computer Applications and Software, vol. 37, pp.78-81, May 2020.
7. Zhang Ruizhe, Zhou Kai, Ye Kuan, et al. Design of Lightweight Solar UAV for Long-distance Power Line Patrol. Power Electronics Technology, vol. 56, pp.46-52, March 2022.

Analysis of common problems of steam engine technical supervision in thermal power plant

Yongqiang Che [1], Yisen Chen [1] and Haizhen Lv [2]

[1] State Grid Shandong Electric Power Research Institute, Jinan 250003, China
[2] Shandong Zhongshiyitong Group Co., Ltd, Jinan 250003, China.

Abstract. Through the work of steam turbine technical supervision, the paper focuses on the construction of technical supervision system, daily management, equipment management, operation indicators, countermeasures management, hidden danger investigation, network-related safety and personnel training and other aspects, and finds out the existing common problems and analyzes them.

Keywords: Large vibration; overspeed hidden danger; high metal temperature; system leakage.

1. Introduction

Technical supervision is an important means to ensure the safe and stable operation of units. As one of the three major majors in thermal power plant, the safe and stable operation of steam engine equipment is very important for the whole thermal power plant[1]. Every summer, due to the high temperature of the weather and the high load of the power grid, the unit is required to operate at full load, which puts forward higher requirements for unit operation. Therefore, regular organization of technical supervision and inspection every summer has become a routine operation, known as the summer kurtosis supervision and inspection. Through summer supervision and inspection of kurtosis, hidden dangers and risks in unit operation can be found, and experience can be provided for other units.

2. Vibration of steam turbine

The vibration of most units is good, but the vibration of many units is too large or even exceeds the alarm value, and some of them have affected the normal load adjustment of units. During the peak summer, some power plants have taken some temporary measures to ensure the normal operation of the units in view of the large vibration problem. If the vibration of the unit is relatively stable and does not continue to increase, it is generally recommended to maintain the operation of the unit first, but monitoring should be strengthened to control the temperature of lubricating oil and the speed of load lifting of the unit[2]. Vibration professionals can be contacted for on-site monitoring and analysis, and preventive and treatment measures can be formulated in advance, which can be put into practice when there is an opportunity to stop maintenance. The vibration of the following units is too large, even exceeding the alarm value, which should be paid attention to:

The vibration of generator excitation end bearing of No. 2 machine in a factory gives an accidental alarm. The DCS screen and historical data show that the bearing vibration of generator excitation end (bearing #9) of unit #2 tends to increase with the increase of load. When the load reaches more than 90%, the bearing vibration value reaches about 80um, and the alarm value has exceeded 79um. The vibration is closely related to the excitation current and has time delay. It is suspected that the vibration is related to the thermal deformation of the generator. Power plants can strengthen the operation monitoring, use the downtime opportunity, eliminate the abnormal vibration of bearings, to ensure the safe operation of equipment.

The two units in the first phase of a factory have the problem of large 9Y vibration, and the vibration value is closely related to the load of the unit. When the load of the unit is close to full load, the vibration reaches the maximum, exceeding 140μm (operation regulation: normal operation is below 150μm, and tripping exceeds 250μm). # 2 unit vibration generator turn-to-turn short circuit, was the main cause of the processing of balancing have made three times in the same place, after each processing of balancing vibration situation has improved significantly, but after a period of time after operation and rev. Stop many times, the vibration value and slowly up to the state before the counterweight, analysis and unit start-stop often have a certain relationship, this problem has been included in the unit overhaul in September, Prepare to return the generator for disposal. Unit #1 also SUFFERED FROM LARGE 9Y vibration recently, and the on-line inspection device did not find the generator inter-turn short circuit problem in the unit, and the vibration cause needs to be further analyzed.

During the operation of the sequence valve of Unit #4 in a factory, the vibration value of bearing bush #1 and #2 in the adjustment of some load sections is fluctuated. The problem mainly occurs in the low load section, where the GV3 opening is 0-33%, and the vibration amplitude tends to increase to the maximum of 120um. Under different load segments and different opening states, the rotor of high regulating gate GV3 of unit #4 is destabilized by steam flow disturbance and the vibration increases when the sequence valve is running. It is suggested to strengthen the observation and trend of #1 and #2 bearing vibration of #4 unit, organize discussion and analysis in time, and formulate corresponding adjustment and emergency measures; When lifting load, closely monitor bearing vibration and shaft temperature, by timely adjusting the opening of the door, control the vibration of any bearing does not exceed the alarm value (127μm); Check and adjust the flow of high and high pressure cylinder.

When the load of unit #7 in a factory is 661MW, the shaft vibration of bearing #1 is 118μm; According to the historical curve, the vibration value of the bearing increases with the increase of load. When the load reaches 900MW, the shaft vibration is 150μm. It is understood that the #1 bearing vibration of this type of unit is relatively common, the main reason is that the #1 bearing is at the end of the shafting, the load is light, and the vibration is easy to occur. It is suggested to organize vibration professionals to carry out on-site monitoring and diagnosis, identify the cause of vibration, and put forward the treatment measures.

3. Steam turbine overspeed risk

Steam turbine valve jamming, tone wave (or off), valve rigor poor motivation (especially oil), oil leak, spring break or valve pin fracture problems still outstanding, which affect the normal regulating function of the steam turbine, load fluctuations or down load operation, serious and even lead to the unit not stop events[3].

In the movable test of the middle joint door of Machine #1 in a factory, the ICV2 valve was stuck around 96% and could not be closed. After replacing the adjusting servo valve, the remote test shows that the ICV2 of machine #1 is fully open and normal. When closing, the instruction is 87.9%, and the feedback is unchanged at 94.7%. It is still stuck and cannot be closed. The replacement of the adjustable servo valve has no effect, and the analysis is that there is a problem in the valve or oil circuit.

The switching valve of main oil cooler of Unit #6 in a factory is stuck. The switching valve of the main oil cooler almost does not switch during normal operation, and the switching valve action frequency is low. After a long period of operation of the unit, the switching valve may be stuck. It is recommended to maintain the current operating state and avoid valve switching operation. It is suggested that the disassembly and overhaul of the main oil cooler switching valve be included in the overhaul project of Unit #6, and the overhaul and maintenance work should be completed as soon as possible to ensure the safety of the system.

References are cited in the text just by square brackets [1]. (If square brackets are not available, slashes may be used instead, e.g. /2/.) Two or more references at a time may be put in one set of brackets [3, 4]. The references are to be numbered in the order in which they are cited in the text and are to be listed at the end of the contribution under a heading References, see our example below.

4. Abnormal bearing metal temperature

The metal temperature of most of the steam turbine bearings is within the normal range, but there are still some units with high bearing temperature, large temperature deviation of each thrust tile, high bearing oil return temperature, especially the case that the temperature difference between each thrust tile exceeds 10℃ is relatively more. High bearing metal temperature is related to bearing load, bearing lubricating oil inlet and return[4], bearing installation quality and other factors. Power plants should closely monitor the changing trend of bearing temperature, conduct statistical analysis on relevant influencing factors, and formulate corresponding treatment measures and preventive measures.

The two temperature measurement points of the front upper part of the thrust bearing of unit #2 in a factory are 66°C and 77°C respectively, and the left and right temperature measurement points of the rear upper part are 73°C and 65°C respectively. The temperature difference between the two front and upper parts is 11℃, and the temperature level between the thrust tiles is not uniform. It is suggested that the power plant should take advantage of the maintenance opportunity to check the contact and self-positioning of each thrust tile, grind each tile to ensure that the load of each tile is uniform and has good self-positioning performance, and overhaul and assemble the thrust bearing in strict accordance with the requirements of maintenance technology and regulations.

When the load of unit #6 in a factory is 243.8MW, the #2 watt temperature is 92.0°C, the #3 watt temperature is 62.2°C, the #4 watt temperature is 94.6°C, and the #3 bearing bush temperature is much lower than that of #2 and #4 bearings. When the load of unit #7 is 232.7MW, the #2 w temperature is 87.0°C, the #3 W temperature is 59.0°C, the #4 W temperature is 96.1°C, and the #3 bearing bush temperature is much lower than that of #2 and #4 bearings. Shafting load is uneven, #2 and #4 bearing load is too large, #3 bearing load is too light, multiple units in the province due to #3 load is too light caused by oil film vibration caused by vibration. The power plant has the opportunity to take measures to redistribute the shafting load, and it is recommended to raise the #3 bearing to increase the bearing capacity.

The temperature of the #4 bearing of the #1 steam turbine in a factory is high. After referring to the operation history curve of the #1 machine, it is found that the oil pressure of the #4 bearing in the local jacking shaft of the #1 machine is zero in normal operation, and the temperature rises abnormally during the shutdown process. It is suspected that there are problems related to the oil pipeline of the #4 bearing jacking shaft. The pipe inside the jacking oil to the #4 bearing box is broken or slipknot is loosened, resulting in pressure relief. It is suggested to use the mediation opportunity to check the oil pipeline of #4 bearing jacking shaft, check the leakage point and deal with it, ensure the normal oil pressure of jacking shaft, and ensure the normal establishment of bearing oil film when the jacking shaft oil system is put into operation in the process of starting and stopping.

The load of unit #1 in a factory is 250MW, and the temperature level difference between thrust tiles on the working face is 16.4°C. The load of unit #2 is 227MW, and the temperature level difference between the thrust tiles on the working face is 18.4°C. It is recommended to analyze the equipment defects together with the design and assembly department of the manufacturer. Add assessment clause in maintenance outsourcing contract.

5. Heater Terminal Difference

In many units, the heater end difference, especially the hydrophobic end difference is large, which affects the unit economy[5]. For example: the difference between the upper and lower ends of Unit #4 #2 in a factory is 9.26℃. The high loading and hydrophobic temperature of unit #1 #3 in a factory is 173.72℃, and the lower end difference is 21℃. The difference between the lower end of the #3 high additive of unit #5 in a factory reaches 21℃, far exceeding the design value. The #3 high additive is operating under bad condition, with many leakage times and high pipe blocking rate. The unit has been running for many years, and the heat transfer performance has decreased. It is suggested to control the water level of #3 high water tank well in operation, and replace the high water tank when conditions permit.

It is suggested to strengthen the tracking and monitoring of the heater end difference and establish the heater ledger under typical working conditions. For the large difference of the hydrophobic end, the water level of the heater should be checked first, and the adjustment should be strengthened in operation to ensure enough hydrophobic water level. At the same time, the running exhaust system of the heater, especially the low - add, is checked to ensure the smooth exhaust. Strengthen the monitoring of the operation of the heater, find leakage in time, and deal with it in time. Replace the high pressure heater if possible.

6. System Leak

All power plants attach great importance to the control of valve leakage, but the valve leakage problem is still common, especially the high and low pressure bypass valves of units, water supply flow recirculation valves and other parts of the leakage phenomenon is serious, the details are as follows:

1) No. 2 machine of a factory is not strict to the secondary cooling auxiliary steam check door, some steam leakage phenomenon, directly affect the unit economy. It is suggested that the power plant take advantage of the maintenance opportunity to repair or replace the secondary cooling to auxiliary steam check door to improve the unit economy.

2) The drainage temperature from main steam to small engine main pipe of Unit #5 in a factory is 307°C, and there is valve internal leakage. Frequent start and stop of the unit, long-term maintenance of the valve, valve quality problems. Valve leakage will reduce the economic efficiency of the unit, and at the same time, there are huge safety risks, which need to be paid attention to. It is suggested that the power plant should try to choose valves with better quality level on the basis of full investigation to solve the problem of valve leakage.

3) The temperature measured after recirculating valve of steam pump A of Machine #6 in A factory is high (98°C), which indicates that there is slight internal leakage. The recirculation valve of steam pump of machine A operates under bad condition, resulting in scour of spool seat. It is suggested to use the mediation opportunity of Unit #6 to replace the valve core and seat of the steam pump recirculation valve of Unit #6 A or consider upgrading the recirculation door.

4) A factory #6 machine #2 bearing shaft seal leakage is large; The low pressure cylinder shaft end of unit #4 and #5 May have air intake. The low pressure shaft seal steam supply pressure of unit #5 reaches 71kPa, which exceeds the normal low pressure shaft seal steam supply pressure, but no steam leakage phenomenon is found at the shaft end. Turbine shaft end seal wear, sealing effect is poor. It is suggested to overhaul the shaft seal of low pressure cylinder of unit #4 and #5, and adopt the sealing form and material with better sealing effect.

5) A factory #8 machine B steam feed pump recirculating pneumatic door leakage, measuring point temperature 102°C. Valve leakage is mainly caused by the valve core, seat sealing line is not complete, the seal is not strict. It is suggested to take advantage of the overhaul opportunity of the unit to overhaul the hydrophobic pneumatic door of the main steam pipeline to eliminate the internal leakage problem. #8 Engine A steam pump recirculating electric regulating door for disassembly inspection, check the valve sealing line, thoroughly eliminate the problem of internal leakage.

6) Internal leakage of closed water cooler A of Unit #5 in A factory. The closed water cooler in power plant is a multi-layer dense plate heat exchanger with large sealed cooling area and many potential dew points. Leakage may occur after long-term operation. It is recommended to switch the closed water to B cooler for operation, and organize maintenance personnel to carefully check the dew point and repair and restore as soon as possible.

7) When the load of machine #1 in a factory is 344MW, the feed water flow at the inlet of steam pump is 1322T /h, the feed water flow at the inlet of economizer is 1036T /h, and the recirculation valve opening of feed water pump is 0. The internal leakage of this valve is serious. Internal leakage of recirculating valve of feed water pump is a common problem. The unit frequently participates in depth peak regulating, start and stop peak regulating, and the recirculating valve of feed water pump often operates, especially the scouring under small opening has the greatest influence on the tightness of the valve. Some power plants install pneumatic shut-off ball valve before the recirculation valve of the feed water pump to share the scouring under small opening. It is suggested that the

power plant should investigate the modification experience in this aspect.
8) There is leakage in the low-pressure cylinder water temperature reducing regulating valve of Machine #2 in a factory. When the bypass water temperature reducing regulating valve is closed to 0% in the process of cylinder cutting, there is still water spraying. It is judged that there is leakage in the valve, resulting in the waste of warm water and affecting the normal use and regulation. When the valve leaves the factory, the spool seat does not meet the standard, or the abnormal flushing in the process of operation leads to a large amount of internal leakage. It is suggested to repair or replace the water jet temperature reducing regulating valve of the low pressure cylinder of machine #2 combined with the shutdown opportunity to reduce the internal leakage and improve the reliability of the equipment.
The valve leakage in the thermal system directly loses the effective energy of the high quality working medium, which has an obvious influence on the coal consumption of the unit. Therefore, it is necessary to strengthen the control of valve leakage. At present, the power plant has been attached great importance to the problem, has taken many effective measures, such as strengthening the valve before and after the temperature monitoring, replace more safety and reliable long-life valve, improve the quality of valve repair, valve leakage problem has improved, but the valve type selection, operation adjustment habits, there are large difference between valve leakage management, governance effect is difference is bigger also.
Power plants should make use of the opportunities for major and minor repairs and downtime to strengthen the control of valve internal leakage, make good selection of new valves and quality control of acceptance, strengthen the supervision of process quality in the valve maintenance process, and ensure that the valve can meet the quality standards of flexible switch without sticking, tight closure and leakage after maintenance. It is suggested that each factory establish a special management ledger of valve leakage, timely grasp the valve leakage situation, planned and step management; At the same time, each factory should strengthen communication and exchange, especially in the valve selection, maintenance, operation and monitoring aspects of mutual learning and reference, common improvement.

7. Cooling Effect

The cooling tower with multiple units has the problem of uneven water pouring, which reduces the heat exchange performance of the cooling tower[7], affects the water temperature of the tower, and thus reduces the output of the unit. Such as:
1) There is water column in the #4 cooling tower of a factory, and the water is not even. The reason is that the packing or water distribution pipe in the cooling tower is damaged. It is recommended to use the overhaul opportunity to check the packing and water distribution pipe in the cooling tower, and # 4 cooling tower can also be retrofitted according to the retrofit experience of the other three cooling towers.
2) In the normal operation of the cooling tower of Unit #4 in a factory, there are many places where the circulating water column falls. The outlet water temperature in the tower is about 2°C higher than that of other units, and the height of the cooling tower is 2-3°C higher than that of unit #3, which affects the economy of unit #4. The cooling tower of unit #4 adopts ceramic packing, which is seriously weathered and collapsed and has the problem of uneven water shower. Dehydrator, water distribution pipe, nozzle and other equipment have different degrees of aging, damage, deformation problems, affect the efficiency of steam water removal, water film formation will cause poor evaporation efficiency. Treatment suggestions: combined with the comprehensive treatment and transformation of the cooling water tower of the technical reform project, upgrade and replace the packing (can be replaced with the same PVC material as the packing material of the #3 machine cooling water tower); Check the dehydrator, water distribution pipe, nozzle and other equipment, replace the damaged parts, replace the efficient nozzle.
3) There is a lot of silt in the bottom pool of cooling tower of Unit #3 in a factory, and the packing has scaling, sinking and aging phenomenon. Treatment suggestion: Take advantage of the unit maintenance opportunity to thoroughly inspect the cooling tower components of Unit 3, repair the damaged packing, splashing device, water remover, and remove the scale.
4) The cooling water tower of Unit #2 in a factory has the problem of large water column and nozzle falling off. Reason analysis: the unit runs all year round, and the poor quality of the circulating water causes the aging and falling off of the spraying equipment of the cooling tower. Rectification suggestions: check and replace the damaged parts according to the shutdown opportunity; Check other spraying devices and fillers of the cooling tower to eliminate hidden dangers.
5) The outlet water temperature of cooling tower of Unit #7 in a factory is 2°C higher than that of unit #6, and the outlet water density of cooling tower of unit #7 is not uniform. Reason analysis: packing of cooling tower of #7 unit falls off, bamboo lattice rots and piles up, which affects the ventilation of water tower and reduces the heat exchange effect. Rectification suggestions: Include in the maintenance plan, and replace the core packing of the cooling tower.

8. Summary

Through inspection, it is found that all power plants attach great importance to the technical supervision of steam engines, with a sound supervision system, complete personnel configuration and clear division of responsibilities. It can actively carry out various supervision work according to the requirements of turbine technical supervision work regulations, and give full play to the role of advanced supervision, risk control and advanced technology promotion in the aspects of unit stability, economic operation and energy saving transformation. The power plants pay more attention to the formulation and implementation of anti-accident measures, but the daily management, equipment maintenance and operation still need to be strengthened and improved, and the technical supervision of each plant needs to be further strengthened.

References

1. Yu Yingli, Han Yi, Cai Bin, et al. Current situation and development of boiler technology supervision in electric power industry. Power & Energy. 2017,38(06), p. 809-811.
2. Guo Jia, He Xinrong, Tan Rui, et al. Analysis and Treatment of Abnormal Shaft Vibration in 600 MW Supercritical Steam Turbine Generator Unit. Northeast Electric Power Technology. 2022,43(06), p.40-43, 48.
3. Zhang Xing. Cause Analysis and Treatment on Control Instruction Fluctuation of Medium-pressure Regulating Valve. Zhejiang Electric Power. 2016,35(02), p. 45-47.
4. Wang Fei, Xie Weiyang. Abnormal Analysis and treatment of low-pressure bearing temperature of 1000MW steam turbine. Turbine Technology. 2021,63(06), p. 455-457.
5. Zhan Hailong. Cause Analysis on Big Difference of Heater Lower End of Regenerative System in 330 MW Unit of Thermal Power Plant. Internal Combustion Engine & Parts. 2016, (12), p. 62-63.
6. Zhang Heng. Causes and countermeasures of pipe valve leakage in thermal system of thermal power plant. Inner Mongolia Coal Economy. 2021, (18), p. 168-169.
7. Wang Yongquan, Cai Lanyan, Ye Xiaozhen. Cooling tower performance research and upgrading. Energy and Environment. 2021, (01), p. 45-46, 48.

https://doi.org/10.1051/e3sconf/202235801032

Study on reservoir physical properties of Nadanhada terrane

Guohui Qu *, Xiunan Li, Xuebin Tian, Zilu Zhang, Weizhong He, Bowen Li, Yikun Liu, Fengjiao Wang, Anqi Shen

Key Laboratory of Enhanced Oil Recovery (Northeast Petroleum University), Ministry of Education, Daqing 163318, China

Abstract. The Nadanhada terrane is basically a large tectonic sedimentary hybrid, which is an accretionary complex belt formed by the subduction of the Pacific plate to the Eurasian plate. Through field geological survey, paleomagnetic analysis, paleontological analysis and microscopic observation of core slices, its structural and petrological characteristics were studied. The results show that the accretionary complex is mainly composed of 7 parts, including undeformed sedimentary rocks, shear deformed sedimentary rocks, shear deformed volcanic rocks, turbidites, deformed siliceous rocks, limestone blocks and basic ultrabasic rocks. Shear deformation structures of deformed sedimentary rocks and volcanic rocks are developed. The sediment grain size of the Lower Cretaceous reservoirs in each basin in the northern region is relatively coarse, the sorting is moderately deviated, and the rounding is poor, mainly secondary to angular. The grain size of the lower Cretaceous sediments of each basin in the southern region is relatively fine, with coarse, medium and fine sandstones, but mainly medium fine sandstones, with medium sorting preference and sub angular to sub round roundness.

Key words. Nadanhada terrane; Reservoir; Petrological characteristics

1. Introduction

The Nadanhada (Wanda Mountain) terrain is located at the eastern edge of Heilongjiang Province. It starts from Dahe Town of Baoqing in the west, ends at the edge of the Wusuli River in the east, starts from Dongfanghong Town of Hulin County in the south, and ends at the edge of Heilongjiang in Fuyuan County in the north. It borders Russia in the east and north. The whole area is a strip about 80km wide from east to west and 240km long from south to north. The tectonic location is in the Nadanhada area east of the Yuejin Mountain fault and north of the Dunhua Mishan fault zone [1,2]. In addition, marine Mesozoic strata are also developed in the eastern part of the Nadanhada region and the eastern part of Sikhot Alin in Russia, which can be compared with the Nadanhada terrane [3, 4].

2. Nadanhada terrane structure

Paleolatitude data (Table 1): in the Middle Triassic (T2), the Nadanhada terrane was near the equator (8.0 °), and then gradually drifted northward. From the Middle Triassic (T2) to the Neogene (N), the Nadanhada terrane drifted northward, 37.8 ° (=45.8-8.0 °), about 4150km (110km × 37.8 °), the average drift speed is about 17.3km/Ma. The current position that has been drifting from the equator [5].

The Nadanhada terrane is basically a large tectonic sedimentary hybrid, which is an accretive complex belt formed by the subduction of the Pacific plate to the Eurasian plate [6]. In the process of field petroleum geological survey, it is found that the Wandashan accretionary complex is mainly composed of 7 parts, including undeformed sedimentary rocks, shear deformed sedimentary rocks, shear deformed volcanic rocks, turbidites, deformed siliceous rocks, limestone blocks and basic ultrabasic rocks. Undeformed sedimentary rocks are mainly composed of mudstone, siliceous rock and siltstone, and locally contain manganese. Pillow basalt formed by rapid condensation of submarine volcanic eruption is also developed, representing the deep-sea sedimentary environment (Figure 1).

* Corresponding author: quguohui1001@126.com

a Nadanhada terrane, flysch formation, deep-sea turbidite deposit, gray black mudstone mixed with gray siliceous rock, dark gray siltstone

b Nadanhada terrane, pillow basalt, ellipsoidal in shape, with almond shaped structure locally developed

Figure 1. Evidence of accretionary complex in the Nadanhada terrane

The shear deformation structure of deformed sedimentary rocks and volcanic rocks is developed (Fig. 2). It is preliminarily speculated that it may be related to the subduction of the ancient Pacific plate and is the main component of the accretion wedge. Turbidite is composed of medium thin fine siltstone and mudstone, mainly composed of A-B-C and B-C sections, and is the product of trench sedimentation. The deformed siliceous rocks are thin bedded, and fold structures are very developed. Sharp folds are formed in some areas, which may be related to the compression of the ancient Pacific plate from west to east. Limestone blocks occur in clastic rocks in blocks, with typical features of exotic rocks, and are the products of tectonic involvement in the process of oceanic crust subduction. The basic ultrabasic rocks are composed of peridotite, pyroxenite, cumulate gabbro, layered gabbro, diabase and pillow basalt, which have typical characteristics of ophiolite and represent oceanic crust fragments, while basalt, siliceous rock and limestone constitute oceanic island/seamount.

a Nadanhada terrane, basalt, developed cleavage, Guanmenzuizi Village

b Nadanhada terrane, Hongqiling, siliceous rock, sharp edge fold

Figure 2. Evidence of shear deformation structure development in Nadanhada terrane

Previous studies have shown that, according to the late Carboniferous Permian limestone blocks contained in the melange, it can be speculated that this period could be a shallow sea shelf environment on the continental margin. In the Middle Late Triassic, the area gradually expanded into an ocean basin, mainly developed radiolarian siliceous rock, argillaceous rock and ophiolite formation. The Nadanhada area is still a deep-sea radiolarian bearing siliceous rock formation (containing deep-sea manganese nodules) from the early Jurassic to the early Late Jurassic. The unconformity above it covers a set of continental - marine - continental - marine sediments from the late Late Jurassic to the Early Cretaceous, uniformly distributed in the Jiamusi block and Nadanhada area, and extended to the Far East [7].

The Nadanhada terrane was spliced on the Jiamusi terrane along the Dahezhen fault in the middle of the Late Jurassic, which has an important influence on the evolution of the eastern basin group in Heilongjiang Province. Firstly, due to the collage of Nadanhada terrane, the basement of Qianjin Depression in the east of Sanjiang Basin and Suibin Depression in the west is obviously different. The basement of Qianjin Depression is marine siliceous rock series and diabase of early and middle Jurassic, which is similar to the lithological characteristics of Nadanhada terrane; Secondly, the sedimentary filling sequence is also obviously different. The seismic profile interpretation results show that the Qianjin depression began to deposit

https://doi.org/10.1051/e3sconf/202235801032

from the Dongshan Formation in the late Early Cretaceous, while the Suibin depression began to deposit from the Suibin Formation in the late Jurassic.

Table 1 Nadanhada Paleomagnetic Data Sheet

Number	Geologic era	Sampling location		Remanence direction				Ancient latitude/°
		longitude /°	latitude /°	D	I	a95	K	
1	N	133.4	46.7	38.4	64.1			45.8
2	K_1	133.5	46.8	19.6	65.0			47.0
3	J_3-K_1	134.2	47.3	61.3	52.7	2.9	88.8	33.3
4	J_3	133.1	46.6	26.6	35.8	4.6	51.8	19.8
5	T_3	133.5	47.1	25.9	10.6	8.1	18.9	5.4
6	T_2	133.1	47.0		15.6			8.0

3. Petrological characteristics

3.1 Mineral composition characteristics and sandstone composition classification

During the field petroleum geological survey of the basin groups in the north and east of Songliao Basin in Northeast China, a large number of reservoir samples were collected, thin sections were ground, and microscopic observations and identification were carried out. The lower Cretaceous reservoir is mainly lithic arkose (Fig. 3), followed by feldspathic sandstone (IV), feldspathic lithic sandstone (VI) and lithic sandstone (VII). Volcanic rocks are also developed in some areas of the Cretaceous system.

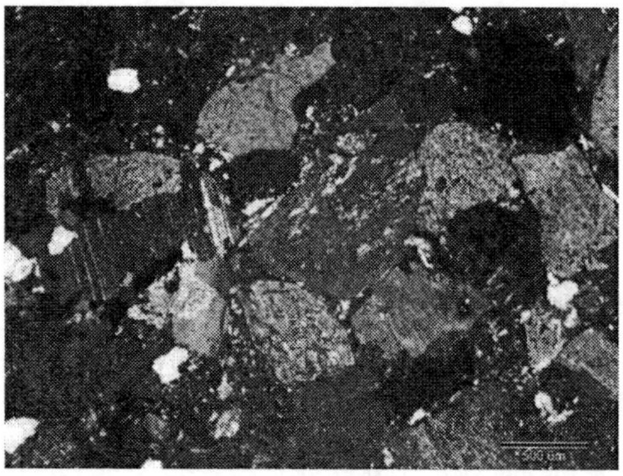

Figure 3. Cuttings Arkose, K_1, Jixi Basin, Chengzihe Cutting Profile, (+)

3.2 Structural characteristics and classification

3.2.1 Particle size characteristics and particle size designation.

The grain size and sorting property of clastic rocks are the indicators of the capacity and efficiency of transportation agents. The size of clastic particles directly determines the type and nature of rocks, and is an important basis for the classification and naming of clastic rocks. Refer to the current Petroleum and Natural Gas Industry Standard of the People's Republic of China - Rock Slice Identification. The coarse clastic rocks of small and medium-sized basin groups in the north and east of Songliao Basin are divided into conglomerate, megasandstone, coarse sandstone, medium sandstone, fine sandstone, very fine sandstone, coarse siltstone and fine siltstone. The grain size of reservoir sediment is related to its sedimentary environment. In the environment with strong hydrodynamic force, the grain size of sediment is relatively coarse; In the environment with weak hydrodynamic force, the sediment grain size is fine. In the northern Dasanjiang area of the study area, Boli Basin, Jixi Basin and Ning'an Basin are in the sedimentary center, with fine reservoir grain size, mainly developing medium fine sandstone; Hulin basin, Sanjiang basin, Tangyuan fault depression and Fangzheng fault depression are located at the edge of Dasanjiang basin, with coarse reservoir grain size and medium coarse sandstone. In the south of the study area, Tonghua Basin and Hongmiaozi Basin are closer to the sedimentary center of the Daliuhe Basin (including the current Liuhe Basin, Tonghua Basin and Hongmiaozi Basin) at that time, and their reservoir granularity is smaller than that of the lower Cretaceous clastic rocks in Liuhe Basin. Coarse sandstones are developed in Shuangyang and Liuhe basins, while medium fine sandstones are mainly developed in Tonghua and Hongmiaozi basins.

3.2.2 Sorting and rounding.

Generally speaking, sandstone with good roundness has good sorting property, even pore space size, many pore coordination numbers, wide throat connecting pores and good connectivity, and small pore throat ratio; The sandstone with poor roundness is also poor in sorting. The reservoir with good clastic particle sorting has large pore space and uniform distribution, single pore type, small pore throat ratio, few fillings in pores, good connectivity between pores, high porosity and permeability; On the contrary, for the reservoir with poor sorting, the particle size distribution is uneven, the fine particles occupy part of the pore space, the argillaceous matrix fills part of the pores, and blocks the throat of part of the pores, changing the geometric shape of the clastic rock in which the pore structure is determined by the skeleton particles, making the pore space of the reservoir complex, with large differences in pore space size, uneven distribution, and large pore throat ratio, The connectivity between pores is poor, and the porosity, permeability and reservoir capacity are poor. The clastic particles of the Lower

Cretaceous reservoir in the study area are mainly medium to coarse, with medium sorting, and the roundness is mostly angular~sub round (Plate 4, Figure 5). On the whole, the sediment grain size of the Lower Cretaceous reservoirs in each basin in the northern region is relatively coarse, with medium deviation in sorting, poor roundness, and mainly secondary to angular. However, the structural maturity of the Lower Cretaceous reservoir in Boli Basin is relatively high, with good selection and sub prismatic to sub round roundness, which indicates that the sediments in the northern basin are mostly proximal sediments with short transport distance, and Boli Basin was the sedimentary center of the Dasanjiang Basin at that time. This research result is the same as that of sedimentary facies. The research on the survey and evaluation of the national oil and gas resources strategic selection area (2004-2009) shows that during the sedimentary period of Chengzihe Formation and Muling Formation in the middle of Early Cretaceous, the northern part of the study area is a unified offshore basin (Dasanjiang Basin), and the sedimentary center is near the Boli Basin.

Figure 4. Poor sorting and roundness, Tongdi Well 1, K_1h, 267.77~267.83m, (+)

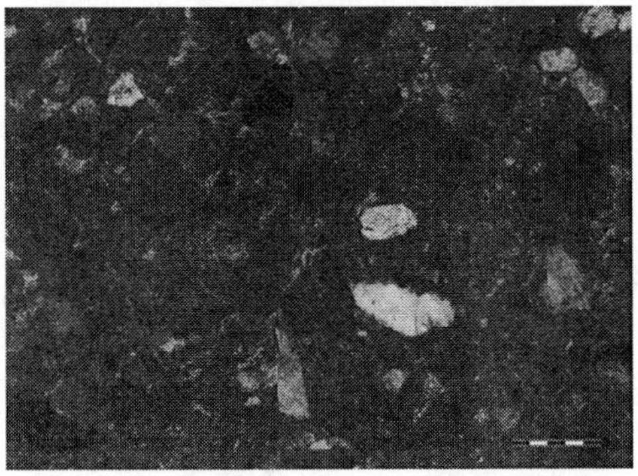

Figure 5. Grain roundness is sub round, with high shale content. Well Tongdi 1, K_1h, 256.26~256.37m, (+)

The grain size of the lower Cretaceous sediments in each basin in the southern region is relatively fine, with coarse, medium and fine sandstones, but mainly medium fine sandstones, with medium sorting preference and sub angular to sub circular roundness. For example, the lower Cretaceous sediments in Hongmiaozi Basin and Tonghua Basin have fine grain size, good sorting and roundness, and parallel bedding is developed (Figure 6), which indicates that a large sedimentary basin (Daliuhe Basin) may be developed in the southern region at that time, Later, it was uplifted, denuded and disintegrated to form small and medium-sized residual basins. The coarse clastic rocks at the edge of the basin were denuded, and the fine clastic rocks at the center of the basin were preserved. According to the erosion amount calculated by Ro, the erosion amount of the Lower Cretaceous in Tonghua Basin at that time was more than 1800m [8].

Figure 6. Parallel bedding, Tonghua Basin, K_1

4. Conclusion

The accretionary complex is mainly composed of 7 parts, including undeformed sedimentary rock, shear deformed sedimentary rock, shear deformed volcanic rock, turbidite, deformed siliceous rock, limestone block and basic ultrabasic rock. Shear deformation structures of deformed sedimentary rocks and volcanic rocks are developed.

The sediment grain size of the Lower Cretaceous reservoirs in each basin in the northern region is relatively coarse, the sorting is moderately deviated, and the rounding is poor, mainly secondary to angular.

The grain size of the lower Cretaceous sediments of each basin in the southern region is relatively fine, with coarse, medium and fine sandstones, but mainly medium fine sandstones, with medium sorting preference and sub angular to sub round roundness.

Acknowledgments

This work was financially supported by Heilongjiang Provincial Undergraduate University Central Support Local University Reform and Development Fund(2020YQ17).

References

1. G.H. Qu, N. Liu, Y.K. Liu Y.L. Meng, Z.L. Zhang, W.Z. He(2021). Simulation Study of Hydrocarbon Generation Characteristics in the Siliceous Rock.

CHEMISTRY AND TECHNOLOGY OF FUELS AND OILS, vol.56,no.6,p. 985-993.

2. G.H. Qu, N. Liu, Y.L. Meng and X.N. Li(2019). Research progress on migration and accretion of Nadanhada terrane. International Conference on Oil & Gas Engineering and Geological Sciences, vol.384,no.1,p.12120.
3. G.H. Qu, N. Liu, Y.L. Meng and X.N. Li(2019). Advances in tectonic genesis of Nadanhada terrane. International Conference on Oil & Gas Engineering and Geological Sciences, vol.384,no.1,p.12121.
4. G.H. Qu, N. Liu, Y.L. Meng and X.N. Li(2019). Evaluation of Source Rocks in Sanjiang Basin of Nadanhada Terrane, vol.6,no.5,p.276-279.
5. P.W. Li, S.H. Zhang, N.H. Shen. The contrast of paleomagnetic results between Nadanhada area, Hei Longjiang province and Mino area, Japan, and its significance. Journal of Changchun Institute of Geology, Vol.27, no.1,1997, pp. 63-67.
6. H. Xu, Y.Q. Liu, H.W. Kuang , N. Peng, W. An, N.S. Wang. Characteristics and tectonic significance of Wandashan accretive complex in Northeast China. Abstracts of papers on sedimentology and unconventional resources at the 2015 National Sedimentology Conference. 2015:318-319.
7. Z.Q. Feng, Y.H. Wang, S.M. Lei, Z.H. Feng, Exploratory techniques and their advancement of deep volcanic gas reservoirs in the Songliao basin, Natural gas industry.vol.27, no.8, 2007,pp. 9-12.
8. Y.L. Meng, W.Q. Shen, X.G. Zhou, S.Z. Li, D.D. Wang, W.H. Zhang, G.H. Qu ,C.X. Cui. Characteristics of the Lower Cretaceous source rocks and shale gas exploration potential of eastern basin group, NE China. vol.37, no.6, pp. 893-902.

Study on Advanced Anchor - grouting Support of Mining Roadway in Close Lower Coal Seam

Guozhi Lu [1], Ping Ni [1,*], Xin Li [2], Jinhua Niu [2] and Tao Wang [2]

[1] College of Energy and Mining Engineering, Shandong University of Science and Technology, Qindao, China
[2] Shandong Zeming Energy Technology Co., Ltd, Taian, China

Abstract. Aiming at the problems of difficult support of mining roadway under goaf of the close distance coal seam, obvious large deformation, and long-term continuous deformation of surrounding rock, an advanced bolt-grouting support system is proposed. The system can effectively solve the problem of roadway support from the design system, construction system, and monitoring and evaluation system, and realize the intelligent evaluation of bolt-grouting support. Based on the actual engineering background of 29204 working face in Dongqu Coal Mine, the supporting material and scheme, grouting parameters, and construction technology are studied and analyzed. Finally, the supporting scheme is adjusted and optimized timely by combining with the intelligent evaluation system of anchor grouting. The field practice results show that the advanced bolting-grouting support system has good applicability to roadway deformation under different conditions. The maximum subsidence of the roof in the test section is reduced by 73.48 % compared with the original support scheme, and the maximum stress of the anchor cable is reduced by 50.68 %. The research results provide a set of safe and effective support system and feasible technical ideas and methods to solve the problem of surrounding rock control of mining roadway under goaf of the close distance coal seam.

1. Introduction

As the main energy in China, coal is of great significance for the development of China's production [1, 2]. In recent years, with the increasing scope and breadth of coal mining, complex geological conditions and high ground stress environment make the control of roadway surrounding rock more difficult. Especially for close coal seam, the traditional support method is difficult to effectively control the stability of the roadway [3, 4]. Therefore, in order to solve the problem of roadway surrounding rock control, a set of safe and effective support technology is urgently needed.

In recent years, domestic and foreign scholars have carried out a lot of research work on roadway support in close coal seams. In advance support of coal seam mining under goaf, aiming at the difficulty of top coal support under goaf in thick coal seam slicing mining, Chai et al. [5] put forward the anchor support mode based on shed support, 'short dense anchor cable reinforcement+anchor suspension', which provided a certain reference for roadway support under goaf. Wang and Fang et al. [6, 7] proposed a variety of combined support technologies for the deformation of roadway surrounding rock under the goaf of close coal seam group, which effectively ensured the stability of roadway surrounding rock and provided a method for solving the problem of roadway surrounding rock control under the goaf. In the study of advanced bolting and grouting support, Li [8] used the method of combining numerical simulation with field practice to study and analyze the theory and technology of grouting anchor cable advanced support in mining roadway, which provided a certain reference for advanced bolting and grouting support. Ning et al. [9] analyzed and verified the advance control range and grouting reinforcement technology and timing according to the distribution and evolution of mine pressure in secondary mining roadway and the characteristics of measured anchor cable tension value, which provided a method for supporting failure caused by the broken surrounding rock. Yao et al. [10] improved the traditional passive advance support technology and proposed the method of grouting anchor cable advance active support, which provided a certain reference for the advance anchor grouting support technology of mining roadway. Professor Lu [11] optimized the timing of grouting bolt support by combining field measurement and numerical calculation results, which provided a method for grouting parameter design. In the aspect of advanced support monitoring and management, Zhang et al. [12] carried out non-destructive testing and early warning analysis of bolt support quality in the deep roadway and proposed early warning technology of roadway surrounding rock stability based on axial bolt load detection.

In summary, many scholars at home and abroad have carried out a lot of research work on the problem of roadway support. However, for the Xishan mining area, the high stress environment and the upper coal seam goaf cause the surrounding rock of the roadway to show obvious large deformation and long-term continuous

* Corresponding author: 1906385617@qq.com

deformation characteristics, and the traditional support system is frequently unstable. Therefore, the traditional support method must be improved. Taking Dongqu Mine in Xishan Mining Area as the research background, this paper puts forward a kind of advanced bolting and grouting support system, and studies the design system, construction system, and monitoring and evaluation system. Through field industrial test, the application effect of advanced bolting and grouting support system is verified, which provides a certain reference for solving the problem of serious deformation of mining roadway under close coal seam goaf.

2. Engineering overview

Dongqu Mine is located in the Gujiao mining area of Xishan Coal and Electricity. The 29204 working face is located in the second mining area of +860 level, which belongs to 9# coal seam, the average depth is 240 m, the dip angle is 0-5°, and the average thickness is 2.67 m. The rock properties of the roof and floor of the working face are mainly sandy mudstone, as shown in Fig. 1. The average distance between 8# and 9# coal seams is 4.35 m, belonging to the close coal seam.

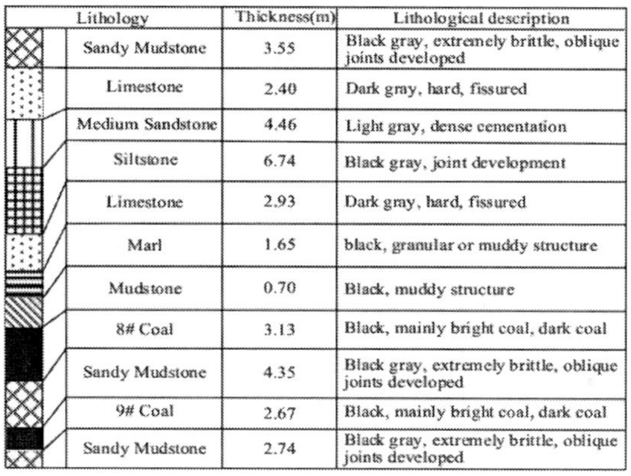

Figure 1. Roof and floor situation of working face

In recent years, affected by 8# coal mined-out area, roadway stress concentration, roof crushing, and 'π' beam deformation occurred in 29204 working face of Dongqu Coal Mine. Therefore, it is necessary to develop a set of advanced bolting and grouting support system suitable for the site conditions of the Xishan mining area.

3. Advance bolting and grouting support system

Advanced bolting and grouting support system is a dynamic system which is based on ' grouting anchor cable + on-line monitoring of mine compaction ' and intelligent evaluation method of bolting and grouting. The advanced bolting and grouting support system mainly includes three aspects: design system, construction system and monitoring and evaluation system.

Among them, the design system mainly adopts the method of combining engineering practice with numerical simulation to study and determine relevant parameters from three aspects of supporting materials, supporting scheme and grouting parameters. The construction system is mainly through the construction technology design and site construction supervision to ensure the design effect; the monitoring and evaluation system is mainly divided into two parts: one is to supervise and evaluate the situation after the site construction to ensure that the construction requirements are met; the second is to monitor the field implementation effect in the process of advancing the working face, and evaluate the implementation effect in real time through the self-developed intelligent evaluation system of anchor injection. When the site does not meet the construction requirements, timely adjustment and optimization of construction technology; when the field implementation effect is not ideal, the design system is optimized in time to determine the best scheme to achieve the optimal control effect.

4. Field practice of advanced bolt-grouting support

4.1 Selection of supporting materials

In view of the roof strata of 29204 working face roadway in Dongqu Coal Mine, the field sampling was first conducted and the standard specimens were made. Then the relevant indoor mechanical tests are carried out to determine the physical and mechanical properties and composition of the rock. Related parameters are shown in Table 1.

Table 1. Parameter table of physical and mechanical properties of roof rock

Name	Uniaxial compressive strength/ MPa	Elastic modulus /MPa	Uniaxial tensile strength/ MPa	Cohesion /MPa	Internal friction angle /°
Sandy mudstone	19.57	5671	4.48	3.86	58.5

Combined with the actual field detection, the coal seam spacing is 4.35m. Combined with the physical and mechanical properties of roof strata, the hollow grouting anchor cable of SKZ22-1/1860 was selected, and ZHM-III Gu'anfeng grouting reinforcement material was selected.

The length of the anchor cable is calculated by the limit equilibrium theory in elastic-plastic mechanics. The specific calculation formula is as follows [13]:

$$R = a\left[\frac{(K_1\gamma H + K_2C\cdot\cot\varphi)(1-\sin\varphi)}{P_i + K_2C\cdot\cot\varphi}\right]^{\frac{1-\sin\varphi}{2\sin\varphi}} \qquad (1)$$

In the formula, K_1 is the repeated mining coefficient; γ is coal density, kN/m^3; h is the depth of coal seam, m; P_i is support resistance, MPa; a is the equivalent radius of the roadway, the value can be obtained by the circumcircle of

 https://doi.org/10.1051/e3sconf/202235801033

the roadway, m; C is cohesion, MPa; φ is the internal friction angle, °; K_2 is the correction coefficient of mechanical parameters of coal and rock mass.

After calculation, the plastic radius of the 29204 track trough is 3.44 m. Considering the roof lithology characteristics, the length of the grouting anchor cable is finally determined to be 4.3 m.

4.2 Determination of support scheme

The determination of support scheme should be based on field engineering practice. At the same time, the characteristics of coal and rock strata, the adaptability of support control technology and the structural coordination of support conditions should be analyzed. Reasonable design should be carried out for the row spacing between anchor cables, the position of anchor cables, the arrangement of anchor cables, the grouting method and sequence.

In view of the problems existing in the track groove of 29204 working face in Dongqu Coal Mine, the preliminary design scheme of field support is as follows: two hollow grouting anchor cables are arranged in each row at the top of the roadway, 1 000 mm apart from the two sides, 1 600 mm apart and 2 000 mm apart. The anchorage force is not less than 200 kN, and the bearing capacity is not less than 350 kN by using 250 mm×250 mm×17 mm high strength bracket and supporting special lock. The anchor cables are all arranged vertically to the rock surface. The interval grouting method is adopted. One row of observation anchor cables is set in each two rows. The grouting time of observation anchor cables is determined according to the deformation of the roadway in the advance section. If the roof in the advance section is reliable and without large deformation, in order to ensure the timely collapse of the working face corner, the observation anchor cables are not grouted. The support scheme is carried out on the basis of the original support.

4.3 Determination of grouting pressure and time

4.3.1 Model establishment and scheme.

This paper uses COMSOL numerical simulation software to study and analyze the grouting pressure and time. The model size is set to 20 m×20 m, and a grouting hole is arranged in the middle part. Darcy 's law is used to solve the model. The grouting pressure and time range are obtained according to the field experience, and the specific simulation scheme is shown in Table 2. The mesh of the model is shown in Fig. 2, and the model parameters are shown in Table 3.

Table 2. Scheme of numerical simulation

Grouting parameters	Simulation scheme									
Grouting pressure/MPa	0.5	1	1.5	2	2.5	3	3.5	4	4.5	5
Grouting time/min	1	2	3	4	5	6	7	8	9	10

Table 3. The parameter of the model

Name	slurry viscosity/Pa·s	slurry density /kg/m³	Permeability of surrounding rock/m²	Gap width /mm	Porosity
Grouting parameters	19.57	5671	4.48	0.5	0.012

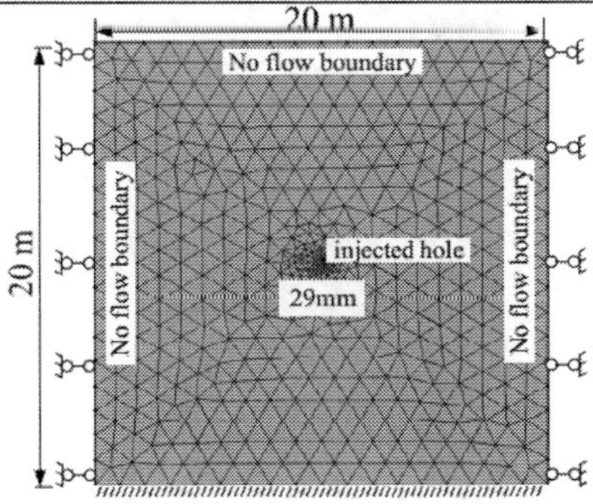

Figure 2. Model mesh generation and boundary conditions

4.3.2 Analysis of simulation results

(1) Grouting pressure

Fixed slurry diffusion time is 10 min, according to the grouting pressure simulation scheme shown in Table 2. The simulation results are selected when the grouting pressure is 1 MPa, 2 MPa, 3 MPa, 4 MPa and 5 MPa, as shown in Fig. 3. As can be seen from Fig. 3, under the same grouting time, the slurry diffusion area increases with the increase of grouting pressure. In order to study its evolution law more carefully, the slurry diffusion area is quantitatively analyzed, and the results are shown in Fig. 4. It can be seen from the figure that at the moment of grouting, the diffusion range of slurry increases significantly, by about 63% compared with that without grouting, considering the breakage of lower coal roof strata caused by upper coal mining. But the growth ratio generally increases first and then decreases. When the grouting pressure is less than 4 MPa, the growth ratio of the slurry diffusion range is greater than 10%. After more than 4 MPa, the slurry diffusion range basically does not increase, the growth rate is less than 10%. Considering that when the grouting pressure is too large, the integrity of the surrounding rock will be further reduced, so the appropriate grouting pressure is determined to be 4 MPa.

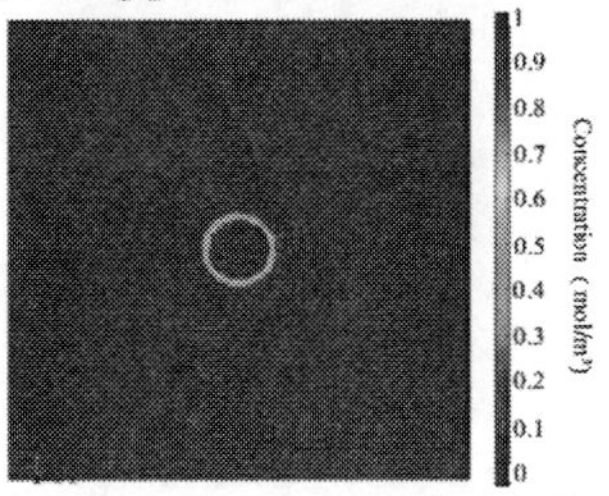

(a) Grouting pressure is 1 MPa

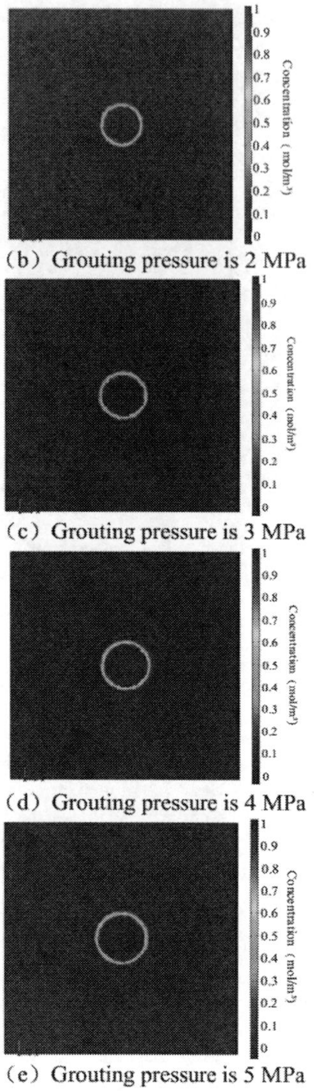

Figure 3. Partial simulation results

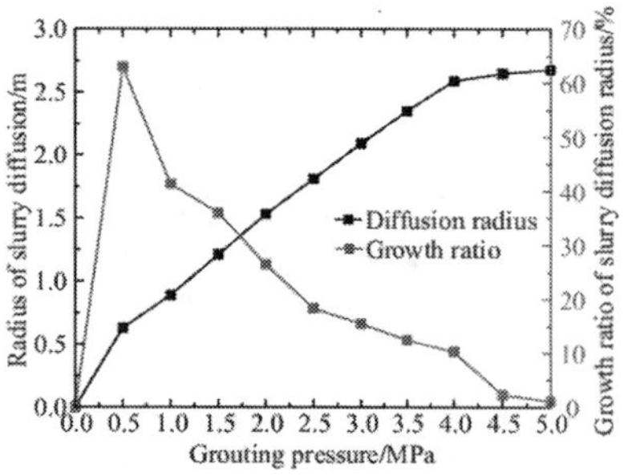

Figure 4. Slurry diffusion effect at different grouting pressure

(2) grouting time

Based on the above analysis of grouting pressure, the fixed grouting pressure is 4 MPa, and the simulation analysis is carried out according to the grouting time simulation scheme shown in Table 2. The results are shown in Fig. 5.

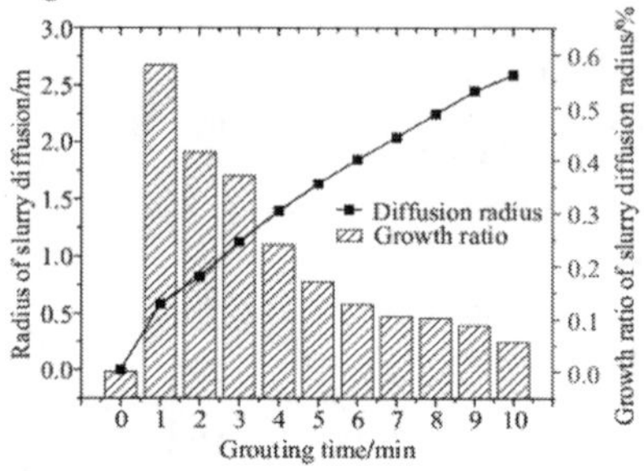

Figure 5. Slurry diffusion effect at different grouting time

It can be seen from the diagram that under the action of grouting pressure, the slurry diffusion radius increases gradually, but it also shows a trend of increasing first and then decreasing. When the grouting time is less than 8 min, the growth rate of slurry diffusion radius is more than 10%, but the overall growth rate decreases gradually. When the grouting time was more than 8 min, the growth rate was less than 10%, showing a rapid downward trend, and the diffusion effect was poor. So the best grouting time is 8 min.

4.4 Time of grouting

Pulverized coal detection was carried out at 6 m, 20 m, 28 m, 41 m, 60 m, 82 m and 105 m ahead of the working face in the track trough of 29204 working face. Taking into account the length of the hollow grouting anchor cable, the average amount of coal powder in the 1~5 m drilling range of different distances in the advanced working face is selected for monitoring and analysis, as shown in Fig. 6.

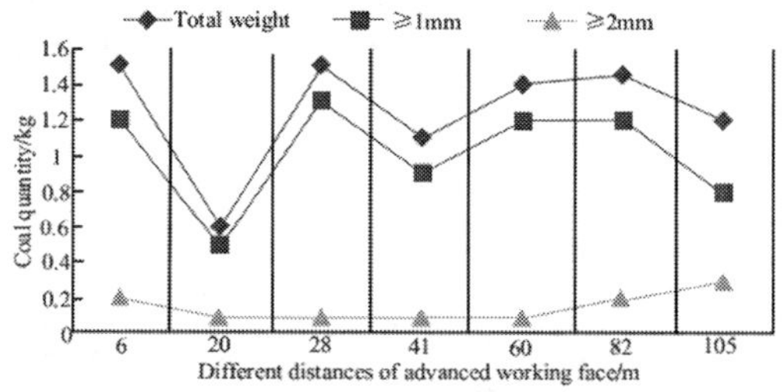

Figure 6. Statistical curve of average coal powder quantity

The amount of coal powder in different drilling depth ranges from 20 m to 28 m in advance working face generally increases, and some boreholes appear stuck drilling and drilling difficulties, indicating that the pressure in this area increases, which may have a greater damage to surrounding rock. At the same time, considering the advancing speed of the working face, the convenience of grouting construction and the solidification time of grouting slurry, it is suggested that the grouting reinforcement time should be selected at about 20 m ahead of the working face.

5. Construction Technology Design and Monitoring Effect Analysis

5.1 Site Construction Technology and Design Supervision

The range of 570~670 m in advance working face is selected as the field construction test area based on the analysis of the current roadway stress in the track roadway of 29204 working face. A station is set every 10 m to ensure the correctness of the monitoring data.

In the field construction, the anchor hole position is determined according to the design and layout requirements of the grouting anchor cable, and the drilling bit with a diameter of 29 mm is used for drilling. Then the grouting anchor cable is pushed into the fundus, and the drug roll is stirred for not less than 10 s after the push. After the stirring is completed, the bolt machine should maintain the thrust of not less than 2 min to ensure the grouting effect. When grouting, using from inside to outside grouting sequence, the speed is slow. After the grouting is filled, the grouting pump can be closed, and the grouting waits for 2~3 min until the grouting is filled again, and the grouting mouth is blocked with screws in time to prevent the slurry from flowing out. In the process of grouting, grouting pressure should be observed at all times to prevent leakage and running.

5.2 Site supervision and monitoring

From Fig. 7, it can be seen that the deformation of surrounding rock and the stress of anchor cable in 570~600 m roadway are relatively stable, indicating that the support effect is better. However, the deformation of the roadway is suddenly too fast between 600~610 m, and the stress of the anchor cable is suddenly increased, and exceeds the warning value set by the mine, indicating that the original support scheme cannot effectively control the deformation of the surrounding rock of roadway. The field analysis shows that the surrounding rock of roadway in this area is broken and belongs to the broken area of the surrounding rock, so it is necessary to adjust the support scheme. Combined with the intelligent evaluation platform of anchor grouting support, the original support scheme is analyzed and predicted. The anchor cable row spacing is changed to 1 000 mm, and the grouting method is used in each row in turn.

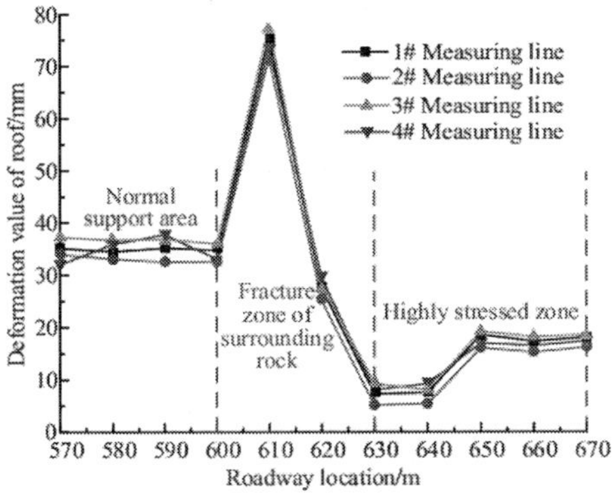

(a) Deformation value of Roof

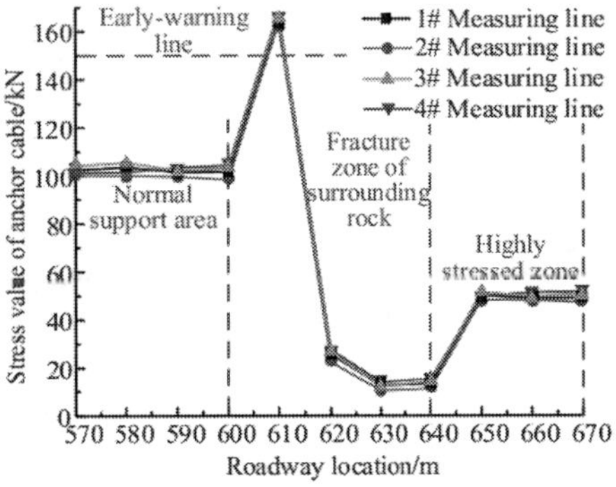

(b) Stress value of anchor cable

Figure 7. Monitoring results and scheme optimization

It can be seen from Fig. 11 that the deformation of the roadway decreases rapidly, and the stress value of the anchor cable is also reduced to below the warning value, and tends to be stable. With the continuous advancement of the working face, within the range of 630m-640m, it is found that the roof of the roadway has no subsidence, and the suspended roof greater than 5m appears after the mining of the working face, which affects the safety production of the working face. The field analysis shows that the surrounding rock of the roadway in this area belongs to high stress area, so it is necessary to adjust the support scheme appropriately. Combined with the intelligent evaluation platform of anchor grouting support, the support scheme is analyzed and predicted, and the grouting sequence is changed to interval grouting.

After the optimization of the support scheme, it can be seen from Fig. 7 that the deformation of the roadway increases slightly, but finally tends to be stable. The stress value of the anchor cable also shows the same trend, and the final values of the two also meet the requirements of safe production of working face.

Compared with the original support scheme, the overall field implementation shows that the maximum subsidence of the roof is reduced by 73.48% and the maximum stress of anchor cable is reduced by 50.68% compared with the original support scheme, and the support effect is improved significantly. It can be seen that the intelligent evaluation platform of anchor-grouting support proposed in this paper can accurately optimize the design of support scheme, so as to achieve good field application effect.

6. Conclusion

This paper puts forward a kind of advanced bolting and grouting support system, and carries out engineering application of the working face track channel, and achieves the following results:

1) An advanced bolting and grouting support system and intelligent evaluation of bolting and grouting is proposed. The system can effectively solve the problem of roadway support from the design system, construction system and monitoring and evaluation system.

2) Aiming at the track trough of 29204 working face in Dongqu Coal Mine, the grouting pressure, time and opportunity in the grouting construction technology were

studied and analyzed, respectively. By detecting the amount of pulverized coal in the same drilling depth in different distance boreholes ahead of the working face, the grouting opportunity was determined to be 20m.
3) The results show that the maximum subsidence of roof in the test section is reduced by 73.48% compared with the original support scheme, and the maximum stress of anchor cable is reduced by 50.68%, indicating that the design of support system can effectively improve the support effect of roadway surrounding rock.

Acknowledgments

This work was financially supported by the National Natural Science Foundation of China (No.52174121).

References

1. H.P. Xie, L.X. Wu, D.Z. Zheng. Prediction on the energy consumption and coal demand of China in 2025[J]. Journal of China Coal Society, 2019, 44(07): 1949-1960.
2. H.P. Kang, G. Xu, B.M. Wang, et al. Forty years development and prospects of underground coal mining and strata control technologies in China[J]. Journal of Mining and Strata Control Engineering, 2019, 1(1): 013501
3. L. Li, S.J. Li. Evolution rule of overlying strata structure in repeat mining of shallow close distance seams based on Schwarz alternating procedure[J]. Journal of Mining and Strata Control Engineering, 2021, 3(2): 023515.
4. G.M. Yuan, T. He. Feasibility discrimination method of upward mining in close-distance coal seam based on data statistics[J]. Journal of Mining and Strata Control Engineering, 2021, 3(3): 033526.
5. J. Chai, Z.C. Han, Y. Qiao, et al. Top coal stability of long-span opening under goaf in layered mining[J]. Journal of Mining & Safety Engineering, 2022, 39(2): 282-291.
6. L.F. Wang, Z.C. Chang, Z.B. Yang, et al. Combined support technology of roadway under mined gob of ultra-distance seams in deep mine[J]. Journal of Mining & Safety Engineering, 2018, 35(4): 686-692.
7. B.M. Fang. Support technology for gateway under goaf in deep coal seams[J]. Safety in Coal Mines, 2021, 52(12): 121-127.
8. L.H. Li. Study on the application of advance support technology with grouting and anchor cable in mining roadway[D]. Xuzhou: China University of Mining and Technology, 2020.
9. S. Ning. Advance rapid control technology of broken surrounding rock in secondary dynamic pressre roadway of hilt working face[J]. Coal Engineering, 2022, 54(3): 56-61.
10. Q.L. Yao, X.H. Wang, Z. Xia, et al. Key technology and application of active forepoling for longwall coal mining in coal mine[J]. Journal of Mining & Safety Engineering, 2020, 37(2): 289-297.
11. Y.L. Lu, X.Y. Feng, K.Z. Wu, et al. Development and application of online intelligent digital monitoring system for grouting in underground roadway of coal mine[J]. Journal of Mining & Safety Engineering, 2022, 39(2): 347-358.
12. Zhang Houquan, Miao Xiexing, Zhang Guimin, et al. Non-destructive testing and pre-warning analysis on the quality of bolt support in deep roadways of mining districts[J]. International Journal of Mining Science and Technology, 2017, 27(6): 989-998.
13. T. Tian. Research on deformation law of surrounding rock and supporting parameters of roadway under short distance goaf in dongqu coal mine[D]. China University of Mining and Technology, 2020.

Research on the in-situ stress inversion method of mine based on GA-BP algorithm

Zhongcheng Qin, Xiaohe Li*, Yongle Liu, and Guanghui Gao

College of Energy and Mining Engineering, Shandong University of Science and Technology, Qingdao, China

Abstract. In-situ stress is the basis for designing mine roadway supports. Firstly, based on the GA-BP algorithm, establish a numerical analysis model, apply orthogonal tests, to establish the relationships between in-situ stress and lateral pressure coefficient, different rock parameters. Then, the optimization model is built by setting the sum of squares difference between measured and calculated in-situ stresses as optimization targets. Finally, using the model to solve the in-situ stress. Results show that the relative error between GA-BP algorithm inversion results and measured values is 2.9% averagely, while the relative error between BP algorithm inversion results and measured values is 4.4% averagely. Inversion methods can provide reference for similar mine in-situ stress inversion.

1. Introduction

The in-situ stress acts on mine rock distortion and damage fundamentally [1], its magnitude and distribution have a great influence on mine exploit, channel design and support parameters [2-4]. The borehole stress relief method could measure in-situ stress, but the actual results are inferior in terms of dispersion and reflecting local in-situ stress only due to impacts of site, funding and testing techniques [5].

Scholars have done lots of research on nonlinear optimization algorithms for in-situ stress inversion. Yang [6] et al. obtained in-situ stress field in complex geological bodies using numerical analysis and genetic programming algorithm. Meng Wei [7] et al. took geothermal temperature into account in tress inversion, obtained in-situ stress of the geological body. Zhao Chen [8] et al. improved lateral pressure coefficient method by considering factors like strata stripping, rock mechanical character, and faults.

Analysis shows that, the in-situ stress inversion has strong dependence on the accuracy of measurement points. In this paper, Genetic Algorithm (GA) and BP neural network are used to invert the in-situ stress in coal mine.

2. Study of in-situ stress inversion method

2.1 Measured in-situ stress acquisition

During in-situ stress measurement, points should lay out in stable rock layers of appropriate thickness, avoid interfering with tunnel construction or other production processes.

Actual measurements of in-situ stress get three main stress magnitudes, azimuth and dip angles. The data with higher dispersion and errors will be cut out to prepare for inversion.

2.2 Creation of in-situ stress influence function

In-situ stress consists of self-gravity and geological movement of the rock mass. The tectonic stress is related to rock properties and geological tectonic. This paper mainly consider the self-weight stress and structural stress.

For lithology, the elastic modulus E and Poisson's ratio μ are regarded as inversion objectives, and rock masses divided into n zones to discuss separately.

In a block dominated by a self-weight stress field, the vertical stress at any depth H below the surface is equal to the gravity of the overlying rock per unit area[9].

$$\sigma_z = \sum_{i=1}^{m} \gamma_i H_i \tag{1}$$

Where, σ_z vertical stress, MPa; γ_i the weight of the ith layer of rock, kN·m^{-3}; H_i the thickness of the ith layer of rock, m.

Therefore, the rock weight γ is chosen as one of the in-situ stress inversion factors.

Tectonic stress field is controlled by horizontal tectonic motion. In-situ stress field changed by considering ground uplifting by tectonic movement and denudation. Thus, the lateral pressure coefficient is set as the inverse factor.

Establish in-situ stress inversion objective function as:

$$\sigma_k = f_k\left(\gamma_i, E_i, \mu_i, \lambda\right)\left(i = \left(1, n\right), k = \left(1, 3\right)\right) \tag{2}$$

Where, σ_k the kth principal stress, $f_k(\cdot)$ the kth principal stress function.

Analysis shows that basic ideas of in-situ stress inversion are concluded as follows: based on easy-to-get

* Corresponding author: lixiaohe0425@163.com

rock parameters, with tools in numerical simulation and orthogonal tests, establish GA-BP algorithm model.

3. GA-BP algorithm inversion analysis

BP neural networks were proposed by Rumelhart, which training multilayer feedforward networks by error back propagation[10], has merits of high nonlinearity, robustness and parallel processing capability.

Based on the literature[11], the output equation of each node of the hidden layer is calculated:

$$z_{dk} = f\left(net_{dk}\right) = f\left(\sum_{i=0}^{n} v_{ki}x_{ki} + \theta_k\right)(k = 0, q) \quad (3)$$

Calculate the formula for each node of the output layer:

$$o_{dk} = f\left(net_{di}\right) = f\left(\sum_{k=0}^{q} w_{ik}z_{dk} + \theta_j\right)(j = 0, m) \quad (4)$$

According to the study, the excitation function is determined Sigmoid function.

Output layer node weight adjustment formula:

$$\Delta w_{jk} = \eta\sum_{p=1}^{p}\left(-\frac{\partial E_p}{\partial net_{pj}} \times \frac{\partial net_{pj}}{\partial w_{jk}}\right) = \eta\sum_{p=1}^{p}\delta_{pj}z_{pk} \quad (5)$$

Put the learning rate of stochastic gradient descent as η(0.1~0.3) and the signal error as $\delta_{pj} = (t_{pj} - y_{pj})\cdot y_{pj}\cdot(1 - y_{pj})$.

The formula for adjusting the weights of each neuron in the hidden layer:

$$\Delta v_{ki} = \eta\sum_{p=1}^{p}\left(-\frac{\partial E_p}{\partial net_{pk}} \times \frac{\partial net_{pk}}{\partial w_{ki}}\right) = \eta\sum_{p=1}^{p}\delta_{pk}x_{pi} \quad (6)$$

Make the implied layer signal error $\delta_{pk} = \left(\sum_{j=1}^{m}\delta_{pj}w_{jk}\right)\times z_{pk}\left(1 - z_{pk}\right)$.

Genetic algorithms (GA) achieve self-adaptive search for optimal solutions by simulating selection, crossover, and mutation actions in biological genetic and evolutionary[12].The basic components of a genetic algorithm can be defined as:

$$SGA = \left(C, E, P_0, M, \varphi, \tau, \psi, s\right) \quad (7)$$

Where C individual coding method, E individual fitness evaluation function, P_0 initial population size, M population size, φ selection operator, τ crossover operator, ψ variation operator, S genetic algorithm termination condition.

GA algorithm operates on the threshold, weights of BP neural network layers to obtain the optimal solution and then use it as the new weight threshold of BP neural network to complete optimization.The GA-BP neural network model flow chart is as follows.

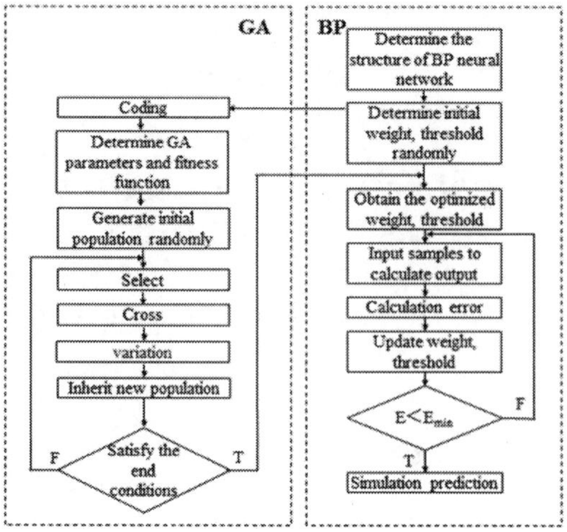

Figure 1. GA-BP neural network model flow chart.

4. Instance Analysis

In order to obtain the coincidence of inversion of in-situ stress using GA-BP algorithm, this paper takes the three-mining area of a mine as an example for analysis.

The average thickness of loose layer in the three mining area is 224m, the average thickness of bedrock is 60.3m, and the maximum mining depth has reached 800m; the dip angle of coal seam is 8° ~18° , the average dip angle is 13° ; the coal thickness is 3.7~5.1m, the average coal thickness is 4.4m; to the north is the trackway of the three-mining area and the working face to be mined, to the south is the fault YF16 and 3303 working face mining void area, to the west is the 4# magmatic rock engulfment area, and to the east is the three-mining area tape downhill. The roadway deformation is serious during 3301 working face chute boring, the coal and rock is identified to have medium impact tendency, the threat of impact power disaster is high.

4.1 In-situ stress measurement

In order to optimize the design of underground roadway support and scientific prevention of rock burst, the research group carried out in-situ stress measurement.

K1 locates in the upper track lane of -534 three-mining area, with less lane lay-out at same level. k2 locates in -740 three-mining area lower track road, belongs to deep roadway, which is affected little by mining. K3 locates in -602 five-mining area track downhill refuge, without mining influence. The relevant parameters of the boreholes are shown in Tab.1, the stress relief curves and cores are shown in Fig.2, the measured results are shown in Tab.2.

 https://doi.org/10.1051/e3sconf/202235801034

Table 1. Construction drilling parameters.

Points	Location	Depth /m	Drilling parameters				
			Guide hole depth /m	Small hole depth /m	Set of core length /m	Azimuth /°	Dip /°
1#	Upper track lane of 3-mining area	534	10.28	0.36	0.42	251	10.5
2#	Lower track lane of 3-mining area	740	10.24	0.35	0.41	135	10.0
3#	Downhill escape chamber of 5-mining area	602	10.26	0.38	0.44	216	12.0

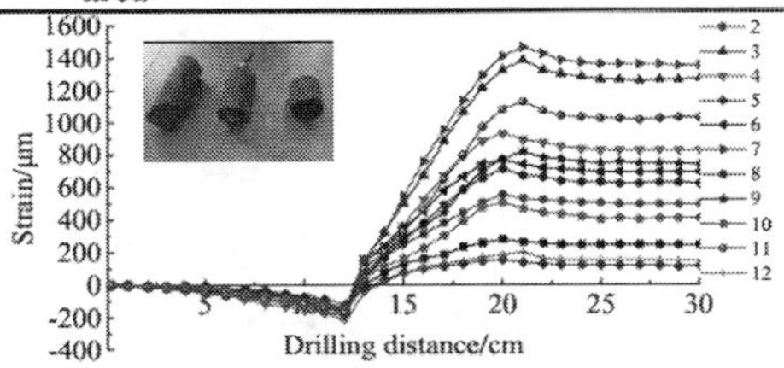

a.1# Stress relief curve.

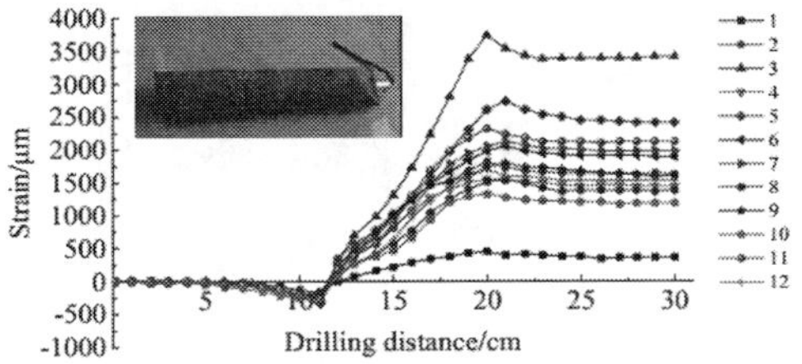

b.2# Stress relief curve.

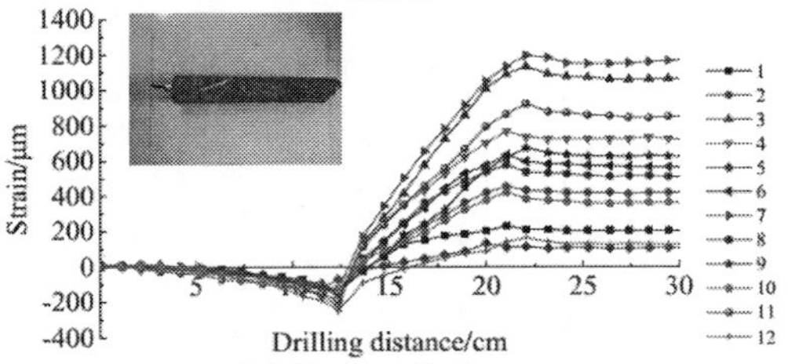

c.3# Stress relief curve.

Figure 2. Stress relief curve.

Table 2. Measurement results of in-situ stress.

Points	Principal stress	Measurement/MPa	Azimuth/°	Dip/°	σ_v	σ_i/σ
1#	σ_1	21.94	256.74	7.43	15.05	1.46
	σ_2	9.88	341.14	-36.80		0.66
	σ_3	8.55	176.42	-52.20		0.59
2#	σ_1	38.19	125.32	8.73	20.43	1.87
	σ_2	20.93	67.85	-74.07		1.02
	σ_3	15.43	213.25	-13.23		0.76
3#	σ_1	25.78	221.22	9.62	15.64	1.65
	σ_2	11.69	303.67	-37.77		0.75
	σ_3	10.40	143.12	-50.59		0.66

4.2 Establishment of mining area influence function

To obtain training and test samples, numerical analysis models were established and in-situ stress inversion analysis was carried out based on actual measurement data, shown in Figure 3.

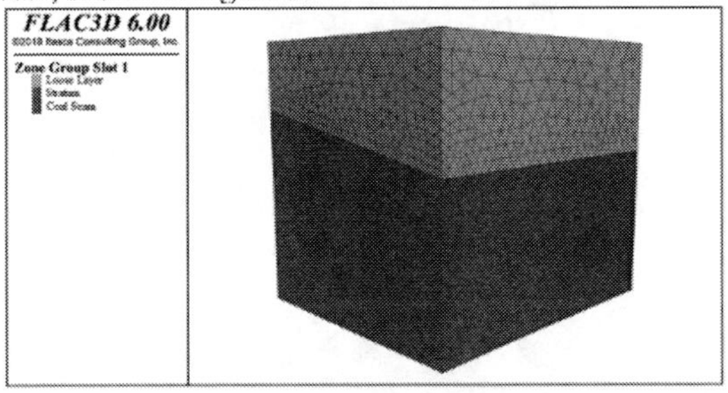

Figure 3. Numerical analysis model.

Considering that there are obvious differences in mechanical properties of each layer, the volumetric weight, elastic modulus and Poisson's ratio of the overlying loose layer, bedrock layer and coal seam are indicated by γ_1, E_1, μ_1; γ_2, E_2, μ_2; γ_3, E_3, μ_3. Clarify the value of the lateral pressure coefficient in this mining area is 0.6~2.3.

Table 3. Values of rock mechanics parameters in each partition.

Zone	Volume weight γ/kN·m^{-3}	Poisson μ	Elasticity modulus E/GPa
Loose layer	18-23	0.35-0.43	0.55-1.23
Stratum	25-29	0.27-0.35	16.88-17.96
Coal seam	13-18	0.18-0.22	95.81-131.74

10 factors of loose seam, coal seam, bedrock seam parameters and lateral pressure coefficient were selected for orthogonal test, capacitance γi set to 6 levels and the others to 3 levels for totally 36 experiments, Tab. 4 shows part of calculation scheme.

Table 4. Partial orthogonal analysis calculation scheme of in-situ stress inversion(Input samples).

No.	λ	Loose layer			Stratum			Coal seam		
		E_1	μ_1	γ_1	E_2	μ_2	γ_2	E_3	μ_3	γ_3
1	0.6	0.55	0.35	18	95.81	0.27	25	16.88	0.18	13
3	0.6	0.55	0.43	19	113.78	0.31	28	17.96	0.18	18
5	0.6	0.89	0.39	21	95.81	0.27	28	16.88	0.22	17
7	0.6	0.89	0.43	22	131.78	0.35	29	17.96	0.20	13
9	0.6	1.23	0.43	21	95.81	0.35	30	17.42	0.18	15
11	0.6	1.23	0.35	21	131.74	0.31	29	16.88	0.22	14
13	1.5	0.55	0.39	19	95.81	0.35	29	16.88	0.18	16
15	1.5	0.55	0.43	22	131.74	0.31	25	17.42	0.22	17
17	1.5	0.89	0.43	18	95.81	0.31	27	16.88	0.20	15
19	1.5	0.89	0.39	21	131.74	0.31	28	17.42	0.18	13
21	1.5	1.23	0.35	22	95.81	0.27	26	17.42	0.18	18
23	1.5	1.23	0.35	19	113.78	0.31	30	16.88	0.20	17
25	2.3	0.55	0.35	23	95.81	0.31	30	17.96	0.22	13
27	2.3	0.55	0.43	21	113.78	0.27	27	16.88	0.22	18
29	2.3	0.89	0.35	20	95.81	0.31	29	17.42	0.20	18
31	2.3	0.89	0.39	22	113.78	0.27	30	16.88	0.18	14
33	2.3	1.23	0.35	18	113.78	0.35	28	17.42	0.22	16
35	2.3	1.23	0.39	19	131.74	0.27	27	17.42	0.20	13

To get training samples, every set of parameters in Tab. 4 was brought into model and FLAC applied for calculation to get 36 sets of principal stresses for 3 measurement points. Taking parameters as input and corresponds to the principal stresses as output, some training samples corresponding to the orthogonal analysis calculation scheme are shown in Tab. 5.

Table 5. Partial algorithm training samples(Output samples).

		Output								
		1#			2#			3#		
μ_3	γ_3	σ_1	σ_2	σ_3	σ_1	σ_2	σ_3	σ_1	σ_2	σ_3
0.18	13	21.28	6.88	5.58	32.38	12.39	7.08	21.90	5.87	5.35
0.18	18	21.84	8.36	7.06	33.56	12.49	7.18	22.65	7.52	6.72
0.22	17	9.63	7.82	6.52	34.62	11.73	6.42	23.32	7.12	6.81
0.20	13	23.44	9.22	7.92	35.34	15.13	9.82	24.30	9.91	8.91
0.18	15	22.92	9.06	7.76	35.03	14.94	9.63	23.85	9.55	9.45
0.22	14	22.91	7.70	6.40	34.81	13.24	7.93	23.85	8.34	7.76
0.18	16	21.86	15.54	14.24	33.74	23.94	18.63	22.72	11.16	10.05
0.22	17	23.70	13.59	12.29	34.52	23.94	18.63	24.04	11.91	11.32
0.20	15	21.32	12.29	10.99	32.8	21.01	15.7	22.05	8.34	8.25
0.18	13	22.91	12.92	11.62	34.58	23.59	18.28	23.70	11.26	10.46
0.18	18	23.39	12.49	11.19	34.65	22.38	17.07	24.06	10.66	10.12
0.20	17	21.86	11.71	10.41	33.93	22.68	17.37	22.77	10.00	9.31
0.22	13	30.94	14.96	13.66	50.44	25.95	20.64	33.28	16.82	16.68
0.22	18	29.69	14.45	13.15	45.26	24.08	18.77	29.30	15.50	15.49
0.20	18	28.68	13.82	12.52	44.77	23.88	18.57	28.97	14.93	14.42
0.18	14	29.88	14.71	13.41	47.85	25.44	20.13	30.75	16.31	16.16
0.22	16	25.93	13.64	12.34	44.97	23.91	18.6	28.20	13.05	13.91
0.20	13	26.34	13.92	12.62	42.77	24.28	18.97	26.72	14.55	14.42

4.3 Local in-situ stress inversion of mining zone

4.3.1 Network learning training.

In this paper, a two-layer BP neural network model is used, with 10 neurons in the input layer, 9 neurons in the output layer, 1 implicit layer, 21 nodes in the implicit layer selected by experience and trial-and-error method, and 429 thresholds for the network connection weights. a flexible BP algorithm is chosen, and the maximum number of network iterations is set to 1000 and the target fit difference is 0.001.

Genetic algorithm optimization weight threshold parameters are set as follows: population size 50, selection operator 0.08, crossover probability 0.4, variation probability 0.1, and maximum genetic generation 100 generations. The neural network is floating-point encoded with different inter-layer connection weights and implied layer thresholds. Random traversal sampling is used for the choice operator and advanced recombination operator is used for the crossover operator.

The BP and GA-BP algorithms were trained for 36 sets of samples respectively, to make the results representative, divided into training set (30 sets) and test set (6 sets), each set performed 20 times of network training, and the results were taken for comparison with in-situ stress actual data.

Comparing R^2 for 20 training sessions of GA-BP and BP, the R^2 of GA-BP model is 0.9778 and BP model is 0.8668. It can be seen that GA optimizes BP's initial weights and thresholds, which improves the network convergence success rate and certain extent. So the GA-BP model is significantly better than the BP model.

4.3.2 Error analysis and algorithm comparison of inversion results.

To verify the reliable of results, Tab.6 shows error analysis of measured in-situ stresses with GA-BP, BP algorithm inversion values. Fig. 4 shows comparison of relative errors between GA-BP and BP.

Table 6. Comparison between GA-BP and BP algorithm inversion values and measured values

Points	Sample	σ_1 Measured	σ_1 Inversion value GA-BP	σ_1 Inversion value BP	σ_2 Measured	σ_2 Inversion value GA-BP	σ_2 Inversion value BP	σ_3 Measured	σ_3 Inversion value GA-BP	σ_3 Inversion value BP
1#	1	21.90	22.19	21.53	9.88	9.52	10.37	8.55	9.01	8.96
	2		21.01	22.51		10.00	9.34		8.39	8.98
	3		22.49	22.18		10.15	10.39		8.67	7.79
	4		22.78	21.16		9.89	10.03		8.42	8.17
	5		22.97	20.67		10.18	10.29		8.39	8.42
	6		22.39	21.29		10.4	10.39		8.39	8.98
2#	1	38.19	39.39	37.45	20.93	20.42	19.90	15.43	14.72	16.45
	2		39.34	37.47		22.00	19.80		14.61	15.80
	3		39.35	36.45		20.38	19.75		14.99	14.50
	4		38.78	36.45		21.26	19.90		15.02	14.83
	5		39.36	36.48		21.81	21.26		15.13	16.26
	6		39.36	37.47		19.90	21.26		14.69	14.41
3#	1	25.78	26.23	24.30	11.69	12.23	10.60	10.40	10.16	10.71
	2		25.97	24.10		11.24	11.49		10.28	9.39
	3		26.20	26.65		12.43	11.30		10.22	11.08
	4		26.68	26.29		12.42	11.40		10.28	9.64
	5		25.90	25.96		11.28	11.17		9.98	9.92
	6		27.30	27.47		11.27	12.18		9.98	10.98

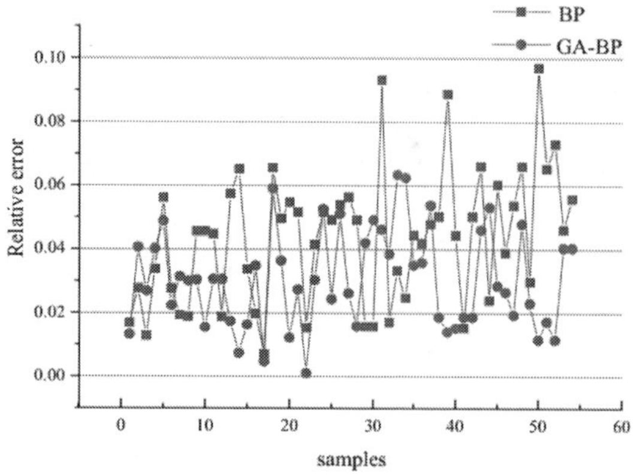

Figure 4. Comparison of relative error between GA-BP, BP algorithm and actual measured values.

Evidently, relative errors between three principal stresses obtained by inversion using GA-BP and measured principal stresses ranged from 0.1%~6.3%, the relative errors of BP ranged from 3.4%~9.7%. The mean relative error of BP algorithm is 0.044 and the MSE is 0.720, while the mean relative error of the GA-BP algorithm is 0.029 and the MSE is 0.431. The mean squared error of the GA-BP algorithm is smaller than BP algorithm, which show that the GA-BP neural network is better than the BP. The inversion results of GA-BP algorithm are close to measured values, which has a better agreement.

5. Conclusion

The inversion parameters selected based on the actual works , with reasonable classification of rock, considering mixed parameters of lateral pressure coefficient, capacity, elastic modulus and Poisson's ratio, to get more approximate inversion results.

The three-dimensional fast Lagrangian method is used to calculate the sample structure of the algorithm, and the orthogonal test design scheme is used to simulate the engineering practice more reasonably and reduce the numerical simulation conditions.

The GA-BP algorithm is used to invert the in-situ stress of the mine. Combining the self-learning and adaptive characteristics of the neural network with the global parallel search method of the genetic algorithm, the distribution characteristics of the underground stress field can be simulated efficiently and truly, which plays an important role in the design and construction of coal mine engineering and disaster prevention.

Acknowledgments

This work was financially supported by ZR2021ME248 fund and ZR2020KE030 fund.

References

1. Qian Minggao, Shi Pingwu, Xu Jialin, et al. Mine pressure and strata control, Jiangsu: China University of Mining and Technology Press, 2010, PP. 41-46.
2. KANG Hongpu, SI Linpo, ZHANG Xiao. Characteristics of underground in-situ stress distribution in shallow coal mines and its applications. Journal of China Coal Society, Vol.41, No.6, PP.1332-1340.
3. KANG Hongpu, WU Zhigang, GAO Fuqiang, et al. Effect of geological structures on in-situ stress distribution in underground coal mines. Chinese Journal of Rock Mechanics and Engineering, Vol.31, Supp.1, PP.2674-2680.
4. WANG Jinan, LI Fei. Review of inverse optimal algorithm of in-situ stress field and new achievement. Journal of China University of Mining & Technology, Vol.44, No.2, PP.189-205.
5. HE Jiangda, XIE Hongqiang, WANG qizhi, et al. Inversion analysis of initial geostress in dam site of Guandi Hydropower Project. Chinese Journal of Geotechnical Engineering, Vol.31, No.2, PP.166-171.
6. YANG Zhiqiang, GAO Qian, ZHAI Shuhua, et al. Intelligent inversion method of in-situ stress field for a complicated engineering geological body. Journal of Harbin Institute of Technology, Vol.48, No.4, PP.154-160.
7. MENG Wei, HE Chuan, ZHANG Junbo, et al. Inversion analysis of the initial geostress field of rock masses under high geo-temperature and high geostress. Chinese Journal of Rock Mechanics and Engineering, Vol.39, No.4, PP.749-760.
8. ZHAO Chen, XIAO Ming, CHEN Juntao, Inversion analysis method for in-situ stress field under complex geological conditions. J. Huazhong Univ. of Sci. & Tech. (Natural Science Edition), Vol.45, No.8, PP.87-92.
9. SHEN Mingrong, CHEN Jianfeng, et al. rock mechanics, Shanghai: Tongji University, 2015, PP.129-142.
10. HUANG Li. BP neural network algorithm improvement and application research, Chongqing Normal University, 2008.
11. Li Guoyong, YANG Lijuan. Neural Fuzzy Predictive Control and its MATLAB Implementation, Beijing: Publishing House of Electronics Industry, PP.85-102.
12. ZHAO Tongbin. Back analysis of roadway displacement based on genetic algorithm, Shandong University of Science and Technology, 2004.

Study on the Stratigraphic Framework and Main Control Factors of Hydrocarbon Formation in Thin Reservoirs of Carboniferous Clastic Rocks in Tahe Oilfield

Feng Deng [1], Li Sun [2,*], Youjuan He [1], Huaxing Hou [2], Zhi Yang [2]

[1] China Petroleum & Chemical Corporation Northwest Oilfield Branch Exploration and Development Research Institute, Urumqi, China
[2] Beijing Huimeng Qihang Petroleum Technology Co., Ltd, Beijing, China

Abstract. The Carboniferous Karashayi Formation in Tahe Oilfield is one of the most promising formations in the field of clastic rocks in Tahe Oilfield, which is rich in hydrocarbon resources, but the stratigraphic framework is unclear, which restricts the exploration of hydrocarbon formation. The stratigraphic features and seismic section analysis of the area combined with oil and gas distribution are used to analyze the stratigraphic framework and reservoir control factors of the Karashayi Formation. The results show that the lower part of the Carboniferous sedimentary strata is overburdened and the top part is denuded. The Karashayi Formation can be divided into four tertiary sequences and seven system domains. The oil and gas reservoirs are mainly developed near the interface of the sequences. Good transport conditions and stratigraphic unconformity cusp blocking can also make the marine intrusion system domain a favorable location for oil and gas accumulation. The Carboniferous Karashayi Formation is located on a convex overburden background and is a composite reservoir controlled by a stratigraphic unconformity and stratigraphic sharp extinction, which is influenced by tectonics, fractures, and lithology.

Keywords: Stratigraphic Framework, Reservoir-forming Principal Control Factors, Carboniferous, Tahe oil field

1. Introduction

Large stratigraphic reservoirs have been the mainstay of China's oil and gas reserves growth in recent years, with proven reserves accounting for 78% of China's new oil and gas discoveries in the same period, and have become an important area for increasing oil and gas reserves and production [1-2]. The Karashayi Formation of the Carboniferous System in the Tahe Oilfield is one of the most promising formations in the field of clastic rocks in the Tahe Oilfield and is rich in hydrocarbon resources. Exploration practice shows that the formation conditions of the Karashayi Formation lithology-stratigraphic reservoir group are complex, and the lack of understanding of its formation conditions and oil and gas enrichment laws has seriously restricted the next oil and gas exploration deployment and planning decisions in the basin. Based on the comprehensive analysis of the basic reservoir formation conditions of the Upper Karashayi Formation, the stratigraphic stratigraphic and reservoir seismic stratigraphic theories are used as the guidance, and the stratigraphic stratigraphic framework is constructed by combining the previous research results and geological and logging data; and under the control of the stratigraphic framework, three levels of stratigraphic sequences are delineated, the stratigraphic characteristics are analyzed, and the reservoir formation characteristics are summarized. The research understanding is of great significance for the exploration of oil and gas reservoirs in the Carboniferous clastic lithology-stratigraphy of Tarim Land.

2. Regional Geological Overview

The Tarim Basin is a large composite basin formed by the superposition of the Paleozoic Craton Basin and the Middle Cenozoic Foreland Basin, which has been developed for a long time on a shallow metamorphic crystalline basement in the Pre-Earthquake period in western China. During its long geological history, it has undergone many complex tectonic movements, forming seven primary tectonic units: the Tarbei Uplift, the Central Uplift, the Tanan Uplift, and the Kucha Depression, the Northern Depression, the Southwest Depression, and the Southeast Depression. The study area is located in the Tahe oil field. The study area, the Tahe oil field, is located in the south-central part of the Akkukul Bulge of the Shaya Uplift within the Tabei Uplift. The study area is located in the middle-south part of the Akkul Bulge, within the Shaya Uplift of the Tabei Uplift, with

* Corresponding author: 63027743@qq.com

the Ruta Fault to the north, the Halahatang Depression to the west, the Caohu Depression to the east, and the Manager Depression to the south. (Figure 1-a).

The Akkul Bulge was a stable carbonate platform in the Cambrian-Middle Ordovician; the base was formed in the Pre-Auranian; the Late Ordovician-Devonian basin was uplifted; the basin was strongly uplifted in the Carboniferous-Permian, resulting in the absence of the Upper Carboniferous and Permian in most of the basin; the Triassic-Cretaceous stable The Tarim Basin was controlled by the collision of the Indian and Eurasian continental plates during the Himalayan (Cenozoic), and several foreland basins, foreland rebellions, and large strike-slip fracture activities developed, while the northern part was strongly subsided under the influence of the Kucha foreland basin, forming the Akkul Bulge. The main oil-bearing systems in the study area are the Carboniferous, Triassic, and Jurassic systems, which have all been denuded to different degrees during the geological and historical periods by multiple phases of tectonic uplift The study area is the Santa Woodworking Zone. The study area of this paper is the most abundant block of clastic oil and gas reservoirs in the Tahe Oilfield, and a variety of reservoir types are developed, including lithologic reservoirs, stratigraphic reservoirs, and stratigraphic-lithologic composite reservoirs. The Karashayi Formation in the Santa woodworking area is integrated with the underlying Bachu Formation bimodal tuffs and has an unconformable contact with the overlying Permian volcanic rocks (Figure 1b). Previous studies have pointed out two sedimentary phases in the study area, the deltaic and tidal ping, and the source system is mainly NNE oriented source.

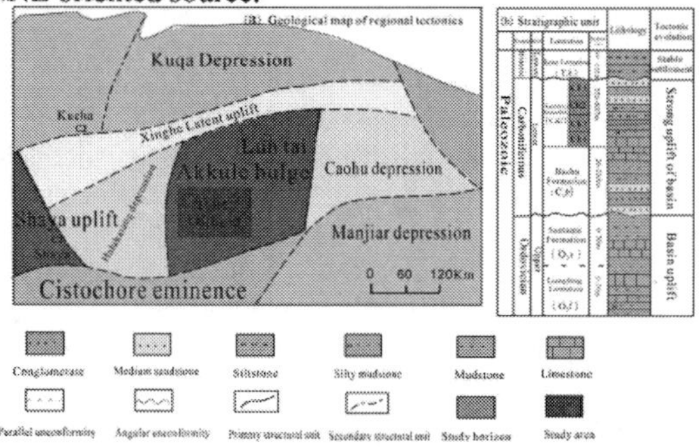

Figure 1. Tectonic Location Map and Paleozoic Integrated Stratigraphic Column Map of Tarhe Oil Field in Tarim Basin

3. Carboniferous Clastic Rocks Thin Reservoir Sequence Frame Delineation

3.1 Stratigraphic Feature

The Carboniferous strata were severely denuded by the uplift of tectonic activities such as Haixi and Indo-Yanshan, and the study area as a whole was denuded in the northern part by regional tectonic activities and sea-level rise and fall, while the southern part was relatively well preserved. Longitudinally, the lower and top lithologic sections of the Carboniferous are partially missing, with the depositional characteristics of overburdened lower and denuded top. The lower part of the Carboniferous is partially missing and has the characteristics of deposition of the overburdened lower part and denuded top part. The overall Taher Oilfield Carboniferous Karashayi Formation can be divided into the lower mudstone section and the upper sand mudstone section: the upper mudstone section is mainly dark gray, brownish-gray, and tan mudstone with stable thickness in the plan; the sand mudstone section is a large set of brown mudstone, tan mudstone interbedded with sandstone and conglomerate bearing sandstone (Figure 1). Based on the internal lithological characteristics, they are divided into four subsections from CK1 to CK4. Figure 2 shows a comprehensive column diagram of a single well in study area B. The unconformity surface and the important water recession surface are used as the interface for stratigraphic division, and four tertiary stratigraphic sequences were divided in this time: SQ1 (middle mudstone section-upper mudstone section), SQ2 (CK4), SQ3 (CK3-CK2), and SQ4 (CK1) (Fig. 2), and seven system domains (Fig. 2).

The SQ1 sequence is mainly mudstone sediments with a small amount of chert, and the sand content gradually increases from the bottom to the top, which is a set of stable mudstone stratigraphy, and there is no oil and gas display in this sequence. The resistivity curve at the boundary between the HST and TST system domains has increased significantly, and the natural gamma curve value has decreased significantly. The SQ3 sequence as a whole is characterized by thin sand-mud interbeds. the TST system domain is mainly a deltaic foreland deposit, and the lithology is mainly grey medium sandstone, conglomerate sandstone with mudstone and siltstone interbeds, with thin sand layers at the bottom and increasing mudstone upwards. the HST system domain is a deltaic plain deposit, and the lithology is a grey conglomerate, conglomerate sandstone and sandstone with grey mudstone interbeds. The conglomerate is lenticular in shape. The SQ4 sequence succeeds the SQ3 sand-mud thin interbedded sedimentary features, mainly yellow fine sandstone, siltstone, and mudstone interbedded. The logging curve has obvious abrupt change characteristics, upward resistivity value decreases, natural gamma curve value increases, this sequence contains a large amount of oil and gas display around the unconformity surface.

The various tertiary sequences of the Karashayi Formation, consisting of the HST controlled by fluvial action and the TST controlled by tidal action, lack the LST. several layers of the S60-1 well contain significant hydrocarbon resources, and the hydrocarbons are mainly distributed around the interface of the sequence and the unconformity.

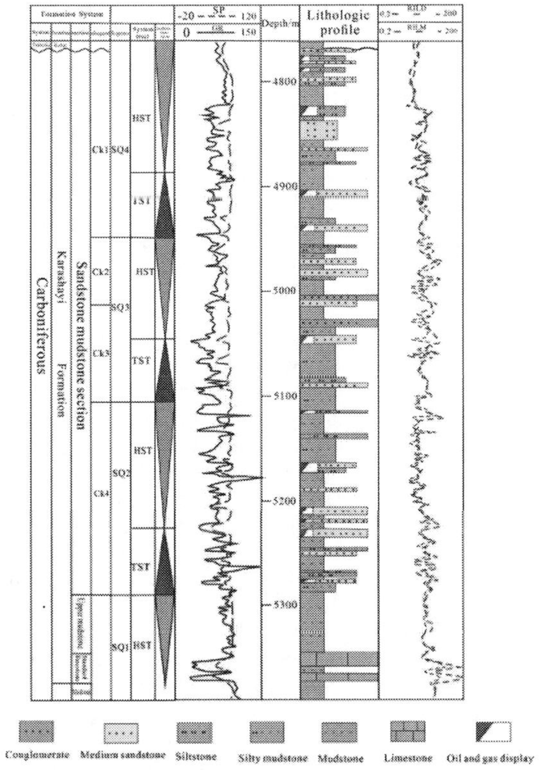

Figure 2. Single-well histograms of wells

4. Comparative Stratigraphy of Continuous Wells

Based on a single well stratigraphic delineation, a continuous well comparison of stratigraphic stratigraphy was conducted to establish a north-south (pro-source) and east-west (pro-source) stratigraphic grid in the study area to provide a framework for studying the distribution of reservoir sands.

4.1 Cis-source Continuous Well Profile.

The subsections have good continuity from the cis-source continuous well profile. Among them, the CK1 formation is gradually thinned from south to north and the northern CK1 formation is denuded. Clear stratigraphic cusps of extinction can be seen on the seismic profile (Figure 4). The longitudinal sands are stacked on top of each other, and have short lateral extension distances and poor connectivity; among them, the CK4-CK3 sands are developed and the CK2-CK1 sands are relatively undeveloped. There is little inter-well variation and strong contrast, indicating relatively stable deposition in this area during the depositional period of the Karashayi Formation. From the seismic profile, T50 (thick mudstone of the Triassic Ke Tul Formation), and T56 (bimodal tuff of the underlying Carboniferous Batu Formation) have a clear reflection interface. The upper mudstone section on the seismic profile is a strong amplitude parallel continuous reflection feature, which is easy to identify. CK2, CK3, and CK4 on the seismic profile can be seen as obvious precessional features. The top of CK1 on the seismic profile shows a continuous strong axis feature, the stratigraphy is stripped in the northern part of the work area, and the sharp extinction feature of the stratigraphy on the seismic is clear (Figure 4). Several layers have oil and gas distribution, and the oil and gas reservoir is rich near the unconformity stripping surface of CK1.

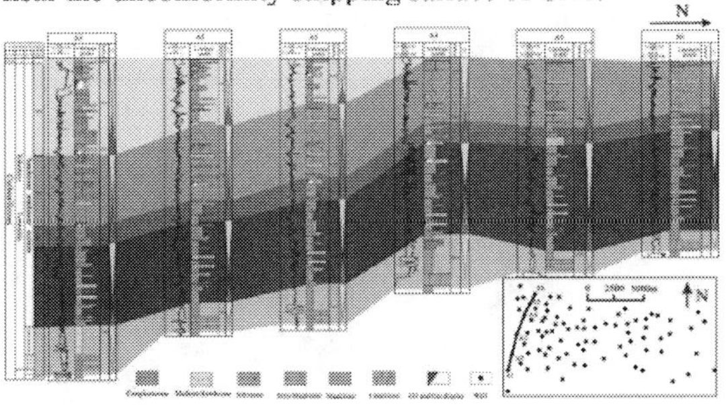

Figure 3. Cis-source continuous well profile (A1-A2-A3-A4-A5-A6)

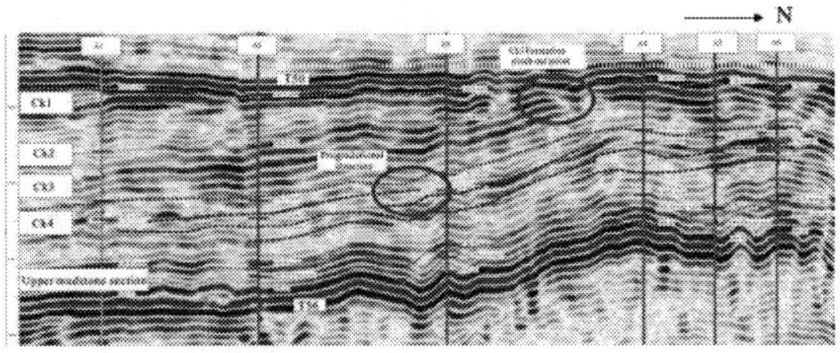

Figure 4. Seismic profile of transwell A1-A2-A3-A4-A5-A6

4.2 Vertical Source Continuous Good Profile.

From the vertical source continuous well profile, the continuity of each small layer in the western part of the study area is relatively good, while the continuity in the eastern part is poor, and the continuity of each small layer in the vertical source direction is good, and the thickness of the stratum does not change much. The sand body in the study area changes from west to east in a clear pattern, and from west to east the sand body gradually does not develop; vertically the western CK4-CK sand body is developed, and multiple sets of sand bodies are superimposed on each other, and the CK2-CK1 sand gradually thins; the eastern part of the work area is dominated by mudstone deposits, occasionally interspersed with thin sand bodies (Figure 5). The upper mudstone section on the seismic profile is characterized by strong amplitude parallel continuous reflections, and the top of CK1 is characterized by continuous strong axes (Figure 6). Overall, the gas reservoir is developed in the western oil region and underdeveloped in the eastern oil and gas reservoir.

https://doi.org/10.1051/e3sconf/202235801035

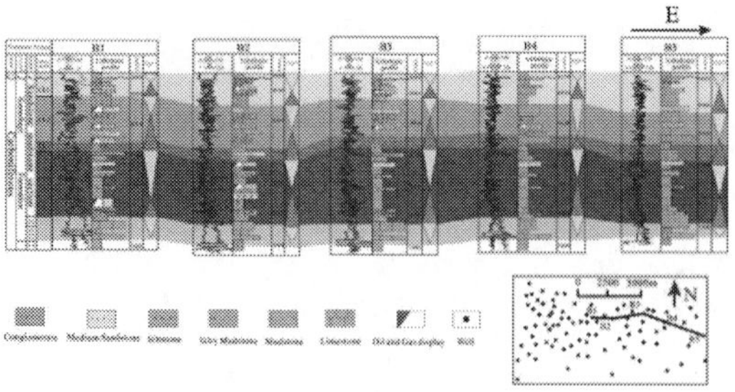

Figure 5. Vertical source continuous good profiles (B1-B2-B3-B4-B5)

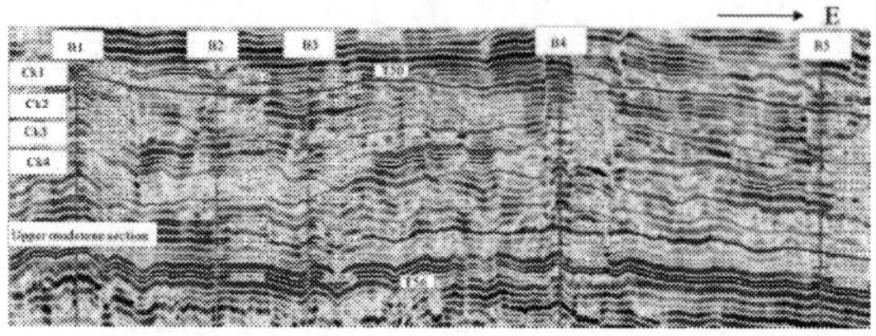

Figure 6. Overwell B1-B2-B3-B4-B5 seismic profiles

5. Main Control Factors of Oil and Gas Formation

The Taher oil field is located in the south-central part of the Akkul Bulge of the Shaya Uplift. Under the influence of multi-phase tectonic movements, the strata have been stripped and overburdened many times, and multiple phases of unconformity and fractures have been developed. Meanwhile, the Cambrian-upper Ordovician marine hydrocarbon source rocks in the Manage depression, which produce hydrocarbons under large burial depth and high-pressure conditions, continuously supply hydrocarbons to the Carboniferous Karashayi Formation through deep and large fractures [17]. It has favorable conditions for the formation of stratigraphic oil and gas reservoirs. The multi-phase stratigraphic unconformity of the Carboniferous Karashayi Formation plays a dominant role in the formation of stratigraphic traps and hydrocarbon reservoirs, and combined with tectonic and lithologic factors, it forms a complex situation of the formation characteristics of the Carboniferous hydrocarbon reservoirs.

5.1 Reservoir Formation Control Factors

5.1.1 Stratigraphical Ringwork.

The Carboniferous Karashayi Formation reservoirs are mainly braided river deltaic sands, river and small deltaic phase estuarine dams, and branch river sands, which are thin in a single layer thickness and interbedded with mudstones in a lenticular form. The upper part of the clastic rocks of the Karashayi Formation is eroded (the stripping line is shown in Figure 7), and the capping layer that plays a major role in capping the Carboniferous reservoirs is a large set of dense mudstone of the overlying Lower Triassic Kotur Formation, while its reservoir development is mostly lenticular and interbedded with mudstone, and the mudstone developed between the sands also plays a good capping role, forming a multiple oil and gas system of thin reservoirs of the Carboniferous clastic rocks. The lower sandstone section of the Carboniferous Batu Formation is distributed in the southwest of the work area and extinguished to the northeast, forming a better reservoir-cover combination with the mudstone of the lower mudstone section overlying it. At the same time, the salt body of the Batu Formation also seals the oil and gas to a certain extent.

5.1.2 Fracture-disintegration Three-dimensional Transport System.

Carboniferous oil and gas distribution is closely related to fractures, which are the main channel for the upward transport of oil and gas in the Lower Ordovician. During the Carboniferous-Permian period, the intense uplift of the basin caused the denudation of most of the Silurian-Devonian strata, and the Carboniferous system was unconformable above the Ordovician system, forming unconformable surfaces. Oil and gas were transported upward through such unconformities and deep major fractures. Multiple phases of tectonic movements have occurred since the Permian, and the resulting multi-phase fractures are the conduits for the diffusion of hydrocarbons within the Carboniferous, which eventually accumulate in tectonic traps and closures to form reservoirs (Figure 7).

5.1.3 Deep Supply.

There are no hydrocarbon source rocks in the Carboniferous System of the Tahe Oilfield, and hydrocarbons come from the Cambrian-Superior Ordovician marine hydrocarbon source rocks in the Manage Depression. The marine hydrocarbon source rocks produce hydrocarbons under the conditions of large burial depth and high pressure, and continuously supply hydrocarbons to the Carboniferous Karashayi Formation through deep and large fractures, which is one of the conditions for the enrichment of Carboniferous hydrocarbons in the Tahe area.

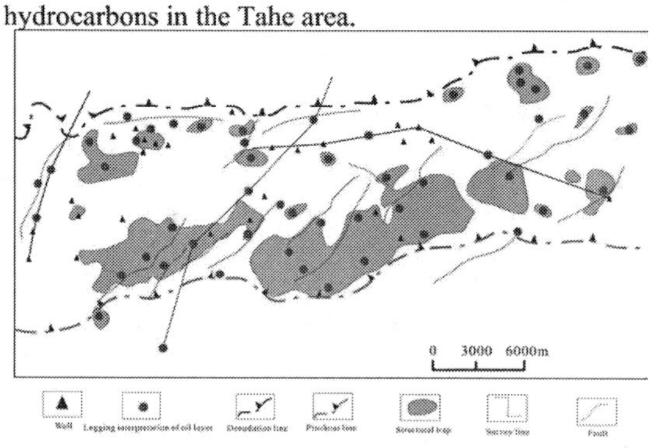

Figure 7. Overlay of CK1 oil and gas showing well site plan distribution and tectonic trap fracture distribution in the Karashayi Formation

5.2 Hydrocarbon formation model

The hydrocarbon formation mode of the Carboniferous Karashayi Formation is source-storage separation type-overburden-truncation composite formation mode (Figure 8).

(1)The stratigraphic cusp extinguishes upper dip masking and forms a trap background. The Carboniferous strata were subjected to multiple stages of denudation, and the clastic-based Karashayi Formation is integrated with the underlying Bachu Formation bimodal tuff and is in unconformable contact with the dense mudstone of the overlying Lower Triassic Kotur Formation. The resulting numerous stratigraphic unconformities provide good trap conditions for the oil reservoir. In addition, the upper part of the clastic rocks of the Carboniferous Karashayi Group is eroded, and when the overlying Triassic volcanic rocks are covered by poorly permeable strata, good trap conditions are formed. From the north-south direction, the Karashayi Formation is cut off to the north and makes unconformable contact with the Triassic volcanic rocks. well S47 CK1 oil group is cut off from the Triassic volcanic rocks and the oil and gas are blocked by the Triassic volcanic rocks.

(2)Lower growth and upper storage, long-distance cross-layer transport, and aggregation. Carboniferous hydrocarbons are supplied from the lower Ordovician suture hole reservoir, transported into the Carboniferous strata through deep and large fractures and unconformities, and formed into Carboniferous clastic thin reservoir strata hydrocarbon reservoirs through the multi-phase fracture transport formed at a later stage and the occlusion of the stratigraphic sharp extinction zone trap background.

The comprehensive analysis concludes that the Carboniferous Karashayi Formation is located in a complex stratigraphic reservoir controlled by a stratigraphic unconformity and stratigraphic sharp extinction, which is influenced by tectonics, fractures, and lithology, on a convex overburden background.

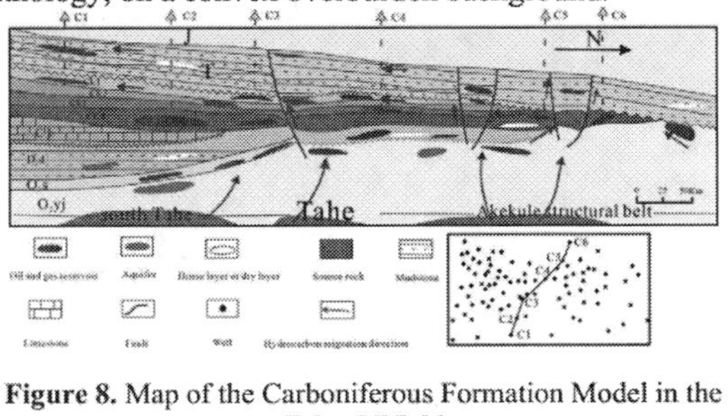

Figure 8. Map of the Carboniferous Formation Model in the Tahe Oilfield

6. Conclusion

(1) The lower and top lithologic sections of the Carboniferous are partially missing due to regional tectonic activity and sea-level rise. On the whole, Carboniferous sedimentation is characterized by overburden in the lower part and denudation in the top part. The interface between the unconformity and the important water recession surface is used as the interface for stratigraphic division, and four tertiary stratigraphic sequences are divided: SQ1, SQ2, SQ3, and SQ4.

(2) The distribution of stratigraphic sequences in the study area has a certain pattern, which is generally characterized by thinness in the middle and thickness from the north to the south, and strip-like spreading from the east to the west. The CK1 stratum in the northern part of the work area is stripped, and the stratigraphic sharp extinction feature is clear on the seismic.

(3) The Carboniferous Karashayi Formation is located on the overburdened background of convexity and is a composite reservoir controlled by a stratigraphic unconformity and stratigraphic extinction that is influenced by tectonics, fractures, and lithology.

References

1. Jia Chengzuo. Several important issues on current oil and gas exploration in China[J]. Journal of Petroleum, 2012,33(S1):6-13.
2. Zheng and R., Yin W., Hu Zongquan, et al. Main control factors and exploration directions for hydrocarbon enrichment in the clastic field of central and western China[J]. Petroleum and Natural Gas Geology, 2010,31(6):753-762
3. Jia Chengzuo, Tectonic features, and hydrocarbons in the Tarim Basin, China.1997 [M] Beijing Petroleum Industry Press.
4. Jia Chengzuo, Wei Guoqi. Tectonic features and hydrocarbon-bearing properties of the Tarim Basin. [J] Science Bulletin, 2002, 47(Suppl.): 1-8
5. Zheng M.L., Wang Y., Jin Z.J., et al. Stacked evolution and hydrocarbon accumulation in the Tarim Basin[J]. Petroleum and Natural Gas Geology, 2014, 35(6): 925-934.
6. Ma Q.Y., Lv H.T., Jiang H.S., et al. Scheme of tectonic unit delineation in the Tarim Basin platform-basin area[J]. Offshore Oil and Gas Geology, 2015, 20(01):1-9.
7. He Dengfa, Jia Chengzuo, Desheng, et al. Formation and evolution of the Tarim multi-rotational stacked basin[J]. Petroleum and Natural Gas Geology, 2005(01): 64-77.
8. Zhong DK, Qi LX, Liu KN, et al. Study on the sedimentary phases of the Carboniferous Karashayi Formation in the southern part of the Tahe Oilfield[J]. Journal of Sedimentology, 2009, 27(4): 614-621.
9. Liu Zhanhong, Chen R, Song Chengbing, et al. A comprehensive study on the spinodal stratigraphy and stratigraphic sequence of the Carboniferous Karashayi Formation in the Tarim Basin[J] Earth Sciences - Journal of China University of Geosciences, 2012, 37(5): 1051-1065
10. Genetic relationship between tensor-torsional fracture zone and basin margin orogenic zone in the basin hinterland and control of hydrocarbon formation—a Jurassic system in the East Rim of the West Depression of 1 well in Junggar Basin as an

https://doi.org/10.1051/e3sconf/202235801035

example[J]. Fracture Block Oil and Gas Field, 2021,28(6):805-809,822

11. Du Weiwei, Shi Yuanyuan, Hong Caijun. Fine evaluation of the favorable zone for exploration of the Carboniferous Karashayi Formation in Tahe Oilfield[J] Petroleum Experimental Geology, 2017, 39(4): 580-586
12. Zhao Linfeng. Identification and distribution characteristics of faults in the Rotai and Sentamu oil fields [D]. China University of Petroleum (East China), 2018
13. ZhangZheng Xiaoli. Fracture characteristics of the Bachu uplift in the Tarim Basin and the migration and evolution pattern of the ancient uplift [D]. Zhejiang University, 2016.
14. Wu B, He Dengfa, Sun Fangyuan, et al. Tectonic I stratigraphic features and evolution of the Caohu Depression in the northeastern Tarim Basin[J]. Frontiers of Marine Geology, 2016, 32(9): 17-25.
15. Pang Xiangqi, Chen Junqing, Li Sumei, et al. Source of crude oil from very large marine fields in Tarim Basin: evidence from deep low-TOC hydrocarbon source rocks and evaluation of relative contribution[J]. Journal of Petroleum, 2018, 39(1):23-41.
16. Kuang L.C., Zhi D.M., Xiao J., et al. Formation conditions and exploration directions of large lithologic-stratigraphic reservoirs in the Upper Permian Upper Urho Formation of the Junggar Basin[J]. Journal of Petroleum, 2022

Study of urbanization trend prediction in Nanchang based on CA-Markov model

Shuhao Li

School of Civil and Surveying Engineering, Jiangxi University of Technology, Ganzhou, Jiangxi 341000, China.

Abstract. In order to promote the healthy development of Nanchang city in the future and maintain the virtuous cycle of urban ecosystem, this paper uses CA-Markov model to simulate and forecast the land use of Nanchang city in 2030 and explore the urbanization process of Nanchang city. Firstly, the land use changes in Nanchang from 2000 to 2020 are described, followed by land use simulations for Nanchang in 2020 as a verification of accuracy, and land use projections for Nanchang in 2030 based on accuracy requirements. The results show that:(1) the rate of urban expansion will continue to be maintained or even gradually increase from 2020 to 2030 under the expansion model based on the period 2000-2020. (2) The main factor in the change of land use types in the built-up area of Nanchang over the past two decades is socio-economic factors, with natural factors taking second place. (3) The area of land development intensity in Nanchang will reach its peak in 2030, and the expansion of the municipal districts will gradually slowdown in favour of Nanchang and Xinjian counties, which have a more extensive land area and are close to the central urban area.

Keywords: Nanchang; CA-Markov model; urbanization; land use change; forecasting study

1. Introduction

As a complex ecosystem combining nature-economy-society, the city is the main place for human survival and development, where society, economy and nature interact with each other and influence each other.[1-3] With the rapid development of the world economy, China is experiencing a rapid rate of large-scale urbanisation unprecedented in human history. At the same time, from the perspective of urban development around the world, large-scale urbanisation is an inevitable trend resulting from a robust process of socio-economic development and a natural process of geospatial aggregation of non-agricultural industries and populations in the context of high industrialisation [4-6]. At the same time, the process of urbanisation is accelerating, human demand for natural resources is increasing, the ecological environment in which it is located is gradually being destroyed, and urban ecological problems are becoming increasingly prominent. The current situation of urban land use is gradually changing rapidly and the changes are characterised by different spatial and temporal distributions, so the study of the urbanisation process is of great importance for sustainable urban development.[7]

With the development of remote sensing technology, RS and GIS technology has the
significant advantage of low cost and high efficiency in the study of urban land use change, so the macro control of urbanization process through remote sensing method has become a new research hot spot[8-9] . At the same time, the promotion of the concepts of green cities and sustainable development has led to a growing interest in land-use change modelling by scholars and people from all walks of life.[9-10] CA-Markov models are gradually being applied to the prediction of land use change, compared to the traditional GIS-based prediction of land use types. CA-Markov models can support multi-factor and multi-objective forecasting, and the generated results can be visualized into maps, which can facilitate targeted analysis of land use change.[11] Zhu Lei et al. (2020) made a sensitivity analysis of different parameters based on CA-Markov model for arable land in the oasis area of the Manas River basin[12] ; Wang Lixia et al. (2020) used the Weihe River region as the study area, calculated interannual NDVI values based on MODIS data, and used the CA-Markov model to simulate the spatial and temporal variation of NDVI in the region using the rank of NDVI as the metacell type. [13]; Ruci Wang and Hao Hou et al. (2019) monitored and simulated changes in farmland dynamics in Hangzhou from 1995 to 2035[14] . Nanchang, the capital city of Jiangxi Province, is the core city of the Poyang Lake City Cluster, the economic, political and cultural centre of Jiangxi Province, and one of the important central cities in the middle reaches of the Yangtze River. Since the new century, its development has been a great achievement, actively driving the local regional economy while the rate of urbanisation is accelerating and environmental problems are becoming increasingly serious[1]. Monitoring the urbanisation process in Nanchang and predicting and evaluating its

future land use changes are of great significance, and the results can be used as a basis for decision making on the sustainable development of Nanchang city, which is needed to promote the healthy development of the city in the future, and can also serve as a theoretical guarantee for maintaining a virtuous cycle in the urban ecosystem.
This study will take Nanchang City as an example, based on the Landsat Global 30m Fine Classification of Land Cover Types 2000-2020, carry out monitoring of the urbanisation process in Nanchang City over the past two decades and analyse its intrinsic driving forces, simulate and forecast the land use of Nanchang City in 2030 through the Land Suitability Atlas and CA-Markov model, and explore the The results of this study will provide theoretical support for the future development of Nanchang City and provide a scientific basis for the formulation of sustainable urban development policies for the harmonious co-existence of people and the environment.

2. Data and methods

2.1 Study area

Nanchang (Fig. 1), located in the lower reaches of the Ganjiang and Fuxiang rivers, is situated slightly north of the central part of Jiangxi province. It is bordered by Dongxiang County in Fuzhou City and Yugan County in Shangrao City to the east, Fengxin County and Gao'an City in Yichun City to the west, Linchuan County in Fuzhou City and Fengcheng County in Yichun City to the south, Yongxiu County and Duchang County in Jiujiang City to the north, and Poyang Lake, the largest freshwater lake in China, to the northeast.[16] . The topography of the city is mainly plain terrain, its geographical location of bearing east and west, connecting north and south, and the superposition of multiple policy locations have injected a constant impetus to the development of Nanchang over the past 20 years, with the development of Honggu Tan New Area, the construction of Poyang Lake Ecological Economic Zone, the construction of Changjiu Integration and a series of other policy measures implemented one after another, resulting in a significant increase in the economic level of Nanchang, a continuous expansion of the population, a faster pace of urbanisation and a continuous increase in the area of urban land. The pace of urbanization has been accelerated, and the area of urban construction land has been increasing.

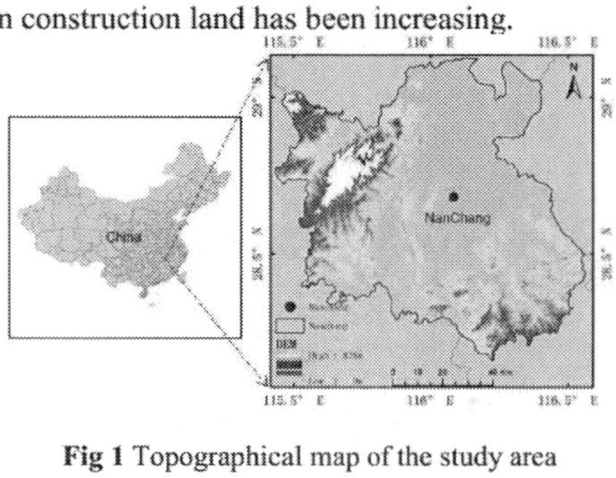

Fig 1 Topographical map of the study area

2.2 Research Data

The land use data used in this study was obtained from the Earth Big Data Science Project Data Sharing Service Platform (https://data.casearth.cn/), in which the global 30m land cover fine classification products for 2000, 2010 and 2020 were obtained, and Nanchang City was selected for the classification products to be downloaded; and according to the help document therein, the The land use names corresponding to the corresponding codes were then extracted and reclassified using ArcGIS 10.2 software, and the final land use results data were presented as five primary classifications of forest land, cropland, wetland, building land and water bodies for the final land use change analysis and the prediction and simulation of land use results in Nanchang in 2030.

2.3 Model and accuracy validation

The models currently used for the simulation of land use types in built-up areas are mainly the CLUE-S model, SD model and CA model, among which the CA model has good results in simulating the spatial and temporal processes of urban built-up area systems due to its top-down dynamic simulation. The CA-Markov model, on the other hand, combines the ability of the CA model to simulate spatial changes in complex systems with the Markov model's ability to make long-term predictions, enhancing the accuracy of the landscape type transfer matrix and achieving good results in analysing and modelling land use, and has therefore also been widely used in recent years in land use cover[17-18] .
In order to verify the validity of the CA-Markov model simulation in Nanchang City, model validation is needed first. Kappa coefficients are often used to assess the accuracy of the classification of remotely sensed data and to calculate the consistency of two maps[14] . Kappa coefficients quantify the changes in quantity, location and integrated information during landscape change in terms of spatial location and quantity, and are calculated as follows[19-20]

$$\text{Kappa}=\frac{P_0-P_c}{1-P_c}$$

where P_0 is the correct simulated proportion. P_c is the expected value of the correct simulated proportion in the random case. kappa coefficient is usually between 0 and 1. When the kappa coefficient is greater than 0.75, the agreement between the simulated results and the true distribution is high, and when the kappa coefficient is between 0.4 and 0.75, the agreement is more general. When the kappa coefficient is less than 0.4, the agreement is poor. Previous research has shown that Map comparison Kit 3 software can provide both quantitative and positional errors to validate the model.

3. Analysis of results

3.1 Land use distribution

The resultant data after reclassification of the land use result data for the three periods 2000, 2010 and 2020 are shown in Figure 2 below, and the study area was classified into five feature types, namely wetlands, cropland, forest land, water bodies and building land. The reclassification results were subjected to a transformation matrix analysis to obtain the conversion of the land use areas of the various feature types in the study area for the two adjacent periods.

From 2000 to 2010 (Table 1), the expansion of building land increased from 503.18km^2 in 2000 to 797.47km^2, a growth rate of 58%, reaching the peak of the town's growth rate in 20 years. The increase in the area of building land was mainly due to the conversion of feature types such as arable land, woodland and water bodies, with 275.56km^2 of arable land, or 34.55%, and 14.59km^2 of woodland, or 1.83%. It shows that in the early 20th century, Nanchang city expanded from the centre outwards, encroaching on large areas of arable land, while forest land was also damaged to a large extent. In addition, there were varying degrees of conversion between the areas of the three land use types: wetlands, arable land and water bodies. During this period, 3.75km^2 of arable land was converted to wetland type, accounting for 10.36% of the total wetland area in 2010, and 2.28km^2 of water bodies were converted to wetland, accounting for 6.29% of the total wetland area in 2010. Moreover, the total area of woodland and wetland were both on a downward trend during this period, while that of cropland, wetland and building land was on an upward trend, indicating that the general trend of land use type change was from natural state land use types such as woodland and wetland to artificially constructed land use types such as cropland and building land, indicating a sharp increase in urbanisation.

From 2010 to 2020 (Table 2), the area of built-up land increases by 200km^2, from 797.47km^2 in 2010 to 997.87km^2, a growth rate of 25%, which is also a reflection of the increased urbanisation process. The increase in the area of building land was mainly converted from feature types such as arable land and water bodies, with the area of arable land being 184.64km^2, accounting for 18.50% of the area of building land in 2010, and the area of water bodies being 6.33km^2, accounting for 0.63% of the area of building land in 2010. Overall, the areas of arable land and forest land all show a clear downward trend, while the areas of building land, water bodies and wetlands show an upward trend during the period 2010-2020. The total area of cultivated land decreases by 189.16km^2 and the total area of forested land decreases by 75.86km^2. The slowdown in the rate of growth of urban areas and the increase in the area of water bodies and wetlands indicate that government departments have become aware of environmental issues during this period and have begun to slow down the damage caused by human activities and gradually increase the area of land use types in their natural state.

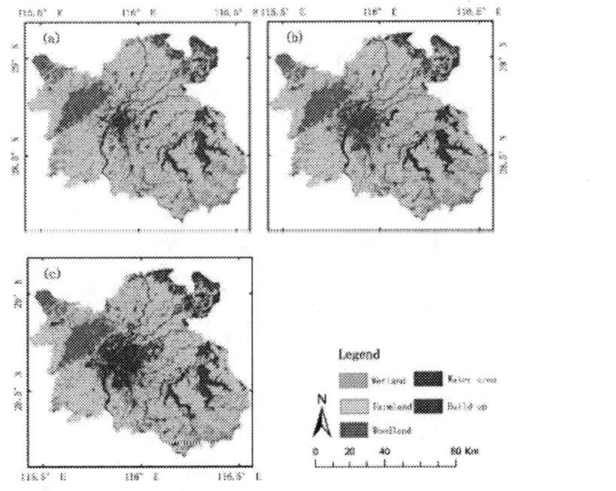

Fig 2 Land use distribution in Nanchang from 2000 to 2020: (a) Land use distribution in Nanchang in 2000; (b) Land use distribution in Nanchang in 2010; (c) Land use distribution in Nanchang in 2020

Table 1 Change in land use type transfer in Nanchang from 2000 to 2010

	Arable land	**Woodland**	**Wetlands**	**Building sites**	**Water bodies**
Arable land	5379.86	25.33	3.75	275.56	60.30
	98.42%	3.11%	10.36%	34.55%	5.36%
Woodland	39.02	787.39	0.00	14.59	0.67
	0.71%	96.70%	0.00%	1.83%	0.06%
Wetlands	0.02	0.00	30.16	0.00	0.01
	0.00%	0.00%	83.34%	0.00%	0.00%
Building sites	0.00	0.00	0.00	503.18	0.00
	0.00%	0.00%	0.00%	63.10%	0.00%
Water bodies	47.14	1.56	2.28	4.14	1047.40
	0.86%	0.19%	6.29%	0.52%	94.58%
Total area	5466.04	814.28	36.19	797.47	1107.47

Table 2 Change in land use type transfer in Nanchang from 2010 to 2020

	Arable land	**Woodland**	**Wetlands**	**Building sites**	**Water bodies**
Arable land	5176.25	18.70	7.57	184.64	78.87
	98.09%	2.53%	15.82%	18.50%	6.80%
Woodland	80.31	719.62	0.00	9.42	4.94
	1.52%	97.45%	0.00%	0.94%	0.43%
Wetlands	0.00	0.00	36.19	0.00	0.00
	0.00%	0.00%	75.62%	0.00%	0.00%
Building sites	0.00	0.00	0.00	797.47	0.00
	0.00%	0.00%	0.00%	79.92%	0.00%
Water bodies	20.32	0.11	4.10	6.33	1076.61
	0.39%	0.02%	8.56%	0.63%	92.78%
Total area	5276.88	738.43	47.85	997.87	1160.43

3.2 CA-Markov model prediction results

Figure 3 shows the prediction results of land use types based on the CA-Markov model, and the land use results for 2020 and 2030 were predicted in the study. The year 2020 is used as the evaluation data for accuracy verification, and the Kappa coefficients are calculated with the result data of land use of Nanchang City in 2020 extracted through remote sensing data, and the simulation results of Nanchang City in 2020 are verified through the software Map comparison Kit 3 to obtain the Kappa histograms of 0.81, 0.92, 0.83, the Kappa location and

Kappa cumulative statistics, indicating that the selected impact factor dataset has met the accuracy requirements of the prediction. This image factor dataset was used for the simulation of the land use status in 2030 to obtain Figure 3-b.

The land use results for 2020 were analysed in a transformation matrix with the land use types predicted for 2030 by the CA-Markov model to obtain the conversion of land use areas for various feature types in the study area for the adjacent years 2020-2030 (Table 3). The results of the transformation matrix analysis show that the area of built-up land still shows an increasing trend and the rate of increase is greater compared to the previous decade, with a total increase of 668.13km2. The increased area of built-up land is mainly converted from the former arable land, accounting for 33.51% of the urban area in 2030. This indicates that the increase in town area during that period was mainly through the conversion of arable land resources, with a large amount of arable land being developed into building land.

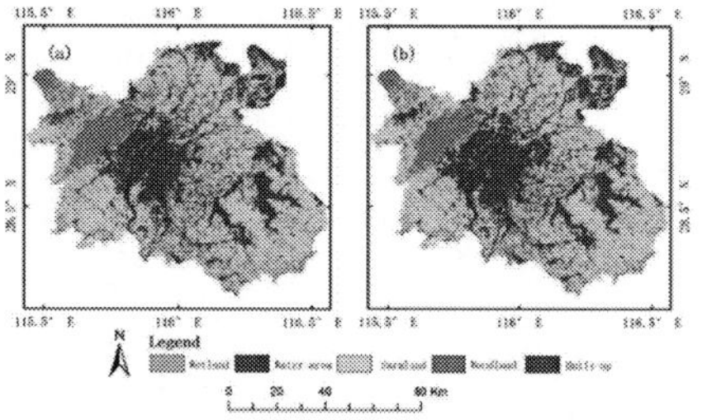

Fig 3 CA-Markov model land use type projection results: (a) 2020 land use projection results; (b) 2030 land use projection results

Table 3 Change in land use type transfer from 2020 to 2030 in Nanchang

	Arable land	Woodland	Wetlands	Building sites	Water bodies
Arable land	4499.19	1.46	0.00	719.88	56.36
	98.26%	0.22%	0.00 %	43.21%	4.68 %
Woodland	79.25	659.14	0.00	0.00	0.03
	1.73%	98.30%	0.00 %	0.00%	0.00 %
Wetlands	0.48	0.00	47.37	0.00	0.00
	0.01%	0.00%	46.44 %	0.00%	0.00 %
Building sites	0.00	9.97	38.10	946.12	3.68
	0.00%	1.49%	37.35 %	56.79%	0.31 %
Water bodies	0.00	0.00	16.54	0.00	1143.89
	0.00%	0.00%	16.21 %	0.00%	95.01 %
Total area	4578.92	670.57	102.01	1666.00	1203.96

4. Discussion

Combined with the results of the transformation matrix, it can be seen that the urbanisation process in Nanchang City reached its peak around 2010. From Figure 2, it can be seen that the expansion of the built-up area was dominated by the northwest and due north directions, and gradually changed from single-direction expansion to multi-directional joint development, and the expansion directions during the study period were mainly concentrated in the due north and northwest directions, with the northeast and northwest directions in the time period of 2005-2010 The expansion of the town during the period 2010-2020 was mainly in the north-west, north-east and due-south directions, while the expansion in other directions was not significant. Due to the large number of lakes within Nanchang and the significant influence of topography, the expansion of Nanchang's built-up area shows a clear directional pattern, indicating that urban expansion in Nanchang is influenced by both anthropogenic activities and natural conditions.

The CA-Markov model was used to simulate the development of the built-up area of Nanchang in 2030, based on the river and lake systems, slope, topographic relief and the distance of traffic routes, etc. The results of the study show that the built-up area of Nanchang is mostly located around river and lake systems, major traffic routes and areas with gentle slope and topographic relief, with the highest probability of the initial type of arable land being transformed into land for construction The probability of the initial type of arable land being converted into building land is the highest. At the same time, the simulation results show that the intensity of land development in Nanchang will peak in 2030, and the expansion of the municipal districts will gradually slow down to Nanchang County and Xinjian County, where the land area is more extensive and close to the central city.

5. Conclusion

This paper uses the Landsat Global 30m Land Cover Fine Classification product for 2000-2020, reclassifies it to obtain land use type data for Nanchang in 2000, 2010 and 2020, and simulates and predicts the land use type of Nanchang in 2030 by CA-Markov model based on the land use data of the three periods. The land use result data for 2030 was obtained. Based on the above land use types in Nanchang for different periods, the spatial and temporal dynamics of the built-up area in the study area for different periods were monitored and the underlying driving forces were analysed through a transfer matrix.

The results of the study show that the main factor in the change of land use types in the built-up area of Nanchang over the past two decades has been socio-economic factors, with natural factors taking second place. Socio-economic development is an important factor contributing to the expansion of the built-up area, and is also the main driving force behind the gradual increase in the expansion rate of the built-up area. When the economy of an area develops at a high rate, it will have a polarising effect, gathering capital and population, which will further drive the expansion of the built-up area. There is a clear spatial and temporal divergence in the development of the built-up area of Nanchang, which is due to the differences in geographical characteristics between its different regions, creating the existence of a geographical divergence in expansion. The simulation results show that the rate of urban expansion will continue to be maintained or even gradually increase from 2020 to 2030 based on the expansion pattern from 2000 to 2020. Therefore, the future development of the built-up area of Nanchang should strengthen the intensive use of land, improve the government's intervention mechanism, protect the ecological environment and promote the coordinated development of urban and rural areas.

References

1. Zhou Jiangwen. Natural ecological evaluation and driving force analysis of Nanchang City [D]. Donghua University of Technology, 2020. DOI:10.27145/d.cnki.ghddc.2020.000084.
2. Jay Gao, Yansui Liu, Yifu Chen, Land cover changes during agrarian restructuring in Northeast China, Applied Geography 26 (2006) 312- 322
3. Jieying Xiao, Yanjun Shen, Jingfeng Ge, Ryutaro Tateishi, Changyuan Tang,Yanqing Liang, Zhiying Huang, Evaluating urban expansion and land use change in Shijiazhuang, China, by using GIS and remote sensing, Landscape and Urban Planning 75 (2006) 69-80.
4. Ma Ting. Spatial and temporal characteristics of urbanization in China from the perspective of night-light remote sensing [J]. Journal of Geoinformation Science,2019,21(01):59-67.
5. Wang X.X., Liu J.Y., Zhuang D.D., Wang L.M. Spatial and temporal characteristics of spatial morphological changes in Chinese megacities[J]. Journal of Geography,2005(03):392-400.
6. Zhou Chunshan, Ye Changdong. Spatial growth characteristics of Chinese megacities and its causes[J]. Journal of Geography,2013,68(06):728-738.
7. Li Yuan, Lin Feng, Yan Zexing. Analysis of urbanization evolution based on Landsat remote sensing image--Xiamen city as an example[J]. Urban Architecture,2019,16(25):138-142.DOI:10.19892/j.cnki.csjz.2019.25.030.
8. Jinxia Lv , Weiguo Jiang , Wenjie Wang, Zhifeng Wu, Yinghui Liu, Xiaoya Wang and Zhuo Li Wetland Loss Identification and Evaluation Based on Landscape and Remote Sensing Indices in Xiong'an New Area, Remote Sensing. 2019, 11, 2834; doi:10.3390/rs11232834
9. Lopez, E., Bocco, G., Mendoza, M., & Duhau, E., Predicting land cover and land use change in the urban fringe a case in Morelia City, Mexico. landscape and Urban Planning, (2001) 55(4), 271-285.
10. Mathan M. & Krishnaveni M., Monitoring spatio-temporal dynamics of urban and peri-urban land transitions using ensemble of remote sensing spectral indices-a case study of Chennai Metropolitan Area, India, Environ Monit Assess (2020) 192: 15
11. Qian Ya. Evolutionary trends and prediction of vegetation cover in Jiangsu Province, 2006-2015[D]. Nanjing University of Information Engineering, 2021. DOI:10.27248/d.cnki.gnjqc.2021.000288.
12. Zhu Lei, Xia Xinxin, Yang Aimin, Jin Han. Parametric sensitivity analysis of CA-Markov model for arable land expansion in the oasis zone of Manas River basin [J/OL]. Arid Zone Research:1-12[2020-0822].
13. Wang Lixia, Zhang Jiawei, Meng Nina, Sui Lichun, Zhang Shuangcheng, Liu Zuo. Simulation and prediction of spatial and temporal variability of NDVI in the Weihe River basin based on CA-Markov[J]. Soil and water conservation research,2020,27(04):206-212.
14. Wang, Ruci, Hao Hou and Yuji Murayama 2018. Scenario-Based Simulation of Tianjin City Using a Cellular Automata-Markov Model. Sustainability 10: 2633.
15. Wang Yanjie. Study on the evolutionary characteristics of wetlands and their driving forces in Nanchang City [D]. Jiangxi Normal University, 2020. DOI:10.27178/d.cnki.gjxsu.2020.000709.
16. Ye Bin. Research on the spatial and temporal evolution characteristics of urban heat island effect in Nanchang based on Landsat satellite images[D]. Guilin University of Technology,2019. doi:10.27050/d.cnki.gglgc.2019.000002.
17. Li X. Research on remote sensing monitoring and simulation of urban built-up areas in Nanchang [D]. Donghua University of Technology,2017.
18. Schucknecht, Anne, Matschullat, Jörg, Erasmi, Stefan. Spatial and temporal variability of vegetation status in Paraíba, Northeastern Brazil(2012).DOI:10.1109/IGARSS.2012.6351643
19. Zhang X. Simulation and prediction of wetland landscape pattern dynamics in the south coast of Hangzhou Bay based on CA-Markov model[D]. Zhejiang University, 2013

Design of Underwater Grinding Device for Nuclear Power Plant

Guangzheng Tian [1], Hongwei Sha[1], Jianpeng Fu [3,*], Jianhua Zhou [2], Chao Li [2]

[1] Daya Bay Nuclear Power Operations and Management Co.,Ltd. ,Guangdong ,China
[2] China Nuclear Power Operations Co.,Ltd .,Guangdong ,China
[3] China Nuclear Power Technology Research Institute Beijing Division, Beijing, China

Abstract. In order to solve the problems of difficulty in cleaning the inner wall of the hammer-shaped component in the RIC system of the nuclear power plant and the low surface accuracy after cleaning, a profiling cleaning system was designed. First, the working principle of profiling cleaning is introduced to ensure that the surface accuracy after cleaning meets the installation requirements. Secondly, in order to ensure the normal operation of the profiling airbag, a mechanical model between the air supply volume of the air pump and the force on the conical inner wall is derived. Finally, in order to accurately control the output torque of the motor, the PID method is used to control the output torque of the motor. The system can increase the success rate of cleaning the inner wall of the hammer-shaped component in the RIC system and improve the efficiency of nuclear power plants.

Keywords: Copying cleaning system, Nuclear power plant, Air bag stress model, PID control.

1. Introduction

After the failure of thermocouple in nuclear power plant, it is necessary to clean the foreign matter in the inner wall of the hammer assembly in RIC system, including the treatment of burrs and the recovery of impurities, so that the hole size and precision of the hammer assembly can meet the requirements of assembly. Moreover, the treatment of burrs and the recovery of impurities are related to the effective implementation of this project. At present, the main cleaning methods for the inner wall of the pipeline are rigid contact treatment, sand blasting treatment and dry ice cleaning. Rigid contact treatment mainly refers to electric drill treatment and grinding wheel treatment in daily life and chemical industry application. Its advantages are significant effect, suitable for many scenes, high cleaning efficiency. The disadvantage is also obvious, the contact between the electric drill and the oxide layer is rigid to rigid contact, in the cleaning process is easy to appear stuck, affecting the cleaning efficiency; At the same time, its copying ability is poor, it is difficult to achieve high-precision cleaning, and the impurities after treatment are not easy to recycle, the workload will increase; In addition, it is not suitable for flammable and explosive, underwater and other special scenarios[1]. Sandblasting treatment can put the workpiece surface rust, oil and so on all impurities dirt completely removed, at the same time, the workpiece surface sandblasting will form the surface of the hair surface, is conducive to coating and bonding. The use of sand blasting treatment for machining parts burr cleaning, that is, through the treatment of fine particles to clean the surface of high precision cleaning, while having a high profile ability. However, as far as this project is concerned, the dirt after sandblasting treatment is too much, which is easy to react with boric acid solution, causing pollution to the boric acid pool, and the impurities are difficult to recover [2][3]. Dry ice cleaning mainly uses high-density dry ice particles to impact the surface of the object to be cleaned, but this cleaning method cannot be used in a small clean space, so the above three impurity treatment methods are not applicable to this project[4][5]. Based on the comparison of the above two cleaning methods for impurities in the inner wall of hammer shaped components, it is not difficult to see that the traditional cleaning methods can't meet the cleaning requirements of this project, the portable device is too rigid, and the device with high efficiency is easy to pollute the boric acid pool. Therefore, in order to reasonably and effectively solve the hammer assembly wall of foreign matter and oxide layer cleaning, this project plans to design a new scheme. In this scheme, a set of airbag copying cleaning device and grinding bit are designed to take into account the rigidity of the inner wall of hammer assembly and the effective recovery of impurities after cleaning treatment.

2. Mechanical Mechanism Design of Copying Cleaning System

The main function of the copying cleaning structure is to clean the inner wall of the conical component. The inner wall view and partial section view of the conical

* Corresponding author: fujianpeng@cgnpc.com.cn

component are shown in Figure. 1. It can be seen from Figure. 1. In order to make the accuracy of the inner wall holes meet the installation requirements after cleaning, the overall schematic diagram of the copying cleaning mechanism designed in this paper is shown in Figure. 2:

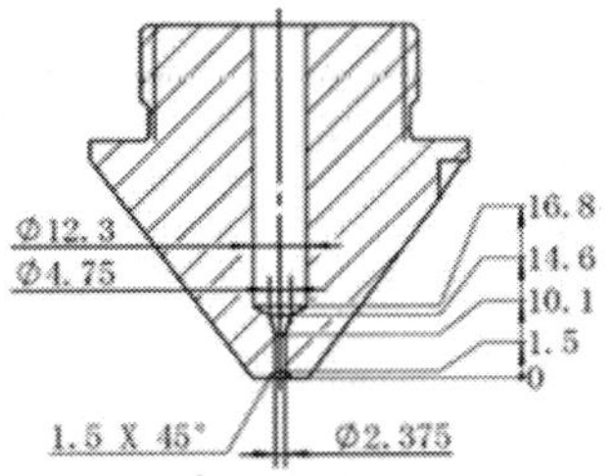

Fig. 1. Taper component wall

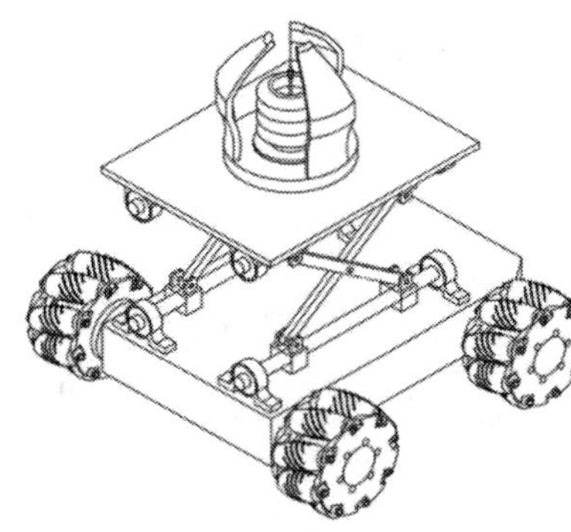

Fig. 2. Schematic diagram of copying cleaning system

Figure. 1 shows that the minimum inner wall diameter of the conical assembly is only 2.375mm, so when cleaning the inner wall, the centering mechanism is needed to ensure that the cleaning tentacle smoothly enters the inner wall hole under the action of the lifting platform. Centering mechanism adopts floating automatic centering fixture, with its floating characteristics, the cleaning tentacle automatically and accurately centering.

Section view of copying grinding mechanism is shown in Figure.3, which is mainly composed of grinding net 2, airbag 3, sealing gasket 4, bellows 5, drainage hole 6, chassis 7 rotating disk 8, charging chamber 9, charging pipeline 10, sealing ring 11, motor 12, motor bracket 13 and connecting ventilation needle 14. When to clean inner hole, the first 14, abrasive mesh 2, airbag will connected pores 3 processing into the hole, after filling line to infuse air plenum chamber 9 and 10 again by the top allied interface even vent needle 14, 141 and 142 from gas plenum chamber bottom allied interface to airbags within 3 bag 3 drives the lapping network expansion, 2 At this time, the motor 12 is started, the output shaft of the motor is connected with the air needle 14 of the United States, and even the ventilation needle 14 is also connected with the rotary disk 8. The air bag 3 and the grinding net 2 are fixed on the rotary disk 8, rotating with the rotary disk 8, so as to complete the grinding of the processed surface.

In the process of grinding process, to ensure that debris does not escape into the water, to avoid such as nuclear power plant spent fuel pools of water quality requirements higher application site contamination, the invention designs the liquid extract this step, the water inlet of the pump to connect to port 6 of the present invention, as grinding treatment processing hole cleaning, Remove clean debris at any time along with the liquid in bellows 5.

During processing, sealing gaskets are pressed against the surface of the workpiece to improve pumping efficiency and avoid debris scattering in the water. At the same time, because of the elasticity of the bellows, the sealing gasket can always fit the surface to be machined.

After the processing is completed, the gas in the air bag is discharged through the ventilation needle 14, the inflation chamber 9, and the inflation pipeline 10. At this time, the air bag is reduced, and the processing parts can be removed to process the next hole. Airbag inflation and contraction are shown in Figure. 4.

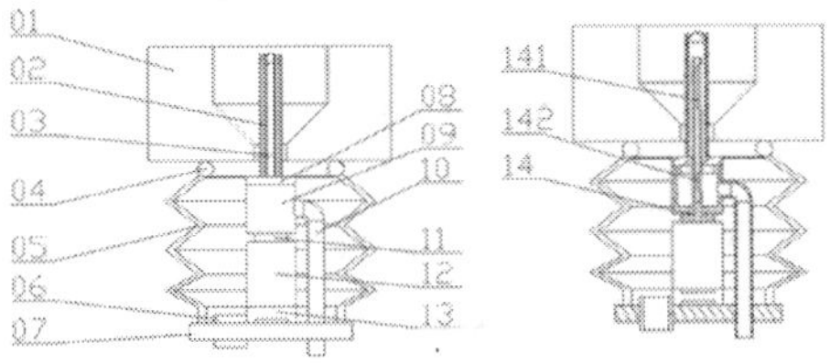

Fig. 3. Section view of grinding mechanism

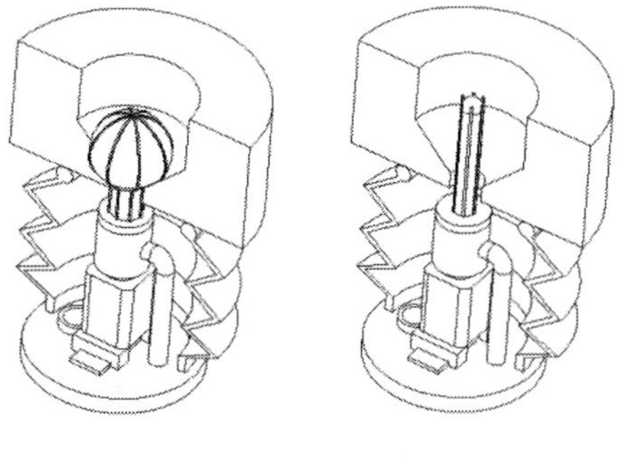

Fig. 4. Schematic diagram of airbag inflation/contraction

3. Design Criteria for Each Subequipment

In order to ensure that the catheter end plug cleaning device can successfully complete the cleaning work without being damaged due to excessive feed or unable to stick to the inner wall of the catheter end plug due to too small feed, it is necessary to model the catheter end plug cleaning device. The force of the catheter end plug is shown in Figure. 5:

 https://doi.org/10.1051/e3sconf/202235801037

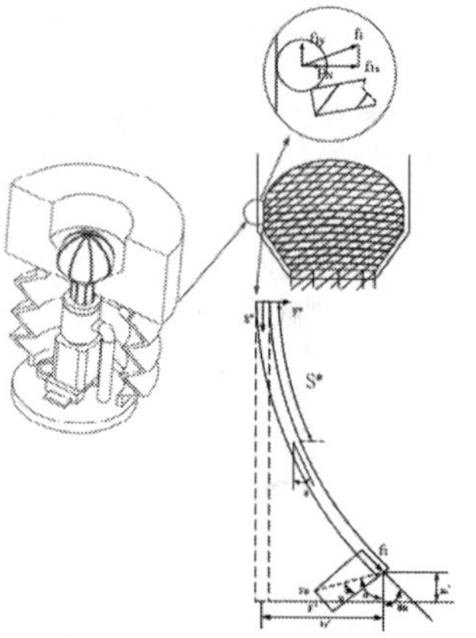

Fig. 5. Catheter end plug cleaning device is stressed

When establishing the mechanical model of the pipe end plug cleaning device, it can be assumed that the dirt is a rigid ball and the grinding net is a flexible beam. During the cleaning process, the grinding net is in contact with the dirt. At this time, the grinding net exerts a positive pressure on the dirt, thus generating a force to cut the dirt. According to the bending degree of lapping net, cleaning state can be divided into three kinds: large deflection, small deflection and critical state. When the bending degree of lapping mesh is small, the force analysis of flexible lapping mesh can be assumed as the cantilever beam model. When the deformation of the flexible lapping net is large, the flexible beam model with large deflection is used for analysis [6].

Differential equation of deflection line of beam structure:

$$EI\frac{d\theta}{ds}=M \quad (1)$$

In the above formula, E is the elastic modulus; I is the moment of inertia; $d\theta/ds$ is the curvature of the beam; $M=-F'x=-Fx$,where, $F'x$ is the positive pressure of the dirt on the grinding net, and x is the axial distance of the coordinate system.

Under small deflection, the distance is the same as the axial distance, and the deflection Angle is the same as the slope, so $d\theta/ds$ can be approximated by d^2y/dx^2 . Therefore,

$$EI\frac{d^2y}{dx^2}=-F'x \quad (2)$$

Then the terminal position shape variable is:

$$\begin{cases} y=\dfrac{F'L^3}{3EI} \\ \theta=\dfrac{F'L^2}{2EI} \end{cases} \quad (3)$$

In the above formula, L is the length of grinding mesh; y is the displacement of grinding mesh.

When the deformation of the flexible lapping net is large, it obeies the flexible beam model, at this stage, the cantilever beam model cannot be used. The deflection differential equation of the beam needs to be solved:

$$\begin{cases} \dfrac{d\theta}{ds^*}=\dfrac{E}{EI}[\cos\alpha(\delta_y^*-y^*)+ \\ \sin\alpha(L-\delta_x^*-x^*)] \\ \dfrac{dx^*}{ds^*}=\cos\theta,\dfrac{dy^*}{ds^*}\sin\theta \end{cases} \quad (4)$$

Dimensionless:

$$\begin{cases} \dfrac{d\theta}{ds}=\beta f \\ [\cos\alpha(\delta_y-y)+\sin\alpha(1-\delta_x-x)] \\ \dfrac{dx}{ds}=\cos\theta,\dfrac{dy}{ds}=\sin\theta \end{cases} \quad (5)$$

After the positive pressure generated by the grinding net on dirt is obtained, the amount of air pump output to the air bag can be calculated by $p=F/s$.

4. General Safety Criteria

The electrical part of the conduit end plug cleaning device consists of four parts: the motor control system that drives the conduit end plug cleaning device to clean, the motor control system that drives the lifting of the conduit end plug cleaning device, the control system that drives the negative pressure recovery, and the motor control system that drives the lifting platform. Figure. 6 and Figure. 7 shows the electrical control system. In case of accidents, multiple protective measures should be taken to recover the foreign bodies generated and foreign bodies are strictly prohibited to fall into the boric acid pool[10].

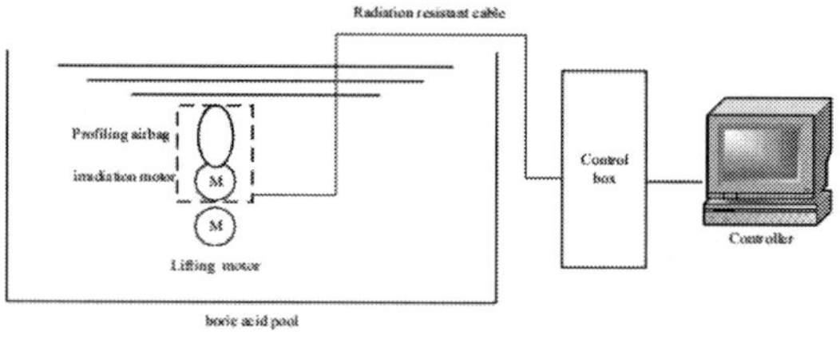

Fig. 6. Electrical diagram of pipe end plug cleaning device

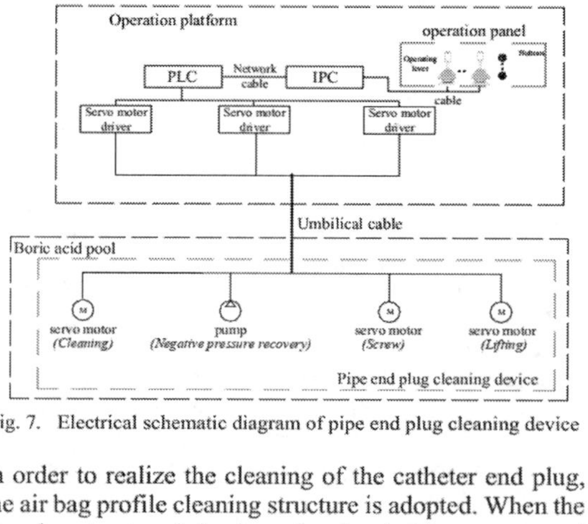

Fig. 7. Electrical schematic diagram of pipe end plug cleaning device

In order to realize the cleaning of the catheter end plug, the air bag profile cleaning structure is adopted. When the cleaning structure is in place, the cleaning face opens, and the catheter end plug cleaning device obtains the torsion force provided by the strong irradiation resistant servo motor through the transmission parts, so as to realize the cleaning of the catheter end plug inside. At the same time, the negative pressure recovery unit is activated to drain clean dirt from the inside of the catheter end plug. Considering the operating space narrow and strong radiation environment, use will be easily affected by the irradiation motor drive is installed in the control box outside the pool boric acid solution[7], strong resistance to radiation servo motor under the pool boric acid solution to guarantee the feasibility, among them, strong resistance to radiation between servo motor with the corresponding driver by strong resistance to irradiation on multi-core cable connection

In order to realize the free rise and fall of the pipe end plug cleaning device, a strong radiation resistant servo motor drives the cleaning device, so that it can obtain longitudinal movement power to achieve the lifting function of the pipe end plug.

In order to ensure that the catheter end plug cleaning device can work normally and safely, the logic shown in Figure. 8 is used to control it. The specific process is as follows:

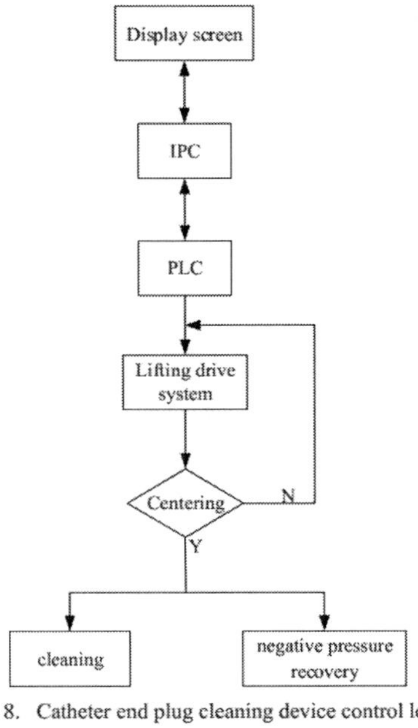

Fig. 8. Catheter end plug cleaning device control logic

According to the impurity information feedback from the display screen, the operator gives instructions for cleaning the pipe end plug to the industrial computer through the operating rod and button.

Industrial computer analysis of the control panel issued instructions to PLC control instructions, drive cutting fork mechanism lifting and complete alignment.

After the alignment is completed, the completed signal is transmitted to PLC. PLC drives the catheter end plug cleaning device to lift. After lifting in place, the cleaning face opens to clean the dirt. At the same time, the negative pressure recovery device drains the dirt from the inside of the catheter end plug. In order to achieve the best cleaning effect, fuzzy PID algorithm is adopted to control the output torque of the highly irradiation-resistant servo motor [8].

In order to ensure the smooth cleaning process of the inner wall of the pipe end plug, it is necessary to control the output torque of the motor. In order to ensure the robustness of the control process, the fuzzy PID algorithm is used to control the output torque of the motor reasonably.

Due to the narrow working environment of the pipe end plug cleaning device and the strong corrosion of the boric acid solution pool, the torque of the irradiation-resistant servo motor can not be obtained directly by using the torque sensor. However, according to the relationship between the motor current and torque, the motor torque can be measured indirectly through the motor current. Among them, the motor current and torque have a positive change relationship. During the operation of the motor, a large current means a large torque. By controlling the size of the current, the size of the torque can be controlled, indicating that the current ring is also a torque ring.

As shown in FIG. 10, the fuzzy PID control algorithm is used to control the torque of the motor and realize the cleaning of the pipe end plug. The fuzzy controller adopts the structure of two inputs and three outputs, in which two inputs are the deviation and the change rate between the measured value and the set value of the motor torque. The output is the modified values of PID parameters K_p、K_I、K_D. The input quantity enters into the fuzzizer first, and is transformed into the corresponding theory domain after certain fuzzification rules. Then the control quantity is deduced according to the database and the corresponding inference rules. Finally, the control quantity is defuzzified, so as to output the corresponding three parameters. The method is the same as "mobile platform equipment". The measured torque can be converted from the current to the measured torque based on the current-mechanical relationship conversion model [9][10].

$$M = kI \quad (6)$$

In the above formula, M is the motor torque; I is the motor current; K is a constant.

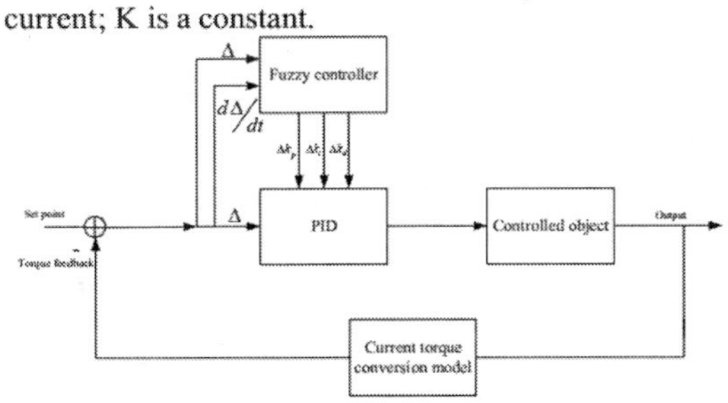

Fig. 9. Flow chart of torque control based on fuzzy PID

5. Conclusion

The inner wall aperture of the hammer component in RIC system is small and difficult to clean. In this study, a set of copying cleaning mechanism is designed for the cleaning of the conical inner wall. The copying cleaning mechanism can fit the grinding net and the conical inner

wall well, and ensure that the inner wall can meet the requirements of installation accuracy after cleaning. At the same time, a set of negative pressure recovery system is designed to ensure that the debris can be extracted from the boric acid solution in time in order to cause pollution by falling into the boric acid solution.
In order to ensure that the air bag can clean normally and the conical wall is not damaged due to excessive force, the relationship between the force on the conical wall and the air pump output to the air bag is established in this paper. In order to ensure that the output torque of the motor can normally complete the recovery of debris and the cleaning of the inner wall, this paper introduces the process of using PID to control the negative pressure recovery system and the output torque of the motor of copying cleaning structure.
The application of profile-copying cleaning in the process of inner wall cleaning of hammer components in RIC system of nuclear power has successfully solved the problem that the conical inner wall is difficult to clean and the debris cannot be recovered after cleaning, greatly improving the success rate of inner wall cleaning, enhancing the competitiveness of enterprises and maximizing the production benefit.

References

1. YM.Gu, "Easy Positioning and Convenient Adjustable Electric Hand Drill," Science Initiation, no.07,pp.12-13.
2. ZX.Lu,Q.Liu and WM.Ge, "The Development of Sandblasting Equipments on Pipeline Nodes," Manufacturing Automation, vol.38,no.04,pp.30-35, 2016.
3. QF.Li and GC.Wang, "Effect of Straight Diameter of Nozzle on Cleaning Efficiency of Sand Blasting," Materials Protection,no.51-52,pp.89-90,2001.
4. XM.Duan and YF.Liu, "Analyses of the Advantage and Profits of Dry Ice Cleaning for Tyre Mould," Cleaning World,no.007,pp.44-46,2018.
5. "Economical, Efficient and Time-saving Dry Ice Cleaning Application," Shanghai Coatings, vol.046, no.005, pp.44-45,2008.
6. HQ.Xu, "Design and Analysis of The Flexible Beam of a Rotational Gyroscope," M.S. thesis, Harbin Institute of Technology,Harbin,China,2017.
7. L.Liu, KM.Bi,TC.Liu et al., "Irradiation Test Study of Silicon Steel Sheets for High Temperature Irradiation-Resistant Motor," Instrumentation, vol.28, no.04,pp. 87-90+74,2021.
8. JB. Zhou, JS.Cheng and JH.Duan, "Implementation of timely Dynamic Fixing Technology of Pontoon Bridge Based on PID Algorithm," Technology Innovation and Application,vol.11,no.32,pp.19-22, 2021.
9. LB.Li,EW.Qi,WX.Lv and Sl.Hu, "Design and Implementation of Cryopump Control System," Cryogenics,no.04,pp. 19-22+34,2021.
10. B.Zhang,"Application of PID Algorithm in Control System," Information Recording Materials, vol.22, no.07,pp.197-199,2021.

 https://doi.org/10.1051/e3sconf/202235801038

Microbial degradation for the remediation of soils contaminated by crude oil in Peru: A systematic review of the literature

Favio Jimenez, Ronny Maldonado

School of Industrial Engineering, University of Lima, Lima, Peru

Abstract. In this research, a systematic literature review is carried out whose objective is to identify, compare and analyze the main fundamentals to be considered in order to face the problematic of remediation of soils contaminated by crude oil in order to establish and achieve a high efficiency of this. This work was elaborated through a comprehensive analysis of the various suitable scientific articles that provide a more comprehensive basis in relation to what is discussed in this research, taking as a basis the results that were optimal in the remediation of soils in terms of their various experiments made that were successful in the realization. Therefore, finally, it is obtained that the various technologies used for the remediation of the contaminated soil should not be treated only at the laboratory level but also at the field level due to the fact that there is a great gap in the treatment of the soil due to the conditions and variables that change significantly, thus reducing their remedial application in contaminated soil, presenting a minimal environmental impact and high investment costs in large-scale projects.

1 Introduction

In Peru, oil activity in hydrocarbon extraction began in 1971 with the discovery of the first oil wells in the Peruvian Amazon, motivating the construction of the North Peruvian oil pipeline, whose extension of its infrastructure reaches approximately 1,106 km of pipes that crosses the Amazon and Peruvian Andes to carry crude oil to the limit with the Pacific Ocean [1].

Crude oil, considered the "black gold" in the Peruvian Amazon, represents 85% of the energy consumption in the country, added together with those from natural gas and coal; given its importance in the country, Perupetro made a statistical estimate related to the oil consumed, which reaches around 1,577 million barrels since 1994 being a very important source over the years due to the constant energy demand it has. However, although it estimates a production of 1,577 million barrels, oil production is actually 763 million barrels; this amount being only a national production of 48%, making Peru a country without energy autonomy.

Given the magnitude of the problem, a subordinate indicator of an idealistic image regarding the reality of oil was evidenced, generating a controversial context about extractive activities, the legal framework and the stakeholders involved [1].

This is mainly due to the fact that in this part of Peru, is considered an area of oil exploitation this is due to the often emergency situations of oil spills or incidents of the companies in charge of extractive activities. One of the sources of contamination are hydrocarbons and oils derived from petroleum that generate harmful damage mainly to soils, waters, the structure of ecosystems and bioprocesses [2]. Evidencing accidental signs whose concern about environmental sustainability is latent since, consequently, they cause damage to microorganisms, plants, animals and humans. An example of these is the contamination of water and soil through the underground injection of oil wells for oil extraction, the soil affected by the splash caused by extraction, accidental spills in water and soil, the smoke generated by the refineries, the water used in the oil and petroleum processes that are discarded from there without an adequate form of treatment.

This study is based on the collection of data from the oil lots in the north-east of Peru, specifically, the Peruvian jungle; where as a result of their activities in the pipeline systems, tank cleaning, oil-water separators and transfers of crude oil to storage tanks, large amounts of pollutants are generated [3]. This type of oil spill has an imminent and profound effect on the environment that can seep into the soil to the water table and groundwater, contaminating it to the point of making it irremediable for its recovery [3]. We can review this in Fig 1 where a segmented graph is shown with the main causes generated by the contamination of Amazonian soil by crude oil, according to the National Human Rights Coordinator (CNDDHH).

According to another study carried out by the same entity, between 2000 and 2019 around 474 oil spills were identified in the Peruvian Amazon. 65.4% of said spills were due to corrosion of the pipelines, operational failures or unsafe conditions of the facilities, 28.8% caused by human actions, being among the main actions various attacks or sabotage caused by people outside the facilities, and 5.8% due to natural causes [1], as show in Fig 2.

https://doi.org/10.1051/e3sconf/202235801038

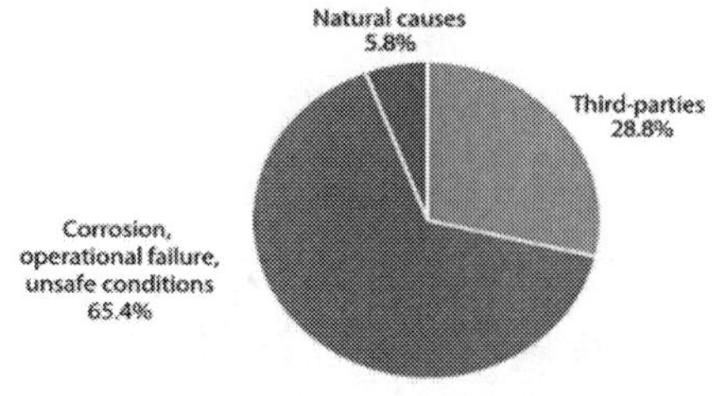

Fig. 1. Percentage of spills by cause in Amazon batches and the ONP (2000 -2019). Font: OEFA y OSINERGMIN.

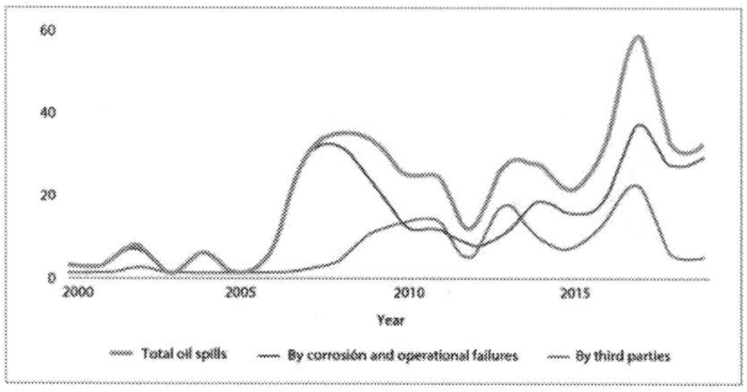

Fig. 2. Total number of spills in the Amazon lots and the ONP (2000-2019). Font: OEFA y OSINERGMIN.

Therefore, the initiative taken for this research work is aimed at the use of a bioremediation process of contaminated soils, opting for microbial remediation as a more efficient process, which consists of the use of bacterial strains under the study of its functional structure. It was shown that its microbial enzymatic activity is responsible for the biodegradation of contaminants, being useful for the development of this research [4]. This allows, through the bioremediation cycle, the bacterial strains to reduce the concentration of hydrocarbons and petroleum-derived oils; being an agent treating the recovery and regeneration of contaminated soil, favoring the restoration of the ecosystem and the biota of the site where the greatest environmental concern can be registered [5]. Therefore, we will emphasize addressing this issue in our research; for which we present the following main problem: What type of research on microbial degradation for the remediation of soils contaminated by crude oil, have been carried out in recent years in America, Asia and Africa.

For this we must deepen the research on the chosen bioremediation methods; Although all types of bioremediation carry out the elimination of the toxins present in the soil using living beings, there are differences between them, which consist of the use of the hydrocarbon degrading agent; within these are Bioaugmentation and Biostimulation. Bioaugmentation is responsible for introducing exogenous microorganisms that are capable of degrading hydrocarbons in the soil [6].

Biostimulation in which nutrients are added to the affected soil in order to stimulate the growth of microorganisms and thus increase their population. These two methods will be reviewed and evaluated according to their degree of impact for our research.

In effect, due to the uncertainty of the achievement of this type of bioremediation compared to the rest of the types, it generates that the general objective is to analyze the effectiveness obtained from the methods used in investigations that deal with the remediation of soils contaminated by oil. oil in America, Asia and Africa in recent years, through a systematic review of the literature.

In this research, the main fundamentals that should be considered for the development of the experimentation of the different types of bioremediation were identified, compared and analyzed; as well as the reports for months of the experimentation of the contaminated soil for its bioremediation. The articles reviewed on the experiences shown in bioremediation, the phases of bioremediation, and their effectiveness of use in each type of bioremediation, were identified through a systematic review of the literature. The results of this research were obtained based on the analysis of the various scientific articles and technical documents published between 2015 and 2021.

Given the diversity of the articles reviewed, in this research a general concept of treatment of contaminated soil is not used to limit the search, but the concepts of treatment of contaminated soil are accepted according to each author who has dealt with the topic addressed in the articles consulted.

This scientific article was developed based on the methodology of systematic review of the literature that was carried out with the following contributions: (1) analyze the laboratory studies of contaminated soils and how they define the development of the type of bioremediation used, (2) analyze the main stages of the experimentation of the various types of bioremediation used (3) and contrast the optimal type of bioremediation according to its level of remediation on the contaminated soil.

2 Methodology

2.1. Type of study

This research was carried out using a qualitative approach, this is because the research aims to present results of the systematic review of the literature about soil contamination by crude oil; with the purpose of having an impact on this research topic and its subtopics that are the subject of study by the scientific community. The type of research is non-experimental, despite the fact that a scientific assessment was made for a literary purpose, information gathering and the evaluation of scientific articles according to the impact or significance that it has.

Finally, this research is of longitudinal design where relevant information was collected from the scientific articles in English and Spanish consulted on the remediation of soils contaminated by crude oil since 2015 and its development to the present, including the various research subtopics.

2.2 Process

The methodology for the systematic review of scientific literature is developed based on the information collected from articles, scientific journals, etc., indexed in bibliographic repositories with scientific contribution, related to science, technology and medicine such as Scielo, Springer, Hindawi, Redalyc, ScienceDirect, Scopus and Web of Science (WOS). Likewise, this research begins with the definition of the topic, subject of the present; then, the selection criteria of the scientific articles are developed and the scientific systematic review is carried out; the same that is displayed in the search, selection, processing and technical review. Finally, the conclusions reached in the systematic review carried out are provided.

2.2.1. Classification of scientific articles to be included in the study

The classification method used is according to the following criteria that we will show below.

- Being in an indexed academic journal.
- Written in English or Spanish.
- Being in the quartiles Q1, Q2 or Q3.
- Contain less than thirty-five pages
- Proven results with a specific methodology.
- Have been published between 2015 and the present.
- Deal with the remediation of soils contaminated by crude oil in general and the microbial degradation of contaminated soils.
- Have experimental studies at the laboratory and field level.
- Be referred to microbial degradation in soils affected by crude oil.

Then, in the article we proceeded to use the bibliographic reference management tool called Mendeley, a tool in which the abstracts of each article could be read to see if they meet or not the requirements specified as search criteria and selection of scientific articles to be collected for systematic review of the literature. Finally, after reviewing the scientific articles, only twenty remained, which were subsequently evaluated to see the quality of their content and their compliance with the main objective of ensuring the reliability and integrity of the information shown in the review. systematics of the final literature. For this, criteria were used that allowed reaching a quality measurement of each scientific article; assigning them a classification and possible qualification score of "Excellent" = 2, "Fair" = 1 and "Poor" = 0.

That is why only those investigations whose quality score was higher than 12 (75% of the perfect score) could be accepted to subsequently collect and synthesize the data extracted from said selected scientific articles. Then, the following will show Fig 3 with the classified articles and Fig 4 with the details of these articles, previously classified.

Qualifying criteria	SA1	SA2	SA3	SA4	SA5	SA6	SA7	SA8	SA9	SA10	SA11	SA12	SA13	SA14	SA15	SA16	SA17
The summary presents the research, the problems, objectives and the methodology developed, as well as the most relevant findings and conclusions.	2.00	2.00	2.00	2.00	2.00	2.00	2.00	2.00	2.00	2.00	2.00	2.00	2.00	2.00	2.00	2.00	2.00
It describes the problems that give rise to the research and bases the need to solve them.	2.00	2.00	2.00	2.00	2.00	2.00	2.00	2.00	2.00	2.00	2.00	2.00	2.00	2.00	2.00	2.00	2.00
Clearly defines the purpose of the research.	2.00	2.00	2.00	2.00	1.00	2.00	2.00	2.00	2.00	2.00	2.00	2.00	2.00	2.00	2.00	2.00	2.00
The introduction describes the knowledge framework and is based on relevant bibliographic references. Identifies the variables and dimensions of the research.	1.00	2.00	2.00	2.00	2.00	2.00	2.00	1.00	2.00	2.00	2.00	2.00	1.00	1.00	1.00	2.00	2.00
The methodology is sufficiently described, detailing the methods and techniques used.	2.00	2.00	1.00	2.00	2.00	2.00	2.00	2.00	2.00	2.00	2.00	2.00	2.00	2.00	2.00	1.00	2.00
The results are fully exposed.	2.00	2.00	2.00	2.00	2.00	2.00	2.00	1.00	2.00	1.00	2.00	2.00	2.00	2.00	2.00	2.00	1.00
The discussion explains the behavior of the results in light of their trends and their comparison with the results of other similar research.	1.00	2.00	1.00	2.00	1.00	1.00	2.00	2.00	1.00	2.00	2.00	2.00	1.00	2.00	1.00	2.00	2.00
TOTAL	12.00	14.00	13.00	14.00	12.00	13.00	14.00	12.00	13.00	13.00	14.00	14.00	12.00	13.00	12.00	13.00	13.00

Fig 3. Qualification of the selected scientific articles.

Nº SA	Title	Author(s)
SA1	Efectos ambientales de los hidrocarburos. Una revisión.	Ortiz Diaz, D. (2019)
SA2	Microbial degradation of total petroleum hydrocarbon in crude oil polluted soil ameliorated with agro-wastes	Agbor, R. B., Antai, S. P., & Nkanang, A. J. (2018).
SA3	Bioremediation of hydrocarbon degradation in a petroleumcontaminated soil and microbial population and activity determination	Wu, M., Li, W., Dick, W. A., Ye, X., Chen, K., Kost, D., & Chen, L. (2017).
SA4	Bioremediation of Contaminated Soil. Introduction to Petroleum Biotechnology	Speight, J. G., & El-Gendy, N. S. (2018).
SA5	Biodegradation of diesel oil and n-alkanes (C18, C20, and C22) by a novel strain Acinetobacter sp. K-6 in unsaturated soil.	Chaudhary, D. K., Bajagain, R., Jeong, S. W., & Kim, J. (2020).
SA6	Biostimulation and bioaugmentation of native microbial community accelerated bioremediation of oil refinery sludge	Roy, A., Dutta, A., Pal, S., Gupta, A., Sarkar, J., Chatterjee, A., & Kazy, S. K. (2018).
SA7	Morphological, biochemical and molecular identification of petroleum hydrocarbons biodegradation bacteria isolated from oil polluted soil in Dhahran, Saud Arabia	Al-Dhabaan, F. A. (2018).
SA8	Bioremediation potential of natural polyphenol rich green wastes: A review of current research and recommendations for future directions.	Kuppusamy, S., Thavamani, P., Megharaj, M., & Naidu, R. (2015).
SA9	Environmental Nanotechnology: Applications of Nanoparticles for Bioremediation	Bhandari G. (2018)
SA10	A Comprehensive Review of Aliphatic Hydrocarbon Biodegradation by Bacteria	Abbasian, F., Lockington, R., Mallavarapu, M., & Naidu, R. (2015).
SA11	Identification and overcome of limitations of weathered oil hydrocarbons bioremediation by an adapted Bacillus sorensis strain	Ouatha, M., Al-Kaabi, N., Al-Ghouti, M., & Zouari, N. (2019).
SA12	Synergistic degradation of crude oil by indigenous bacterial consortium and exogenous fungus Scedosporium boydii	Yuan, X., Zhang, X., Chen, X., Kong, D., Liu, X., & Shen, S. (2018).
SA13	Microbial degradation of petroleum hydrocarbons	Varjani, S. J. (2017).
SA14	Abiotic factors controlling bioavailability and bioaccessibility of polycyclic aromatic hydrocarbons in soil: Putting together a bigger picture.	Yu, L., Duan, L., Naidu, R., & Semple, K. T. (2018).
SA15	A comprehensive guide of remediation technologies for oil contaminated soil - Present works and future directions.	Lim, M. W., Lau, E. V., & Poh, P. E. (2016).
SA16	A new look on factors affecting microbial degradation of petroleum hydrocarbon pollutants	Varjani, S. J., & Upasani, V. N. (2017).
SA17	Bioremediation of oily sludge polluted soil employing a novel strain of Pseudomonas aeruginosa and phytotoxicity of petroleum hydrocarbons for seed germination	Varjani, S., Upasani, V. N., & Pandey, A. (2020).

Fig 4. Scientific articles selected for the study.

3 Results and discussion

This section explains the results of the systematic literature review on the remediation of soil contaminated by crude oil through microbial degradation as an efficient method. The level of implementation of bioremediation in soil treatment is defined, the main aspects that influence the implementation of bioremediation are distinguished and the main stages of experimentation of the types of bioremediation that are used are determined, being crucial for future research by professionals and researchers in the environmental sector.

The search of the authors of these scientific articles is based on demonstrating the effectiveness of their bioremediation methods used using bacteria or a different agent, since each of these will grant a different appreciation regarding the use of the implementation.

3.1 Methodologies used in scientific articles

Some of the methodologies observed in the selected articles are the amplitude of the characterization of soil contaminated by crude oil, the selection of the microbial community in the soil treated under in vitro conditions and the compilation of the technologies used in the different scientific studies that will be part of the execution of the purpose of this study research. Initially, the contaminated soils are removed through a Dutch-type auger, at a depth of 0 to 25 cm deep, which are homogenized in labeled plastic buckets (PB) where they are placed to be taken to a laboratory for study. These PBs that contain the contaminated soils were thoroughly mixed and left to rest for a period of fourteen days; time necessary for the autochthonous microorganisms to acclimatize to the new soil condition [7].

While for [8] the subtracted contaminated soil sample it must be collected at a depth of 30 cm to reduce the pertinent intervention by animals or man. After being collected, these samples are kept in cold gel packs and adjusted to a temperature of 4°C, after arriving at the laboratory. The experiences of the laboratory tests are intended to remove the hydrocarbons present in the soil, observe the conditions of the soil contaminated by crude oil (physicochemical-microbiological analysis of the contaminated soil), obtain a study of native bacterial strains that will be isolated and placed tested to measure its aggressive degrading capacity against hydrocarbons from crude oil in contaminated soil. These samples were transferred to the laboratory at rest, according to [9], the samples must be analyzed physicochemically since it positively favors bioremediation as a traditional technique since factors such as: Bioavailability that alludes to the rate of degradation whose dependence it is the microbial metabolism that they have, it is related to the presence of water in required quantities that must be in contact with the contaminant; as well as also allow the growth of the microbial community and the provision of electron acceptor such as oxygen that allows microbial respiration. In addition, the study of the molecular structure of the samples that are an important chemical factor since their study affects the analysis of their composition to determine their capacity for hydrocarbon biodegradation.

On the other hand, it must be taken into account that regarding the samples analyzed microbiologically speaking, their observation at the level of oxidation, degradation, transformation and the complete mineralization process of the contaminants adhered to the soil are important [9]; but, given the above, a factor must also be taken into account that is transcendental with what has been studied, which is the environmental factor of the soil; for which, the reviewed samples have a specific particularity, a temperature, a pH, oxygen, humidity and availability of nutrients, adhered to it [9].

Having carried out the study of the soil to be treated, the current state of the extracted sample was verified, so for [10] it is necessary to indicate the composition of the crude oil that is at rest in the affected areas, being a negative impact on the environment. This composition can vary depending on the age of the affected oil field and its location, as well as the depth of the oil well and the polluting soil around it. Crude oil is divided into four broad fractions: Saturated (Aliphatic), Aromatic (Annulated Hydrocarbons), Resins and Asphaltenes. It is important to know this type of fractions because in the study of the extracted soil the percentage of each fraction is observed and based on this, verify what type of bacterial strain is its greatest degrading agent, as can be seen in Fig. 5.

Crude oil (Fractions)	Bacterial strain	References
Aliphatics	Acinetobacter Alcanivorax sp. Azoarcus sp. Bacilo sp. Brevibacterium Desulfosarcina sp. Desulfococcus sp. Marinobacter sp. Micrococos sp. Ochrobactrum sp. Oleispira sp. Pseudomonas sp. Rhodococcus sp. Stenotrophomonas sp. Thalassolituus sp. Aspergilo sp. Candida sp. Penicillium sp.	Mittal y Singh (2009) y Foght (2008) Harayama y col. (2004) y Brooijmans et al. (2009) Widdel y Rabus (2001) Ghazali et al. (2004) y Das y Mukherjee (2007) Leahy y Colwell (1990) Jaekel et al. (2013) Jaekel y col. (2013) Yakimov et al. (2007) Roy et al. (2002) y Ghazali et al. (2004) Varjani et al. (2015) Harayama et al.(2004) y Brooijmans et al. (2009) Mittal y Singh (2009), Rocha et al. (2011), Sajna et al. (2015), Varjani et al.(2015), Varjani y Upasani (2016c) Abbasian y col. (2015) Varjani et al. (2015) Brooijmans et al. (2009) Wilkes y col. (2016) Leahy y Colwell (1990) Atlas (1981) Sajna y col. (2015)
Monoaromatics	Pseudozima sp. Acinetobacter sp. Archaeoglobus fulgidus Aromatoleum aromaticum Bacilo sp. Halomonas sp. Pseudomonas sp. Rhodococcus sp. Sphingobacterium sp.	Batista y col. (2006) Wilkes y col. (2016) Janbandhu y Fulekar (2011) Widdel y Rabus (2001) Mittal y Singh (2009) y Meckenstock et al. (2016) Leahy y Colwell (1990) y Salleh et al. (2003) Janbandhu y Fulekar (2011)
Polyaromatics	Achromobacter insolitus Bacilo sp. Cicloclasticus sp. Phanaerochaete chrysporium Pseudomonas sp. Vibrio sp. Penicillium janthinellum	Janbandhu y Fulekar (2011) Mittal y Singh (2009) Harayama et al. (2004) Salleh et al. (2003) Widdel y Rabus (2001) y Meckenstock et al. (2016) Widdel y Rabus (2001) Boonchan et al. (2000)
Resins	Pseudomonas sp. Family member Vibrionaceae, Members of the family Enterobacteriaceae, Moraxella sp.	Leahy y Colwell (1990) Chandra et al. (2013)

Fig 5. Components – fractions of crude oil which are degraded by baterial strains.

In the process of the laboratory study, the isolation of bacteria that are attached to contaminated soils was developed. [11] explained that what is indicated to proceed in the phase of enrichment and isolation of hydrocarbon-degrading bacteria is the use of the selective enrichment technique and the Bushnell-Hass broth, which includes the increase in the strengthening and multiplication of the degrading microbial community, while the growth of other microorganisms that do not support the study is partially or totally inhibited and the crude oil of the contaminated sample is diluted. Said selected microbial community was gradually isolated and placed in nutritive agars to give a control and maintenance routine to the pure bacterial strains. These pure bacterial strains were put under observation for their bacterial identification and characterization process; being decisive to choose if it is the most suitable degrading agent with the study and how much it can increase its performance with the respective stimulants. According to [12], a morphological characterization process must be developed in which its shape, color, transparency and what type of differential behavior it has in its colony must be verified, as well as the biochemical characterization process that is fundamental in the study, as shown in Fig 6. This allows various metabolic characteristics of bacterial strains to be observed and can be verified through rapid techniques such as the oxidase test, catalase test, urease test, indole test, Simmons citrate test, methyl red test (MR) and hydrolysis of gelatin.

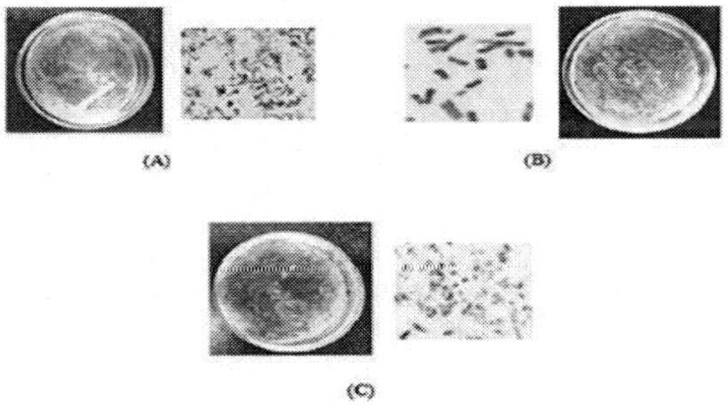

Fig. 6. Morphological characters and Gram stain for bacterial strains A, B and C isolated from Khurais oil. Font: Al-Dhabaan, F. A. (2018).

In various studies, it was possible to demonstrate the various bioremediation technologies used. For [8], in the field of bioremediation, biological alternatives were proposed that contributed to the oxidation, degradation, transformation and complete mineralization of the contaminating agents adhered to the soil. These bioremediation technologies are used according to ecosystem type criteria that include climatic variables in the affected areas at the water table, type and structure of soil, concentration of the type of hydrocarbon, environmental impact, investment budget, among others.

The technologies used by professionals in the environmental sector are mentioned below.

In this study, given the selected scientific articles, bioaugmentation and biostimulation were chosen as the most appropriate bioremediation technologies to be used in the soil to be treated. As can be seen in Fig 7, it explains [8] that in developing his experimentation with four bioremediation techniques (Natural Attenuation, Bioaugmentation, Biostimulation and Bioaugmentation + Biostimulation) and separating his samples with different nutrients such as selected native Bacteria (BA), Nitrate (N), Phosphorus (P) or combined (NP), found that one of the samples had a significant performance; as a result of the practice of the Bioaugmentation + biostimulation technique (BA + NP) through which an increase of 75% was observed compared to the rest of the samples that have between 35% and 63% of hydrocarbon degradation, being the largest with respect to the rest of the proposed samples.

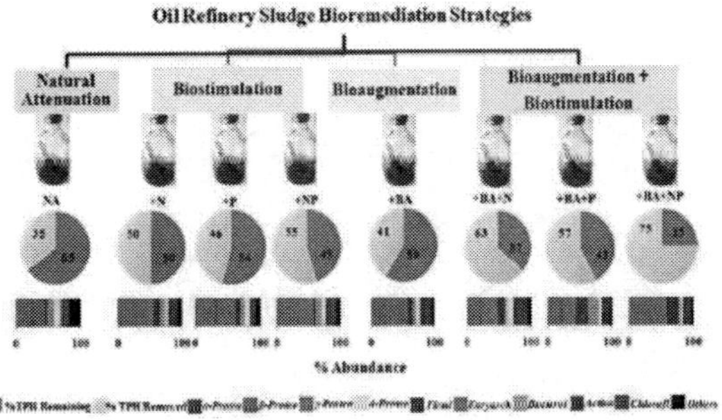

Fig. 7. Comparison of the methods used for the experimentation of the remediation of contaminated soil at the laboratory level. Font: Roy, A. et al. (2018).

According to the above figure, these authors considered that this improvement is substantial and stable due to the increase in the yield of the Bacillus, Archrobacter, Rhodobacter and Pseudomonas strain bacteria used, adding nutrients such as NP that, in effect, promoted the activation of sulfate reducers, the generation of fermentative abundances, increased yield of hydrocarbon degraders and high activity of methanogenic populations that are within the contaminated soil microbiome [8].

On the other hand, according to [13], who developed a pyro-sequencing study to verify changes in bacterial diversity and the proportions of their populations in soils contaminated by crude oil. He argues that this type of treatment with various concentrations of crude oil, showed severe effects on bacterial communities and their biodiversity in soils, clear detail was observed in several bacterial communities, such as Fibrobacterales, Thormoanaerobacterales Nautiales, Enterobacteriales, Thermotogales, Ustilaginales and Plocamiales that had a high susceptibility in the treatment and disappeared in the contaminated soils with low concentrations of crude oil of 0.5% and 2.5%. Being able to show that, not only should an appropriate selection technique be developed for the bacterial communities to later be subjected to bioaugmentation, but also biostimulation should be added to maximize the performance of the bacterial communities so that they can resist these concentrations, making the degradability and its ability to resist high levels of contaminants greater.

After synthesizing the importance of bioaugmentation + biostimulation in the research and with the intention of corroborating the effectiveness of this treatment [14], he carried out the germination of a Vigna radiata plant, which under normal conditions has a germination rate of 95.23% at fifth day of experimentation, while in the case of soil contaminated by crude oil, it was observed that its germination rate is 4.76% on the tenth day of experimentation. Having evidenced this significant difference in the germination rate, the bacterial inoculation supplementation was developed together with other nutrients that help the Vigna radiata seeds to have a germination rate of 80.95% on the fifth day of experimentation, considering the good progressive and rapid growth, demonstrating that the contaminated land of the extracted sample has managed to largely eliminate the concentration of crude oil present, for which this treated land became fertile again, having great relevance in agricultural and environmental sustainability, providing a way to agricultural application on land contaminated by crude oil after the bioremediation process.

Finally, in the present investigation carried out based on the selected scientific articles, it was shown an effective experimental process with respect to their studies, specifically, in biostimulation + bioaugmentation techniques, having as a key point the sum of these two techniques as a more complete method for bioremediation in situ, being viable as an associated technology with a great impact on the environment in the areas surrounding the area contaminated by crude oil.

4 Conclusions

From the systematic review of the literature, it was possible to verify that the remediation of soils contaminated by crude oil has been a relevant research topic in the last six years. As a result of these investigations, it has been possible to find and verify the effectiveness of the use of bacterial strains in the recovery and restoration of soils contaminated by crude oil.

When the compilation and revision of the scientific articles was carried out, it was possible to appreciate that a large part of these reflected positive experiences, obtained in the laboratory. Leaving in evidence that although bioremediation applications were developed using various tools with optimal results in their experimentation; In reality, they had a great gap when putting them into practice in the field or affected area, not being able to obtain the expected results because the conditions of the variables that condition the process of degradation of the contaminated soil are not the same.

However, the importance of laboratory-scale studies is not denied, since their development is essential since it can minimize the various problems associated with their application of In-Situ bioremediation and achieve a significant improvement in the use of the method [15].

On the other hand, the behavior of the crude oil fractions is unique to each contaminated area as well as the impact generated in them; that is why exhaustive evaluations are carried out, to determine what would be the type of appropriate technique with which soil remediation will be addressed [16]. Instead, the influence of abiotic factors in direct relation to soil structure must be understood intrinsically. Defining the series of parameters involved is a complex task that makes it difficult to develop a global procedural method regarding the various factors that control the bioavailability and bioaccessibility of crude oil in contaminated soil [17].

Each bioremediation technology used has its own characteristics and the cleaning methods applied will not be established as universal methods but as alternatives for the bioremediation of soils contaminated by crude oil [16]. Based on this, the mixed design of bioremediation technologies is proposed that allows the integration of two or more bioremediation technologies given that they have a higher performance and effectiveness in soil restoration, as in the case of Bioaugmentation + Biostimulation that were used in other investigations [18].

As part of the research, new proposals were observed to cover the remediation of contaminated soils, one of them being nanoremediation, which is an approach under development and has a promising perspective in improving bioremediation [19], incorporating the different types of food, agro-industrial and tree waste materials; as important sources of polyphenols in the successful removal of environmental pollutants [20]; and as a last proposal, genetic engineering, which proposes the genetic modification of the microorganisms or organisms used, although the application of this proposal is controversial and with little acceptance, due to the fact that the real impact that these microorganisms or organisms could have on the the ecosystem.

Finally, the great challenge in the future for researchers is to achieve a broader coverage of their knowledge regarding the microbial populations that degrade crude oil hydrocarbons [21], deeply understanding that symbiotic relationships and their metabolizing pathways implant in the complicated system of the contaminated soil. The proposal of this will help to expand and extrapolate the knowledge obtained in pilot scale experimentations in large complex field systems.

References

1. León, A., & Zúñiga, M. (2020). The shadow of oil. A report on oil spills in the Peruvian Amazon from 2000 to 2019.
2. Zamora, A., & Ramos, J. y. (2012). Effect of oil pollution on chemical and microbiological properties of a savanna soil. Bioagro, 5-12.
3. Cardenas, P. G. (2017). Biodegradation of total petroleum hydrocarbons by biostimulation with Cachaza and Guano de Islas in soils of the Conchan-Petroperu Refinery, Lima, 2017. Lima: Cesar Vallejo University.
4. Plaza, G., Cábala, M., Torres, N., Velásquez, M., Corbalán, E., & Rodriguez, C. (2001). Biorremediación en suelos contaminados con hidrocarburos. Avances en Energías Renovables y Medio Ambientes.
5. Oualha, M., Al-Kaabi, N., Al-Ghouti, M., & Zouari, N. (2019). Identification and overcome of limitations of weathered oil hydrocarbons bioremediation by an adapted Bacillus sorensis strain. Journal of Environmental Management, 250, 109455. doi:10.1016/j.jenvman.2019.109455
6. Wu, M., Li, W., Dick, W. A., Ye, X., Chen, K., Kost, D., & Chen, L. (2017). Bioremediation of hydrocarbon degradation in a petroleum-contaminated soil and microbial population and

activity determination. Chemosphere, 169, 124–130. doi:10.1016/j.chemosphere.2016.11.059

7. Agbor, R. B., Antai, S. P., & Nkanang, A. J. (2018). Microbial degradation of total petroleum hydrocarbon in crude oil polluted soil ameliorated with agro-wastes. Global Journal of Earth and Environmental, 3(1), 1-7.
8. Roy, A., Dutta, A., Pal, S., Gupta, A., Sarkar, J., Chatterjee, A., Kazy, S. K. (2018). Biostimulation and bioaugmentation of native microbial community accelerated bioremediation of oil refinery sludge. Bioresource Technology, 253, 22–32. doi:10.1016/j.biortech.2018.01.004
9. Ortiz, D. (2019). Efectos ambientales de los hidrocarburos. Una revisión. (Doctoral dissertation, Universidad Santiago de Cali).
10. Varjani, S. J. (2017). Microbial degradation of petroleum hydrocarbons. Bioresource Technology, 223, 277–286. doi:10.1016/j.biortech.2016.10.037
11. Chaudhary, D. K., Bajagain, R., Jeong, S. W., & Kim, J. (2020). Biodegradation of diesel oil and n-alkanes (C 18, C 20, and C 22) by a novel strain Acinetobacter sp. K-6 in unsaturated soil. Environmental Engineering Research, 25(3), 290-298.
12. Al-Dhabaan, F. A. (2018). Morphological, biochemical and molecular identification of petroleum hydrocarbons biodegradation bacteria isolated from oil polluted soil in Dhahran, Saud Arabia. Saudi Journal of Biological Sciences. doi:10.1016/j.sjbs.2018.05.029
13. Abbasian, F., Lockington, R., Mallavarapu, M., & Naidu, R. (2015). A Comprehensive Review of Aliphatic Hydrocarbon Biodegradation by Bacteria. Applied Biochemistry and Biotechnology, 176(3), 670–699. doi:10.1007/s12010-015-1603-5
14. Varjani, S., Upasani, V. N., & Pandey, A. (2020). Bioremediation of oily sludge polluted soil employing a novel strain of Pseudomonas aeruginosa and phytotoxicity of petroleum hydrocarbons for seed germination. Science of The Total Environment, 139766. doi: 10.1016/j.scitotenv.2020.139766
15. Varjani, S. J., & Upasani, V. N. (2017). A new look on factors affecting microbial degradation of petroleum hydrocarbon pollutants. International Biodeterioration & Biodegradation, 120, 71–83. doi:10.1016/j.ibiod.2017.02.006
16. Lim, M. W., Lau, E. V., & Poh, P. E. (2016). A comprehensive guide of remediation technologies for oil contaminated soil — Present works and future directions. Marine Pollution Bulletin, 109(1), 14–45. doi:10.1016/j.marpolbul.2016.04.023
17. Yu, L., Duan, L., Naidu, R., & Semple, K. T. (2018). Abiotic factors controlling bioavailability and bioaccessibility of polycyclic aromatic hydrocarbons in soil: Putting together a bigger picture. Science of The Total Environment, 613-614, 1140–1153. doi:10.1016/j.scitotenv.2017.09.025
18. Yuan, X., Zhang, X., Chen, X., Kong, D., Liu, X., & Shen, S. (2018). Synergistic degradation of crude oil by indigenous bacterial consortium and exogenous fungus Scedosporium boydii. Bioresource Technology, 264, 190–197. doi:10.1016/j.biortech.2018.05.072
19. Bhandari G. (2018) Environmental Nanotechnology: Applications of Nanoparticles for Bioremediation. In: Prasad R., Aranda E. (eds) Approaches in Bioremediation. Nanotechnology in the Life Sciences. Springer, Cham. https://doi.org/10.1007/978-3-030-02369-0_13
20. Kuppusamy, S., Thavamani, P., Megharaj, M., & Naidu, R. (2015). Bioremediation potential of natural polyphenol rich green wastes: A review of current research and recommendations for future directions. Environmental Technology & Innovation, 4, 17–28. doi:10.1016/j.eti.2015.04.001
21. Speight, J. G., & El-Gendy, N. S. (2018). Bioremediation of Contaminated Soil. Introduction to Petroleum Biotechnology, 361–417. doi:10.1016/b978-0-12-805151-1.00010-2

Research on capacity allocation optimization of a wind-photovoltaic -hybrid-battery power generation system with multi-energy complementary

Yu Lei, Jianjun Xu *, Lichao Pan, Dikang Sun

School of Electrical Engineering and Information, Northeast Petroleum University, Daqing, Heilongjiang 163318, China

Abstract: The output of complementary energy is the core of power generation system planning, and researching its configuration is the basis for realizing safe, reliable, economical and stable operation of the system throughout total life cycle. This paper comprehensively considers the constraints of power supply reliability and battery energy storage operation, and proposes a capacity optimization method for wind-photovoltaic-hydro-storage system aiming at life cycle cost and unit cost of electricity. Take the Northwest Region as an example, the optimization problem was solved by HOMER software, the optimal capacity allocation result was obtained. Compared with other solutions, the life cycle cost was reduced by 12.1%, the unit cost of electricity was reduced by 11.9%, and the load power shortage rate probability was relatively reduced. The result prove that the complementary system helps to reduce power generation cost and improve power supply stability.

Key words: complementary energy; wind-photovoltaic-hydro-storage; capacity configuration; HOMER software; life cycle cost.

1. Introduction

Since the output power of wind and solar resources is greatly affected by weather factors, no single energy source can provide stable and reliable power. In order to improve the reliability of power supply, it is usually necessary to combine and configure two new energy sources to form complementary output[1-3]. In the northwest region, the hydropower, wind energy and solar energy have multiple complementarities. The abundance and dryness of water power and the complementary wind energy and solar energy of day and night form seasonal complementary advantage[4-6]. Using hydropower with low operating cost and rapid start-stop, taking into account the tracking load and bidirectional charging characteristics of energy storage devices, it provides key technologies for adjusting the output power characteristics of wind and solar output and forming a multi-energy complementary power generation[7-10].

This paper will consider the factors of economy, reliability and technical feasibility, establish the capacity configuration optimization of the wind-photovoltaic-hydro-storage multi-energy complementary generation system with the system life cycle cost and unit cost of electricity as the optimization object, taking into account the reliability of power supply and the operation constraints of the battery. For the wind-solar hybrid power generation system, the multi-energy hybrid optimization model is built and simulated by the HOMER software developed by the United States National Renewable Energy Laboratory. After optimization, the capacity configuration results that meet different constraints are obtained, and the optimal results are analyzed.

2. Structure of Wind-photovoltaic-hybrid-battery Multi-energy Complementary Generation System

As shown in Figure 1, the main power generation part includes photovoltaic array and wind turbine, hydroelectric generator and battery in the system to assist in generating electricity. The system initially sets the wind and solar utilization to the limitation, that is, the photovoltaic array and wind turbine work at the maximum power point. Firstly the system provide electric energy to the load, and quickly charge the battery when the electrical energy provided by the system has excess energy. If the electric energy generated by the main power source is not enough to the load demand of the user, the auxiliary power supply is supplied by the backup power source; if the electric energy of the user is not met, the electric energy is purchased from the grid to meet the load demand of the complementary system.

* Corresponding author: 18709138654@163.com

https://doi.org/10.1051/e3sconf/202235801039

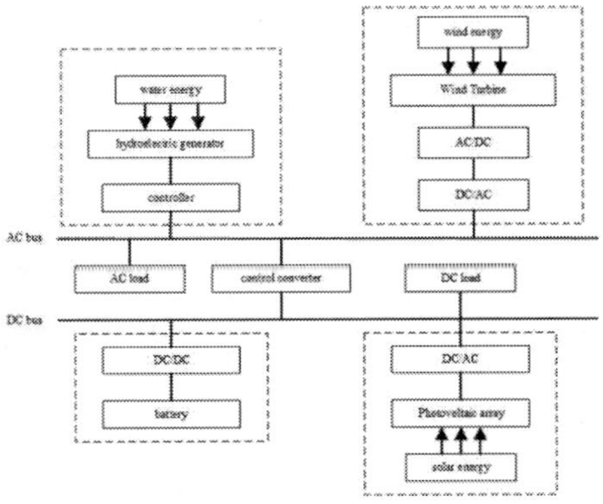

Fig. 1 Structure of wind-photovoltaic-hybrid-battery multi-energy complementary generation system

3. Capacity Configuration Optimization Model

3.1 Objective Function

In the planning of the complementary power generation system, the operation investment cost should be reduced as much as possible on the premise of ensuring the feasibility and reliability of the system[11]. Therefore, considering the system status and actual requirements, choose the life cycle cost and unit power generation cost of the overall system as the optimization goals to meet the safe, stable, reliable and economical power demand of users, and finally the optimal capacity allocation of photovoltaic cells, wind energy, hydropower and energy storage. Select the wind power capacity N_{wind}, photovoltaic capacity N_{pv}, hydropower capacity N_H and battery capacity N_{bat} in wind-photovoltaic-hybrid-battery complementary generation system as optimization variables. Variable as:

$$N = \left[N_{wind} \quad N_{pv} \quad N_H \quad N_{bat} \right] \tag{1}$$

3.1.1 Life Cycle Cost

The life cycle cost is the net present value of the system, which refers to all costs incurred during the overall system life cycle due to installation and operation, minus the present value of the earning revenue during the cycle. The cost mainly include initial investment cost, replacement cost, operation and maintenance cost, cost of fuel and pollution, and cost of purchasing electricity from the grid when electricity is insufficient. Since this system only considers renewable energy power supply, excluding the cost of fuel and pollution. Since the research has not considered the connection with the power grid and has not generated funds with the power grid, the present value of the power grid income is temporarily recorded as zero income.

Therefore, the life cycle cost of the research system can be expressed as:

$$C_{NPC} = C_{I.invest} + C_{R.reset} + C_{O\&M} - C_{income} \tag{2}$$

1) Equipment Investment Cost

$$C_{I.invest} = C_{I.wind}N_{wind} + C_{I.pv}N_{pv} + C_{I.bat}N_{bat} + C_{I.H}N_H \tag{3}$$

where $C_{I.wind}$ is the initial investment costs of wind turbines; $C_{I.pv}$ is the initial investment costs of photovoltaic arrays; $C_{I.H}$ is the initial investment costs of battery packs; $C_{I.bat}$ is the initial investment costs of hydroelectric generator.

2) Equipment Replacement Cost

$$C_{R.reset} = \sum_{1}^{k=n} \left(C_{R.wind}N_{wind} + C_{R.pv}N_{pv} + C_{R.bat}N_{bat} + C_{R.H}N_H \right) * \frac{1}{(1+i)^{(k-1)*Y}} \tag{4}$$

where n is reset time, take its value as the ratio of the system life cycle Y_{system} to the equipment life Y_A and round up, A is any device in the system; $C_{R.wind}$, $C_{R.pv}$, $C_{R.H}$, $C_{R.bat}$ are the replacement investment costs of wind turbines, photovoltaic arrays, hydroelectric generators, and battery packs, respectively; i is the annual real discount rate, can be calculated by $i = \frac{i' - f}{1+f}$, where i' is nominal discount rate, f is expected inflation rate.

3) Equipment Operation and Maintenance Cost

$$C_{O\&M} = \sum_{1}^{k=n} \left(C_{O\&M.wind}N_{wind} + C_{O\&M.pv}N_{pv} + C_{O\&M.bat}N_{bat} + C_{O\&M.H}N_H \right) * \frac{1}{(1+i)^{(k-1)}} \tag{5}$$

where $C_{O\&M.wind}$ is operation and maintenance costs for wind turbines; $C_{O\&M.pv}$ is operation and maintenance costs for photovoltaic arrays; $C_{O\&M.H}$ is operation and maintenance costs for hydroelectric generators; $C_{O\&M.bat}$ is operation and maintenance costs for battery packs.

3.1.2 Unit Cost of Electricity

The unit cost of electricity characterizes the average cost of a system to generate 1kWh of active electrical energy, can be expressed as:

$$C_{COE} = \frac{C_{annual}}{E_{ELE}} \tag{6}$$

where

$$C_{annual} = \frac{i(1+i)^N}{(1+i)^N - 1} C_{NPC} \tag{7}$$

where C_{annual} is the total cost of annual power generation conversion of the system; E_{ELE} is annual total power load of the system; N is system life cycle.

3.2 Power Supply Reliability Permit Model

Due to the periodic characteristics of new energy power generation, ensuring the safety and reliability of the power supply system is a basic requirement. The evaluation criteria of power supply reliability include load power shortage rate and energy waste rate. The loss of power supply probability $LPSP$ represents the probability of a shortage of system capacity demand, and the loss of renewable energy rate $LRER$ represents the ratio of the annual curtailment of wind power and photovoltaics to the total annual wind power generation. When optimizing the system capacity, give the two equal weights to record a new index power deviation PD [12], which can be expressed as:

$$PD = 0.5LPSP + 0.5LRER \tag{8}$$

1) The Loss of Power Supply Probability

$$LPSP = \frac{\sum_{t=1}^{8760} P_{lack}}{\sum_{t=1}^{8760} P_{load}} \tag{9}$$

where

$$P_{lack} = P_{load} - P_{pv} - P_{wind} - P_H - P_{bat} \tag{10}$$

where P_{lack} is the system of lacks power at time t ; P_{load} , P_{pv} , P_{wind} , P_H , P_{bat} are the output power of the load, the output power of the photovoltaic array, the output power of wind turbines, the output power of the hydroelectric generators and the output power of the battery.

2) The Loss of Renewable Energy Rate

$$LRER = \frac{\sum_{t=1}^{8760} \left(P_{loss.pv} * \Delta t + P_{loss.wind} * \Delta t\right)}{\sum_{t=1}^{8760} \left(P_{pv} * \Delta t + P_{wind} * \Delta t\right)} \tag{11}$$

where $P_{loss.pv}$, $P_{loss.wind}$ are the load-loss power of photovoltaic and wind power when the load is not met respectively; P_{pv} , P_{wind} are the photovoltaic and wind power outputs at time t under ideal conditions respectively.

3.3 Constraint Condition

3.3.1 Power Balancing Constraints of the System

When renewable energy generation is sufficient to support the load,

$$P_{pv} + P_{wind} + P_H = P_{load} + P_{bat} + P_{loss} \tag{12}$$

When renewable energy generation is insufficient to the load demand,

$$P_{load} = P_{pv} + P_{wind} + P_H + P_{bat} + P_{lack} \tag{13}$$

3.3.2 Every Equipment Output Constraint

For any kind of power output at time t, it is subject to constraints:

$$0 \le P_i(t) \le P_i^{\max} \tag{14}$$

where $P_i(t)$ is the actual output power value of a certain power supply at time t; $P_i^{\max}$ is the maximum output power value of each power supply.

3.3.3 Battery Operating Constraints

The use of battery energy storage needs to ensure its service life and operating limitations, and the battery needs to consider two constraints during the charging and discharging process:

$$SOC_{\min} \le SOC \le SOC_{\max} \tag{15}$$

$$\begin{cases} P_{ch}(t) \le 0.2E_{bat} / \Delta t \\ P_{dch}(t) \le 0.2E_{bat} / \Delta t \end{cases} \tag{16}$$

where SOC is battery charge and discharge status; $P_{ch}(t)$ is charging power per hour; $P_{dch}(t)$ is discharge power per hour; E_{bat} is the available power of the battery at the moment. The use time of the battery is limited by the charging and discharging power. Excessive charging and discharging rate will detract from its use time. Therefore, the upper limit of the charging power per hour cannot exceed $SOC/5$, Δt which is taken as 1h[12].

3.3.4 System Power Supply Reliability Constraints

$$PD \le \lambda_{\max} \tag{17}$$

where $\lambda_{\max}$ is the maximum acceptable power deviation for the power supply system.

4. Case Analysis

4.1 Simulation Basic Data

4.1.1 Basic Components

Optimization and sensitivity analysis of HOMER allows to evaluate the economic and technical feasibility of a large number of technology options, taking into account changes in technology cost and energy availability[9]. Therefore, it is necessary to consider the geographical conditions of the study site, natural resource conditions, meteorological data, local power grid data, and required loads to determine the type, capacity and power of various equipment in the current power generation system, so as to ensure that multi-energy complementary power generation system remain as optimal as possible. In this paper, the economic parameters of related equipment are set as shown in Table 1 by referring to papers and related materials and considering the premise of parallel economy and reliability.

Tab. 1 The economic parameters of each equipment in the system

Device parameters	Component type				
	Photovoltaic	Wind Turbines	Hydroelectric generator	Battery	Control converter
Equipment capacity (kW)	1	10	100	1kWh	50
Initial investment cost (¥)	5000	48000	105500	1500	2100
Replacement cost (¥)	4800	45000	75000	1190	2000
Operation and maintenance cost (years/¥)	100	1400	8000	100	120
Service life (years)	25	20	30	15	15

The design life of the research system is assumed to be 25 years. It can be seen from the table that the photovoltaic cell and hydroelectric generator do not need to be replaced during the overall system life cycle. In addition to the parameters described in the table above, other parameters need to be set as follows:(1) When photovoltaic cells generate electricity, there will be a demand for land occupation. This article sets the upper limitation of the allowable photovoltaic installed capacity to 500 kW. (2) The nominal number of 10kW wind turbines is set to 0,10, 20, 30, 40. (3) The 100kW hydroelectric generator is limited by the head H and flow Q, the parameters are set in the system. (4) The battery used is a 1 kWh lithium-ion battery and the SOC limitation is 30% to 100%. Usually, the maximum battery is 90% in practical applications. (5) The rectification efficiency and inverter efficiency of the converter are both set to 95% and the capacity variable of the converter is set to 0/50/100 kW.

4.1.2 Resource Data Setting

In the HOMER software, after determining the latitude and longitude coordinates of a certain place, you can directly download the wind energy resources, solar energy resources and water resources data of the research area from the meteorological resources website database.

1) solar and wind data

The research takes a region in northern China (34°15'N, 108°54'E) as the research object. According to the geographic coordinates of the region, solar and wind energy data are downloaded from the homer energy website or obtained from the National Meteorological Information Center website. The required historical meteorological data for 2016-2020 and take the monthly average of the data. Table 2 shows the monthly average radiance S, the corresponding clarity index k and the wind speed data at a height of 10m used for meteorological measurement. According to the statistical results, it is obvious that the wind energy and solar energy resources are relatively abundant.

Tab. 2 Meteorological data at the study site

Month	Clearness index(k)	solar radiation S (kWh/m^2/day)	Speed v (m/s)
January	0.543	5.460	6.320
February	0.557	5.780	5.190
March	0.569	5.980	5.690
April	0.612	6.240	3.670
May	0.655	6.320	4.380
June	0.690	6.420	2.620
July	0.692	6.530	3.210
August	0.681	6.750	3.450
September	0.622	6.430	5.390
October	0.601	6.230	3.530
November	0.527	5.230	6.450
December	0.517	5.120	6.340

Enter the resource data results obtained by statistics in the corresponding resource tab of HOMER. In order to simulate the output curve as accurately as possible, many factors need to be considered. Therefore, 4, 6, and 8 (kWh/m2/day) are used for the average solar radiation in the selected base year. Similarly, the average annual wind speed in the wind power resource components is 4, 6, and 8 (m/s). When the average radiance value of the year is 6, the distribution curve of solar energy resources is shown in Figure 2; when the average wind speed value is 8, the distribution curve of wind energy resources is shown in Figure 3. The selection of annual average is related to the simulation results.

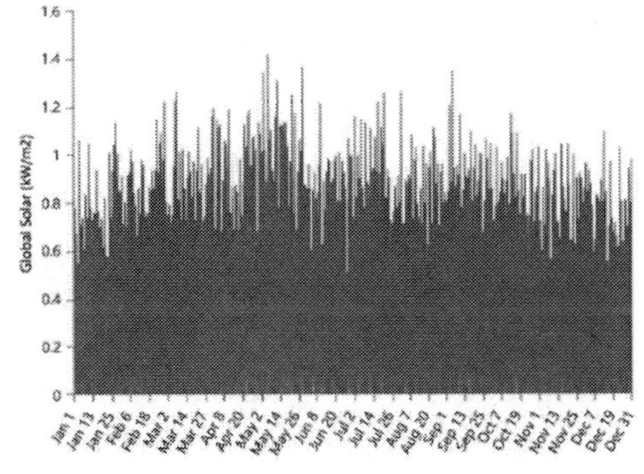

Fig. 2 The local hourly solar radiation data for the whole year after separation

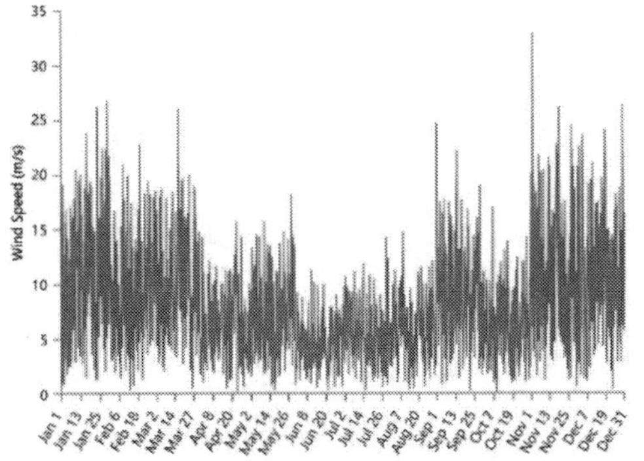

Fig. 3 Local hourly wind speed data throughout the year after separation

2) water resource data

The local water resources are abundant throughout the year. The annual average water flow is about 172.5L/s, and the maximum flow could reach 200L/s. The monthly average water flow in this area is shown in Table 3 through long-term observation.

Tab. 3 The average monthly water flow in the local area throughout the year

month	1	2	3	4	5	6
annual average stream flow L/s	190	195	200	200	190	185
month	7	8	9	10	11	12
annual average stream flow L/s	160	125	135	145	160	185

3） load data

Assuming that only AC load is considered, according to the existing load model in the software, combined with the load demand of a certain region, the load model of this study can be obtained. By entering the corresponding information into the corresponding load tabs related to the load demand and adding daily variability and time variability, a load distribution of 8760 hours in 12 months of the year could be generated, as shown in Figure 4. The model is a load based on "Community Load", which conforms to the normal law of residential electricity consumption. The peak of the load curve appears at 18:00-21:00, and the trough period is in the early morning.

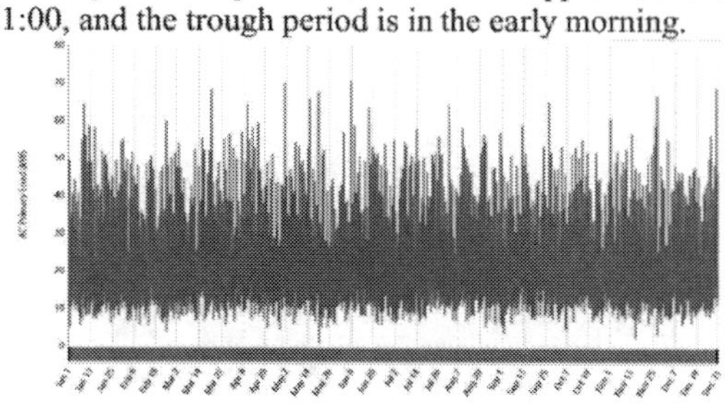

Fig. 4 Annual load data in the study area

4.2 System modeling and simulation result analysis based on homer software

Compared with other optimization software, the advantage of HOMER software developed by the National Renewable Energy Laboratory is that it can quickly and accurately help users to obtain configuration results. During the simulation process, the software will internally judge whether the design is feasible and compare various configuration results. In this paper, the simulation and optimization are carried by HOMER software, the optimization goal is to minimize the life cycle cost of the power generation system and the lowest unit power generation cost. According to the conditions of the selected research area, a model of a hybrid power generation system with wind-solar-water-storage is established. The model established in the software is shown in Figure 5. As can be seen from the figure, the main power generation parts of wind turbine and photovoltaic cell, including hydroelectric generator, are arranged on the AC side, and the battery is connected to the DC bus. The two are connected by an AC-DC converter to facilitate energy flow conversion and battery charging and discharging.

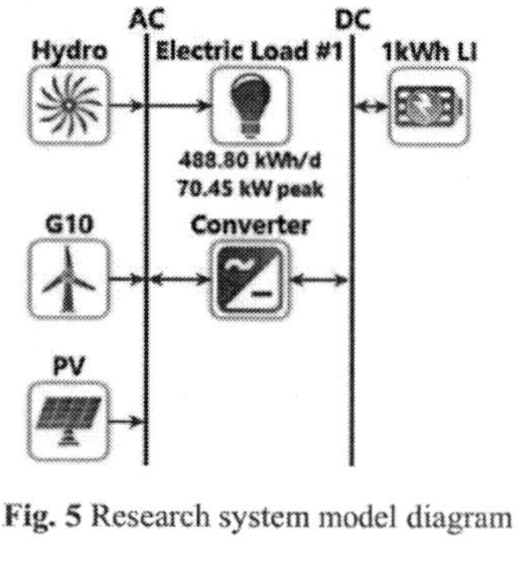

Fig. 5 Research system model diagram

The basic service life of the system is set to 25 years, the expected expansion rate is 2%, the nominal discount rate is 8%, and the maximum allowable power outage probability is 5%, resulting in an actual discount rate of 5.88%. The control strategies are selected as load following strategy (LC) and cyclic charging (CC) strategy. The parameters of the design selection and various resource and load data are entered into the model.

4.2.1 Analysis of Simulation Results

Press the "calculate" button to simulate according to the set basic data and constraints. The simulation outputs an operable capacity configuration plan, and the obtained configurations are arranged in order of life cycle cost from low to high. A part of the configuration scheme of the optimized configuration results is shown in Figure 6. It can be clearly seen from the figure that the smaller total net present value of the system, the smaller power generation cost; on the contrary, the greater load power shortage rate will be, that is, the required operating capacity that the system can provide will not meet the actual operating capacity.

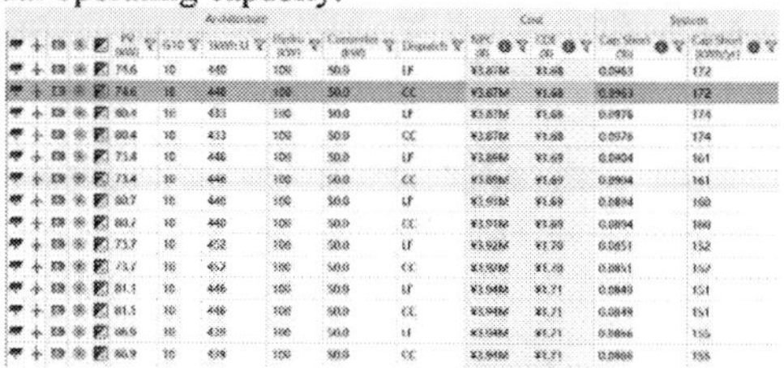

Fig. 6 Partial capacity configuration results

In different combinations in the optimal configuration plan, the electric energy is selected to be provided by one or both of the three energy sources of wind energy, solar energy and water energy, or the electric energy is provided by the three together. The optimal allocation results of different combination schemes are shown in Figure 7, and the specific allocation and investment costs of each scheme are shown in Table 4.

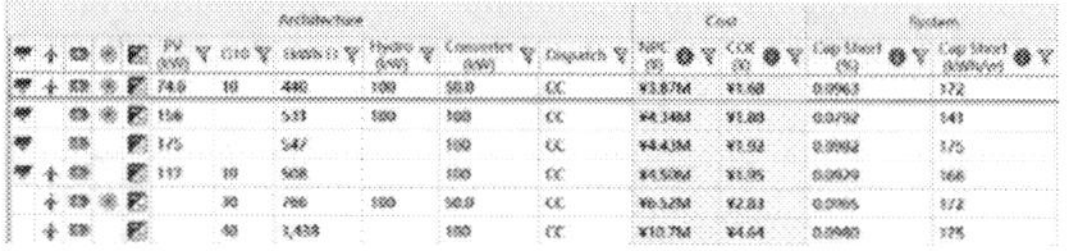

Fig. 7 Optimal configuration results for each scenario

Table 4. The specific allocation and investment costs of each scheme are shown

Configuration	1	2	3	4	5	6
PV (kW)	74.6	156	175	117	/	/
Wind Turbines (kW)	100	/	/	100	300	400
Hydroelectric generator (kW)	100	100	/	/	100	/
Battery (kWh)	440	531	547	508	766	1438
Converter (kW)	50	100	100	100	50	100
Life cycle cost (mil)	3.87	4.34	4.43	4.50	6.52	10.7
Unit cost of electricity (￥)	1.68	1.88	1.92	1.95	2.83	4.64
Capacity shortage fraction (%)	0.096	0.079	0.098	0.092	0.096	0.098
Capacity Shortage(kW)	172	141	175	166	172	175

It can be seen from Figure 7 and Table 4 that the configuration schemes listed in the software are obtained after the optimized operation results of the given capacity configuration are in line with the power supply reliability settings. Therefore, from the configuration results, when the life cycle cost is 3.87 million yuan and the unit power generation cost is the lowest 1.68 yuan/kWh, the optimal solution is the solution 1 in the table. Compared with other solutions, the whole life cycle cost is at least reduced 12.1%, the unit cost of power supply decreased by 11.9%. The specific allocation of capacity is as follow: wind turbine capacity is 100kW, hydroelectric generator capacity is 100kW, photovoltaic cell capacity is 74.6kW, battery capacity is 440kWh, and bidirectional converter capacity is 50kW.

At this time, the annual load power shortage rate of the optimal configuration result can be obtained through the formula $LPSP = \frac{E_{CS}}{E_{demand}}$. Among them, E_{CS} is the annual capacity shortage; E_{demand} is the total electricity demand. The calculated result is 0.096%. According to the HOMER software, the comparison curve between the total annual load demand and the annual load shortage of 8760 hours a year can be generated, as shown in Figure 8. It can be seen from the figure that there is a significant capacity shortage during the peak load months.

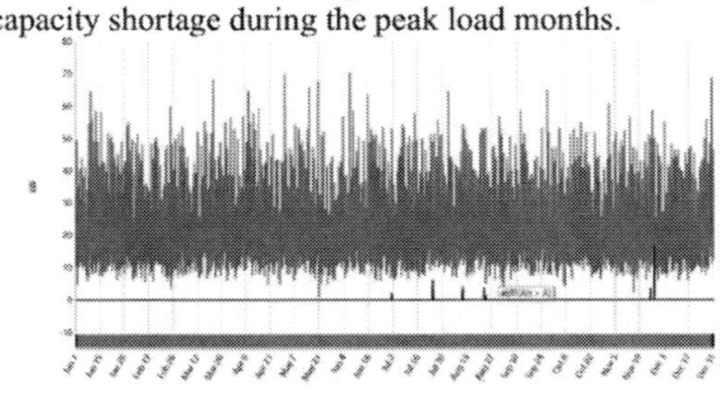

Fig. 8 The comparison curve of the total annual load demand and the annual load shortage

When the wind-solar-hybrid-battery power generation system is optimally configured, according to the data in Table 5, it can be seen that in the determined configuration results, wind power accounts for 49.6% occupying the backbone of annual power generation. The sources of renewable energy power generation for 12 months of the year are shown in Figure 9. It can be seen that the output of hydroelectric generator is mainly concentrated from December to June; In April, June and July (the wind speed is small), the output is relatively small.

Tab. 4 System renewable energy power generation data

Renewable energy	power generation (kWh/year)	Proportion of power generation (%)
Photovoltaic	131143	17.5
Wind turbine	372806	49.6
Hydroelectric generator	247401	32.9
Total	751350	100

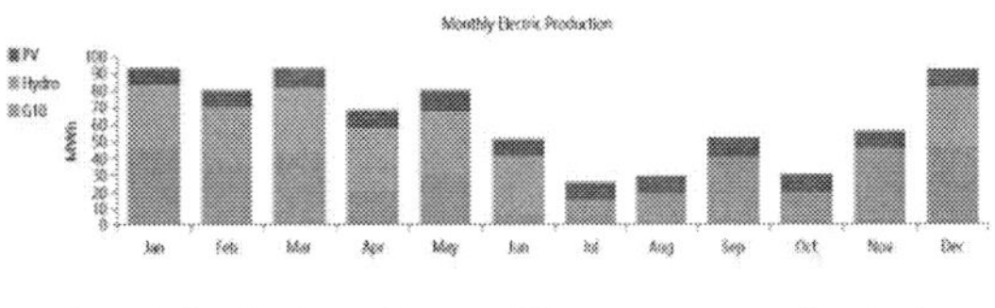

Fig. 9 Distribution of renewable energy generation in the system

5. Conclusions

In this paper, the HOMER software for optimizing the power generation system is used to establish a wind-solar-hydro-storage power generation system with multi-energy complementary in the rich resources research area. According to the selected parameter data of each equipment, the data of renewable energy resources in a certain place and the load demand, the constraints such as the relevant parameters of the research system, sensitive variables and the range of capacity variables of each part are determined. Based on the relevant conditions, the simulation of the whole life cycle of the research object in hours is carried out. Under the premise of the lowest life cycle cost, the optimal configuration combination of wind-solar-water-storage that meets the requirements of designing a complementary power generation system is obtained, so costs and load shortage rates are significantly optimized. Therefore, it is proved that the wind-solar-water-battery system containing a variety of renewable energy has broad development prospect and contributes to the sustainable development of the energy system in the future.

References

1. FAISAL M, HANNAN M A, KER P J,et al. Review of Energy Storage System Technologies in Microgrid Applications: Issues and Challenges [J]. Ieee Access, 2018, 6: 35143-64.
2. ZHANG Y, LIAN J, MA C,et al. Optimal sizing of the grid-connected hybrid system integrating hydropower, photovoltaic, and wind considering cascade reservoir connection and photovoltaic-wind complementarity [J]. Journal of Cleaner Production, 2020, 274.
3. WANG K Y, LUO X J, WU L, et al. Optimal Dispatch of Wind-hydro-thermal Power System With Priority Given to Clean Energy [J]. Proceedings of the CSEE (Natural Science Edition), 2013, 33(13):27-35.
4. XIAO B, ZHANG Y, HAN J,et al. A multi-energy complementary coordinated dispatch method for integrated system of wind-photovoltaic-hydro-thermal-energy storage [J]. International Transactions on Electrical Energy Systems, 2019, 29(7).
5. JIANG H, DU E, ZHU G, et al. Review and Prospect of Seasonal Energy Storage for Power System with High Proportion of Renewable Energy[J].

Automation of Electric Power Systems (Natural Science Edition), 2020, 44(19):194-207.

6. QU X, WU M, LI Q, et al. Review on Comprehensive Evaluation of Multi-energy Complementary Integrated Energy Systems[J]. Electric Power (Natural Science Edition), 2021, 54(11):153-63.
7. XU L, RUAN X, ZHANG B, et al. An Improved Optimal Sizing Method for Wind-solar-battery Hybrid Power System[J]. Proceedings of the Chinese Society for Electrical Engineering (Natural Science Edition), 2012, 32(25):88-98+14.
8. LI X, ZOU F, LI X, et al. Multi-objective Hierarchical Optimal Configuration Method for Microsources of Stand-alone Microgrid[J]. High Voltage Apparatus (Natural Science Edition), 2018, 54(10):170-8.
9. YU D, ZHANG J, WANG X, et al. Optimal Capacity Configuration of Grid-connected Wind-PV-storage Hybrid Power Generation System[J]. Proceedings of the CSU-EPSA (Natural Science Edition), 2019, 31(10):59-65.
10. GONG Q, FANG J, QIAO H,et al. Optimal Allocation of Energy Storage System Considering Price-Based Demand Response and Dynamic Characteristics of VRB in Wind-PV-ES Hybrid Microgrid [J]. Processes, 2019, 7(8).
11. GE S, LI J, LIU H,et al. Research on Operation-Planning Double-Layer Optimization Design Method for Multi-Energy Microgrid Considering Reliability [J]. Applied Sciences-Basel, 2018, 8(11).
12. TAN Y, LYU Z, LI S. Hybrid micro-grid energy optimization and sensitivity analysis of influencing factors[J]. Journal of Guangxi University (Natural Science Edition) (Natural Science Edition), 2015, 40(05):1210-9.

A Short-term Wind Power Prediction Framework based on Two-layer Decomposition and the Combination of Ensemble Model and Deep Network

Feichao Di [1, 2, 3]

[1] CCCC Wuhan Harbor Engineering Design and Research Institute Co., Ltd, Hubei Wuhan, 430040, China
[2] Hubei Provincial Key Laboratory of New Materials and Maintenance and Reinforcement Technology for Offshore Structures, China
[3] Transportation Industry Transportation Infrastructure Intelligent Manufacturing Technology R&D Center, China

Abstract. The time series of wind power is influenced by many external factors, showing strong volatility and randomness. Aiming at the problem of low prediction accuracy of wind power time series, this paper proposes a short-term wind power prediction framework based on two-layer decomposition and the combination of ensemble model and deep network, which is composed of complete ensemble empirical mode decomposition (CEEMD), sample entropy (SE), stacking ensemble, linear regression (LR), variational mode decomposition (VMD), long short term memory (LSTM) and multi-layer perceptron (MLP). Firstly, CEEMD is used to decompose the time series of wind power into different modes and then SE is used for reconstruction. Secondly, different models are applied to predict the different reconstruction components and select the optimal model. Subsequently, VMD is used to decompose the partially decomposed reconstruction components and a combined prediction model of stacking ensemble and LSTM is established. Finally, in order to further improve the prediction accuracy, MLP is applied to correct the error and the corrected error is superimposed with the prediction results and other reconstruction components to obtain the final predicted value. The simulation results show that the accuracy and effectiveness of this model is superior to the traditional model and the prediction accuracy of short-term wind power time series is improved effectively.

Key words: Wind power prediction; complete ensemble empirical mode decomposition; variational mode decomposition; stacking ensemble; long short term memory; multi-layer perceptron.

1. Introduction

With the development and continuous progress of society, people's demand and requirements for energy continue to increase. Traditional fossil fuels are far from meeting people's needs. The emergence of new energy can not only solve the problem of energy shortage, but also effectively alleviate the environmental issues such as the greenhouse effect and air pollution. Wind energy is the most widely used new energy with the largest application prospects. It is of great significance to use wind energy reasonably and realize wind power generation. However, due to the uncertainty and volatility of wind, when wind power generation is carried out, it will bring a severe test to the safe and stable operation of the power grid and the safety and reliability of the grid connection of the power system. In order to reduce the impact of wind power generation on the power system, it is of great significance to achieve accurate prediction of wind power in wind farms [1-4].

At present, domestic and foreign researchers have mainly divided the methods of wind power prediction into physical method [5-8], statistical method [9-13] and combined prediction method [14-16]. The physical method is based on the wind speed, wind direction, longitude, temperature, air pressure and other impact factors to realize wind power prediction. Due to the large number of impact factors, it is usually necessary to perform feature selection; the statistical method is based on the time series of historical wind power, using methods such as data mining and deep feature analysis to realize wind power forecast; the combined forecasting method is currently the most popular method, which often manifested as the ensemble of multiple models such as the combination of shallow model and deep model, to improve the prediction accuracy to a certain extent. Currently, support vector machine [17-18], neural networks [19-21] and deep networks [22-26] are often used in wind power prediction. It proposes an ultra-short-term wind power forecasting method combining multi-cluster algorithm and hierarchical clustering algorithm, which clustering historical power sequences and historical meteorological sequences and establishing particle swarm optimization-back propagation (PSO-BP) neural network

 https://doi.org/10.1051/e3sconf/202235801040

prediction model separately to obtain the final prediction result in reference [27]. It reports a LSTM prediction model based on multiple meteorological features and error correction of the final prediction result to obtain the final prediction value in reference [28]. In reference [29], Ying-Yi Hong proposes a deep feature extraction of wind power time series by convolutional network (CNN) and then establishes an LSTM prediction model to realize the prediction of wind power. Reference [30] proposes a new short-term wind power prediction method consisting of partial direct prediction and iterative prediction through the combination of chaotic time series analysis and singular spectrum analysis (SSA).

2. Experimental Principle and Methods

2.1 COMPLETE ENSEMBLE EMPIRICAL MODE DECOMPOSOTION(CEEMD)

Complete ensemble empirical mode decomposition (CEEMD) is an improvement of empirical mode decomposition (EMD) and integrated empirical mode decomposition (EEMD). EMD can decompose the fluctuation and change trend of the original signal, making the signal tends to be stable. CEEMD adds a set of positive and negative white noise to the original signal, which can not only suppress the EMD mode aliasing problem, but also make the residual noise maintain on a very small scale. The main steps are as follows:

Step 1: Add n sets of positive and negative white noises $m^+, m^-$to the original signal x(t) to form the composite signals P(t) and N(t).

$$\mathrm{P(t)} = \mathrm{x(t)} + m^+$$
$$\mathrm{N(t)} = \mathrm{x(t)} + m^-$$

Step 2: Make empirical mode decomposition on the composite signals, where the jth IFM component of the ith signal is expressed as c_{ij}.

Step 3: Calculate the average value of all inherent modal function (IMF) components to obtain the decomposition result of CEEMD, as shown in Equation 1.

$$C_j = \frac{1}{2n}\sum_{i=1}^{2n} c_{ij} \quad (1)$$

2.2 SAMPLE ENTROPY(SE)

Sample entropy (SE) is a method of measuring the complexity of time series. The larger the entropy value, the higher the complexity of the time series. The SE does not depend on the data length and has better consistency. The calculation of the sample entropy value is shown in Equation 2.

$$\mathrm{SampEn(k, r, L)} = -\ln \frac{B^{k+1}(r)}{B^k(r)} \quad (2)$$

Where k is the dimension, r is the threshold, L is the length, and $B^k(r)$ is the probability of matching k points when the threshold is r.

2.3 VARIATIONAL MODE DECOMPOSITION(VMD)

Variational mode decomposition (VMD) is a new signal decomposition method. Its purpose is to decompose a signal into multiple discrete modes $u_k(\mathrm{k} = 1,2,3, \ldots. \mathrm{K})$. For each mode, the state u_k surrounds the center frequency ω_k and the VMD method mainly includes the solution and construction of the variational problem. To decompose the original time series into a series of modal functions with limited bandwidth, the main process is as follows:

Step 1: Hilbert transform is performed for each mode u_k to obtain a single-sided spectrum.

Step 2: For each mode u_k, according to the estimated center frequency, adjust the spectrum to the corresponding base band.

Step 3: Using the Gaussian smoothness of the demodulated signal to estimate the bandwidth of each mode, the constraint variation problem is expressed as Equation 3.

$$\min\{\sum_{k=1}^{K} \left\|\partial_t\left[\left(\delta(t)+\frac{j}{\pi t}\right)u_k(t)\right]e^{-j\omega_k t}\right\|_2^2\} \quad (3)$$

$$\mathrm{s.t.} \sum_{k=1}^{K} u_k = f(t)$$

Step 4: Introduce the second penalty factor C and Lagrange multiplier θ to transform the constrained optimization problem into an unconstrained optimization problem, as shown in Equation 4.

$$\mathrm{L}(u_k, w_k, \theta) = \mathrm{C}\{\sum_{k=1}^{K} \left\|\partial_t\left[\left(\delta(t)+\frac{j}{\pi t}\right)u_k(t)\right]e^{-j\omega_k t}\right\|_2^2\} + \|f(t) - \sum_{k=1}^{K} u_k\|_2^2 + \langle\theta(t), f(t) - \sum_{k=1}^{K} u_k\rangle \quad (4)$$

Step 5: Solve the problem by using the alternating direction multiplier algorithm (ADMM), and get the updated formulas of u_k and w_k.The specific expressions are shown in Equation 5 and Equation 6.

$$\hat{u}_k^{n+1} = \frac{\hat{f}(\omega) - \sum_{k=1}^{K}\hat{u}_k(\omega) + \frac{\hat{\theta}(\omega)}{2}}{1+2C(\omega-\omega_k)} \quad (5)$$

$$\omega_k^{n+1} = \frac{\int_0^\infty \omega|\hat{u}_k(\omega)|^2 d\omega}{\int_0^\infty |\hat{u}_k(\omega)|^2 d\omega} \quad (6)$$

2.4 STACKING ENSEMBLE

The Stacking ensemble model is currently the most widely used model in data mining competitions. It is a method of classification and regression using the fusion of mu ltiple models. The key lies in the selection of the base model and the number of cross-validation. The selection requires that the effects of the base models are close and the correlation is low. The main flows of the Stacking model are as follows:

Step 1: Divide the dataset into training set and test set.

Step 2: Cross-validate the training set to obtain the prediction result of the validation set and the test set.

Step 3: Calculate the weighted average of the test set prediction results, the same operation is performed according to the number of base models, until a new training set and test set are constructed.

2.5 LONG SHORT TERM MEMORY(LSTM)

Long short term memory (LSTM) is an improvement of the recurrent neural network. It is an improved network proposed to solve the gradient disappearance or gradient explosion of the recurrent neural network (RNN). The LSTM consists of an input gate, a forget gate, and an output gate, as shown in Figure 1.

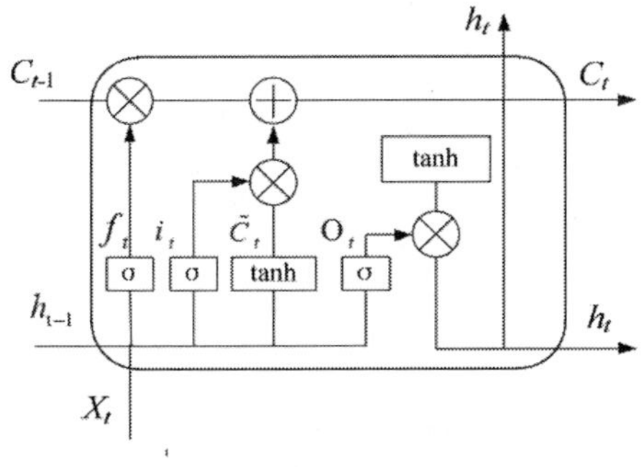

Figure 1. The block diagram of LSTM

Among the three gates, the forget gate determines which information will be discarded, as shown in Equation 7.

$$f_t = \sigma(w_f \cdot [h_{t-1}, X_t] + b_f) \quad (7)$$

The input gate determines which information is retained, and receives new input at the same time, and updates the control parameter C_t, as shown in Equations 8,9,10.

$$i_t = \sigma(w_i \cdot [h_{t-1}, X_t] + b_i) \quad (8)$$

$$\widetilde{C}_t = tanh(w_c \cdot [h_{t-1}, X_t] + b_c) \quad (9)$$

$$C_t = f_t * C_{t-1} + i_t * \widetilde{C}_t \quad (10)$$

The output gate generates the output of the LSTM at the current time according to the control parameter C_t, as shown in Equations 11 and 12.

$$O_t = \sigma(w_o \cdot [h_{t-1}, X_t] + b_o) \quad (11)$$

$$h_t = O_t * \tanh(C_t) \quad (12)$$

Where X_t is the input, h_t is the output, h_{t-1} is the output at the last time, C_{t-1} is the control parameter at the last time, C_t is the latest control parameter, i_t is the information to be saved.

2.6 MULTILAYER PERCEPTRON(MLP)

Multi-layer perceptron (MLP) is a simple multi-layer neural network, the hierarchy of MLP is a directed graph. Each layer is fully connected to the next layer, and the output of the neuron in the previous layer is the input of the neuron in the next layer. The key to MLP lies in the number of hidden layers, the number of neurons in the hidden layer, and the connection weights and offsets, which are often adjusted by grid search and other methods.

3. Establishment Of Prediction Model

3.1 DATA PREPROCESSING

In order to eliminate the dimensional influence between the data, we can make the data be normalized. After the original data is normalized, the data will be in the same dimension. When the input of the model is the same dimension, the operating efficiency of the model will be greatly improved.

The equation for data normalization is shown in the Equation 13:

$$X^* = \frac{x - x_{min}}{x_{max} - x_{min}} \quad (13)$$

Where X_{max} is the maximum value of the sample data, X_{min} is the minimum value of the sample data.

3.2 ONE-LAYER DECOMPOSITION

Due to the large volatility of wind power time series, in order to improve the prediction accuracy, many signal decomposition techniques are widely used in wind power prediction. At present, EMD, EEMD, WT, VMD and so on are widely used. In the one-layer decomposition, this paper uses complete ensemble empirical model decomposition (CEEMD), which can effectively suppress the modal aliasing problem of EMD and make the residual component always keep at a very small level. Also, the original wind power data can be completely decomposed without loss. Calculating the sample entropy value (SE) of each IMF and reconstructing the components which the SE are closed to form a new reconstructed component. The process of one-layer decomposition is shown in Figure 2.

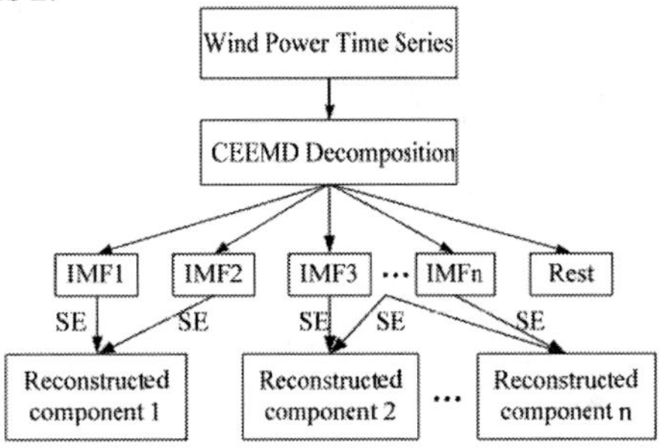

Figure 2. The structure diagram of CEEMD + SE

After obtaining different reconstructed components, Analyzing the reconstructed components and selecting different models for prediction according to the characteristics of the different components, Finally, we select the optimal model .

3.3 TWO-LAYER DECOMPOSITION

When the volatility of one of the reconstructed components is large, a two-layer decomposition is performed on the basis of the one-layer decomposition. The method of two-layer decomposition in this paper adopts the variational mode decomposition (VMD) technology. After the two-layer decomposition, the reconstructed component obtains k modes and the mode number k is determined by the center frequency ω_k of each mode.

For each different mode, the combined model of stacking and LSTM is used to make predictions and the final prediction results are accumulated to obtain the wind power prediction results. Among them, the selection of the basic model of the stacking model are random forest (Random Forest), Xgboost and LightGBM three ensemble models, making full use of the characteristics of the three basic models and constructing new features as the input of the LSTM model to predict wind power, which is shown in Figure 3. All prediction results are superimposed to obtain the final prediction value. Since certain prediction errors will occur, in order to further improve the prediction accuracy, the prediction errors are corrected by the MLP model. The specific process of the

https://doi.org/10.1051/e3sconf/202235801040

two-layer decomposition of the reconstructed components is shown in Figure 4.

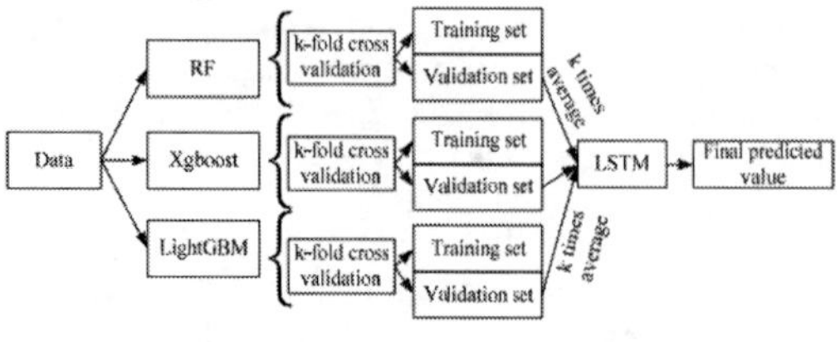

Figure 3. The flowchart of Stacking ensemble model

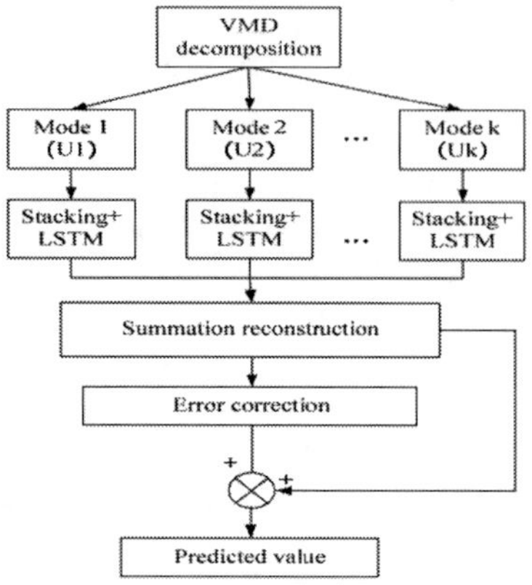

Figure 4. The flowchart of two-layer decomposition

A short-term wind power prediction framework based on two-layer decomposition and the combination of ensemble model and deep network is shown in Figure 5.

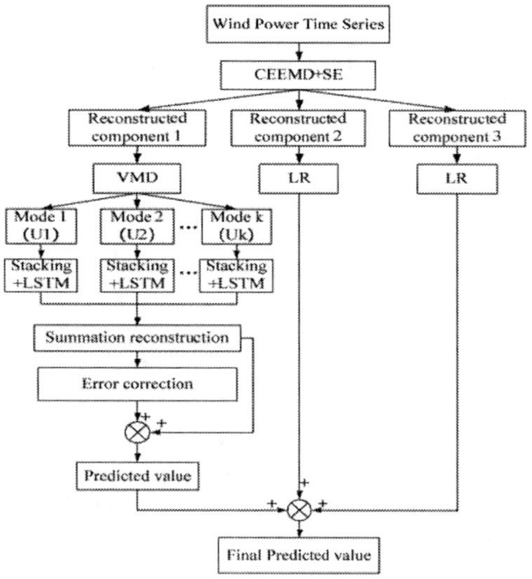

Figure 5. short-term wind power prediction model based on two-layer decomposition and the combination of ensemble model and deep network

The main steps of the short-term wind power prediction model based on two-layer decomposition and the combination of ensemble model and deep network are:

Step 1: Perform CEEMD decomposition and SE reconstruction on the original wind power time series to form reconstructed component 1, reconstructed component 2, reconstructed component 3, and the residual component can be directly ignored because of its small magnitude.

Step 2: For reconstructed component 2 and reconstructed component 3, analyzing their data characteristics and selecting different models for comparison and prediction, we decide to use linear regression (LR) model on the reconstructed component 2 and reconstructed component 3.

Step 3: Because the volatility of reconstructed component 1 is relatively large, VMD is used for two-layer decomposition and the combined prediction models of stacking and LSTM are used for each mode, and error correction is performed using MLP to obtain the final predicted value of reconstructed component 1.

Step 4: Accumulate the predicted results of the reconstructed components 1, 2 and 3 to obtain the final predicted value of wind power.

3.4 D. EVALUATION FUNCTION

The evaluations function of the prediction model selected in this paper are:

a. Root mean square error:

$$\mathrm{RMSE} = \sqrt{\frac{1}{m}\sum_{i=1}^{m}(y_i - \hat{y}_i)^2}$$

b. Mean absolute error:

$$\mathrm{MAE} = \frac{1}{m}\sum_{i=1}^{m}|y_i - \hat{y}_i|$$

c. R-Square:

$$R^2 = 1 - \frac{\sum_{i=1}^{m}(y_i - \hat{y}_i)^2}{\sum_{i=1}^{m}(y_i - \bar{y})^2}$$

Where y_i is the actual value of the wind power at the i-th moment, $\hat{y}_i$ i is the predicted value of the wind power at the i-th moment, and $\bar{y}$ is the average value of the m wind power values.

4. Case studies

In this paper, we use a wind farm data in the UK for short-term wind power prediction. In order to prove the validity of the model, this paper selects a total of 4,490 wind power data from the wind farm at 1:00 on March 1, 2018 to 8:00 on April 14, 2018.

In the experiment, the sampling interval between the experimental data points is 10 minutes and the time series of the wind power obtained is shown in Figure 6. Among them, the first 3995 data points are used as the training set and the last 495 data points are used as the test set. The previous 5 wind power data predict the wind power at the next moment to achieve rolling prediction.

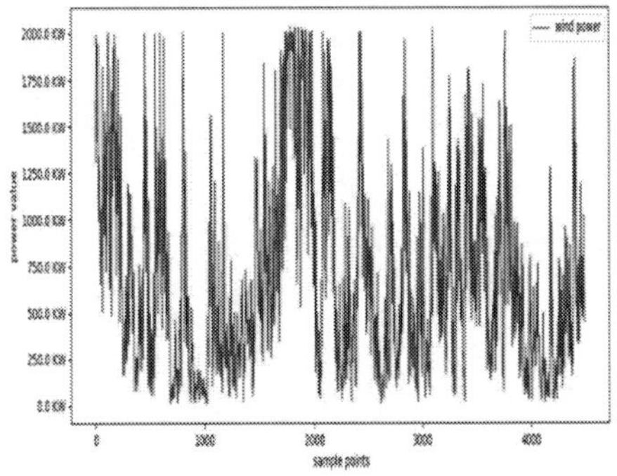

Figure 6. Time series diagram of wind power

It can be seen from Figure 6 that the time series of wind power has obvious fluctuation, so CEEMD decomposition is performed on it, and the specific decomposition results are shown in Figure 7.

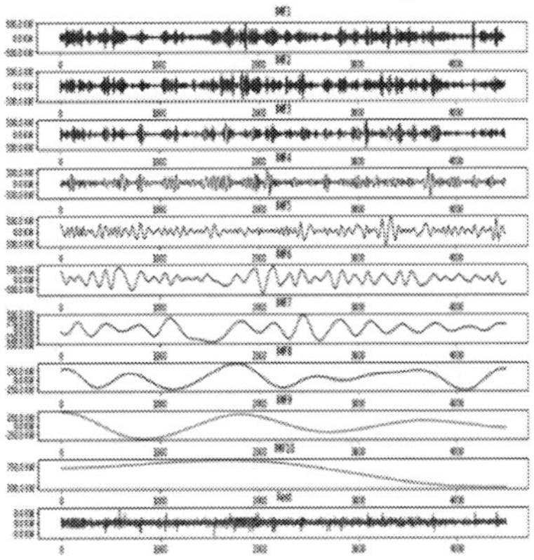

Figure 7. The graph of CEEMD decomposition results

Because that the residual component Rest is always on a very small scale, only the sample entropy values of the decomposed components (IMF1-IMF10) need to be calculated, the sample entropy threshold r is set to 0.2, the length L is set to 2, and each component The sample entropy values are shown in Table 1.

Table 1. Sample entropy of each component

	IM	IM	IM	IM	IM	IM	IM	IM	IM	IM
	F1	F2	F3	F4	F5	F6	F7	F8	F9	F10
S	5.7	5.9	5.7	5.4	4.7	3.9	2.5	1.3	1.0	0.1
E	57	43	00	02	97	77	90	00	53	00

It can be drawn from Table 1 that the sample entropy values of IMF1-6 are close, so the components are superimposed to obtain the reconstructed component 1; IMF6-9 is used as the reconstructed component 2 and IMF10 is used as the reconstructed component 3.

For the reconstructed component 1, the graph is shown in Figure 8.

It can be seen from Figure 8 that the time series of wind power still exhibits a certain degree of non-stationarity, so VMD decomposition is performed on the reconstructed component 1 and the center frequency method is used to determine the number of modes k. The general value range of k is 3~ 8. The center frequency when k takes different values is shown in Table 2.

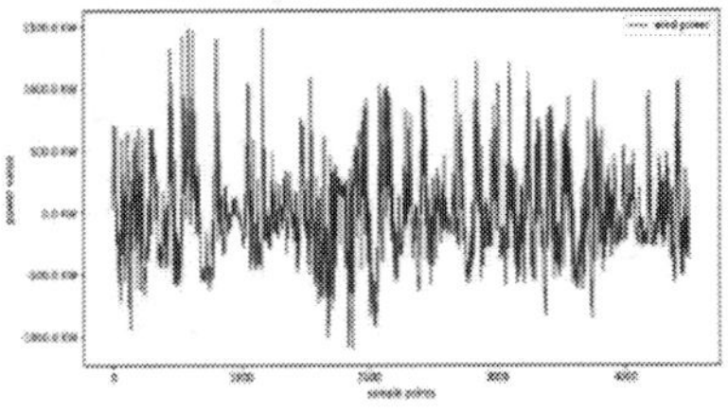

Figure 8. The graph of reconstruction component 1

Table 2. Center frequencies for different values of k

	u1	u2	u3	u4	u5	u6	u7
K=3	9.147	48.973	139.975	-	-	-	-
K=4	7.787	31.679	78.593	156.950	-	-	-
K=5	7.720	30.964	76.114	150.410	263.141	-	-
K=6	7.673	30.466	74.291	145.673	235.520	354.132	-
K=7	7.202	25.636	54.285	93.459	156.300	251.171	359.550

It can be seen from the center frequency in Table 2 that when k=7, the center frequency appears 25.636 and 54.285, the two center frequencies are relatively close and the number of modes of VMD decomposition can be determined to be 6.As is shown
in Figure7, Comparing with the six modes before decomposition, the wind power is decomposed two times, which reduces the fluctuation of wind power and has better stability. The results of VMD decomposition are shown in Figure 9.

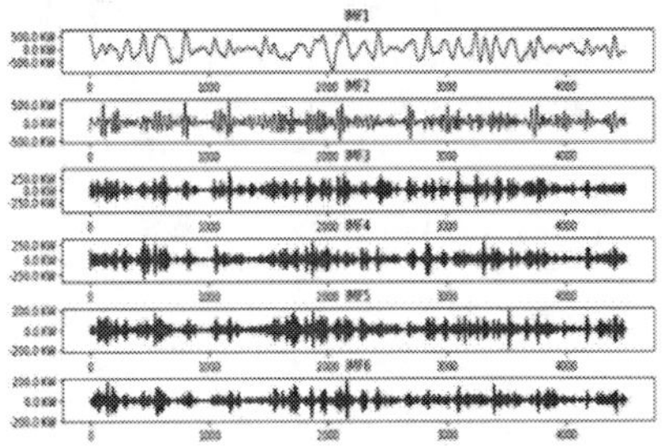

Figure 9. The graph of VMD decomposition results

When training and predicting each mode, we build a combined prediction model of stacking and LSTM. The three base models are random forest (RF), xgboost and LightGBM. Among them, the parameters of each model are determined by grid search, where the max depth of three base models is 4, the learning rate is 0.1, LSTM is a 4-layer network structure, the number of neurons in each layer is 50, dropout is 0.1, the optimizer is adam and The number of iterations is 50. when MLP is used for error correction, the number of MLP network layers is 4 and the number of neurons in each layer is 100, 100, 100, and 50 respectively.

Table 3 lists the predicted evaluation indexes of four different models and Figure 10 shows the predicted results of the four models.

https://doi.org/10.1051/e3sconf/202235801040

Table 3. Evaluation indicators of the four models

Model	RMSE(KW)	MAE(KW)	R-Square
LSTM	155.71	97.57	0.687
stacking-LSTM	153.42	96.62	0.801
VMD-stacking-LSTM	62.35	43.21	0.949
VMD-stacking-LSTM-MLP	**12.21**	**8.87**	**0.998**

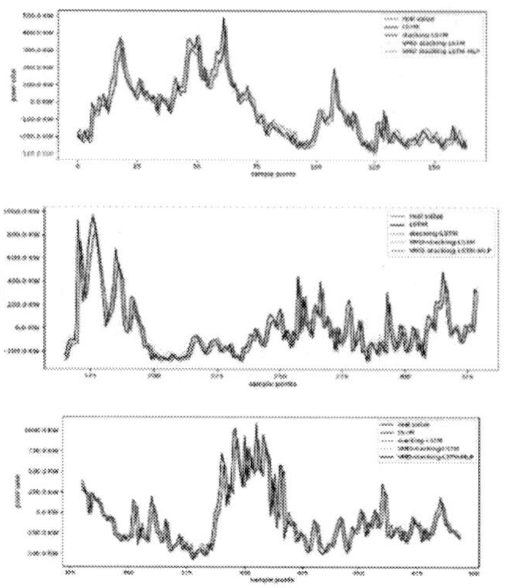

Figure 10. The graph of four model prediction curves

According to the predicted results in Figure 10 and Table 3, we can conclude that for the reconstructed component 1, the four models can both predict the wind power at the next moment. Compared with the one-layer decomposition and the two-layer decomposition, the MSE is reduced by 91.07 KW and MAE are reduced by 53.41KW. In terms of predicted accuracy, VMD-stacking-LSTM-MLP has the highest predicted accuracy and R-Square achieves 0.998, indicating that this model has better fitting ability and better predicted effect than other models.

For the reconstructed component 2 and the reconstructed component 3, the graphs of the two reconstructed components are shown in Figure 11 and Figure 12.

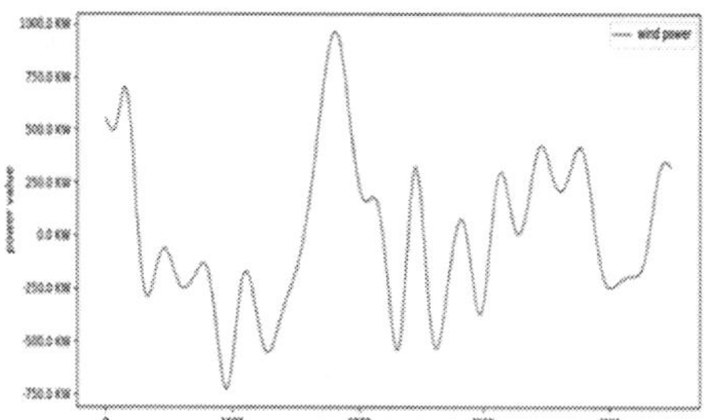

Figure 11. The graph of reconstruction component 2 curve

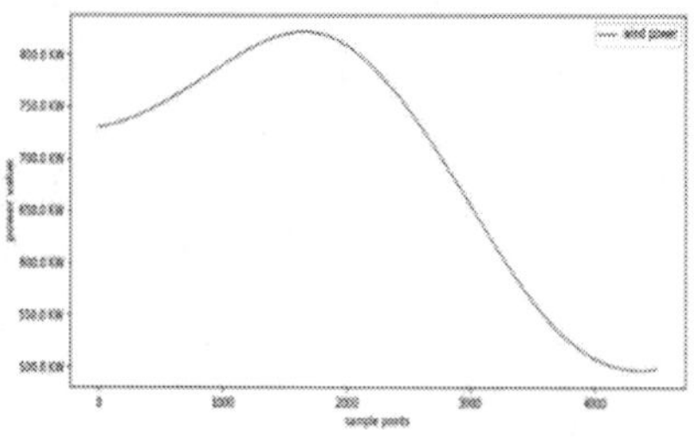

Figure 12. The graph of reconstruction component 3 curve

As can be seen from Figure 11 and Figure 12, the curve relationship between the reconstructed component 2 and the reconstructed component 3 has a certain linear relationship, because of its potential data characteristics, a linear regression (LR) model is selected for prediction and the predicted effect is the best. As shown in Figure 13, Figure 14.

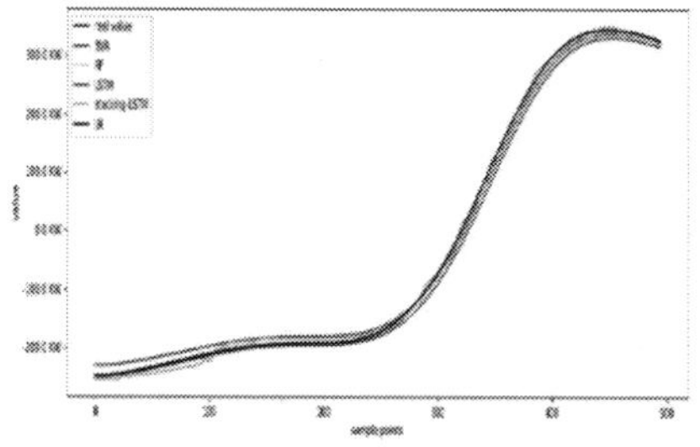

Figure 13. The graph of reconstruction component 2 prediction curve

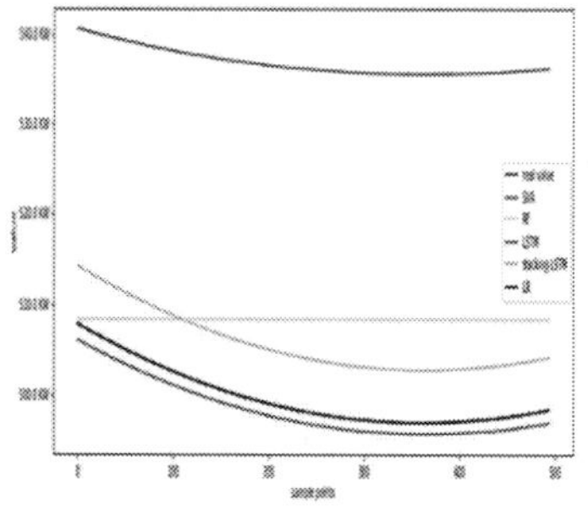

Figure 14. The graph of reconstruction component 3 prediction curve

It can be drawn from Figure 13 and Figure 14, compared with other prediction models, the predicted effect of the linear regression (LR) model on the reconstructed component 2 and the reconstructed component 3 is almost completely fitted with the true values, which shows the predicted effect of the model is the best.

Because of the CEEMD decomposition is a complete decomposition and almost no loss. After the predicted values of each reconstructed component are obtained, they are accumulated to obtain the final predicted value of wind power. Table 4 shows the predicted indexes of the 7 models. The graphs of the final prediction results of the 7 models are shown in Figure 15. Among them, Model 1 uses the SVR model for prediction directly; Model 2 uses the LSTM model for prediction directly; Model 3 uses the LSTM model for prediction after one –layer decomposition; Model 4 uses a combination of stacking and LSTM for prediction after one-layer decomposition of CEEMD; Model 5 uses stacking and LSTM for combined prediction after one-layer decomposition of VMD; Model 6 uses stacking and LSTM combined prediction after two-layer decomposition; Model 7 uses LR, stacking, LSTM and MLP after two-layer decomposition combination forecast.

Tab 4. Evaluation indicators of the seven models

Model	RMSE(KW)	MAE(KW)	R-Square
Model 1	168.31	124.79	0.753
Model 2	165.75	115.64	0.768
Model 3	161.24	100.62	0.773
Model 4	153.42	98.68	0.795
Model 5	81.08	54.78	0.942
Model 6	62.34	43.22	0.966
Model 7	**13.15**	**9.44**	**0.998**

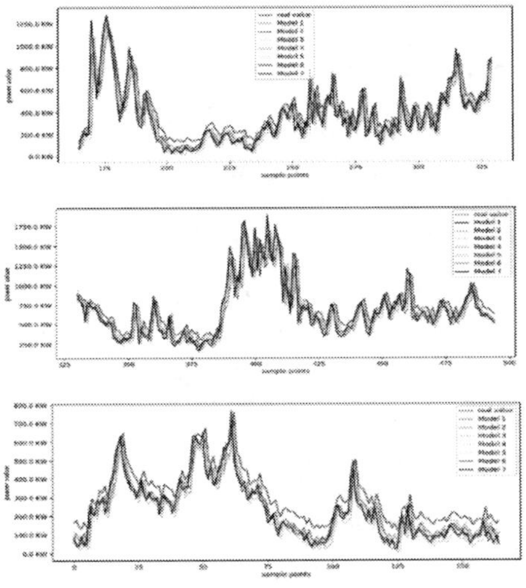

Figure 15. The graph of seven model prediction curves

Analyzing Table 4 and Figure 15, we can get:
(1) Compared with the SVR model, The RMSE and MAE of the LSTM model is reduced by 2.56KW and 9.15KW respectively.
(2) After one-layer decomposition of CEEMD, RMSE and MAE are reduced by 4.51KW and 15.02KW respectively and R-Square is increased to 0.773.
(3) After one-layer decomposition of CEEMD, using stacking integration and LSTM combined prediction, both RMSE and MAE are reduced, which indicates that the combined prediction model has a strong fitting ability than the single model.
(4) The second decomposition of VMD based on the first decomposition of CEEMD greatly improves the prediction accuracy of the model. RMSE and MAE are reduced by 91.08KW and 55.46KW respectively and R-Square is also greatly improved to 0.966.
(5) After MLP correction of the error, RMSE and MAE were reduced to 13.15KW and 9.44KW respectively and R-Square is increased to 0.998, which indicates that the correction of the error can improve the prediction accuracy of the model.
Therefore, compared with other models, the prediction model proposed in this paper can better predict the short-term wind power changes and the predicted accuracy of the model has also been greatly improved, which verifies the feasibility and effectiveness of the proposed method.

5. Conclusion

This paper proposes a short-term wind power prediction based on the two-layer decomposition and the combination of ensemble model and deep network and draws the following conclusions:
a. Signal decomposition techniques such as CEEMD, VMD can effectively improve the predicted accuracy of the model.
b. The predicted results obtained by the combined prediction model are significantly better than the single prediction model.
c. Perform feature analysis on the data to find the most suitable model for prediction, which can not only get better predicted results, but also speed up the operation efficiency of the model.
d. When it is found that the prediction error shows a certain regularity, an appropriate model can be used to train and correct the error, which can further improve the predicted accuracy of the model.
At present, the extensive use of ensemble models and combined forecasting models, the combined use of shallow models and deep models and the correction of errors are the main development directions of forecasting models. In the future, we can do major research in these three areas.

References

1. Liu Shuai. Research on Short-Term Power Forecasting Methods for Wind Farms [D]. Beijing: North China Electric Power University, 2019.
2. Liang Zhifeng, Wang Yan, Feng Shuanglei, et al. Ultra-short term forecasting method of wind power based on fluctuation law mining [J]. Power System Technology, 2020, 25 (10): 1-10.
3. Peng Xiaosheng, Xiong Lei, Wen Jinyu, et al. Summary of improved short-term and ultra-short-term power prediction accuracy methods for wind power clusters [J]. Chinese Journal of Electrical Engineering, 2016, 36 (23): 6315-6326 + 6596.
4. Lu Peng, Ye Lin, Tang Yong, et al. Multi-time scale active power optimization scheduling strategy for wind power cluster based on model predictive control

[J]. Proceedings of the CSEE, 2019, 39 (22): 6572-6583.

5. Wang Yining, Xie Da, Wang Xitian, et al. Prediction of wind turbine network interaction based on PCA-LSTM model [J]. Chinese Journal of Electrical Engineering, 2019, 39(14): 4070-4081.
6. Song Jiakang, Peng Yonggang, Cai Hongda, et al. Study on short-term wind power prediction considering multi-site NWP and atypical characteristics[J]. Power System Technology, 2018, 42 (10): 3234-3242.
7. J. Yan, H. Zhang, Y. Liu, et al. Forecasting the High Penetration of Wind Power on Multiple Scales Using Multi-to-Multi Mapping [J]. *Transactions on Power Systems*, 2018, 3276-3284.
8. Wang Youjia, Lu Zongxiang, Qiao Ying, et al. Statistical upscaling prediction of regional wind power based on feature clustering [J]. Power System Technology, 2017, 41 (05): 1383-1389.
9. Wu Xiaomei, Lin Xiang, Xie Xuquan, et al. Short-term wind power prediction based on VMD-PE and optimized correlation vector machine [J]. Journal of Solar Energy, 2018, 39 (11): 3277-3285.
10. Zhu Qiaomu, Li Hongyi, Wang Ziqi, et al. Ultra-short-term prediction of wind farm power generation based on long-short-term memory network [J]. Power System Technology, 2017, 41 (12): 3797-3802.
11. Mu Gang, Yang Xiuyu, Yan Qian, et al. Prediction method of field group continuous power characteristics based on the evolution trend of wind farm group convergence [J]. Chinese Journal of Electrical Engineering, 2018, 38 (S1): 32-38.
12. S.Tian, Y. Fu, P. Ling, et al. Wind Power Forecasting Based on ARIMA-LGARCH Model[J]. 2018 International Conference on Power System Technology, Guangzhou, 2018, pp. 1285-1289.
13. E. Yatiyana, S. Rajakaruna, A. Ghosh. Wind speed and direction forecasting for wind power generation using ARIMA model[J]. 2017 Australasian Universities Power Engineering Conference, 2017, pp. 1-6.
14. Yin Hao, Ou Zuhong, Chen De, et al. Ultra-short-term wind power prediction based on quadratic mode decomposition and cascading deep learning[J]. Power System Technology, 2020 (02): 445-453.
15. Hao CHEN, Fangxing LI, Yurong WANG. Wind power forecasting based on outlier smooth transition autoregressive GARCH model[J]. Journal of Modern Power Systems and Clean Energy, 2018, 6(3).
16. Yu Hua, Lu Jiping, Zeng Yanting, et al. Nonlinear combined prediction of wind power based on different optimization criteria and generalized regression neural network [J]. High Voltage Technology, 2019, 45 (03): 1002-1008.
17. Y. Zhang, H. Sun, Y. Guo, Wind Power Prediction Based on PSO-SVR and Grey Combination Model[J], IEEE Access, 2019, pp. 136254-136267.
18. Li Yanqing, Yuan Yanwu, Guo Tong. Ultra-short-term wind power combination prediction based on AMD-ICSA-SVM [J]. Power System Protection and Control, 2017, 45 (14): 113-120.
19. H. J. Lu and G. W. Chang. Wind Power Forecast by Using Improved Radial Basis Function Neural Network, 2018 IEEE Power & Energy Society General Meeting, Portland, OR, 2018, pp. 1-5.
20. Ding Ming, Zhang Chao, Wang Bo. Short-term prediction and error correction of wind power based on the power rate fluctuation process [J]. Automation of Electric Power Systems, 2019, 43 (03): 2-12.
21. J. Varanasi, M. M. Tripathi. Artificial neural network based wind power forecasting in belgium, 2016 IEEE 7th Power India International Conference, 2016, pp. 1-6.
22. Jinhua Zhang, Jie Yan, David Infield, et al. Short-term forecasting and uncertainty analysis of wind turbine power based on long short-term memory network and Gaussian mixture model[J]. Applied Energy, 2019, pp. 229-244.
23. X. Peng, D. Deng, J. Wen, et al. A very short term wind power forecasting approach based on numerical weather prediction and error correction method[C]. 2016 China International Conference on Electricity Distribution , 2016, pp. 1-4.
24. Xiong Yindi, Liu Kaipei, Qin Liang, et al. Short-term wind power prediction method based on dynamic weather classification of time series data [J]. Power System Technology, 2019, 43 (09): 3353-3359.
25. B. Khorramdel, C. Y, Chung, N. Safari, G, C. D. Price, A Fuzzy Adaptive Probabilistic Wind Power Prediction Framework Using Diffusion Kernel Density Estimators, IEEE Transactions on Power Systems, 2018, pp. 7109-7121.
26. T. Ouyang, H. Huang, Y. He, Ramp events forecasting based on long-term wind power prediction and correction, IET Renewable Power Generation, 2019, pp. 2793-2801.
27. Peng Chenyu, Chen Ning, Gao. Ultra-short-term wind power prediction method combining multiple clustering and hierarchical clustering [J]. Automation of Electric Power Systems: 1-8 [2019-11-10].
28. Shuang Han, Yan-hui Qiao, Jie Yan, et al. Mid-to-long term wind and photovoltaic power generation prediction based on copula function and long short-term memory network[J]. Applied Energy, 2019, pp, 181-191.
29. Ying-Yi Hong, Christian Lian Paulo P. Rioflorido. A hybrid deep learning-based neural network for 24-h ahead wind power forecasting[J]. Applied Energy, 2019, pp. 530-539.
30. N. Safari, C. Y. Chung, G.C.D. Price, Novel Multi-Step Short-Term Wind Power Prediction Framework Based on Chaotic Time Series Analysis and Singular Spectrum Analysis, IEEE Transactions on Power Systems, 2018, pp. 590-601.

Feasibility Analysis of Vacuum Pipeline Magnetic Levitation Energy Storage System Based on Existing Magnetic Levitation Transportation Technology

Ziming Fan [1,2], Jun Yang [1,2], Hua Xun [1,2]

[1] Inner Mongolia Power (Group) Co., Ltd., Inner Mongolia Power Research Institute Branch, Hohhot 010020, China
[2] Inner Mongolia Enterprise Key Laboratory of High Voltage and Insulation Technology, Hohhot 010020, China

Abstract: This paper is mainly summarized the research progress of maglev transportation technology. The vacuum pipeline magnetic levitation energy storage system is constructed based on the existing four types of magnetic levitation as technical prototypes, and the four schemes are formed: as followed: electromagnetic suspension (EMS) medium and low speed magnetic levitation type vacuum pipeline energy storage system, EMS high speed magnetic levitation type vacuum pipeline energy storage system, electric suspension (EDS) high speed magnetic levitation type vacuum pipeline energy storage system and super high-speed rail type vacuum pipeline energy storage system. The energy storage and energy storage cost of these four energy storage systems are analyzed to study their energy storage feasibility.

Keywords: Energy storage system; vacuum pipeline; magnetic levitation.

1. Introduction

In order to effectively solve the problems of large-scale grid connection, scheduling and consumption of renewable energy power in China[1, 2], and alleviate the increasingly serious phenomenon of wind and light abandonment based on China's west to east power transmission channel, energy storage technology has been rapidly developed and applied. According to different principles, energy storage technology can be divided into three categories of mechanical energy storage, electrochemical energy storage and electrical energy storage[3-6].

At present, the energy storage methods with mature technology are pumped storage and lithium battery energy storage. However, pumped storage is not easy to locate and lithium battery energy storage capacity is small, and the other energy storage methods are in the stage of demonstration operation or research and development. In addition, all current energy storage technologies are difficult to integrate energy storage characteristics such as large capacity, easy location, low cost, high efficiency and fast response. Therefore, new large-scale electric energy storage technology should be further explored.

As a kind of mechanical energy storage comparing with the mature pumped storage, flywheel energy storage, and air energy storage based on abandoned mines, vacuum pipeline magnetic levitation technology exhibits lots of advantages such as flexible, convenient location, fast response, large monomer capacity, simple control and high efficiency[7-9]. In general, as a new mechanical large-scale energy storage technology, vacuum pipeline magnetic levitation can effectively integrate the advantages of large capacity, easy location and fast response, and the research, development and application about which can provide new ideas for large-scale grid connection, dispatching and consumption of renewable energy power in China[10-13].

The vacuum pipeline magnetic levitation energy storage technology is to combine the advantages of magnetic levitation transportation technology and vacuum pipeline technology through controlling the acceleration and deceleration of the linear motor/generator in the energy storage system, which can realize the mutual conversion between the electric energy of the power grid and the mechanical energy of the energy storage system, and complete the large-scale energy storage application with the characteristics of large capacity, easy location, low cost, high efficiency, fast response, etc.

2. Key technical features of existing maglev transportation technology

According to the different ways of magnetic levitation force generation the maglev train can be divided into two types of electromagnetic suspension (EMS) and electric suspension (EDS). EMS maglev train mainly uses the electromagnetic attraction between electromagnet and track to realize the suspension of vehicle body. Compared with EMS, EDS maglev train realizes the suspension of the vehicle body through the repulsive force generated by the relative movement between the on-board magnet and the track.

So far, more mature transportation technologies include EMS medium and low speed maglev train technology, EMS high speed maglev train technology and EDS high speed maglev train technology. As main application of the maglev transportation technology, super high speed railway is expected to achieve a breakthrough in 20 years[14-19].

The research of EMS maglev train was carried out earlier and received more attentions due to its research results were more mature. After decades of research, EMS maglev train has moved from the initial test line research to the construction of commercial operation line[20-22]. At present, EMS maglev trains put into commercial operation include TR maglev trains in Germany as shown in Fig.1(a), HSST low-speed maglev trains in Japan as shown in Fig.1(b)and CMS series low-speed maglev trains in China as shown in Fig.1(c). In 2001, a commercial operation line using German TR maglev train was built between Shanghai longyang road and pudong airport, which was put into operation in 2004. The line has a total length of about 30 km and a maximum operation speed of 431 km/h. The HSST low-speed maglev train developed by Japan was also put into commercial operation on the commuter line (Eastern hilly line) of the aichi world expo in 2005, with a maximum running speed of 100 km/h. The University of national defense science and technology has carried out research on the application of magnetic levitation technology to rail transit since the 1980s. In 1995, it successfully developed a full-size single bogie system of maglev train that can be stably suspended. In 2001 and 2009, 204 m long and 1.5 km long medium and low-speed maglev test lines were built in Changsha and Tangshan respectively, as shown in Fig.1(d). At the same time, CMS03 test vehicle, CMS03A engineering vehicle and CMS04 practical engineering vehicle have also been successfully developed, and the coupling operation of vehicles has been realized. Among them, CMS04 maglev train can run at a maximum speed of 110 km/h on the straight-line track.

Fig.1 Maglev train

(a) TR maglev trains in Germany (b) HSST low-speed maglev trains in Japan (c) CMS series low-speed maglev trains in China (d) Changsha maglev line in China

After more than 30 years of painstaking research and nearly 10 years of trial operation and testing, the low-speed maglev train technology has gradually matured in China. At present, China's first medium and low-speed maglev transportation demonstration line has begun to construct in Beijing, which is with fully independent intellectual property rights. The medium and low-speed maglev transportation is expected to be opened to traffic by the end of 2016, which marks that China's maglev train technology has entered the commercial operation stage .

The key technical parameters of four different typical maglev transportation are summarized in Table 1. The four typical maglev lines are Changsha maglev line, Shanghai maglev line, Xingan maglev line and Dubai super high speed railway line.

Table.1 The key technical parameters of existing maglev train technology

Maglev line	Speed (km/h)	Load ratio (t/m)	Total length (km)	Cost ratio (HMkm)	Suspension mode	Application characteristics
EMS medium and low speed maglev (Changsha maglev line in China)	100	≈2	18.55	≈2.3	Electromagnetic levitation	Low load, Low speed, Suitable for short distance passenger transport in the city
EMS Type High Speed Maglev (Shanghai maglev line in China)	430	≈2.06	30	≈3	Electromagnetic levitation	Low load, High speed, Suitable for intercity and long-distance passenger transport
EDS type high speed maglev (low-speed maglev trains in Japan)	603	≈3.52	286	≈12.3	Superconducting electric levitation	High load, High speed, High cost, Suitable for intercity and long-distance passenger transport
EDS type high speed maglev (Dubai super high speed rail line)	1200	≈1	10	≈8.8	Permanent magnet electric suspension	Low load, High speed, Suitable for long-distance passenger transport, the technology is at the research stage

3. Feasibility analysis of energy storage based on existing maglev rail transit technology

In this section, the feasibility of energy storage is studied by analyzing the energy storage and energy storage cost of these energy storage systems. The vacuum pipeline magnetic levitation energy storage system will be constructed based on the four existing magnetic levitation technology prototypes as listed in Table 1. The four magnetic levitation vacuum pipeline energy storage systems are EMS medium and low-speed magnetic levitation type vacuum pipeline energy storage system, EMS high-speed magnetic levitation type vacuum pipeline energy storage system, EDS high-speed magnetic levitation type vacuum pipeline energy storage system and super high-speed rail type vacuum pipeline energy storage system respectively.

Assuming that the above four energy storage systems are all operating in an ideal state. The ideal state of the vacuum pipe is always in a full vacuum state, and the electromagnetic resistance is ignored. That means the electromagnetic resistance, pneumatic friction resistance and vacuum maintenance loss of the energy storage system are all 0. The efficiency of linear induction motor and synchronous linear motor in maglev transportation technology are 64% and 91% respectively. Therefore, the mutual conversion efficiency between the electric energy of the power grid and the mechanical energy of the energy storage system based on the linear induction motor and the synchronous linear motor is 40.96% and 82.81%, respectively. In addition, the energy storage calculation is performed by making the vehicle rails of the four energy storage systems equal in length, for the purpose of comparison and analysis with the following text.

Four kinds of vacuum pipeline magnetic levitation energy storage systems are calculated and analyzed based on existing magnetic levitation transportation technology. The results are summarized in Table 2. At present, the largest China Fengning pumped storage power station under construction in the world has a total installed capacity of 3.6 million kilowatts, a total investment of 19.2 billion, and the energy storage system cost of about 5200 ¥/kW. It should be informed that the phase I project of Fengning pumped storage power station in China has an installed capacity of 1.8 million yuan, an investment of 9.9 billion yuan and a tax of 360 million yuan. It is estimated that the annual power generation can reach 2.2 billion yuan and the daily power generation is about 6 million yuan. If the planned operation period is calculated as 50 years, plus various operating costs of the power station within 50 years, the energy storage cost of Fengning pumped storage power station is about 0.1 ¥/kW·h.

Table 2 Analysis table of four energy storage systems based on existing maglev technology

Programme index	EMS medium and low speed maglev vacuum pipeline energy storage	EMS high speed magnetic levitation vacuum pipeline energy storage	EDS high speed magnetic levitation vacuum pipeline energy storage	Super high speed rail type vacuum pipeline energy storage
Load ratio (t/m)	2	2.06	3.52	1
Cost ratio (hundred million/m)	≈2.3	≈3	≈12.3	≈8.8
Energy storage efficiency	40.96%	82.81%	82.81%	82.81%
Total length of rail (km)	3×10^5	3×10^5	3×10^5	3×10^5
Total load (kg)	6×10^5	6.18×10^5	1.056×10^6	3×10^5
Maximum kinetic energy of the system (J)	2.3148×10^{11}	1.2246×10^5	4.0714×10^{12}	3.2150×10^{12}
Single energy storage (kW·h)	2.6337×10^4	2.8169×10^5	9.3716×10^5	7.3954×10^5
Total system cost (hundred million)	>730	>940	>3730	>2680
cost(¥/kW·h)	>151.87	>18.29	>21.77	>19.86

According to the analysis in Table 2, the maximum annual power generation of the vacuum pipeline magnetic levitation energy storage scheme based on the four existing magnetic levitation transportation technologies is 340 million, which is far less than the annual power generation of Fengning pumped storage power station of 2.236 billion. The cost of energy storage is the lowest, which is far higher than the energy storage cost of Fengning pumped storage power station. In general, the vacuum pipeline energy storage system based on the existing maglev transportation technology has no advantages in energy storage capacity and energy storage cost compared with pumped storage. Therefore, the existing maglev transportation technology needs to be further improved to meet the applications in maglev energy storage.

4. Conclusion and Prospect

There are three main reasons why the existing magnetic levitation transportation technology is difficult to be used for electric energy storage. Firstly, the existing magnetic levitation transportation technology is passenger carrying transportation, and its operating speed is generally not high, load is relatively small, maximum kinetic energy is not high, so its energy storage capacity is not high. Secondly, the motor efficiency of the vacuum pipeline energy storage system based on the existing magnetic levitation transportation technology is not high, indicating that the energy loss in the process of mutual conversion between the electric energy of the grid and the mechanical energy of the maglev vehicle is large, so the energy storage efficiency is low. Thirdly, the cost ratio of the existing maglev transportation technology is generally high, so it is obviously unreasonable to apply it in the energy storage industry.

In short, the existing technology needs to be improved in three aspects of energy storage capacity, energy storage efficiency and economy. All in all, magnetic levitation energy storage technology has great potential in electric energy storage.

Acknowledgments

This research was financially supported by 2022 Science and technology project of Inner Mongolia Power (Group) Co.,Ltd (Grant No. 2022-20).

References

1. B. YUAN, Y.Z. ZHANG, G. LU, et al.Research on key issues of energy storage development and application in power systems[J]. Electric Power, 2019, 52(3): 1-8.
2. [2]Consulting Group of SGCC to Prospects of New technologies in power systems. An analysis of prospects for application of largescale energy storage technology in power systems[J]. Automation of Electric Power Systems, 2013, 37(1): 3–8.
3. [3]H.S. CHEN, H. LI, W.T. MA. Research progress of energy storage technology in China in 2021[J]. Energy Storage Science and Technology, 2022, 11(3): 1056-1076.
4. [4]P. LIAO, Z. YAO, J. LEMMON. Research progress and prospect of battery energy storage technology[J]. Energy Storage Science and Technology, 2020, 9(3): 670-678.
5. [5]C.X. WANG, Q.H. LI, X.J. LEI. Methodology for analyzing the value of energy storage to power system frequency control context of high shares of renewable energy[J]. Electric Power, 2016, 49(10): 148-152.
6. D. HAN, Z.H. ZHAO, B.Z. YAN, et al. Status and prospect of China's pumped storage development in 2021[J], Water Power, 2022, 48(5), 1-5.
7. W.C. TU, W.Y. LI, Q. ZHANG, et al. Engineering application of flywheel energy storage in power system[J]. Energy Storage Science and Technology, 2020, 9(3): 869-877
8. W.J. LI, G.M. ZHANG, X.W. Wang, et al. Integration design of high-temperature superconducting bearing and electromagnetic thrust bearing for flywheel energy storage system[J]. Transactions of china electrotechnical society, 2020, 35(1): 10-18.
9. S.H. YU, W.Y. GUO, Y.P. TENG, et al. A review of the structures and control strategies for flywheel bearings [J]. Energy Storage Science and Technology, 2021, 10(5): 1631-1642.
10. Y.J. CAO, F.M. XIA, H.L. ZHU, et al. Application and prospect of superconducting energy storage in new energy power system [J]. 2021, 10: 1-7.
11. R.V. HOLLA. Energy storage methods-superconducting magnetic energy storage-a review[J]. Journal of Undergraduate Research, 2015, 5(1): 49-54.
12. W.Y. GUO, J.Y. ZHANG, Z.F. ZHANG, et al. Research status and application prospect of superconducting energy storage system[J]. Science and technology review, 2016, 34(23): 68-80.
13. C.Y. TIAN, C.H. ZHANG, K. LI, et al. Micro grid composite energy storage technology with compressed air energy storage and its cost analysis[J]. Aotumation of electric power systems, 2015, 39(10): 36-41.
14. .Z.G. DENG, H.T. LI. Recent development of high-temperature superconducting maglev[J]. Materials china ,2017, 36(05): 329-334.
15. K. GLATZEL, G. Khurdok, D. Rogg. The development of the magnetically suspended transportation system in the federal republic of germany[J]. IEEE Transactions on Vehicular Technology, 1980, 29(1): 3-17.
16. S. Suzuki, M. Kawashima, Y. Hosoda, et al. HSST-03 system[J]. IEEE Transactions on Magnetics, 1984, 20(5): 1675-1677.
17. J.Y. XIONG, Z.G. DENG. Research progress of high-speed maglev rail transit[J]. Journal of Traffic and Transportation Engineering, 2021, 21(1): 177-198.
18. J.X. Wang. Introduction to the principle and application of maglev train[J]. Scientific and technological innovation, 2019, (15): 38-39.

19. Y. ZHANG, C. WU. The maglev technology development route[J]. Technology and market 2017, 24(6): 101-104
20. Y.S. LI, A. XU. Magnetic levitation technology development in Germany[J]. Urban Mass Transition, 2001, (2): 64-68.
21. C.W. FENG, X. FANG, H.M. LI, et al. Technological development of high speed maglev system based on low vacuum pipeline[J]. Strategic Study of CAE, 2018, (6): 105-110.
22. Z.L. YU, K.H. REN, T.W. XU. Development trend of high speed rail transit[J]. Equipment Manufacturing Technology, 2020, (3): 230-232.

Synthesis and performance of a small molecular quaternary ammonium salt inhibitors

Jun Jia *, Chaoyang Xu, Chao Hao, Hongwei Dong

CNPC Chuanqing Drilling Engineering Company Ltd., Xi'an, 710018, China

Abstract. In this work, the quaternary ammonium salt small molecule inhibitors with melamine and sodium chloroacetate as raw materials was synthesized and it has certain inhibition on the hydration swelling of bentonite. Through linear swelling test, mud ball test and drilling fluid performance evaluation, its inhibition performance in clay is evaluated. The results show that the linear swelling rate of bentonite in 0.5% MAA-3 solution is 57.28% at 90 min, which is similar to that in 4.0% KCl solution.

Keywords: Synthesis; Hydration; Swelling; Performance

1. Introduction

From the drilling practice at home and abroad in recent decades, the commonly used drilling fluid systems for drilling are water-based drilling fluid and oil-based drilling fluid [1-2]. With the development of oil exploration and development to the deep and offshore, we often encounter complex situations such as tripping resistance, sticking, low ROP, wellbore instability, lost circulation and formation damage [3-4]. The traditional method to solve these problems is to use oil-based and synthetic based drilling fluids [5-6]. However, the development of water-based drilling fluid has been promoted by the strict environmental restrictions and the challenges of deep-water drilling [7-9]. When drilling with water-based drilling fluid, a series of problems such as wellbore instability, bit mud package and borehole purification caused by the hydration swelling and dispersion of shale increase the difficulty of drilling, thus putting forward higher requirements for drilling fluid treatment agents, especially inhibitors [10].

Secondly, in recent years, the severe energy tension, the rapid growth of energy prices and the large-scale development of shale gas in the United States have made shale gas resources widely valued all over the world [11]. Shale gas mainly exists in mudstone, shale and other formations with hydrocarbon generation capacity in both adsorption and free states [12]. Although shale gas resources in China have great potential, shale gas resources are deeply buried and exploration and development started late [13]. In the process of shale gas exploitation, there are still major problems such as ultra-low permeability of shale gas reservoir, reservoir protection, prevention of shale rock deformation and so on. Therefore, the development of shale inhibitors with good inhibition performance, little damage to the reservoir and little environmental pollution will also play a certain role in promoting the exploitation of shale gas [14]. Amine inhibitors have been widely concerned by researchers in recent years because of their successful application in foreign water-based drilling fluids. The common goal of researchers is to optimize the molecular structure of amines and their derivatives, and finally develop amines small molecule inhibitors with good inhibition effect, good compatibility and little harm to the environment [15-16]. Melamine and sodium chloroacetate were used as raw materials to synthesize amine inhibitors, which provided some experimental basis for the further development of amine small molecule inhibitors.

2. Experimental Materials and Methods

2.1 Materials and Reagents

Melamine was purchased from Tianjin kemio Chemical Reagent Co., Ltd. Sodium chloroacetate was purchased from Beijing 57601 organic chemical plant. KCl and Na_2CO_3 were purchased from Tianjin Zhiyuan chemical reagent factory. KD-03, PAM and modified starch (MS) were purchased from Yangzhou Runda Oilfield Chemical Co., Ltd. Calcium bentonite was purchased from Xi'an permanent Chemical Co., Ltd. Sodium bentonite was purchased from Qingtongxia Changqing bentonite Co., Ltd.

2.2 Synthesis of inhibitors

According to the molar ratio of functional groups (acid base), add melamine and sodium chloroacetate into the flask, use water as solvent, heat and reflux for 1H. The reaction is shown in the figure below.

* Corresponding author:

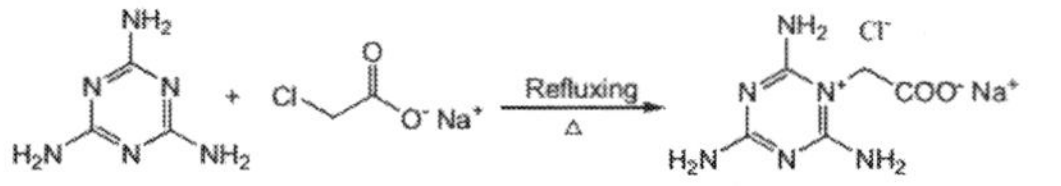

Figure 1. Synthesis of quaternary ammonium salt

Divide the solution into two parts, one with NaOH to adjust pH=7 and the other without NaOH. See Table 5-3 for the name of the product.

Table 1. Naming of reaction products of melamine and sodium chloroacetate

$n_{melamine}$: $n_{Sodium\ chloroacetate}$	pH	Name
1:1	Acidic	MAA-1
	Neutral	MAA-2
1:2	Acidic	MAA-3
	Neutral	MAA-4
1:3	Acidic	MAA-5
	Neutral	MAA-6
1:4	Acidic	MAA-7
	Neutral	MAA-8
1:5	Acidic	MAA-9
	Neutral	MAA-10
1:6	Acidic	MAA-11
	Neutral	MAA-12
1:7	Acidic	MAA-13
	Neutral	MAA-14
1:8	Acidic	MAA-15
	Neutral	MAA-16
1:9	Acidic	MAA-17
	Neutral	MAA-18

2.3 Linear Swelling Experiment

According to SY/T 6335-1997《evaluation method of shale inhibitors for drilling fluids》, the effect of clay swelling inhibitor on linear swelling rate of bentonite was evaluated. And the swelling rate was calculated according to Formula (1-1):

$$S_r = \frac{R_0}{\Delta L} \times 100 \quad (1)$$

Formula: S_r is the linear swelling rate of bentonite, %. R_0 is the swelling of bentonite, mm. ΔL is core height, mm.

2.4 Mud ball test

At room temperature, the mixture of bentonite and tap water was mixed in a ratio of 1.2:1 by mass. Then the mixture was made into small clay balls of about 10 g/piece, which were immersed in the same volume of different kinds of inhibitor hydrosol or tap water for 24 h, respectively. The surface morphology of the expanded clay balls was used to evaluate the inhibition performance of the clay stabilizers.

2.5 Evaluation of drilling fluid performance

The KD-03, MS and PAM were prepared by adding 0.5% KD-03, 1.0% MS and 0.04% PAM into the medium pre-hydrated bentonite base mud. And then the selected inhibitors were added respectively, the AV, PV, YP, YP/PV, F_L, ρ and μ were measured after high speed stirring.

3. Results and Discussion

3.1 Determination of linear swelling rate of Bentonite

The effect of the synthesized inhibitor on the linear swelling rate of bentonite is evaluated. The experimental results are shown in Figures 2 to 10.

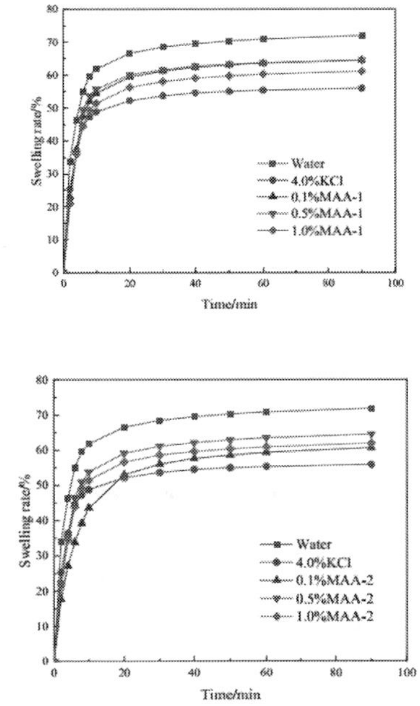

Figure 2. Effect of MAA-1 and MAA-2 with different concentrations on linear swelling rate of bentonite

It can be seen from Figure 2 that MAA-1 and MAA-2 of different concentrations have a certain inhibitory effect on the hydration swelling of bentonite, the mass fraction of inhibitor has little effect on the linear swelling rate of bentonite, and the inhibitory effect of MAA-1 and MAA-2 on bentonite is similar. At 90 min, the linear swelling rates of bentonite in 1.0% MAA-1 and 1.0% MAA-2 solutions were 61.06% and 61.97% respectively.

 https://doi.org/10.1051/e3sconf/202235801042

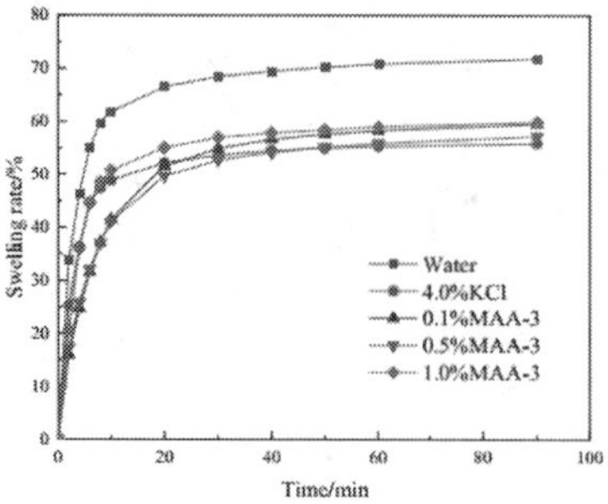

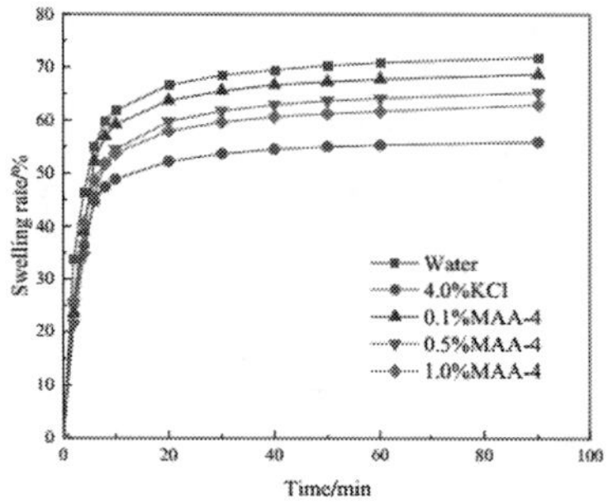

Figure 3. Effect of MAA-3 and MAA-4 with different concentrations on linear swelling rate of bentonite

It can be seen from Figure 3 that MAA-3 has a stronger inhibitory effect on the hydration swelling of bentonite than MAA-4. In 0.1%, 0.5% and 1.0% MAA-3 solution, the linear swelling rate of bentonite is basically the same, which is similar to that in 4.0% KCl solution. At 90 min, the linear swelling rate of bentonite in 0.5% MAA-3 solution is 57.28%; At 90 min, the linear swelling rate of bentonite in 1.0% MAA-4 solution is 62.90%, and its inhibition on the hydration swelling and dispersion of bentonite is better than that in 0.5% and 0.1% solutions, but worse than that in 4.0% KCl solution.

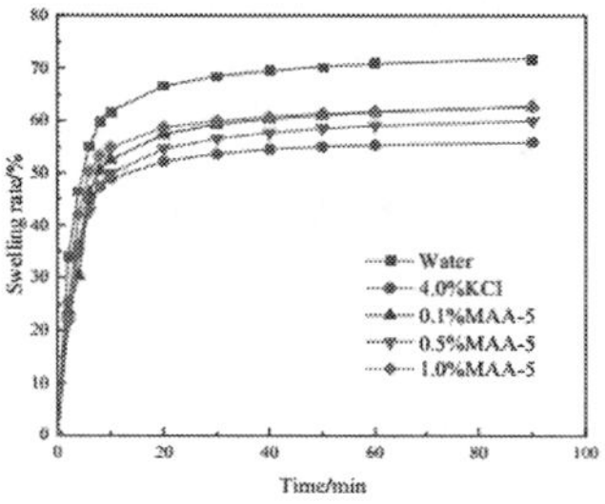

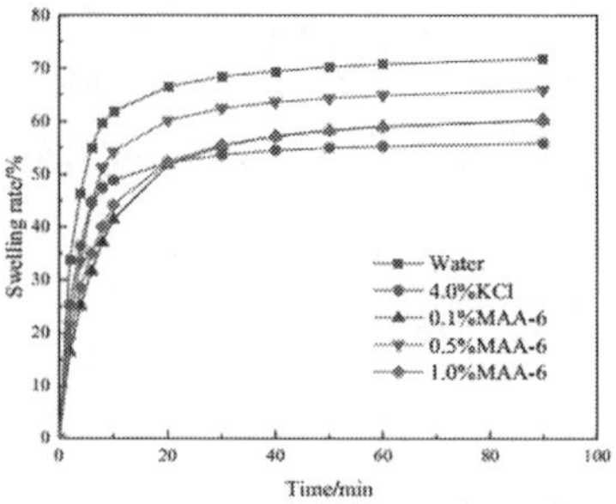

Figure 4. Effect of MAA-5 and MAA-6 with different concentrations on linear swelling rate of bentonite

It can be seen from Figure 4 that MAA-5 and MAA-6 have similar effects on the linear swelling rate of bentonite, which is worse than that in 4.0% KCl solution. The linear swelling rate of bentonite in 0.1%, 0.5% and 1.0% MAA-5 solution are similar. At 90 min, the linear swelling rate of bentonite in 0.5% MAA-5 solution is 59.97%. The linear swelling rates of bentonite in 0.1% and 1.0% MAA-6 solutions are basically the same, which are 60.27% and 60.33% respectively at 90 min. The inhibition of bentonite is slightly better than that of 0.5% MAA-6.

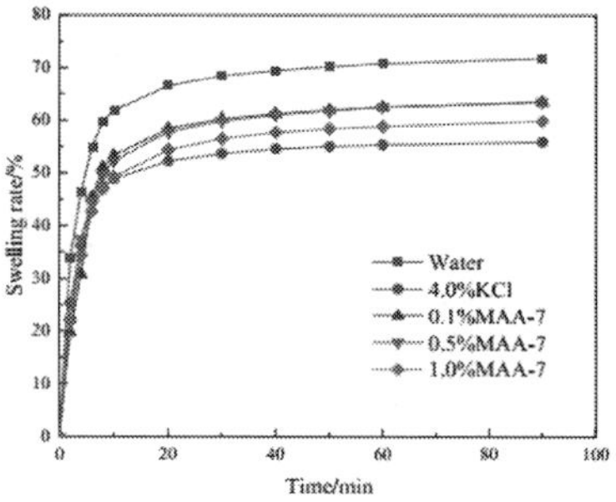

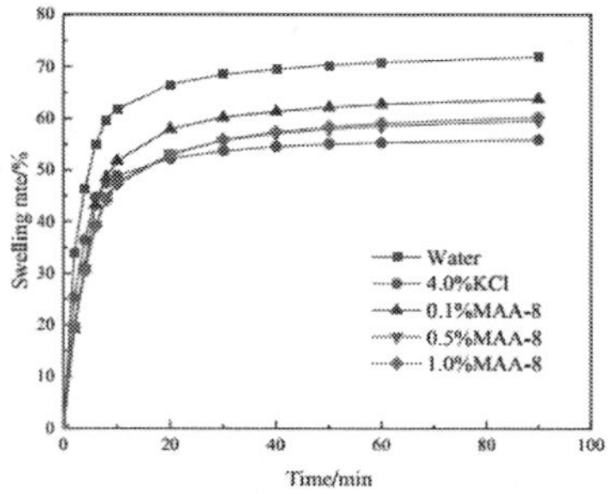

Figure 5. Effect of MAA-7 and MAA-8 with different concentrations on linear swelling rate of bentonite

It can be seen from Figure 5 that MAA-7 and MAA-8 have similar inhibitory effects on the hydration swelling and dispersion of bentonite. With the increase of inhibitor concentration, the linear swelling rate of bentonite in inhibitor solution decreases, but both are larger than that in 4.0%KCl solution. At 90 min, the linear swelling rate of bentonite in 1.0% MAA-7 solution is 59.86%, and that in 1.0% MAA-8 solution is 59.57%.

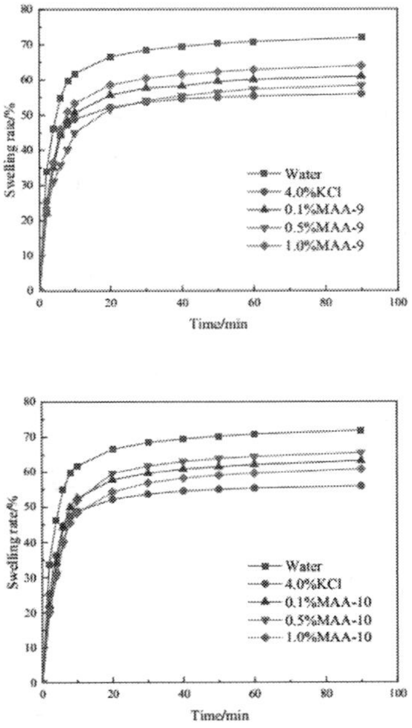

Figure 6. Effect of MAA-9 and MAA-10 with different concentrations on linear swelling rate of bentonite

It can be seen from Figure 6 that MAA-9 and MAA-10 have no significant difference in the inhibition effect on the hydration swelling and dispersion of bentonite, and the linear swelling rate of bentonite in different concentrations of inhibitor solution is greater than that in 4.0%KCl solution. At 90 min, the linear swelling rate of bentonite in 0.5% MAA-9 solution is 58.31%, and that in 1.0% MAA-10 solution is 60.86%.

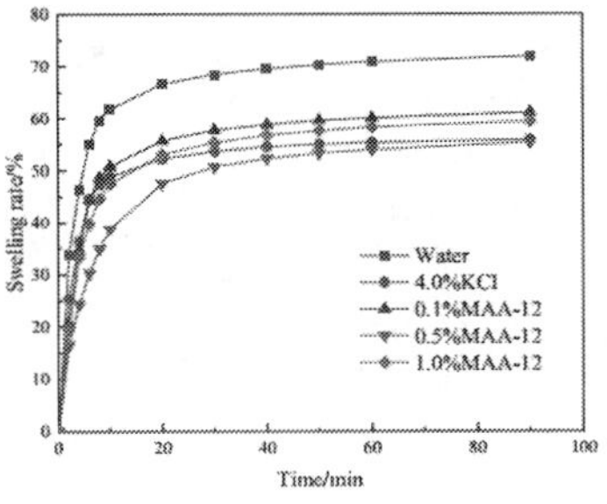

Figure 7. Effect of MAA-11 and MAA-12 with different concentrations on linear swelling rate of bentonite

It can be seen from Figure 7 that MAA-12 has slightly better inhibition on the hydration swelling and dispersion of bentonite than MAA-11. At 90 min, the linear swelling rate of bentonite in 0.5% MAA-12 solution is 55.37%, which is similar to that in 4.0% KCl solution. The linear swelling rate of bentonite in 0.5% MAA-11 solution is 60.12%, which is slightly larger than that in 4.0% KCl solution.

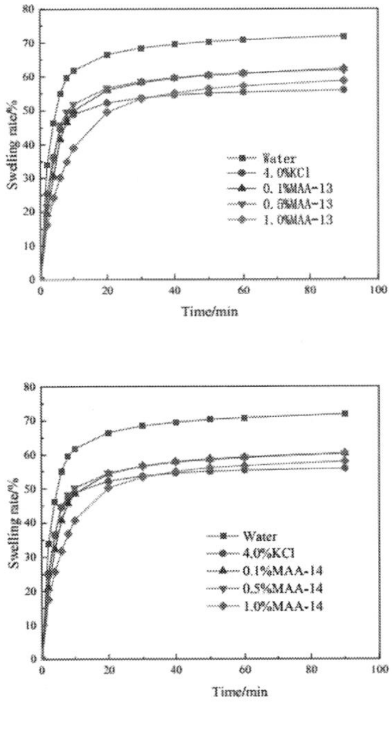

Figure 8. Effect of MAA-13 and MAA-14 with different concentrations on linear swelling rate of bentonite

It can be seen from Figure 8 that MAA-13 and MAA-14 with different concentrations have basically the same inhibitory effect on the hydration swelling and dispersion of bentonite, and their inhibitory effects on the hydration swelling and dispersion of bentonite are similar. The linear swelling rate of bentonite in 0.1% and 0.5% inhibitor solution is similar. The linear swelling rate of bentonite in 1.0% inhibitor solution is slightly smaller than that in 0.1% and 0.5%, but slightly larger than that in 4.0% KCl solution. At 90 min, the linear swelling rate of bentonite in 1.0% MAA-13 solution is 58.67%, and that in 1.0% MAA-14 solution is 58.08%.

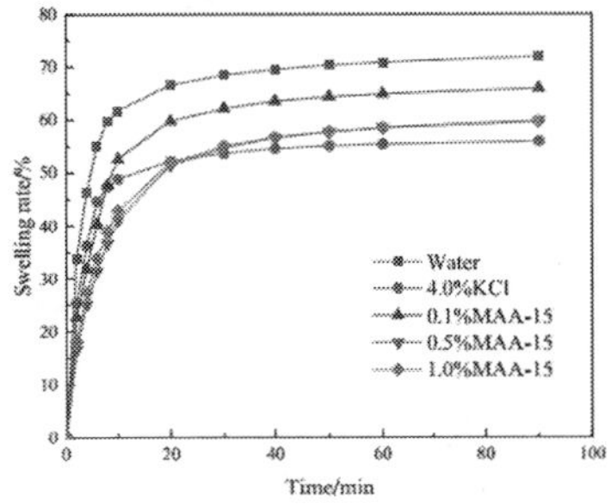

 https://doi.org/10.1051/e3sconf/202235801042

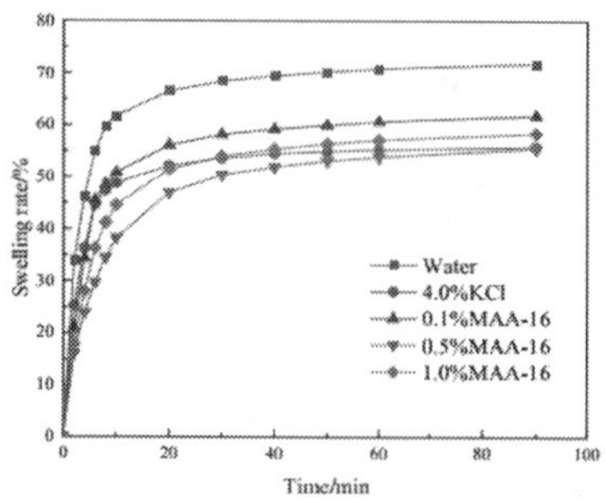

Figure 9. Effect of MAA-15 and MAA-16 with different concentrations on linear swelling rate of bentonite

It can be seen from Figure 9 that MAA-16 has a slightly stronger inhibition effect on the hydration swelling and dispersion of bentonite than MAA-15. At 90 min, the linear swelling rate of bentonite in 1.0% MAA-15 solution is 59.58%, which is slightly larger than that in 4.0% KCl solution. 55.57% in 0.5% MAA-16 solution, which is similar to that in 4.0% KCl solution.

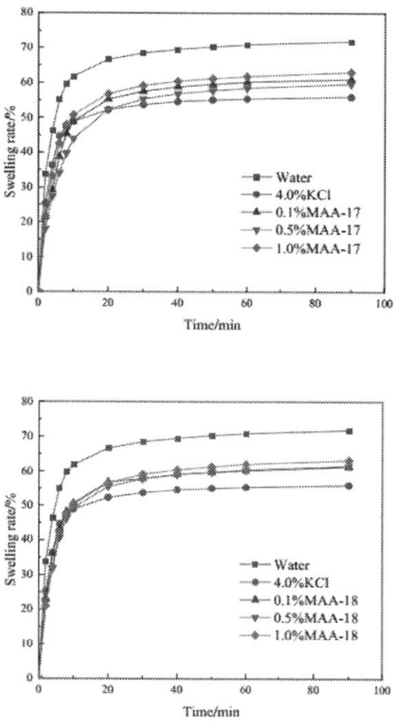

Figure 10. Effect of MAA-17 and MAA-18 with different concentrations on linear swelling rate of bentonite

It can be seen from Figure 10 that MAA-17 and MAA-18 have similar inhibitory effects on the hydration swelling and dispersion of bentonite, and the linear swelling rate of bentonite in 0.1%, 0.5% and 1.0% inhibitor solution is not different, which is slightly larger than that in 4.0% KCl solution. At 90 min, the linear swelling rate of bentonite in 0.5% maa-17 solution is 59.57%, and that in 0.5% MAA-18 solution is 59.61%.

To sum up, the synthesized inhibitor has a certain inhibitory effect on the hydration swelling of bentonite, and the mass fraction of the inhibitor and the acidity and alkalinity of the inhibitor have little effect on the linear swelling rate of bentonite. Among them, 0.5% MAA-3 and 0.5% MAA-12 have better inhibition effect on the hydration swelling of bentonite.

3.2 Clay ball experiment

The results of the mud ball experiment are shown in Figure 11. It can be seen from Figure 11 that with the extension of time, the water absorption of the mud ball continues to increase. After soaking in tap water for 72 h, the surface of the mud ball is loose and its volume increases, and its hydration swelling and dispersion are the most obvious, as shown in Figure 11 (a). The hydration swelling degree of the mud balls soaked in 0.5% MAA-3 and 0.5% MAA-12 aqueous solutions is slightly smaller, and there are more small cracks on the surface of the mud balls, and the cracks of the mud balls in 0.5% MAA-12 aqueous solution are more obvious than those in 0.5% MAA-3 aqueous solution, as shown in (b) and (c) in Figure 11. MAA-3 and MAA -12 have certain inhibitory effects on the hydration swelling and dispersion of bentonite.

(a) Tap water

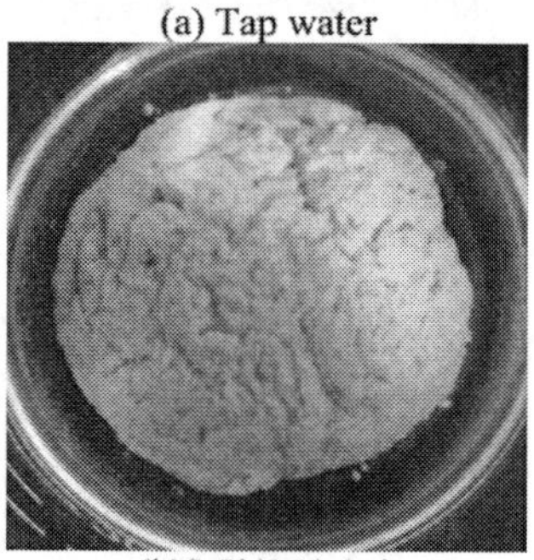

(b)0.5%MAA-3

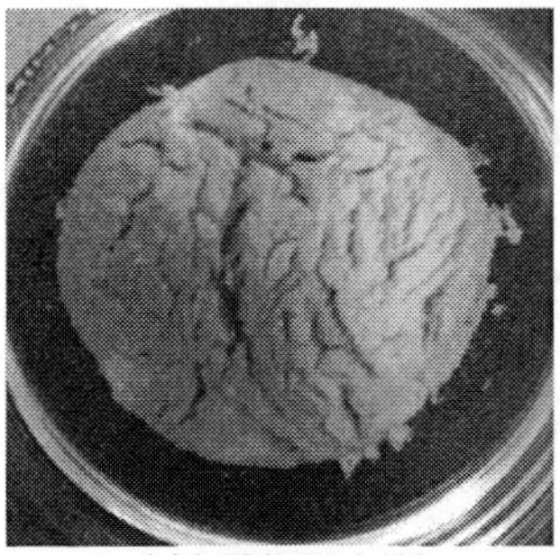

(c)0.5%MAA-12

Figure 11. Appearance of mud balls soaked in different aqueous solutions for 72 h

3.3 Evaluation of drilling fluid performance

Add the selected melamine and sodium chloroacetate reactants to the polysaccharide treated mud, PAM treated mud and modified starch treated mud respectively. See Table 2 to Table 4 for the performance parameters of the drilling fluid at room temperature. From Tables 2 to 4, it can be seen that the AV, YP and YP/PV of polysaccharide drilling fluid increased significantly after adding the screened melamine and sodium chloroacetate reactants. μ decrease, F_L increases. AV, PV and YP of PAM drilling fluid decreased, while F_L increased, μ almost unchanged. AV, YP and YP/PV of MS drilling fluid increased. Melamine and sodium chloroacetate reactants have a prominent impact on the rheological properties of polysaccharide drilling fluid, PAM drilling fluid and MS drilling fluid. The AV, PV and YP of PAM drilling fluid are significantly reduced. The changes of rheological parameters of polysaccharide drilling fluid and MS drilling fluid are opposite to those of PAM drilling fluid. The reason may be that after adding a certain amount of PAM to the base mud, the drilling fluid will produce flocculation. 0.04% PAM is the maximum amount of drilling fluid that does not produce flocculation. After adding the screened melamine and sodium chloroacetate reactant to the PAM drilling fluid, the electrostatic interaction between cationic amine and clay enhances the force between clays, making clay particles gather together and produce flocculation. Therefore, the viscosity of drilling fluid decreases, F_L increases. MAA-3 and MAA-12 have good compatibility with KD-03 and MS.

Table 2 Effects of quaternary ammonium salt inhibitors on the properties of polysaccharide drilling fluid

Treated pulp	AV /(mPa•s)	PV /(mPa•s)	YP /Pa	YP/PV Pa/(mPa•s)	F_L /mL	μ
KD-03 03	4.9	2.8	2.1462	0.7665	11.0	0.0875
KD-03 +0.5%MAA-3	6.9	2.3	4.7012	2.0440	14.6	0.0524
KD-03 +0.5%MAA-12	7.1	2.4	4.8034	2.0014	15.5	0.0699

Table 3 Effects of quaternary ammonium salt inhibitors on the properties of PAM treated pulp

Treated pulp	AV /(mPa•s)	PV /(mPa•s)	YP /Pa	YP/PV Pa/(mPa•s)	F_L /mL	μ
PAM	8.0	4.2	3.8836	0.9247	12.0	0.0699
PAM +0.5% MAA-3	2.5	0.5	2.0440	4.0880	17.0	0.0524
PAM +0.5% MAA-12	3.8	0.5	3.3215	6.6430	16.2	0.0699

Table 4 Effect of quaternary ammonium salt inhibitor on properties of MS treated pulp

Treated pulp	AV /(mPa•s)	PV /(mPa•s)	YP /Pa	YP/PV Pa/(mPa•s)	F_L /mL	μ
MS	6.0	4.0	2.0440	0.5110	7.5	0.0699
MS+0.5% MAA-3	8.9	3.6	5.4166	1.5046	7.5	0.0875
MS+0.5% MAA-12	7.9	1.6	6.4386	4.0241	11.5	0.0612

4. Conclusion

The experimental results of linear swelling rate of bentonite show that 0.5% MAA-3 and 0.5% MAA-12 in the reactant of melamine and sodium chloroacetate have a good inhibitory effect on bentonite. The linear swelling rate of bentonite is basically the same as that of adding 4.0% KCl. The clay ball experiment shows that the addition of 0.5% MAA-3 has a certain inhibitory effect on the hydration expansion of the clay ball. The above experimental results show that quaternary ammonium salt inhibitor has a certain inhibition on the hydration swelling and dispersion of bentonite.

Acknowledgments

We are very grateful to Cai Dan for her help in these experiments. We also thank the work of Modern Analysis and Testing Center of Xi`an Shiyou University.

References

1. Huang X.B., Sun. J.S., Jin. J.F., Lv. K.H., Li. H.C. Use of silicone quaternary ammonium salt for hydrophobic surface modification to inhibit shale hydration and mechanism study[J]. Journal of Molecular Liquids, 2021, 341.

2. Lv. K.H., Liu. J., Jin. J.F. Synthesis of a novel cationic hydrophobic shale inhibitor with preferable wellbore stability[J]. Colloids and Surfaces A: Physicochemical and Engineering Aspects, 2022, 637.
3. Ye. Z.Q., Ye. Z.P., Huang. L., Bai. Y., Li. L.L. Preparation and Application of a New Crosslinked Polyammonium as a Shale Inhibitor[J]. Journal of Applied Biomaterials & Functional Materials, 2018, 16(1), pp. 119-124.
4. Ítalo. G.M.S., Bertolino. L.C., Lucas. E.F. Correlation between clay type and performance of swelling inhibitors based on polyetherdiamine in aqueous fluids[J]. Journal of Applied Polymer Science. 2019, 136(25), pp. 47661.
5. Zhu. Y.Y., Wang. W.S., Wu. X.M., Zhang. H.C., Xu. J., Yan. J., Cao. L.L., Ran. H.Q., Zhang. J.C. Main technical innovations of Songke Well No.2 Drilling Project[J]. China Geology, 2018, 1(2), pp. 187-201.
6. Chen. G., Yan. J., Li. L.L., Zhang. J, Gu. X.F. and Song. H. Preparation and performance of amine-tartaric salt as potential clay swelling inhibitor[J]. App Clay Sci. 2017, 138, pp. 12–16.
7. Bai. X., Wang. H., Luo. Y., Zheng. X., Zhang. X., Zhou. S., Pu. X. The structure and application of amine-terminated hyperbranched polymer shale inhibitor for water-based drilling fluid[J]. Appl. Polym. Sci. 2017, pp. 134.
8. Gu. X.F., Zhang. J.J., Zhang. J., Chen. G., Ma. C., Zhang. Z.F. Stabilization of montmorillonite by ammoniated lignosulfonates and its use in water-based drilling fluid[J]. Sci. Adv. Mater. 2017, 9, pp. 928–933.
9. Zhang. J., Zhang. Q., Chen. G. Mechanism study and application of modified polysaccharides as drilling fluid additive with high inhabitability[J]. Technol. Develop. Chem. Ind. 2013, 42, pp. 1–5.
10. Song. Z.F., Zhang. L., Huang. L., Liu. X.X., Zhang. J., Du W.C., Chen. G. Preparation and application of a novel polyammonium as potent shale hydration inhibitor[J]. Journal of Macromolecular Science, Part A, 2020, 57(5), pp. 326-331.
11. Chen. G., Zhang. J., Yang. N.W., Ma. Y.F. The evaluation of sodium hydroxymethyl lignosulfonate as an ecofriendly drilling fluid additive[J]. Pet. Sci. Technol. 2014, 32, pp. 1816–1823.
12. Zhang. R.J., Gao. L., Duan. W.G., Hu. W.M., Du. W.C., Gu. X.F., Zhang. J., Chen G. The application of ferric chloride-lignin sulfonate as shale inhibitor in water-based drilling fluid[J]. Molecules, 2019, 24(23), pp. 4331.
13. Wang. Q.D., Michal. S., Gu. X.F., Miao. Z.P., Du. W.C., Zhang. J., Chen. G. Lubricity and rheological properties of highly dispersed graphite in clay-water-based drilling fluids[J]. Materials, 2022, 15(3), pp. 1083-1083.
14. Du. W.C., Pu. X.L., Sun. J.S., Luo. X., Zhang. Y.N., Li. L. Synthesis and evaluation of a novel monomeric amine as sodium montmorillonite swelling inhibitor[J]. Adsorption Science & Technology, 2018, 36(1), pp. 655-668.
15. Vryzas. Z., Kelessidis. V.C. Nano-based drilling fluids: A review[J]. Energies 2017, 10, pp. 540.
16. Xuan. Y., Jiang. G.C., Li. Y.Y., et al. Inhibiting effect of dopamine adsorption and polymerization on hydrated swelling of montmorillonite[J]. Colloids and Surfaces A, 2013, 422, pp. 50–60.

Research of a modified nano-depressure and augmented injection agent

Qianhong Pan [1,*], Yanhong Shen [1], Xiuhua Liu [1], Kun Zhao [1,2], Jialu Du [1], Yali Li [1,3]

[1] Xi'an ChangQing Petrochemical Corporation Co.Ltd., Xi'an, 710018, China
[2] No. 8 Oil Production Factory, Changqing Oilfield Company, CNPC, Xi'an, 716000, China
[3] Xi'an Key Laboratory of Tight oil (Shale oil) Development, Xi'an Shiyou University, Xi'an, 710065, China

Abstract. In this paper, the research and application of nanomaterials in oilfield chemical engineering was briefly described with emphasis on the nano-silica in enhanced oil recovery, depressure and augmented injection. A study on the modification of nano-silica was carried out indoors, and a modified nano-depressure and augmented injection agent was prepared by using a coupling agent for surface grafting modification, followed by introducing functional monomers for in-situ modification. The effects of silica dosage, coupling agent dosage, dispersant and regulator ratio, and modification time on the modified product were investigated. It was analyzed by infrared spectroscopy, thermogravimetric analysis, nanoparticle size analyzer, transmission electron microscope, atomic force microscope, etc. A series of experiments were conducted to evaluate the temperature and salt resistance, wettability, as well as depressure and augmented injection performance of the modified products.

Key words: Oilfield chemical engineering; nanosilica modification; enhanced oil recovery; depressure and augmented injection

1. Introduction

Nanomaterials are defined as materials with at least one dimension in the three-dimensional structure ranging from 1 to 100 nm. They have many unique properties such as surface effects, volume effects, quantum size effects, macroscopic quantum tunneling effects, and mechanical properties effects. In oilfield chemical engineering, it mainly involves nanomaterials in many different forms, including nanoparticles, nanofilms, nanocrystals, nanofibers, and nanofluids. In drilling and completion construction, as a kind of additive, nanomaterials play a great role in reducing the liquid filter loss, enhancing the walled performance of drilling fluid, improving the pore structure of cementing cement ring and the construction quality. After adding nanomaterials to the drilling fluid, it was found that its lubrication was improved and the construction friction was significantly reduced. In addition, the rheology, lubrication, mud cake wall-building and crack sealing properties of drilling fluid were greatly improved, while the pore penetration of cement ring and the thickening time of the cement slurry were reduced. The introduction of nanomaterials in the fracturing fluid system can improve its rheology and sand suspension, form a more quality crack network system, enhance the fracturing construction effect, and improve the efficiency of increasing production and transformation.[1-4] It can meet the current effective development of low permeability reservoirs and unconventional oil and gas. The application of nanomaterials in oil recovery has become a research topic, especially in the nanomaterials to change rock wettability, reduce the oil-water interface tension, flood oil, pressure depressure and augmented injection. With a large number of unsaturated functional groups on their surface, nanomaterials can efficiently catalyze the organics and bacteria in oilfield sewage so as to make the emulsification separation of oil and water through oil absorption and demulsification. In addition, the integral aerogel material can be constructed based on the structural unit of nanomaterials, and the dual function of demulsification and oil-water separation can be realized.[5-10]

The conventional resistance reduction, depressure and augmented injection measures adopted in the low permeability oilfield are mainly divided into two categories: reservoir modification and rock surface modification. Among them, the reservoir modification mainly includes fracturing and acidification technology. Rock surface modification uses surfactant to form an adsorption layer on the pore surface of the oil layer, which increases the injection by reducing the interface tension of solid and liquid. Field tests also show that fracturing can cause the formation flow, water rush, causing irreversible damage to the formation, and the short validity period of acidification can easily cause cumulative damage and

* Corresponding author: 314505704@qq.com

pollution to the formation. The adsorption force of surfactant on the rock surface is weak and short, and its application effect is also greatly limited.[11-14]

In recent years, more and more researchers pay attention to nano depressure and augmented injection technology, its technology is to bring nanoparticles into the reservoir through specific dispersion, and adsorb and deposit in the appropriate position, change the wettability and roughness of the micropore wall, form the surface microstructure, so that the micropore wall produces a large sliding length, thus playing the purpose of depressure and augmented injection. Needle from oil field water injection process field, this paper proposed a kind of modified nanomaterials with a water phase system for depressure and augmented injection. The silica dosage, coupling agent dosage, dispersant and regulator ratio, and modification time on the modified product were investigated, and the optimum synthesis method and parameters were determined.

2. Materials and methods

2.1 Materials

Materials used for the study such as ethanol (Sinopharm Co., Ltd.), hydrochloric acid (Tianjin Damao Chemical Co., Ltd.), coupling agent (Nanjing New Materials Co., Ltd.), silica (Shanghai Kay Chemicals Co., Ltd.), functional monomer (Shanghai Kay Chemicals Co., Ltd.) were purchased, respectively. Dispersants, stabilizers were synthesized in the laboratory.

The resulting modified silica was characterized by Fourier Transform Infrared Spectrometer (Nicolet Summit, Thermo Fisher Scientific Co., Ltd., USA); TGA-DSC thermal analyzer (TG-SDTA851, Mettler-Toledo Co., Ltd., Switzerland); laser particle size distribution analyzer (NANO ZS90, Malvern Instruments Co., Ltd., UK); transmission electron microscope (HT7800, Hitachi Co., Ltd., Japan); atomic force microscope (SPM-9700HT, Shimadzu Co., Ltd., Japan); scanning electron microscope (Nova NANOSEM450, FEI Co., Ltd., USA); contact angle measuring instrument (DSA30, Clues Co., Ltd., Germany). The HA-2 integrated oil flooding device was used to evaluate the depressure and augmented injection performance through the core flow experiment. The properties of anti-calcium sulfate scale and anti-swelling percentage were evaluated according to the formulas in Standard Q/SY 126 and SY/T 5971.[15-19]

2.2 Preparation of nanosilica

A certain amount of nanosilica was slowly added to the dispersant and stirred at a certain temperature to form solution 1. In addition, the coupling agent and the modifier were added to a certain amount of the regulator, and the solution 2 was formed by stirring at 25 °C, and then the solution 2 was slowly added to the solution 1. The reaction was carried out at a certain temperature for 6 h to obtain a modified nano-depressure and augmented injection agent with water-phase system.

3. Results and discussion

3.1 Nanosilica selection and dosage determination

The solid content of the nanosilica was determined using the standard GB 12005.2. A certain amount of nanosilica was added to the dispersant to form a uniform liquid, and then its particle size was determined to gradually increase the amount of nanosilica until the particle size was greater than 100 nm.

Table.1 Solid content of nanosilica

Order number	Sample number	Appearance	Solid content (%)
1	1#	White emulsion	22
2	2#	White emulsion	62
3	3#	White emulsion	57
4	4#	Conviscous homogeneous liquid	8
5	5#	Conviscous homogeneous liquid	20
6	6#	White powder	98
7	7#	Homogeneous liquid	48

Table.2 Determination of the dosage of nanosilica

Order number	Sample number	Nanosilica dosage (%)			
		4	8	16	32
		Particle size corresponding to the dosage of nanosilica (nm)			
1	1#	231	323	434	567
2	2#	121	145	156	189
3	3#	168	201	218	248
4	4#	136	219	232	276
5	5#	120	160	204	289
6	6#	130	143	156	183
7	7#	32	31	36	87

As shown in from Table 1 and Table 2, it can be found that the particle size began to increase when the dosage of nanosilica was 32%. Therefore, the orthogonal trials were designed with the dosage of nanosilica of 4%, 8%, 16%, and 32%.

3.2 Determination of the dosage of the coupling agent

When the dosage of nanosilica was 16%, the reaction time was 4 h and the temperature was 60 °C, and the effect of the coupling agent dosage on the grafting percent was explored to determine the selection range of orthogonal trials. According to Table 3, the grafting percent of nanosilica was maximum at the coupling agent dosage of 0.8%. When the coupling content was further increased, the grafting percent decreased slightly, due to the large amount of the coupling agent hydrolysis will partially self-polymerization, enhanced competitive reaction, and decreased grafting percent. Therefore, in the orthogonal trials, the dosage of the coupling agent was determined as 0.1%, 0.2%, 0.4% and 0.8%.

Table.3 Determination of the range of coupling agent dosage

Order number	Coupling agent dosage (%)	Grafting percent
1	0.1	0.0476
2	0.2	0.0521
3	0.4	0.0618
4	0.8	0.0763
5	2	0.0431
6	4	0.0354
7	8	0.0321
8	10	0.0336

3.3 Modifier selection and dosage determination

The performance of functional monomers was evaluated using standard Q / SY 126 and the experimental results are shown in Table 4.

Table.4 Selection of modifiers

Order number	Sample number	$CaCO_3$ prevention percent (%)	$CaSO_4$ prevention percent (%)	$BaSO_4$ prevention percent (%)
1	1#	63.71	4.91	40.63
2	2#	86.69	98.75	2.74
3	3#	83.87	9.72	0
4	4#	87.90	98.33	/
5	5#	43.95	78.89	8.21
6	6#	/	0.52	/
7	7#	59.27	31.87	0.1
8	8#	70.65	96.67	1.16
9	9#	70.44	87.29	2.21
10	10#	59.96	97.71	38.32
11	11#	73.79	97.92	4.84
12	12#	83.44	0	0

Based on the results of Table 4, and according to its functional group structure, and the comprehensive performance, it is determined to use 2# as the optimal modifier. When the dosage of nanosilica was 16%, the reaction time was 4 h and the temperature was 60 °C, and the effect of the ratio of dispersant and modifier on the grafting percent was explored to determine the selection range of orthogonal trials. As shown from Table 5, the ratio of dispersant and modifier was 4:1, the grafting percent of nanosilica was maximum. Subsequently, the graft percent decreased when the ratio of dispersant and modifier increased. Therefore, in the orthogonal trials, the ratio of dispersant and modifier was 1:1, 2:1, 3:1 and 4:1.

Table.5 Determination of the ratio of dispersant and modifier

Order number	The ratio of dispersant and modifier	Grafting percent
1	1：1	0.0234
2	2：1	0.0317
3	3：1	0.0402
4	4：1	0.0533
5	5：1	0.0421
6	6：1	0.0414
7	7：1	0.0401

3.4 Investigation of the optimal modification condition

Orthogonal trials with the four factors and four level were designed so as to investigate the effect of nanosilica dosage (factor A), coupling dosage (factor B), the ratio of dispersant and modifier (factor C), reaction time (factor D) on the particle size of nanosilica, as shown in Table 6.

Table.6 Orthogonal table

Level	Experimental factor			
	Nanosilica dosage (%)	Coupling agent dosage (%)	The ratio of dispersant and modifier	reaction time (h)
	Factor A	Factor B	Factor C	Factor D
1	4	0.1	1：1	2
2	8	0.2	2：1	4
3	16	0.4	3：1	6
4	32	0.8	4：1	8

Table.7 Orthogonal scheme and the structural intuitive analysis

Order number	Experimental factor				the particle size (nm)
	Nanosilica dosage (%)	Coupling agent dosage (%)	The ratio of dispersant and modifier	reaction time (h)	
	Factor A	Factor B	Factor C	Factor D	
1	1 (4)	1 (0.1)	1 (1：1)	1 (2)	1419
2	1 (4)	2 (0.2)	2 (2：1)	2 (4)	1233
3	1 (4)	3 (0.4)	3 (3：1)	3 (6)	667
4	1 (4)	4 (0.8)	4 (4：1)	4 (8)	774
5	2 (8)	1 (0.1)	2 (2：1)	3 (6)	877
6	2 (8)	2 (0.2)	1 (1：1)	4 (8)	642
7	2 (8)	3 (0.4)	4 (4：1)	1 (2)	585
8	2 (8)	4 (0.8)	3 (3：1)	2 (4)	108
9	3 (16)	1 (0.1)	3 (3：1)	4 (8)	270
10	3 (16)	2 (0.2)	4 (4：1)	3 (6)	416
11	3 (16)	3 (0.4)	1 (1：1)	2 (4)	342
12	3 (16)	4 (0.8)	2 (2：1)	1 (2)	523
13	4 (32)	1 (0.1)	4 (4：1)	2 (4)	85
14	4 (32)	2 (0.2)	3 (3：1)	1 (2)	226
15	4 (32)	3 (0.4)	2 (2：1)	4 (8)	63
16	4 (32)	4 (0.8)	1 (1：1)	3 (6)	68
Level K_1	4093	2651	2471	2753	
Level K_2	2212	2517	2696	1768	
Level K_3	1551	1657	1271	2028	
Level K_4	442	1473	1860	1749	
k_1	1023	663	618	688	
k_2	553	629	674	442	
k_3	388	414	318	507	
k_4	111	368	465	437	
R	913	295	356	251	

As shown in Table 7, it was found that the minimum particle size was 63 nm and the optimal combination was selected as $A_4B_4C_3D_3$. Depending on the magnitude of the

 https://doi.org/10.1051/e3sconf/202235801043

error, the importance of each factor to the particle size of modified nanomaterials can be judged, the dosage of nanosilica (A)> the ratio of dispersant and modifier (C)> the dosage of coupling agent (B)> reaction time (D).

3.5 Performance evaluation of nanosilica

3.5.1 Characterization of nanosilica

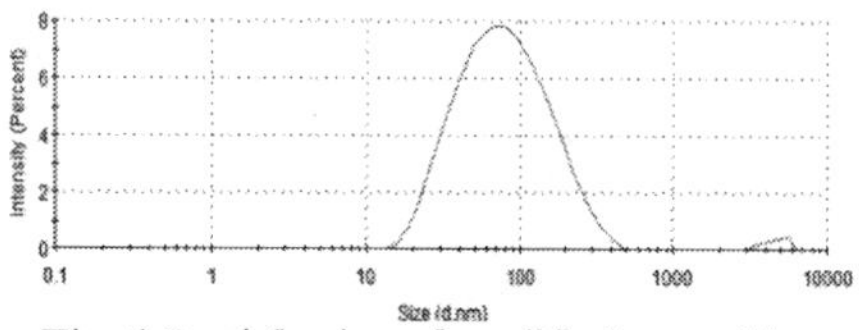

Fig. 1 Particle size of modified nanosilica

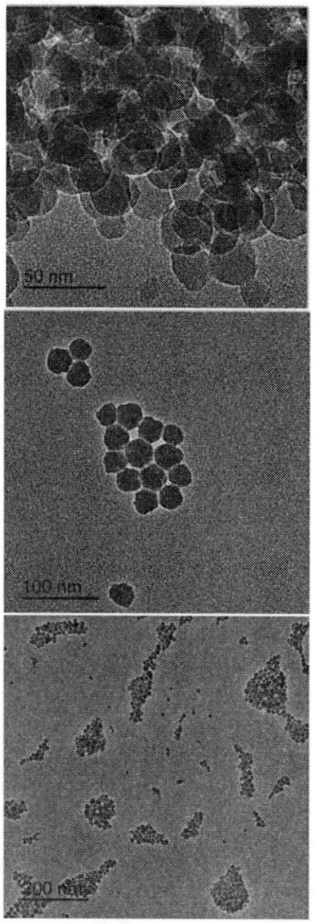

Fig. 2 TEM of modified nanosilica

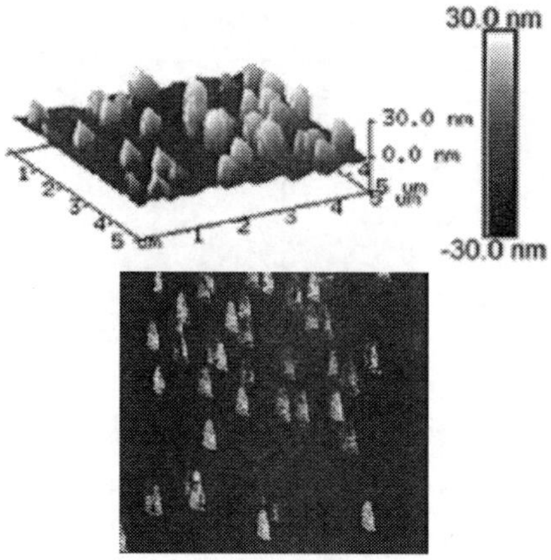

Fig. 3 AFM of modified nanosilica

As shown in Fig. 1, Fig. 2 and Fig. 3, it is clear that the particle size of the modified nanomaterials is less than 100 nm.

3.5.2 Determination of salt resistance

Modified nanosilica was added to 40 g/L salt solution at 60 ℃ in comparison with conventional modified nanosilica and unmodified silica, and the salt resistance result was shown in Fig. 4. It can be found that the modified nanosilica has a high salt resistance ability.

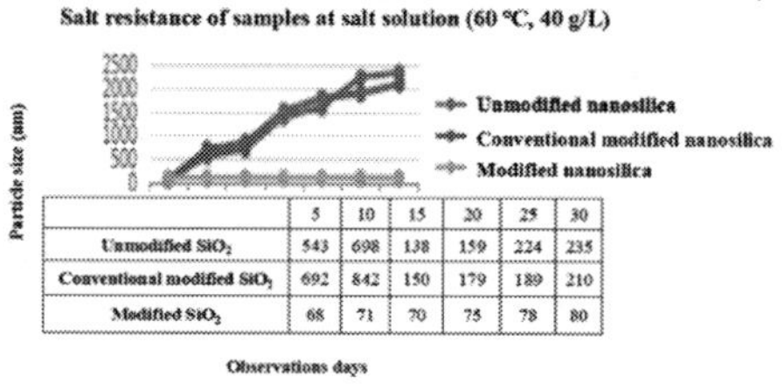

Fig.4 Salt resistance of modified nanosilica

3.5.3 Functional evaluation

The evaluation results of anti-scale and anti-swelling performance were shown in Table 8 and Table 9. It can be found that the modified nanosilica enhance the prevention percent and anti-swelling percentage obviously.

Table.8 Anti-scale performance of modified nanosilica

Sample number	$CaCO_3$ prevention percent % (100 mg/L)	$CaSO_4$ prevention percent % (100 mg/L)	$BaSO_4$ prevention percent % (100 mg/L)
Unmodified nanosilica	1.2	1.6	0
Conventional modified nanosilica	0.3	0.5	0
Modified nanosilica	68	98	10

Table.9 Anti-swelling performance of modified nanosilica

Sample number	Anti-swelling percentage % (0.5%)
Unmodified nanosilica	3.5
Conventional modified nanosilica	2.7
Modified nanosilica	14

3.5.4 Evaluation of wettability

With 40 g/L salt solution as solvent, a certain amount of 0.5% sample solution was prepared. The slide was immersed in the sample solution for 4 h, and then removed and dried, measuring the wettability of distilled water on its surface, as shown in Fig. 6(a-b). The contact angle of Fig. 6(a) was determined as 10°, indicating its hydrophily after the immersion of the unmodified nanosilica solution. The contact angle of Fig. 6(b) was determined as 102°, indicating its hydrophobicity after the immersion of the modified nanosilica solution.

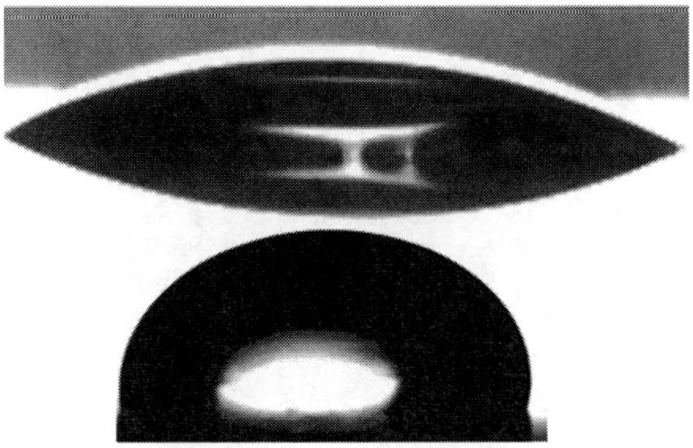

(a) Unmodified nanosilica (b) Modified nanosilica

Fig. 6 Contact angles of unmodified nanosilica and modified nanosilica

3.5.5 Performance evaluation of depressure and augmented injection

Take a cleaned core and measure the air permeability. The saturated salt solution was then loaded into the displacement device as shown in the Fig. 10, and a certain "PV" was continuously displaced with salt solution at a certain flow rate "Q" to ensure that the system was stable, and then record the pressure difference "ΔP_0" before and after. Change the flow rate "Q" , record the pressure difference before and after and calculate the permeability "K_{0W}". In the constant current mode, a modified nanosilica agent was used for displacement, and the pressure difference "ΔP_1" before and after was recorded, and different flow rates "Q" were changed, the pressure difference before and after was recorded, and the phase permeability "K_{0Y}" was calculated. After the modified nanosilica was displaced to a certain "PV", it is replaced by salt solution, and the pressure difference "ΔP_2" before and after. Change the flow rate "Q" , record the pressure difference before and after and calculate the permeability "K0Z". The following formula to calculate the depressurization rate (A) and the change of permeability (K).[20-23]

$$A = \frac{\Delta P_0 - \Delta P_2}{\Delta P_0} \times 100\%$$

$$K = \frac{\Delta K_0 - \Delta K_{0Z}}{\Delta K_0} \times 100\%$$

Three cores were collected to determine the depressure and augmented injection performance of the modified nanosilica. The evaluation results were as follows. The Table10 showed that the modified nanosilica had a considerable depressure effect.

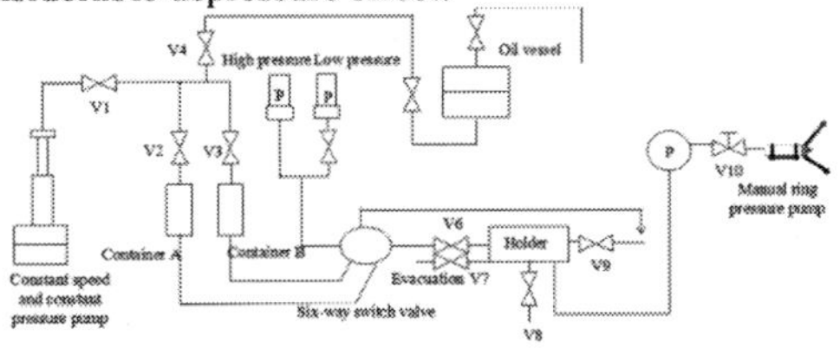

Fig. 7 Flow chart of the depressure and augmented injection and displacement device

Table.10 Performance of of depressure and augmented injection based on modified nanosilica

Core number	Core air permeability(mD)	Reduced pressure rate% (0.5%)	Rate of change in permeability% (0.5%)	Sample name
1#	3.2	-0.6	-12.3	Unmodified nanosilica
2#	4.1	1.4	0.1	Conventional modified nanosilica
3#	3.8	13	2.45	Modified nanosilica

4. Conclusions

The development of nanofluids in hypotonic and extreme low permeability reservoirs has a broad development prospect, which can be in the direction of simple process, controllable synthetic structure and long-term stable dispersion. In this paper, a modified nanomaterials were prepared by coupling agent grafting modification and surface modification technology. When the reaction temperature was 60 °C, the modification time was 6 h, the amount of nanosilica was 16%, the coupling agent was 0.8%, and the ratio of dispersant and modifier was 3:1, the particle size of the modified nanosilica is 63 nm, and the optimal modification effect was achieved. Indoor displacement experiment showed that it has a considerable effect of depressure and augmented injection.

References

1. Cui CH, Li XJ, Zhang YZ, et al. Application of novel surfactant system with low content in lowpermeability reservoir[J]. Advances in Fine Petrochemicals, 2004, 5(1): 7-9.
2. Liu K. The effect of cationic and zwitterionic surfactants[J]. Journal of Petrochemical Universities, 2013, 26(5): 55-59.
3. Ma HY, Pan Y, Yang SC. Study on determining methods of anionic surfactants[J]. Contemporary Chemical Industry, 2013, 42(3): 373-375.
4. Wang H, Ren XJ, Li DY, et al. Experimental research on decreasing in jection pressure with surfactant inLangdong oil field[J]. Liaoning Chemical Industry, 2010, 39(10): 1007-1009.
5. Fang XH. The behaviors of several kinds of surfactant system and effects on decreasing displacement pressure[J]. Petroleum Geology & Oilfield Development in Daqing, 2002, 21(2): 62-63.
6. Tang JZ, Fang XH. Research and application of decreasing injection pressure by surfactant in low-permeability reservoir[J]. Chemical Industry Times, 2004, 18(1): 51-54.
7. Qu JK, Zhu YY, Zhou GY. Application of efficiency and effectiveness on surfactants flooding system[J]. Science &. Technology in Chemical Industry, 2004, 12(1): 1-4.
8. Yang Y, Liu YB, Pu WF, et al. The research progression and prospect of a new generation of Geminis surfactants[J]. Petroleum Geology and Recovery Efficiency, 2005, 12(6): 67-70.
9. Tan ZL, Han D, Yang PH. The properties and application prospect in enhanced oil recovery of Gemini surfactants[J]. Oilfield Chemistry, 2003, 20(2): 187-191.
10. Wu YH, Gong H, Xia KJ, et al. Applicability of a nano-silica microemulsion for depressurization and injection of low permeability reservoirs[J]. Drilling & Production Technology, 2020, 43(2): 111-114.
11. Guo Y. Discussion on the technical difficulties and development countermeasures of low-permeability oilfield exploitation[J]. Chemical Industry Management, 2017 (24): 121.
12. Shu SL, Guo YC. Study and application of seepage mechanism of special low-permeability reservoir[J]. Oil and gas geology and recovery, 2016,23(5): 58-64.
13. Qu ZQ, Qi N, Wang ZQ, et al. New progress in acidification transformation of low permeability oil layer[J]. Oil and gas geology and recovery rate, 2006, 6: 93-96.
14. Yin Y, Jia JF. Ananionic complex microemulsion system is preferred and effective fruit[J]. Applied chemical industry, 2018, 47 (6): 1105-1108.
15. Gao M, Song KP. New sulfobetaine suitable for low permeable sandstone oil layer surfactant studies[J]. Oilfield chemistry, 2008, 3: 265-267.
16. Feng AZ, Zhang JQ, Jiang P, et al. Experimental study of depressurization and augmented injection by high concentration surfactant system for low permeability reservoir[J]. Oilfield Chemistry, 2011, 28(1): 69-73.
17. Zhang J, Zhang G, Ge J, et al. Laboratory studies of depressurization with a high concentration of surfactant in low-permeability reservoirs[J]. Journal of Dispersion Science & Technology, 2012, 33(11): 1589-1595.
18. Zhang X, Liu D, Li L, et al. Synergistic effects of surfactants on depressurization and augmented injection in high salinity low-permeability reservoirs: formula development and mechanism study[J]. Colloids and Surfaces A: Physicochemical and Engineering Aspects, 2021.
19. Xiong S, Shi L, Cheng S. Study of depressure and augmented injection agents for low permeability oilfields[J]. Journal of Petrochemical Universities, 2014.
20. Chen G, Lai N. Study on the adaptability of depressurization and injection of MD film in low permeability reservoirs[J]. IOP Conference Series: Earth and Environmental Science, 2019.
21. Hua Y, Li F, Li W. Decompression and augmented injection theories research of nanometer polesilicon in oilfield water flood development[J]. Natural Science Journal of Harbin Normal University, 2005.
22. Wang K, Wang C, Sun L, et al. Experimental research on decompression and augmented injection effect by using active SiO2 nano-powder in middle and low permeability cores[J]. IEEE, 2010:1-4.
23. Zhang X, Liu D, Li L, et al. Synergistic effects of surfactants on depressurization and augmented injection in high salinity low-permeability reservoirs: formula development and mechanism study[J]. Colloids and Surfaces A: Physicochemical and Engineering Aspects, 2021.

 https://doi.org/10.1051/e3sconf/202235801044

Research on Land Use Dynamic Change Based on RS Technique in The West Coast New Area of Qingdao

Leibo Wang [1], Chao Zhang [1, 2,*], Shuang Wu [3]

[1] School of Civil Engineering, Shandong Jiaotong University, Jinan.250000, P.R.China
[2] College of Geography and Environment, Shandong Normal University, Jinan. 250000, P.R. China
[3] School of Land Resources and Surveying, Shandong Agriculture and Engineering University, Jinan.250000, P.R. China

Abstract: The West Coast New Area of Qingdao has witnessed a fast-developing economy and accelerated urbanization since its establishment as a national new area, which also leads to the rapid change of land utilization status. In such a context, ways of quickly monitoring land utilization changes and leveraging land resources more efficiently and intensively are practical problem that requires urgent settlement. Focusing on Landsat8 OLI images in 2013, 2017 and 2021 of the West Coast New Area of Qingdao, data of land utilization changes in three periods were first extracted using ENVI5.3 software. And then, land utilization was dynamically monitored with the post classification comparison method. Finally, the driving force was quantitatively analyzed from the perspectives of socio-economic development, population change, and policy in combination with document literature collected, which can provide data supports for land planning of the West Coast New Area of Qingdao with suggestions proposed.

Keywords: The West Coast New Area of Qingdao; land utilization; dynamic monitoring; driving force analysis

1. Introduction

1.1 Research Background and Significance

As a new regional economic function area, the West Coast New Area of Qingdao is a representative in the new wave of opening up, reform and development. Its urban planning and land utilization changes are quite distinctive from those of traditional cities. The West Coast New Area of Qingdao is studied in this research in an attempt to investigate the characteristics and laws of land utilization changes from the perspective of temporal- spatial changes. All in all, the research findings will contribute to exploring the development planning of national new areas in the future [1], in addition to providing elementary data support for the study on national new areas. The study is conducted to provide scientific decision-making basis for land utilization and urban planning quickly and accurately.

1.2 Research Contents

Research contents are presented below:

(i) Extraction of land utilization classification data of the West Coast New Area of Qingdao in 2013, 2017, and 2021 are extracted.Images of the West Coast New Area of Qingdao in 2013, 2017 and 2021 were pre- treated via radiometric calibration, atmospheric correction, and image clipping. Based on this, land utilization types were divided for monitored classification with the maximum likelihood method through combining land utilization conditions and image characteristics of the West Coast New Area of Qingdao.

(ii) Study on the characteristics of land utilization changes in the West Coast New Area of Qingdao.Land utilization transfer matrices from 2013 to 2017 and from 2017 to 2021 were obtained from the land utilization classification results of the West Coast New Area of Qingdao in 2013, 2017 and 2021. And the dynamic change laws, including the law of land type quantity, change speed, and mutual conversion between different types of land utilization in the West Coast New Area of Qingdao, were analyzed in accordance with the land utilization change information displayed in the transfer matrix. After that, the overall information and law of land utilization change in the West Coast New Area of Qingdao from 2013 to 2021 were obtained completely.

(iii) Study on the driving force mechanism of land utilization change in the West Coast New Area.Some practical problems can be found in land utilization and urban construction in the West Coast New Area of Qingdao through analyzing the characteristics of land utilization change and driving factors behind the change from the perspectives of population mobility, policy

* Corresponding author: naonao81@126.com

https://doi.org/10.1051/e3sconf/202235801044

implementation, and socio-economic development. In response these problems, solutions and suggestions are put forward by means of referring to open statistical data together with the characteristics of the the West Coast New Area of Qingdao.

2. Research on Land Utilization Classification

2.1 Acquisition of Remote Sensing Images

Image data utilized in this paper were acquired from geospatial data cloud, with LandSat8 OLI image data products with cloud amount not exceeding 5% selected. Images were acquired on Aug. 23th, 2013, Sept. 3rd, 2017 and Aug. 10th, 2021, respectively, which are images in summer.

2.2 Pre-treatment of Remote Sensing Image

After radiation calibration and atmospheric correction preprocessing of the three phase images, from the spectral curves of various land types, the spectral curves of the ground objects displayed on the remote sensing images after atmospheric correction become smooth, close to the spectral curve characteristics of the actual ground objects. As shown in Figure 1, 2 and 3 below, the image quality has been improved to a certain extent, laying a foundation for land use type classification.

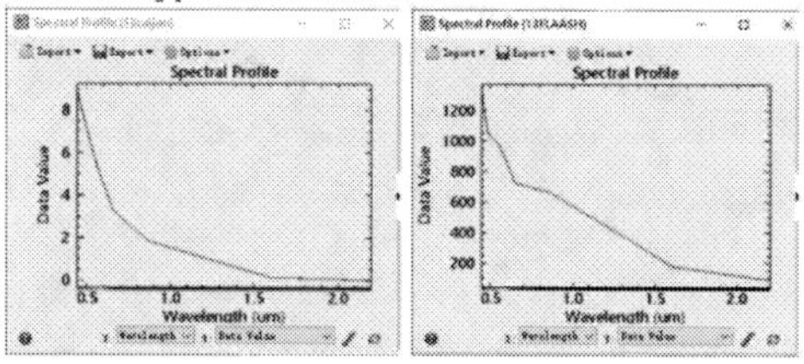

Fig.1 Spectral reflectance of water area before and after atmospheric correction

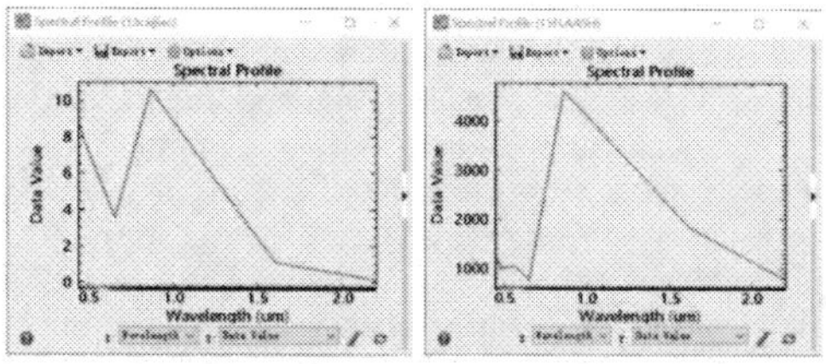

Fig.2 Spectral curves of agricultural and forestry land before and after atmospheric correction

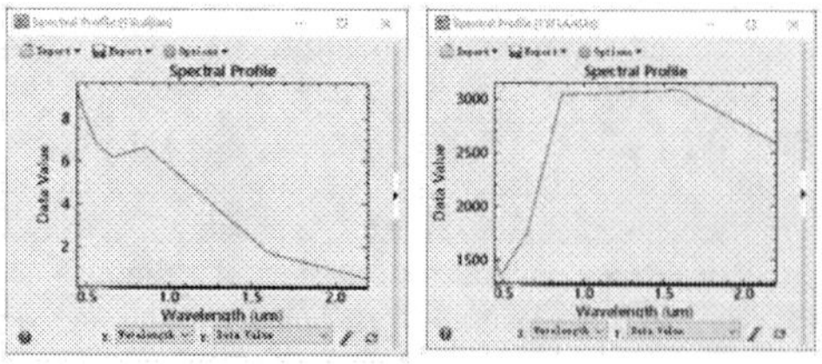

Fig.3 Spectral curve of building land before and after atmospheric correction

2.3 Extraction of Information on Land Utilization Change

To begin with, interpretation marks of water area, agricultural and forestry land, and building land are determined. Samples of the type of ground objects can be directly determined through visual interpretation in the pre-treated remote sensing image. Then the classification of images is supervised using the likelihood method based on the training samples selected at an earlier stage. Meanwhile, the classified land utilization patches are treated upon classification, such as clustering and merging. Further, the supervised classification results can be evaluated in terms of accuracy using the confusion matrix, as shown in Table 1:

Table 1. Classification accuracy evaluation

	2013	2017	2021
Classification Accuracy	98.2792%	98.3764%	98.6642%
Kappa coefficient	0.9744	0.9765	0.9822

As can be observed from Table 2.1, the overall classification accuracy and kappa coefficient of the remote sensing images of the West Coast New Area of Qingdao in 2013 after supervised classification are 98.2792% and 0.9744, respectively; the overall classification accuracy of images and the kappa coefficient in 2017 are 98.3764% and 0.9765, respectively; the overall classification accuracy and the kappa coefficient in 2021 are 98.6642% and 0.9822, respectively. Apparently, the accuracy of the supervised classification results in 2013, 2017 and 2021 is in conformity with the accuracy requirements of land utilization classification. Results of land utilization classification in Phase III are obtained, as shown in Fig. 4, Fig. 5and Fig. 6:

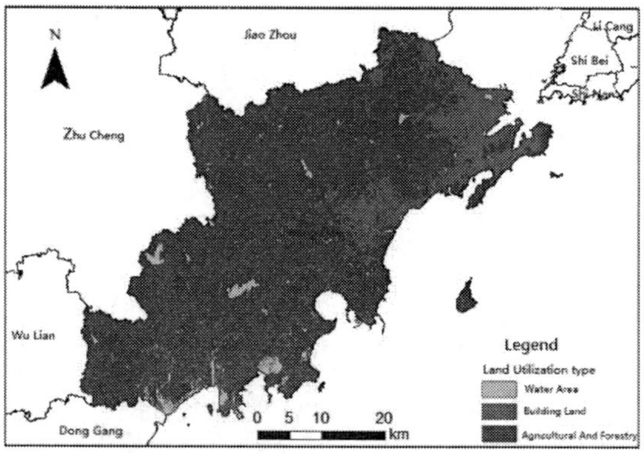

Fig.4 2013 land utilization classification map

 https://doi.org/10.1051/e3sconf/202235801044

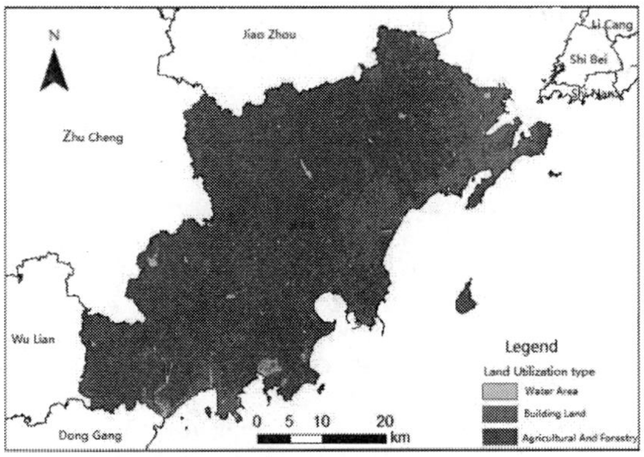

Fig.5 2017 land utilization classification map

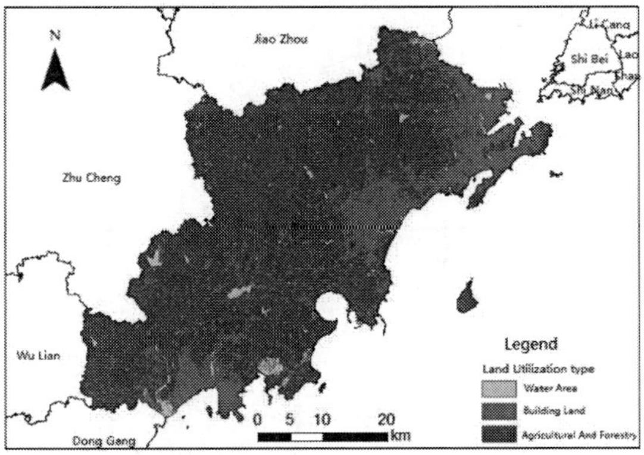

Fig.6 2021 land utilization classification map

3. Analysis of Changes in Land Utilization

3.1 Data Statistics of Land Utilization Types

The areas of different land Utilization types in different years are obtained using ENVI5.3 software, the classified statistical tool. And the area transfer matrix of land utilization types can be generated according to the area changes of different land utilization types, which presents the mutual conversion of various land utilization types in the study period. The transfer matrix of the land utilization area is shown in Table 2 and Table 3.

Table 2. 2013-2017 land utilizationArea Transfer Matrix (km^2)

2013 / 2017	Water Area	Building Land	Agricultural and Forestry	Total
Water Area	26.7822	0.9387	3.3498	31.0707
Building Land	18.2331	339.2181	136.0638	493.5150
Agricultural and Forestry	13.5549	77.9472	1447.3188	1538.8209
Total	58.5702	418.104	1586.7324	2063.4066
Change	-27.4995	75.4110	-47.9115	

Table 3. 2017-2021 land utilizationArea Transfer Matrix (km^2)

2017 / 2021	Water Area	Building Land	Agricultural and Forestry	Total
Water Area	26.2809	13.0311	9.8208	49.1328
Building Land	3.8097	418.0428	181.143	602.9955
Agricultural and Forestry	0.9801	75.4722	1347.8571	1411.2783
Total	31.0707	493.515	1838.8209	2063.4066
Change	18.0621	109.4805	-127.5426	

The water area of the West Coast New Area of Qingdao was decreased by 27.5 square kilometers, with the area of building land increased by 75.42 square kilometers, and the area of agricultural and forestry land decreased by 47.9 square kilometers from 2013 to 2017. By comparison, the water area of the West Coast New Area of Qingdao was increased by 18.06 square kilometers, with the area of building land increased by 109.48 square kilometers, and the area of agricultural and forestry land decreased by 127.54 square kilometers from 2017 to 2021. The area of building land in the West Coast New Area of Qingdao was on the rise from 2013 to 2021, whereas the areas of agricultural and forestry lands were decreased due to the rapid economic development and accelerated urbanization process of the New Area. according to data analysis. As a result, large areas of other land types have been converted into building lands. But the area of water area is presented in a trend of first decreasing and then increasing, which was decreased by 27.5 square kilometers from 2013 to 2017, with a decline of 47% and recovered by 18.06 square kilometers from 2017 to 2021.

3.2 Analysis of Transferred Land Utilization Type

The transfer of different types and years is counted according to the classification results of land utilization types. Results are shown in Table 4 and Table 5:

Table 4. Dynamic change table from 2013 to 2017

Land Change Type	Conversion Area（km^2）
Water Area To Building Land	18.2331
Water Area To Agricultural And Forestry	13.5549
Building Land To Water Area	0.9387
Building Land To Agricultural And Forestry	77.9472
Agricultural And Forestry To Water Area	3.3498
Agricultural And Forestry To Building Land	136.0638

Table 5. Dynamic change table from 2017 to 2021

Land Change Type	Conversion Area（km^2）
Water Area To Building Land	3.8097
Water Area To Agricultural And Forestry	0.9801
Building Land To Water Area	13.0311
Building Land To Agricultural And Forestry	62.4411
Agricultural And Forestry To Water Area	9.9208
Agricultural And Forestry To Building Land	181.143

https://doi.org/10.1051/e3sconf/202235801044

The largest conversion area is from agricultural and forestry land to building land from 2013 to 2017, reaching 136.0638 square kilometers, followed by the conversion from building land to agricultural and forestry lands, covering 77.9472 square kilometers, according to data analysis results of Table 4. Evidently, mutual conversion between agricultural and forestry land and building lands is a primary characteristic of land changes in the West Coast New Area of Qingdao from 2013 to 2017. The transferred amount from the water area to other area types is far greater than that from the water area to other area types, resulting in the declined area of water area.

The largest conversion area is, equally, from agricultural and forestry land to building land from 2017 to 2021, reaching 181.143 square kilometers, followed by the conversion from building land to agricultural and forestry lands, covering 62.4411 square kilometers, according to data analysis results of Table 5. It can be seen that the land change of the West Coast New Area of Qingdao from 2017 to 2021 is also dominated by the mutual conversion between agricultural and forestry land and building land.

3.3 Summary and Analysis of Changes in Land Utilization Type

The overall change trend is the conversion of other land types such as agricultural and forestry land and water to building land, with an obvious change in the coastal and rural-urban integration areas. From the spatial perspective, there is a great change of land utilization in the central and southwestern areas. To be specific, an increase has occurred in the central building land due to the change of the development center of the West Coast New Area of Qingdao, leading to a gradual rise of building land area in the southwest in 8 years. And land utilization type is change slightly in the northern and eastern regions because of relatively stable land utilization structure with high development degree of lands such as the economic and technological development zone, bonded port area and Xuejiadao area in the east.

4. Analysis on the Driving Force of Land Utilization Change

According to land utilization change, main factors driving the changes in land utilization in the West Coast New Area of Qingdao are presented below:

Demographic factors: Tremendous foreign population inflows have been attracted by the favorable development prospects, the incentive policy of investment promotion and talent introduction, and the livable climatic conditions since the West Coast New Area of Qingdao was formally established in 2014. And the rising area of building land area is a primary land utilization change resulted from population growth[2]. For example, the area of building land was increased by 184.9 square kilometers from 2013 to 2021, indicating a change significantly higher than other land types. Therefore, it can be seen that the land utilization structure is more significantly changed with population factors.

Policy factors: The overall pattern of land utilization change in the West Coast New Area of Qingdao is affected by policy factor, leading to an increasing trend of urban building land area. With the increase of building land, the water area was once occupied and destroyed in a large area. In response to it, the government of the West Coast New Area has attached great importance to ecological protection through issuing a series of relevant protection measures and policies since 2018. And remarkable results have been achieved in the ecological protection of the West Coast New Area of Qingdao by the end of September 2021.

Socio-economic development: The West Coast New Area of Qingdao has enjoyed a steady growth in economic development year by year since its establishment as a national new area, with GDP accounted for 30.9% of the city's GDP[2] in 2021. Moreover, the building land area was increased by 184.9 square kilometers in the West Coast New Area of Qingdao from 2013 to 2021. Hence, socio-economic development is a significant factor driving the land utilization change of the West Coast New Area of Qingdao.

5. Conclusion

Land utilization types are classified in the study based on the phase III images of the West Coast New Area of Qingdao from 2013, 2017 and 2021 before analyzing the conversion between different land utilization types in different periods. The results show that agricultural and forestry land and water areas being conversed into building land is a major conversion structure of land utilization in the West Coast New Area of Qingdao at this stage. It is an inevitable cause of urban expansion. Serious challenges, however, pose severe challenges to local ecology along with the rapid development of the city. In such a context, undeveloped building lands should be checked in the subsequent development process to promote the efficiency of building land utilization. Moreover, ecological protection red line must be obeyed to protect cultivated land, thereby achieving the balance between economic development and ecological protection.

References

1. Xu Weifang,Chen min. Detection and driving forces of land use change in Beijing based on Landsat images [J]. Beijing surveying and mapping, 2018(01):58-61.
2. Xue yawei,Zhang jian,Yun lexin. Research on industrial life Cycle and Evolution of National New District -- Based on the data of West Coast New District[J]. Technical Economics and Management studies,2018(03):120-124.
3. Qiao Haiyang,Ma dandan,Wang tian. Analysis of land use change and driving force in Feicheng city[J]. Beijing surveying and mapping, 2020(03):376-380.
4. Wang huili. Study on land use change and its driving force during urbanization in Zhengzhou City[D]. Henan university,2017.
5. Daniel, C.B., Mathew, S. & Subbarayan, S. GIS-Based Study on the Association Between Road

 https://doi.org/10.1051/e3sconf/202235801044

Centrality and Socio-demographic Parameters: a Case Study. Journal of Geovisualization & Spatial Analysis, 2022, 6(1):1.

https://doi.org/10.1007/s41651-021-00095-6.

6. Fu wei,Zhou baotong,Tian yu. Study on land use change and driving force in Rongchang District of Chongqing[J]. Jiangsu Agricultural Sciences, 2020(06): 214-222.
7. Chen ruiqin. Driving forces of land use change in Qingdao City [J]. Land and Natural resources Research, 2008(4):2.
8. Hecht, R., Behnisch, M. & Herold, H. Innovative Approaches, Tools and Visualization Techniques for Analysing Land Use Structures and Dynamics of Cities and Regions (Editorial). Journal of Geovisualization & Spatial Analysis, 2020, 4(2): 19. https://doi.org/10.1007/s41651-020-00060-9.
9. Li xindong,Wang lina,Guo weili. Monitoring and analysis of urban spatial expansion in Beihai City based on RS and GIS [J]. Mapping and spatial geographic information, 2016, 39(1):3.
10. Jin mingyi. Research on urban spatial expansion of Harbin based on RS and GIS [J]. Mapping and spatial geographic information, 2016, 39(11):4.

Development and Application of Educational Mini CNC Milling Machines

Xin'an Zhou, Junbing Pan

Shaanxi Institute of Technology, Xi'an 710300, China

Abstract. Industrial CNC milling machine is huge in volume, complex in structure, poor in system openness and expensive in price, which increases the cost of training relevant professional and technical personnel in enterprises and schools, and restricts the further development of CNC milling machine. In order to solve this problem, based on the existing domestic CNC milling machine and previous research results, this paper designs a teaching type micro CNC milling machine with high integration, low cost, convenient handling, simple structure and high flexibility, and carries out processing experiments on the micro CNC milling machine. The experimental results show that when the spindle speed is 5000 r/min, the feed rate is 2000 mm/min and the engraving cutter is 1 mm, the machining error of the teaching type micro CNC milling machine is 0.1 ~ 0.2 mm, can well reflect the shape of the object to be processed, can meet the students' understanding and learning of CNC machine tools, and meet the needs of carving enthusiasts, in order to make contributions to the further development of CNC milling machines in China.

Keywords: Micromachine, CNC machine, machining, CNC milling machine.

1. Introduction

The appearance of CNC machine tools has accelerated the pace of modern machining technology [1,2]. Since the 20th century, CNC equipment has been widely applied in industrial production, and the advanced level of CNC machine tools has become an important symbol for evaluating the level of a country's manufacturing industry [3,4]. At present, with the rapid industrial growth in China, the number of traditional CNC milling machines and miniature CNC milling machines is increasing year by year [5], which has further improved the quality requirements of machine tool maintenance and repair personnel. Industrial CNC milling machines feature complex structures [6-8], poor openness of the system, and high prices, which increase the training cost of enterprises and schools for related professionals and restrict the further development of CNC milling machines. In addition, existing mini CNC milling machines are most widely used to perform precision machining, and they are so expensive to run that have hindered the popularization of machine tools. Therefore, it is urgent to design a type of mini CNC milling machines with simple structure and affordable prices, to facilitate operations of disassembly, learning and civilian promotion.

To address the above problems, based on current CNC milling machines and prior research results in the country, the present study designs an educational mini CNC milling machine that is characterized by high integration [9], low cost, convenient handling, simple structure and high flexibility, with the purpose of contributing to the training of CNC machining professionals in our country, while at the same time fulfilling the needs of engraving enthusiasts, thereby pushing forward in-depth development of CNC milling machines domestically.

2. Design of machine tool structures

2.1 Machine tool structure.

It is a 3-axis CNC milling machine and it shares basically the same structure with industrial CNC milling machines. It operates on the X, Y and Z axes and is composed of a transmission system, workbench, guide rail, column, base, control system, etc., as shown in Figure 1. Its related parameters are shown in Table 1.

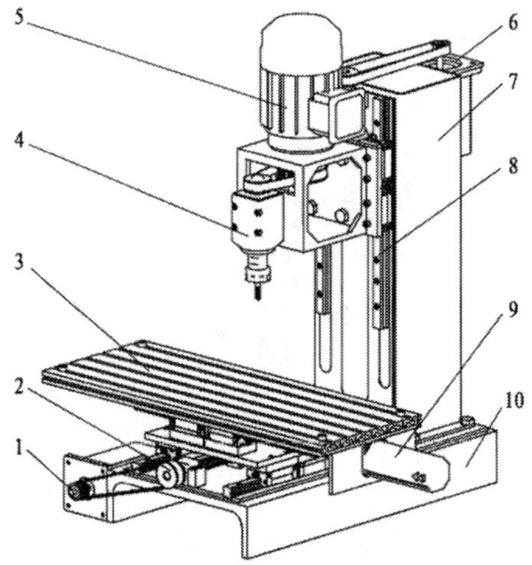

1. Y-axis motor; 2. Y-axis drive pulley; 3. Workbench; 4. Z-axis cutter installation device;
5. Z-axis motor; 6. Spindle driving motor; 7. Column; 8. Guide rail; 9 X-axis motor ; 10. base

Fig. 1 Teaching Micro NC milling machine

2.2 Working Principles.

The educational mini CNC milling machine moves along the X, Y, and Z axes driven by the axis motors. That is, the motors drive the workbench to move forward and back, left and right and lift and rotate the tool. During operations, fix a workpiece on the workbench first, and prepare for clamping and process transmission. Then, find the origin of the tool relative to the workpiece using the tool setting method of industrial CNC machining. Finally, complete the machining process of the workpiece manually or by driving the X, Y and Z axis motions through the numerical control system.

Table 1 Three Scheme comparing

project	parameter
Overall dimension (length × wide × Height) /mm	700×600×750
Three axis stroke /mm	400×200×200
Spindle power /W	750
Maximum spindle speed / ($r \cdot min^{-1}$)	10000
Maximum feed speed of spindle / ($mm \cdot min^{-1}$)	12000
Maximum tool diameter /mm	16
Cooling mode	Air cooling
Error /μm	≤ 10
control system	XC09MC
Processing materials	Acrylic, wood, hard aluminum, steel (soft), etc

The numerical control system can precisely control the position of all three axes (X, Y, Z), as well as precisely controlling the rotational speed of the Z-axis where the cutting tool is located. In this CNC milling machine, the control system adopts the internationally recognized G and M codes.

3. Control system

This mini CNC milling machine does not have high levels of precision or large cutting load, but its open-loop control system has strengths of a simple structure, low costs, being easy to debug and maintain, etc. Incorporating the design requirements of this miniature machine tool into comprehensive considerations, the present study decides to adopt an open-loop control electric spindle in the spindle system, and an open-loop stepping motor system in the feeding system. Figure 2 is a block diagram of the open-loop control system.

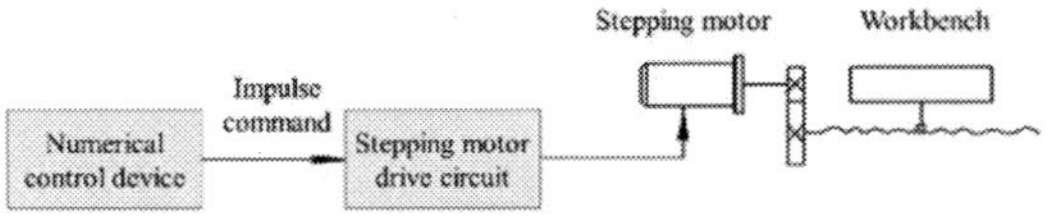

Fig. 2 Block diagram of open loop control system

This mini CNC milling machine for educational settings adopts XC09MC CNC system. It can control the machine tool to realize milling, tapping, boring, drilling and other machining processes. Using a high-performance 32-bit microprocessor, the CNC system can realize real-time multi-task control technology and hardware interpolation technology, with high interpolation accuracy up to 0.001 mm. In addition, the system has the following advantages:

(1) A simple operational interface comes equipped with 5 shortcut function keys (from F1 to F5) and is easy to operate;

(2) It can recognize the G and M codes, is compatible with FANUC instructions, highly adaptable and easy to learn;

(3) Program files can be transferred via USB, which greatly improves the efficiency of the machine tool;

(4) The interface is available in two languages: Chinese and English, which lowers the threshold for entry-level learners;

(5) It can support multiple operation authorities, which facilitates equipment management, and can lock the system within a time limit, which improves system security;

(6) It has good compatibility and can support processing codes written by third-party software.

4. Finite element design

A frame model is created in SolidWorks, as shown in Figure 3. The material of the frame is Q235. The model is meshed and force is exerted. The Z-axis device and the counterweight device are shown in simplified diagrams, with a gravitational force of 500 N and 300 N, respectively. The lower end of the base is fixed on the level ground, and a simulation is carried out in the Simulation module of SolidWorks. The result is shown in Figure 4.

 https://doi.org/10.1051/e3sconf/202235801045

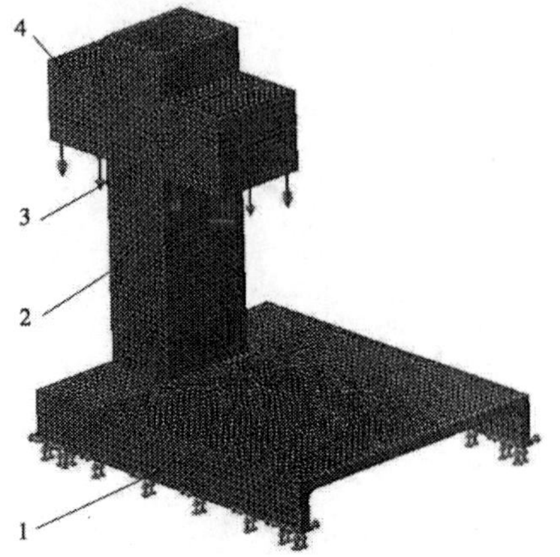

1. Base; 2. Column; 3. Z-axis device;
4. Counterweight device

Fig. 3 Frame model

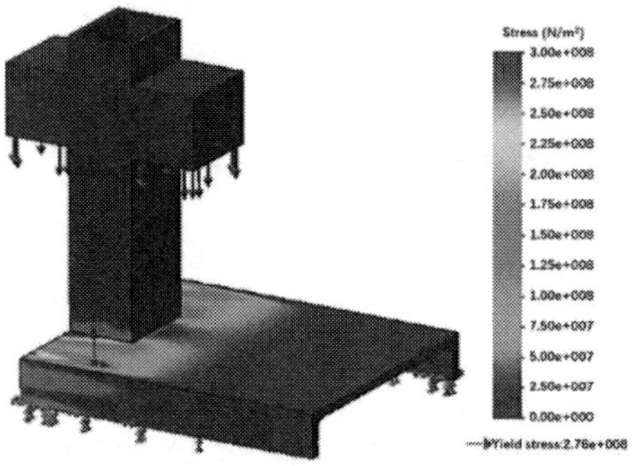

Fig. 4 Stress cloud

It can be seen from Figure 4 that the yield stress of the frame is 2.76×10^8 N. When the actual stress is greater than the yield stress, the frame will deform dramatically and the maximum value of the deformation can reach up to 0.21mm, which reduces the accuracy of the machine tool. The maximum deformation will occur on the left and right sides of the beam column joint. Under the action of the counterweight, no large deformation takes place in the frame, so the frame structure is reasonably set and can meet the working requirements of the miniature CNC milling machine. If overloading is considered, a reinforcing rib can be added to the beam column joint to improve rigidity there.

Fig. 5 Test site

Fig. 6 Machining shape

Fig. 7 Digital model

5. Experimental studies

5.1 Materials and methods.

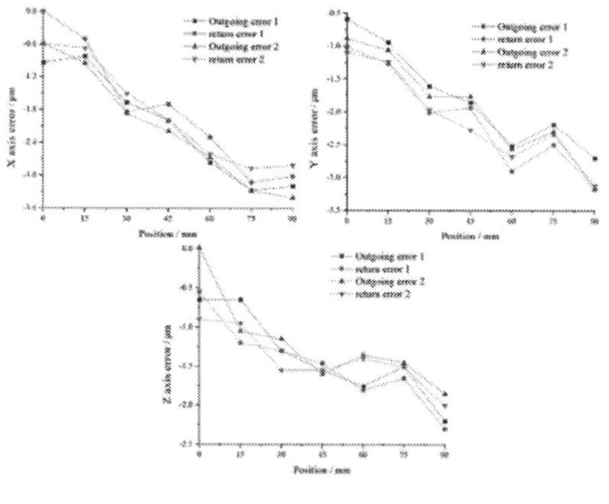

Fig. 8 Deviation diagram of measurement results and set stroke

In order to test the performance of the educational mini CNC milling machine, prototype testing was conducted in a training center of Shaanxi Institute of Technology on March 30, 2022. The testing site is shown in Figure 5. In this experiment, the school badge of Shaanxi Institute of Technology was taken as a case study, as shown in Figure 6. Considering the complexity of the pattern, in order to improve coding efficiency and accuracy, a digital model of the pattern was created before the test on a 100×100mm square in UG, as shown in

Figure 7. Using the CNC machining module in UG, a 1 mm engraving tool was selected and the spindle speed was set at 5000 r/min; the feed rate was set at 2000 mm/min to reduce cutting resistance, and a 3-axis machine tool path for processing the pattern was generated. The simulation was verified, after which the tool path was post-processed to obtain the corresponding machining program. Input the program into the CNC machine tool control system using a U disk to complete preparations for pre-machining.

In the stage of clamping, a 100×100×30 mm workpiece was fixed on the workbench. Tool setting operations was completed through the photoelectric edge finder SOE-20S, and a machining coordinate system was established with regard to the machine tools. Switching to the automatic mode, machining of the workpiece was completed. After that, the Contura G2 3-coordinate measuring machine was employed to detect errors in the shape and size of the machined pattern. The errors measured were between 0.1 and 0.2 mm. At the same time, the error analysis experiment of each axis of the machine tool is carried out, and the experimental results are shown in Figure 8.

5.2 Test result analysis.

As can be seen from Figure 8, the X, Y, and Z axes of the mini CNC milling machine developed in this project shall reach a positioning accuracy up to ±8 μm, and a repeated positioning accuracy of ±5 μm. Compared with the experiment data, the teaching prototype has reached the design requirements raised by the project in terms of the positioning accuracy and repeated positioning accuracy of the X, Y, and Z axes.

5.3 Error analysis.

Although the miniature CNC prototype machine meets the design criteria from the perspective of the test results, it should still reach high levels of machining accuracy.

There is still a lot of room for improvement in machining. The main accuracy errors are listed below:

(1) There are installation errors for various machine tool components, which lower the accuracy of workbench movements and ultimately affect the experiment results pertaining to the accuracy of the machine tool.

(2) The open-loop control method is adopted in this mini CNC milling machine, so it cannot automatically correct deviations. There is only sequential and one-way control between the input and output ends and no reverse connections. If the characteristics and parameters of the components of the system are relatively stable, and external interference rarely shows up, open-loop control can still maintain accuracy to some degree. However, due to random errors and external interference, the motion unit is bound to have certain errors.

(3) There will be errors in the machining accuracy of the machine tool because of the backlash of each mechanical moving part of the machine tool.

(4) Measurement of the experimental environment (noise and vibration) will also cause certain positioning errors.

6. Discussion and Conclusion

While the industrialization process in our country sees rapid advances, the number of CNC milling machines has increased year by year, leading to greater demand for related professionals and technical personnel. However, traditional CNC milling machines are very large, have complex structures and can be quite costly, which increases the training cost of enterprises and schools, and restricts further development of CNC milling machines. To address the above problems, based on current CNC milling machines and prior research results in the country, the present study has developed an educational miniature CNC milling machine that is characterized by high integration, low cost, convenient handling, simple structure and high flexibility, and carried out machining tests on this mini CNC milling machine for educational settings. The test results show that when the spindle speed was 5000 r/min, the feed rate was 2000 mm/min, and the machining error was 0.1~0.2 mm when machining with a 1 mm engraving knife, positioning accuracy up to ±8 μm, this mini CNC milling machine could well reflect the shape of the machined object, which could meet students' requirements for learning the structures and working principles of CNC milling machines while also satisfying the needs of engraving enthusiasts. In colleges and universities, enterprise-level training units and scientific research institutes are capable of replacing industrial CNC milling machines with this educational mini CNC machine as teaching and training equipment with the purpose of vastly improving the teaching effect. It is expected that the development of this educational mini CNC milling machine can promote the training of CNC machining technicians in China and play a positive role in promoting "Made in China 2025".

Acknowledgments

Thank you for the support of the fund for the key research project of and Fund for ideological and political research and practice of school curriculum in China's "double high school plan" (SGKCSZ2020-794).

References

1. LOU S J, Chen J C. In-process surface recognition of a CNC milling machine using the fuzzy nets method. Computers & Industrial Engineering, Vol. 33 (1997) No. 10, p. 401-404.
2. LU Z N, Yoneyama Takeshi. Micro Cutting in the Micro Lathe Turning System. International Journal of Machine Tools & Manufacture. Vol. 39 (1999) No. 7, p.1171-1183.
3. LI H T, LAI X M, LI C F, et al. Meso-scale Milling Machine Tool and Its Dynamic Performance Analysis. Journal of Shanghai Jiao Tong university, Vol. 42 (2008) No. 7, p. 1041-1045 .
4. Koren Y, Heisel U, Jovane F, et al. Design of Reconfigurable Manufacturing Systems. CIRP

Annals -Manufacturing Technology, Vol. 48 (1999) No. 2, p. 527-540 .

5. Milos Glavonjic, Dragan Milutinovic, Sasa Zivanovic, et al. Desktop 3-axis parallel kinematic milling machine. The International Journal of Advanced Manufacturing Technology, Vol. 46 (2010) No. 1, p. 51-60.
6. B. Malleswara Swamil, K. Sunil, Ratna Kumar. Design and structural analysis of CNC vertical milling machine bed. Swami et al, International Journal of Advanced Engineering Technology. 2012, p. 97-100.
7. Migbar Assefa. The Effect of Tapers, Apertures and Ribs On the structural Dynamics of Machine Tool Column Using Finite Element Method. Canadian Journal on Mechanical Sciences & Engineering, Vol. 2 (2011) No. 8, p.185-94.
8. JIANG T, Chiredast M. A System Approach to Structural Topology Optimization: Designing Optimal Connections. Journal of Mechanical Design. Vol. 119 (1997) No.1, p. 40-47.
9. M. F. Zaeh, Th. Oertli, Finite element modeling of ball screw feed drive systems. CIRP Annals-Manufacturing Technology, Vol. 53 (2004) No.1, p. 287-294.

E3S Web of Conferences **358**, 01046 (2022) https://doi.org/10.1051/e3sconf/202235801046
GESD 2022

Introduction to Marine Data Source Analysis and Sharing

Shanshan Geng, Zhenmin Liu and Xiaoxiao Shi *

National Marine Data and Information Service, Tianjin, China

Abstract. You should leave 8 mm of space above the abstract and 10 mm after the abstract. The heading Abstract should be typed in bold 9-point Times. The body of the abstract should be typed in normal 9-point Arial in a single paragraph, immediately following the heading. The text should be set to 1 line spacing. The abstract should be centred across the page, indented 17 mm from the left and right page margins and justified. It should not normally exceed 200 words.

1. Introduction

The development of marine science has roughly gone through three stages: the first is the theoretical traction stage due to the lack of observational data, the scientific research at this stage mainly relies on the establishment of scientific theories to explain ocean phenomena. The second is the observational traction stage, with vigorous development of different observation means and instruments and equipment, human footprints have advanced from the offshore to the deep sea, and established a three-dimensional observation network covering the world's oceans. The third is the data traction stage. With the continuous enrichment of observational technologies, data sources are becoming more extensive and diverse, and the amount of data has increased to the PB level [1]. It provides a strong foundation for scientific research, social economy and national security in the field.

The ocean data described in this paper [2]mainly refers to the data obtained from various types of observations and simulations conducted in the marine natural environment, including the dynamic elements of the background field of the ocean environment and the ecological elements of the ocean environment.

2. Ocean Data Sources

There are a wide range of marine data sources. According to the means and methods of production, this paper summarizes the marine data sources into on-site observation data, satellite telemetry data and reanalysis assimilation data. The relationship between the three is inseparable and complementary to each other. On-site observation mainly refers to shore-based, sea/subsea-based and air-based: shore-based refers to the coastal, island, offshore platform and other fixed ocean observation platform, the coastal waters of the marine hydro-meteorological environment, marine ecological environment for observation and monitoring, to obtain a long time series of marine information; sea/subsea base refers to the most basic mobile observation platform, such as marine research vessels, to obtain marine hydrometeorology, marine biochemistry, seafloor topography and geomorphology, seafloor sediments, marine geophysical fields and other elements, which are in the dominant position in marine data sources; space-based in this species refers to unpowered aerostats flying in the stratosphere, including zero-pressure high-altitude balloons, large overpressure balloons, small overpressure balloons, infrared hot air balloons, etc., generally used to measure the physical and chemical properties of the atmosphere above the near-surface layer, and high-altitude balloons are mainly used in oceanographic surveys to observe meteorological elements. Satellite telemetry data refers to all kinds of marine and atmospheric remote sensing products reflected by altimeters, radiometers, and scatterometers carried by marine satellites and meteorological satellites, which have the advantages of wide coverage, fast speed, high efficiency, and not limited by national boundaries and geography compared with shore-based, sea/subsea-based and air-based observation means. In addition, the reanalysis data based on the dynamic ocean-atmosphere model and assimilation technology effectively make up for the uneven spatial and temporal distribution of the data, and are irreplaceable fundamental information for gaining insight into the ocean-atmosphere motion at multiple spatial and temporal scales and exploring the mechanisms of sea-air interaction.

2.1 In-situ observation data

2.1.1 Special domestic marine survey,

The first-hand data obtained from the survey is the most direct and objective reflection of various marine phenomena and distribution patterns, and is the basic resource for conducting various types of marine scientific research and applications. Since China independently launched marine survey in 1958, it has acquired a huge

* Corresponding author: shannuist@126.com

amount of marine survey data, and successively carried out major national projects, polar expeditions and ocean resource and environment surveys. The results obtained have gained important progress in related fields and expanded the national influence. China's marine scientific investigation has undergone a gradual metamorphosis from none to have, the region from offshore to ocean, the ship from backward to advanced, equipment from reliance on imports to independent development, investigation and research personnel from less to more, the investigation environment from difficult to superior.

2.1.2 Domestic marine observation/monitoring business.

China's modern independent operational marine observation began in the early 20th century. In 1958, 119 marine hydrometeorological observation stations were set up along the coast of the country. Then, the ice survey began in the 1960s, and the voluntary ship observation began in the 1970s. The buoy observation began in the early 1980s. As of 1997, there were 524 various coastal observation stations in China, and 248 marine pollution monitoring stations nationwide [3]. In 2002, with the construction of China's marine environment monitoring system, the operational observation system was further developed, and the collection frequency and transmission time frame of observation data reached the minute level. After decades of construction and development, China's ocean observation/monitoring data sources continue to expand, management systems continue to improve, data types, quantities and timeliness grow rapidly, and data processing capabilities and technologies continue to improve. In terms of ocean observation, a global three-dimensional ocean observation network consisting of ocean stations, buoys, volunteer ships, high-frequency ground wave radar, X-band radar and standard ocean observation sections, combined with offshore oil and gas platforms, satellite/aerial remote sensing and emergency mobile platforms, has been built. In terms of marine monitoring, it has covered marine environmental quality, marine ecological status, marine environmental supervision, public welfare services, marine ecological and environmental risks, marine environmental carrying capacity monitoring and early warning, etc. The monitoring scope covers the waters under China's jurisdiction and the four Fukushima nuclear radiation monitoring areas in the western Pacific. In terms of seafloor observation, the series of ocean observation instruments and land are connected using submarine optical cable, which realizes real-time data return of ocean observation instruments and realizes all-weather, multi-parameter, long-term, real-time and continuous observation. It is a new observation platform for exploring marine scientific research and the current development needs of marine science, marine ecological research, national security analysis and natural disaster prevention.

2.1.3 The International Ocean Observation Cooperation Program,

The Program started in the 1960s, is carried out about every 10 years. The International Geophysical Year from 1956 to 1959 was the first large - scale international joint operation in the history of ocean observation. The Decade of International Ocean Expeditions from 1971 to 1980, went through a medium-sized plan from 1980 to 1985, and developed into a global change from [4]. Since then, countries and organizations still carry out ocean observation plans on a global scale through cooperation and other means such as Argo, WOD, DBCP, ICOADS and other international organizations/plans. It has realized real-time update and release of observation data and information products such as ocean temperature and salinity, ocean currents, meteorology, geology and topography for global users.

Table 1. Major marine international programmes

No.	Plan Name	Country or Organization	Time	China's Contribution
1	*WOCE*	National Oceanographic Data Center	1990~2002	A large number of marine survey data obtained from multiple voyages such as the multidisciplinary comprehensive survey of the Western Pacific, the Sino-Japanese Kuroshio cooperative survey, and the Southern Ocean scientific survey were exchanged as WOCE data.
2	ARGO	CORIOLIS Operational Oceanography(USGODA)	1995 to present	China's Argo Ocean Observation Project has built the basic platform of China's Argo Ocean Observation Network for the first time in the northwestern Pacific Ocean adjacent to my country, creating a long-term, high-resolution and large-scale real-time monitoring of temperature and salinity in the water depth of 0-2000 meters in my country. Monitor business.
3	TOGA-COARE	World Climate Research Program (WCRP)	1985 to present	Eight TOGA expeditions were conducted between 1985 and 1990. Three expeditions were conducted between 1992 and 1993, contributing a large number of observations to the entire program.
4	JOGFS	JGOFS International Program Office	1990~2000	From 1991 to 1995, the key project voyages of the "East China Sea Shelf Edge Ocean Flux Research" were carried out . From then to 1999, more than 80 voyages were carried out in the Taiwan Strait, East China Sea, West Philippine Sea and other waters, all of which were included in the international JGOFS. The plan has effectively promoted the study of ocean fluxes at the edge of the international shelf.
5	GEOTRACES	GEOTRACES International Data Assembly Centre(GDAC)	2010~present	China's first GEOTRACES cross-sectional survey voyage was led by Xiamen University from April to June 2019, and the mission was successfully completed by the Xiamen University "Jiageng" scientific research ship [4], demonstrating the comprehensive strength of China's marine scientific research.
6	ETOPO 5	National Geophysical Data Center	-	The horizontal resolution of the global grid is 5', and the horizontal resolution can reach 30" in 48 areas near the United States. The product information includes global topographic and bathymetric data, digitized land shore boundaries and gridded gravity anomaly fields.
7	GEBCO	IHO& IOC	-	Currently, two datasets are available for download and sharing, namely the 30" GEBCO_08 Grid dataset and the 1' GEBCO One Minute Grid dataset.

2.2 Satellite remote sensing data

The United States successfully launched the first meteorological satellite in 1960 [5], and launched the first ocean satellite again in 1978. Subsequently, active remote sensing technology corresponding fast algorithms and inversion methods have formed a situation of rapid development, and remote sensing ocean and atmospheric science application models are becoming increasingly mature. Based on the development of the aerospace industry, the quality of satellite remote sensing data has been greatly improved. This chapter mainly introduces several types of commonly used satellite data.

2.2.1 *Altimeter data are usually used to measure the sea surface height, effective wave height, sea surface topography and other element data, and then analyze the ocean dynamic process and mesoscale ocean phenomena [6].*

In 1991, ESA European Space Agency launched the ERS-1 satellite. The radar altimeter remote sensor was installed on the satellite to provide users with elements including the geoid, ocean gravity field, polar sea ice area, etc. The measurement accuracy of the altimeter was 3cm. In 1992, NASA and the French National Center for Space Research jointly launched the TOPEX/Poseidon satellite, which was equipped with an active radar altimeter to detect sea level. The resolution can reach 2.2km, and the standard deviation is between ±14cm [7]. Then in 2001, NASA and the French National Space Research Center jointly launched Jason-1 with an accuracy of 3.0 cm, and other indicators were consistent with the T/P satellite. At the end of 2001, EN-VISAT-1 (Environmental satellite) satellite was launched, equipped with a 13.5GHz/Ku-band radar altimeter, and its horizontal resolution under calm sea conditions was greatly improved compared with T/P satellites and Jason-1 satellites, Up to 1.7km. Countries such as Europe and the United States have successively launched a series of altimeter satellites, including ENVISAT (2002~2012), Jason-2 (2008-present), Jason-3 (2016-present), Saral/AltiKa (2013-present), Sentinel-3A (2016 to present) [8]et al. In 2011, my country launched the HY-2A, which is equipped with a dual-frequency radar altimeter, and uses a pulse-limited working method to measure sea surface height, effective wave height and sea surface wind speed. The altimeter has a high measurement accuracy of 4cm under the condition of an effective wave height of 20m, and a high accuracy of 2cm under the condition of an effective wave height of 4m.

2.2.2 *Radiometer data is a sensor made according to passive remote sensing theory.*

The commonly used radiometer is the Advangced Very High Resolution Radiometer, which is the main detection instrument of the NOAA series of satellites, with a resolution of 1.1km and an imaging period of 6d in this range. It can be used to invert sea surface temperature, chlorophyll concentration, surface vegetation coverage and the distribution of fog in the atmosphere, etc. The inversion results have high resolution and high reliability. NOAA provides two high-resolution daily average sea surface temperature analysis products, with a spatial resolution of 0.25° × 0.25° and a temporal resolution of 1 day, covering the global ocean and the time range from 1981 to the present. The Moderate Resolution Imaging Spectroradiometer (MODIS) is an important remote sensing instrument developed by NASA. It is a passive imaging spectroradiometer that can acquire images of targets such as land and ocean temperature, primary productivity, land surface coverage, clouds, aerosols, water vapor, and fire. MODIS provides image data products with high radiometric resolution of all surfaces in the world, sunlight reflection and day and night thermal radiation, and the image resolution is between 0.25 and 1 km. Basic objectives of MODIS measurements include: land and ocean surface temperature and ground fire, ocean coloration, water sediment and chlorophyll, global vegetation mapping and change detection, cloud characterization, aerosol concentrations and properties, atmospheric temperature and humidity detection, Snow Cover and Characterization, Ocean Circulation. Commonly used microwave radiometers include TMI, AMSR-E, AMSR-2, MWRI carried by my country's FY-3 satellite, and RAD carried by HY-2 satellite.

2.2.3 *The scatterometer data is an active microwave radar dedicated to monitoring global sea surface winds. All-weather, high-resolution global ocean near-surface wind data can be obtained using satellite-borne scatterometers.*

In 1991, ESA launched the ESA-1 satellite (European Space Agency Remote Sensing Satellite), which was equipped with a scatterometer with a working frequency of the C-band , which was used to observe the basic elements of ocean dynamics , and to retrieve wind speed and direction . The spatial resolution of wind field data products is 25 km × 25 km. In 1999 , NASA launched Quick-SCAT (quick scatter satellite) satellite, carrying the Ku-band Sea The Winds scatterometer, a satellite wind data product with a spatial resolution of 25 km × 25 km, was discontinued in 2009 due to antenna failure. In 2006, ESA launched the Metop-A meteorological satellite, which carries the ASCAT microwave scatterometer operating in the C-band. The current release of ASCAT wind field products includes ocean wind fields at 25 km and 12.5 km resolution, and nearshore ocean wind field products at 12.5 km resolution, with a spatial coverage of the global ocean. In 2011, China launched the "Ocean No. 2" (HY -2A) satellite, which is China's first marine dynamic environment monitoring satellite. The HSCAT microwave scatterometer on board is mainly used for global sea surface wind vector observation. There are four types of field data products: L1B-level data products, L2A-level data products, L2B-level data products, and L3B-level data products.

2.2.4 *Synthetic Aperture Radar (SAR) [9]data,*

which is an active microwave remote sensing imaging radar, is usually used to observe the direction spectrum of sea surface wind, internal waves, and ocean waves [10], and can also monitor sea ice movement and marine oil spills. Since the frequency band of synthetic aperture radar is the microwave band, the instrument can work normally even in the dark, with the advantages of all-day, all-weather and penetrate all features imaging[11], its use is very wide.

Spaceborne radar has developed rapidly in the 1990s. So far, SAR has been carried on some launched satellites, such as Seasat SAR, Almaz SAR, JERS-1 SAR, and ERS-1/2 SAR. In 1999, ESA launched the Envisat-1 satellite loaded with ASAR. In 2001, Canada launched Radar Satellite 2 with full polarization measurement capability, and LightSAR launched in 2002 is a practical imaging radar with L-band multi-polarization and interferometry and scanning modes. In 2006, Japan launched the ALOS satellite, and the PALSAR radar sensor carried is a multi-polarization, multi-mode radar system. Provide rich data sources for the development of digital earth .

2.3 Reanalysis data

The lack of high-quality, high-resolution, long-term and continuous global observations due to the heterogeneous spatial and temporal distribution of global ocean-atmosphere observations has posed great difficulties and challenges to the in-depth understanding of global ocean-atmosphere changes, cognition of climate evolution patterns and unveiling of phenomenal mechanisms. Based on the ocean-atmosphere dynamic model, scholars use data assimilation technology to assimilate historical observation data into the ocean-atmosphere dynamic model, and obtain a class of product datasets [12]that effectively solve the above problems. This chapter mainly introduces the dynamic data used for the driving field of the ocean-atmosphere model , which is based on the reanalysis data of the observational data.

Table 2. Major reanalysis data products in the marine sector

No.	Data name	Country or Organization	Basic Information
1	NCEP/NCAR [13]	NOAA	Carry out quality control and assimilation analysis and processing of global meteorological observation data from 1948 to the present, including ground, ships, sounding balloons, radio soundings, aircraft, and satellites. The monthly average reanalysis data is obtained four times a day, with a spatial resolution of 2.5°×2.5°, covering the period from 1948 to the present, and the coverage area is global. According to the data content, it is divided into isobaric surface data, ground data and flux data.
2	ICOADS Information	NCDC	Data sources include observation data from ships, fixed-point buoys, drifting buoys and ocean station data. Covering a variety of marine hydrology and meteorological element variables: sea surface temperature, air temperature, air pressure, humidity, meridional wind, zonal wind and cloud cover, etc., the space covers the world.
3	ECMWF	ECMWF [13]	Provide weather observation data and model forecast products as well as related data for global sea-air model calculation [14]. The data set provided by ECMWF contains a variety of variables, with a grid resolution of 2.5°×2.5°, 144 points in the latitude and 73 points in the meridian, and the data time resolution is four times a day.
4	SODA [14]	University of Maryland	The assimilation time of SODA 2.2.4 is more than 100 years, the spatial range is 0.25°E~359.75°E, 75.25°S~89.25°N, the horizontal resolution is 0.5°×0.5°, and the vertical direction is divided into 40 unequally spaced layers, From 10m to 250m, the time resolution is monthly average, and the elements include temperature, salinity, zonal flow velocity, meridional flow, sea surface height, zonal wind and meridional wind.
5	GECCO [16]	UH	Developed on the basis of ECCO products released by MITgcm [14_15]unequally spaced layers, ranging from about 10m to 500m, Product containing 6 variables: temperature, salinity, zonal flow velocity, meridional flow velocity, vertical flow and sea surface height.
6	HYCOM [14]	U.S. Naval Research Laboratory	Product [17]elements include water temperature, salinity, zonal mobility, meridional mobility, and sea surface height. The time span of the reanalysis data product is 1992~2012, the time resolution is 1 day, the latitude range is 80.48°S~80.48°N, the horizontal resolution is 1/12°, and the vertical direction is 40 layers with unequal distances.
7	OFES	The Earth Simulator Center, Japan	The average horizontal resolution of the product is about 12km, and the output variables are re-interpolated to a standard grid of 1°×1°. The vertical resolution is 50 layers, and the time resolution is the daily average.
8	GLORYS [18]	CMEMS	The time span of the product is from 1993 to 2015, and the time resolution is the daily average, including 10 variables, namely water temperature, salinity, and zonal flow velocity.
9	FIOCOM	First Institute of Oceanography	The horizontal resolution is 0.1°×0.1°, the vertical is divided into 54 layers, the depth range is 1m~5316m, and the time resolution is 1d, including 5 variables: temperature, salinity, meridional current velocity, zonal current velocity and sea surface height.
10	CRA	China Meteorological Administration	The product has a horizontal resolution of 34km and a time resolution of 6hr. It is divided into 64 layers vertically, and the height range is from the ground to 55km.

3. Ocean big data sharing

The progress of marine science depends on the effective use and full sharing of scientific data. Faced with the huge amount of big data generated every day, countries have invested a lot of manpower and material resources to actively promote the sharing of marine scientific data. This chapter mainly summarizes the information released by the data sharing platform to provide reference for scientific researchers in need.

3.1 Domestic marine data sharing

Domestically, China has actively advocated information sharing in marine data sharing work, fully exploited marine information resources, made very important contributions in relevant international cooperation and programs, effectively promoted the development of marine scientific research, and increasingly reflected the status of a great power. The national marine authorities and relevant sea-related ministries have released a large amount of marine data information for the use of the society through various ways and with the help of national scientific data centers and characteristic sharing platforms.

Table 3. Domestic Marine Data Sharing

No.	Shared Agency	URL	Share Content
1	National Marine Data Information Center	http://www.nmdis.org.cn/	Through the National Oceanographic Data Center, it provides external data sharing services, and the shared information includes marine multidisciplinary measured data, analysis and forecast products, and thematic information.
2	National Satellite Ocean Application Service	http://osdds.nsoas.org.cn	Three types of satellite data products are shared and released: HY-1A, HY-1B and HY-1C series of marine aqua satellite products, HY-2A, HY-2B and CFOSAT series of marine dynamic environmental satellite products, and high score SAR satellite products.
3	National Marine Environmental Forecasting Center	http://www.nmefc.cn/	Provide various marine forecast products, provide real-time analysis product maps of waves and sea temperature in the Northwest Pacific Ocean, and provide special numerical forecast products for the North and South Pole.
4	Polar Research Institute of China Polar Research Institute of China	http://www.chinare.org.cn/data	The coverage of polar science data resources includes: Antarctic Zhongshan Station Area, Antarctic Great Wall Station Area, Dome A, the highest point of the Antarctic Ice Sheet, Emory Ice Shelf, Grove Mountains, Arctic Ocean Bering Sea, Chukchi Sea (Taiwan), Canada my country's important polar expedition areas such as the sea basin and the Arctic New region
5	First Institute of Oceanography, MNR	https://www.fio.org.cn/	The China Ocean Sample Museum is affiliated to the China Ocean Mineral Resources Research and Development Association and is operated by the First Oceanographic Research Institute of the Ministry of Natural Resources, providing deep-sea sample application services.
6	Second Institute of Oceanography, MNR	https://www.sio.org.cn/	Operate and maintain the Hangzhou Global Ocean Argo System Field Scientific Observation and Research Station, and provide Argo real-time data, Argo grid data sets and other products through anonymous ftp; Inquire.
7	Third Institute of Oceanography, MNR	http://www.tio.org.cn/	The No. 3 Marine Research Institute builds and maintains a marine microbial strain preservation and management center, provides strain inquiry and sharing services, accepts the entrustment of various units and individuals across the country to preserve strains, and charges for strain identification matters.
8	South China Sea Institute of Oceanology, CAS	http://www.scsio.csdb.cn/	Provide measurement data of physics, chemistry, biology, geology, geophysics and other disciplines obtained from on-site ocean observations, satellite remote sensing, ocean remote sensing, ocean model simulation and assimilation data, as well as various data products, etc.
9	The Institute of Oceanology, CAS	http://msdc.qdio.ac.cn/data/	Released the COMS global ocean science data set, collected the global ocean observation data of WOD and China's independent survey since 1900, including 13 elements such as sea temperature, salinity, pH, dissolved oxygen, CO_2 partial pressure; XBT, CTD, Argo, Glider, buoy and other 11 types of instrument data.
10	Guangzhou Marine Geological Survey (GMGS)	http://www.gmgs.cgs.gov.cn/	Relying on the geological cloud to carry out data sharing, the sharing includes geological survey business data and geological scientific research data. Geological survey business data includes five categories: basic geology, mineral geology, hydraulic environment geology, physical and chemical remote sensing, and geological drilling. The scientific research data includes geological research projects and field observation stations. The forms of sharing include space vector data, pictures and photos, audio and video, pdf files, software systems, databases, patents, methods, models, etc.
11	Qingdao Institute of marine geology	http://www.qimg.cgs.gov.cn/	

3.2 Foreign Ocean Data Sharing

In foreign countries, various international organizations are committed to the sharing of global marine data, formulating corresponding data policies and management methods, and upgrading marine data sharing to development strategies and national policies to provide data support for various marine fields.

Table 4. Oceanographic data sharing by foreign countries and international organizations

No.	Country or Organization	Shared Agency	URL	Share Content
1	U.S.	NODC	ftp://ftp.nodc.noaa.gov	The WOD observation data include: temperature, salinity, dissolved oxygen, phosphate, nitrate, silicate, chlorophyll, alkalinity, pH, pCO_2, etc.; the WOA atlas contains objective analytical fields for the 1° square region of world ocean temperature, salinity, dissolved oxygen, dissolved oxygen satiation, nutrient salinity and statistical products used to generate WOA.
2	U.S.	NGDC	http://www.ngdc.noaa	Providing interactive access to geophysical, geological, surveying and environmental data and information products, indexed by map.
3	U.S.	NCEI	https://www.ncei.noaa.gov/	The shared data includes 16 categories. The data update frequency is related to the data type , and can be divided into real-time update, delayed update and one-time update. For important data, NCEI adopts a delayed public release and sharing method.
4	Japan	JMA	http://www.j	Publish short-term, one-week and long-term weather

No.	Country or Organization	Shared Agency	URL	Share Content
			odc.go.jp/	forecasts, as well as typhoon, snowstorm forecasts and advisories. The published data include meteorology, balloon environment and climate, oceanography, earthquakes and other disciplines, and commonly used high-resolution daily average sea surface temperature grid data.
5	Japan	JODC	http://www.jodc.go.jp/	Providing global coverage of basic oceanographic elements such as temperature, salinity, ocean currents, tides, tidal currents, geomagnetism, gravity and water depth.
6	Japan	JAMSTEC	http://www.jamstec.go.jp	Published data include document reports, observation data (Argo, tropical moored buoy array, tropical subsurface moored ADCP database, paleoclimate data, ocean chemistry, etc.), geoscience data (crust structure, seafloor seismic data, etc.), analysis forecasts and models, etc.
7	South Korea	KMA	http://web.kma.go.kr/eng/index.jsp	Provide geophysical data on the Korean Peninsula and its surrounding areas such as sea, air and earthquakes
8	South Korea	KODC	www.nifs.go.kr/kodc/	The released data are obtained by South Korea's independent observations, including data from oceanographic stations around South Korea and cross-sectional observation data.
9	International Organizations	WDC	http://www.wdc-mare.org/	Provide professional data such as environmental oceanography, marine geology, and marine biology in the field of global change and earth system research.
10	International Organizations	ISA	https://www.isa.org.jm/	Sharing data related to marine mineral resources obtained by various institutions around the world, including mining location map, navigation, side scan, CTD, water depth, electrical method, annual report and other data. The data in this database can be downloaded and authorized for use
11	International Organizations	IOC	http://ioc-unesco.org/	Providing ocean science, ocean observations, ocean data and information exchange, and ocean services such as tsunami warnings.
12	International Organizations	IODE	http://www.iode.org	Publishing links to data nodes, marine environmental data and information resources independently operated and maintained by working groups and countries, including observational data, marine meteorological products, information services, etc. These websites mainly provide data and information through open download or registered download serve.
13	International Programs	GTSPP	https://www.nodc.noaa.gov/GTSPP/	The data type of GTSPP is mainly temperature-salt profile data. The data time range is from 1990 to the present. The data is downloaded through the web page and ftp, which is delayed data, and the update frequency is monthly.
14	International Programs	GOOS	http://www.gsf.de/UNEP/goos.Html	Collecting and analyzing all-weather continuous observation data in various sea areas of the world's oceans, including various data sent by the World Meteorological Monitoring Network, the Global Joint Ocean Service System, the Global Sea Level Observing System, and the Oceanographic Data System of the Drifting Buoy Observation Network.
15	International Programs	Argo	http://www.jcommops.org/board?t=Argo	Argo data sharing has been released through the following platforms. The acquisition of foreign Argo data is mainly based on traditional FTP download. JCOMM has released JCOMMOPS, which enables interactive search and retrieval of Argo data.

4. Conclusion

This paper introduces a variety of marine data sources, summarizes the data situation of domestic and foreign marine on-site observation/monitoring data, satellite remote sensing data and reanalysis product data, and also elaborates the basic information of major domestic platforms and international common marine data websites from the perspective of data sharing and usage. Due to the wide variety of marine data, the information listed in this paper is only a few commonly used at present. At the same time, due to the limitation of space, many good marine data cannot be listed one by one. It is hoped that this article will serve as a primer to help readers understand marine data from multiple perspectives. On the basis of the information provided in this article, readers will be able to further explore and utilize Internet resources, and give full play to the scientific value of marine data.

Acknowledgments

This work was financially supported by National Key R&D Program of China (No.2021YFC3101600) fund.

References

1. QIAN Chengcheng, CHEN Ge. Big Data Science for Ocean: Present and Future [J].Subject and Field,2018,33(08):884-891.DOI:10.16418/j.issn.1000-3045.2018.08.018.
2. HOU Xueyan, HONG Yang, ZHANG Jianmin,et al. Marine big data: concept, applications and platform construction [J]. Marine Science Bulletin,2017,36(04):361-369.
3. Huang Dongmei, Zhou Guolian, et al. Big Data Science for Ocean [D]. ShangHai Scientific&Technical Publishers.
4. XU Yanping. Review of standardized data archiving and management of international ocean observation programs [J]. Marine Sciences, 2021, 45(07): 127-136.
5. Lu Daren,Wang Pucai,Qiu Jinhuan,et al. An Overview on the Research Progress of Atmospheric

Remote Sensin and Satellite Meteorology[J].Chinese Journal of Atmospherie Sciences, 2003(04): 552-566.

6. Song Tingting, Fan Wei, Wu Yumei,2013. Review on the fishery analysis with satellite remote sensing sea surface height data[J]. Marine Science Bulletin,32(4):474-480.
7. JIANG Xingwei, SONG Qingtao. Satellite Microwave Measurements of the Global Oceans and Future Missions [J]. Science,2010,28(3):105-111.
8. YANG Lei,Zhou Xinghua,Xu Quanjun et al. Research status of satellite altimeter calibration[J]. Journal of Remote Sensing,2019,23(3):392-407.
9. Zhang Xi,Zhang Jie,Ji Yonggang. Sea Ice Detection From SAR Images of the Liaodong Bay Based on Texture Analysis [J]. Advances Marine Science,2008,26(3):782-790.
10. Li Ying,Li Guannan,Cui Cui.Research progress of oil spill detection based on the spaceborne SAR[J]. Marine Science Bulletin,(3),2017,36(03):241-249.
11. Hua Lei,Li Jingwen.Study of Jamming Technology to Synthetic Aperture Radar[J]. Telemetry,Tracking and Command,2003(02):14-18.
12. Wang Hui, Liu Guimei, Wan Liying. Review on the Data Assimilation into Marine Ecosystem Model [J]. Advances in Earth Science, 2007, 22(10): 989-996.
13. Shi Maochong, Gao Guoping, Bao Xianwen. Method for Oceanographic Investigation [M]. China Ocean University Press, 2008.
14. Wang Shihong,Zhao Yiding,Yin Xunqiang,at el.Current Status of Global Ocean Reanalysis Datasets[J].Advances in Earth Science,2018,33(8):794-807.
15. Marotzke J,Giering R,Zhang KQ,et al.Construction of the adjoint MIT ocean general circulation model and application to Atlantic heat transport sensitivity[J].Journal of Geophysical Research Oceans,1999,104(C12):29 529-29 547 .
16. Kohl A,Stammer D.Decadal sea level changes in the 50-year GECCO Ocean Synthesis[J].Journal of Physical Oceanography,2008,21(9):1 876-1 890 .
17. Bleck R,Benjamin S G.Regional weather prediction with a model combining terrain-following and isentropic coordinates.Part I:Model description[J]. Monthly Weather Review,1993,121(6):1770.
18. Parent L, Ferry N, Garric G, et al. GLORYS2: A Global Ocean Reanalysis Simulation of the Period 1992-Present [Z]. Geophysical Reearch Abstracts, 2011.

Drilling practice Zhongzihao gas exploration well in the thrust belt of the western margin

Hao Wang, Yan Wei *

Drilling & Production Engineering Technology Institute, CNPC Chuanqing Drilling Engineering Company Ltd., Xi'an, Shaanxi, 710018, China

Abstract: The geological structure of Zhong-7 well and Zhong-14 well is the Hengshanpu section of the thrust belt in the western margin of the Ordos Basin, and there have been few construction wells in this area in the past two years. During the construction of the two wells, the formation stability of the entire well section was extremely poor, especially Shiqianfeng. The following strata collapsed seriously and frequently encountered blockages during tripping. During the construction, measures such as optimizing the system, enhancing plugging, improving suppression, wall patching in complex well sections, and fiber cleaning were adopted to obtain valuable experience for the later construction of the Zhongzi gas exploration well in the thrust belt of the western margin.

Keywords: Thrust zone; Gas exploration; Collapse

1. Construction of Zhongzi Well

1.1 Construction of Zhong-7 well

The designed well depth of Zhong-7 well is 3350m, geological structure: Hengshanpu section of thrust belt in the western margin of the Ordos Basin, the completion of drilling the stratum Zhuoshan Formation, the well was opened at 18:00 on March 22nd, surface well depth: 723.48 meters, on April 2nd start at 00:00, the drilling was completed at 00:00 on April 22nd, the drilling depth was 3350 meters, the drilling fluid density was $1.28g/cm^3$, the funnel viscosity was 65s, the medium pressure water loss was 3.8mL, and the pH was 9. Among them, on April 16nd, after drilling through the Shihezi Formation, the middle test showed that the well wall was basically stable, and the average well diameter expansion rate was 7.85%. After drilling, the electrical measurement was down to 2501 meters (stone box) and encountered resistance. The instrument was pulled out to drive the well. When the well was drilled, it returned to Shiqianfeng, and the stone box dropped a lot. It took 14.2 days for the double-stone layer to be drilled. On April 28nd, the electrical measurement was completed, and the electrical measurement showed that the diameter of Shiqianfeng and the stone box was seriously enlarged. It was difficult to drill holes in Shiqianfeng and Shihezi before running the casing, and the turntable was severely held back, making it impossible to drill holes.

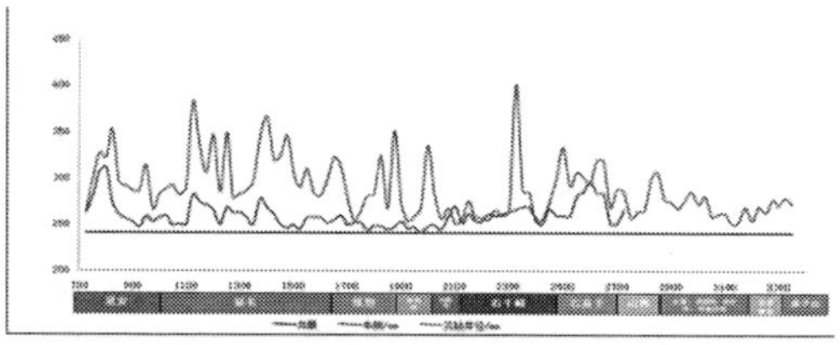

Fig. 1 Diameter diagram of mid-term and completion-drilling electrical logging in Zhong-7 well

* Corresponding author: zjwyan@cnpc.com.cn

https://doi.org/10.1051/e3sconf/202235801047

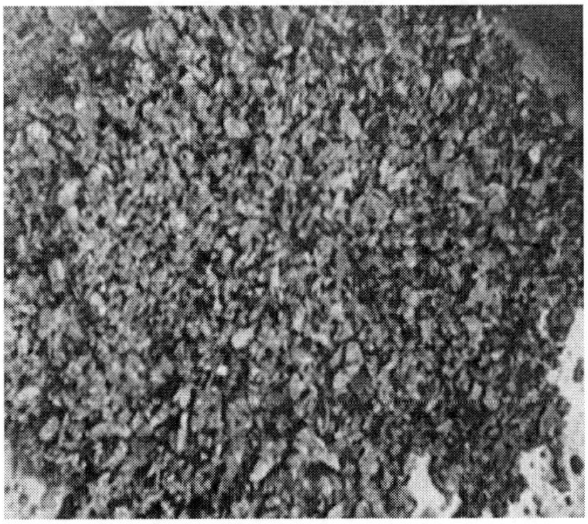

Fig. 2 Return drop during well drilling

1.2 Construction of Zhong-14 well

The designed well depth of Zhong-14 well is 3285m. Geological structure: Hengshanpu section of thrust belt in the western margin of the Ordos Basin. The well was opened at 21:00 on August 3nd, and the surface layer was drilled to a depth of 777m. The second was opened at 23:00 on August 9nd, and was drilled to a depth of 3357m (Shanxi) at 4:00 on August 24nd. After 6.5 days of drilling, the drilling fluid density was 1.36g/cm^3, the funnel viscosity was 68s, the medium pressure water loss was 4.2mL, and the pH was 9. During tripping, Shiqianfeng and the stone box encountered serious obstructions. When the pump was turned on, the hole was drawn backwards. When drilling through the well, the pump was held back by the pump, and a large number of double-rock layers fell out, making it impossible to proceed to the next step of construction.

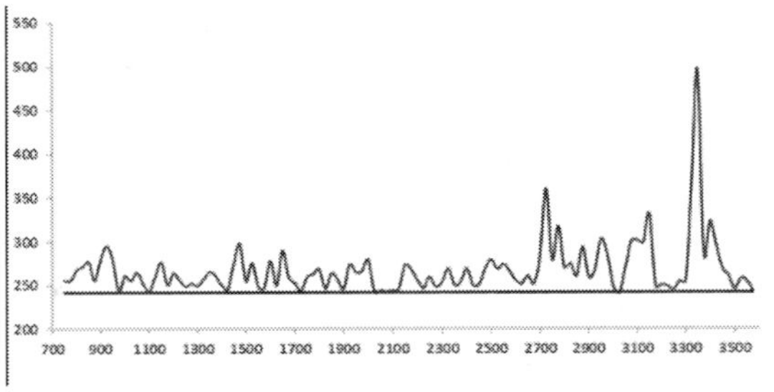

Fig. 3 Completion electrical logging diameter of Zhong-14 well

Fig. 4 Blocks returned from the pass-through of Zhong-14 well

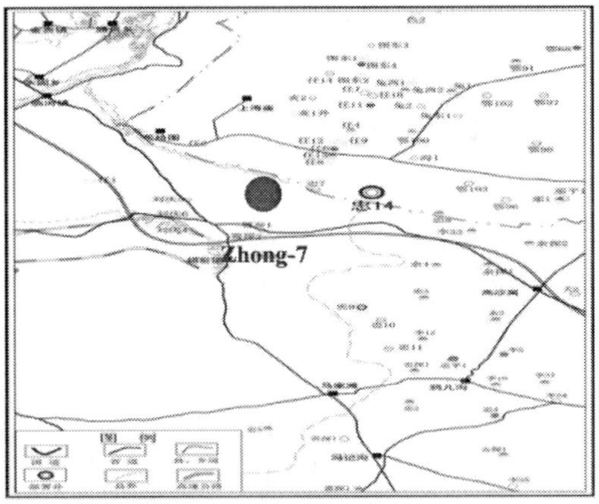

Fig. 5 Geographical location of Zhong-7 and Zhong-14wells

 https://doi.org/10.1051/e3sconf/202235801047

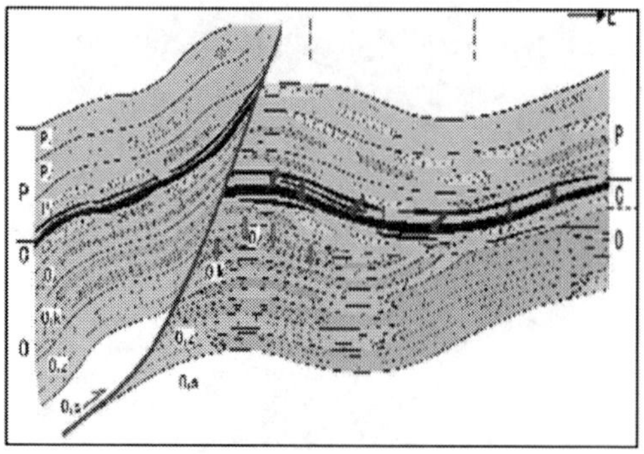

Fig. 6 The structure of the thrust belt in the western margin

2. Stratigraphic characteristics of the thrust belt in the western margin

There are faults in the thrust belt of the western margin from north to south, the stratum is broken[1], and the dip angle is large. Although there is no fault in Zhongzi well, there is in-situ stress concentration, and physical collapse prevention alone cannot balance the formation collapse pressure. The completion density of Zhong-7 well is 1.28 g/cm^3, and the double-stone layer collapses seriously. Zhong-14 well has a density before core-tripping. 1.36g/cm^3, the double stone layer still collapsed.

3. On-site construction drilling fluid technology

3.1 Optimize system performance while improving system inhibition

The compound salt system was used in the construction of the second opening of Zhong-7 well. The lower part of the Yanchang Formation was transferred to the system. After the system was transferred, the compound salt was added by 5-7%. When drilling to the Zhifang Formation, the pulping was serious, and the extended block was lost during the tripping and returning[2]. It shows that the mudstone in the upper stratum has strong dispersive slurrying property, and the system has insufficient inhibition. Adding 5% KCl to the system, the slurrying is obviously improved, and the block loss of the Yanchang Formation is reduced[3].

Fig. 7 Blocks of Yanchang Formation in Zhong-7 well are easily dispersed and collapsed after being soaked in water

After the second opening of Zhong-4 well, the system was switched. After the system was switched, 7% KCl was added, and the KCl content of the Yanchang group was increased to 10-12%. The slurry production was relieved to a certain extent, but still could not be completely suppressed. The wellbore stability was improved, and the return sand pattern rules.

Table 1 Performance of the second opening of Zhong-7 and Zhong-14 wells Liujiagou

Parameter	Hashtag	Anding-Zhiluo	Yan'an	Yanchang	Zhifang	Heshanggou	Liujiagou
Density	Zhong-7	1.01	1.02	1.08	1.09	1.12	1.16
	Zhong-14	1.10	1.15	1.20	1.23	1.27	1.30
Viscosity	Zhong-7	33	34	50	68	46	48
	Zhong-14	46	44	48	50	50	55
Dehydration	Zhong-7	18	15	4.4	4.8	5.6	5.2
	Zhong-14	5.2	5.0	5.4	5.0	5.2	4.4

Table 2 Comparison of well diameters of Zhong-7 and Zhong-14 wells in the second opening and Liujiagou

Hashtag	Well section	Formation	Average well diameter expansion rate	Maximum well diameter expansion rate
Zhong-7	750-2700	Erkai-Liujiagou	8.06	23.46
Zhong-14	725-2025	Erkai-Liujiagou	24.91	58.59

From the comparison of Table 1 and Table 2, it can be seen that the transformation of the system in advance, the enhancement of performance, and the enhancement of inhibition can effectively alleviate the collapse of the upper water-sensitive formation[4].

3.2 Using wall solidifying agent to enhance system blocking, improve mud cake and reduce water loss

During the construction of Zhong-7 well, 2% clay slurry was added to the system after the double-stone layer. After Zhong-14 well entered Shiqianfeng, the content of clay slurry was increased to 3%. During the drilling of the well, 1% of wall-reinforcing agent was added to reduce water loss and the quality of mud cake was Better[5], the large drop-out blocks will be gradually reduced when the well is opened in the later stage.

Table 3 Performance comparison before and after the addition of wall cement in Zhong-14 well

	Density (g/cm^3)	Funnel viscosity (s)	Medium pressure water loss (mL)	High temperature and high pressure water loss (mL)
Before wall reinforcement	1.38	136	3.6	14.8
After wall reinforcement	1.38	152	2.7-3.2	13.6

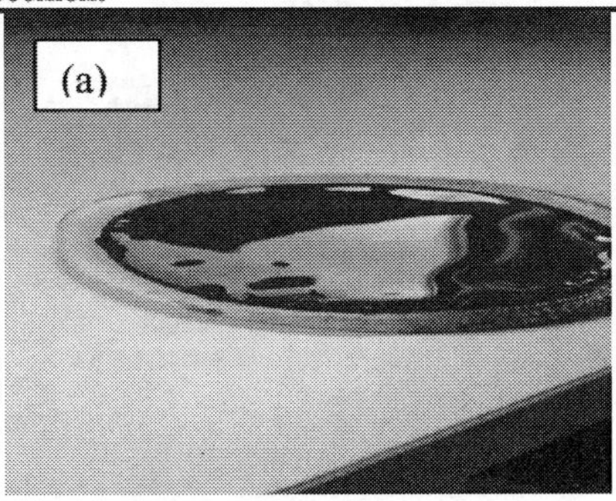
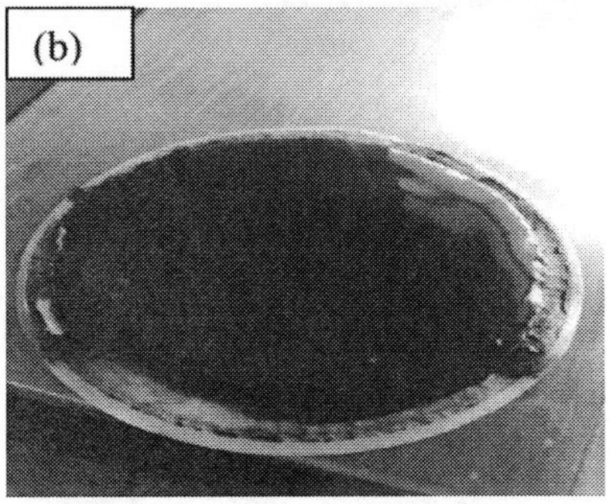

Fig. 8 Medium pressure loss cement cake(a) and high pressure loss of cement cake(b)

3.3 Using thick pulp, heavy pulp, heavy thick pulp, and fiber cleaner to circulate wells during reaming, and carry large droplets in the well

Table 4 Blocks dropped in the well carried by different technical measures in Zhong-14 well

Measure	Performance	Effect
Heavy pulp	On the basis of the original pulp, the density is raised to 1.60-1.65g/cm^3	4 times of cleaning, the first 2 times can bring out some large pieces, and the effect is not obvious in the later period
Thick paste	Viscosity:230-280s	After cleaning 6 times, it can bring out large pieces each time, about 0.5 tons of pieces can be brought out each time, and more pieces can be brought out when used continuously
Heavy and thick pulp	Density:1.55-1.60g/cm^3 Viscosity:160-180s	Use 2 times, the effect of carrying the block is not obvious
Fiber sweeping pulp	Original pulp + 0.13% fiber cleaning agent	Large droplets are not carried out much, but flaky and fine-grained sand is brought out

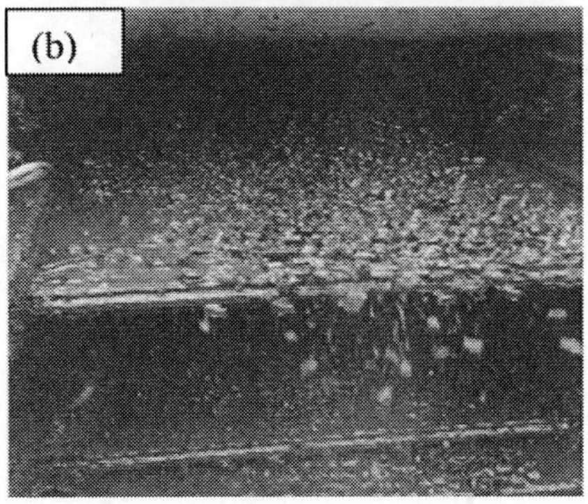

Fig. 9. Heavy pulp belt out of block(a) and thick slurry brings out lumps(b)

https://doi.org/10.1051/e3sconf/202235801047

3.4 The ductile cement retaining wall in the difficult section of drilling

The electrical survey of Zhong-7 well showed that the double-stone layer collapsed seriously, and the turntable jumped seriously when the well was drilled, and it was difficult to ream the hole[6]. A total of two wall protection cements were applied; the double-stone layer of Zhong-14 well was drilled 6.5 days after drilling, the blockage was serious, and the well was opened. When the turntable jumped frequently, a large number of double-stone layers dropped out, and the cement wall was injected twice on the double-stone layer[7]. The turntable still jumped seriously after the first cement injection was completed and the plug was removed, and the returned cement blocks were high in hardness, large in size, and frequent[8-10]. Hold down the turntable. After the second cement injection, the drill plug is basically normal. After the plug is drilled, it is short and normal, and then the drill is drilled and cored[11, 12].

Table 5 Cement wall protection of Zhong-7 and Zhong-14 wells

Hashtag	Number of wall repairs	Drill plug situation	Wall effect	Remark
Zhong-7	the first time	Plug section 2390-2546m; plug length 156m	The tripping in the wall-filling well section is normal	Continue to scribble, 2597-2636m (stone box) is difficult to scribble, hold down the turntable, lift up and down and encounter resistance, hit cement wall again
	the second time	Plug section 2505-2741m; plug length 236m	Short up and down after eye stroke is normal	After the cement plug is drilled, the lower well section is normal, and the drilling and casing are normal after the well is drilled to the bottom
Zhong-14	the first time	Plug section 2822-3148m; plug length 326m	During the plugging process, the cement block is large, and the turntable is seriously held back	Continue reaming to the bottom of the hole (3357m), the turntable jumps seriously during the reaming process, and the drilling is still abnormal after short tripping, and the cement wall is continued during tripping (3000-3300m)
	the second time	Plug section 3158-3306m; plug length 148m	After the cement plug is swept, the eye-scraping turntable is still throbbing	Continue reaming after the plug is removed, the turntable still jumps, and it is normal after repeated reaming. Before coring, the drill is short tripped for 2 times. After normal, the drilling and coring are normal

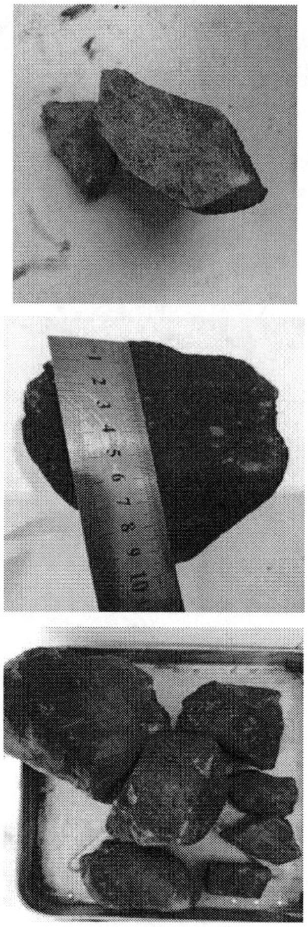

Fig. 10 The cement block returned after the first cement injection in Zhong-14 well

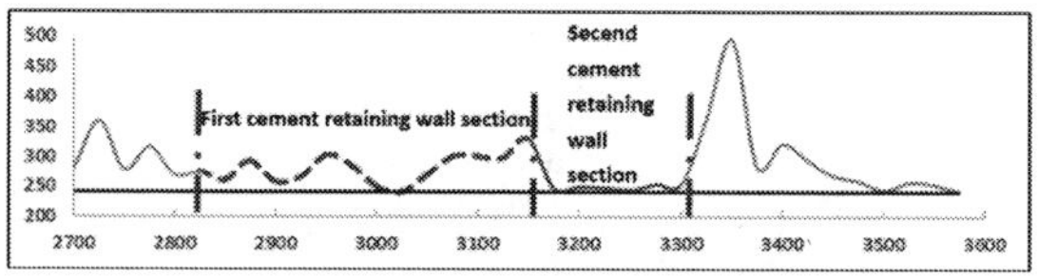

Fig. 11 Shiqianfeng of Zhong-14 well bottom hole diameter

4. Knowledge and advice

There is in-situ stress concentration in the geological structural formation of the thrust belt in the western margin, and the stress release period varies after the borehole is drilled. In this area, the upper strata have good water-sensitivity and pulping ability, and the collapse of the upper strata can be weakened by the measures of turning the system in advance, improving the inhibition, controlling the water loss and increasing the density. Cement solid wall technology has a good effect of stabilizing the well wall for complex well sections, but the wall solidification timing and cement formulation still need to be further optimized and improved. The wellbore structure in this area still needs to be further optimized. For the current two-opening structure, it can be optimized to three-opening. Before the core is taken, the next layer of technical casing is used to seal the complex formation above the stone box.

References

1. Jing C.Q, Zhang H.L, Hou Y.K. Application of leakage plugging technology in liujiagou layer of exploration well[J]. Chemical Engineering and Equipment, 2021(08): 78-79.
2. Wang S.G. Thinking about the application of geological logging in shale gas exploration wells[J]. Chemical Management, 2018(19): 224.
3. Qin C, Chen X.Y, Cheng Z.Y. Integrated processing technology for salvage and reaming of large-diameter collapses in downholes[J]. Petroleum Drilling Technology, 2020, 48(06): 61-64.
4. Hu Y.K, Ma X, Ma H.Y. Analysis of key technologies for borehole stability in deep exploration wells[J]. Technical Supervision of Petroleum Industry, 2021, 37(10): 20-24.
5. Wang Y.D, Du Z.D, Kang H.Y, et al. Research and application of high-precision 3D exploration technology in the thrust belt in the western margin of the Ordos Basin [C]. Proceedings of the China Petroleum Society 2021 Geophysical Technology Seminar Proceedings., 2021: 156-159.
6. He D.F, Shao D.B, Kai B, et al. Structural style and trap distribution in the Majiatan area on the western margin of the Ordos Basin[J]. Journal of the Earth, 2019, 40(01): 219-235.
7. Miao S.J, Yao L, Li B.X. Geology-geophysical characteristics and geothermal exploration of the Pingliang section of the thrust belt in the western margin of the Ordos Basin[J]. Geology of Gansu, 2017, 26(04): 51-56.
8. Zhang Y.L, Guo Y.R, Gao J.R, et al. Analysis of Ordovician paleo-tectonic features and exploration prospects in the thrust belt in the western margin of the Ordos Basin[C]. Collection of abstracts of the 8th China Petroleum System and Oil and Gas Reservoir Academic Conference., 2015: 113.
9. Luo Q. Characteristics of the faults in the thrust belt in the Majiatan area on the western margin of the Ordos Basin and its reservoir-controlling model[J]. Journal of the Earth, 2008(05): 619-627.
10. Wang Z.C, Wang Y.X. Majiatan detachment thrust structure in the western margin of Ordos[J]. Petroleum and Natural Gas Geology, 1996(03): 221-225.
11. Wu H.T, Xu X, Wang L, et al. Mud purification technology on construction site[J]. Construction Technology, 2020, 49(S1): 1690-1691.
12. Jia X.P. Practice and understanding of drilling fluid construction in X-P1 well[J]. Western Prospecting Engineering, 2020, 32(06): 36-38+42.

https://doi.org/10.1051/e3sconf/202235802001

Hyperspectral characteristics of altered rock and soil features and compositional correlation analysis

Zhiming Liu *, Xiangcui Liu, Tong Zhang

Institute of Chemical Defense, Academy of Military Science, Beijing, China

Abstract. Hyperspectral characteristics of ground features are the basic data of modern remote sensing. In this paper, reflectance spectrum detection was carried out on the altered rock and soil features from mountainous environment. The analysis between reflectance spectrum and its related compositions was studied on the rock and soil samples collected. The typical hyperspectral characteristics and their influencing factors of the altered rock and soil features were proposed. The relevant conclusions could provide spectral information reference for remote sensing analysis of altered rock and soil features and environmental evaluation of the northern mountainous environment in China.

1. Introduction

Chengde area of Hebei province in China is a typical transition zone from mountain area to the plain area. This area has a continental monsoon climate with a long winter under low temperature and a short spring with dry and windy weather every year. In summer, the sunlight is strong, and the temperature difference between day and night is very large, meanwhile the precipitation is less. These lead to the easily weathering alteration of rocks and turn them into sands and soils. According to the field investigation, the parent rocks in the mountainous area are mostly granite and metamorphic rocks, and aeolian and residual materials are the main parent soils of this area. The typical ground features include altered rock, sandy soil and loess.

Mastering the spectral characteristics of different rock and soil features is the basis of mineral information extraction using remote sensing data, especially the spectral characteristics of altered features under different physical and chemical environments [1-3]. It is of great value to carry out the analysis of hyperspectral characteristics of altered rock and soil features with their related mineral compositions.

2. Reflectance spectrum detection and material composition analysis of rock and soil features

The spectral measurement of rock and soil features such as rock, sandy soil and loess was carried out under the condition of cloudless weather and fine sunlight illumination in daytime. The portable spectrometer of Spectral Evolution's SR 3500 was used for spectral detection with the white board used for calibration.

X-ray diffractometry (XRD) was used to study the major material compositions of typical ground samples collected. The XRD instrument used was the D8 Advance of Bruker, Germany. The test condition was atmospheric and without sample treatment. The composition and structure of samples were analyzed at the same time.

2.1 Reflectance spectrum measurement of rock and soil features

The typical rock and soil features in mountainous environment were altered rock fragments, sandy soil on the surface and loess under the surface. The typical ground object samples of rock, sandy soil and loess collected were shown as follows.

Rock

Sandy soil

* Corresponding author: liuzhiming1979@126.com

Loess

Figure 1. Typical ground samples of rock, sandy soil and loess

The portable spectrometer of SR-3500 was used to conduct reflection spectrum detection on rock, sand, and loess samples respectively by using the probe with a built-in halogen light source in the mode of close up measurement. The spectrum rang was from 350nm to 2500nm. Reflectance spectra of ground samples were shown in Figure 2.

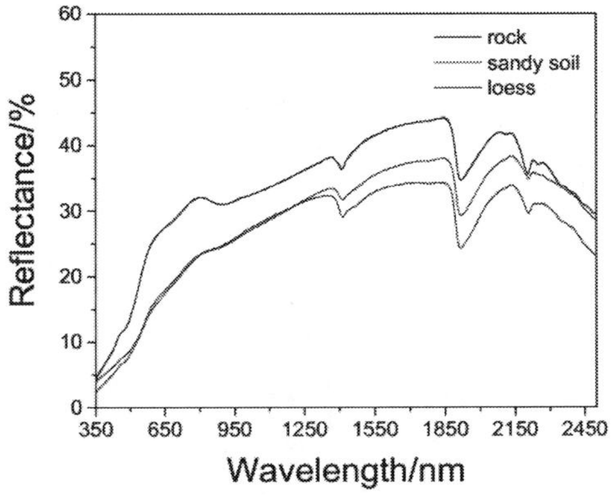

Figure 2. Reflectance spectrum detection of ground samples in the room

In the spectral range of 350nm to 2500nm, there were mainly four spectral features in the reflectance spectra of ground samples. It included the gradually rising of reflectance from 350nm to 780nm, the near-infrared plateau between 780nm and 2500nm, the weak absorption around 870nm, the series of characteristic absorption around 1410nm, 1920nm and 2200nm. The reflectance spectra of typical ground features revealed the law of reflectance changing with wavelength. In order to analyze the influence of the material compositions of the samples on their spectral characteristics, the main material compositions of collected samples were analyzed as follows.

2.2 Main material compositions of rock and soil features

Quantitative X-ray diffraction analyses of main material compositions of the rock and soil features were carried out. There were about quartz 35 wt%, potassium feldspar 16 wt%, albite 30 wt% and clay 19 wt% in the rock samples. There were about quartz 20 wt%, potassium feldspar 19 wt%, albite 25 wt%, clay 36 wt% in the sandy soil samples. And there were about quartz 40 wt%, potassium feldspar 17 wt%, albite 28 wt%, calcite 4 wt% and clay 11 wt% in the loess samples. The stongest diffraction angle located around 26.775 degrees in these samples, which belonged to quartz in their common compositions. The content of clay in sandy soil samples was higher than it in the rock samples. The additional clay in sandy soil samples might come from the altered rock samples on the ground. And the content of calcite in loess samples might come from the biomineralization of microorganisms.

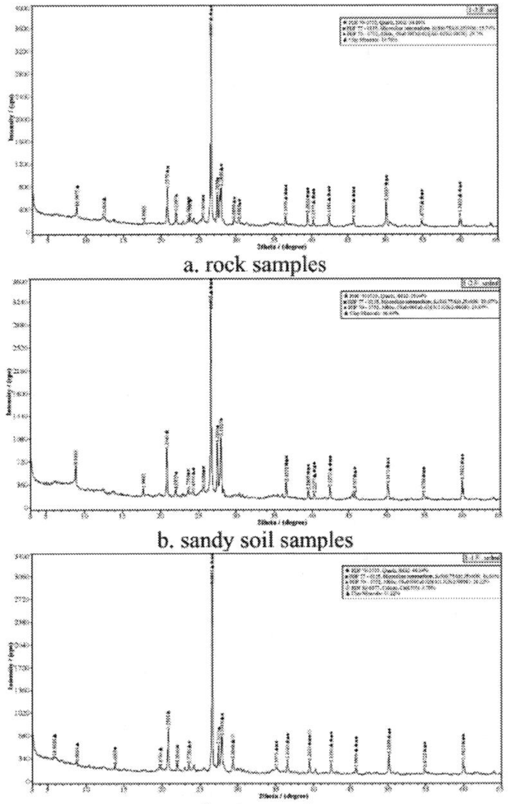

a. rock samples

b. sandy soil samples

c. loess samples

Figure 3. X-ray diffraction spectra of the collected ground samples

3. Correlation analysis of reflectance spectrum of rock and soil features with its relevant material compositions

The main material compositions of rock, sand and loess included quartz, albite, clay and so on [4]. The common minerals in clay were montmorillonite, illite and so on. In order to analyze the influence of the main constituent minerals of rock and soil samples on its reflectance spectrum, a spectral detection of the constituent minerals was conducted with the portable spectrometer SR-3500 in the room. Reflection spectra of the mineral powders of quartz, albite, montmorillonite and illite were shown as follows.

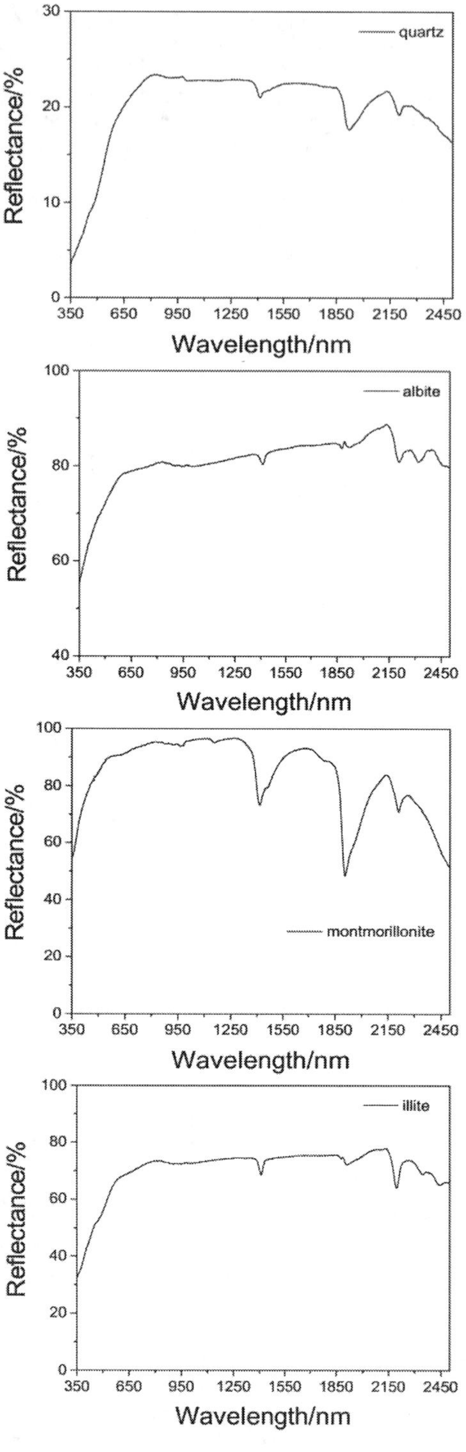

Figure 4. Reflectance spectra of the constituent mineral powders

In the spectral range from 350nm to 2500nm, the reflectance spectra of mineral powders of quartz, albite and illite were very similar to those of the collected ground samples, including the gradually rising of reflectance from 350nm to 780nm, the near-infrared plateau between 780nm and 2500nm, the series of characteristic absorption around 1413nm, 1920nm and 2204nm [5]. However, the absorption of montmorillonite around 1413nm, 1908nm and 2211nm were very strong compared with those of other constituent minerals.

According to the material analysis of the main compositions of rock and soil features, and based on the above spectral analysis, the spectral characteristics and relevant influencing factors of typical rock and soil features in the northern mountainous area were proposed in the following table.

Table 1. Spectral characteristics of typical rock and soil features and their influencing factors

Type of characteristics	Spectral characteristics of rock	Spectral characteristics of sandy soil	Spectral characteristics of loess	Influencing factor
The rising of reflectance	350-799 nm	380-820 nm	380-810 nm	Si-O bond
Near infrared reflectance	31-45%	24-39%	24-34%	Rock particle size
Weak absorption	898nm	870nm	870nm	Fe^{3+}absorption
Characteristic absorption I	1413 nm	1414 nm	1414 nm	Hydroxide radical
Characteristic absorption II	1917 nm	1919 nm	1917 nm	Interlayer water
Characteristic absorption III	2205 nm	2207 nm	2208 nm	Hydroxide radical and Al/Mg-OH

4. Conclusion

In this paper, the altered rock and soil features of Chengde area of Hebei province in China were selected as the subject of hyperspectral analysis and its compositional correlation analysis. By collecting rock and soil samples in the environmental background, the detection of reflectance spectrum and analysis of the main material compositions of the ground samples were carried out. Combining the analysis of the main material compositions, the hyperspectral characteristics of typical altered rock and soil features in the northern mountainous environment and their influence factors were proposed. The relevant conclusions could provide spectral information reference for remote sensing analysis of altered rock and soil features and environmental evaluation of the northern mountainous environment in China.

Acknowledgments

This work was financially supported by relevant scientific research fund.

References

1. K. Stephan, S. Schröder, M. Baque, et al, Multi-spectral investigation of volcanic deposits and their alteration processes on Vulcano/Italy, EPSC Abstracts, 13(2019) EPSC-DPS2019-2062.
2. H.T. Basavarajappa, N.E. Maruthi, Petrochemical characteristics and hyperspectral signatures on corundum bearing precambrian lithounits of Varuna area, Mysuru district, Karnataka, India, IJCRT 6 (2018) 998-1009.
3. Q. H. Yang, X. J. Wu, Z. H. Xiao, Technology research on mineral extraction by using CASI/SASI airborne hyperspectral data. Journal Geomechanics, 21(2015) 241-250. (in Chinese)
4. Y. D. Zhao, T. T. Pi, Spectral prediction model of soil total nitrogen content of clay loam soil in Beijing, Trans. Chin. Soc. Agric. Mach., 47(2016) 144-149. (in Chinese)
5. H. H. Han, G. L. Ren, Z. Zhang, et al, Spectral characteristics of typical altered rocks and minerals from Fangshankou area in Beishan, Northwestern Geology 51(2018) 264-275. (in Chinese)

Design of Flexible Interactive Platform System between Self-owned Power Plant and Power Grid

Lingkai Zhu [1,*], Weishuai Wang [2], Haijing Zhang [2], Qian Wang [1], Wei Zheng [1], Ziwei Zhong [1]

[1] Shandong Electric Power Research Institute, Jinan, China
[2] State Grid Shandong Electric Power Company., Ltd, Jinan, China

Abstract. In response to the call of the state to deepen power system reform, strengthen and standardize the supervision and management of coal-fired captive power plants, and implement regulation from the load side to effectively ensure the power supply and demand balance of the power grid, a flexible interactive platform system between captive power plants and the power grid is developed. The platform software has the functions of self owned power plant file management, resource adjustable potential analysis, operation state detection, energy consumption analysis, transaction management, etc., and the response time to power grid interactive instructions is less than 10 seconds.

1. Introduction

The province's power supply continues to be tight and the balance between supply and demand is uncertain. The traditional power supply and demand balance is to regulate power generation to meet the load demand of power users. With the shortage of power coal supply, the uncertainty of call support outside the province, the increase of temporary fault repair and output reduction of thermal power units in the province, and the improvement of new energy penetration, it is urgent to implement relevant regulation from the load side to effectively ensure the balance of power supply and demand of the power grid[1]. Enterprise owned power plant is an important part of China's power generation industry. It will also become a member of the power market reform and participate in the market competition[2]. The self owned power plant has both power generation resources and a large number of loads, but it has been in the state of spontaneous self use for a long time. Its potential as an available resource for the balance of power supply and demand has not been tapped[3]. The promulgation of the guiding opinions on strengthening and standardizing the supervision and management of coal-fired captive power plants is conducive to scientifically guiding and standardizing the development of captive power plants, promoting the equal management of captive power plants and public power plants, realizing fair competition, giving better play to the role of captive power plants, earnestly maintaining the order of power market and steadily promoting power reform[4].

The development direction of supply-demand interaction between captive power plants and power market is clear. However, through investigation, it is found that due to the flexible interaction between captive power plants and market-oriented supply-demand regulation, the method system of design and operation optimization is not perfect, and the actual operation effect of most demonstration projects has not reached the expected goal[5]. The reasons are mainly reflected in several aspects: the theoretical research on the behavior characteristics and flexibility evaluation methods of self owned power plants is still insufficient, and there is a lack of fine operation technical characteristic model of multi type self owned power plants, resulting in the lack of planning means; Lack of evaluation and Analysis on the characteristics of power generation and consumption behavior of self owned power plants, lack of analysis on on-line monitoring technology and external relevant data, resulting in unclear objectives for flexibility improvement; The research on the coordinated optimization operation method of market mechanism is not enough; Insufficient policy guidance and low willingness of users of self owned power plants to participate in market-oriented supply and demand interactive services.

The platform built in this paper studies the operation characteristics, evaluation and market mechanism of self owned power plants, and improves the results of theoretical and technical research in combination with demonstration projects, in order to form an operable and popularized evaluation system and market mechanism.

2. Technical research content supported by the platform

The platform is developed on the basis of relevant theoretical research of captive power plants. The theoretical basis is as follows: from the perspective of energy interaction and auxiliary service interaction, study the interactive regulation strategy of multiple types of captive power plants in different application scenarios; Study the security protection strategies of boundary

* Corresponding author: zhulingkai@woyoxin.com

security, terminal security and network security for the flexible interaction between the self owned power plant and the power grid when the power grid control command is issued and the state information of the self owned power plant is uploaded; From the perspective of operation monitoring specification of self owned power plant under the framework of technical supervision and online monitoring system, the standardization technology of technical supervision of self owned power plant is studied. Based on the above research, the flexible interactive platform system between self owned power plant and power grid has the functions of self owned power plant file management, resource adjustable potential analysis, operation state detection, energy consumption analysis, transaction management, etc. the system response time to power grid interactive instructions is less than 10 seconds.

Table 1. The platform relies on the theoretical research content.

Number	Technical research content	Concrete content
1	interactive regulation strategy of multi type captive power plants	1.Energy interactive regulation strategy 2.Interactive adjustment strategy of ancillary services
2	flexible interactive safety protection strategy between self owned power plant and power grid	1.Establish border security protection strategy 2.Establish terminal security protection strategy 3.Establish network security protection strategy
3	technical Supervision standardization technology of self owned power plant	1.Establish the Technical Supervision Specification and on-line monitoring specification system of self owned power plants 2.Specification framework for on-line monitoring of captive power plants 3.Technical regulatory framework for captive power plants

Research on interactive regulation strategy of multi type captive power plants from the perspective of marketization and economy, comprehensively consider the general characteristics, dynamic multi priority and power regulation tolerance of captive power plants, the energy interaction mode between captive power plants and power grids in the scenarios of power generation right transaction and new energy consumption, and the regulation strategy of power interactive transaction; Based on the principle of marketization and the normal operation environment of self owned power plants, the interaction mode of auxiliary services and power interactive regulation strategy between various types of self owned power plants and power grid in peak shaving scenario, frequency modulation scenario and other scenarios are studied.

Research on flexible interactive security protection strategy between captive power plant and power grid boundary security protection measures at three levels of terminal credibility, transmission reliability and boundary controllability under the access of public network and power private network; Research the terminal security protection measures of terminal and master station based on digital certificate authentication technology; Research on network security protection measures based on encryption algorithm, identity authentication and other technologies.

Research on the standardization technology of technical supervision of self owned power plants in the form and specific content of technical supervision of self owned power plants, as well as supporting technical policies and management specifications related to power reform; This paper studies the specifications for operation monitoring of self owned power plants based on the architecture of on-line monitoring system.

3. Overall system architecture design

The platform software has the functions of self owned power plant file management, resource adjustable potential analysis, operation state detection, energy consumption analysis, transaction management, etc. the response time to power grid interactive instructions is less than 10 seconds.

The distribution map of self owned power plants of Shandong direct dispatching on the homepage of the platform includes the distribution of self owned power plants of cities in Shandong Province. The main adjustment parameters are displayed after hovering the mouse; The real-time statistical function displays the power generation and consumption data of current operating units, active power, load rate, thermal power ratio and energy consumption power of self owned power plants in Shandong Province in real time; The main indicators of the whole network show the power generation of the whole network yesterday, the power generation of the whole network today and the active power of the whole network at present; The comparison of the regulatory potential of each group shows that each group has the maximum daily upward potential and the maximum daily downward potential; The adjustable potential of the whole network in the current month shows the adjustable potential of the whole network in the current month.

The condition monitoring can monitor the relevant performance of the direct dispatching self owned power plant and its relevant production units in Shandong Province. The relevant parameters of steam turbine include main steam system, double reduction system, feedwater thermal control system, steam turbine safety monitoring thermal control system, etc; Relevant

https://doi.org/10.1051/e3sconf/202235802002

parameters of energy consumption system display the actual power generation and power consumption relationship diagram of the power plant.

Energy consumption analysis shows the coal consumption deviation of each unit through histogram and radar chart, and analyzes the influence of boiler efficiency related parameters, steam consumption rate related parameters and plant utilization rate related parameters of each unit on coal consumption level by level.

Adjustable potential analysis mainly includes the display of power generation potential calculation indicators, power consumption potential calculation indicators, power generation Statistics (day, month and year), consumption statistics (day, month and year), adjustable potential parameter configuration, etc.

The transaction management module mainly realizes the identification of comprehensive utilization of self owned power plants and the display of medium and long-term transaction information. Among them, the comprehensive utilization identification function is the real-time monitoring of comprehensive utilization fuel blending of each unit in the self owned power plant, monthly fuel blending and display of important indicators of the unit. Medium and long-term transaction management includes declaration information module and transaction information module.

File management mainly includes the boiler equipment account, steam turbine equipment account, boiler and steam turbine alarm parameters and data upload and download of each unit (operation procedures, thermal calculation instructions, etc).

The overall directory structure of the platform is shown in Figure 1.

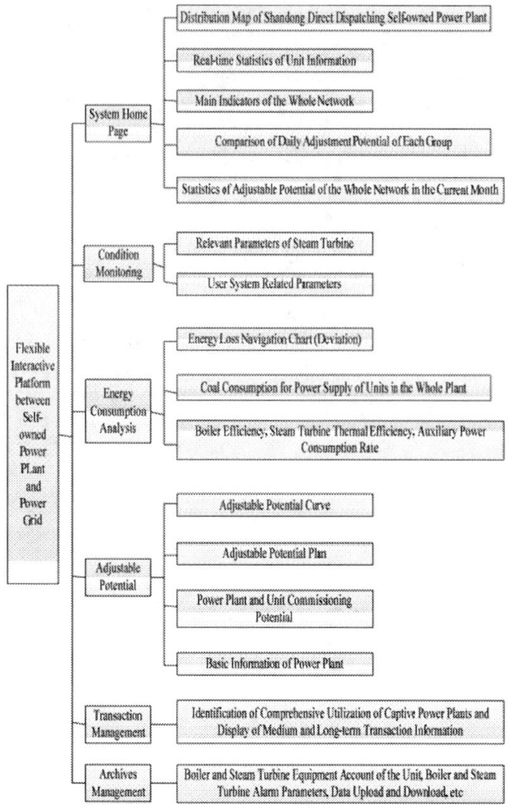

Figure 1. Overall directory structure of system platform

4. Conclusion

This paper introduces a flexible interactive platform system between self owned power plant and power grid. The system has the functions of self owned power plant file management, resource adjustable potential analysis, operation state detection, energy consumption analysis, transaction management and so on. The flexible interactive technology of self owned power plants for market-oriented supply and demand regulation is the key to promote the future development of power reform. Through the research and development of the operation method of self owned power plants, the analysis method of power generation and consumption regulation potential of self owned power plants, the interactive market mechanism between self owned power plants and power grids and the flexible interactive platform between self owned power plants and power grids, it will help to master the key technologies of flexible interactive application of self owned power plants for market-oriented supply and demand regulation, and continue to maintain and consolidate its leading position in the field of power reform.

Acknowledgements

This work is supported by the science and technology project of Shandong Electric Power Research Institute (ZY 2020-05), which studies the calculation method and platform application of minimum operation mode of thermal power plant based on power grid peak shaving dispatching

References

1. Wanyu Liu, Huaqiang Li, Hongli Zhang, etc. Transmission network expansion planning considering flexible supply-demand balance[J]. Power system automation, 208,42(5):56-63.
2. Shifu Zou. Design and implementation of data analysis and processing subsystem of on-line monitoring system in self owned power plant [D]. Chengdu: University of Electronic Science and Technology of China, 2011.
3. Shijun Liu, Qing Cai, Yuhui Bi. Research and judgment on restrictive policies and development direction of enterprise owned thermal power units [J]. Baosteel Technology, 2016,(5):6-12.
4. Shuang Weng. Self owned power plant: how to achieve Nirvana rebirth [J]. China Electric Power Enterprise Management, 2018,(11):22-24.
5. Yue Zhang, Xianzhong Dai, Jinli Lv, etc. Analysis on relevant policies and development problems of self owned power plants under energy transformation [J]. Electrical technology and economy, 2020,(4):65-66.

Chinese medicines used as oilfield chemicals: A case of resource utilization of waste medicines

Yuxi Zhang

The High School Affiliated to Shaanxi Normal University, Xi'an Shaanxi, 710065 China

Abstract: In order to explore the resource utilization of expired traditional Chinese medicine, in this study, three traditional Chinese medicines, LQC, BLG and HXZ were used to evaluate their inhibition effects on clay expansion rate and corrosion inhibition rate in acid solution by weight loss method and linear expansion test. The corrosion inhibition results reveal that the highest corrosion inhibition rate of HXZ at 1000mg/L is 67.65% in 3.0% HCl solution. Linear expansion experiments show that LQC and BLG have stronger inhibition effect when 1.2% of them are added to the drilling fluid, which is better than 4.0% KCl generally used as a reference to bentonite expansion inhibitor. HXZ can decrease the tg obviously in drilling fluid, which means it has a potential use as lubricants in drilling fluid.

Key words: Chinese medicines; drilling fluid; corrosion inhibition.

1. Introduction

The outbreak of the novel coronavirus pneumonia (COVID-19) has been posing a huge threat on human health and social stability since the year of 2019[1-3]. Chinese medicines like LQC, BLG and HXZ play important roles in the treatment and prevention of COVID-19 which is classified as an epidemic according to Chinese medical theory. LQC is a popular Chinese medicine to cure epidemics[4,5]. Previous studies have indicated that LQC may inhibit the biological activity of various viruses such as Middle East Respiratory Syndrome Coronavirus (MERS-CoV) and Severe Acute Respiratory Syndrome Coronavirus (SARS-CoV)[6,7]. In order to prevent infection of COVID-19, many Chinese households stockpiled related flu drugs[8]. However, most drugs expired without ever being used, which can be a waste of resources. Under such circumstances, reusing expired medicines would be beneficial to environmental protection and might contribute to different industries, including the treatment of drilling fluid in oilfields.

In order to explore the resource utilization of expired traditional Chinese medicine, in this work, we screened the use of Chinese medicines in oil field chemistry. The corrosion inhibition effect on metal, the expansion inhibition effect on bentonite and the performance in drilling fluid of three traditional Chinese medicines LQC, BLG and HXZ were evaluated to provide possible industrial utilization of waste drugs.

2. Experiment Details

2.1 Materials

The manufacturer of LQC is Shijiazhuang Yiling Pharmaceutical Co., Ltd., BLG is produced by Guangzhou Baiyunshan Pharmaceutical Co., Ltd. and HXZ was purchased from Yunnan Baiyao Group Co., Ltd.. All three drugs expired at the experiment. Sodium bentonite (technical grade mineralogical composition: 70% montmorillonite, 20% Quartz, 10% Feldspar) and calcium bentonite (technical grade mineralogical composition: 60% montmorillonite, 25% quartz, 15% feldspar) were purchased from Xi'an Chanqing Chemical Co., Ltd. China. $MgCl2·6H2O$, $AlCl3·6H2O$ and NaOH were purchased from Tianjin Shengao Chemical Reagent Co., Ltd. China. HCl (36-38%), petroleum ether (Tian'an of Fuyu Fine Chemical Co., Ltd., 99.9%), acetone (CH3COCH3, Anyang Chemical Reagent Co., Ltd., 99.5%) and anhydrous ethanol (C2H5OH, 99.9%) were used directly. Using analytical grade HCl (36%) as raw material, 3.0% HCl solution was prepared by distilled water dilution method. The Q235 low carbon steel used in this experiment was purchased from Shanghai Mechanical and Electrical Equipment Co., Ltd., and the chemical composition of the low carbon steel was as follows: C: 0.16%, Si: 0.26%, Mn: 0.15%, P: $\leq$ 0.015%, S: $\leq$ 0.09% and Fe. First of all, the mild steel specimens of the required size were rubbed with sandpaper of different grades (400,600,800 and 1000) to remove rust particles, then they were cleaned with petroleum ether to remove oil from the steel surface, and finally were cleared with anhydrous ethanol and dried with cold air.

2.2 Gravimetric measurements

The gravimetric measurements were conducted in 3.0% HCl solution with two types of mild steel at different concentrations of Chinese medicine (0, 100, 200, 300, 500, 700, 1000mg/L) to give reproducible results.All the acid solutions were exposed toair to remove impurities, then cleaned, dried and weighed accurately. The corrosion of mild steel in 3.0% HCl solution was measured after 4 hours at 303 K. The inhibition efficiencyη(%) and corrosion rate Cvwere calculated based on Eq (1) and Eq (2)[9-10],

$$\eta(\%) = 100\times(M_0-M_1)/M_0 \quad (1)$$

$$C_V = (M_0-M_1)/St \quad (2)$$

where M0 and M1 are the weight losses of mild steel under uninhibited and inhibited conditions, respectively, S(m2) is the steel area, and t is the duration of immersion.

2.3 Drilling Fluid Evaluation

To measure the physical and chemical properties of drilling fluid, calcium bentonite (4%) and sodium carbonate (0.2%) were added to tap water (350 mL), stirred for 30 minutes and left to set for 16 hours at 298K. The rheological, filtrational and lubricational properties of drilling fluid, such as AV (apparent viscosity), PV (plastic viscosity), YP (yield point), FL (API filtration) and Tg (friction coefficient), were evaluated using a viscometer (ZNN-D6S, Hengtaida Co., Ltd., Qingdao, China) and medium pressure filtration instrument (GJSS-B12K, Hengtaida Co., Ltd., Qingdao, China), according to the formulas in Chinese National Standard GB/T 16783.1-2006[11].

2.4 Linear Swelling Evaluation

Apart from corrosion inhibition capability of drugs and properties of drilling fluid, linear expansion should also be taken into consideration in application of oilfield industry. The effect of shale inhibitor on the linear expansion rate of bentonite was evaluated according to the oil and gas industry standard SY/T6335-1997 "Evaluation method of Shale Inhibitors for drilling fluids"[12]. The calculation formula of linear expansion ratio of bentonite is as follows:

$$S_r = \frac{R_0}{\Delta L}\times 100\% \quad (3)$$

Where Sr is the linear expansion rate of bentonite; R0 is the expansion amount of bentonite, (in) mm; ΔL is core height, (in) mm.

3. Results and Discussions

3.1 Gravimetric measurements

Gravimetric measurements were used to evaluate the corrosion inhibition performance of LQC, BLG and HXZ in 3.0% HCl solution. Fig.1 shows the hanging of the flakes with the addition of LQC at concentrations of 0, 100, 200, 500, 700 and 1000mg/L. It can be seen from Fig.1 that as the concentration increases the colour of 3.0% HCl solution deepens, and bubbles are produced on the surface of the steel flakes soon, indicating that the reaction between HCl solution and Fe in Q235 steel flakes releases H_2.

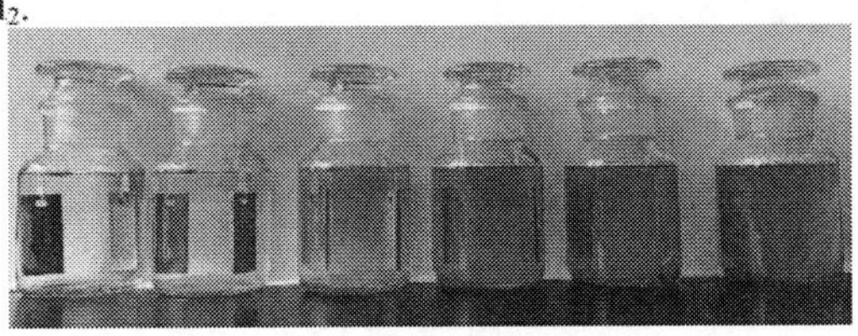

Fig.1 Weight loss hangings at different concentrations of LQC

In 3.0% HCl solution, LQC, BLG and HXZ with a concentration of 100, 200, 500, 700, and 1000mg/L were added respectively. The results after immersed for 4 h at 303K were shown in Fig.2, which indicates that the corrosion inhibition rate increases with the concentration, and the maximum corrosion inhibition rate of HXZ, LQC and BLG are 67.65%, 51.79% and 37.61% respectively at a concentration of 1000mg/L.

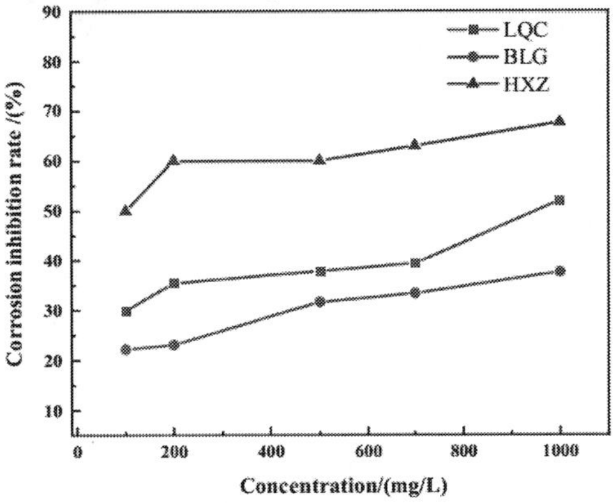

Fig.2 Experimental results of weight loss method

3.2 Linear Swelling Experiment

Since HXZ has the highest corrosion inhibition rate among the three drugs, its corrosion inhibition rates at different temperatures 303, 313, 323, and 333 K were measured, as shown in Fig.3. The corrosion inhibition rate of HXZ gradually decreases with the rise in temperature. The maximum corrosion inhibition rate of 1000mg/L HXZ is 67.65% at temperature 303K. This indicates that the ambient temperature can affect the corrosion of metal surfaces.

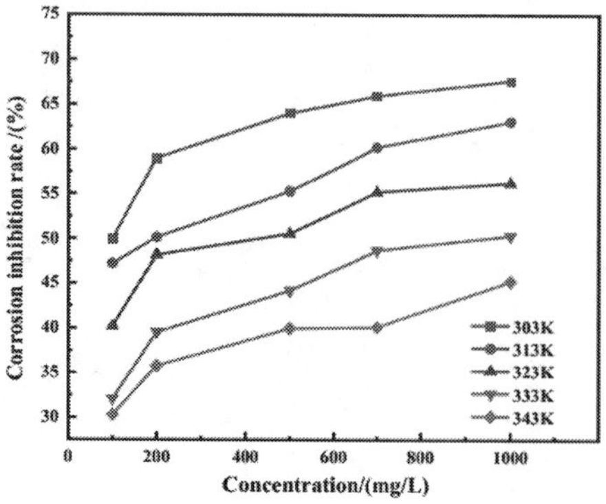

Fig.3 Corrosion inhibition rate of HXZ at different temperatures

3.3 Linear Swelling Experiment

The effects of concentration of the LQC and BLG on the clay hydration swelling inhibition were evaluated, and the results were shown in Fig.4. The expansion rate of the clay shows an increasing trend as time increases. The 1.2% LQC and BLG solution have higher inhibition effect on the hydration swelling of the clay than 4% KCl, which is a common reference in expansion inhibition. The linear expansion rate after 120 minutes is only 3.39% for LQC and 22.69% for BLG, which are much lower than that of 4% KCl.

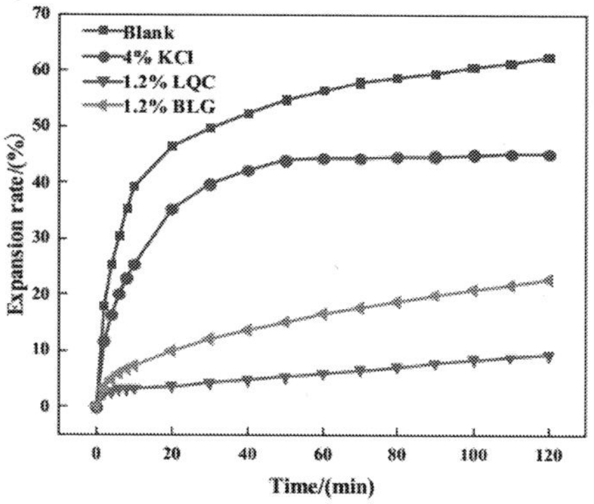

Fig.4 Swelling of bentonite in the suspension.

3.4 Performance in Drilling Fluid

Drilling is the primary link between the development of oil and gas fields. Drilling fluid is critical in drilling engineering and also needs to have good rheology and inhibition effect. According to the drilling fluid performance evaluation method in 2.3, the effects of HXZ on the drilling fluid performance at concentration 0.1%, 0.3%, 0.5%, 0.7% and 1.0% at 298K were evaluated, and the results were listed in Table 1. The AV of the drilling fluid dones not change, PV increases to 2.00, but YP decreas to 0 with the addition of 0.1-0.5% HXZ at room temperature compared to blank mud, while the filtration loss increases significantly. It also can be found the tg decrease obviously, which means HXZ has a potential use as lubricants in drilling fluid.

Table 1 Effect of adding different mass fractions of patchouli in slurry on performance

Dosage	AV /mPa•s	PV /mPa•s	YP /Pa	YP/PV /m•s	Φ600	Φ300	F_L /mL (API)	tg
4%Calcareous slurry	2.00	1.50	0.50	0.33	4.0	2.5	33.0	0.2468
4%Calcareous slurry +0.1%HXZ	2.00	2.00	0.00	0.00	4.0	2.0	35.0	0.0524
4%Calcareous slurry +0.3% HXZ	2.00	2.00	0.00	0.00	4.0	2.0	36.0	0.0815
4%Calcareous slurry +0.5% HXZ	2.00	2.00	0.00	0.00	4.0	2.0	34.0	0.1139
4%Calcareous slurry +0.7% HXZ	2.00	1.50	0.50	0.33	4.0	2.5	34.0	0.1405
4%Calcareous slurry +1.0% HXZ	2.00	1.50	0.50	0.33	4.0	2.5	35.0	0.1589

4. Conclusions

The performances of three traditional Chinese medicines LQC, BLG and HXZ in drilling fluids were evaluated. The highest corrosion inhibition rate of HXZ, measured by the weight loss method, was 67.65% in 3.0% HCl solution. Linear swelling rate experiments indicates that HXZ has some inhibition effects on the hydration swelling of bentonite, which is much lower than that of 4% KCl. It also can be found the tg of HXZ modified drilling fluid decrease obviously, which means HXZ has a potential use as lubricants in drilling fluid. The three of the expired Chinese medicines can be used as oilfield chemicals successfully, providing new ideas for the waste medicines reuse.

Acknowledgments

Special thanks to Xu Zhongying for her help in my experiment, and the work of Modern Analysis and Testing Center of Xi'an Shiyou University.

References

1. Ren J, Zhang A H, Wang X J. Traditional Chinese medicine for COVID-19 treatment[J]. Pharmacological research, 2020, 155: 104743.
2. Ren W, Liang P, Ma Y, et al. Research progress of traditional Chinese medicine against COVID-19[J]. Biomedicine & Pharmacotherapy, 2021, 137: 111310.
3. Xu J, Zhang Y. Traditional Chinese medicine treatment of COVID-19[J]. Complementary therapies in clinical practice, 2020, 39: 101165.
4. Dai Y J, Wan S Y, Gong S S, et al. Recent advances of traditional Chinese medicine on the prevention and treatment of COVID-19[J]. Chinese Journal of Natural Medicines, 2020, 18(12): 881-889.
5. Sun C Y, Sun Y L, Li X M. The role of Chinese medicine in COVID-19 pneumonia: a systematic

review and meta-analysis[J]. The American Journal of Emergency Medicine, 2020, 38(10): 2153-2159.

6. Peng A Y, Chen S. Traditional Chinese medicine works: A politicised scientific debate in the COVID-19 pandemic[J]. Asian Journal of Communication, 2021, 31(5): 421-435.
7. Noor R. A comparative review of pathogenesis and host innate immunity evasion strategies among the severe acute respiratory syndrome coronavirus 2 (SARS-CoV-2), severe acute respiratory syndrome coronavirus (SARS-CoV) and the Middle East respiratory syndrome coronavirus (MERS-CoV)[J]. Archives of microbiology, 2021, 203(5): 1943-1951.
8. Chen K, Chen H. Traditional Chinese medicine for combating COVID-19[J]. Frontiers of Medicine, 2020, 14(5): 529-532.
9. Hsissou R, Benzidia B, Rehioui M, et al. Anticorrosive property of hexafunctional epoxy polymer HGTMDAE for E24 carbon steel corrosion in 1.0 M HCl: gravimetric, electrochemical, surface morphology and molecular dynamic simulations[J]. Polymer Bulletin, 2020, 77(7): 3577-3601.
10. Solomon M M, Umoren S A. Electrochemical and gravimetric measurements of inhibition of aluminum corrosion by poly (methacrylic acid) in H2SO4 solution and synergistic effect of iodide ions[J]. Measurement, 2015, 76: 104-116.
11. Ni W J, Wang W L, Wang Q C, et al. Modification and application of waste shaddock peel as a green additive for water-based drilling fluid. J. Biobased Mater. Bioenergy 2021, 15, 380–384.
12. Jing Y, Zhang J, Hu W, et al. Preparation and Evaluation of Ammonium-Succinic Salts as Shale Swelling Inhibitor and Its Application in Water-Based Drilling Fluids[J]. Russian Journal of Physical Chemistry B, 2021, 15(1): 102-108.

https://doi.org/10.1051/e3sconf/202235802004

Performance test of optimal denitration efficiency of a 350MW unit under the background of new power system

Ziwei Zhong *, Lingkai Zhu, Wei Zheng, Zhiqiang Gong, Panfeng Shang, Junshan Guo, Jun Liu,Yanpeng Zhang

Electric Power Research Institute of State Grid Shandong electric power company, Jinan, China

Abstract. With the establishment of the goal of "carbon peaking and carbon neutralization", a new power system based on new energy is proposed. With the rapid development of ultra-high voltage external power and new energy in Shandong power grid, the difficulty of system peak regulation is increasing. At present, the peak regulation is mainly based on direct regulating thermal power units. The capacity of pumped storage power stations is limited and there are no gas-fired generating units, which also leads to serious shortage of other flexible regulation resources that Shandong power grid can call. In 2022, Shandong Province plans to implement the renewable energy multiplication plan, and it is expected that the installed capacity of new energy will account for 38%. With the large-scale access of new energy, the "double high" characteristics of the power system are highlighted, and the technical characteristics and stability mechanism of the power grid will undergo profound changes. In the new power system with new energy as the main body, the role of thermal power generation has gradually changed from the electric power source to the regulating power source. Under the variable load conditions, find the optimal efficiency conditions and provide necessary data for the optimal denitration efficiency operation.

key word: Deep peak shaving; Optimal denitrification efficiency.

1. Introduction

Now, after a 350MW heating unit is completed and put into operation, it can meet the environmental protection parameter of denitration ultra-low emission 50mg/m3 formulated by the state[1].There are problems such as high ammonia escape rate, uneven NOx distribution at the outlet and low denitration efficiency of SCR device.The chemical reaction formula of denitration process is as follows:

$$4NO+4NH_3+O_2 \rightarrow 4N_2+6H_2O$$

$$2NO_2+4NH_3+O_2 \rightarrow 3N_2+6H_2O$$

The selective catalytic reduction method is selected for denitration process, that is, in the SCR reactor of the tail flue, the flue gas reacts with the injected NH3 to generate nitrogen and water. The selected reductant NH3 is generated by hydrolysis of liquid ammonia and supplied to the denitration unit area[2]. In the mixer, it is mixed with the air provided by the dilution fan and diluted to below 5% volume safety concentration.

2. Establishment of model of out of stock system and field out of stock optimization test

The main purpose of establishing the denitration model is to test whether the flow field design of the denitration device is reasonable. By optimizing some key positions and parts in the denitration flow field, such as the inlet section of the ammonia injection grid, the inlet section of the catalyst layer, and the baffle, etc., a uniform flue gas distribution is established in the SCR reactor before the ammonia injection mixing. At the same time, the performance of the static mixer is tested and optimized.
The study of denitration flow model will include the flue from the outlet of boiler economizer to the inlet of air preheater, including: ammonia injection and mixing system at the inlet flue of reactor, ammonia injection grid system and SCR reactor with rectifier. In the denitration device, the flow field that has not been optimized by numerical simulation calculation has poor uniformity of ammonia nitrogen mixing at the inlet, low denitration efficiency, and ultimately leads to high ammonia escape[3].
Therefore, for the denitration device of Baoding northwest suburb thermal power plant, verify whether it

* Corresponding author: dkyzhongziwei@163.com

meets the uniform distribution from the flow field design to ensure uniform flow field distribution, uniform velocity distribution and uniform ammonia nitrogen mixing.

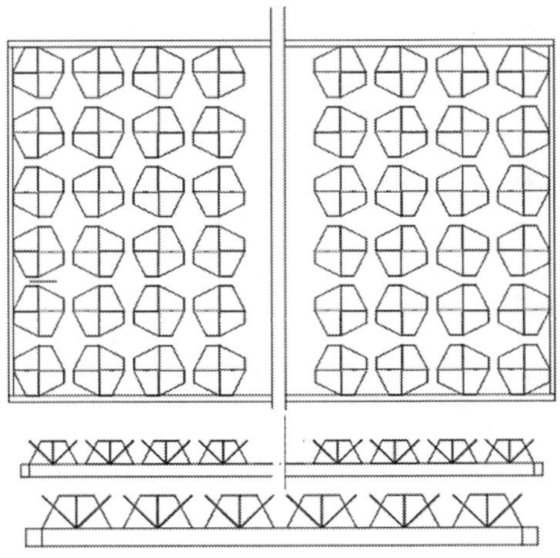

Figure 1. Schematic diagram of AIG blade in the model

2.1 Porous media model

In porous media, momentum, energy and mass transfer phenomena exist widely, especially the study of the fluid movement process in porous media has far-reaching practical value. Porous media model is also widely used in power plants, and the simulation of catalyst layer in this model can also be realized through porous media model. The flow of fluid in porous media channels is various and difficult to describe by simple methods[4]. It is also difficult to describe the resistance and pressure drop. Therefore, appropriate restrictions on the porous media model can play an effective role in modeling, so as to better understand the momentum mechanism of the fluid passing through the partially or completely filled porous media channel . The porous media model established this time has the following limitations:

(1) Most of the research results of numerical calculation are tests carried out under unsteady conditions. In this case, there will be many uncontrollable situations. However, at present, there are very few numerical studies on normal fluid flow, so it is difficult to have accurate numerical reference. In fact, the velocity of the fluid passing through the medium in this model does not change, which means that the transition time of the fluid passing through the medium is not accurately reflected.

(2) In the porous media model, when the fluid is turbulent, fluent will solve the standard conservation equation of turbulent flow by default. Based on this, the solid medium has no influence on the generation and dissipation speed of turbulence[5].

The momentum equation of porous media can be expressed by the following formula, and the additional momentum source term is composed of the following two parts:

$$S_i = \sum_{j=1}^{3} D_{ij}\mu v_j + \sum_{j=1}^{3} C_{ij}\frac{1}{2}\rho\left|v_j\right|\cdot v_j \qquad (1)$$

Where is the momentum source term of the direction and D and C are the prescribed matrices.

For simple homogeneous porous media:

$$S_i = \frac{\mu}{\alpha}v_i + C_2\frac{1}{2}\rho\left|v_j\right|v_j \qquad (2)$$

Wherein: αIs the permeability factor and C2 is the internal resistance factor.

The catalyst used in the test unit is a plate type, which can be regarded as an isotropic porous medium composed of parallel channels[6]. Its internal resistance coefficient in the flue gas flow direction can be obtained from the test or actual measurement.According to the design requirements, the resistance of $\Delta p_y = 150Pa$ ($\Delta p_y \leq 200Pa$) is set in the flue gas flow direction in this simulation to obtain the internal resistance coefficient of the catalyst during calculation:

$$K_y = \frac{1}{2}\rho v_y{}^2 / \Delta p_y \qquad (3)$$

2.2 Field out of stock optimization test

Due to the limitation of load dispatching, it is determined through negotiation by many parties that the continuous operation time at full load from 9:00 to 19:00 shall be the test time, and the test shall not be less than 3 days. The coal quality shall be confirmed by both parties. During the performance assessment test of the unit denitration system, the average load of the unit is 350MW, which is required to meet the requirements of the assessment test and achieve stable and continuous operation. See table 2 for the main operating parameters of the unit during the test. During the test, the 100% load period is from 9:00 to 19:00 every day for 10 consecutive hours. During the test, the inlet NOx concentration range is 200-330mg/Nm^3 (6% O_2) (the NOx concentration during the test is calculated as NO_2), which basically meets the test conditions. During the test, the coal mill shall not be switched and the combustion adjustment shall not be conducted.

The instruments and materials required during the test are shown in the following table:

Table 1 Required instruments and materials

Serial number:	Instrument name:	Model and number	unit	quantity
1	flue gas analyzer	MGA5+	set	2
2	Electronic micropressometer	SWEMA 3000	platform	1
3	GASMET Fourier infrared analyzer (including SO_3, HF, HCl, NO_2, etc.)	FT-IR DX4000	platform	1
4	Total composition flue gas analyzer	U23	platform	2
5	Oxygen analyzer	Kane KM900	platform	2
6	Precision electronic balance	CPA225D	platform	2
7	Digital thermometer	51II	set	2
8	Dust tester	3012H	set	4
9	Thermocouple	Fluke F52	set	3

Table 2 Main operating parameters of the unit during the test

project	7/27	7/28	7/29
Actual power	344.44	342.50	349.41
Main steam pressure	24.12	23.96	24.30
Steam pressure of reheat hot section	3.91	3.94	3.92
Main steam flow	1065.47	1044.29	1115.56
Outlet steam temperature of high temperature superheater	562.66	567.12	567.58
Outlet temperature of high temperature reheater	562.98	566.64	567.53
Fuel quantity after calibration	150.05	148.05	154.57
Outlet flue gas temperature of air preheater a	149.11	148.31	149.97
Outlet flue gas temperature of air preheater B	149.82	149.54	146.01
Oxygen content of flue gas at outlet of air preheater a	3.05	2.68	2.83
Oxygen content of flue gas at outlet of air preheater B	3.32	3.15	3.11

3. Conclusion

The flue gas parameters were tested at the inlet of SCR reactor. The flue gas analyzer at the outlet of the reactor is used to measure the content of NOx and O_2 at the inlet and outlet of the reactor. At the same time, the NOx and O_2 data of CEMS dial are recorded, and the display correction coefficient of NOx and O_2 dial is obtained through comparison. Denitration efficiency is calculated as follows:

$$\eta = (C_{NOx\ in} - C_{NOx\ out}) / C_{NOx\ in} * 100\% \qquad (3\text{-}1)$$

The denitration system meets the reading of CEMS at the inlet and outlet of denitration as required by the assessment, and the CEMS reading is corrected according to the measured flue gas parameters of side a and the calibration coefficient of CEMS reading at the inlet and outlet of side B, and the test results are obtained. The final result is that the average denitrification efficiency of side a and side B is 92.85% and 92.33% respectively at 100% load, and the outlet NOx (calculated as NO_2) concentration is 16.03mg/nm^3 (6% O_2) and 18.52mg/nm^3 (6% O_2) respectively.Through comparison under different load conditions, it can be concluded that the denitrification performance is the best when the unit burns design coal and operates under 100% load, the NOx concentration distribution at the outlet section is relatively uniform, and the ammonia escape is also low.

Acknowledgements

This work is supported by the science and technology project of Electric Power Research Institute of State Grid Shandong electric power company(ZY-2022-15), which studies Research on peak capacity improvement of thermal power unit based on electric heating coordination optimization under guaranteed supply demand.

References

1. Zhanshan Wang, Libo Pan, Yunting Li. Assessment of air quality benefits from the national pollution control policy of thermal power plants in China: A numerical simulation[J]. Atmospheric Environment, 2015, 106: 288-304.
2. Rai P, Gautam N, Chandra H.Investigating Electrostatic Precipitator Design Parameters for Efficient Control of Particulate Matter in Thermal Power Plant: A Case Study[J].Journal of the Institution of Engineers, 2018, 99(2): 303-310.
3. J. K. Tewari.Preventive techniques of pollution control-the reliability and safety in core sectors including thermal power plant installations and economic evaluation[C]. Energy Conversion Engineering Conference, 1997.
4. Tsukamoto S, Namihira T, Wang D.Pollution control of actual flue gas using pulsed power at a thermal power plant[J].2001, 134(4): 28-35.
5. Fluhr R, Aviv D, Galun E.Present Situation on Pollution Control of Thermal Power Plant and

Objective on the Tenth Five-year Plan in China[J].Environmental Protection, 2003.

6. Fluhr R, Aviv D, Galun E. Present Situation on Pollution Control of Thermal Power Plant and Objective on the Tenth Five-year Plan in China[J]. Environmental Protection, 2003.

Evaluation of operation regulation ability of a power plant unit in heating season

Jinxu Lao *, Hu Zhang, Lingkai Zhu, Yangpeng Zhang, Haizhen Lv

State grid shandong electric power research institute, Jinan,China

*Corresponding author e-mail: 1686844029@qq.com

Abstract. In order to further improve the operation level of thermal power units directly regulated by the power grid, the regulation capacity of a power plant unit is comprehensively analyzed and evaluated from the aspects of heating capacity, minimum startup mode of the whole plant, heating state and load capacity, etc., to provide support for the accurate dispatching of the unit.

Keywords: Unit heating, Peak shaving,Minimum output.

1. Equipment introduction

1.1 Turbine Equipment Overview

No. 1 and No. 2 steam turbines are N135-13.24/535/535 ultra-high pressure, single intermediate reheat, single shaft, double cylinder, double exhaust and condensing steam turbines. Unit 1 and unit 2 will complete the high back pressure heating[1] transformation in 2019 and 2021 respectively, with the designed circulating water volume of 6600 t/h and the exhaust pressure of 54kPa.

No. 5 and No. 6 steam turbines are C330-16.7/0.8/538/538 reactive, single shaft, primary intermediate reheat, high and medium pressure cylinder combined, double cylinder double exhaust, extraction and condensing steam turbines. The heating extraction position of unit 5 and unit 6 is the middle exhaust, the design extraction pressure is 0.8MPa, the rated extraction flow of single unit is 280 t/h, and the maximum extraction flow is 300 t/h; The industrial extraction position is the cold reheat and auxiliary steam header, and the designed extraction flow of the cold reheat is 50 t/h. Unit 5 will complete zero output [2-3] heating transformation of low-pressure cylinder in 2020. After the transformation, the maximum heating extraction flow of the middle exhaust is 640 t/h.

1.2 Heat supply

The heating and steam extraction pipes of units 5 and 6 are combined into one pipe in the steam turbine plant and then sent to the initial heating stations of phase I and phase II respectively.

There are two circuits of circulating water return in the heat supply network, which are first combined in the plant area and then divided into two circuits and sent to the condensers of units 1 and 2 for the first heating, absorbing the exhaust heat of the low-pressure cylinder and heating the circulating water temperature to about 75 ℃. The heated circulating water in the heat supply network is first combined and then divided into two circuits and sent to the initial heating station for the second heating.

There are 2 initial heating stations, and the steam source is from the exhaust and supply of No. 5 and No. 6 units. Three steam water heaters, three water-water heat exchangers and one exhaust steam heater are installed in the initial station of phase I; There are 3 circulating pumps in the heat supply network, including 1 electric pump and 2 steam pumps. The steam source of the small steam turbine is from the heating extraction header of unit 5 and unit 6. The phase II initial station is equipped with 3 steam water heaters, 3 water heat exchangers, 1 exhaust steam heater, 3 heat network circulating pumps, 2 electric pumps and 1 steam pump. The steam source of the small steam turbine is from the heating extraction header of unit 5 and unit 6.

1.3 Heating capacity

According to the statistics of the initial stage, the middle stage and the final stage of heating, the initial stage is from November 15 to December 14 every year, the middle stage is from December 15 to February 14, and the final stage is from February 15 to March 15. According to the statistics of heating extraction capacity, heating capacity, thermal power ratio, power generation capacity, power supply capacity, converted heating area and other indicators of a single unit and the whole plant, the converted extraction flow is calculated according to the heating heat index 40 W/㎡ converted heating area and the high back pressure unit. The above thermal power

* Corresponding author: 1686844029@qq.com

parameters are calculated according to the operating hours of the unit. In the heating season from 2019 to 2020, the heating capacity of residents is 148.27×10^{5}GJ, and the heating area of residents is 351.65×10^5 ㎡; In the heating season from 2020 to 2021, the heating capacity of residents is 442.37×10^{5}GJ, and the heating area of residents is 1057.86×10^5 ㎡; In the heating season from 2021 to 2022, the heating capacity of residents is 502.73×10^{5}GJ, and the heating area of residents is 1202.18×10^5 ㎡.

According to the rated external heating capacity of the power plant, the allowance of replaceable heating area of the unit is analyzed: Unit 1 and unit 2 are high back pressure heating units, which have reached the rated heating capacity and are irreplaceable; Unit 5 is divided into two operation modes: cylinder cutting and non cylinder cutting. Under the cylinder cutting operation mode of unit 5, the available heat supply area margin of the whole plant is 1355.85×10^5 ㎡, and under the non cylinder cutting operation mode, the available heat supply area margin of the whole plant is 778.97×10^5 ㎡.The alternative heating area is shown in Table 1.

Table 1 Alternative heating area of each unit and the whole plant

Generator Set	Rated heating extraction flow/ t/h	Rated heating area/ ×10^5 ㎡	Actual heating extraction flow/t/h	Actual heating quantity/ GJ	Actual heating area/ ×10^5 ㎡	Alternative heating area allowance/ ×10^5 ㎡
1	High back pressure heating unit has reached the rated heating capacity					0
2	High back pressure heating unit has reached the rated heating capacity					0
5	No cylinder cutting280/ Cylinder cutting576.68	No cylinder cutting544.44/Cylinder cutting1121.32	145.87	1157489.06	276.79	No cylinder cutting267.79/ Cylinder cutting844.53
6	280	544.44	21.65	138491.71	33.12	511.32
Whole plant (except unit 1 and 2)	No cylinder cutting560/ Cylinder cutting856.68	No cylinder cutting1088.88/ Cylinder cutting1665.76	132.81	1295980.77	309.91	No cylinder cutting778.97/ Cylinder cutting1355.85

2. Operation analysis

2.1 Analysis of operation during Spring Festival

During the Spring Festival, the actual average people's livelihood heating flow is 778.71t/h, and the actual average industrial steam supply flow is 0.30 t/h. The minimum startup [4] mode of people's livelihood heating and comprehensive heating should actually be one large and two small. The power plant reviewed and reported that it was two small and the cylinder of unit 5 was cut off for operation.

He analysis is as follows: in the middle period of heating, the power plant reported that the additional demand of 400 t/h people's heating flow for unit 1 and unit 2 under high back pressure operation exceeded the rated heating extraction flow (280 t/h) of unit 5 and unit 6. However, the maximum heating extraction flow of unit 5 under cylinder cutting mode was 640 t/h, which could meet the demand reported by the power plant, Therefore, before the heating season, the minimum start-up mode of the people's livelihood heating and comprehensive heating in the middle of the heating season is the two small units plus the cylinder cutting operation of unit 5; The actual average heating steam extraction volume of unit 5 during the Spring Festival is 227.02 t/h, while unit 6 basically has no heating steam extraction. The actual heating volume required by unit 1 and unit 2 under high back pressure operation is less than the rated heating steam extraction flow (280 t/h) of unit 5 and unit 6. Therefore, according to the actual heating volume, the minimum start-up mode during the Spring Festival shall be one large and two small, and it is not required to cut off the cylinder of unit 5 for operation.

2.2 Analysis of electric load adjustment capability

Unit 1 and Unit 2 are high back pressure heating units, which operate stably in the heating season and basically have no power load adjustment capacity. Unit 5 has passed the minimum technical output verification test [5]. During the test, the heating steam extraction flow is 218.83t/h, the industrial steam extraction flow is 18.16t/h, and the minimum technical output is 128.48MW (the verification requirement is 132MW); Unit 5 is a cylinder cutting unit, and there are two operation modes: non cylinder cutting and cylinder cutting. However, the average heating steam extraction flow of unit 5 at the beginning, middle and end of heating is lower than the required value of cylinder cutting (since the minimum technical output test after flexible transformation is verified according to 40% rated output, the required value of cylinder cutting is set as 278.49t/h under 40% boiler evaporation). To sum up, the electric load adjustment capacity of units 5 and 6 is as follows:

At the initial stage of heating, the adjustable output range of Unit 5 for people's livelihood heating is 132MW～314.18MW, and the adjustable output range of comprehensive heating is 132MW～ 308.32MW; The adjustable output range of livelihood heating of unit 6 is 170.59MW～306.20MW, and the adjustable output range of comprehensive heating is 173.13MW～299.79MW.

In the medium term of heating, the adjustable output range of Unit 5 livelihood heating is 132MW～272.04MW, and the adjustable output range of comprehensive heating is 132MW～266.94MW; The adjustable output range of livelihood heating of unit 6 is 165MW～330MW, and the adjustable output range of comprehensive heating is 160.36MW～325.36MW.

At the end of the heating period, the adjustable output range of Unit 5 livelihood heating is 132MW～294.81MW, and the adjustable output range of comprehensive heating is 132MW～ 293.92MW; The adjustable output range of livelihood heating of unit 6 is 165MW～330MW, and the adjustable output range of comprehensive heating is 165MW～330MW.

Table 2 Statistics of deep peak shaving and derating

period	Generator Set	Rated capacity of unit (MW)	Minimum output of unit (MW)	Minimum adjustable output of comprehensive heating (MW)	Maximum adjustable output of comprehensive heating (MW)	Whether it has deep adjustment capability	Deep regulation load (MW)	Reduced output load (MW)
Initial stage of heating	1	145	73	98.34	98.34	no	High back pressure heating	
	2	145	73	99.45	99.45	no		
	5	330	165	132	308.32	yes	33	21.68
	6	330	165	173.13	299.79	no	0	30.21
Mid heating period	1	145	73	98.45	98.45	no	High back pressure heating	
	2	145	73	103.66	103.66	no		
	5	330	165	132	266.94	yes	33	63.06
	6	330	165	160.36	325.36	yes	4.64	4.64
End of heating	1	145	73	99.59	99.59	no	High back pressure heating	
	2	145	73	98.40	98.40	no		
	5	330	165	132	293.92	yes	33	36.08
	6	330	165	165	330	no	0	0

2.3 Analysis of electric load adjustment ability after thermal load distribution of the whole plant

Distribution principle: according to the approved minimum operation mode, the heat supply is distributed manually, and the high back pressure units are distributed preferentially. The adjustable range of the remaining heat supply bearing units is calculated, and the sum is the adjustable output range of the whole plant.

According to the actual heat supply, the minimum start-up mode of the people's livelihood and comprehensive heating at the beginning, middle and end of heating is one large and two small, i.e. (1, 2, 5 or 6). Since unit 5 has passed the minimum technical output test after flexible transformation, there are two combinations of the minimum start-up mode (1, 2, 5) and (1, 2, 6). Unit 1 and unit 2 are high back pressure heating units, with average loads of 98.69MW and 101.71MW respectively. When the heat load of the whole plant is distributed [6], the maximum and minimum adjustable output shall be taken as the average value of the whole heating season, and the heating steam extraction shall be taken as the average value of the whole heating season. To sum up, the power load adjustment capacity of single unit and the whole plant after the thermal load of unit 1 and unit 2 is preferentially distributed is as follows:

At the initial stage of heating, the adjustable output range of Unit 5 for people's livelihood heating is 132MW～302.03MW, and the adjustable output range of comprehensive heating is 132MW～289.70MW; The adjustable output range of livelihood heating of unit 6 is 165.75MW～302.03MW, and the adjustable output range of comprehensive heating is 176.01MW～289.70MW. When the minimum startup mode is (1,2,5), the adjustable output range of the whole plant's livelihood heating is 332.40MW～502.43MW, and the adjustable output range of comprehensive heating is 332.40MW～ 490.10MW; When the minimum startup mode is (1,2,6), the adjustable output range of the people's livelihood heating of the whole plant is 366.15MW ～ 502.43MW, and the adjustable output range of comprehensive heating is 376.41MW～490.10MW.

In the middle of heating, the minimum output under the average heating steam extraction of unit 5 calculated according to the design heating working diagram is lower than the minimum technical output test verification value (132MW), and the minimum adjustable output is taken as the calculated value of the heating working diagram. At this time, the adjustable output of the two combinations (1, 2, 5) (1, 2, 6) is the same value. The adjustable output range of Unit 5 or Unit 6 unit for civil heating is 114.67MW～270.78MW, and the adjustable output range of comprehensive heating is 123.67MW～259.96MW. The adjustable output range of people's livelihood heating of the whole plant is 315.07MW ～471.18MW, and the adjustable output range of comprehensive heating is 324.07MW ～ 460.36MW.

At the end of heating, the adjustable output range of Unit 5 livelihood heating is 132MW～330MW, and the adjustable output range of comprehensive heating is 132MW-330MW; The adjustable output range of livelihood heating of unit 6 is 165MW～330MW, and the adjustable output range of comprehensive heating is 165MW～330MW. When the minimum startup mode is (1,2,5), the adjustable output range of the whole plant's livelihood heating is 332.40MW～530.40MW, and the adjustable output range of comprehensive heating is 332.40MW～530.40MW; When the minimum startup mode is (1, 2 and 6), the adjustable output range of the whole plant's livelihood heating is 365.40MW ～ 530.40MW, and the adjustable output range of comprehensive heating is 365.40MW～530.40MW.

3. Analysis of output reduction

3.1 Analysis of actual average heat supply decreasing output

No. 1 and No. 2 units are high back pressure heating units with stable operation. The average load in the heating season from 2021 to 2022 is 98.69 MW and 101.71 MW respectively, which deviates from the blocked capacity applied by the power plant by 1.31 MW and 1.71 MW. The output reduction application is reasonable. The upper limit of the theoretically calculated adjustable output under the actual average heating extraction steam flow and average industrial extraction steam flow of unit 5 during the heating season from 2021 to 2022 is 283.17MW, which is 43.17MW different from the applied blocked capacity of the power plant, and the application for output reduction is unreasonable.

3.2 Analysis on verification test of reduced output

The unit shall be subject to the verification test of output reduction and analyzed according to the test results. Analysis basis: 1) during the test, whether the main steam flow reaches the main steam flow of the unit under TRL (high back pressure) condition; 2) Whether the heating extraction steam flow is reduced.

No. 1 and No. 2 units are high back pressure heat supply units, which operate in the mode of heat and constant power. During the test, the main steam flow did not reach the main steam flow under TRL condition. The unit still has the space for output increase of 11.03MW and 10.16MW. The application for output reduction is unreasonable. During the test of unit 5, the output was not increased by adjusting the heat supply; When the heating extraction steam flow is unchanged, the main steam flow of the unit does not reach the main steam flow under TRL condition, and the unit still has 24.59MW of output increase space, so the application for output reduction is unreasonable.The analysis results of output reduction verification test are shown in Table 3.

Table 3 Verification test results of output reduction in heating season from 2021 to 2022

Generator Set	Unit load during test（MW）	Main steam flow during test（t/h）	Main steam flow under TRL condition（t/h）	Test heating extraction steam flow（t/h）	Industrial extraction steam flow（t/h）	Power plant application adjustable upper limit（MW）	Theoretical maximum output under test steam extraction（MW）	Output up regulation space（MW）	Whether the output reduction application is reasonable
1	109.27	345.5	420	-	0	100	120.30	11.03	unreasonable
2	109.95	363.97	420	-	0	100	120.11	10.16	unreasonable
5	280.18	889.42	980.8	47.36	16.59	240	304.77	24.59	unreasonable

4. Conclusion

This paper analyzes the alternative heating area margin of a power plant from its heating capacity and actual heating situation, analyzes whether the minimum startup mode is reasonable according to the estimated heating capacity of the power plant, verifies its electric load adjustment capacity in this mode, and then verifies its electric load adjustment capacity after the thermal load distribution of the whole plant. According to the actual average heat supply of the unit in the heating season and the verification test of output reduction, analyze whether the application for output reduction of the unit is reasonable, and provide a basis for the reasonable dispatching of the power grid dispatching department.

Acknowledgements

This work was supported by the science and technology project of State grid shandong electric power research institute (ZY-2022-16), which studied the peak shaving capacity of different types of cogeneration units in Shandong power grid.

References

1. Jiang Tang,Xuedong Wang,Yuzhu Zhao, et al.Analysis of Performance Indicators and Peak Regulation Capacity of CondensingUnit After High Back Pressure Retrofit for Heating[J].Power Generation Technology,2018,39(05):455-461.
2. Haisheng Yang, Tuo Zhang, GuangTong Tang, et al. Influence of Zero-Output Technology of Low-Pressure Cylinder on Deep Peak Regulation Performance of Heating Unit and Compensation Standard for Peak Regulation[J]. Journal of Engineering for Thermal Energy and Power, 2020,35(06):268-273.
3. Qinpeng Zhang, Xuedong Wang, Feng Li, et al. Analysis of Heating Capacity and Peak-regulating Capacity of 330 MW Steam Turbine Unit With Low-pressure Cylinder off Operation[J]. Shandong Electric Power, 2020,47(12):72-76.
4. Lihong Zhang, Fusuo Liu,Wei Li, et al. Minimum startup method of a DC near-field thermal power unit based on a unit voltage supporting effect[J].Power System Protection and Control ,2020,48(23):141-147.
5. Xuehui Zhang, Xingsen Yang, Gang Xin, et al.Experimental study on deep peak regulation operation of coal－fired thermal power unit[J].Clean Coal Technology, 2022,28(04):144-150.
6. Lingkai Zhu, Wei Zheng, Junshan Guo, et al. Application of Heat Load Distribution in Improving Peak-shaving Ability of Plant[J]. Power System Engineering, 2019,35(06):43-46+49.

Water pollution assessment and health risk analysis of Puzhehei Lake

Ziyuan Wang, Xingyue Liu, Songlin Zhou, Wei Li*

School of geography and ecotourism, Southwest Forestry University, Kunming 650224, China

Abstract. Puzhehei Lake is located in Wenshan Prefecture, Yunnan Province. Water pollution assessment is of great significance for the pollution control of the lake and the sustainable development of the basin. This study evaluates the water pollution and eutrophication status of three sampling points (upstream, middle and downstream) of Puzhehei Lake based on single factor evaluation method, comprehensive pollution index method and improved comprehensive pollution index method based on entropy weight method, and analyzes the water health risk. The results show that the water quality of the three sampling points is sub-class V using the single factor evaluation method. The water quality evaluation result based on the traditional comprehensive pollution index method shows that the water quality of the three sampling points belongs to heavy pollution. By contrast, the improved comprehensive pollution index method based on entropy weight method shows that the water quality of Leshui Cave belongs to class V, while the water quality of Puzhehei Village and Karst Wetland Park belongs to class V. According to the comprehensive evaluation method, the water quality of the three sampling points before and after correction is ranked from good to bad as follows: Leshui Cave, Karst Wetland Park and Puzhehei Village. The water bodies of Leshui Cave and Puzhehei Village are moderately nutritious, and the water bodies of Karst Wetland Park are oligotrophic. The chemical carcinogens health risk in water at the three sampling points exceeded the standard seriously, and the risk ranking was as follows: Karst Wetland Park > Leshui Cave > Puzhehei Village, in which Cr exceeded the standard seriously. The non carcinogenic health risk are low, and the risk ranking is Puzhehei Village > Leshui Cave > Karst Wetland Park. In the future water quality control, we should focus on the strict control of TN and TP in rivers entering the lake.

Key words: Puzhehei Lake; Entropy weight method; Comprehensive pollution index; Eutrophication; Water health risk.

1. Introduction

The lake ecosystem plays a variety of ecological service functions such as providing drinking water resources, flood control and irrigation, and climate regulation[1]. In recent years, the unreasonable development and utilization of lake resources by humans has caused many problems such as lake area shrinkage, water pollution, and biodiversity loss[2-3]. Since the quality of lake water quality is closely related to the health and safety of regional drinking water, the National Academy of Sciences (NAS) and the United States Environmental Protection Agency (USEPA) took the lead in introducing a health risk assessment model in the 1980s, which provided quantitative information on the relationship between environmental pollution and human health[4-6].
Puzhehei Lake is the largest karst lake in southeastern Yunnan. Its unique topography and lake ecosystem have important research value, and it is also a source of local drinking water[7]. In recent years, the fragility of the karst structure and unreasonable human activities have led to the pollution and destruction of the water quality of Puzhehei Lake, and its water environment protection is facing severe challenges[8].
On the basis of relevant domestic and foreign research, we used the improved comprehensive index based on entropy weight method to evaluate the water pollution status of the upper, middle and lower reaches of Puzhehei Lake, used the health risk assessment model to evaluate the lake. The aim of the present study is provide a theoretical basis for water pollution control and water environment health protection in the study area.

2. Methods

2.1 Sampling

Three sampling points were set to the upper, middle and lower reaches of Puzhehei Lake, namely Leshui Cave, Puzhehei Village and Karst Wetland Park, sampling on

* Corresponding author: ww0592@gmail.com

December 6, 2020. The water was sampled with a 500ml capacity sampling bottle, and two sets of replicates were set for each sampling point. During sampling, choose to take samples at a depth of 0.1~0.2m in the lake when the water level is stable, and avoid shaking the water body. After sampling, tighten the bottle cap to avoid oxidation reaction, and store it in an incubator at about 4°C away from light, and then send it to the testing station for analysis on December 8, 2020. A total of 11 water quality indicators including COD, TN, NH3-N, TP, Cu, Zn, Pb, Cd, Cr, As and Hg were selected for testing.

2.2 Comprehensive index based on entropy weight method

The traditional comprehensive pollution index ignores the weight of each pollution index in the evaluation system. We used the entropy weight method to optimize the comprehensive pollution index, and eliminated the interference of other factors as much as possible[9-10]. The larger the entropy weight, the greater the influence of the index on the evaluation results.

The steps for calculating the weight of each indicator are to first normalize each index value (xij) detected by the three sampling points, then calculate the entropy value Ei of the index (k=lnm), and finally calculate the information weight coefficient ωi (Formula 1-4).

$$p(x_{ij}) = \frac{x_{ij}}{\sum_{j=1}^{n} x_{ij}} (i = 1,2,\ldots,m; j = 1,2,\ldots,n) \tag{1}$$

$$E_i = -k\sum_{j=1}^{n} p(x_{ij}) \ln p(x_{ij}) \tag{2}$$

$$d_i = 1 - E_i \tag{3}$$

$$\omega_i = \frac{d_i}{\sum_{i=1}^{m} d_i} (i = 1,2,\ldots,m) \tag{4}$$

In the formula, p(xij) is the normalized standard value of each index, Ei is the information entropy, and ωi is the weight value of the pollution index i.

Therefore, the improved comprehensive index based on entropy weight method has the following formula:

$$P' = \sum_{i=1}^{m} \omega_i P_i \tag{5}$$

2.3 Health risk assessment model

At present, the health risk analysis of water environment often adopts the health risk assessment model recommended by USEPA, which is divided into chemical carcinogen and non-carcinogen health risk assessment models according to the type of chemicals. The specific formula is referenced from [11].

3. Results

3.1 Water quality evaluation

The weights of Zn, Pb, and Cd were the highest, all greater than 0.10, the weights of other indicators ranged from 0.06 to 0.09, and the weights of Cu and As were the lowest. The comprehensive index evaluation results showed that the index of Puzhehei Village is the highest, followed by Karst Wetland Park, and Leshui Cave is the lowest. It can be seen that the water quality in the upper reaches was relatively the best, the lower reaches was the second, and the middle reaches was the worst. The results of water quality grade evaluation showed that Leshui Cave (upstream) belongs to Class V water, while Puzhehei Village and Karst Wetland Park (middle and downstream) was inferior to Class V (Table 1-3).

Table 1. Weight of each pollution index of Puzhehei Lake

Index	COD	TN	NH_3-N	TP	Cu	Zn	Pb	Cd	Cr	As	Hg
ω_i	0.0698	0.0808	0.0797	0.0717	0.0676	0.1348	0.1408	0.1292	0.0724	0.0676	0.0856

Table 2. Evaluation standard of comprehensive index method based on entropy method

Water quality grade	I	II	III	IV	V
Value	0.27	0.71	1.00	2.12	2.62

Table 3. Evaluation results of traditional and modified comprehensive pollution index method

Sampling point		P'	Average value	Water quality grade
Leshui Cave (the upper reach)	Sampling 1	3.03	2.49	Grade V
	Sampling 2	1.95		
Puzhehei Village (the middle reach)	Sampling 1	5.13	4.84	Inferior Grade V
	Sampling 2	4.54		
Puzhehei National Wetland Park (the lower reach)	Sampling 1	3.34	2.67	Inferior Grade V
	Sampling 2	2.00		

3.2 Water health risk assessment

The overall health risks of chemical carcinogens in the water bodies of the three sampling points in Puzhehei Lake exceeded the maximum risk level (5.0×10-5·a-1) accepted by the International Commission on Radiation Protection, and the order was: Karst Wetland Park > Leshui Cave > Puzhehei Village. The health risks of Cd and As chemical carcinogens were lower than the maximum risk level accepted by the International Commission on Radiation Protection, but slightly higher than the risk level acceptable to USEPA.

The health risks of non-carcinogens at the three sampling points were all lower than the negligible risk level (1×10-8 ·a-1) recommended by the Netherlands Construction and Environment Agency and the negligible risk level (1×10-7·a-1) recommended by the Royal Society of England. The order was: Puzhehei Village > Leshui Cave > Karst Wetland Park. The health risks of each indicator of non-carcinogens from high to low were: Hg, NH3-N, Pb, Cu, Zn.

In general, the total health risk ranking of the three sampling points in Puzhehei Lake was as follows: Karst Wetland Park > Leshui Cave > Puzhehhe Village, and the total health risk values were very high. The health risk of chemical carcinogens occupied an absolute proportion in the total health risk, and the risk contribution rate was very high. Among them, the health risk of Cr was the highest, which played a decisive role in the overall health risk of the lake. In the future health protection of drinking water in Puzhehei Lake, the discharge of Cr into the lake should be strictly controlled, and the treatment of such chemical pollutants should be strengthened to ensure the safety of drinking water in the lake area (Table 4).

Table 4. Water health risk of Puzhehei Lake

Sampling point		Chemical carcinogens								Non-carcinogen											$R_{总}$ (×10⁻³)
		Cd (×10⁻⁶)		Cr (×10⁻³)		As (×10⁻⁶)		R^c (×10⁻³)		NH₃-N(×10⁻¹⁰)		Cu (×10⁻¹¹)		Zn (×10⁻¹²)		Pb (×10⁻¹⁰)		Hg (×10⁻⁹)		R^n (×10⁻⁹)	
		R^a_{ig}	Average	R^a_{ig}	Average	R^a_{ig}	Average			R^b_{jg}	Average	R^b_{jg}	Average	R^b_{jg}	Average	R^b_{jg}	Average	R^b_{jg}	Average		
Leshui Cave (the upper reach)	Sampling 1	9.86	5.75	5.86	5.64	6.73	6.73	5.65		6.36	6.38	8.98	8.98	1.50	1.50	3.14	3.83	2.99	2.24	3.36	5.65
	Sampling 2	1.64		5.42		6.73				6.39		8.98		1.50		4.52		1.50			
Puzhehei Village (the middle reach)	Sampling 1	7.94	10.3	3.53	4.30	6.73	6.73	4.31		5.86	4.93	8.98	8.98	19.5	1.87	3.88	1.96	7.48	6.73	7.53	4.31
	Sampling 2	12.6		5.07		6.73				4.01		8.98		18.0		0.0321		5.99			
Puzhehei National Wetland Park (the lower reach)	Sampling 1	0.0274	0.0274	4.45	5.94	6.73	6.73	5.95		1.90	1.88	8.98	8.98	1.50	1.50	0.0321	0.0321	4.49	2.99	3.28	5.95
	Sampling 2	0.0274		7.43		6.73				1.87		8.98		1.50		0.0321		1.50			

4. Discussion

TN and NH3-N, TN and Pb, NH3-N and Pb, TP and Cd, Zn and Hg were significantly positively correlated, and the correlation coefficients were 0.965, 0.944, 0.908, 0.816 and 0.879, respectively, among which TN and NH3-N, TN and Pb, NH3-N and Pb have the strongest correlation. It can be seen that there was a significant correlation between TN, NH3-N and Pb, indicating that these three pollution indicators may be the main factors that cause water quality changes in Puzhehei Lake[12]. In addition, there was no significant correlation between other factors. But But interestingly, except for COD, there was a certain negative relationship between Cr and other 8 factors, which can be further analyzed in future research.

Table 5. Pearson correlation matrix between pollution indexes of Puzhehei Lake

	COD	TN	NH_3-N	TP	Zn	Pb	Cd	Cr	Hg
COD	1	0.078	0.222	0.44	-0.07	-0.146	0.644	0.114	-0.031
TN		1	0.965**	0.71	0.075	0.944**	0.294	-0.288	-0.066
NH_3-N			1	0.795	0.209	0.908*	0.5	-0.342	0.095
TP				1	0.577	0.52	0.816*	-0.518	0.438
Zn					1	0.036	0.681	-0.578	0.879*
Pb						1	0.126	-0.325	-0.033
Cd							1	-0.325	0.581
Cr								1	-0.796
Hg									1

Note: **significant correlation at the 0.01 level (two-tailed), * significant at the 0.05 level (two-tailed).

In the above conclusions, the middle reaches represented by Puzhehei Village had the worst water quality and the most serious water pollution, which was related to the population growth and the development of tourism service industry in this region in recent years. Specifically, Puzhehei Village is one of the important areas for the development of tourism in Puzhehei Lake, and its large lakeside belt has been converted into construction land, which has caused the water area in the middle reaches of Puzhehei Lake to shrink seriously [13]. At the same time, according to the sixth national census, the populations of Yuezhe Town, Shuanglongying Town and Badaoshao Yi Township where the three sampling points of Puzhehei Lake are located are 33,223, 75,190, and 31,345 respectively. This leads to relatively more non-point source pollution in this area, more domestic wastewater entering the lake, and relatively poor water quality. In response to the problem of water pollution in Puzhehei, Qiubei County of Yunnan Province has formulated the "Overall Plan for the Protection and Management of Puzhehei Lake Basin" to investigate the pollution sources of the lake. So far, 268 households and 308 fish ponds in the basin have been renovated. In addition, in view of the excessive health risks of chemical carcinogens in Puzhehei Lake water body, we should focus on the treatment of pollutants such as Cr and Cd in the future to ensure water safety in the basin.

Acknowledgments

This work was financially supported by the National Natural Science Foundation of China (31760175, 31460158) and Yunnan Provincial Department of Education Scientific Research Fund (2022Y563).

References

1. Guo W, Huo S, Xi B, et al. Heavy metal contamination in sediments from typical lakes in the five geographic regions of China: distribution, bioavailability, and risk. Ecological Engineering 81: 243-255(2015).
2. Gao C, Zhang TL. Eutrophication in a Chinese context: Understanding various physical and socio-economic aspects. Ambio 39: 385-393(2010).
3. Liu J, Yang W. Water sustainability for China and beyond. Science 337(6095): 649-650(2012).
4. Zheng JN, Xu M, Zheng XW, et al. Ecosystem Evolutionary Trajectory of Lake Xiliang over the Past Century Driven by Eutrophication. Environmental Science 43(5): 2518-2526(2022).
5. US National Research Council. Risk assessment in the federal government managing the process. Washington DC: National Academy Press 19-20(1983).
6. US Environment Protection Agency. Risk assessment guidance for superfund volume I human health evaluation manual (part A). Washington DC: Office of Emergency and Remedial Response 1-7(1989).
7. Liu YG, Zhao LQ, Wang Y, et al. Ecological Environment Governance and Restoration in Puzhehei Watershed. Beijing: China Forestry Publishing House (2016).
8. Zhan NC, Wang Y, Liu YG, et al. Distribution characteristics of total phosphorous in water-sediment-vegetable system of typical karst lakeside wetland in southwest Yunnan. Journal of northwest A&F university (Nat. Sci. Ed) 46(3): 50-60(2018).
9. Jiang B, Sun M, Ji YK, et al. Fuzzy Comprehensive Evaluation of Surface Water Quality Based on Entropy Method. Journal of Irrigation and Drainage 37(1): 47-49(2018).
10. Zhang J, Shu LC, Zhang C, et al. Comprehensive Index Based Entropy and Weight for Groundwater Quality Evaluation. Water Resources and Power 28(8): 30-32(2010).
11. US Environment Protection Agency. Superfund public health evaluation manual. Washington DC: USEPA 47(1986).
12. Zhang Y. Water environment quality assessment and eutrophication risk in Fenghe Basin. Xi' an: Chang' an University (2020).
13. Wang Y, Liu YG, Liang QF, et al. Variation of Puzhehei Lake Area in Dry Season from 1977 to 2014. Wetland science 14(4): 471-476(2014).

Construction of dual-carbon goal realization path based on fractal theory

Hai Chi, Xincheng Wang

Jinan Preschool Education College, Jinan 250307, China

Abstract: The amount of CO_2 emissions is related to people's life, social progress and national development. To seek a balance between environmental protection and social progress is a problem that scholars have paid close attention to. This paper firstly studies the data of carbon emissions from 1997 to 2019, and conducts relevant processing on the data. Using fractal theory to predict the processed data, the prediction results show that my country's carbon emissions will continue to grow. Finally, corresponding countermeasures are put forward for the realization of the dual-carbon goal.

Key words: Dual-carbon; fractal; realization path.

1. Introduction:

Since the Industrial Revolution, the process of industrialization and modernization in the world has been accelerating, and environmental problems arising from production and life have gradually emerged. In recent years, the global economic growth and ecological environment have faced major challenges, and there is an urgent need to reach an international consensus on environmental protection and green development [1]. As early as 2015, President Xi proposed five development concepts of "innovation, coordination, greenness, openness and sharing", and announced at the United Nations General Assembly in 2020 that China will strive to achieve "carbon peak" by 2020 and "carbon peak" by 2060. carbon neutrality" goal. At present, we are in a critical period of realizing the "Dual-Carbon Goal", and it is of great practical significance to study the realization path of the "Dual-Carbon Goal". Many scholars have studied the realization path of the "two-carbon goal" from different perspectives [2-4], and have achieved rich research results. The term Fractal was first proposed by American mathematician Mandelbrot to describe the more complex, irregular and unstable phenomena in nature [5]. In Euclidean geometry, the dimensions are all integers, that is, points are 0-dimensional, lines are 1-dimensional, finite planes are 2-dimensional, and solid geometry is 3-dimensional. But it was later found that this geometry could not explain many irregular graphic laws in life. Mandelbrot explained this: the whole can be regarded as the enlargement of the part in a certain proportion, that is, the part and the whole have a similar shape in a certain proportion. In 1975, Mandelbrot proposed fractal geometry and fractal theory successively, he believed that the dimension is continuous, not necessarily an integer. Dimension is an important feature quantity of geometric objects, which contains many information and properties of graphics. Fractal dimension is a very important concept in fractal theory and application. It is the most important parameter to measure the complexity and irregularity of fractal graphics, objects or fractal bodies. rules and vice versa. Fractal dimension shows that the fractal body has self-similarity between the part and the whole in one aspect of shape, structure, information, function, time, energy, etc., and quantitatively describes the degree of self-similarity. In geometric mathematics, the fractal dimension reflects the effectiveness of the space occupied by the fractal body. Fractal dimension expands the meaning of dimension, and the change of space dimension can be either discrete or continuous. Fractal dimension has a wide range of applications in data prediction [6], information recognition [7] and other fields.

This study uses fractal dimension to forecast the national energy consumption, and according to the forecast results, proposes the realization path of the "dual carbon goal".

2. Fractal model and prediction results

A fractal distribution can be defined by the following power exponential distribution:

$$N = \frac{C}{r^D} \tag{1}$$

Among them, r is the characteristic length such as time, length, etc.; N represents the number of observation targets related to r, which can be price, speed, $C0_2$ emission, etc.; C is the undetermined coefficient, which can be calculated by other values; D represents with fractal dimension. In this study, r represents the year, such as the first year, r1=1; the second year, r2=2. N is the national CO_2 emissions in the year. D can be calculated from any two data (N_a, r_a) and (N_b, r_b) on a straight line,

$$D = [\ln(N_a) - \ln(N_b)] / [\ln(r_b) - \ln(r_a)] \quad (2)$$

we can get (3) from (1)

$$C = N_a r^D \quad (3)$$

It should be pointed out that in the research of this paper, all variables involved in the formula are positive numbers. Due to the nonlinear relationship between the data in this study, it is necessary to use the concept of variable dimension fractal for calculation, so as to predict the emission of CO_2, and then put forward corresponding suggestions for the realization of the "dual carbon goal".

This study uses the series transformation of cumulative sum proposed by Fu [6] to process the data, so that the processed data is more in line with the fractal distribution model, so that the future data can be reasonably predicted. The method is as follows:

The first step is to plot the original data in the double log coordinate axis. Obviously, these data without obvious regularity generally do not satisfy the fractal distribution. N can form a basic sequence, namely:

$$N_a = \{N_1, N_2, N_3, \ldots\ldots\}(a = 1,2,3\ldots.) \quad (4)$$

Other sequences can be constructed based on the basic sequence, the first-order cumulative sum sequence S_1, where $S_{11}=N_1$, $S_{12}=N_1+N_2$, $S_{13}=N_1+N_2+N_3$. By analogy, higher-order cumulative sum sequences can be constructed.

$$S1a = \{N_1, N_1 + N_2, N_1 + N_2 + N_3, \ldots\ldots\}(a = 1,2,3,\ldots\ldots a) \quad (5)$$

$$S2a = \{S_{11}, S_{11} + S_{12}, S_{11} + S_{12} + S_{13}, \ldots\ldots\}(a = 1,2,3,\ldots\ldots a) \quad (6)$$

$$S3a = \{S_{21}, S_{21} + S_{22}, S_{21} + S_{22} + S_{23}, \ldots\ldots\}(a = 1,2,3,\ldots\ldots a) \quad (7)$$

The processed data are shown in Table 1:

Table 1 original data and processed data

Year	Original	First-order	Two-order	Three-order
1997	**2923.85556**	**2923.85556**	**2923.86**	**2923.86**
1998	**2886.345645**	**5810.201206**	**8734.056766**	**11657.92**
1999	**2878.639872**	**8688.841078**	**17422.89784**	**29080.81**
2000	**3003.426725**	**11692.2678**	**29115.16565**	**58195.98**
2001	**3250.146167**	**14942.41397**	**44057.57962**	**102253.6**
2002	**3472.133954**	**18414.54792**	**62472.12754**	**164725.7**
2003	**4084.320775**	**22498.8687**	**84970.99624**	**249696.7**
2004	**4695.809704**	**27194.6784**	**112165.6746**	**361862.4**
2005	**5398.279906**	**32592.95831**	**144758.633**	**506621**
2006	**6008.710384**	**38601.66869**	**183360.3016**	**689981.3**
2007	**6546.296649**	**45147.96534**	**228508.267**	**918489.6**
2008	**6761.019282**	**51908.98462**	**280417.2516**	**1198907**
2009	**7333.675089**	**59242.65971**	**339659.9113**	**1538567**
2010	**7904.547464**	**67147.20718**	**406807.1185**	**1945374**
2011	**8741.562054**	**75888.76923**	**482695.8877**	**2428070**
2012	**9080.54839**	**84969.31762**	**567665.2054**	**2995735**
2013	**9534.235131**	**94503.55275**	**662168.7581**	**3657904**
2014	**9451.281631**	**103954.8344**	**766123.5925**	**4424027**
2015	**9253.502801**	**113208.3372**	**879331.9297**	**5303359**
2016	**9256.251069**	**122464.5883**	**1001796.518**	**6305156**
2017	**9408.170796**	**131872.7591**	**1133669.277**	**7438825**
2018	**9621.119877**	**141493.8789**	**1275163.156**	**8713988**
2019	**9794.756397**	**151288.6353**	**1426451.791**	**10140440**

The second step is to establish a fractal model according to the data of the selected order.

Each data point is plotted in the double log coordinates, and compared with the fractal distribution model, the sequence with the best fitting degree is selected, and the corresponding fractal dimension is calculated. Then make predictions about the next data.

Table 2 Fractal dimension of C02 emissions in each year

Year	CO_2Emissions	D	D1	D2	D3
1997	2923.85556	-0.0186	-0.9907	-1.5788	-1.9954
1998	2886.345645	-0.0066	-0.9925	-1.7031	-2.2544
1999	2878.639872	-0.1475	-1.032	-1.7849	-2.4115
2000	3003.426725	-0.3538	-1.0992	-1.8564	-2.5259
2001	3250.146167	-0.3624	-1.146	-1.9154	-2.6153
2002	3472.133954	-1.0534	-1.2995	-1.9954	-2.6984
2003	4084.320775	-1.0448	-1.4196	-2.0794	-2.7785
2004	4695.809704	-1.1836	-1.5374	-2.1658	-2.8569
2005	5398.279906	-1.0168	-1.6059	-2.2436	-2.9319
2006	6008.710384	-0.8991	-1.6436	-2.3095	-3.0014
2007	6546.296649	-0.3709	-1.6038	-2.3526	-3.0621
2008	6761.019282	-1.0157	-1.651	-2.3945	-3.1164
2009	7333.675089	-1.0115	-1.69	-2.4342	-3.1657
2010	7904.547464	-1.4589	-1.7738	-2.4792	-3.2125
2011	8741.562054	-0.5895	-1.7512	-2.5124	-3.2553
2012	9080.54839	-0.8042	-1.7542	-2.54	-3.2941
2013	9534.235131	0.15289	-1.6676	-2.5512	-3.3269
2014	9451.281631	0.39115	-1.5772	-2.549	-3.353
2015	9253.502801	-0.0058	-1.5322	-2.542	-3.3733
2016	9256.251069	-0.3337	-1.517	-2.5346	-3.3889
2017	9408.170796	-0.4811	-1.5137	-2.5283	-3.401
2018	9621.119877	-0.4024	-1.5057	-2.5222	-3.4105
2019	9794.756397				

In Table 2, D represents the fractal dimension of the original data, and D1, D2, and D3 represent the fractal dimension of the 1st, 2nd, and 3rd order accumulation and processing, respectively. It can be seen that the fractal dimension of the unprocessed data is positive or negative, and fluctuates greatly, and the fractal dimension of the processed data tends to be more stable. In the case of the first-order cumulative sum, we choose D=-1.5057 and C=88.1815. According to the formula, in 2019, the national carbon emissions N=9902.2; in the case of the second-order cumulative sum, D=-2.5222, C=3.9570, then N=10763; in the case of the third-order cumulative sum, D=-3.4105, C=0.254, N=11195. The prediction result of the first-order cumulative sum is closer to the real data, so in the following research, we choose the data after the first-order cumulative sum to predict the national CO2 emissions. The prediction results are shown in Table 3:

Table 3 Carbon emissions projections for the next few years

Year	$C0_2$ emission	C	D
2019	**9902.2**	**88.1816**	**-1.5057**
2020	**10558**	**87.8778**	**-1.5068**
2021	**11228**		

Figure 1 shows the changing trend of national $C0_2$ emissions (including forecasts) from 1997 to 2021. It can be clearly seen from Figure 1 that with the development of my country's economy, CO_2 emissions show an upward trend year by year.

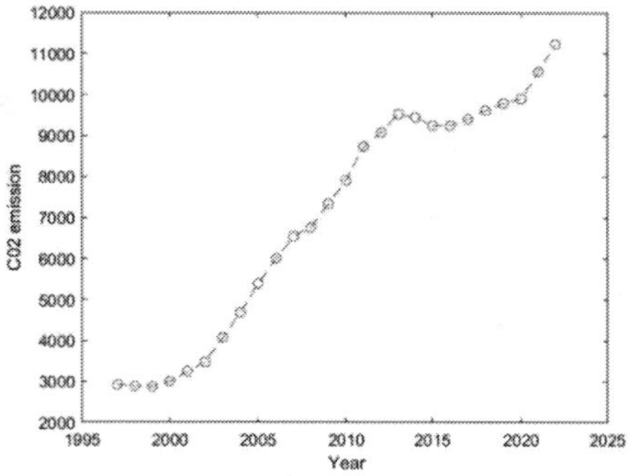

Figure 1 National CO_2 emissions

3. Paths to achieve the "two-carbon" goal

According to the forecast results in the previous section, my country's C02 emissions will continue to grow and remain at a relatively high level for a long time. In order to successfully achieve the "carbon peak" goal in 2030, it is imperative to take measures to reduce carbon emissions.

(1) Vigorously develop clean energy

According to the relevant data of the National Bureau of Statistics, the main fossil fuels in my country are coal, oil and natural gas [3]. In order to reduce our dependence on these highly polluting energy sources, we need to vigorously develop clean energy, for example, support the development of new energy electric vehicles, which can reduce the use of oil. However, the popularity of electric vehicles will also bring new problems. For example, the main power generation method in my country is thermal power generation, and the main fuel used in thermal power generation is coal. Therefore, the new energy electric vehicle reduces the consumption of oil on the surface, but it increases the demand for coal. Therefore, we should increase the proportion of other power generation methods. Our country has abundant water sources and vast land, so we should increase investment in hydropower and wind power.

(2) Transformation and upgrading of key enterprises

In addition to vigorously developing clean energy and reducing people's dependence on coal and oil in their daily lives, it is also an important measure to achieve the goal of "double carbon" by rectifying the phenomenon that some high-polluting enterprises do not meet the pollutant discharge standards. Industry is the cornerstone of a country's prosperity, but the environmental pollution caused by the production process of industrial enterprises does not conform to the principle of sustainable development. Therefore, it is necessary to promote the transformation and upgrading of these enterprises, so that the product skills produced by industrial enterprises can meet the needs of social development and conform to the principle of sustainable development.

(3) Develop high-tech energy-saving technologies

The research and development of energy-saving technology mainly has two aspects, one is independent research and development, and the other is imported from abroad. Introducing technology from abroad is simple, convenient and economical, but in the long run, it is not in our national interest. Therefore, we need to set up our own R&D team, which can not only drive industrial upgrading, but also increase domestic demand and promote the realization of domestic circulation. Energy-saving technologies can be manifested in improving energy utilization, for example, the same coal can generate more electricity; it can also be manifested in energy substitution, developing new energy sources to replace traditional fossil fuels, thereby reducing CO2 emissions and achieving the "double carbon" goal. .

4. Summary

This study uses fractal theory to predict my country's C02 emissions, and according to the prediction results, analyzes the current situation of my country's C02 emissions. In order to successfully achieve the goal of "carbon peaking" in 2030, it is imperative to save energy and reduce emissions, transform the energy structure, and promote the transformation and upgrading of related industries. At the end of the article, this study proposes measures to achieve the "two-carbon" goal.

References

1. Zhu XK, Gong BL. Risk Challenges and Path Choices for Realizing the "Double Carbon" Goal under the Background of High-quality Development [J]. Governance Research, 2022, 38(03): 13-23+2+124.
2. Zhang LP, Luo JH. Research on the realization path of "two-carbon" goal under ESG concept [J]. Theoretical Observation, 2022(03):63-66.
3. Liu Jiang, He YJ. Exploring the Policy Coordination Path to Realize the "Dual-Carbon" Goal [N]. Chinese Journal of Social Sciences, 2022-05-11(007).
4. Ju JD. The optimal path to achieve the goal of "dual carbon" [J]. Tsinghua Financial Review, 2021(12): 2.
5. Zhang JZ. Fractals [M]. Beijing: Tsinghua University Press, 1995.

6. Fu YH, Fu AJ. Using fractal method to predict oil stock price and index [J]. China Ocean Platform, 2002(06): 42-46.
7. Guan Yongming, Wang Gang, Luo KB, Zou Bo, Lu Liang, Shi YL. Network public opinion monitoring method based on multi-fractal situation optimization [J]. Computer Integrated Manufacturing System, 2022, 28(04): 1258-1266.

Study on Numerical Simulation of Fire Flooding Reservoir

Zhencheng Peng, Mingzhong Li *, Chenwei Liu, Fanyu Meng and Shun Zhang

School of Petroleum Engineering, China University of Petroleum, Qingdao, China

Abstract. Fire flooding, also known as in-situ combustion technology, is a mining technology that can greatly improve the recovery of heavy oil reservoir. The composition of the output of the fire flooding oil well is very different from that of other heavy oil thermal recovery wells, and shows regular changes with different stages of fire flooding. These changes have a great impact on the efficient lifting of the fire flooding oil well. Numerical simulation can be used to simulate the combustion process of fire flooding and the heat and mass transfer process in the reservoir, so as to simulate the changes of the components of the output with different stages of fire flooding, and provide guidance for the design of fire flooding lifting. The results show that it is feasible to use reservoir numerical simulation software to simulate the production performance and the change of production composition of in-situ combustion oil wells.

1. Introduction

In-situ combustion is the continuous injection of combustion-supporting gas such as air or oxygen into the oil layer through the injection well, and the continuous combustion of crude oil in the oil layer by ignition or spontaneous combustion. The heavy components in the heavy oil are cracked into light oil at high temperature, and the injected gas, light oil, combustion-generated gas and water vapor drive the crude oil to flow to the production well. The composition of the output of the fire flooding oil well is very different from that of other heavy oil thermal recovery wells, and shows regular changes with different stages of fire flooding. These changes have a great impact on the efficient lifting of the fire flooding oil well.

Kuo C H proposed a mathematical model, introducing two temperature fronts, heat flow front and combustion front [1]. The two-dimensional numerical simulation proposed by Smith J W et al. can be used to predict the effect of fire flooding in closed well pattern [2]. A fully implicit and universal mathematical model for thermal recovery was proposed by Grabowski J W [3]. The model can be used for both fire flooding and steam flooding, and can deal with one-dimensional, two-dimensional and three-dimensional problems. University of Calgary scholar R. G. Moore defines crude oil as light oil, medium oil, and heavy oil based on the number of carbon atoms. His model has seven reaction equations [4]. Stanford University scholar A. R. Kvoscek obtained the correlation equation of heavy oil combustion reaction by numerical simulation through RTO results combined with DSC heat release curve. In this method, the fuel is assumed to be two coke substances [5].

Through the investigation of the numerical simulation of fire flooding, it can be seen that the numerical simulation can simulate the combustion process of fire flooding and the heat and mass transfer process in the reservoir, so as to simulate the change of the composition of the output with different stages of fire flooding, and provide guidance for the design of fire flooding lifting. Based on these studies, it is believed that it is feasible to simulate the oil well production performance and the change of production composition by reservoir numerical simulation software.

2. Numerical Simulation of Fire Flooding Reservoir Development Based on Mechanism Model

The reservoir in the block is a reservoir after long-term fracturing and thermal recovery. The basic parameters of the model are set with reference to the field data of the oilfield, which makes the simulation closer to the field reality and more intuitive and accurate. Describe how the fire flooding production performance changes.

2.1 Establishment of Mechanism Model

The reservoir in the block is a reservoir after long-term fracturing and thermal recovery. The basic parameters of the model are set with reference to the field data of the oilfield, which makes the simulation closer to the field reality and more intuitive and accurate. Describe how the fire flooding production performance changes.

* Corresponding author: Pengzc1998@163.com

 https://doi.org/10.1051/e3sconf/202235802008

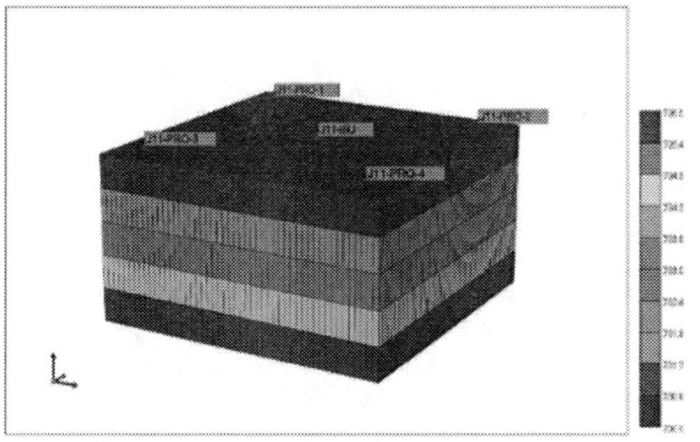

Figure 1. Three-dimensional structural model.

In order to accurately characterize the fire flooding process, it is necessary to consider both the accuracy and reliability of the model and the operation speed of the model. It is necessary to reasonably plan the grid and unit selection of the model. Select a similar proportion to establish a conceptual model corresponding to the actual. The length and width of the model are set to 300 m, the grid blocks are 50 in the horizontal direction, the grid step is 60 m, and there are 7 layers of grids in the vertical direction, with a total of 17500 grid nodes (50 × 50 × 7). The top depth is set to 700 m and the total thickness of oil layer is 7 m. The three-dimensional structure of the model is shown in Figure 1.

2.2 Establishment of Component Model

The model mainly considers 4 phases and 7 components, in which 4 phases are mainly gas phase, water phase, oil phase and solid phase. The 7 components are mainly heavy oil, oxygen, methane, water, carbon dioxide, inert gas (carbon monoxide and nitrogen) and coke.

There are three main chemical reactions involved in the formation: heavy oil is decomposed into light components (methane) and coke, heavy oil combustion, coke combustion.

2.3 Establishment of Fluid Model

The variation curve of reservoir crude oil viscosity with temperature is shown in Figure 2. It belongs to heavy oil under the condition of reservoir temperature, and the viscosity of crude oil decreases exponentially with the increase of temperature, which indicates that the viscosity of crude oil is sensitive to temperature, so it can be explained that it is feasible and inevitable to exploit by fire flooding.

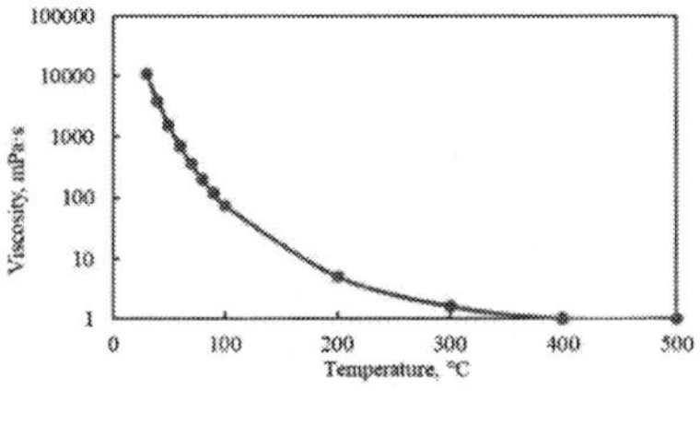

Figure 2. Variation curve of crude oil viscosity with temperature.

Relative permeability is one of the main factors affecting the relative flow of underground fluid. The oil-water relative permeability curve, oil-gas relative permeability curve and oil-gas-water three-phase relative permeability used in the model are shown in Figure 3 to Figure 5. Through relative permeability curve and absolute permeability, the effective permeability of each phase fluid can be calculated when oil, gas and water coexist. Effective permeability is an important parameter in fire flooding simulation.

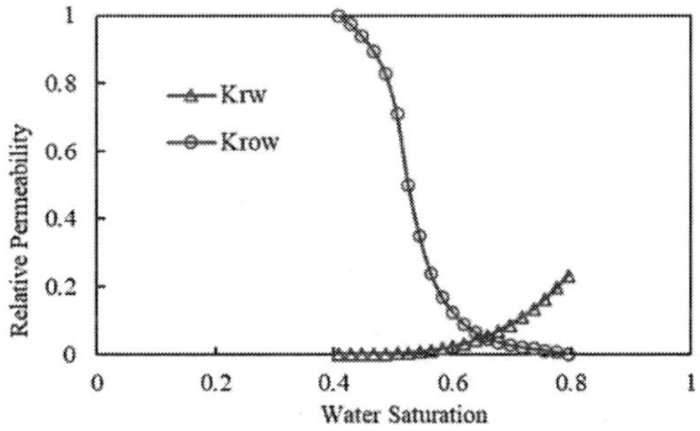

Figure 3. Oil and water relative permeability curve.

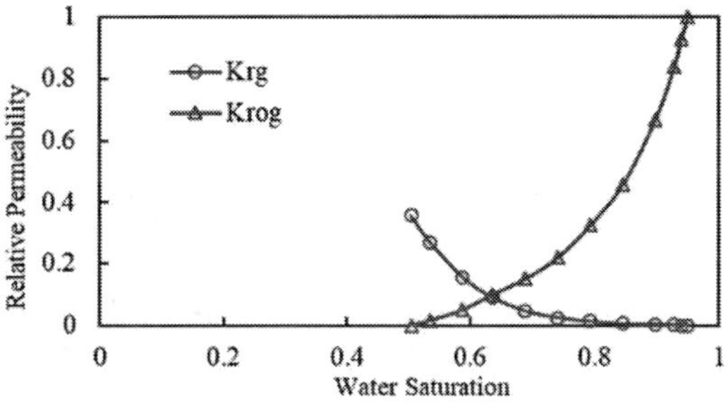

Figure 4. Oil and gas relative permeability curve.

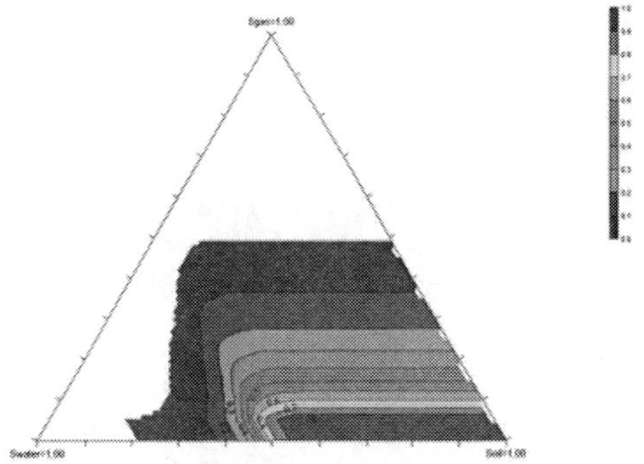

Figure 5. Three-phase relative permeability of oil, gas and water.

2.4 Establishment of Production Dynamic Model

In the model, a well group is used for fire flooding development, the well layout scheme is selected by the inverse five-point method, and the well spacing is 160 m. The perforation scheme adopts the scheme of perforating

 https://doi.org/10.1051/e3sconf/202235802008

the upper two layers. Table 1 is the operating parameters of injection-production wells.

Table 1. Operating parameters of injection-production wells.

Property	Parameter
Bottom hole flowing pressure, MPa	1
Gas injection pressure, MPa	16
Gas injection speed, MPa	48000
Gas injection type	Air

The operating parameters of production wells and gas injection wells are shown in the table. Commissioning began on January 1, 2020, with simulated production lasting eight years until December 1, 2027.

3. Production change of fire flooding reservoir

3.1 Change Law of Fire Flooding Productivity

In the process of oil layer combustion, it is necessary to grasp the position of fire line at any time. According to the radial distance of the fire line, the gas injection intensity at different stages can be reasonably adjusted and corresponding control measures can be taken to make the combustion zone advance uniformly and stably. In the numerical simulation results can be seen in the model at different times each section of the temperature distribution, according to the characteristics of the fire drive front, the highest temperature part can be seen as the combustion front, so you can intuitively find the combustion front in the model, so that the production process of the fire drive stage line division. According to the production characteristics of each stage, the production stage of in-situ combustion can be divided into the following four stages: combustion establishment stage, combustion effect stage, thermal efficiency oil displacement stage and fire stage.

At the same time, there is usually an oil wall in the process of fire flooding. The oil wall is the area where the oil saturation increases within a certain range in the porous medium in a certain period of time. The area has a large oil saturation, and the seepage resistance is large. It can play a certain plugging role and can block the high permeability channel in the process of fire flooding, thereby improving the sweep efficiency. After passing through this area, the pressure will decrease greatly, and this area is the main area of pressure drop.

The fire flooding after steam injection is different from the primary development of fire flooding in the original reservoir. Its characteristic is that the secondary water body caused by steam injection provides a horizontal channel and also improves the speed of fire line advance. The secondary water body leads to a large amount of drainage in the early stage of production, and several stages of water content fluctuation appear in the middle stage. Under the drive of exhaust gas, the secondary water body is continuously expelled. When the fire line advances beyond a certain range, only bound water is left near the production well. At this time, the water content decreases and the daily oil production increases.

3.2 Fire Drive Exhaust Composition

According to the survey results and numerical simulation results, the main components of the fire flooding exhaust gas are CO_2, N_2, and CH_4 from more to less, and there is almost no generation of O_2, H_2S, and CO. By drawing the gas production rate of these four components, the exhaust gas composition characteristics of the four stages of fire flooding are determined.

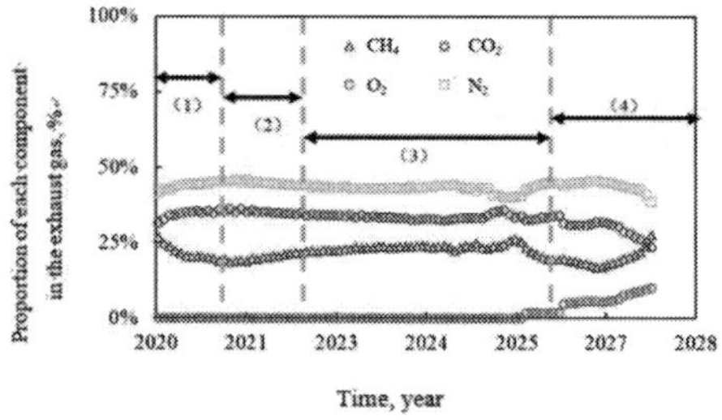

Figure 6. Composition of fire drive exhaust gas.

In the first three stages, the CH_4 production is high. At this time, the combustion front of the fire flooding is normally advanced, and the combustion reaction is fully carried out. The methane production increases rapidly after ignition, and begins to decrease after ignition is completed. This is due to the initial stage of ignition. The combustion is intense, the heat generated is large, the temperature rises rapidly, and more crude oil is cracked. When the fire line continues to advance to the production well, reaching the high temperature effective stage and the thermal effect oil displacement stage, the spread range of the fire flooding becomes larger, the high temperature area becomes larger, and the reaction speed is proportional to the area of the reaction. The combustion reaction is faster, so the yield of small molecule saturated hydrocarbons such as CH_4 is also greater. The gas production rate of CO_2 and N_2 components was found to be basically consistent with the oil production rate, indicating that the output of nitrogen and carbon dioxide changed with the change of reaction rate.

4. Conclusion

In this paper, a three-dimensional injection-production geological model is established using reservoir numerical simulation software.

(1) According to the simulation results of the fire flooding mechanism model, the fire flooding production in this block is divided into the following four stages: combustion establishment stage, combustion effective stage, thermal flooding stage and fire stage. Among them,

the reservoir pressure increases first and then decreases with the progress of fire flooding, and the reservoir pressure and daily oil production are the highest in the thermal oil displacement stage.

(2) The gas-liquid ratio remained in the range of 1000-2000 m^3/m^3 after the establishment of fire flooding. In the combustion effect stage, the composition of tail gas was N_2 (47.33 %), CO_2 (30.06 %) and CH_4 (27.97 %). In the thermal effect oil displacement stage, the composition of tail gas was N_2 (43.51 %), CO_2 (33.59 %) and CH_4 (22.90 %). There was almost no formation of O_2, H_2S and CO in the two stages.

Acknowledgments

This work was financially supported by National Natural Science Foundation of China (NO.2021ZX20210354) fund.

References

1. Kuo C H, Convective heat transfer model for underground combustion, Society of Petroleum Engineers Journal, 1968, 8(04): 323-324.
2. Smith J T and Farouq Ali S M, Simulation of In-situ Combustion in a Two-dimensional System, Society of Petroleum Engineers, 1971.
3. Grabowski J W, Vinsome P K and Lin R C, A fully implicit general purpose finite difference thermal model for in-situ combustion and steam, Society of Petroleum Engineers, 1979.
4. Barzin Y, Moore R G and Mehta S A, A Comprehensive Kinetics Model for Light Oil Oxidation/Combustion Reactions under High Pressure Air Injection Process, Society of Petroleum Engineers, 2013.
5. Kovscek A, Castanier L M and Gerritsen M, Improved Predictability of In-Situ-Combustion Enhanced Oil Recovery, Spe Reservoir Evaluation & Engineering, 2013, 16(2): 172-182.

https://doi.org/10.1051/e3sconf/202235802009

Research on Working Condition Diagnosis of Submersible Screw Pump

Changhui Feng, Chenwei Liu *, Mingzhong Li, Xu Zeng and Mingming Tian

School of Petroleum Engineering, China University of Petroleum, Qingdao, China

Abstract. As the development of each oil field enters the middle and late stage, sand production, wax deposition and the continuous increase of well depth, electric submersible screw pump wells frequently break down, and hidden troubles of equipment can not be found in time, which makes oil wells produce with faults. It increases the wear and tear of equipment, seriously affects the production of oil wells, and causes unnecessary economic losses. Therefore, improving the accuracy of the working condition diagnosis of the electric submersible screw pump production system plays a vital role in improving the overall management level of the electric submersible screw pump well. This can not only enhance the managers' correct understanding of this kind of oil well system, but also comprehensively monitor the working state of the well. It can find the fault as soon as possible, analyze the cause of the fault, and avoid the accident by adjusting the production parameters, so that the downhole screw pump can work safely, reliably and effectively, and maximize the maintenance free period and economic benefits of the electric submersible screw pump well.

1. Introduction

In addition to normal working conditions, screw pump wells also have many other abnormal working conditions, which are regarded as fault conditions. Although there are many kinds of faults in screw pump wells, generally speaking, faults can be divided into surface faults and downhole faults. For the submersible screw pump production system, since the submersible motor and reducer are placed underground, their faults are mainly concentrated in the underground. At the same time, the submersible motor is connected to the screw pump rotor through the motor protector and flexible shaft, so there is no rod breaking and rod tube eccentric wear fault. There are five typical working conditions of submersible screw pump, including flexible shaft breaking, wax deposition, pump jamming, pump leakage and normal operation.

Cui Jinbang and others from North China Bureau, based on understanding the testing method of screw pump well working conditions, researched and designed the working condition test technology of screw pump well mainly consists of working condition test and replay diagnosis [1]. Wang Haiwen of the University of petroleum and others proposed a working condition diagnosis technology of the bare rod stress method by analyzing the stress on the bare rod of the screw pump oil production system, developed the bare rod stress test system, and analyzed some commonly used working condition diagnosis models of screw pump wells by studying the relationship between the working conditions of screw pump wells and relevant data, and compiled the corresponding working condition diagnosis software [2]. Liu Wenbo of Southwest Petroleum Institute has developed a test method that can measure the stress on the polished rod of screw pump well System, and the corresponding auxiliary process is given [3]. Gong Junfeng of Shengli Oilfield realized that there are few working condition diagnosis methods for screw pump wells, and developed a working condition test model for screw pump, which includes two modules - working condition diagnosis and optimization design. In working condition diagnosis, various methods such as current method and liquid production method are used to diagnose oil wells [4]. After analyzing the production status of oil wells, Wu Xiaodong and others from the University of petroleum selected five characterization parameters of the production status of oil wells, such as production, dynamic liquid level and working current. Combined with the two judgment theories of threshold diagnosis and logic diagnosis, they formulated the rules for deriving the working status of oil wells from the state variables and studied a new diagnostic method for working conditions of screw pump wells [5]. Jose F. Correa et al. Proposed a new working condition analysis method of screw pump well, which combines rough set theory with expert system. This method can not only judge the working condition of screw pump well, but also give the optimization scheme of fault well, which is convenient for managers to adjust the fault well [6].

* Corresponding author: 1016880373@qq.com

2. State diagnosis model based on artificial neural network

By analyzing the fault diagnosis of screw pump well, combining the advantages of wavelet packet decomposition theory in extracting signal features and artificial neural network in pattern recognition, a diagnosis method based on wavelet packet extraction of fault features and artificial neural network for online fault recognition is proposed.

2.1 Wavelet packet algorithm

The decomposition algorithm and reconstruction algorithm of wavelet packet are shown in the following formulas:

$$g_j^n(t)=\sum_l d_l^{j,n} u_n\left(2^j t-l\right) \tag{1}$$

Wavelet packet decomposition algorithm:

$$d_l^{j,2n}=\sum_k a_{k-2l} d_k^{j+1,n} \tag{2}$$

$$d_l^{j,2n+1}=\sum_k b_{k-2l} d_k^{j+1} \tag{3}$$

Wavelet packet reconstruction algorithm:

$$d_l^{j,2n+1}=\sum_k h_{l-2k} d_k^{j,2n}+g_{l-2k} d_k^{j,2n+1} \tag{4}$$

It can be seen from the above formula that the wavelet packet transform can decompose both the low-frequency part and the high-frequency part of the image. It has a stronger adaptability and a wider application range, so it is more suitable for various image processing.

2.2 Feature extraction of fault signal based on wavelet packet

Wavelet packet decomposition can extract fault features in a more refined frequency band. The following takes the decomposition of three-layer wavelet packet as an example to illustrate the decomposition process of wavelet packet.

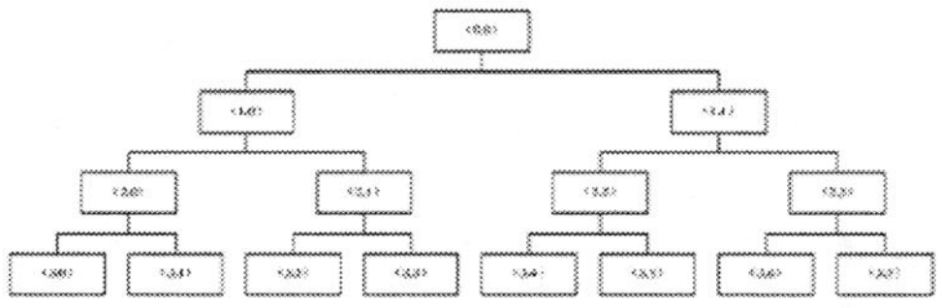

Figure 1. Three layer wavelet packet decomposition tree

The node (I, J) represents the jth node (I = 0, 1, 2,...) of the i-layer, and each node represents certain signal characteristics. Where node (0, 0) represents the original signal s, node (1, 0) represents the low-frequency coefficients of the first layer of wavelet packet decomposition, node (1, 1) represents the high-frequency coefficients of the first layer of wavelet packet decomposition, node (2, 0) represents the low-frequency coefficients of the second layer of wavelet packet decomposition, and so on.

2.3 Artificial Neural Network Model

Artificial neural network constructs an artificial information processing system by simulating biological neural network processing methods, and connects neurons (processing units) based on specific relationships, thus forming a relatively complex network structure. When a large number of samples are given, the rules in the samples are automatically learned and summarized through the learning of the samples, and the relationship weights connecting each neuron are constantly modified, so that the network tends to be stable. Finally, according to the obtained network, a new input mode is given, and a corresponding output mode can be deduced.

(1) Artificial neuron model

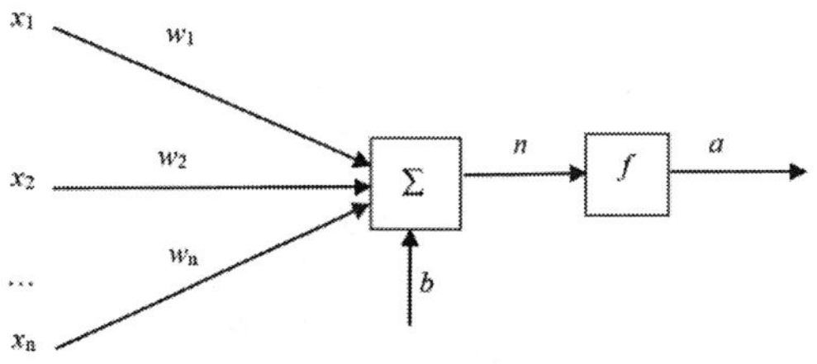

Figure 2. Basic model of neuron

https://doi.org/10.1051/e3sconf/202235802009

It can be seen from the above figure that a neuron is a unit model with multiple inputs and single output. For a typical neuron model, it can be divided into five parts: neuron input, weight threshold, summation unit, activation function and neuron output.

a. Threshold function

$$f(x)=\begin{cases}1 & x\geq 0\\ 0 & x<0\end{cases} \tag{5}$$

b. Linear function

$$f(x)=kx \tag{6}$$

c. Logarithmic sigmoid function

$$f(x)=\frac{1}{1+e^{x}} \tag{7}$$

d. Tangent sigmoid function

$$f(x)=th(x)=\frac{e^{x}-e^{-x}}{e^{x}+e^{-x}} \tag{8}$$

(2) Topological structure of neural network

By organizing a large number of basic neurons in a certain structure, a neural network can be obtained. According to the combination relationship and action mode of each neuron, neural networks with different characteristics can be obtained. Generally speaking, it can be divided into two types: interconnected type and layered type. The layered type can be further divided into forward type, inter layer interconnection forward type and feedback forward type according to whether there is feedback or not and the type of feedback; Interconnected neural networks can be divided into two types according to the connection relationship of neurons: local interconnection and full interconnection.

3. A diagnosis model of working conditions based on logical discrimination

According to the analysis of the causes and characteristics of various abnormal working conditions of electric submersible screw pump, the logical relationship between various working conditions and operating parameter States is summarized, and the inference mechanism of various working conditions is designed according to the logical relationship between operating parameter States and operating conditions.

(1) Flexible shaft / tubing disconnection detection

If the output is zero and the current is ultra-low, it is judged that the flexible rod is broken; If the output is zero and the current is too high, it is judged that the pump is jammed; If the output is zero and the current is normal, it is judged that the oil pipe has fallen off; In all three cases, it is recommended to shut down the well for workover.

(2) Detection of pump card / stator swelling / wax deposition

If the current is too high and the output is zero, it is judged that the pump is stuck, and it is recommended to shut down the well for workover. If the current is too high and the output is not zero, it is judged that the stator is swollen or waxed; The current is on the high side, the pump inlet temperature does not change, and the wellhead temperature increases. It is judged that the stator is swollen or waxed; In the above two cases, it is recommended to run at high speed for 10 minutes first, and then reduce to the original speed. If the front and rear currents do not change much, the working condition is stator swelling, and vice versa; If the working condition is wax deposition, wax removal shall be adopted; If the working condition is stator swelling, reduce the speed and run for a period of time; if there is no improvement, stop the pump for inspection; If the current is only on the high side, it may be motor failure.

(3) Screw pump / oil pipe leakage detection

If the output decreases and the working current of the screw pump is low, the leakage of the screw pump should be analyzed. Calculate the volumetric efficiency according to the identification results of screw pump speed and wellhead flow data, analyze the current volumetric efficiency in combination with the pressure difference between the inlet and outlet of the pump and the characteristic curve of the screw pump, and understand the internal leakage of the screw pump.

Analyze the rotational speed identification results of the screw pump to obtain the volumetric efficiency of the screw pump, and classify the leakage according to the ratio between the calculated volumetric efficiency and the volumetric efficiency in the characteristic curve. If the ratio between the calculated volumetric efficiency and the volumetric efficiency in the characteristic curve is greater than or equal to 85%, it is diagnosed as a micro leakage of the screw pump, and it is recommended to continue working at this time; If the ratio between the calculated volumetric efficiency and the volumetric efficiency in the characteristic curve is greater than 60% and less than 85%, it is diagnosed as a slight leakage of the screw pump. At this time, it is recommended to reduce the speed and continue to operate; If the ratio between the calculated volumetric efficiency and the volumetric efficiency in the characteristic curve is greater than or equal to 40% and less than or equal to 60%, it is diagnosed as moderate leakage of the screw pump. At this time, it is recommended to reduce the speed and continue to operate, and observe the effect. If the situation does not improve, it is recommended to stop the pump; If the ratio between the calculated volumetric efficiency and the volumetric efficiency in the characteristic curve is less than 40%, it is diagnosed that the screw pump has serious leakage. At this time, it is recommended to stop the pump immediately and consider replacing the pump. If only the production is reduced, it is judged as tubing leakage.

References

1. Cui Jinbang, Wang Yansong, Zhang Shengli, et al Research on working condition test technology of screw pump [J] Petroleum drilling and production technology, 1999, 21 (6): 4.
2. Wang Haiwen, Chen Lei, Lei Jian Working condition diagnosis technology of smooth rod stress method in screw pump well [J] Petroleum drilling and production technology, 2003, 25 (5): 4.
3. Liu Wenbo Research and application of working condition diagnosis technology of screw pump [D] Southwest Petroleum Institute, 2004.
4. Gong Junfeng Optimization design and working condition diagnosis system of surface driven screw pump well [J] Petroleum machinery, 2005, 33 (007): 21-22.
5. Wu Xiaodong, Wang Shizhan, Wang Zhishen Fault diagnosis method of surface driven screw pump oil production system [J] Petroleum drilling and production technology, 2003, 25 (005): 82-84.
6. Correa J F , Carvalho P , Filardi V L , et al. Fuzzy Analysis for Progressing Cavity Pump.

A Trait-based Investigation of Fungal Decomposition with Machine Learning

Shiyi Du [1], Yiran Zhao [1], Bingwei Tian [2]

[1] Department of Computer Science (Software Engineering), Sichuan University, Chengdu 610207, China
[2] Institute for Disaster Management and Reconstruction, Sichuan University, Chengdu 610207, China

Abstract. Fungi are of great functional significance in terrestrial ecosystems as the main decomposers. To better understand their decomposing process and population coexistence, we first describe and quantify the decomposition rate, focusing on three traits of interest selected by machine learning algorithm: moisture tolerance, hyper extension rate, and hyphal density and obtain, and use a Ternary Linear Regression Decomposition Model (TLRDM) to quantify the decomposition rate. Then, to incorporate the interactions, we build an Interactive Decomposition Model (IDM) and creatively employ a Three-player Logistic-based Competition Population Model (TPLCM). Based on logistic growth, we formulate a differential equation group, fit the curves of this unsolvable equation group to obtain a function of population density versus time and compare the decomposition rates of three populations under interactive and non-interactive conditions, followed by analyzing the impact of the communications on decomposing ability. We obtain the population combinations that can coexist in certain climates. Furthermore, we include environmental factors, conducting a sensitivity analysis to describe how short-term and long-term climate changes affect our models.

Keywords: Fungal traits; machining learning; regression; logistic growth; sobol.

1. Introduction

Fungi as decomposers matter a lot in the process of the carbon cycle on the earth. They release organic carbon dioxide and other compounds when they break down dead organisms or waste products, like ground litter (dead plant material) and woody fibers [1]. Carbon can cycle quickly through this biological pathway, especially in terrestrial ecosystems. Overall, an estimated 1,000 to 100,000 million tons of carbon move through the biological pathway each year. Visually, a ton is about the weight of a small car or an elephant [2].

While there are many reports on the decomposition of ground litter at home and abroad with the background of ecosystem nutrient cycling and global climate change, those reports are mainly attributed to the two aspects of macroscopic and substrate chemical composition changes. There are very few reports on how the ground litter is decomposed and utilized by soil microorganisms microscopically, especially concerning the inter-population relationships of fungi [3][4][5][6][7]. Thus, a Three-player Logistic-based Competition Population Model is mainly proposed in this work to better describe inter-population effect of fungi on decomposition rate and quantify its both short and long term trend.

2. Methodology

2.1 Data Pre-processing

The datasets from [8][9] include 34 fungal species with their trait data and climate conditions for each location. We integrated the collected data and normalized it. Then, a PCA method was implemented to figure out the most important traits: moisture tolerance, hyphal extension rate and hyphal density.

2.2 Non-interactive Decomposition Model

The moisture tolerance and the hyphal extension rate play a crucial role in the decomposition of fungi. In this section, we formulated a Ternary Linear Regressive Decomposition Model to provide a soluble explanation about how moisture tolerance and hyphal extension rate affect the decomposition rate.

The relationship between decomposition rate and the moisture tolerance is general linear. The moisture tolerance Mt and the hyphal extension rate H are used as explanatory variables to describe the decomposition rates D. The relationship between the explanatory variables and the explained variable D can be expressed as

$$\hat{D_i} = \hat{\beta_0} + \hat{\beta_1} Mt_i + \hat{\beta_2} H_i + \hat{\beta_3} Den_i + e_i \qquad (1)$$

where the parameters β0, β1, β2, β3 are the partial regression coefficients, which indicate the impact of degree of moisture tolerance and hyphal extension rate on the decomposition rate.

According to the (1), we fitted the model with a ternary linear regression function and obtained a solution

$$D = 3.785 - 0.213Mt + 1.664H - 0.664Den \quad (2)$$

Then we tested its goodness of fit and illustrated the result in Fig.1.

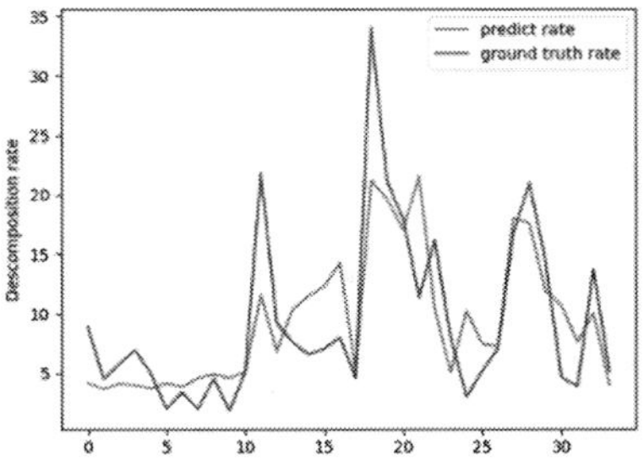

Figure 1. Decomposition rate on the 122th day.

We did the variance analysis of our model. As is shown in Table 1, the significance sig value is small, which means the chance of making mistakes is low [10]. Therefore, the performance of our regression model is not a coincidence but is statistically significant.

Table 1. ANALYSIS OF VARIANCE

Model	Sum of Square	Mean Square	F	Sig.
Regression	1016.901	338.977	14.100	0.000
Residual	721.182	24.039	-	-
Total	1738.083	-	-	-

2.3 Interactive Decomposition Model

We propose a Three-player Logistic-based Competition Population Model (TLCPM) that is closer to the natural environment. Based on the non-interactive decomposition model, the interactive decomposition model (IDM) comprises more internal and external factors. The factors include population density, interactions, environmental and climatic changes.

Here, we modeled an environment formed by three specific populations. Suppose there are three populations A, B, C. According to the logistic model, we can formulate the competition population model as (3).

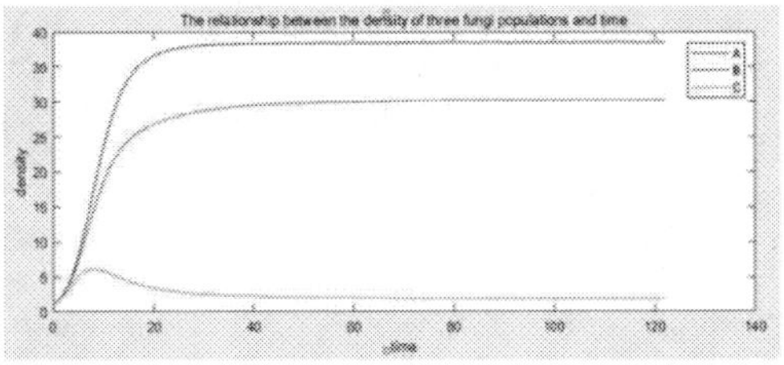

Figure 2. The relationship between density and time.

$$\begin{cases} \frac{dx}{dt} = r_1 x\left(1 - \frac{x}{N_1} - \alpha_1 \frac{y}{N_2} - \beta_1 \frac{z}{N_3}\right) \\ \frac{dy}{dt} = r_2 y\left(1 - \frac{y}{N_2} - \alpha_2 \frac{x}{N_1} - \beta_2 \frac{z}{N_3}\right) \\ \frac{dz}{dt} = r_3 z\left(1 - \frac{z}{N_3} - \alpha_3 \frac{x}{N_1} - \beta_3 \frac{y}{N_2}\right) \end{cases} \quad (3)$$

Although no analytical solutions of (2) can be obtained, we get three curves shown in Fig. 2 that illustrate the relationships between the density of three fungi populations during a 122-day period of time.

Note that the curves rise before 40 days and flat after 40 days, which means that the population reproduces rapidly before 40 days and tends to balance after 40 days. We fitted the curves of the logistic function piecewise and formulated the relationships between decomposition rate and time when t ⩽ 40 in (4).

$$Den(t) = \begin{cases} 5.5 \times 10^{-6}t^3 - 7.9 \times 10^{-4}t^2 + 3.3 \times 10^{-2}t - 2.5 \times 10^{-2} \\ 6.3 \times 10^{-6}t^3 - 7.0 \times 10^{-4}t^2 + 2.5 \times 10^{-2}t - 1.3 \times 10^{-2} \\ 7.4 \times 10^{-6}t^3 - 4.9 \times 10^{-4}t^2 + 8.4 \times 10^{-3}t - 1.2 \times 10^{-2} \end{cases} \quad (4)$$

In the first problem, we employed a ternary linear regressive model to quantify the decomposing rate. However, one explanatory variable, density, varies with time in this problem, so we redefined (2) to (5).

$$D = 3.785 - 0.213Mt + 1.664H - 0.664Den(t) \quad (5)$$

By applying the time-related density function Den(t) to the Equ.4, we obtained function of decomposition rate of A, B, C respectively under interactions. We set the time variable t = 122 and made a comparison with non-interactive results in problem 1. The figures are listed in Table 2. To better illustrate, we draw two bar charts in Fig.3.

Table 2. DECOMPOSITION RATES OF NON-INTERACTIVE AND INTERACTIVE ENVIRONMENTS

Species	Non-interactive (%)	Interactive (%)	Growth rate (%)
A	6.950	3.910	-43.7
B	2.000	4.910	145.5
C	7.61	10.759	41.4
Total	16.560	19.580	18.2

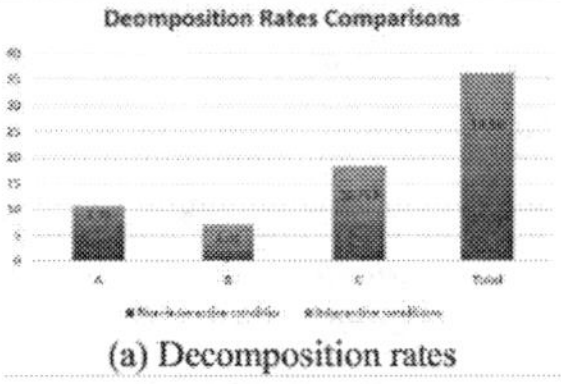

(a) Decomposition rates

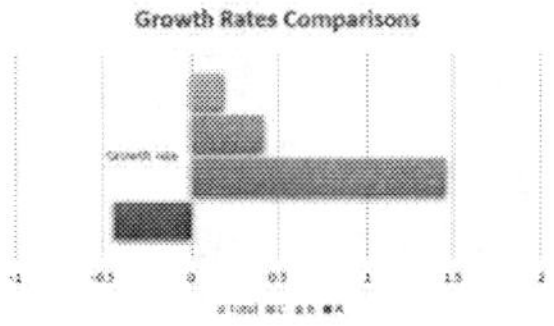

(b) Growth rates under interactions

Figure 3. Decomposition rates between non-interacitve and interactive conditions.

From Table 2 and Fig.3, we can see that in the interactive environment, the decomposition rate of A reduces significantly, indicating that A is suppressed in population competition. The decomposition rates of B and C increase to varying degrees so that these two species have advantages in this competition. For each population, it may gain advantages in the competition and thus decompose more woody fibers (conducive to population development); Or at a disadvantage and unable to decompose enough woody fibers (suppress the population development). The added decomposition rate is slightly larger in the interactive environment than in the non-interactive environment. This trend shows that in our experimental environment, population competition is conducive to the full utilization of environmental resources.

3. Analysis

3.1 Trends Analysis Towards Environmental Changes

Our logistic-based competition population model is a dynamic system, which are affected by short-term environmental fluctuations.

The short-term impacts are caused by rapid fluctuations in the environment, such as extreme weather. To describe the short-term trend, we implemented a sensitivity analysis to TLCPM.

We adopted Sobol, a variance-based sensitivity analysis to (4) in order to determine which trait, among decomposition rate, density of space and hyphal extension rate, is the most sensitive to short-term trend.

We consider the first-order sensitivity index, which measures the contributions of a single input-change to the output variance. Besides, the total sensitivity, which measures the overall contribution is also in our consideration. The results are in Table 3.

We observe that the sensitivity indicators are low, except the hyphal extension rate. Accordingly, we conclude that moisture and density factors are not sensitive to abrupt climate changes.

By contrast, the hyphal extension rate is sensitive to even slight changes. Therefore, rapid environmental fluctuations that impact the hyphal extension rate are more likely to vastly change the decomposition rate and affect the survival and reproduction of fungi.

Table 3. SOBOL ANALYSIS OF FUNGI TRAITS

Parameter	S1	S1_conf	ST	ST_conf
Mt	0.000575	0.00189	0.000463	0.000041
H	**0.996342**	0.077642	**0.996087**	0.058175
Den(t)	0.005076	0.005937	0.004571	0.003765

a. S1, ST represent the first-order and second-order sensitivity respectively.
b. S1_conf, ST_conf represent the 95% confidence.

3.2 Analysis of Dominant Populations Under Varied Environments

Different populations differ in their ability to adapt to certain environment. Some may benefit from the competition and win, while others are inhibited in their reproductive

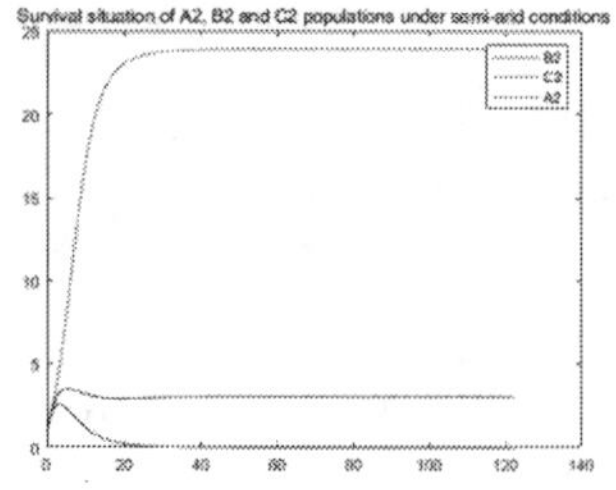

Figure 4. Survival numbers of three populations under semi-arid climate.

survival or even extinguish.

We randomly selected three populations in each typical climate (arid, semi-arid, temperate, arboreal, tropical rainforest) and randomly divided these 15 selected populations into three groups. We noticed that five groups of mixed fungi populations share almost the same trend. So, we choose semi-arid climate condition as a representative, and specifically focus on the

(a) temperature-hyphal rate relationship (b) Matric pot-hyphal rate relationship

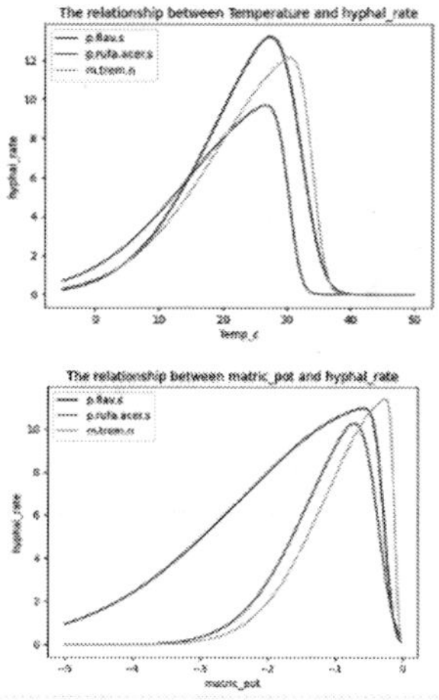

Figure 5. The relationship between temperature, matric pot and hyper rate.

analysis of the results under semi-arid climate. Fig. 4 shows the survival numbers of three testing groups A2 (p.flav.s), B2(m.trem.n), C2(p.rufa.acer.s) under semi-arid conditions.

According Fig.4, the temperature curves of the three populations have similar trends and their adaptation to different temperatures is comparable. The moisture curves, on the other hand, are more different in their adaptability to a more arid environment. A2 is well adapted to the arid climate, while B2 population is the weakest in its adaptability to the arid climate. The analysis verified the reasonableness of our previous figure of the survival situation.

In the process of inter-population competition among A2, B2, C3, according to Fig. 5, the B2 population, due to its population characteristics, has difficulty in adapting to the relatively arid climate of the semi-arid regions, and lacks a strong survival ability under the limited environmental resources. Besides, it perished in the inter-population competition.

4. Discussions And Furture Directions

In conclusion, hyper extension rate and moisture tolerance are two traits of interest in fungi decomposition. Inter-population interactions can be either beneficial or harmful, according to fungal characteristics. Among all traits, the hyphal extension rate is most sensitive to short-term environmental fluctuations. However, the interactions between species are complex, and there are many factors related to it. Although the factors we chose are typical, our model still cannot fully describe the relationships between populations. A more applicable model in real life scenario should be developed.

Acknowledgments

This research is supported by the Strategic Priority Research Program of the Chinese Academy of Science, Grant No. XDA23090502.

References

1. Tedersoo L, Bahram M, Põlme S, et al. 2014 Global diversity and geography of soil fungi Science 346(6213).
2. P. Baldrian 2017 Forest microbiome: Diversity, complexity and dynamics. FEMS Microbiol.Rev. 41, 109–130.
3. 1. Treseder, K. K. & Lennon, J. T. 2015 Fungal traits that drive ecosystem dynamics on land. Microbiol. Mol. Biol. Rev. 9, 243–262.
4. Van Veen J A, Ladd J N, Frissel M J 1984 Modelling C and N turnover through the microbial biomass in soil Plant Soil 76: 257- 274.
5. Revell, L. J. 2010 Phylogenetic signal and linear regression on species data. Methods Ecol. Evol. 1, 319–329.
6. W. R. Wieder, A. S. Grandy, C. M. Kallenbach, P. G. Taylor, G. B. Bonan 2015 Representing life in the Earth system with soil microbial functional traits in the MIMICS model. Geosci. Model Dev. 8, 1789–1808.
7. C. A. Aguilar-Trigueros et al., Branching out: Towards a trait-based understanding of fungal ecology. Fungal Biol. Rev. 29, 34–41 (2015).
8. Maynard D S, Bradford M A, Covey K R, et al. 2019 Consistent trade-offs in fungal trait expression across broad spatial scales Nat. Microbiol. 4: 846–853.
9. Lustenhouwer N, Maynard D S, Bradford M A, et al. 2020 A trait-based understanding of wood decomposition by fungi Proceedings of the National Academy of Sciences of the United States 117(21): 11551-11558.
10. Felipe D. Sepúlveda, Freddy Lucay, Jorge F. González, Luis A. Cisternas, Edelmira D. Gálvez 2017 A methodology for the conceptual design of flotation circuits by combining group contribution, local/global sensitivity analysis, and reverse simulation International Journal of Mineral Processing 164: 56-66.

Application of depth learning in prediction of glutenite reservoir

Meng Dong

Research Institute of Petroleum Exploration and Development of Daqing Oilfield Company Limited, 163113, Daqing, Heilongjiang, China

Abstract. Sedimentary facies of gravel reservoir has the characteristics of multiple sources and short flow. Sand body thickness changes rapidly in plane and space. Model-based multi-attribute regression seismic reservoir prediction method often results in unreasonable inter-well prediction results because of insufficient generalization ability of the model. For this reason, a deep learning reservoir prediction method based on fault blocks is proposed. Within a fault block range (regardless of wells outside the fault block), the seismic attributes extracted from the sliding time window of the isochronous sedimentary surface are used as input, and the sandstone thickness data of the target layer of the well point are used as expected output to train the depth neural network model of the multi-layer perceptor. Then, the trained model is used to predict the sandstone thickness distribution in the well distribution area. The application of actual data in A oilfield shows that the predictive coincidence rate of sandstone thickness in known wells is more than 95%, and the drilling occurrence rate of sandstone in horizontal section after completion of a posterior horizontal well is more than 80%. This method has high prediction accuracy and good generalization ability.

Keywords- Glutenite reservoir; reservoir prediction; depth learning; multi-layer perceptron

1. Introduction

Seismic attribute analysis technology is completely faithful to seismic data and has the characteristics of simplicity and efficiency. It is a very effective means of post stack seismic reservoir prediction. At present, hundreds of seismic attributes can be extracted and used, including time, amplitude, frequency, phase and absorption attenuation. In terms of quantitative reservoir prediction using seismic attributes, one is to use single attributes to characterize reservoir physical properties. The common method is to establish mathematical models of attributes and storage physical properties [1]. The other is to use multi-attribute comprehensive analysis to calculate with multiple attributes representing different reservoir characteristics, and establish linear or nonlinear models of multi-attribute and some reservoir characteristics, Reflect some reservoir characteristics with its operation results [2]. For sandy conglomerate reservoirs with complex geological conditions and rapid sandstone phase change, the model-based multi-attribute regression seismic reservoir prediction method often leads to unreasonable inter well prediction results because of the insufficient generalization ability of the model.

In recent years, deep learning technology has been gradually popularized and applied in geophysical exploration fields such as seismic data processing and interpretation and seismic inversion. The model of deep learning technology training is completely data-driven and can mine the deep relationship between input data and expected output. It often has the characteristics of nonlinearity, high resolution and de modeling [3]. Multilayer perceptron is a classical deep learning network model. The network takes neurons as the minimum unit. Its network structure includes input layer, output layer and multiple hidden layers. Through iterative training, the input information of multiple nodes can be nonlinearly mapped to a single output node. In the field of seismic reservoir prediction, multi-layer perceptron depth learning network has been widely used, but no application experiment has been carried out in the prediction of sandy conglomerate reservoir with rapid changes in sand body thickness in plane and space [4].

In this paper, the multi-layer perceptron depth learning network is used to study the reservoir prediction of sandy conglomerate reservoir in fault block reservoir based on seismic attributes. In view of the large difference of models between fault blocks caused by the large difference of stratum buried depth, dip angle and lithology among fault blocks, a deep learning reservoir prediction method based on fault blocks is proposed. Aiming at the problem that the reservoir changes greatly in plane and space caused by rapid reservoir phase change, a seismic attribute extraction technology based on isochronous sedimentary surface sliding time window is proposed, which replaces the determined time window with uncertain time window to eliminate the influence of subjective factors on attribute extraction. Taking the sandstone thickness data of the target layer of the well point as the expected output, the multi-layer perceptron depth neural network model is trained, and then the trained model is used to predict the sandstone thickness distribution characteristics in the well layout area. The application of actual data in a oilfield shows that the prediction coincidence rate of sandstone thickness in known wells is more than 95%, and the drilling encounter rate of sandstone in horizontal section after the completion of posterior horizontal wells is more than 80%. It is verified that this method has high prediction accuracy and good generalization ability.

2. Geological characteristics

The study area is a faulted anticline structure developed under the control of early developed reverse normal faults in the process of basement uplift. The target layer is divided into three oil formations from top to bottom: oil formation I, oil formation II and oil formation III. the lithology is mainly grayish green tuffaceous siltstone and tuffaceous sandy conglomerate, and the overlying grayish green shore shallow lake mudstone and tuffaceous siltstone are the main caprock. The formation thickness of oil formation I of the main oil layer is 65.0 ~ 93.2m, with an average of 81.2m, and the formation thickness of small layer is 16.4 ~ 30.3m, with an average of 23.9m. The formation development is relatively stable.

The east of the main oil layer I oil formation is dominated by multi-source sediments. The sedimentary type is mainly fan delta front sedimentation, and the provenance comes from the northwest. It mainly develops fan delta front underwater distributary channel, inter channel, front sheet sand and estuary bar microfacies. The electrical characteristics of underwater distributary channel in the front of fan delta are that the natural potential is wavy, nearly straight, and some are box and bell shaped; The electrical characteristic curve of the estuarine bar in the front of the fan delta is a medium high amplitude bell shape and slightly toothed box shape. The natural potential curve between distributary channels in the front of fan delta is straight. The sedimentary type is mainly shallow lacustrine sedimentary environment. With the gradual shallowing of water body and the slowing down of fault activity, a large-scale progradational delta began to appear in the north, extending from north to south. The sedimentary period is dominated by underwater distributary channel and estuarine bar sedimentation, with good reservoir physical properties and good oil-bearing property.

The average effective thickness of three wells in the northeast of the study area is only 13.0m, and the vertical wells cannot be effectively used under the existing economic conditions. To design the development of horizontal wells, it is necessary to carry out high-precision reservoir prediction research in the target area. According to the geological analysis of dense well pattern in the Development Zone, the lithology transformation between wells is fast and the thickness difference of multi well sandstone is large. The correlation coefficient between sandstone thickness and 43 extracted seismic attributes is low (the maximum correlation coefficient is 0.56). The conventional model-based multi-attribute regression reservoir prediction technology has low coincidence rate of sandstone thickness on wells and poor inter well prediction effect.

3. Deep neural network reservoir prediction method

In recent years, deep neural network has become a research hotspot in reservoir prediction, which puts forward a new direction based on data-driven for the development of reservoir prediction technology. Multilayer perceptron neural network is a basic type of deep neural network and a common method to realize "multilayer nonlinear transformation". The multi-layer perceptron neural network is expressed as a multi hidden layer neural network. The increase of hidden layer enhances the expression ability of the model, and its complexity increases greatly [5].

The depth neural network takes the seismic attribute as the input and the well point sandstone thickness data as the expected output. In order to achieve good prediction effect, the input data need to have consistent distribution characteristics. In fault block reservoir, there are great differences in reservoir characteristics between fault blocks. In view of this problem, a reservoir prediction based on fault block is proposed. Taking 46 development wells in the fault block as training data, the hidden characteristics of sandstone development (based on multi seismic attributes) are extracted at the depth of the development area, and the results are applied to the design well area. In terms of seismic attribute extraction, aiming at the problem of rapid reservoir phase change, a seismic attribute extraction technology based on isochronous sedimentary surface sliding time window is proposed, which replaces the determined time window with uncertain time window to eliminate the influence of subjective factors on attribute extraction.

3.1 Sliding window attribute extraction technology

Attribute extraction often adopts layer slicing technology. According to sedimentology theory, geological isochronous slicing can obtain high resolution. However, the seismic event events are not completely isochronous. The reason for the time crossing is that on the one hand, there is not enough seismic resolution for a given sedimentary sequence; The other is irregular cumulative thickness or abnormal impedance. There is a strong dependence between seismic event axes and the frequency of seismic data. Some event axes do not change depending on the change of frequency, but some event axes show different contact relationship with the change of frequency. Seismic sedimentology believes that the event axis on the original seismic profile is not completely isochronous, and its isochronity is controlled by seismic frequency. After frequency division of the seismic profile, the occurrence and location of the event axis are consistent on different frequency division profiles, and its isochronity is more reliable [6].

The isochronous interface is identified according to seismic sedimentology, and 43 seismic attributes are extracted by sliding time window. Each attribute is extracted by 40 different time windows, and a total of 1720 (43 * 40) attributes are extracted. The data set will be constructed with the sandstone thickness value at the well point as the label and the seismic attributes after regularization as the training set. The purpose of regularization is to meet the distribution range requirements of neural network algorithm for input data (Fig. 1).

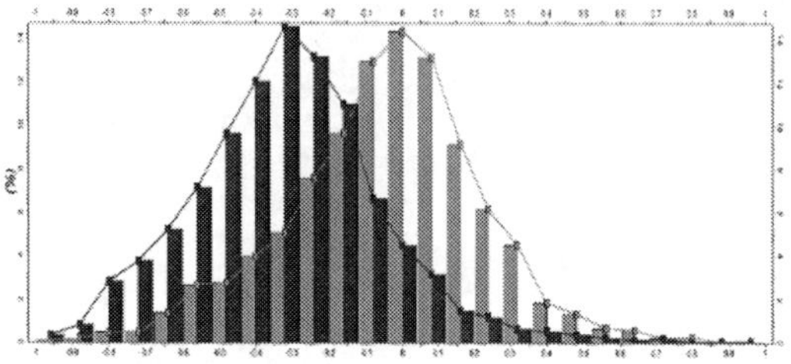

Fig. 1 Histogram after regularization of seismic attributes

3.2 Multilayer perceptron neural network

Multilayer perceptron neural network can be divided into three types: input layer, hidden layer and output layer. All layers are fully connected, and neurons in the same layer are not connected and independent of each other. An important feature of multi-layer perceptron is multi-layer. The first layer is called input layer, the last layer is called output layer, and the middle layer is called hidden layer. Multilayer perceptron neural network does not specify the number of hidden layers, so the appropriate number of hidden layers can be selected according to their respective needs. And there is no limit to the number of neurons in the output layer [7].

Firstly, the neural network structure of multi-layer perceptron is defined, and the seismic attribute at the input well point of the input layer is used as the feature vector. Then, the training and optimization of the model are realized by selecting the activation function and debugging the super parameters. In the process of model training, it is necessary to define the loss function and the optimizer that reduces the loss function the fastest.

The calculation of loss function is a forward propagation process. Forward propagation refers to calculating and storing the intermediate variables of the model in sequence from the input layer to the output layer of the neural network, and finally outputting the results [7]. In this paper, the mean square deviation function (RMSE) is used to evaluate the prediction ability of the model. The calculation formula of RMSE is:

$$\mathrm{RMSE} = \sqrt{\sum_{n=1}^{N} (y_i - p_i)^2 \times \frac{1}{N}}$$

Where, y_i and p_i is the value of the actual porosity and predicted porosity of the sequence, and N is the number of samples.

The optimization process of the optimizer is a back propagation process. Back propagation refers to the method of calculating the parameter gradient of neural network. In general, according to the chain rule in calculus, back propagation successively calculates and stores the gradient of intermediate variables and parameters of the objective function related to each layer of the neural network from the output layer to the input layer.

In the optimization of supervised process, Adam optimization algorithm with momentum is selected for adaptive learning rate. The modified algorithm combines the characteristics that adagrad algorithm is better than sparse gradient and rmsprop algorithm is better than non-stationary target, which ensures the accuracy and stability of gradient descent process and reduces turbulence, and increases the update range of sparse parameters, And reducing the update amplitude of frequency parameters. For example, in network training, with the increase of network depth, some parameters deviate from the set range and need to be debugged. At this time, Adam algorithm can first calculate the exponential weighted average of the gradient at the same time of each training, and then use the obtained gradient value to update the parameters [8].

4. Application effect analysis of horizontal well

Firstly, the model training effect is evaluated [9]. The evaluation of neural network training effect is divided into two parts: one is the accuracy of training data, and the other is the coincidence rate of test data (cross validation Implementation). Through analysis, it is considered that the loss function decreases with the increase of iteration times, the error value of test data fluctuates with the increase of iteration times, and the loss function and test error tend to be consistent with the increase of iteration times (Fig. 2).

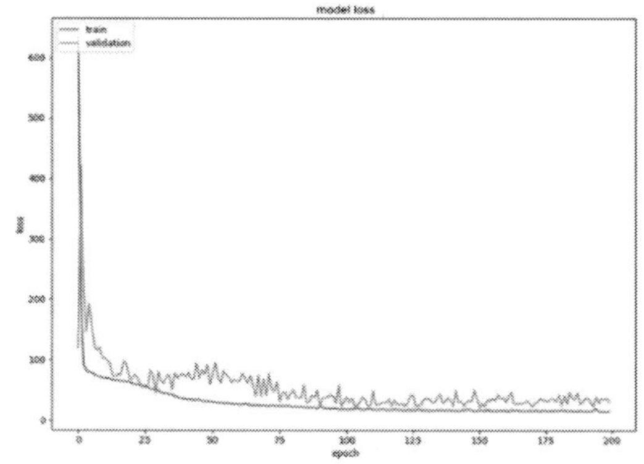

Fig. 2 Loss function variation diagram

The prediction results of sandstone thickness are compared and analyzed. The coincidence rate of well points is increased by 30%, from 59% to 92%. According to the distribution characteristics of sandstone, the thickness of sandstone is distributed in strips, and the predicted sand body distribution is consistent with the sedimentary characteristics of fan delta front. The predicted sand body strip trend is consistent with the provenance direction and matches the regional geological understanding. The prediction results of sandstone distribution in the design well area are consistent with the sandstone distribution law in the development area, indicating that the depth learning model has learned the characteristics of sand body distribution in the development area and applied to the sandstone thickness prediction in the design well area.

The core observation in the study area confirmed that the natural fractures of the reservoir in the study area are not developed. According to the analysis of micro resistivity scanning imaging logging data, the fractures are mainly drilling induced fractures and stress release fractures. Through XMAC data analysis, it is mainly affected by in-situ stress. The anisotropy of the target reservoir is strong, and the direction of the maximum principal stress is 45 °～85°, average 61 °。

According to the quantitative prediction results of sand body distribution in the target layer, according to the following principles:

(1) The structures and faults of the main target strata in the horizontal well area are determined, and the formation dip angle changes little;

(2) The design horizontal section is basically oriented along the direction of principal stress, which is convenient for fracturing and later water injection to supplement energy;

(3) Considering the fracturing scale, the vertical distance between the horizontal well trajectory and the fault shall not be less than 150m;

(4) The target point a of horizontal well shall be at the high point of the structure as far as possible, the drilling shall be from high to low as far as possible, and the well trajectory shall be drilled for a large distance in the oil layer as far as possible;

(5) The length of horizontal section is designed based on the principle of maximizing the controlled reserves of single well.

According to the prediction results of sandstone thickness reservoir in the well layout area, three horizontal wells are designed and drilled, with an average horizontal section length of 800m, a sandstone penetration rate of 85% and a reservoir penetration rate of 69%, and good drilling results are achieved.

5. Conclusion

(1) Using its powerful feature mining ability, deep learning technology can deeply extract the geological features contained in seismic attributes

(2) The deep learning prediction method for sandy conglomerate reservoir in fault block oilfield has good effect, high coincidence rate and matches with regional geological understanding

(3) Sliding time window seismic attribute extraction technology can eliminate the influence of human factors on deep learning feature extraction. It is an attempt to drive data.

References

1. ZHAO W J, YANG W Y, ZHAO W.The review of technical progress in application of seismic reservoir and hydrocarbon prediction [J].Progress in Geophysics, 2014(5):2337-2346.(in Chinese).
2. LIU Zhen ping,LIU Jia qi.Seismic-controlled nonlinear extrapolation of well parameters using neural networks [J].Geophysics,1998,63(6):2035-2041.
3. ZHAO J L, LI G, MA P S,et al. The application of network technology to petroleum logging interpretation. Progress in Geophysics (in Chinese),2010,25(5):1744-1751.
4. SUN Y H, LIU Y.Prediction of S-wave velocity based on GRU neural network. Oil Geophysical Prospecting,2020,55(3):484-492.
5. GOODFELLOW I, BENGIO Y, COURVILLE A,et al. Deep learning. Massachusetts Institute of Technology Press, Cambridge,2016.
6. ZHENG Y Z, YE Z H, LIU X E,et al.Research on prediction method of reservoir physical properties based on deep learning[J].Electronics World, 2018,10(4):23-26.
7. ZHANG L L. Analysis of rock physics characteristics of typical reservoir and applied research. Qingdao:China University of Petroleum(East China),2013.
8. HINTON G E, SALAKHUTDINOV R R.Reducing the dimen-sionality of data with neural networks. Science, 2006,313(5786):504-507.
9. GOODFELLOW I, BENGIO Y, COURVILLE A,et al. Deep learning. Massachusetts Institute of Technology Press, Cambridge,2016.

https://doi.org/10.1051/e3sconf/202235802012

Research on Carbon Emission Market Pricing Based on Putty-Clay Vintage Model

Kailin Zhang*, Ailin Zhao, Qibo Yan, Meiqi Sheng, Xiaochun Zhang

School of North China Electric Power University, Beijing, China

Abstract. With the rapid development of economy, the ecological environment problem of global warming is becoming increasingly serious. In order to effectively implement carbon emission reduction, countries have successively established carbon tax system or carbon emission trading market. China has established carbon emissions trading market in seven pilot provinces since 2013, and plans to implement the national carbon emissions trading market in the '14th Five-Year' period. In this process, how to price 'carbon' is the primary problem of China's carbon trading market construction. This paper first reviews the current development status of carbon emissions pricing in China's carbon trading pilot, and then analyzes the theoretical basis and price formation mechanism of carbon emissions trading pricing in the current carbon trading market, and then puts forward the carbon emissions pricing model based on Putty-Clay Vintage model, and puts forward reasonable policy suggestions for improving the current carbon pricing strategy in China.

1. Introduction

With the continuous seriousness of the ecological environment problem of global warming, people are increasingly aware of the indivisible link between the excessive emission of greenhouse gases and climate change. In order to delay this trend, countries have made efforts to actively respond to climate change. As the largest developing country in the world, China actively undertakes the responsibility of emission reduction. At the seventy-fifth session of the United Nations General Assembly, President Xi Jinping solemnly pledged to China's efforts to achieve its carbon neutrality goal by 2060, stabilising China's path to green and low-carbon development as the world's largest developing country in terms of emissions and its task of achieving the carbon neutrality vision quickly in about 30 years after its peak by 2030. It is difficult to achieve carbon peak only by means of administrative orders, and market means are essential.

Carbon market system is an effective economic means to solve the externality of carbon emissions, which can minimize the cost of emission reduction under the total control goal, and is conducive to high-quality economic growth. China began to participate in the international carbon trading market since 2005, when it mainly participated in the clean development mechanism as a seller. In October 2011, the National Development and Reform Commission issued *Notice on pilot work of carbon emission trading*, marking the official launch of China's carbon emissions trading. Beijing, Tianjin, Shanghai, Chongqing, Guangdong, Hubei, Shenzhen and other seven provinces and cities have successively started the pilot work of carbon emissions trading. Since the formal trading in 2013, China has experienced seven performance periods, with 2837 key emission units, 1082 non-performance agencies and more than 10,000 natural persons participating in the carbon trading pilot in seven provinces. As at 31 December 2020, including Fujian Province, which subsequently joined the pilot carbon market, quota spot transactions in eight pilot carbon markets totalled 445 million tons, with turnover of 10.431 million yuan and transaction price of 23.5 yuan/ton. In addition, national certified voluntary emission reductions (CCERs) totalled 268 million tons.

With the continuous progress of pilot carbon market construction, some problems have also been exposed. In the pilot stage, China' s relatively independent carbon trading market has the problems of system independence and regional segmentation. The carbon trading price is relatively low compared with the foreign market, and the trading price of each pilot market is different. The lack of unified and perfect national pricing mechanism restricts the establishment and improvement of China ' s carbon trading market. If carbon trading cannot form a fair, reasonable and effective price, the function of the carbon market will be greatly weakened. In the process of building a unified national carbon market, reasonable pricing of carbon emission rights has become a key issue.

* Corresponding author: zhangkailin_dut@foxmail.com

2. Theoretical basis and price formation mechanism

Energy-saving low-carbon value-added service competitiveness evaluation model

2.1 Theoretical basis

2.1.1 Environment externality

In the early 20th century, economists Marshall and Pigou proposed the externality theorem, that is, the production or consumption activities of one person or group will affect another. The production externality caused by enterprise economic activity to environment is called environmental externality. The production and other economic activities of enterprises need to emit carbon dioxide and other greenhouse gases outward, resulting in pollution to the social environment, which causes economic activities of other enterprises and consumers to cause environmental external diseconomy, affecting their normal production and life activities.

The tragedy of the commons is the most obvious in environmental externalities, which is the phenomenon that limited social resources are overused because of unlimited and undemanded, resulting in the damage to the interests of all people. Carbon dioxide emissions are also facing the same situation. If enterprises only focus on maximizing their own interests and do not curb carbon dioxide emissions, it can be expected that the social environment will be seriously polluted, thereby affecting the production activities of enterprises and the quality of life of the people. The establishment of the carbon market is to add constraints on the use of natural resources, thereby inhibiting excessive emissions of carbon dioxide, carbon price is an effective means to control carbon dioxide emissions.

2.1.2 Coase theorem and property right theory

Coase theorem mainly refers to the external non-economic or inefficient situations can be solved by negotiation between the two parties, and ultimately achieve Pareto optimality. The establishment of the carbon market provides a trading mechanism and venue for carbon emissions. Its purpose is to trade the carbon emission rights based on the enterprise ' s own needs through carbon price changes, so as to maximize social benefits.

For the negative externality problem widely existing in society, Coase believes that as long as the property right is clear, the external uneconomic problem can be well solved. He believes that when the transaction cost is very small, no matter how the rights are defined at the beginning, the most effective allocation of resources can be achieved through free market transactions, namely Pareto optimality. The establishment of property rights trading market is the basis for the implementation of property rights system. The free trade of property rights will guide carbon emission rights resources to the regions or industries with the highest economic efficiency, to the subjects that can create more wealth for the society, and to achieve the optimal allocation of resources. The national carbon trading market that China is striving to build is based on this theory. The establishment of national carbon trading venues will promote the rational allocation of atmospheric resources in China, and promote the continuous improvement of regional economic development level and environmental governance capacity.

2.1.3 Resource scarcity

To trade carbon emission rights assets, carbon emission rights must have commodity attributes, and the value of commodities comes from their scarcity. Atmospheric resources are scarce in the long term, but short-term enterprises do not seem to have any constraints on carbon dioxide emissions. To make atmospheric resources also scarce in the short term, it requires government intervention, artificially delimiting the scarcity of carbon emission rights, and specifying the total amount of carbon emissions in a certain region during a certain period. In this way, the scarcity of carbon emission rights will lead to market competition, and benign competition will achieve the optimal allocation of resources, and ultimately achieve the ultimate goal of energy conservation and emission reduction with low social costs. The advantage of carbon emissions trading is to combine the government's visible hand with the invisible hand of the market, which can not only ensure the total emissions of society, but also introduce flexible market mechanisms to encourage enterprises to conduct trading activities in the carbon emissions trading market according to the principle of cost-effectiveness.

Economic theories such as environmental externality and resource scarcity provide theoretical support for the development of the carbon market, and the embodiment of relevant theories in the carbon market is shown in Figure 1.

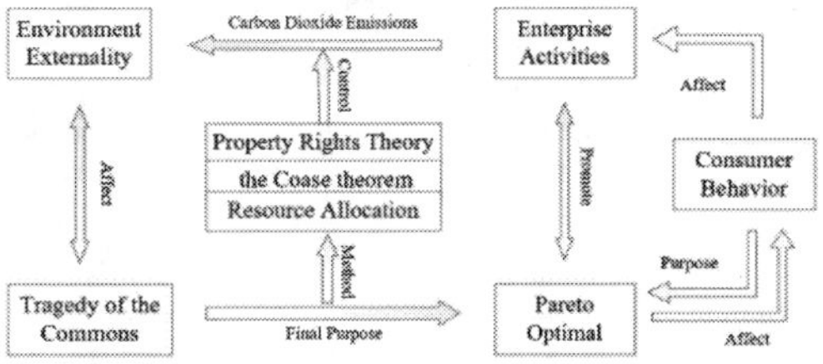

Figure 1. Embodiment of relevant theories in carbon market.

2.2 Price formation mechanism

The mechanism of the carbon emission trading system sets emission limits for the emitters and allows them to perform the contract by trading emission quotas. Carbon trading mechanism is directly from the 'quantity' dimension, trading prices with the 'quantity' restrictions and spontaneous adjustment. Carbon emissions trading system is shown in Figure 2.

E3S Web of Conferences **358**, 02012 (2022) https://doi.org/10.1051/e3sconf/202235802012
GESD 2022

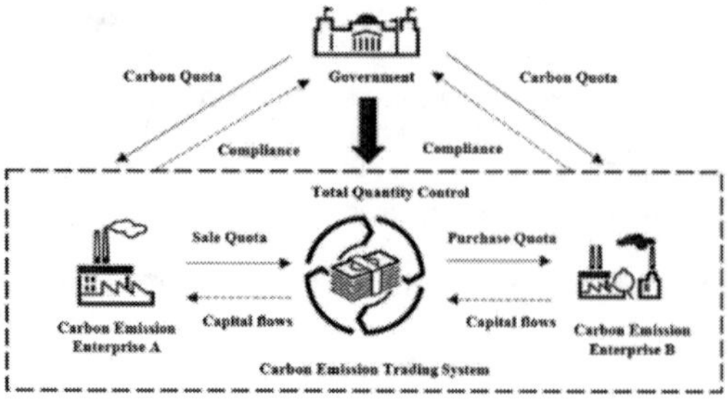

Figure 2. Carbon emission trading system.

3. Pricing Strategy

3.1 Model thought

This paper uses the Putty-Clay Vintage model as a method for evaluating the value of carbon emission rights, based on the intrinsic value of the model. The evaluation principle of the model is that carbon emission rights are regarded as the production factors of enterprises, and the maximum profit of enterprises under certain constraints is calculated. The maximum benefit of carbon emission power is its maximum value to the enterprise, that is, its intrinsic value of carbon emission power. The basic idea of the model is : under the premise of the total allowable carbon emissions set by the state, each region sets its own emission reduction targets and decomposes them into various enterprises. Each enterprise should carry out its own production and operation activities within the emission quota set by the government. Therefore, the emission reduction cost of enterprises can be incorporated into the total cost of enterprises, and the profit function is used to obtain the maximum profit that enterprises can obtain under the constraint of emission quota index.

3.2 Model construction

This paper uses the Putty-Clay Vintage model as a method for evaluating the value of carbon emission rights, based on the intrinsic value of the model. The evaluation principle of the model is that carbon emission rights are regarded as the production factors of enterprises, and the maximum profit of enterprises under certain constraints is calculated. The maximum benefit of carbon emission power is its maximum value to the enterprise, that is, its intrinsic value of carbon emission power. The basic idea of the model is : under the premise of the total allowable carbon emissions set by the state, each region sets its own emission reduction targets and decomposes them into various enterprises. Each enterprise should carry out its own production and operation activities within the emission quota set by the government. Therefore, the emission reduction cost of enterprises can be incorporated into the total cost of enterprises, and the profit function is used to obtain the maximum profit that enterprises can obtain under the constraint of emission quota index.

3.2.1 Construct profit function

Under the limitation of the total amount of carbon emission rights, the output of an enterprise is limited by the carbon emission rights. Therefore, with the increase of the production output of an enterprise, the cost of carbon emission increases gradually. When constructing the profit function, we include the cost of carbon emission into the variable cost accounting of the enterprise.

$$\prod = S - VC - FC = S - (VC_1 + VC_C) - FC \quad (1)$$

In the formula, $\prod$ represents enterprise profit; S represents enterprise sales; VC represents the total variable cost of enterprise; FC represents the fixed cost of enterprise; VC_C represents the variable cost related with carbon emissions; VC_1 represents the variable cost unrelated to carbon emissions.

3.2.2 Decompose the cost of carbon emission rights

Carbon emission rights are mainly composed of three parts: one is the cost of the company's own emission reduction activities, the second is the expenditure required by the company to purchase carbon emission rights in the market, and the third is the cost of fines caused by excess emissions. This article does not study the impact of changes in technology level on the cost of carbon emission rights under long-term conditions. Assuming that the technology level will not change in the short term, the variable cost related to carbon emissions can be expressed as:

$$VC_C = C_r + C_b + C_f \quad (2)$$

In the formula, C_r represents the cost of the enterprises' own emission reduction activities; C_b represents the cost for enterprises to purchase carbon emission quota in the market; C_f represents the the fine caused by excess emissionss.

3.2.3 Constraint

In the actual emission reduction work, the actual carbon emission of the enterprise should be less than or equal to the sum of the carbon emission quota owned by the enterprise, the carbon emission quota purchased by the enterprise and the excess emission of paying fines.

$$e_T \le e_q + e_b + e_f \quad (3)$$

In the formula, e_T represents total actual emissions of enterprises; e_q represents carbon emission quotas owned by enterprises; e_b represents carbon emission quota purchased by enterprises; e_f represents excess emissions subject to fines.

3.2.4 Construct carbon emission pricing model

Integrate profit function and constraints.

$$\prod = S - (VC_1 + C_r + C_b + C_f) - FC \tag{4}$$

$$\text{s.t. } e_T \le e_q + e_b + e_f$$

After the derivation of the profit function on carbon emissions, the marginal profit generated by unit carbon emissions is obtained, that is, the intrinsic value of carbon emission rights, which is the basis for pricing carbon emission rights. The intrinsic value of carbon emissions plus some other factors that may affect the price of carbon emissions adjustment coefficient is the price of carbon emissions.

$$P = \frac{d\Pi}{de_T} + \sigma \tag{5}$$

In the formula, σ represents impact of other factors on price.

4. Conclusion

This paper studies the carbon emissions market pricing. Firstly, this paper reviews the current development of carbon emissions pricing in China's carbon trading pilot, and then analyzes the theoretical basis and price formation mechanism of carbon emissions trading pricing in the current carbon trading market, which lays the foundation for carbon emissions pricing model. Then, a carbon emission pricing model based on Putty-Clay Vintage model is constructed. The objective function of this model is to maximize the profits of enterprises, mainly considering the cost of carbon emission rights composed of three parts, so as to improve the trading activity of the carbon emission market and promote the construction of a unified national carbon emission rights market.

Acknowledgments

This work was financially supported by The National Social Science Fund of China (No. 19ZDA081) and The Fundamental Research Funds for the Central Universities (No. 2020MS067).

References

1. Li Yi. An empirical study on the external influencing factors of carbon emission trading pricing [J], Price: Theory & Practice, 2020(06):146-149.
2. Wang Ke, Liu Yongyan. China's carbon market: reviews and prospects for 2020 [J], Journal of Beijing Institute of Technology(Social Sciences Edition), 2020,22(02):10-19.
3. Men Ming, Zhang Qiuli. A review of carbon emissions trading production model and discussion about its applications in China[J], International Business, 2010(06):44-51.
4. Wu Huijuan, Zhang Guangzhong. Causes of China's low carbon trading price: theoretical models and empirical analysis [J], Modernization of Management, 2020,40(06):75-81.
5. Pan Lu. Research on the valuation method of carbon emission rights[J], Appraisal Journal of China, 2020(03):39-43.
6. Shen Hongtao, Huang Nan, Liu Lang.A Study of the Micro-effect and Mechanism of the Carbon Emission Trading Scheme[J], Journal of Xiamen University (Arts & Social Sciences), 2017(01):13-22.
7. Wang Liping, Wang Zhijia. Research progress and prospect of carbon emission trading system[J], Review of Economic Research, 2019(19):49-62.
8. Yang Hua. Research on the influencing factors of carbon emission trading price in China[D], Xi'an University of Science & Technology, 2020.

Research on Economy of Clean Energy Power Generation Technology Based on CORE Method

Ailin Zhao [1,*], Kailin Zhang [1], Meiqi Sheng [1], Qibo Yan [1], Weili Ren [2], Baohua Bai [2], Xichao Zhou [2], Xiaochun Zhang [1], Ming Zeng [1]

[1] School of Economics and Management, North China Electric Power University, Beijing, China
[2] Integrated Energy Service Group, Xicheng District, Beijing, China

Abstract. Clean energy power generation has good economic and social benefits for promoting regional development. However, clean energy power generation is in the growth stage, and the difference of investment income is obvious. Therefore, it is of great significance to study the economy of clean energy power generation technology. In view of this, considering the influencing factors on economy of clean energy power generation technology, this paper proposes a technical economy evaluation and cost trend analysis model for clean energy power generation, which includes three modules: input, output and analysis calculation. Finally, taking Africa as an example, under the three scenarios of high utilization hours, medium utilization hours and low utilization hours, the paper analyzes the economy of photovoltaic power generation in Africa, and concludes that the cost of photovoltaic power generation in the northern and southern of Africa has a certain degree of network transmission economy.

1. Introduction

At present, the contradiction between energy demand and energy supply has become an important factor restricting social and economic development. In order to realize the sustainable development of energy, the development of clean energy is of great strategic significance to promote the revolution of energy production and consumption, and to establish a clean, low-carbon, safe and efficient modern energy system. With the development of clean energy power generation technology, countries all over the world are developing clean energy, which has good economic and social benefits for promoting regional development. However, the current clean energy power generation project is in the growth stage, affected by the regional resources, the investment income difference is obvious, so the research on the economy of clean energy power generation technology is very important. Firstly, this paper analyzes the influencing factors of clean energy power generation technology economy. Secondly, based on the basic model framework of LCOE method, a technical and economic evaluation and cost trend analysis model for clean energy power generation is established to realize the quantitative evaluation of technical and economic efficiency of clean energy power generation. Finally, the economy of photovoltaic power generation in Africa is analyzed based on different scenarios.

2. Factors influencing the technical economy of clean energy power generation

The main influencing factors of clean energy power generation cost include the basic parameters of the project, power generation cost and government incentive policies and measures and so on.

(1) Basic parameters

The basic parameters of the project include the life cycle of the project, utilization hours of power generation equipment, benchmark internal rate of return (IRR), financial parameters, etc. For the life cycle, the longer the life cycle is, the higher the ROI of the project will be, and the corresponding internal rate of return will also be improved without considering the renewal cost of the equipment and the same boundary conditions. The utilization hours of power generation equipment and clean energy power generation technology equipment are important indicators affecting power generation economy. Affected by the local resource conditions, equipment quality, management and operation and maintenance level, power grid capacity and so on. The benchmark internal rate of return is the discount rate when the total present value of capital inflow is equal to the total present value of capital flow and the net present value is equal to zero. It is particularly important to analyze clean energy power generation assets, which can be used to compare the operating costs of different clean energy power generation. Financial parameters mainly include loan proportion, loan

* Corresponding author: 120192206949@ncepu.edu.cn

repayment period, depreciation rate, depreciation period, residual value rate of fixed assets, formation rate of fixed assets, proportion of working capital, etc. The impact of financial parameters on the economy of clean energy power generation project is obvious.

(2) Power generation cost

The generation cost of clean energy power generation project includes initial investment cost (including unit), operation and maintenance cost, grid connection cost and other costs. The cost items included in the power generation cost have an internal impact on the project economy. The initial investment cost of clean energy power generation project is the most important factor to determine the total cost of the project, which has the greatest impact on the economy of the project. Generally, it includes the equipment cost, construction cost, financial cost, design and other miscellaneous items, and other expenses of the project.

(3) Policy measures

Policy measures are the external factors of power generation cost, but they will have a crucial impact on the cost. Electricity price policy, fiscal and taxation policy, financial policy, full guarantee purchase policy, green certificate trading mechanism and other policies, as well as development planning and scale objectives have an important impact on the cost of clean energy power generation technology.

3. Economic analysis model of clean energy power generation technology

In addition to the basic parameters, the most important factors affecting the economy of clean energy power generation are power generation cost, utilization level and policy measures. Among them, although the policy measures will not directly affect the cost of power generation, they will affect the basic parameters, grid connection cost, other costs and utilization level that determine the cost of power generation, which has an important impact on the economy of the project. Therefore, this paper takes policy measures as an important external influencing factor model into the model system, and analyzes its impact on the relevant factors and parameters of power generation economy.

3.1 Analysis steps

According to the above factors and components of clean energy power generation economy, based on the basic model architecture of LCOE method, this paper comprehensively considers the advantages and disadvantages of different technical and economic analysis methods and models, and establishes a economics evaluation and cost change trend analysis model (CORE) for clean energy power generation technology, so as to realize the quantitative evaluation of clean energy power generation technology economy. The specific analysis steps are as follows: firstly, the main cost factors such as initial investment cost, construction cost, operation and maintenance cost are determined according to the technology type and project scale. Then determine basic parameters such as lifetime, depreciation, and internal rate of return, and predict operating parameters such as the utilization hours of different types of clean energy power generation equipment based on historical data. Finally, according to the influencing factors of policies and measures, the impact of on grid electricity price, fiscal and tax policies, financial policies and other related parameters are determined, and the CORE value and IRR level of different clean energy power generation projects are calculated.

3.2 Model framework

The core model consists of three modules: input module, output module and analysis module. The input and output parts adopt custom data format, and the analysis module is the core module of the model.

(1) Input module

For a given clean energy power generation project, input the internal factors including technical parameters, economic parameters, operating parameters, and external factors including on grid electricity price, fiscal and tax policies, financial policies, and development planning objectives.

(2) Analysis and calculation module

Analysis and calculation module is the core function module of core model, which is divided into three parts: generation cost calculation module, internal rate of return calculation module and generation cost prediction module.

1) Power generation cost calculation module

According to the technical parameters, economic parameters, operation parameters and other internal influencing factors input by the input module, the cost, profit, cash flow and so on in the whole life cycle of clean energy power generation project can be calculated.

The mathematical formula for calculating the cost of clean energy power generation is as follows:

It is known that the value F of each future period is lower than that of the current period. The discount rate r is used to measure the difference.

$$P = F(1+r)^{-n} \tag{1}$$

$$NPV = \sum_{t=1}^{n} \frac{C_t}{(1+r)^t} - C_0 \tag{2}$$

$$\sum_{n=0}^{N} C_n (1+r)^{-n} = \sum_{n=0}^{N} (A_n P_n + B_n)(1+r)^{-n} \tag{3}$$

$$LCOE = \frac{I_0 + V_R(1+r)^{-n} + \sum_{n=0}^{N}(C_n - B_n)(1+r)^{-n} + R_I + R_E}{\sum_{n=0}^{N} A_n (1+r)^{-n}} \tag{4}$$

Where NPV is the net present value; C_0 is the initial investment; C_t is the cash flow of year t; R is the discount rate; N is the life cycle of the investment project; C_n is the total expenditure; A_n is the electric energy produced; B_n is income from other sources; I_0 is the initial investment; V_R is the system residual value; R_I is the internal factor of risk cost; R_E is the risk cost of external factors.

2) Internal rate of return calculation module

IRR can be obtained by the following formula:

$$IRR(NPV)=\sum_{t=0}^{n}(CI-CO)_t(1+IRR)^{-t}=0 \quad (5)$$

Criterion: if the benchmark discount rate is i_0, if $IRR \geq i_0$, the economic effect of the project is acceptable; If $IRR < i_0$, the economic effect of the project is not acceptable.

The solution process is as follows: firstly, the present value coefficient of annuity $(p/A,IRR,n)=K/R$ is calculated; secondly, look up the table of annuity present value coefficient, find two adjacent coefficients $(p/A,i_1,n)$ and $(p/A,i_2,n)$ and corresponding i_1、i_2, which satisfy $(p/A,i_1,n)>K/R>(p/A,i_2,n)$; finally, the IRR is calculated by interpolation: $(IRR-I)/(i_1-i_2)=[K/R-(p/A,i_1,n)]/[(p/A,i_2,n)-(p/A,i_1,n)]$.

3) Generation cost prediction module

According to the given future forecast period, combined with the basic input parameters of the input module, the future development scale, planning objectives and other external factors, the medium and long-term power generation cost of clean energy power generation project can be calculated, and the future cost change trend is analyzed, and the cost decline curve and cost change roadmap are given. In addition, the core analysis and calculation module of the CORE model also has the function of uncertainty analysis, which can analyze the sensitivity of input parameters for the economy of clean energy power generation technology, and also analyze and calculate the break even hours, break even payback period, etc.

(3) Output module

According to the input parameters of the input module, through the analysis and calculation of the core functions of the calculation module, the economic analysis results and power generation cost prediction results of the clean energy power generation project are output. Including power generation cost (LCOE), IRR, profit margin, etc; medium and long-term power generation cost, cost change trend, etc. In addition, according to the calculation of uncertainty analysis function, the results of break even analysis and sensitivity analysis can be output.

3.3 Example analysis

Taking Africa as an example, the analysis of power generation economy mainly considers the initial investment cost, financial parameters and other basic factors, and is divided into three scenarios: high utilization hours, medium utilization hours and low utilization hours. The specific boundary condition parameters are shown in Table 1.

Table 1. Boundary conditions and scenarios for economic analysis of photovoltaic power generation in Africa

Boundary condition	Northern Africa	Southern Africa
Initial investment cost	1120$/kW	
Operation and maintenance cost	35000$/MW/year	
Effective utilization hours corresponding to resource conditions	1800-2000	1700-1800
Lending rate	11%	
Capital ratio	21%	
Benchmark IRR	8%	
Scenarios	**Northern Africa**	**Southern Africa**
High utilization hours	2000	1800
Medium utilization hours	1900	1750
Low utilization hours	1800	1700

The economic calculation results are as follows:

1) High utilization hours scenario: the cost of photovoltaic power generation in northern Africa is about 0.51 yuan / kWh; The cost of photovoltaic power generation per kilowatt hour in southern Africa is about 0.58 yuan / kWh.

2) Medium utilization hours scenario: the cost of photovoltaic power generation in northern Africa is about 0.54 yuan / kWh; The cost of photovoltaic power generation per kilowatt hour in southern Africa is about 0.61 yuan / kWh.

3) Low utilization hours scenario: the cost of photovoltaic power generation in northern Africa is about 0.58 yuan / kWh; The cost of photovoltaic power generation per kilowatt hour in southern Africa is about 0.63 yuan / kWh.

To sum up, the per kilowatt hour cost of photovoltaic power generation in northern Africa is about 0.51-0.58 yuan / kWh; About 0.58-0.643yuan / kWh in the south.

At present, in Europe, Germany, Italy, Britain and other countries, the cost of photovoltaic power generation is about 0.63-0.80 yuan / kWh. Compared with European photovoltaic power generation, the cost of photovoltaic power generation in Northern and Southern Africa has certain transmission economy, but considering the transmission cost, the economy is not prominent.

4. Conclusion

The economy of clean energy power generation technology is a practical problem to be further studied. Based on the analysis of the economic factors of clean energy power generation technology and the basic model framework of LCOE method, this paper establishes a economics evaluation and cost trend analysis model for clean energy power generation technology. Through setting different scenarios, this paper analyzes the economy of photovoltaic power generation in Africa, and draws the conclusion that the cost of photovoltaic power generation in Northern and Southern Africa has a certain degree of network transmission economy, but the economy is not prominent after considering the transmission cost. This model can more comprehensively compare and evaluate the economy of clean energy power generation.

Acknowledgments

This work was financially supported by State Grid Corporation Technology Project (Research and application of integrated energy system regulation technology of power source, grid load and storage interaction (No. SGFJJY00GHJS1900066)).

References

1. Yang Haishou. Impact of photovoltaic power grid on power grid operation and countermeasures[J]. Application of IC, 2021,38(03):72-73.
2. Mao Aihan, Li Faxiang, Yang Siyuan, et al. Clean energy power generation potential and value in Qinghai Province[J]. Resources Science, 2021, 43(01):104-121.
3. Yan Qingyou, Zhu Mingliang. Empirical research on economy of wind power grid connected based on LCOE method[J]. Journal of Technical Economics & Management, 2017(11):21-25.
4. Shan Ming, Liu Yanqing, Ma Rongjiang, et,al. Comparison of economy and emission performance of different technologies from coal to clean energy in northern rural areas[J]. Environment and Sustainable Development, 2020,45(03):43-49.
5. Liu Xiaochun, Shan Baoguo, Wang Chengjie, et al. Substitution potential model of high proportion of clean energy and analysis of key affecting factors[J]. Power System Technology, 2017,41(09):2755-2761.
6. Liu Ximei, Bai Kai, Deng Chun, et al. Research on levelized cost model of large-scale wind power projects[J]. Renewable Energy Resources, 2016,34(12):1853-1858.
7. Chen Rongrong, Sun Yunlin, Chen Siming, et al. LCOE analysis of grid-connected photovoltaic power generation project[J]. Renewable Energy Resources, 2015,33(05):731-735.
8. Zeng Ming, Lu Wei, Duan Jinhui, et al. Study on cost of solar photovoltaic power generation using double-factors learning curve model[J]. Modern Electric Power, 2012,29(05):72-76.
9. Wang Hengtian, Yang Xiaolong. Economic evaluation, influencing factors and Countermeasures of grid-parity photovoltaic power generation projects[J]. Enterprise Economy, 2021,40(03):96-104.
10. Wu Mingliang. Research on characteristics and economics of solar photothermal power generation technology[J]. Qinghai Electric Power, 2019, 38(02):18-22.

Water quality evaluation of typical Lake Wetland Parks around Dianchi Lake

Ziyuan Wang, Xingyue Liu, Weichao Wang, Wei Li*

School of Geography and Ecotourism, Southwest Forestry University, Kunming 650224, China

Abstract. Dianchi Lake is one of the key management points of China's "three rivers and three lakes basins", and the construction of wetland park plays an important role in purifying Dianchi lake water body and maintaining the health of lake ecosystem. Based on single factor evaluation method and three improved Nemerow pollution index methods, this study evaluates water quality of South Dianchi Wetland Park, Xihua Wetland Park and Laoyu River Wetland Park, with concentrations of total nitrogen (TN), total phosphorous (TP), chemical oxygen demand (COD), zinc (Zn), nickel (Ni), cadmium (Cd), chromium (Cr), lead (Pb) specifically measured and monitored. The results showed that: (1) the pollution levels of COD, TN, TP and Cr in the three wetland parks are relatively high. (2) based on single factor evaluation method and the improved Nemerow pollution index method, the evaluation grade of the water quality of the three wetland parks is inferior to Grade V. On the whole, the water pollution of Laoyu River Wetland Park is the most worrisome, and the water pollution of South Dianchi and Xihua Wetland Park is slightly better. (3) COD and TP are the main types of pollutants entering the lake in the basins where the three wetland parks are located. Therefore, immediate attention should be paid to the management of water quality state of Dianchi Lake, and the treatment of urban, agricultural non-point source and some uncollected point source pollution should be further strengthened.

Key words: Wetland park; Dianchi Lake; water quality evaluation; single factor evaluation method; Nemerow pollution index method; pollution source.

1. Introduction

Water resource is indispensable material resources for the survival and development of all life on earth. Since the 21st century. Wetland protection has become a global environmental problem and one of the themes of international wetland research[1]. At the same time, wetland park is an organic combination of protecting and restoring wetlands, and plays an important role in purifying water bodies and maintaining a virtuous cycle of wetland ecosystem.

Dianchi Lake, located in the southwest of Kunming, is the largest lake in Yunnan Province, known as the "Pearl of the plateau". However, with the rapid development of urban construction, the water pollution problem of Dianchi Lake is becoming serious. In recent years, the rise of the construction of wetland parks around Dianchi Lake not only provides ecological sightseeing and leisure places for the city, but also focuses on improving the water quality of the lake and restoring and giving full play to the key ecosystem functions of wetland around it.

On the basis of relevant research, we use single factor evaluation method and the improved Nemerow pollution index method to evaluate the water quality of Laoyu River Wetland Park, South Dianchi Wetland Park and Xihua Wetland Park, and analyze the pollution sources, in order to provide suggestions and references for the water quality treatment and optimization of wetland parks around Dianchi Lake.

2. Methods

2.1 Sampling

Starting from the three directions of East, West and south, this study selects three typical wetland parks in Dianchi Lake Basin as water quality sampling points (South Dianchi Wetland Park, Xihua Wetland park and Laoyu River Wetland Park). The position information of our sampling points is provided in Table 1. Water sampling was completed on December 6, 2020. The sampling points were selected along the wetland park, and two water samples were collected at each sampling point. In the process of sampling, when the water level was stable, pressed the sampling bottle to sink into the water for sampling at the depth of 10 cm-20 cm, and avoided shaking the water body. After sampling, closed the bottle

* Corresponding author: ww0592@gmail.com

cap immediately to prevent oxidation reaction, then sealed it and kept it away from light[3].

The alkaline persulfate digestion technique was applied to determine total nitrogen (TN) and total phosphorous (TP) of water samples, and Potassium dichromate method was used to determine chemical oxygen demand (COD). Heavy metals in water samples were determined using ultrasound-assisted dispersive liquid-liquid microextraction method[4-5]. The water quality evaluation indexes refer to According to China' s Environmental Quality Standards for Surface Water (GB3838-2002), and the standard concentration value is compared with the class III water quality standard.

Table 1. The position information of sampling points

Sampling points	Longitude	Latitude	Altitude
South Dianchi Wetland Park	102°37′52.64"E	24°41′6.45"N	1892m
Xihua Wetland Park	102°39′33.87"E	24°52′40.55"N	1867m
Laoyu River Wetland Park	102°46′12"E	24°49′33.6"N	1894m

2.2 Single factor evaluation method

Single factor evaluation method compares the measured value of the pollution factors of the evaluated water body with the environmental quality standard value of surface water, and selects the category with the worst water quality among all pollution factors as the category of the measured water quality. The method is the simplest and widely used, but it has the defect that it can not comprehensively evaluate the water body. The calculation formula is shown below.

$$F_i = \frac{C_i}{S_i} \tag{1}$$

F_i: Single factor pollution index of pollution factor i at the sampling point, its value reflects the degree of exceeding the standard of pollutants. C_i: the measured value of pollution factor i at the sampling point (mg/L). S_i: the standard value of pollution factor i at the sampling point(mg/L) [6].

2.3 Nemerow pollution index method

Nemerow pollution index method considers the pollution degree of other index factors in the evaluation system. It is a comprehensive evaluation method. However, the traditional method still does not consider the weight of each pollution factor. Therefore, we use the improved Nemerow pollution index method to evaluate the water quality.

First, determine the weight of each pollution factor, and then calculate the improved Nemerow pollution index. The calculation formula is as follows.

$$r_i = S_{\max} / S_i \tag{2}$$

$$\omega_i = r_i / \sum_{i=1}^{n} r_i \tag{3}$$

$$F'_{i,\max} = \frac{F_{i,\max} + F_{\omega}}{2} \tag{4}$$

$$\overline{F'} = \sum_{i=1}^{n} \omega_i F_i \tag{5}$$

$$P' = \sqrt{\frac{{F'_{i,\max}}^2 + \overline{F'}^2}{2}} \tag{6}$$

S_{max}: the maximum value of standard concentration among all pollution factors. n: the number of pollution factors participating in the evaluation. ω_i: the weight value of the ith pollution factor. $F_{i,max}$ is the maximum value of F_i. F'_{i},max: the weighted maximum value of F_i. F_{ω}: Single factor pollution index of the pollution factor with the largest weight value. $\overline{F'}$: the average value of corrected F_i. P': the improved Nemerow pollution index.

3. 3. Results

In general, the water quality of the three wetland parks was highly polluted in terms of COD, TN, TP and Cr, among which the concentration of COD, TP and Cr exceeded the standard concentration by the highest multiple, and the other four indicators were not polluted, among which the concentration of Zn and Pb was the lowest, followed by Ni and Cd. The metal elements detected in Laoyu River Wetland Park were significantly lower than the other two wetland parks, but the concentration of TP was the highest, the concentration of TN and Cr in South Dianchi Wetland Park was the highest, and the COD of Xihua Wetland Park was significantly higher than that of other wetland parks(Table 2).

Table 2. The concentrations of all index in water samples

Sampling points		COD (mg/L)	TN (mg/L)	TP (mg/L)	Metal element				
					Zn (mg/L)	Ni (mg/L)	Pb (ug/L)	Cd (ug/L)	Cr (mg/L)
South Dianchi Wetland Park	Sampling 1	104	2.99	0.256	0.019	0.002	1.49	4.650	0.415
	Sampling 2	95	2.32	0.33	0.012	0.001	48	2.400	0.660
Xihua Wetland Park	Sampling 1	257	1.59	0.222	0.013	0.001	30	0.400	0.200
	Sampling 2	124	1.77	0.236	<0.001	<0.001	1.12	0.500	0.500
Laoyu River Wetland Park	Sampling 1	76	2.52	1.14	<0.001	<0.001	<0.01	<0.01	0.060
	Sampling 2	99	1.79	1.47	0.02	<0.001	<0.01	<0.01	0.150

Single factor evaluation method was used to evaluate the water pollution level according to the worst pollution index of each water sample. The results showed that the water pollution level of the three wetland parks is inferior to Grade V. The serious exceeding of the detection concentrations of TP, COD and Cr was the reason why the water quality of the three wetland parks was inferior to Grade V under this method. However, the concentration of other pollution indicators was far lower than the standard concentration. This meant that Single factor evaluation method pays too much attention to the worst indicators without comprehensive consideration, and the results may be too conservative(Table 3).

Table 3. Evaluation results of single factor pollution index method

Sampling points	F_i								Grade of pollution
	COD	TN	TP	Zn	Ni	Pb	Cd	Cr	
South Dianchi Wetland Park	4.975	2.655	5.86	0.015	0.075	0.495	0.705	**10.75**	Inferior Grade V
Xihua Wetland Park	**9.525**	1.68	4.58	<0.007	< 0.05	0.311	0.09	7	Inferior Grade V
Laoyu River Wetland Park	4.375	2.155	**26.1**	<0.01	< 0.05	<0.0002	<0.002	2.1	Inferior Grade V

By calculating the weight of each index, the weight value of Cd was the largest, indicating that this index has the greatest impact on the evaluation system. The second was Ni, and COD had the least impact(Table 4). Next, according to the standard, the pollution grade classification standard of the improved Nemerow pollution index method was obtained, and the Nemerow index was calculated. The calculation results can be seen from Table 5. The improved Nemerow pollution index of the water samples from Laoyu River Wetland Park was much higher than that of the other two wetland parks. Therefore, overall, the water pollution of Laoyu River Wetland Park was the most serious, and that of South Dianchi lake and Xihua Wetland Park was slightly better. The pollution assessment results of this method were consistent with single factor evaluation method, that is, The water quality of the three wetland parks was inferior to Grade V.

Table 4. Evaluation results of Nemerow pollution index method

Sampling points		$F'_{i,max}$	$\overline{F'}$	P'	P'_{mean}	Grade of pollution
South Dianchi Wetland Park	Sampling 1	4.615	1.485	3.428	4.202	Inferior Grade V
	Sampling 2	6.840	1.655	4.976		
Xihua Wetland Park	Sampling 1	6.465	0.646	4.594	4.119	Inferior Grade V
	Sampling 2	5.050	1.024	3.644		
Laoyu River Wetland Park	Sampling 1	11.401	1.556	8.136	9.318	Inferior Grade V
	Sampling 2	14.701	2.093	10.500		

Note: P'_{mean}: average of P'.

4. Discussion

The serious exceeding of individual index concentrations in the water quality evaluation system is the main reason for the poor overall water quality of the three wetland parks. We should focus on controlling the emissions of COD, TP and Cr, and attach great importance to the pollution treatment of urban non-point sources and agricultural non-point sources. According to the characteristics of aboveground and underground runoff in Dianchi Lake Basin, build a more scientific urban drainage system, reduce urban non-point sources, stabilize urban flood peaks, regulate the scouring of pollutants by urban rainstorm runoff, and control the entry of non-point source pollutants into the lake in the whole process[7-8]. For agricultural non-point sources, we should actively apply and promote agricultural non-point source pollution prevention and control technologies integrating drip irrigation and sprinkler irrigation technology, slow-release fertilizer technology, precision fertilization technology and solid waste treatment technology, so as to intercept and eliminate agricultural pollutants from the source, process and end point[9].

Compared with the previous data, the concentration of most indicators of the three wetland parks detected in this study decreased. In addition, Studies have found that the concentrations of TN, TP and NH_3-N at the outlet of Dianchi wetland park were significantly lower than those at the inlet[10], indicating that the wetland has a certain purification effect on the water quality of Dianchi Lake. According to the statistics[11], the daily tourist volume of South Dianchi, Xihua and Laoyu River Wetland Park is, 91 ~ 236, 67 ~ 173 and 357 ~ 1537 respectively. Among them, the daily tourist volume of Laoyu River Wetland Park is the largest and that of Xihua Wetland Park is the least. Generally, the number of tourists is inversely correlated with the degree of pollution, which is consistent with the evaluation results of our study. Therefore, we should strengthen the management of

Wetland Park tourism, strictly maintain the daily number of tourists above the environmental carrying capacity, and implement the management of diversion of tourists. In addition, in the future research, we should try to avoid the problem that the detected water samples are affected by the lake shore environment, and carry out the dynamic monitoring of water quality in Dianchi Lake wetland parks for a long time series.

Acknowledgments

This work was financially supported by the National Natural Science Foundation of China (31760175).

References

1. Hu S, Niu Z, Chen Y, et al. Global wetlands: Potential distribution, wetland loss, and status. Science of The Total Environment (586): 319-327(2017).
2. Nan YH, Zhao QL, Tan XM. On the research gap of Wetland Park in Dianchi Lake Basin. INTELLIGENT CITY (3): 205-225(2021).
3. He J, Xu XM, Yang Y, et al. Problems and effects of comprehensive management of water environment in Lake Dianchi. J. Lake Sci. 27(2): 195-199(2015).
4. Patton C.J, Kryskalla J.R. Methods of Analysis by the U.S. Geological Survey National Water Quality Laboratory — Evaluation of alkaline persulfate digestion as an alternative to Kjeldahl digestion for determination of total and dissolved nitrogen and phosphorus in water, U.S.
5. Geological Survey Water Resources Investigations Report 03-4174, Denver, CO (2003).
6. Li Z, Yu G, Song J, Wang Q, Liu M, Yang Y. Study on the determination of heavy metals in
7. water samples with ultrasound-assisted dispersive liquid-liquid microextraction prior to FAAS. Water Science and Technology 67(2): 247-253 (2013).
8. Shen HX, Niu Beibei, Li FQ, et al. Water quality assessment and pollution source analysis of
9. The Dawen river. Safety and environmental engineering 28(5): 176-185(2021).
10. NI YF. Characteristics and Control Progress for Controlling Urban Non-point Source Pollution. Environmental science and management 33(2):53-57(2008).
11. ZHAO JW, SHAN BQ, YIN CQ. Application and progress in engineering technologies for controlling urban non point source pollution. China water & waste wate 23(12): 1-5(2007).
12. GUO HC, HE B, SONG LR, et al. Study on water pollution control and eutrophication control technology in Dian Lake Basin. Beijing: China Environment Publishing House (2017).
13. ZHANG JS, LI L, MA F, et al. Water Quality Monitoring and Evaluation of Dianchi Lake Wetland Park W in Kunming from 2016 to 2020. Yunnan Chemical Technology 48(7): 82-86(2021).
14. FU JM, ZHOU J, GONG XH, et al. Estimation of the Number of Tourists in the Wetland Parks in the Shore of the Dian Lake. Wetland science 19(4): 501-506(2021).

Identification of potential CCUS clusters and its pipeline network optimization in China

Pengtao Wang [1], Xi Wu [2], Gangke Ge [1], Xiaoyan Wang [1,*], Haifeng Wang [1], Yang Zhang [1], Wei Wang [1], Xi Yan [1], Yan Zheng [3], Yongbin Miao [4], Bing Wang [5]

[1] Oil Production Technology Institute, Dagang Oilfield Company, CNPC, Tianjin, 300280, China
[2] CNPC Dagang Oil Company, Tianjin, 300280, China
[3] No.5 Oil Production Plant, Dagang Oilfeild Company, PetroChina, Tianjin, 300280, China
[4] Tianjin Undergroud Gas Storage Company, Petrochina Dagang Oilfield Company, Tianjin, 300280, China
[5] No.1 Oil Production Plant, Dagang Oilfeild Company, PetroChina, Tianjin, 300280, China

Abstract. Climate change is a huge challenge facing the world. Carbon Capture, Utilization and Storage (CCUS) is an important technology that is expected to ensure the continued safe and large-scale use of fossil energy under the pressure of climate change in China. However, the current CCUS project has a large initial investment, and it is difficult to realize the commercialization of CCUS technology. The development of CCUS cluster centers can effectively reduce the overall investment of CCUS projects, which is the development trend of CCUS projects. This study identified 20 coal-fired power plant cluster centers and 6 oil and gas field gathering and transportation centers by utilizing Chinese coal-fired power plant data and oil and gas field data. Finally, the CO_2 pipe network between the cluster centers is planned. This study provides data and research methods for identifying potential CCUS clusters and storage and transportation centers in China. It can provide a theoretical reference for further research on the matching site selection of CO_2 transport pipelines and CO_2 sources and sinks in China.

Keywords: CCUS cluster, Pipeline network optimization, CO_2 transport

1. Introduction

Climate change is a huge challenge facing the world. The United Nations Intergovernmental Panel on Climate Change (IPCC) believes that CCUS technology plays a vital role in the scenario of achieving a 450ppm CO_2 equivalent concentration in 2100 (the target of controlling the temperature rise by 2 °C) [1]. The 450 ppm scenario suggests in IEA statement that implementation of CCUS would reduce required emission reductions by 12%[2]. As the world's largest CO_2 emitter, it actively implements the national strategy for mitigating climate change, and proposes the goal of "strive to achieve carbon peak by 2030 and carbon neutrality by 2060" in China. CCUS is an important technology that is expected to ensure the continued safe and large-scale use of fossil energy in China under the pressure of climate change. Large-scale commercial promotion of CCUS is faced with the development obstacles of high investment and operating costs, especially after the implementation of CCUS in power plants, the levelized power cost will increase by 29%-32%[3].
CCUS project cluster planning and development has many advantages. Amortize CCUS investment and operating costs by utilizing shared infrastructure. Studies have shown that in CO_2-EOR oilfield clusters, the construction of public CO_2 storage and transportation centers can significantly reduce the initial investment[4]. Lowering the entry barriers for participating in CCUS projects, providing opportunities to participate in CCUS for emission sources with small CO_2 emissions, as well as emission sources that do not have CO_2 storage or have limited storage capacity. At the same time, it can also provide sufficient and reliable CO_2 sources for CO_2-EOR and other CO_2 utilization projects[5].
In order to identify potential CCUS clusters and storage and transportation centers in China. The clusters and storage and transportation centers identified in this study can provide a theoretical reference for further research on the matching of CO_2 transportation pipelines and CO_2 sources and sinks in China.

2. Methods and Data

2.1 Methods

The research method of this paper is shown in Figure 1. The identification method of cluster center and gathering center is divided into three modules. The first part is the CO_2 cluster center identification module. It is determined

* Corresponding author: wxiaoy05@petrochina.com.cn

https://doi.org/10.1051/e3sconf/202235802015

that the power plants that meet the implementation of CCUS are screened for suitability of power plant data, imported into ArcGIS software, and the function of Kernel density estimator is used to further confirm the power plant cluster center. In the second part, the suitable oil and gas field screening criteria are used to determine the suitable oil and gas fields for CO2 storage, and the Kernel density estimator function is also used in ArcGIS software to realize the storage and transportation center of the storage site. The third part is the identification module of the shortest CO2 transmission pipeline, which links the power plant with the oil and gas field cluster center, and determines the main pipeline network of CO_2 transportation.

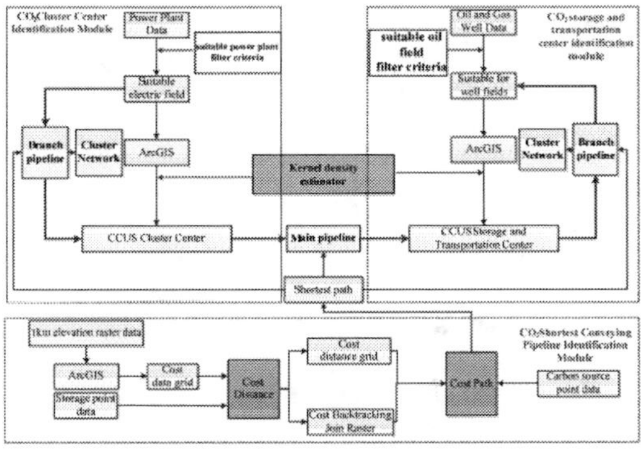

Fig. 1 Power plant and oil and gas field cluster identification method

2.2 Data

Screen the power plants suitable for deployment of CCUS and calculate the CO2 emissions of the power plants. From the surveyed power plant data, obtain the name, address, installed capacity, annual power generation, construction time and steam turbine type of each power plant, and screen out the power plants suitable for CCUS implementation according to the CCUS retrofit suitability specification for coal-fired power plants[2]. The power plant was built after 1995, with an installed capacity of more than 300 MW and within 800 kilometers of the storage basin. After the screening was completed, there were 591 power plants meeting the above rules, with a total installed capacity of about 664 GW.

The reservoir database comes from China Petrochemical Corporation[6]. The reservoir database contains 4386 reservoirs. The database has detailed statistics on oil reservoirs. There is the location of each oil field, including the name, location, and topography of the oil reservoir, as well as the basin and affiliated oil company of each oil reservoir. In terms of recoverable reserves, there are details of the time when the oil field was discovered, the proven recoverable reserves of the reservoir, and the amount of oil that has been extracted.

3. Results

3.1 Identification results of power plant cluster centers

From the power plant cluster results, a total of 20 large coal-fired power plant cluster centers were identified, and these 20 collection centers covered a total of 580 large coal-fired power plants, with a total annual CO_2 emission of about 2.2 billion tons. The total emissions of the 20 coal-fired power plant cluster centers account for about 25% of the country's total emissions from coal-fired power plants. There are 3 cluster centers whose CO_2 capacity exceeds 2% of the total emissions of coal-fired power plants, namely Nanjing, Jiyuan and Shenyang. There are a total of 9 cluster centers with CO_2 capacity exceeding 1% of total emissions, as shown in Figure 2.

According to the analysis of power grid area, the East China Power Grid Regional Collection Center is the largest, which is the Nanjing Gathering and Transportation Center, covering 42 power plants with an annual emission of about 200 million tons. In the Northeast Power Grid region, the Shenyang collection center covers 65 power plants with an annual emission of about 180 million tons of CO_2, making it the largest collection center in the Northeast region. In the North China power grid area, the number of collection centers is at most 7, involving a total of 138 power plants and an annual CO_2 capacity of about 570 million tons. The Shijiazhuang collection center is the largest cluster center in North China. The Central China Power Grid Regional Collection Center has a total of 3 large-scale collection centers. Among them, Jiyuan Collection Center is the largest cluster center in Central China, involving 49 coal-fired power plants, with an annual CO_2 storage capacity of 190 million tons. There are 3 large-scale coal-fired power plant cluster centers in the coverage area of Northwest Power Grid, namely Shuozhou, Changji and Wuzhong cluster centers, involving a total of 82 power plants. The largest coal-fired power plant cluster center is Wuzhong cluster center, involving 41 power plants. The CO_2 capacity can reach 140 million tons.

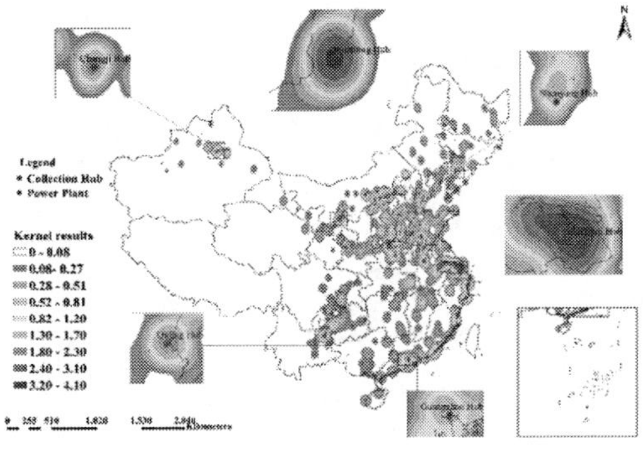

Fig. 2 Power plant cluster center identification results

3.2 Identification results of cluster centers of electric oil and gas fields

From the analysis of oil and gas field clusters, a total of 6 large oil and gas gathering and transportation centers have been identified, as shown in Figure 3. They are Daqing Oil and Gas Gathering and Transportation Center in Songliao Basin, Karamay Oil and Gas Gathering and Transportation Center in Junggar Basin, and Shengli Oil and Gas Gathering and Transportation Center in Bohai Bay Basin. Oil and gas gathering and transportation center in Yanchang in Ordos Basin, Changqing oil and gas gathering and transportation center in Ordos Basin, and Jianghan oil and gas gathering and transportation center in Hanjiang Basin. The six oil and gas gathering and transportation centers cover a total of 81 large oil and gas fields, with a total CO_2 storage capacity of 2.3 billion tons, of which about 190 million tons of CO_2 can be stored through CO_2-GAS, and the rest need to be stored and transported through CO_2-EOR. The sum of the CO_2 storage capacity of the six potential storage and transportation centers is about a quarter of the country's total annual emissions. The potential CO_2 storage and transportation capacity of Shengli, Karamay and Daqing accounts for about 5% of the country's total annual emissions.

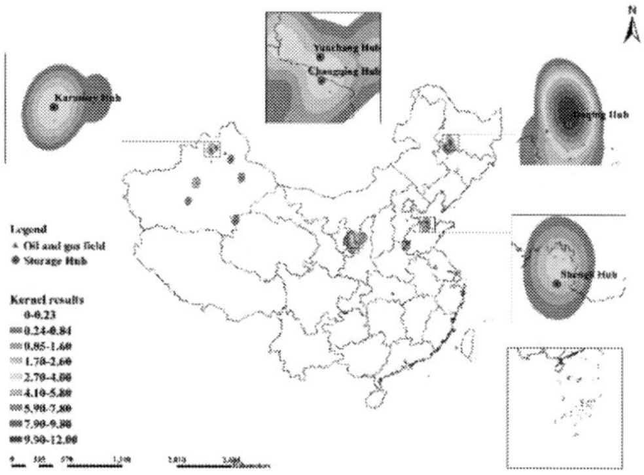

Fig. 3 Identification results of oil and gas field gathering and transportation centers

3.3 Cluster Center Matching

The matching results of the direct shortest distance pipeline network between the coal-fired power plant cluster center and the oil and gas gathering and transportation center are shown in Figure 4. There are a total of 6 matching pipeline paths. The transportation distance of the main network is basically distributed between 200-300km, and the longest between the Shenyang power plant cluster area and the Daqing storage and transportation area is 518km. The gathering and transportation center of Changji coal-fired power plant is matched with the pipeline network of Karamay oil and gas gathering and transportation center. The actual distance of the pipeline network is 213 kilometers, and the transportation volume is expected to be 390 million tons. The CO_2 pipeline network is 170 kilometers long from the gathering and transportation center of the Wuzhong coal-fired power plant to the Yanchang oil and gas gathering and transportation center. The transportation distance from the gathering and transportation center of the Wuzhong coal-fired power plant to the Changqing oil and gas gathering and transportation center is 153 kilometers. From the Shenyang coal-fired power plant cluster center to the Daqing oil and gas gathering and transportation center, the transportation distance is 518 kilometers, and the transportation volume is about 720 million tons of CO_2. The Tianjin coal-fired power plant cluster center is matched with the pipeline network of the oil and gas gathering and transportation center of Shengli Oilfield, with a transportation distance of 217 kilometers and a transportation volume of about 400 million tons of CO_2. From the Pingdingshan coal-fired power plant cluster center to the Jianghan oil and gas gathering and transportation center, the CO_2 transportation pipeline network is 371 kilometers long, and the transportation volume is about 50 million tons. The CO_2-EOR storage capacity of oil and gas fields is limited, and other CO_2 utilization projects need to be added, such as deep saline aquifer storage, CO_2-ECBM, CO_2-EGS, etc.

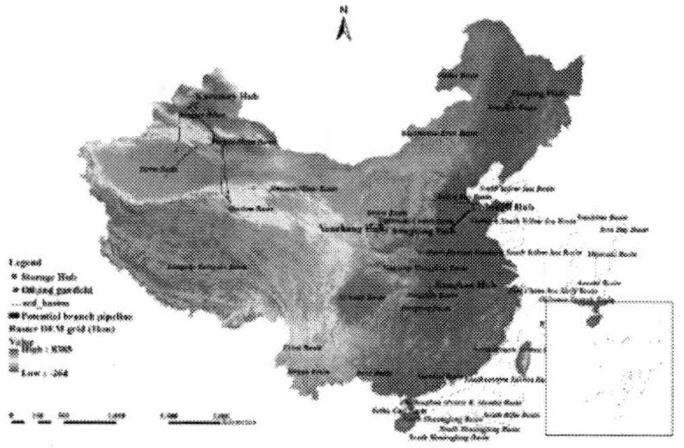

Fig. 4 Identification results of oil and gas field gathering and transportation centers

4. Conclusions

From the power plant cluster center, the total CO_2 emissions in the power plant clusters in Nanjing, Jiyuan and Shenyang all account for more than 2% of the national annual CO_2 emissions. The CCUS clusters with the highest average power plant emissions are distributed in the Ningbo cluster area, with an average annual CO_2 emission of about 8Mt per power plant. From the perspective of oil and gas field storage and transportation centers, the amount of CO_2-EOR storage used in oil and gas field storage and transportation areas is only a quarter of the country's annual CO_2 emissions. In the storage and transportation areas of the Daqing, Karamay and Shengli CCUS oil and gas fields, the combined CO_2 storage potential accounts for 70% of the total CO_2-EOR storage capacity. From the perspective of the proportion of public pipelines, the power plant cluster areas in Tianjin, Jiyuan and Shantou share more than 30% of the pipelines. In the storage and transportation areas of Yanchang and Changqing oil and gas fields, the pipeline sharing ratio is about 25%. From the perspective of the power plant cluster center and the oil and gas field storage and transportation center, Wuzhong power plant cluster center and Yanchang, Changqing oil field storage and transportation center, Shenyang power plant cluster center and Daqing oil field storage and transportation center, Tianjin power plant cluster center and Shengli oil field storage and transportation center, The development of CCUS cluster is more suitable. This is also the area where the CCUS demonstration project currently carried out is located in China.

Based on the above research, this paper puts forward corresponding policy recommendations. On the basis of the existing CCUS demonstration projects, especially the CCUS demonstration projects distributed in Yanchang, Changqing and Shengli Oilfields, the CCUS storage and transportation center is planned to prepare for the future CCUS cluster development and CO_2 transportation network construction. Utilize the advantages of its shared infrastructure, as well as its flexible and reliable operation, to verify the feasibility of reducing the construction and operation costs of CCUS projects. Explore new business models suitable for the development of CCUS clusters in China.

References

1. IPCC. Climate Change 2014: Synthesis Report. Contribution of Working Groups I, II and III to the Fifth Assessment Report of the Intergovernmental Panel on Climate Change [Core Writing Team, R.K. Pachauri and L.A. Meyer (eds.)]. IPCC, Geneva, Switzerland2014.
2. IEA. The potential for equipping China's existing coal fleet with carbon capture and storage. Paris, France: IEA; 2016. p. 34.
3. Viebahn P, Vallentin D, Holler S. Prospects of carbon capture and storage (CCS) in China's power sector - An integrated assessment. Applied Energy. 2015;157:229-44.
4. Kemp AG, Kasim S. The economics of CO2-EOR cluster developments in the UK Central North Sea. Energy Policy. 2013;62:1344-55.
5. IEAGHG. CCUS Hubs and Clusters. 2016.
6. Markit H. Energy & Natural Resources. 2019.

Elastic impedance inversion method based on Quantum Annealing MH algorithm

Chen Zhao, Fengming Jin, Guomeng Han, Shuwen Guo, Pengtao Wang, Delu Wang, Wei Guan

Dagang Oilfield Company, PetroChina, Tianjin, China.

Abstract. Seismic inversion is one of the important methods to realize reservoir prediction in oil and gas field exploration. The seismic stochastic inversion based on Metropolis-Hastings algorithm makes full use of well logging data and improves the vertical resolution of inversion results. However, conventional Metropolis-Hastings algorithm has slow convergence speed and poor stability. In this paper, an elastic impedance inversion method based on optimized Metropolis-Hastings algorithm is established by combining quantum annealing algorithm with Metropolis-Hastings algorithm. The feasibility of the method is verified by model calculation and actual data test.

1. Introduction

Exploration of oil and gas is extremely important to secure energy supplies. As the exploration of oil and gas become more and more extensive, exploration target has been transferred from structural reservoirs to complicated reservoirs. Seismic inversion is an important method to provide efficient and reliable support for the identification of complicated oil and gas reservoirs.

How to describe the complicated reservoir based on seismic inversion has become current research hotspots.

For conventional linear inversion method, it is often difficult to meet the identification requirements of the complicated reservoir because of seismic is band-limited. Then, we study the stochastic inversion which uses seismic data as a soft constraint to improve the vertical resolution of the inversion results. The Metropolis-Hastings (MH) algorithm is widely used in stochastic inversion. However, the MH algorithm has some limits. Due to the limited computing time Settings or inappropriate search strategy, MH algorithm is easy to fall into local extremum for a complex parameter space.

In 2006, Wei Chao realized the inversion of log data based on Quantum annealing. In 2015, Badr Alulaiw and Mrinal K. Sen applied QA algorithm in a seismic pre-stack inversion to Cana field. The idea of Quantum annealing applied to optimization is come from simulated annealing. Get inspiration from it, we also can improve the MH method based on the theory of Quantum annealing.

Connolly (1999) first proposed the concept of elastic impedance. Whitcombe (2002) applied a new normalized form of elastic impedance to improve the stability of parameters extraction. Elastic impedance inversion based on angle partial stack gathers has higher efficiency, higher stability, higher noise immunity and lower dependence on the quality of seismic data than AVO inversion.

In this paper, we propose a new fast stochastic inversion method based on Quantum MH method to deal with the stochastic elastic inversion problems.

2. Quantum MH method

Quantum Annealing (QA) is a global optimization method that is inspired from the mechanics of the quantum fluctuations known as tunneling effect. Compared to Simulated Annealing, Quantum Annealing algorithm converges faster than SA and escapes from being trapped in a local minimum. The construction of Hamiltonian of the system is crucial.

To apply Quantum Annealing to seismic inversion, the kinetic energy of the system H_0 is analogue to the error function ΔE, thus the Hamiltonian H can be rewritten as:

$$\mathrm{H} = \Delta \mathrm{E} + \mathrm{C}\tau \quad (1)$$

Where C is a constant, τ is the transverse field acting on the system.

Then the probability of the quantum jump can be written as:

$$\rho(\mathrm{H}) = (\mathrm{e}^{-H/PT})^{P} \quad (2)$$

Where P is the number of model parameter.

Get inspiration from the Quantum Annealing algorithm, the acceptation probability ω of the Quantum Metropolis-Hastings method can be expressed as:

$$\omega = \begin{cases} 1, & if\ \Delta E \leq 0 \\ \exp[-(\Delta E + \varphi)], & if\ \Delta E > 0 \end{cases} \quad (3)$$

Where φ is a variable which is gradually decreased.

We realize the elastic inversion based on the Quantum Metropolis-Hastings method to test the feasibility of the inversion method. The flow chart of the Quantum Metropolis-Hastings method is shown in Figure

 https://doi.org/10.1051/e3sconf/202235802016

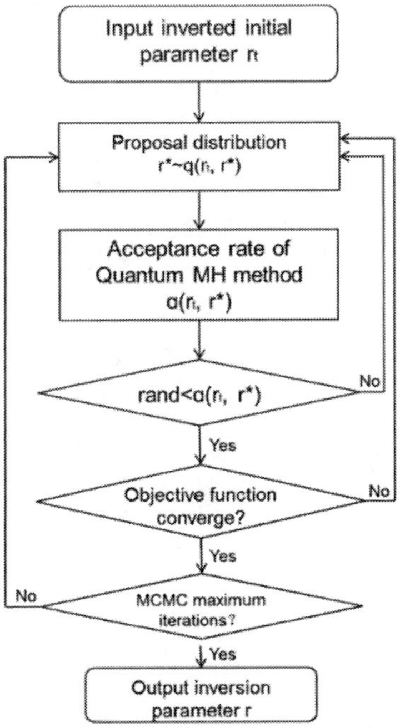

Figure 1 The Process of Quantum MH method

3. Elastic impedance equation

Connolly (1999) first proposed the concept of elastic impedance based on Zoeppritz equation. The elastic impedance can be expressed as:

$$\mathrm{EI}(\theta)=\alpha^{(1+tan^2\theta)}\beta^{-8Ksin^2\theta}\rho^{(1-4Ksin^2\theta)} \quad (4)$$

The changes of the incident angle will affect scale characteristics in the elastic impedance equation. In order to eliminate the influence, Whitcombe (2002) applied a new normalized form of elastic impedance to improve the stability of parameters extraction. Then the elastic impedance can be written as:

$$\mathrm{EI}(\theta)=\alpha_0\rho_0\left[\left(\frac{\alpha}{\alpha_0}\right)^{(1+tan^2\theta)}\left(\frac{\beta}{\beta_0}\right)^{(-8Ksin^2\theta)}\left(\frac{\rho}{\rho_0}\right)^{(1-4Ksin^2\theta)}\right] \quad (5)$$

Where α is P-wave velocity, β is S-wave velocity, ρ is density. θ is incident angle, and $K=\beta^2/\alpha^2$. α_0, β_0, ρ_0 are the average value of P-wave velocity, S-wave velocity, density respectively.

4. The posterior probability density of inverted parameter

Based on Bayes' theorem, the posterior probability density of inverted parameter is expressed as:

$$\mathrm{p}(\mathrm{r}|d)\propto \mathrm{p}(\mathrm{r})\cdot \mathrm{p}(\mathrm{d}|r) \quad (6)$$

Where d is seismic observation data, r is reflection coefficient, $\mathrm{p}(\mathrm{r}|d)$ is posterior probability. $\mathrm{p}(\mathrm{d}|r)$ is likelihood function.

Assuming seismic background noise obeys independent Gaussian distribution with zero-mean and σ_n^2-variance, it can be expressed as:

$$\mathrm{p}(\mathrm{d}|r)=\frac{1}{2\pi\sigma_n^2}\exp\left[-\sum\frac{(d-g(r))^2}{2\sigma_n^2}\right] \quad (7)$$

Where g is the forward operator. n is the sample number of seismic data.

Assuming that the priori information of reflection coefficient obeys Cauchy distribution to stand out weak reflection of the underground media, the priori information is expressed as:

$$\mathrm{p}(\mathrm{r})=\frac{1}{(\pi\sigma_m)^M}\prod_{i=1}^{M}\left[\frac{1}{1+r_i^2/\sigma_m^2}\right] \quad (8)$$

So objective function can be expressed as:

$$\mathrm{J}=\frac{1}{2\sigma_n^2}\sum(d-g(r))^2+\frac{1}{\sigma_m^2}\sum_{i=1}^{M}\ln\left(1+r_i^2/\sigma_m^2\right)+\varepsilon\sum(I-I_0)^2 \quad (9)$$

Where $\varepsilon\sum(I-I_0)^2$ represents the smooth constraint of elastic impedance, I_0 is smooth elastic impedance constraint information constructed according to the log data, ε is elastic impedance smoothing factor

To obtain the posterior probability distribution, we apply Quantum MH method to obtain the inversion results.

5. Case study with 1D data

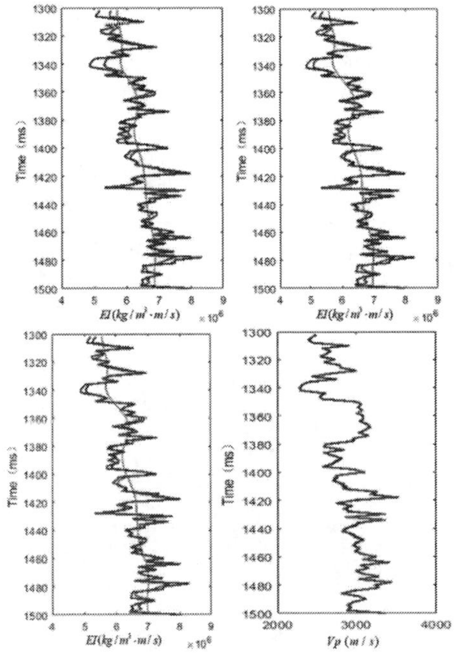

(a) (b) (c) (d)

Figure 2 Elastic impedance inversion results of small angle (a), the middle angle (b), large angle (c) and the extracted P-velocity result (d) by using quantum MH method when SNR=3.

To test the feasibility of the Quantum Metropolis-Hastings method, we carry out the feasibility test of one well. The elastic inversion test is to prove the feasibility and the superiority of the Quantum MH method.

From Figure 2 and Figure 3, We can see that the inversion results match the model data well and convergence and stability are greatly improved for quantum MH algorithm compared with MH algorithm. The factor φ in quantum

 https://doi.org/10.1051/e3sconf/202235802016

MH algorithm can reduce the unnecessary transfer of the current state, so that the Markov chain can converge to the posterior probability density more rapidly and steadily.

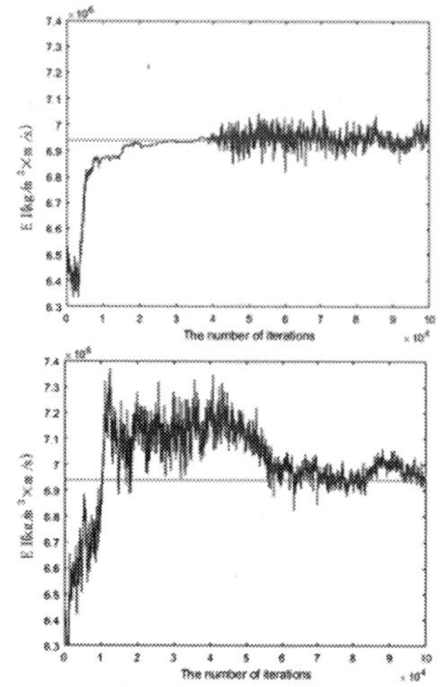

(a (b)

Figure 3 The change of elastic impedance value (the middle angle) along with the change of the number of iterations for an sampling point in quantum MH algorithm (a) and MH algorithm (b).

6. Case study with 2D real data

To test the feasibility of the elastic impendence inversion based on Quantum MH method, we apply this method to 2D real seismic data. The inversion results are shown in Figure 4. The red region represent oil and gas reservoir.

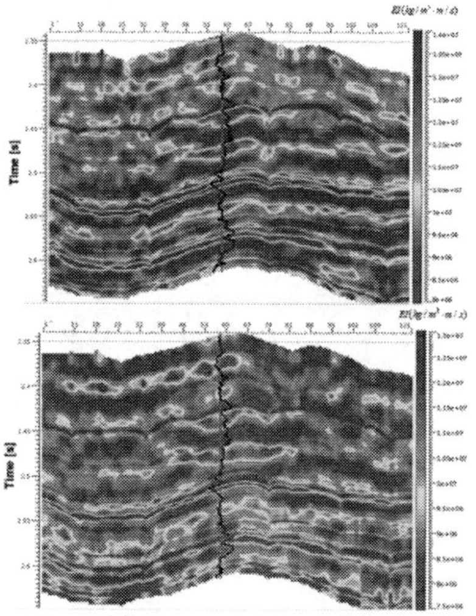

(a) (b)

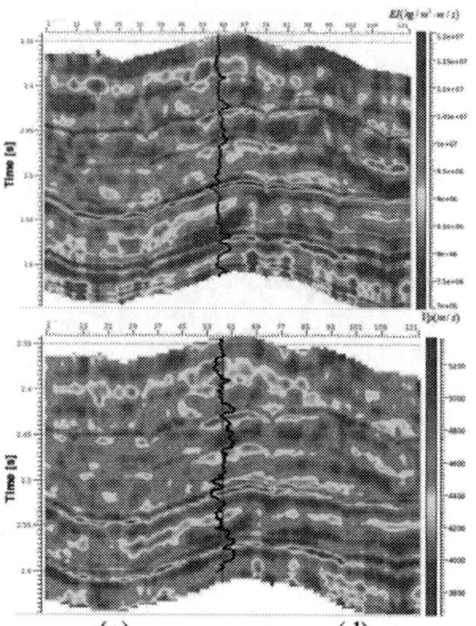

(c) (d)

Figure 4 The elastic inversion result of small angle (a), medium angle (b), large angle (c) and the extracted P-velocity profile (d).

We can see from Figure 4 that the results of the elastic inversion match the logging data well. The elastic inversion method based on quantum MH method is feasible in practical applications.

7. Conclusions

Quantum Metropolis-Hastings method is a feasible inversion method that is inspired from Quantum MH method and the objective function can be quickly convergent. The 1D data test and 2D real data test show that quantum MH algorithm can effectively obtain the elastic inversion results.

Acknowledgments

This work is supported by the Post-doctoral Fund project "The seismic prediction technology and its application for sweet spot in tight sandstone reservoir".

References

1. Francis, A. Limitations of deterministic and advantages of stochastic seismic inversion. CSEG Recorder, 2005, 30: 5-11.
2. WEI C, ZHU P M, WANG J Y. Quantum annealing inversion and its implementation. Chinese Journal of Geophysics, 2006, 49(2): 577-583
3. Song Zhu, Genhai Mao, Guohua Liu; Yuefei Huang. The improvement of MCMC method and its application. Journal of Hydraulic Engineering, 2009, 40(8):1019-1023.
4. Xinpeng Pan, Guangzhi Zhang, and Changlu Sun. Nonlinear elastic impedance inversion for effective pore-fluid bulk modulus. 85th Annual International

https://doi.org/10.1051/e3sconf/202235802016

Meeting, SEG, Expanded Abstracts, 2015, 3446-3450.

5. Xiaodan Wang, Xingyao Yin, Baoli Wang, and Fusen Yang. Pre-stack seismic stochastic inversion with VFSA-based PSO algorithm. 85th Annual International Meeting, SEG, Expanded Abstracts,2015, 3441-3445.

 https://doi.org/10.1051/e3sconf/202235802017

Application of UAV technology in environmental impact assessment of mine reclamation project

Jiaxin He *, Zhuo Zhang, Xinyu Xie

land science and technology, China University of Geosciences (Beijing), 100083 Beijing China

Abstract. It is of great significance to carry out environmental impact assessment on the renovated mining area for the assessment of reclamation benefits. The maturing of UAV technology brings new possibilities for environmental assessment. Aiming at the problem of insufficient environmental impact assessment system of mining area renovation project, this paper constructs the environmental impact assessment system of mining area renovation project, and puts forward a set of scheme to carry out the environmental impact assessment of mining area after reclamation based on the data obtained by UAV platform. The research results have certain value for the benefit assessment of comprehensive renovation and ecological reclamation of mining area

1. Background

The past 10 years have been a period of rapid development of UAV aerial photogrammetry and remote sensing technology. In the 1960s, UAV technology has started. In 2004, there were only three articles on the application direction of UAV remote sensing at the Istanbul International photogrammetry and remote sensing conference; At the 2012 Melbourne International Conference on Photogrammetry and remote sensing, there were more than 50 articles on the application of UAV aerial survey and remote sensing, and 9 sub venues had relevant reports [1]. According to the Research Report on the market prospect and investment opportunities of China's civil UAV industry released by the China Business Research Institute, by 2020, China's UAV output value has exceeded 36billion yuan, with an average annual compound growth rate of 66.88%. It is expected that by the end of 2020, the scale of the domestic UAV industry will reach 45.3 billion yuan, which shows the wide application of UAV technology in China. There are also numerous scientific researches on civil UAVs, involving agriculture, water, industry, cities and other aspects.

2. Environmental assessment index system of mining area reclamation

2.1 Construction principle of index system

2.1.1 Scientificity and relevance

The evaluation index system shall be based on the terrain and geomorphic data obtained by UAV and other local natural environment characteristics to reflect its scientific rationality.The selection of indicators should start from the purpose and impact of ecological restoration, and the construction of indicator system should focus on the impact on the environment according to the characteristics of mining area reclamation.

2.1.2 Systematic

As a relatively complete system, the index system of the ecological restoration project for mining area renovation should not only reflect the impact of the system on the environment, but also reflect the relationship between the evaluation factors within the system. Secondly, the index system should have a clear structure, clear levels, and appropriate details, with clear levels vertically and clear categories horizontally.

2.1.3 Principle of dominance

The general environmental impact assessment indicators of engineering projects include atmosphere, water, sound, ecological environment, social environment and other aspects. The environmental impact assessment indicator system of mining area reclamation projects should start from the identified environmental impact factors, screen out the dominant indicator factors, and eliminate the cross overlap and indicator factors caused by mutual influence.

2.1.4 Stability

The selection of indicators should not be greatly affected by accidental factors, but should select stable and referential indicator factors to provide reference for the environmental impact assessment of similar mining area renovation projects.

* Corresponding author: 26404425@qq.com

2.2 Determination of index system

The basic requirements for the construction of the environmental impact assessment index system of the project are to identify the resource, ecological and environmental impacts generated by the project, preliminarily judge the nature, scope and extent of the impacts, determine the assessment focus, clarify the environmental objectives, and establish a scientific and reasonable assessment index system. Based on the analysis of the environment and characteristics of the mining area, learn from relevant theories and practical experience, follow the principle of stability and dominance, and the index system shown in Table 1 is reasonably constructed.

Table 1. Environmental impact assessment index of mining area reclamation project

Target layer	Criterion layer	Index layer	Evaluation basis*
Natural resources and environment	terrain	Slope stability, height, gradient, aspect, surface roughness, proportion of dangerous rock mass and Deluvial deposit	comparison with that before construction
	atmosphere	Smoke and dust emission, inhalable particles (PM10), SO_2, NO_2	ambient air quality standard
	hydrology	Types of pollutants, number of pollution factors, sewage discharge, receiving water	environmental quality standard for surface water
Ecological environment	land use and cover	Land utilization rate and vegetation coverage	comparison with that before construction
	soil	Soil texture, structure, soil configuration and soil environmental sensitivity	soil environmental quality standard
	biodiversity	Wildlife richness and biodiversity index	regional biodiversity evaluation criteria
	water and soil conservation capacity	Soil loss amount, area index of water and soil loss (land desertification and land salinization)	monitoring and evaluation standards for water and soil conservation of production and construction projects
Social environment	Employment	New jobs and labor income	Comparison with that before construction
	Transportation and social economy	Number of new roads, road traffic operation index and road traffic congestion rate	Comparison with that before construction

*Only typical national standards or industrial standards shall be retained as the evaluation basis, and other relevant national standards and industrial standards shall be combined in practice.

The measurement units of each group of indicator data in the evaluation indicator system are different, so it is impossible to conduct unified analysis and calculation. Therefore, it is necessary to carry out dimensionless processing for each indicator, that is, standardization of indicators, to solve the problem of homogenization of different qualitative indicator values.First, the evaluation indexes are classified and assigned according to a certain range of changes. The range of assignment is 0~1. Then, the values of each index in the index layer are calculated respectively according to the change values after the project construction. The expert scoring method is used to give them different weights, and then the values of each index and the weights are multiplied and added to obtain a comprehensive score.

3. UAV technology and environmental impact assessment

At present, the environmental protection departments in all provinces, cities and regions in China have established stations to serve the environmental impact assessment. However, due to the comprehensiveness and macro nature of the environmental impact assessment, the simple ground investigation can not meet its needs [2].Therefore, the UAV technology has great application potential in the environmental impact assessment. UAV is a new platform that can carry different types of cameras and sensors, such as ordinary digital camera (visible light), lidar scanner, multispectral imager, hyperspectral imager, thermal imager, etc. [3]. Based on the environmental impact assessment indicators of the mining area reclamation project constructed above, this part uses UAV to obtain targeted data. Table 2 describes the UAV acquisition means corresponding to common environmental assessment factors.

 https://doi.org/10.1051/e3sconf/202235802017

Table 2. Applications of different sensor UAVs in environmental impact assessment

Sensor type	Output data	Application in environmental assessment of mining area renovation reclamation
Visible light digital camera	orthophoto, digital surface model	provide terrain, vegetation, biodiversity and land use information
Multispectral sensor	multi band spectral information	provide vegetation index to calculate the biomass of reclaimed vegetation, etc
Hyperspectral sensor	full band spectral information	water quality monitoring around the mining area, plant diseases and insect pests of reclaimed vegetation, growth monitoring of reclaimed crops and chemical gas monitoring in the mining area
Thermal infrared camera	temperature map	wildlife ecological investigation, heat emission monitoring and water temperature monitoring in the mining area
Lidar	point cloud, digital surface model, digital ground model	extraction of slope surface roughness, estimation of biomass in the mining area and calculation of vegetation coverage in the mining area

Visible light digital camera is widely used in surface data acquisition at home and abroad, because it can directly produce orthophoto images and digital surface models, and build three-dimensional models, which is intuitive and convenient. Yutianyu et al. Used UAV visible light and shadow image to identify dead and standing trees, which is of great value to vegetation monitoring, biodiversity protection and forest carbon storage change assessment[4]. In the environmental impact assessment of the Dongzhuang water control project on the Jinghe River in Shaanxi Province, the UAV equipped with visible light digital camera technology was used to carry out the tracking monitoring of the protection areas (sensitive points) involved in the project, and the ecological environment protection measures were timely adjusted by comparing the vegetation types of the protection areas (sensitive points) and their area changes, water area changes, etc. [5]. In the mining area, the UAV equipped with visible light digital camera is used in landform identification and slope analysis. Similarly, the terrain, vegetation, biodiversity and land use information provided by the UAV are the basic data for environmental impact assessment in the mining area.

UAV equipped with multispectral sensors and hyperspectral sensors has a wide range of significance in agricultural production monitoring, and is also used as one of the important data acquisition means in ecological and environmental science research. Zangchuankai et al. Used UAV hyperspectral remote sensing to detect the water environment of rivers and lakes, and extracted water color parameters and water quality parameters to provide reference for the efficient development of water environment assessment[6]. In terms of vegetation survey, multispectral sensors can provide rich vegetation index, which is convenient for judging vegetation growth and nutrition[7]. Therefore, when carrying out the benefit evaluation of mine regulation, the UAV equipped with multispectral sensors and hyperspectral sensors has important application prospects in vegetation investigation and water quality monitoring.

The temperature map generated by the thermal infrared camera is mainly used to investigate the distribution and quantity of wild animals after ecological restoration in the mining area, and to monitor the water temperature and heat emission.In terms of water pollution monitoring, the thermal infrared camera carried by UAV is also widely used, which can directly provide information on the type, distribution and diffusion of pollution sources[8].

Airborne lidar technology is often used to accurately identify the height of vegetation. In the land reclamation of the mining area, chenqiuji and others used the airborne lidar measurement system to quickly and accurately obtain the vegetation parameter information of the mining area[9]. In the environmental impact assessment of mining area renovation projects, lidar can be used to extract the surface roughness of slope, estimate the biomass of mining area, estimate the vegetation coverage of mining area, etc.

4. Results and discussion

In this paper, the environmental impact assessment system is constructed for the mine renovation and ecological restoration projects, and a scheme for carrying out the environmental impact assessment of the mine after reclamation based on the data obtained from the UAV platform is proposed. The results show that UAV has great application potential in environmental impact assessment of mining area reclamation projects. The innovation of technology promotes the progress of science. The continuous development and maturity of UAV technology provide new possibilities for the related work of environmental impact assessment. The combination of traditional ground stations and UAV technology improves the efficiency and accuracy of environmental assessment. How to apply UAV technology to practical projects is worth further exploration and research.

Acknowledgements： This work was supported by Innovation and Entrepreneurship Training Program for College Students of China University of Geosciences (Beijing)202111415080.

References

1. I. Colomina, P. Molina, ISPRS, 92, 79 (2014)
2. Y. Chen, P. Guo, Huanjin Fazhan, 32, 97 (2020)
3. Z. Zhang, Q. Xu, B. Wang, Shengtai Xuebao, 37, 4029 (2017)
4. T. Yu, W. Ni, J. Liu, Z. Zhang, Yaogan Xuebao, 25, 725 (2021)
5. B. Wu, L. Wang, Y. Wang, M. Zhu, Huanjin Yinxiang Pinjia, 43, 51 (2021)
6. C. Zang, F. Shen, Z. Yang, Ziran Ziyuan Yaogan, 33, 45 (2021)

https://doi.org/10.1051/e3sconf/202235802017

7. G. Sun, W. Huang, P. Chen, X. Wang, Nongye Jixie Xuebao, 49, 1 (2018)
8. Y. Hong, H. Yang, Y. Li, H. Zhu, S. Zhao, H. Guo, Zhongguo Huanjing Jiance, 0, 163 (2015)
9. Q. Chen, M. Hang, J. Li, X. Wang, Meitan Kexue Jishu, 48, 113(2020)

 https://doi.org/10.1051/e3sconf/202235802018

Field test practice and understanding of effective utilization technology for dense and refractory reservoir in block A

Zengpeng Xiu, Linhai Qi, Li Liu

1Hulunbeier subsidiary of Daqing oilfield Co., Ltd., Inner Mongolia, Hulunbeer, 021000, China

Abstract. In view of the ultra-low permeability reservoirs in Block A of Hailar Oilfield, it is difficult to establish an effective driving system between oil and water wells due to the influence of many oil-bearing strata, poor reservoir physical properties and large vertical span of oil layers. With the deepening of development, it is more difficult to use reservoirs. In order to realize the effective utilization of low permeability reservoirs in Beier Oilfield, the successful experiences of tight oil and Hailar large-scale fracturing in peripheral oilfields are summarized, and the effective utilization technology of large-scale fracturing supplementary energy development is explored. The large-scale fracturing evaluation of ultra-low permeability reservoirs and the optimization of potential well areas are carried out, so as to fully tap the

1. Introduction

Taking Block A as an example, the buried depth of the oil reservoir in the second member of the south of this area is about 2500 m, the porosity is 9 %, and the permeability is 0.61 MD, which are all low porosity and ultra-low permeability reservoirs. Low permeability reservoirs have large reserves and low oil recovery rate. In recent years, many energy supplement tests have been carried out, but no effective breakthrough has been achieved. In order to solve the problems of large unutilized reserves and low productivity of single well in tight sandstone reservoirs, large-scale fracturing well group tests are carried out in ultra-low permeability tight reservoirs in Hailar Oilfield from the perspective of effective utilization in developed areas, improving oil recovery and revitalizing inefficient assets, so as to implement the feasibility of ultra-low permeability reservoirs. Affected by factors such as strong reservoir heterogeneity, poor physical property, large vertical span of oil layer and water injection, the fracturing effect is quite different. Therefore, it is urgent to carry out reserve evaluation and optimization of ultra-low permeability reservoirs and study on subsequent energy supplement methods[1,2,3]. The following four main results were achieved:

2. Deepening the geological understanding of low permeability tight oil reservoirs

Carry out geological research, deepen the understanding of well selection sweet spot area and effective reservoir, optimize the fracturing scheme of low permeability tight reservoir, and provide theoretical basis for the next large-scale fracturing promotion.

2.1 Geological re-understanding and analysis of favorable and unfavorable conditions for reservoir formation

The favorable conditions for reservoir formation in Block A are as follows: near-source reservoir formation, low structural uplift area, structural-lithologic reservoir formed by the cutting of early sedimentary faults, and sand deposition between rivers. The intersection of multiple rivers leads to differences in adjacent well facies, sand thickness and physical properties. Unfavorable factors : sand body staggered contiguous distribution, vertical multi-stage superimposed, poor lateral connectivity, different layers of thickness difference[4,5].

2.2 Effective thickness review under the guidance of fine facies control characterization to identify reservoir development sweet spot area

The south second member of A well area is mainly controlled by the south and southeast provenance, and mainly develops underwater distributary channel, mouth bar and turbidite sand. Through fine description of sedimentary microfacies, the development scale and distribution law of developed oilfield fans are revealed.
The effective thickness of old wells was re-examined, and the interpretation coincidence rate was improved by correcting the chart. The effective thickness of some wells changed greatly, and the effective thickness distribution of the new interpretation was more in line with the understanding of sand bodies under the guidance of sedimentary microfacies. The dessert area was re-recognized in the development zone. The effective thickness of oil group II was relatively developed in the middle of the fault block, and the effective thickness of oil

groups III and IV was mainly concentrated in the northern and eastern regions of the fault block[6,7,8].

3. Standard and direction of well layer selection for large scale fracturing oil production characteristics

3.1 Optimizing the sweet spot area of tight reservoirs

The project focuses on the research on ultra-low permeability reservoirs in Bell Oilfield, combining the sedimentary characteristics of ultra-low permeability reservoirs, and in accordance with the existing knowledge of the reservoir and the results of geological and logging research, the introduction of facies-controlled geostatistics seismic attribute inversion, and the focus of tight oil development. Reservoir classification evaluation in large-scale fracturing test well area, realizing the evaluation and optimization of high-quality reserves of ultra-low permeability oil reservoirs in Block A.

Table 1. Reservoir classification evaluation in well block A

Comparative projects		reservoir classification: Class I	Class II
single layer thickness(m)		>5.0	5.0-3.0
sedimentary micro		subaqueous distributary channel	sand barrier
reservoir physical property	Porosity (%)	>15.0	15.0-10.0
	Permeability (mD)	>5.0	5.0-1.0
mercury intrusion parameters	Average throat radius(μ m)	>2.0	2.0-1.0
	displacement pressure (MPa)	<0.5	0.5-1.0
	median pressure (MPa)	<3.0	3.0-5.0
	median radius (μ m)	>0.3	0.3-0.2

3.2 Geological analysis of fracturing wells

Referring to the exploration ideas of tight oil and the technology of volume fracturing, secondary fracturing of old wells such as A39-61 in the oil layer A39-61 in the second member of the South around block A was carried out. Good oil-increasing effect, the combined pressure of the third section of Well A15, 6611 cubic meters of fluid and 380 cubic meters of sand were added. Compared with the conventional fracturing scale of this well, the fluid volume increased by 4.4 times and the sand volume increased by 1.7 times. , and achieved 6.6t output through fracture mesh fracturing. The distance between Well A15 and Well A3-69 in Block A is 1200m. The curve shows that the oil layer of Well A15 is buried deep, has poor physical properties, poor oil-bearing properties, and has few oil-bearing intervals. It is believed that Well A15 is located at the edge of the near-shore underwater sector. Analysis of A3- From the perspective of Well 69, the reservoir of the second member of the South is better than that of Well A15, so it is used as the basis for well selection and layer selection[9,10].

Table 2. Comparison of construction parameters between conventional fracturing and fracture network fracturing

Well number	conventional fracturing: liquid measure (m^3)	sediment discharge (m^3)	sand strength	quantity	production (t)	network fracturing: liquid measure (m^3)	sediment discharge (m^3)	sand strength	quantity	production (t)
A15	460	50	2.8	4	0.1	2246	120	6.8	7	6.6
	610	70	3.2	4		2196	140	6.4	7	
	520	60	3.5	4		2169	120	7.1	7	
A55-67	550	60	1.2	5	0.9	2300	60	1.2	7	4.6
	220	25	3.2	4		1960	32	4.1	7	
A39-61	750	80	3	5	1.3	2462	80	3.0	7.5	3.6
A4-1	358.5	30	0.5	2.5	0.6	2650	124.5	1.8	5	5.8
average value	452	49	2.4		0.6	1997	74	3.9		4.9

4. Formation of mature matching technology for tight oil fracturing

According to the characteristics of reservoir development and development status, the fracturing technology and fracturing scale are selected, and the key parameters are defined for the low permeability interbed of sand and mud. The sand production and sand plugging risk are reduced by optimizing the process to ensure the fracturing effect. In view of the problems of difficult sanding, sand stuck pipe string and long flowback period in 2019, several process improvements of tight oil fracturing geological design and engineering design in 2020:

4.1 Geological design optimization of tight oil fracturing

Reduce the number of single-stage fracturing sections and reduce the risk of stuck strings. The number of single-stage strings is controlled at 3-4 segments and the span of single-stage string tools is controlled within 150 m.

The application of hydraulic pump to speed up the flow rate, large-scale fracturing fluid volume, long time into the well, fracturing fluid fracture ability at the same time to consider reducing residual pollution, after the use of hydraulic pump drainage, the daily flow rate of pumping unit increased from 30 m^3 to 100 m^3, strive to flow rate within 30 days can reach 100 %.

4.2 Optimization of tight oil fracturing process design

In view of the difficulties in sanding, sand stuck string and long flowback cycle in 2019, the engineering department puts forward the following five suggestions for fracturing in 2020.

4.2.1 Optimized fracturing fluid for comparative test

Large-scale fracturing fluid consumption, long time into the well, fracturing fluid fracturing ability while considering residual pollution; it is suggested that 1 – 2 clean fracturing fluid tests should be continued in 2020, and guar gum fracturing fluid should be used in wells. The concentration of crosslinking agent and the adding procedure of gel breaker in each stage should be optimized.

4.2.2 Increasing construction displacement and reducing sand plugging risk

The construction displacement is proportional to the seam width. The maximum construction displacement in 2019 is 7 m^3/min, and the pump truck pressure limit is 60 MPa. In order to improve the seam width and reduce the risk of sand plugging, it is suggested that the construction capacity should be improved in 2020, the pump truck pressure limit should be increased to 65 MPa, and the displacement should be increased to 8 m^3/min.

4.2.3 Optimizing the structure of seat pressure string and reducing the risk of pipe sticking

In 2019, 24 layers of 5 wells were constructed. Due to the large scale of construction and long time of well down, the phenomenon of pipe string sticking after pressure is common. In order to reduce the influence of tool string factors on the pipe string, it is recommended to replace the pipe string once in 2020 when the stratification exceeds 3 segments or the tool crossing distance exceeds 100 m.

4.2.4 Speed up drainage at high water cut stage

The average oil breakthrough time after well pressure measures in 2019 is 40 days, and the average production is 237 days, with a cumulative return rate of 82 %. During the operation in 2020, after the fracturing string is lifted, the pump is not lifted. Three wells are optimized to use hydraulic pump to discharge liquid. The daily discharge capacity is about 100 m^3, and the return rate is expected to reach 100 % within 30 days.

4.3 Adopt a variety of monitoring means to ensure fracturing and post-fracturing effects

Two large-scale fracturing wells were selected to carry out tracer monitoring, and tracer was injected into the fracturing interval. Through continuous sampling and analysis of the samples, the water production profile of each fracturing interval and the water production contribution rate during continuous monitoring can be obtained. From the water production contribution rate of each layer curve, it shows that fracturing has transformed the reservoir into a relatively stable seepage environment, and a comprehensive interpretation and evaluation of the fracturing transformation status of the stratified fracturing construction wells can be carried out. Comprehensive evaluation of fracturing effect in Well A2-70 : the total fluid inflow is 4850 m^3, the general flowback is 3117.2 m^3, and the general flowback rate is 64.27 %. The return rate of each section from large to small is as follows: layer 2, 1, 5, 3, 4. See the following table for details. The recovery rate of tracer from high to low is as follows: layer 2, 1, 5, 3, 4.

Table 3. 60-day A2-70 Water Production and Drainage Statement

section / item	Layer 1	Layer 2	Layer 3	Layer 4	Layer 5
liquid production （m^3）	1101.0	1334.0	837.0	539.0	1039.0
Water yield （m^3）	720.5	1130.6	450.4	172.2	643.5
flowback rate （%）	65.4	84.8	53.8	31.9	61.9
Types of indicator	Er	La	Tb	Ce	Dy
Total addition of tracers （g）	4324.0	5240.0	3268.0	2076.0	4076.0
Tracer output （g）	2612.6	4099.6	1633.6	624.6	2333.8
Recovery rate of tracer （%）	60.4	78.2	50.0	30.1	57.3

The tracer test results show that the main water production layer of Well A2 - 71 is the second, first and fifth layers, and the contribution rate of water production during the 60 - day monitoring period is 80 % in total. The contribution rate of water production varies at different times, with small variation in layers 1, 2 and 5 and large variation in layers 3 and 4. The contribution rate increased to paragraphs 2 and 3 ; contribution rate decreased to layers 1,4 and 5 ; fracturing effect is good for layer 1, 2, 5, medium for layer 3, poor for layer 4.The contribution rate of water production was linearly and positively correlated with the total liquid inflow and sand addition. The correlation coefficient between water production contribution rate and total liquid inflow is 0.8552, and the correlation coefficient with sand addition is 0.9984. It shows that the amount of sand is the main factor of fluid volume in each fracturing layer.

5. The subsequent energy supplement of tight oil fracturing wells

There are 14 water injection wells in the test area. At present, there are 4 wells in the pressure-boosting water injection test, and the average daily water injection rate of single well is 7.8 m^3/d. The main block is elastic mining, and the production decreases rapidly. It is necessary to carry out energy enhancement test. One is to plan to carry out large-scale fracturing well CO2 huff and puff, and use the mechanism of CO2 expansion crude oil volume to supplement formation energy. Considering the energy supplement for large-scale fracturing wells, through the principle of carbon dioxide huff and puff replacement, considering the energy depletion of large-scale wells, considering the low liquid production, low daily oil production, insufficient fluid supply and well condition with good casing sealing, carbon dioxide huff and puff is carried out.

6. Conclusion

Through the geological re-understanding of structure, sedimentation, sand body and oil-water distribution law, the optimization of test well area is realized. Through the comprehensive classification and evaluation of reservoirs, the classification and evaluation standard of tight oil reservoir reserves is established, and the optimization of well location and horizon of large-span reservoir in Nantun Formation is completed.

Through the study of rock mechanical properties, block stress field and fracturing mode optimization, combined with the experience of large-scale fracturing in 2019, the optimization of fracturing scale and construction parameters are realized

Conventional fracturing cannot meet the effective production demand of tight reservoirs. According to the integration of 'sand body scale and fracturing scale', the breakthrough of effective production technology of tight reservoirs can be achieved by improving the sand body control degree and reserve production rate through the energy supplement of old wells in the to achieve the goal of improving sand body control degree and reserve utilization rate.

References

1. Yu Gaoming, et al. "Production Mechanism and Development Strategy of Sandstone Bottom Water Reservoir." Journal of Petroleum, 18.02 (1997): 99-107
2. Li Shujun, et al. "Reservoir Reservoir Description in Pingqiao Horizontal Well Area of Sai Oilfield." Journal of Xi'an Petroleum University (NATURAL SCIENCE EDITION) 15.04 (2000): 8-12
3. Wu Shenghe, et al. "Three- dimensional Modeling of Sedimentary Microfacies in Pingqiao Horizontal Well Area of Ansai Oilfield." acta sedimentologica sinica 21.02 (2003): 266-271.
4. Liu Xinju, et al. " Research on evaluation and influencing factors of horizontal well development effect in ultra-low permeability reservoir." Journal of Petroleum and Natural Gas (Journal of Jianghan University of Petroleum) 33.06 (2011): 318-321+325
5. Min Qi, et al. " Numerical simulation study on the overall opening and releasing mode of horizontal wells in Ansset Oilfield." Journal of low permeability oil & gas field 2.01 (1997): 45-49
6. Zhang Li, et al. " Microfracture characteristics of Alshan reservoir in Hanan and its influence on development." Journal of Oil and Gas Geology 22.02 (2001): 158-160+193
7. Ran Xinquan, et al. " Study on Horizontal Well Pattern of Ultra - low Permeability Sandstone Reservoir." Journal of Petroleum 29.01 (1998): 89-92
8. Ling Zongfa, et al. " Horizontal well injection-production well pattern optimization." Journal of petroleum exploration and development 34.01 (2007): 65-72
9. Ou Yangjing, et al. " Optimization design of horizontal section length of horizontal well." xinjiang's oil and gas 4.02 (2008): 43-45+48
10. Hu Yueting, et al. " Optimization Design Method for Horizontal Section Length of Horizontal Well." Journal of Petroleum, 21.04 (2000): 80-86

 https://doi.org/10.1051/e3sconf/202235802019

CO_2 Emission Accounting Model in Nonferrous Mining Industry

Liping Ju*, Yibin Zhu

BGRIMM Technology Group, 100160, Beijing, China

Abstract. The policy of carbon peak in 2030 address the question of how to account CO2 emissions of Nonferrous Mining. The aim of this study is to establish CO2 emissions accounting model, focused on usual activities, e.g. fuel combustion emission, process emission, emission from purchased and exported electricity and heat. And also to discuss explosive and urea emission sources to identify the significant process that contribute the most to climate change. Moreover, the accounting model is chosen to evaluate emissions of one mine in china, and explosive explosion and urea use of CO2 emissions accounting for 4.61% and 3.97%, respectively.

Keywords: CO2 Emission; Nonferrous Mining Industry

1. Introduction

Nonferrous mining industry, as a high energy consumption and high emission industrial sector, is one of the key industries to achieve the 2030 carbon peak in China [1].

In recent years, many scholars have carried out research on greenhouse gases in the Nonferrous mining and mineral industry from the aspects of carbon emission reduction, carbon peak and carbon footprint [2-4]. In 2015, "Guidelines for Accounting Methods and Reporting of Greenhouse Gas Emissions from Mining Enterprises" was issued [5]. In 2016, "General guideline of the greenhouse gas emissions accounting and reporting for industrial enterprises" was implemented [6]. Those two files build accounting models including CO_2 emissions from fuel combustion, process emissions, emissions from purchased electricity and heat, and emissions from exported electricity and heat. In 2019, the 2019 refinement to 2006 IPCC guidelines for national greenhouse gas inventory was adopted at the 49th IPCC plenary session. The guidelines include four aspects, such as energy, industrial process and product use, agriculture, forestry and other land use, waste [7].

The nonferrous mining industry has its own characteristics. Most of the mining machinery is diesel equipment, which involved CO_2 emissions caused by the use of urea. At the same time, explosive explosions produce CO_2. The two emission sources are rarely mentioned. Therefore, this study, which based on four activities, including fuel combustion emission, process emission, emission from purchased and exported electricity and heat, will refine explosive and urea emission sources to identify the significant process that contribute the most to climate change. Finally, the accounting model is chosen to evaluate emissions of one mine in China.

2. Source of carbon dioxide emissions

The CO_2 emission sources in nonferrous mines are shown in Table 1.

Table 1. Statement of sources and types of greenhouse gases

Accounting boundary	Greenhouse gas source	Types of greenhouse gases
Fuel combustion emission	Stationary facility or mobile facility	CO_2
Process emission	Explosive explosion process	CO_2
	The process of using urea for diesel vehicles	CO_2
Emission from purchased electricity and heat	Purchased electricity and heat	CO_2
Emissions from exported electricity and heat	Exported electricity and heat	CO_2

* Corresponding author: juliping2013@126.com

3. CO_2 emissions accounting model

The emissions from nonferrous mines should be estimated using Equation(1).

$$E=E_{\text{fuel}}+E_{\text{process}}+E_{\text{purchased}}+E_{\text{expoted}} \quad (1)$$

E=total amount of CO_2 emitted due to nonferrous mine exporation activities, t CO_2;

E_{fuel}=CO_2 emissions from fuel combustion, t CO_2;

E_{process}=CO_2 emissions from process;

$E_{\text{purchased}}$=CO_2 emissions from purchased electricity and heat;

E_{expoted}=CO_2 emissions from exported electricity and heat.

3.1 CO_2 emissions from fuel combustion

The emissions from nonferrous mines are given by Equation(2).

$$\mathrm{E_{fuel}} = \sum_{i=1}^{n} \left(\mathrm{AD_i} \times \mathrm{CC_i} \times \mathrm{OF_i} \times \frac{44}{12}\right) \quad (2)$$

AD_i=amount of fuel combusted, t or 10^4Nm^3;

CC_i=carbon content by type of fuel, tC/t or tC/10^4Nm^3;

OF_i=the carbon oxidation factor by type of fuel, %;

i=type of fuel.

3.2 CO_2 emissions from process

To calculate the emissions of process, the result must be summed as equation (3).

$$\text{Eprocess=Eindustrial explosives+Eurea} \quad (3)$$

$E_{\text{industrial explosives}}$=$CO_2$ emissions from industrial explosive explosion in mine production;

E_{urea}=CO_2 emissions from urea for diesel vehicle.

3.2.1 CO_2 emissions from industrial explosive

There are three kinds of explosives commonly used in mining enterprises, namely emulsion explosives, ammonium oil explosives and powdery ammonium nitrate explosives. In 2021, the total output of industrial explosives was 4.4151 million tons, of which ammonium oil explosives accounted for 22.67%, emulsion explosives 67.36%, ammonium nitrate explosives 7.91%, and others 2.06%.

When the explosive explodes, CO_2, CO, NO, NO_2, N_2O_3, HCl, H_2S and a small amount of other toxic gases are generated. Gases of five kinds of explosives is shown in table 2[8].

Table 2. Gas composition of five explosive

Name	Gas composition %					
	CO_2	CO	H_2	N_2	NOx	subtotal
1 # ammonium antimony explosive	31.5	2.9	2.6	63.1	0.063	100.16
2# ammonium antimony explosive	29	0	0.3	63.7	2.19	95.19
Hw-1 explosive	28.1	1.8	3.0	65.2	0.027	98.13
Hw-2 explosive	27.8	3.3	3.8	64.6	1.02	100.52
Anfo explosive	22.6	1.7	3.4	68.2	0.09	95.99

Carbon in explosives is oxidized to CO_2 after one and two reactions. As long as the chemical formula of explosives is written, CO_2 emissions can be calculated by carbon element, which is named carbon balance calculation method. There are two main estimation methods.

Firstly, if the carbon content of explosives is known, Estimation of CO_2 emission factors for explosive explosions is shown by equation (4).

$$EF_{\text{industrial explosives}} = C \times \eta \times \frac{44}{12} \quad (4)$$

$EF_{\text{industrial explosives}}$=emission factor of industrial explosives, tCO_2 per explosives;

C=Carbon content in mixed explosives, %;

η=Percentage of carbon converted to carbon dioxide gas. The values should be based on the monitoring data of different explosives, and the default value can be 100 %;

The second, if the chemical formula in explosives and its corresponding mass fraction are provided, Estimation of CO_2 emission factors for explosive explosions is shown by equation (5).

Explosives are a mixture. If the chemical formula of each material composition is $C_{a_i}H_{b_i}N_{c_i}O_{d_i}$ and the proportion of each component is β_i, so $\sum_{i=1}^{n}\beta_i = 1$.

$$EF_{\text{industrial explosives}} = \sum_{i=1}^{n} \frac{44\beta_i a_i}{12a_i+b_i+14c_i+16d_i} \times \eta \quad (5)$$

i= material composition iof explosives;

β_i=mass fraction, %;

η= Percentage of carbon converted to carbon dioxide gas. The values should be based on the monitoring data of different explosives, and the default value can be 100 %;

Equation 6 should be used to estimate the amount of CO_2 emitted from explosion.

$$E_{\text{industrial explosives}} = \sum_{i=1}^{n} AD_{\text{industrial explosives},\ i} \times EF_{\text{industrial explosives},\ i} \quad (6)$$

$E_{industrial\ explosives}$=emissions of industrial explosives, tCO_2;

$AD_{industrial\ explosives,i}$=amount of industrial explosives consumed, t;

$EF_{industrial\ explosives,i}$=emission factor for industrial explosives i, t CO_2 per ton of industrial explosives;

i=industrial explosives.

3.2.2 CO_2 emissions from urea

CO_2 emissions from urea are calculated using equation (7).

$$E_{urea}=\sum_{i=1}^{n} AD \times EF \times P \times \eta \qquad (7)$$

E_{urea}=emissions of urea, tCO_2;

AD_i=amount of urea consumed, t;

EF=emission factor for urea, tCO_2 per ton urea, 0.7333[9];

P=Urea concentration in urea solution, %, 32.5%[9];

η=Decomposition rate of urea; The values should be based on the monitoring data of different diesel vehicle, and the default value can be 100 %.

3.3 CO2 emissions from purchased electricity and heat

The emissions from purchased electricity and heat should be estimated using Equation(8).

$$E_{purchased} = E_{\text{purchased electricity}} + E_{purchased\ heat} = AD_{electricity} \times EF_{electricity} + AD_{heat} \times EF_{heat} \qquad (8)$$

$E_{purchased}$=CO_2 emissions from purchased electricity and heat;

$E_{purchased\ electricity}$=$CO_2$ emissions from purchased electricity;

$E_{purchased\ heat}$=CO_2 emissions from purchased heat;

$AD_{electricity}$= the amount of purchased electricity;

AD_{heat}= the amount of purchased heat;

$EF_{electricity}$=annual average carbon dioxide emission factor of regional power grid;

EF_{heat}= emission factor for heat.

3.4 CO2 emissions from exported electricity and heat

The emissions from exported electricity and heat should be estimated using Equation(9).

$$E_{exported} = E_{\text{exported electricity}} + E_{exported\ heat} = AD_{electricity} \times EF_{electricity} + AD_{heat} \times EF_{heat} \qquad (9)$$

$E_{exported}$=CO_2 emissions from exported electricity and heat;

$E_{exported\ electricity}$=$CO_2$ emissions from exported electricity;

$E_{exported\ heat}$=CO_2 emissions from exported heat;

$AD_{electricity}$= the amount of exported electricity;

AD_{heat}= the amount of exported heat;

$EF_{electricity}$=annual average carbon dioxide emission factor of regional power grid;

EF_{heat}= emission factor for heat.

4. Data and Results

The accounting model is developed on one open-pit mine in China. Ore and rock is transported by diesel vehicle using urea. And ammonium oil explosives should be used in mine. All data are provided by the enterprise. The CO_2 emission are shown in Table 3.

The annual carbon emission is as high as 54382.57 t, and power consumption is the main source of carbon emission, accounting for 57.59 % of total carbon emission. Process emissions, which includs explosive explosion and urea use, are also important components of greenhouse gases, accounting for 4.61% and 3.97%, respectively.

Table 3. CO_2 emissions from one open-pit mine

source		Emissions, t per year	%
fuel combustion emission		18398.97	33.83
process emission	Explosive explosion process	2508.32	4.61
	The process of using urea	2158.68	3.97
purchased electricity		31316.6	57.59
total		54382.57	100

5. Discussion

Fuel combustion caused by transportation is one of the main source of carbon emissions, but transportation is usually commissioned and may not be included in statistics. Therefore, it is suggested that CO_2 emissions within the geographical boundaries should be included in the accounting scope.

CO_2 emissions of explosive explosion and urea use is an important source, which should be included in the CO_2 emission accounting model of Nonferrous Mining Industry.

References

1. Notice on strengthening the management of corporate GHG emission reports, 2021.
2. N. Zhao, Y.L. Song, X.W. Zhang, et al. Jinchuan Science and Technology, 1(2022).
3. L.J. Wang, Z.Q. Shao, H. Xiong,et al. Research of Environmental Sciences, 35(2022).
4. A. LI, Z. Zhou Zhiyuan, Y. Liu Yong, et al. Mining Technology, 3(2021).
5. Circular on Issuing the Third Batch of Greenhouse Gas Accounting Methods and Reporting Guidelines for 10 Industries, 2015.

6. General guideline of the greenhouse gas emissions accounting and reporting for industrial enterprise: GB/T32150-2015(2016).
7. .B.F. Cai, S.L. Zhu. Environmental engineering, 37(2019).
8. C.X. Wu. Blasting Safety, 34(2003).
9. Requirments for carbon dioxide emission accounting and reporting, DB11/T1786-2020(2020).

Study on oil-water interface of Budate fracture reservoir in each block of Suderte Oilfield

Yuhang Liu*

Hulunbeier branch, Daqing Oilfield, Daqing City, Heilongjiang Province, China

Abstract. It is well known that in addition to bound water in the reservoir, formation water mainly exists in the form of edge water and bottom water. Fractured buried hill metamorphic rock reservoirs are generally bottom-water reservoirs. In fractured reservoirs, rock blocks are cut by fractures and oil and water are distributed in fractures. Due to the low capillary pressure of fractures, the oil-water transition zone is also small. In the same fault block, there is relatively uniform oil-water interface. And oil-water interface to judge correctly or not will directly related to the determination of oil-bearing area and effective thickness division and well placement, thus correctly determine the position of the oil-water interface of oil field development is of great significance, through the variation of oil-water interface also presumably or some guidance on the seismic section can't precisely divide the small fault.

Key words: Buried hill oil-water interface.

1. Introduction

Budate reservoir in Suderte oilfield is a fractured and karst cave type reservoir, which is horizontally divided into eight blocks. According to the current research, the bottom water of four blocks has been seen, while the bottom water of other blocks has not been seen yet. Because the geological structure of Suderte oilfield is very complex, the large fault block is cut into many small fault blocks due to the existence of faults, which leads to one or several different oil-water interfaces in the same large fault block. In addition, faults which cannot be determined in seismic profile can also be pushed back by accurate oil-water interface. The most direct and effective method is to determine the oil-water interface by using oil test data. In the early stage, the oil- water interface is mainly determined by exploration and evaluation well, but it has certain limitations to judge the oil-water interface by using a single exploration and evaluation well because of the large control range of exploration and evaluation well and the relatively large perforation horizon span. In this paper, oil-water interface and oil-water interface uplift are determined based on the data of exploration and evaluation of well test and the dynamic changes of water cut of the oil Wells in operation, which provides a certain basis for the division of effective thickness of the oil Wells in later production and whether some slow drilling can be completed.

2. Fault sealing

The Budate group of Suderte oilfield is a metamorphic sandstone reservoir with abundant dual porosity and fractures, and the faults are extremely well developed. The fractures are mainly caused by the unified stress field during the formation of the faults and the derived stress field caused by the relative dislocation of the two plates during the fault action, so the fractures of the Budate group are also extremely well developed. For block-fractured reservoirs, there is the same oil-water interface in the same fault-block reservoir, but the underground structure is very complicated in practice, and there are often several oil-water interfaces in the same fault-block. Suppose in a hydrocarbon migration, after the entire block has the same oil-water interface, then for some reason makes two fault block edge is not closed, for secondary hydrocarbon migration, and as a result of the central fault closed, so in the same block formed different oil-water interface, for example well1,well2,well3,well4 reflect the same in different small fault block reservoir oil-water interface, all have independent oil-water interface; Wells A,B and C pass through different small reservoirs and reflect the transition zone between the two reservoirs. Well 42-1 reflects the oil-water interface of the two reservoirs. In the blocky reservoir, oil and water co-produced in the upper part, while pure oil layer or oil and water co-produced in the lower part with thicker interlayer. Assuming that the central fault is not closed, Wells 2 and B will produce water from the upper part and oil from the

* Corresponding author: h_liuyh@petrochina.com.cn

lower part. Examples of the above situations have occurred in Sudert oilfield. Based on FIG. 1 and FIG. 2, this paper analyzes and speculates the oil-water interface of fractured reservoir of Budate Group in each block of Suderte Oilfield.

3. oil-water interface analysis

3.1 Analysis of Oil-water interface in Budate fault Block bei16:

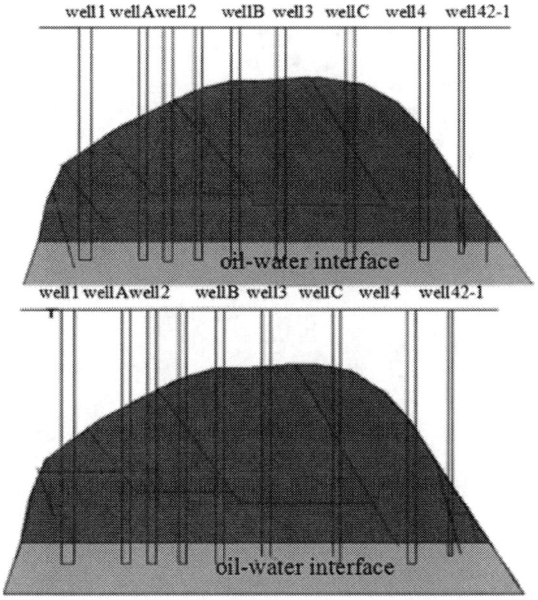

Fig. 1 Oil-water interface after initial hydrocarbon migration

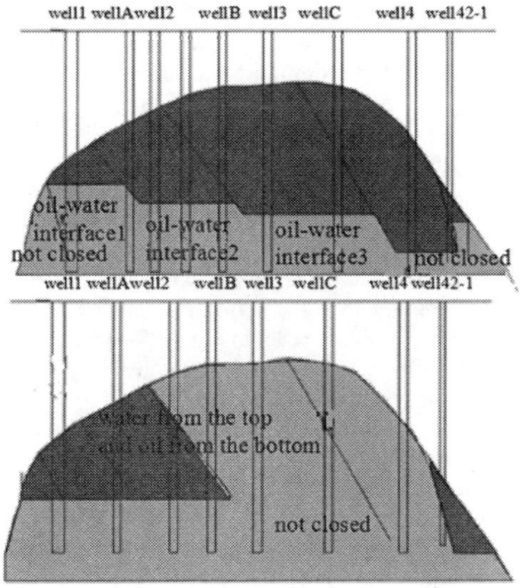

Fig. 2 Oil-water boundary after secondary hydrocarbon migration

3.1.1 Structural characteristics

The top structure of the Xing 'anling Group and Budate Group was explained by suderte structural belt in 2004. The top structure of the Budate group in De1 fault block is a fault block surrounded by faults on three sides. The top structure of the Budate group in De1 fault block is a nearly east-west fault-uplift belt, and the high parts of the uplift are in De1 and De2 The first line of well 2 is northeast, and the high point is about -1075m above sea level. After drilling the development well, the structure of the top surface of budate Group was reinterpreted. De1 fault block is a fault block surrounded by faults on three sides. Within the fault block, an NE to NE chevron fault cut into three small fault blocks, among which the two small fault blocks close to the large fault in the north are closed independent fault blocks, and the southern fault block opens to the southwest. There are three Wells in the north west fault block, De3, Bei1 and Bei2.Within the fault block there is a nearly north-south trending fault ridge, high in the middle, lower to the west and east, and the high point is in the Bei2 well area in the east is a negative structure construction; there is only one well Bei3 in the eastern fault block in the north, and there is a nose-breaking machine in the fault block that always dips southward. The main body of the southern fault block is a nose-breaking structurethat dips southward. According to the drilling data, the BIII term plane structure map is drawn. According to the map,the structural high point of the western fault block in the north is still in the Bei1 well area, which is a broken nose structure dipping south and east, and the eastern wing is a slope dipping east. The NE fault block is an eastward dipping monocline. The southern fault block is a broken nose structure that dips southward.

3.1.2 Distribution of oil and water

Budate buried-hill reservoir in Block De1is considered to be a massive bottom-water reservoir at present. According to the oil test data, it can be seen that the oil and water distribution of fault block De1 is oil-loading and water entering, and the energy of edge and bottom water is sufficient, among which De4 shot 14.5m from 19333.0-1947.5m. The average daily water test was up to 106.8m3.

3.1.3 Development of fractures

According to the coring data of Well De1, the Budate oil layer is extremely well developed, with an average of 14 fractures /m and a maximum of 24 fractures /m(Table 2-2).It can be seen that microspheres in some sections of Budate oil reservoir in De5 are obviously smaller than dual direction finding and have obvious inverse display, and the acoustic time difference curve shows cycle jump phenomenon, the density and neutron curve are abnormally low, and the fractures and holes are abundant.

https://doi.org/10.1051/e3sconf/202235802020

Table 1. Reservoir fracture parameters of Budate Group in well De1

Serial number	Well section (m)		Thickness (m)	Number	Density (crack/m)
1	1777.0	1818.0	41	674	16
2	1818.0	1823.5	5.5	13	2
3	1823.5	1875.0	51.5	637	12
4	1875.0	1881.0	6	42	7
5	1881.0	1891.0	10	82	8
6	1895.5	1906.0	10.5	251	24
Subtotal			124.5	1699	14

3.1.4 Reservoir connectivity

Through the pulse interference well test, the fracture connectivity of De1 fault block is very good. In 2005, the pulse well test was carried out with the self-gusher well De1 as the excited welland the self-gusher well De5 as the reaction well. The distance between the two Wells is 1118 meters 27 Wells of the exciting cycle. The periodic fluctuation indicates that Well De5 is connected with Well De1, and the time delay between them and the agonist well is 73.60h, the force response amplitude is 0.0448685MPa(Figure 2-2).

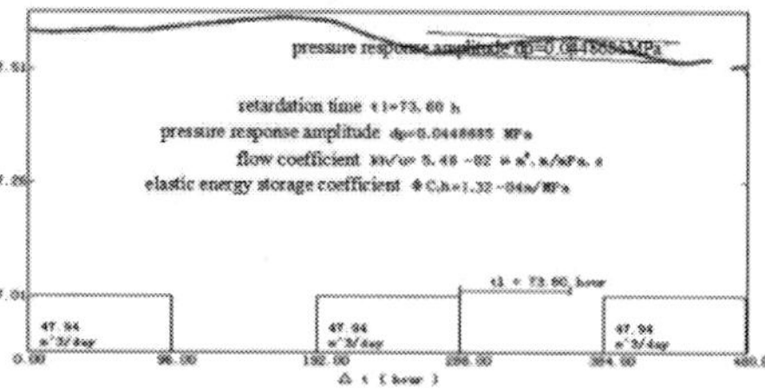

Fig. 3 Interpretation results of pulse interference well test

3.1.5 Production status

In October 2005, all the8 development Wells in fault block De1 were put into operation. Excluding the inter-pumping well De2 well and high-containing well Bei3,the average daily output of liquid was 36.1t, the daily out put of oil was 31.7t ,the water cut was 12.1%,and the flow pressure was13.75MPa.After one year of elastic exploitation, the daily out put of fluid was 33.8t and the daily output of oil was 27.0t, and the flow pressure was still 8.70MPa.

3.1.6 Analysis of Oil-water interface of each fault block

The Budate buried-hill reservoir in blockDe1 is a massive bottom-water reservoir.The Reservoir is divided into three parts by a large fault in the middle.There is no unified oil-water Interface in the whole area, but the same fault block basically has a relatively unified oil-water interface. There is only Bei3 well in the north east fault block, and the water cut of this well is 100% in the initial stage of production. Therefore, it is considered that the oil-water interface of this fault block is above the perforation top limit of this well or there is no oil layer in this fault block. There are three Wells in the north west , and the oil-water interface is about 1915m by comprehensive analysis. A total of 7 Wells in the south (among which the appraisal wells De4 and De6 have been used as water well injection in Xing 'anling of Bei16 main block)were preliminary analyzed the oil-water interface is located between 1980 and 2050m, with an average depth of 2015m. There is obvious dynamic and static discrepancy between De2, which should be further studied in geology (Table 2).

Table 2. Statistical table of oil-water interface of fault block De1

block	well no	perforation top boundary(m)	perforation bottom(m)	initial stage			at present			note
				water content (%)	water top(m)	oil-water interface(m)	water content (%)	water top(m)	oil-water interface(m)	
north east	Bei3	1802	1837	100	more than1802.0	more than1802.0	100	more than1802.0	more than1802.0	more than1802.0
north west	De3	1868.2	1940.8	35.3	more than1940.8	1915	97.5	more than1940.8	1833	interlayer interference
	Bei2	1881	2011.8	50.1	more than 2011.8		78.9	more than 2011.8		
	Bei1	1790.6	1833	0	below1833.0		8.8	more than1833.0		
south	De2	1783.4	1826	0	below1826.0	2015	0	below1826.0	2015	
	De1	1760.4	1880	0	below1880.0		0	below1880.0		
	Bei4	1850	1980	0	below1980.0		0	below1980.0		
	De6	1888.2	2128	0	below1878.0		28.4	below1878.0		

3.1.7 Conclusions and Suggestions

(1) through the current production data can be seen that the north west fault block oil-water interface has been handled by about 1915 m in the early rise to about 1833 m, up nearly 80 m or so, the oil-water boundary is close to or reach Bei1 perforation bottom, at this point if this well for large production will form water cone, causes water to rise faster, so this well should not be adjustable Large parameter production.

(2)De5 when testing in the bottom water layer, at the beginning of the production of the Inter-layer interference has been no water, with declining formation energy, and upper reservoir cannot suppress lower water layer, lead to water cut rising, due to the perforation layer 1 and layer 2 176 m of interlining, in order to control water cut increase, keep the formation energy, suggest to lower 2052.0-2128.0 m for water shutoff.

(3) The perforating top boundary of well De3 is above the oil-water interface, so the water injection intensity must be controlled to prevent the injection water from forming flow check along the fracture, leading to premature and rapid flooding of the oil well. This has been fully taken into account in the injection scheme.

(4) Due to the obvious dynamic and static differences between De2 and Bei3, preliminary analysis suggests that De2 is a single block or on the same fault block as Bei3.In that case,since there is no oil well connected to it, De2

should not be designedas an injection well. If it is designed as a water well, it is necessary to study the fault distanceand water injection at the bottom of the fault . If it is completely disconnected, the surrounding oil wells will not be affected. It is considered that the geology should be further studied.

(5) It is suggested to drill observation wells in favorable areas to study the uplift of the oil-water interface.

(6) once the fractured oil reservoir after water breakthrough in oil and water phase viscosity and pressure formed by the comprehensive water cone or bottom water along the development degree of good inter porosity flow causing crack, the well water cut rising rapidly, but due to the crack of the capillary pressure is small, the gravity differentiation is obvious, so before forming water cone or channeling Wells stop mining, make water cone cannot form or make water cone height To the minimum, so as to control the rate of water rise. It is suggested to adopt cycle production test for high water content well Bei2. According to the production data in November 2006, well B4 has a daily fluid of 30.4t, a daily oil of 5.2t, water content of 82.9%, the efficiency of pump is 38.5%. Therefore, it is suggested to adopt the working system of pumping in three days,comprehensively track the change of water cut and timely adjust the oil recovery cycle to find out the best oil recovery cycle, so as to provide experience for whether it is the feasible to adopt the cycle oil recovery after water is found in other fault blocks.

(7) At present, the oil production rate of the western north fault block is as high as 9.2%, and the water cut has risen to 27.8%. In order to improve the final recovery factor, the oil production rate of the western North fault block should be controlled.

References

1. Lou Zhanghua, Gao Ruiqi, CAI Xiyuan, on groundwater dynamic field evolution and hydrocarbon migration and accumulation in Songliao Basin [J]. Acta sedimentologica sinica, 1997, 15(4):116-120.
2. Kang Yongshang, Pang Xiongqi. Principle and application of fluid dynamic system analysis for hydrocarbon accumulation [J]. Acta sedimentologica sinica, 1998, 16(3):81-84.
3. ZENG Jianhui, Dynamic characteristics of groundwater and its influence on hydrocarbon migration and accumulation in Taibei Sag [J]. Acta sedimentologica sinica, 2000, 18(2):274-278.
4. Tian Shicheng, Chen Yongjin, Zhang Xingguo, et al. Discussion on the hydrodynamic mechanism of reservoir forming dynamic system [J]. Earth science frontiers, 2001, 8(4):330-335.

 https://doi.org/10.1051/e3sconf/202235802021

Analysis of hedge loss factors and protective measures in the western part of Area-A

Xinxin Cui

Technical Management Office, No.6 Operation Area, No.4 Oil Production Plant, Daqing Oilfield Co., Ltd., Daqing, Heilongjiang 163000, China

Abstract: With the continuous development of oilfield, the number of oil and water wells is increasing, and the number of casing loss wells is on the rise. This paper analyzes and researches to further understand the causes of casing loss, and proposes measures to protect casing loss wells in the western part of area A. It summarizes the experience and methods of "six checks, six preventions and six controls" for casing loss protection, which provides direction for preventing casing loss and provides a basis for reasonable development.

Key words: Prevention of nesting; causes of nesting; protective measures.

1. Introduction

The western part of area A belongs to medium and low permeability oil reservoir, with oil-bearing area of 24.83×km^2, geological reserves of 8094.25×10^{4}t and a total of 2318 oil and water wells. With the extension of oil recovery and water injection time and the continuous adjustment of oil field development program, the working condition of casing becomes worse and damaged, which destroys the normal injection and extraction well network system and causes the well network imperfection. In 1996, the first set loss peak appeared in a mining area, and the set loss wells concentrated in A north block in 2013, and the second set loss peak appeared in 2014, and the number of annual set loss wells reached 57.

2. Analysis of the causes of casing damage

2.1 Geological factors

2.1.1 Water Swelling and Creep Effects of Mudstone

When the injected water enters Nen II member, the stress state will change due to the unstable nature of the mudstone, and these changes will cause displacement, expansion and deformation of the mudstone, increasing the amount of load on the outside of the casing. When the external load degree of the casing is greater than the compressive strength, the casing will be extruded and deformed or even wrongly broken. After the oil field is developed with water injection, the injected water in the sandstone permeates through the pores, there is no softening of the rock skeleton, and there is no change in ground stress. While the mudstone absorbs the softening of injection water, its cementing force gradually disappears and creep rate increases. For the un-shooting section, the casing is subject to elliptical stress that cannot be released, forcing the non-uniform horizontal stress to squeeze the outer wall of the casing, and when the equivalent damage load is greater than the casing yield strength, the casing produces elliptical deformation, which eventually causes casing damage.

2.1.2 Rock sliding

In the west of area-A, there are Sa0~SaI interlayer and SaI~SaII interlayer. If the interlayer does not absorb water, the action of the original ground stress keeps the rock formation stable, but the weak interlayer usually has a strong ability to absorb water. In the process of oil field development, when the injection pressure reaches a certain value, the injection water will flee to the interlayer through the fracture and make it absorb water, resulting in the destabilization and sliding of the rock formation, thus causing oil and water well set loss.

2.2 Engineering Factors

2.2.1 Poor quality of cementing injection water run-off

The poor quality of cementing (cementing quality rate less than 80%) provides a channel for the water to escape from the injected wells into Nen II member, causing the mudstone to absorb water and slide, which leads to the loss of cementing.

2.2.2 Scrapping is not a complete tampering

Census 189 scrapped well situation, found 68 wells scrapped when the well has fallen, of which 50 wells have been drilled to update / side-slant wells, scrapped well condition is unclear scurry into the tender second section, there is a set of potential damage.

3. Countermeasures to protect the casing loss well area

The work of casing protection adheres to the work idea of mainly hidden danger investigation and combining prevention and control, and through the work of investigation, prevention and treatment, the casing loss protection countermeasures of "6 investigation, 6 prevention and 6 control" have been summarized to achieve a clear situation of hidden danger and reasonable preventive measures, which effectively control the speed of casing loss and gradually stabilize the situation of casing loss.

3.1 Carry out casing hidden danger survey

3.1.1 Checking the quality of well cementing in the layer section

Using the acoustic variable density curve, the cementing data of 988 layers sections of 247 poly-drive injection wells in the key monitoring area of the casing loss can be viewed. It can be seen from the cementing quality contour map that the cementing quality of the poly-drive wells in the planar B North block casing loss area is the worst, with 21.4% of the whole wells having a quality rate of 80% or more; the cementing quality rate of the longitudinal cement return high ~ Sa I1 and Sa II1 ~ Sa II4 layer sections is low.

3.1.2 Census layer section abnormal water absorption condition

Using isotope information, the current situation combined with history, a census of 1,176 isotope data in the past ten years, focusing on the proportion of water absorption at the top boundary of the injection hole, focusing on the analysis of a single layer of water absorption of more than 30%, stopping the injection layer of water absorption of the potential problems of the well, to take a total of 80 wells treated by means of measurement and adjustment of the amount of lowering, operation adjustment, etc., to reduce the risk of set loss.

3.1.3 Census casing protection packer down into and sealing condition

Census of 985 injection wells found 42 wells were shut down without protective packers due to unplugging, set loss and other reasons. For 16 wells with abnormal pressure to implement the seal of the packer, 2 wells are not sealed, all sealed after the second release.

3.1.4 Census well group injection and extraction relationship

In recent years, the proportion of oil layer parts set loss has increased, the injection and extraction relationship of the water-driven well group was analyzed, and 19 layers sections with injection, no extraction and injection greater than extraction were stopped and controlled after reservoir reawakening, and the water injection intensity was adjusted downward by 1.57 m3/d.m.

3.1.5 Census of previously hedged injection well rupture pressure

After the rupture loss occurred in the standard layer, the top boundary depth of the injection hole in the rupture pressure calculation formula was shifted up to the depth of the standard layer, and the rupture pressure of the ruptured wells in the standard layer was adjusted downward by reducing the calculation depth to reduce the risk of rupture loss. A survey was conducted on 37 injection wells that had been damaged in the Tender II section and had been overhauled and opened, and the rupture pressure of 10 wells was adjusted downward by 0.5MPa.

3.1.6 Census of water injection strength of wells in the hidden area

A survey was conducted on 206 water wells with water injection intensity inside and at the edge of the hidden area. Aiming at 80 wells with water injection intensity greater than 6m3/d.m, the relationship between injection and extraction of well groups and the degree of hidden damage of the set was analyzed and 16 well reductions were implemented, with water injection intensity decreasing by 1.32m3/d.m.

3.2 Carry out casing damage prevention work

3.2.1 Comprehensive enhancement of hedge protection awareness

Adopt the way of "lecture, examination and investigation", carry out training on knowledge of set loss at different levels and with different focus; implement the management of set loss wells, pay attention to the dynamic changes of oil pressure, set pressure and water volume; develop "emergency plan for handling abnormal wells", increase production reduction to reduce the impact of abnormal wells; refine the analysis process of set loss wells, strictly implement the new set loss wells "Weekly analysis system".

3.2.2 Timely circling of hidden hazards

Combined with the four data items of "the distribution of the set loss wells in Nen II member, the degree of set loss, the distribution of the submerged water in Nen II member, and the implementation of the pipe column", seven key areas for monitoring set loss were divided. Combined with the "proportion of lossy wells and the submerged water", we drew up a map of the level of potential loss, and defined three levels of "danger zone, buffer zone and relative safety zone" in each key lossy monitoring area.

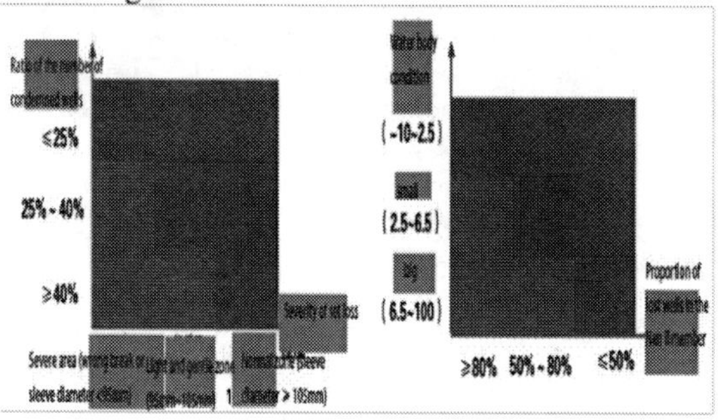

Fig.1 Proportion of lossy wells and the submerged water

3.2.3 Resolutely implement shutdown and control to prevent the spread

After the discovery of the set of damaged wells, resolutely implement the surrounding 300m injection wells all shutdown, and the development of shutdown control program immediately after implementation, until the implementation of the well condition without hidden problems before resuming water injection.

3.2.4 Accelerated investigation to prevent unclear well conditions

Based on the principle of "two timely, a comprehensive" implementation, that is, the timely development of the survey program, timely and effective survey methods, comprehensive coverage of the survey, the implementation of a total of 164 wells well condition survey, electromagnetic flaw detection, timely detection to determine the set loss of 56 wells, reducing the risk of hidden defects.

3.2.5 Balanced set control area boundary pressure difference

Monitor the difference in formation pressure inside and outside the boundary of the hidden area. In order to prevent large changes in the pressure difference between the boundary inside and outside the block, 16 well adjustments to the injection and extraction parameters were implemented to control the pressure difference between the boundary inside and outside the block to within 0.2MPa.

3.2.6 Prevention of water injection on the hidden danger

In recent years, for the history of 156 Sa II 1 ~ 4 injection wells have been stopped to take tests, isotopes and other ways to verify. At the same time, increase the monitoring of injection pressure, water injection, 156 Sa II 1 ~ 4 stop injection layer section to view and analyze 271 isotopes, test results and other information, no water absorption was found to show.

3.3 Casing protection monitoring work

Establishing six attention standards to reduce the risk of hedging losses by laying out a full range of monitoring well points in a flat and vertical manner.
Through the above comprehensive protection measures, the number of casing loss wells decreased from 57 to 37 in 2017 compared with 2014, and the casing loss situation turned better.

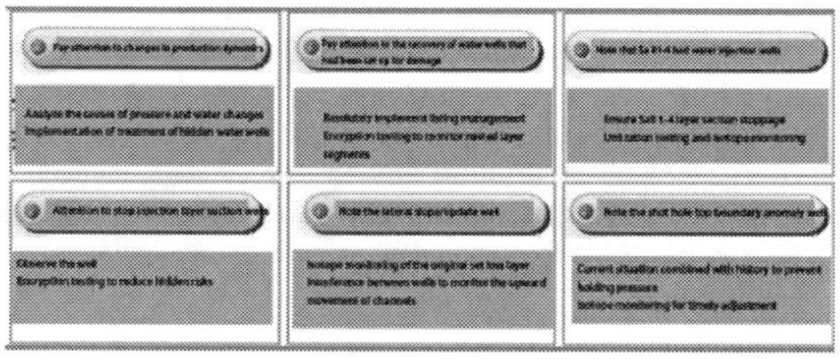

Fig.2 Establishing six attention

4. Conclusion and Awareness

(1) The situation of casing damage in Area A is serious, and the prevention and management capabilities of casing damage wells should be continuously strengthened.
(2) To strengthen the management of casing damage wells, new processes should be introduced and multiple measures should be taken to reduce the frequency of casing damage.
(3) Through continuous monitoring of the well group injection and extraction profile and pressure measurement data, detailed analysis of the injection and extraction relationship can provide a basis for the prevention of damage.
(4) The prevention and control work should be done in time after the occurrence of casing damage to avoid the expansion and spread of the casing damage area.

References

1. Sun Chao, Wang Jun, Liu Chengshuang. Analysis of casing damage factors and preventive measures in Jilin oilfield [J]. Natural Gas and Petroleum. 2011,29(02):43-47.
2. Zhou Ying, Zhang Xinwei, Sun Hongan, et al. Research on the mechanism of oil and water well casing loss and application of comprehensive protection technology [J]. Special Oil and Gas Reservoirs, 2005,12(03):79-82
3. Zhao GuoShi. Analysis of casing damage causes in Songfangtun low-permeability oilfield [J]. Daqing Petroleum Geology and Development, 2014, 23(04):40-41.

Analysis of the causes of rising water content in the late stage of poly-drive in western block III of Area B and study of adjustment methods

Qiong Wu

Geological Research Institute of the Fourth Oil Production Plant, Daqing Oilfield Co., Ltd., Daqing, Heilongjiang 163000, China

Abstract: Seizing the main contradiction of fast water-bearing rebound in western block III, analyzing the reasons of fast water-bearing rebound in response to the contradiction, meanwhile, by understanding the mechanism of polymer oil drive and combining the development effect of three western injection and gathering blocks, finding out that the main problem of slowing down water-bearing rebound is the establishment of pressure system, and therefore the corresponding management measures at the injection and extraction ends respectively.

Key words: Late poly injection; control of water content rise rate; injection pressure; injection and extraction system

1. Introduction

With the time of poly injection, the western poly drive block III in area B has experienced blank water drive, water content decline period, water content low value period, water content rebound early stage, and is currently in the late stage of water content rebound, as can be seen from the proportion of oil production in different stages of the injected poly block, the stage of oil production in the early stage of water content rebound occupies a larger proportion in the whole process of poly drive due to the longer exploitation time, so the comprehensive adjustment of the water content rebound period plays an important role in the whole development of poly drive [1].

2. Block overview

The block is located in the western part of Area B of X Oilfield in City A, starting from the four rows of Area I in the north to the three rows of Area III in the south, bounded by Road S in the east and adjacent to the western transition zone in the west, with an oil-bearing area of 5.93km2.
The block was put into development in August 2011 and was injected with poly in August 2012. The geological reserves are 732.19×10^4 t. The exploited layer is Portuguese I 1~3, the thickness of sandstone is 11.28m, the effective thickness is 8.63m, the permeability is $494 \times 10^{-3} \mu m^2$, the degree of control of poly-drive is 75.4%, and the five-point method and 125m well spacing are used for well placement. As of June 2016, the average integrated water content of the block is 93.67%, which is in the late stage of water content rebound. There are currently 342 oil and water wells in the block, including 166 injection wells and 176 recovery wells.

3. Analysis of the current main problems and their causes

3.1 Rapid return of water content

The monthly water content of the block increased by 0.35 percentage points in the first half of 2016 and the number of wells with a monthly water content increase of greater than 0.5 percentage points was 28 wells, with a rapid rate of water content rebound.

 https://doi.org/10.1051/e3sconf/202235802022

Table 1 Rising water content of wells in Western Block III in the first half of the year

Classification	Number of wells (wells)	June 2016				Monthly water content uplift value (Percent)
		Daily fluid production (t)	Daily oil production (t)	Aqueous (%)	Sunkenness (m)	
≤0.3	49	35.2	2.4	92.84	342	-0.13
0.3~0.5	15	40.0	2.2	94.74	284	0.41
> 0.5	28	40.2	2.1	94.63	427	0.75
Set control area	62	15.4	1.2	90.92	187	0.68
Total	168	30.6	2.0	92.87	324	0.35

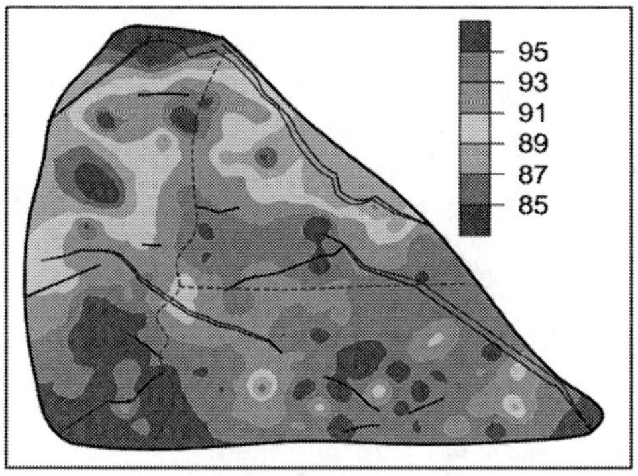

Figure 1 Water-bearing isopleth map of recovered wells in Western Block III

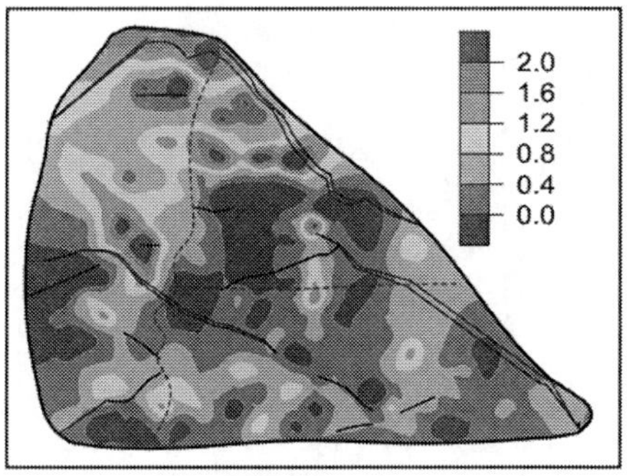

Figure 2 Water-bearing rise isogram for Block III West

3.2 Analysis of the causes of rising water content

The 28 wells with rising water content were analysed well by well for the reasons for rising water content into four categories: inefficient and ineffective circulation, insufficient fluid supply, insignificant profile improvement and high recovery [2].

The development of inefficient and ineffective circulation is characterized by high water content and prominent inter-stratigraphic conflicts, the solution focuses on controlling ineffective circulation and stopping high seepage; the development of undersupplied wells is characterized by low well sinkage, which is manifested by insufficient fluid supply, the solution focuses on enhancing fluid supply by increasing the injection volume; the development of insignificantly improved profiles is characterized by a lack of significant improvement in the ratio of layer activation and thickness activation, the solution focuses on stopping high seepage, adjusting dissection and increasing injection concentration; the development of high recovery is characterized by high cumulative oil production per unit thickness and a large maximum drop in water content, the solution focuses on reducing the injection of high seepage layers to improve the activation of low seepage layers.

4. Adjustment methods

Polymers act as oil repellents in two main ways. Firstly, due to the nature of polymers, they can be adsorbed into capillaries in the rock to seal the water pathway without any significant effect on the permeability of the oil phase, which is due to the hydrophilic and oleophobic characteristics of the phthalimide and completion acid groups. In addition, the high viscosity of the polymer increases the viscosity of the water phase, which can play a role in reducing the flow rate of the water phase, thus playing a role in reducing the water-oil flow rate ratio [3]. The change to high polymer concentration at the end of the injection period improves the viscosity of the polymer, changes the fluid properties and movement within the formation and improves the intra- and inter-formation conflicts in the injector well.

After we understood the oil drive principle of polymer, by comparing the development effect of the three western injection and gathering blocks, the injection and gathering process, the western block III injection pressure is always higher than the other two blocks, the pressure space is kept smaller, it is found that the wells with higher injection pressure, the succession layer is continuously moved during the injection and gathering process, the injection profile is improved, it is recognized that the pressure system has a greater impact on the development effect, for this reason, the pressure system is reasonably adjusted as the focus to establish an effective drive system.

The main means of adjustment for this is to adjust the injection and extraction parameters and increase the injection pressure.

Table 2 Rate of water content rise in each block of Zone B during the aggregation injection phase

Block	Initial water content (%)	Duration of low value period(month)	Maximum water content reduction (Percent)	Monthly water content uplift value (Percent)	
				Early stage of rebound	Late rebound
West I Block	97.26	5	15.37	0.88	0.45
West II Block	95.23	8	13.55	0.55	0.43
West III Block	93.43	13	14.45	0.45	0.35

4.1 Injection side

For the 53 wells with small injection pressure space, we focus on the work idea of ensuring injection and system, firstly, for the 18 wells with pressure space less than 0.2MPa, we implement the reduction of concentration and quantity 17 times to ensure continuous injection; secondly, we strengthen the sticky loss management for wells with pressure space between 0.2~0.5MPa to ensure the injection system, ensure the system fluctuation to the site, pressure change to the site, troubleshooting The "three to the site" principle, at the same time to implement a clear viscous loss of each node, to identify the two major factors affecting viscous loss: the impact of machine pumps and pipeline impact, to this end we actively repair all kinds of pump failure in the first half of the year more than 1,000 times, and at the same time to solve the pipeline viscous loss of high-temperature water cast ball 21 wells, through the above work, the western III Through the above work, the qualification rate of the western III block system was maintained at over 90%.

For the 58 wells with a large injection pressure space, we focused on the idea of increasing the injection pressure and reducing the pressure space, firstly, we implemented 10 wells with a thicker volume for wells with a pressure space between 0.5 and 1.0MPa according to the injection strength, and the average injection pressure per well increased by 0.2MPa after adjustment; for wells with a pressure space greater than 1.0MPa, we implemented 11 wells with a large thicker volume and 11 times, and stopped 2 wells with high seepage, and the average injection pressure per well increased by 0.6MPa after adjustment. The average single well injection pressure increased by 0.6MPa after the adjustment.

4.2 Extraction end

In view of the contradiction between the current development of the western block III, where the water content is rising rapidly and the recovery concentration is rising rapidly, we refined the classification of the extraction wells to explore the potential; at the same time, based on the relationship between the flow pressure and the water content and recovery concentration, we established a chart and concluded that the current reasonable flow pressure limit is 3.0~5.0MPa, for which we made targeted adjustments in conjunction with the injection and extraction conditions.

For the 4 wells with high water content and high concentration of poly, 9 wells were adjusted to control the inefficient and ineffective cycle, and 4 wells were plugged with water from the extraction wells, after which the average daily oil production per well increased by 1.3t and the water content decreased by 3.05 percentage points. For low water-content high poly wells to strengthen the injection and extraction structure adjustment, the implementation plan was adjusted 10 times, and 8 wells were transferred to small reference, after which the water content decreased by 1.50 percentage points and the sinkage increased by 46 m. For low water-content high poly wells to extend the low value of water content to maintain the effect of poly drive, the implementation plan was adjusted 12 times, and 4 wells were transferred to large reference, after which the daily oil production increased by 0.3 t and the sinkage decreased by 39 m.

5. Conclusion

A reasonable injection plan is a prerequisite to ensure the effect of injection, and timely adjustment of the plan is a guarantee to prolong the effect of injection; for the characteristics of late injection should reduce the injection rate, increase the injection concentration, expand the wave volume, and maximize the remaining oil. At the same time,

 https://doi.org/10.1051/e3sconf/202235802022

in order to control the rate of water content recovery to control the pressure system as a key element.

References

1. Zhao Qirui. Analysis of the effect of adjustment of measures during the period of polymer drive water content rebound [J]. Chemical Management, 2014, 5: 85.
2. Wei Yuting. Analysis of the causes of water content rise in the late stage of polymer injection in the second class oil layer in the east zone and measures to control it[J] . Chemical Management, 2016,3:175.
3. Zhang Kunpeng. Application and progress of polymer drive in improving oil and gas recovery[J]. Contemporary Chemistry, 2016,4:860~862.

https://doi.org/10.1051/e3sconf/202235802023

Analysis of Enhanced Oil Recovery Technology in Extra-high Water Cut Period in Oilfield

Dezhang Lin

Institute of Geology, No.6 Oil Production Plant of Daqing Oilfield Company, Zip code: 163000 China

Abstract. Most onshore oil fields in China are in the later stage of development, and the water content is more than 90%, which increases the difficulty of remaining oil mining. In order to improve the oil production, it is necessary to develop a variety of oil recovery technologies, increase the oil field space pressure, weaken the oil viscosity, and improve the recovery factor. Taking L oilfield as an example, the paper briefly expounds the characteristics of extra-high water cut period, and on this basis, puts forward recovery technology, grasps development rules, and realizes technological innovation. The accumulative oil production reaches 2.32 million tons, and the recovery factor increases by 9%, thus providing reference for related workers.

Key words: Oilfield; Extra high water cut stage; Recovery techniques.

1. Preface

Most of the old oilfields in China are in the stage of ultra-high water cut, the distribution and flow of underground fluid in the reservoir change, the overall recovery rate is low, there are a lot of remaining oil in the watered-out layer, and it is difficult to dig. Various contradictions are highlighted in the late development of oilfields, and it is difficult to maintain the output of oilfields by conventional development. Traditional oilfield recovery is mainly carried out by chemical means and fine water flooding. Due to local enrichment, high dispersion and complex distribution of ultra-high water cut remaining oil, recovery technology must be optimized to improve oil recovery and maintain stable oil production.

2. Overview of oil

L Oilfield is divided into 59 faults, most of which are normal faults. In the north, there is a group of east-west normal faults with a fault distance of > 40m. In the south, there are east-south normal faults with a fault distance of 10-20m. According to the fault development characteristics of L oilfield, seismic interpretation data were used to accurately control the fault extension scale and occurrence, establish the contact relationship between geological layer and fault, and carry out 3d fine modeling, as shown in Figure 1.

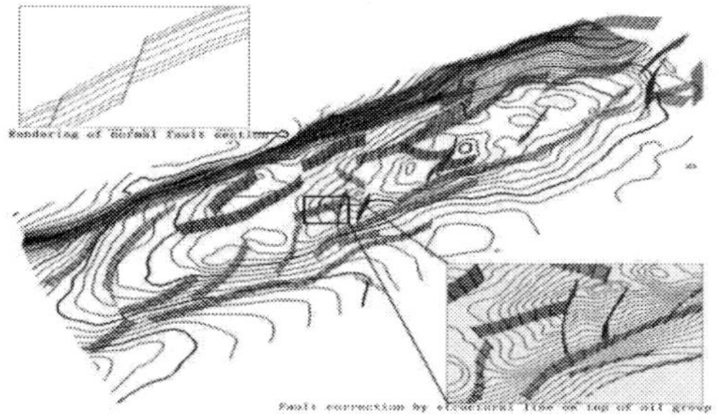

Fig. 1 Fault model of L oilfield

L oilfield is in the middle and late stage of development, with a high degree of well pattern improvement, a relatively complete understanding of geology, and a lot of well point data to ensure the static reliability of reservoir. According to the numerical simulation of L oilfield, the remaining oil distribution is predicted and the remaining oil saturation of main T12-1 small layer is plotted, as shown in Figure 2.

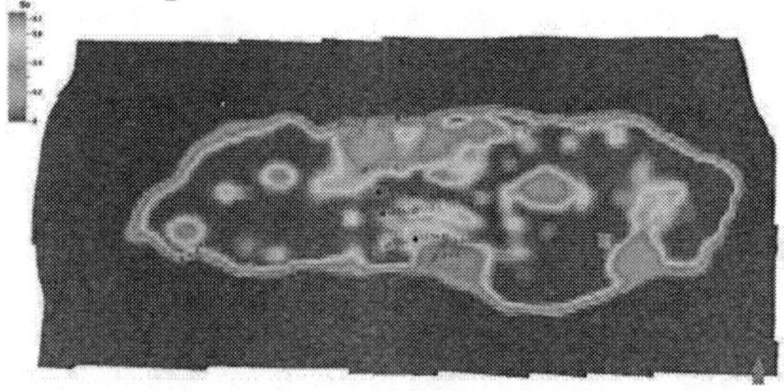

Fig. 2 Remaining oil saturation

L oilfield is in the stage of high water cut development, and the remaining oil is scattered, mostly flake, with high saturation and insufficient water flooding effect. Features are as follows:

1.1 Water sources are complex. In the ultra-high water cut stage of an oil field, the water layer is thick, the water area is wide, and the water sources are complex, including bottom water and edge water. Bottom water is the main reason for the increase of oil water volume, while edge water has little influence. The combined action of the two will improve the top interface of oil layer, increase the floating type of oil layer, and then reduce oil recovery [1].

1.2 Development law changes. Oil field development rules are different in ultra-high water cut stage, including oil production rate, water cut rise rate, liquid production rate and other parameters, which all reflect the changes of underground macro reservoir parameters.

1.3 The well pattern changes greatly. After the ultra-high water cut, the well pattern of the oilfield has been adjusted for several times. In the process of water flooding, the multi-phase fluid interaction between reservoirs is distributed, and the position of enrichment and highly dispersed remaining oil in some places becomes the mining object, concentrated at the top of the thick oil layer.

1.4 Change of potential objects. With the deepening of development, injection-production Wells are adjusted by multiple infilling, and the plane classification of remaining oil varies greatly. The infilling and adjustment of well pattern is to improve the control degree of reserves at first, and to tap the potential of local remaining oil at the later stage.

3. Technical measures to improve oil recovery during extra high water cut period

3.1 Fine water drive technology

L oilfield is in the stage of ultra-high water cut, and water injection is general in the early stage, resulting in high water cut, local high permeability bands and serious underground invalid water circulation. Fine water flooding technology can be adopted to adjust injection-production parameters, change injection-production working system, optimize liquid flow direction and improve water flooding recovery [2]. In oilfield exploitation, streamline data simulation technology is used to fit the dynamic production of the block, with "water injection efficiency" as the main factor, and the water injection volume of injection well is adjusted to avoid invalid circulation of injected water. The specific steps are as follows:

(1) The production history of the block is simulated by using numerical results of streamlines. After full fitting, numerical results of simulated streamlines are called to calculate the e_I single well water injection efficiency and the average water injection rate of E 12 block. The calculation formula is as follows:

$$\bar{e} = \frac{\sum q_o}{\sum q_i} \tag{1}$$

Among them, $\sum q_i$ is the daily water injection volume in the work area(m³/d); $\sum q_o$ is the daily oil production of the working area(m³/d)。

(2)Optimize water injection volume, Take the average water injection rate of $\bar{e}$ block as the boundary, Water injection well if $e_i > \bar{e}$, The injection volume needs to be increased, On the contrary, the amount of water injected is reduced. The calculation method of new injection volume after adjustment is as follows:

$$q_i^{new} = q_i^{old} * (w_i + 1) \tag{2}$$

Among them, q_i^{old} is the original injection volume(m³/d); q_i^{new} is to optimize the injection volume(m³/d); w_i is the weight coefficient, The calculation formula is as follows:

$$e_i > \bar{e}: w_i = \left(\frac{e_i - \bar{e}}{e_{max} - \bar{e}}\right)^a * w_{max} \tag{3}$$

$$e_i < \bar{e}: w_i = \left(\frac{\bar{e} - e_i}{\bar{e} - e_{min}}\right)^a * w_{min} \tag{4}$$

In practical application, the maximum amount of water injection needs to be adjusted due to the limitation of working system and production status, w_{min} Set to -0.5, w_{max} Set to 0.5, The maximum amount of water injection can be adjusted to the original amount 50%。 e_{max} 、 e_{min} is under existing production, Maximum/minimum water injection efficiency of water injection well in work area, a is the optimization index, take 2, The weight coefficient changes gently。

(3)To optimize the fluid production of oil Wells, the adjustment of water injection is the main result. According to the situation of production Wells, the associated injection Wells and distribution coefficient of relevant production Wells are calculated. The adjusted water injection volume is multiplied by the distribution coefficient, and the value of adjusted oil production is obtained after summing up [3]. It is assumed that four injection Wells around Px in the production well contribute to its production.The water volume is respectively Δq_1、Δq_2、Δq_3、Δq_4。 distribution coefficients are, respectively $f_{1\to Px}$、$f_{2\to Px}$、$f_{3\to Px}$、$f_{4\to Px}$, and $\Delta q_i = q_i^{new} - q_i^{old}$, PxThe adjustment formula of liquid production is as follows:

$$\Delta q_{Px} = \Delta q_{P1} f_{1\to Px} + \Delta q_{P2} f_{2\to Px} + \Delta q_{P3} f_{3\to Px} + \Delta q_{P4} f_{4\to Px} \tag{5}$$

Formula of fluid production after adjustment of production well:

$$q_i^{new} = \Delta q_{P1} + q_i^{old} \tag{6}$$

(4)The streamline model prediction is a domestic effect. The optimized liquid production and injection volume are simulated for a period of time, and the water injection efficiency is calculated again. The above steps are repeated to optimize the water injection efficiency.

3.2 Well pattern reconfiguration technology

The oil field is at the stage of ultra-high water cut, and the remaining oil distribution is characterized by "local enrichment and overall dispersion". About 30% of

 https://doi.org/10.1051/e3sconf/202235802023

recoverable reserves remain to be developed. Therefore, in order to improve oil recovery, local infill Wells can be used to develop the remaining oil and improve the development effect. The location of infill Wells is an important problem, especially when oil prices are low and drilling costs are high, so the location of infill Wells must be optimized. In the past, the optimization was mainly based on the index of remaining oil reserves richness, and the formation pressure and reservoir physical properties were ignored because of single factors [4]. During well pattern reorganization, the optimized production potential calculation formula can be adopted to select the infill well location based on the remaining oil. Taking L oilfield as an example, there are 25 production Wells in total and 6 injection Wells have been shut down for a long time. Analysis shows that most of them are located at the edge of the oilfield, and it is speculated that the intrusion of edge water causes oil well flooding and reduces the efficiency of injection Wells. In the adjustment well deployment, the oil well diversion mode is selected to transfer the production Wells and high water content Wells in the periphery of the well area. Some production Wells are longitudinal-flooded and shut down for a long time; some production Wells are water-cut high and in the marginal area; some production Wells are poor in water plugging operation, so they are all transferred.

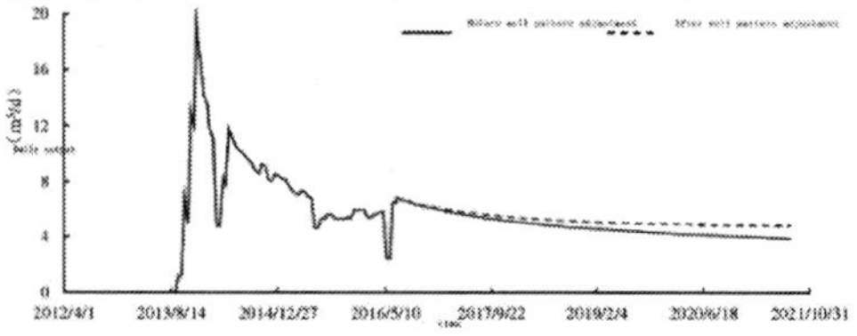

Fig. 3 Comparison of oil production of adjusted well pattern

3.3 Asp flooding technology

In order to improve oil recovery, the "self-efficient oil displacement" composite technology was developed. Details are as follows:

(1) Design and synthesis technology of polymer molecules and surfactants. Based on the strong base alkylbenzene sulfonate surfactant products used for oil displacement, the microscopic oil displacement mechanism was deepened, the molecular performance and structure relationship of surfactants were determined, and the weak base alkylbenzene sulfonate products with high symmetry and high branched degree were developed [5]. For high temperature, high salinity, low permeability oil field conditions, the product can use non-ionic, anionic, amphoteric ionic alkali free surfactants, reduce the amount of composite flooding polymer, increase the technical and economic effects, reduce the cost of 20% and 40% of the polymer amount, achieve good results.

(2) Multi-parameter quantization scheme. Reservoir geological and developing factors affecting the thorough analysis, build a series of strata combination, object of exploitation, the degree of control, well pattern well spacing design standards and technologies such as line, form the best technical effect of asp flooding engineering plan, build in order to enlarge sweep volume efficiency, improve to effective prescription, standard of parameters and design process.

(3) Tracking adjustment technology. Oil field development, in order to ensure the effect of asp flooding, different mining stages in order to enhance the capacity of reservoir flood, used as the key point, follow the "power, adjustable pressure, ability to control water cut, to protect, prevent breakthrough" the adjustment of the tracking principle, realizes the whole control, key adjustment, increase the single well oil recovery, a quantitative targets, advanced prediction analysis, classification, form, active control adjustment. Through this method, the tracking and control standards of plugging, pressure, control and adjustment can be formed to realize accurate control of oilfield adjustment. The compliance rate of implementation plan is > 90%, and the whole oilfield can maintain high injection-production capacity and low water cut stable period.

3.4 Chemical flooding technique

L oilfield is an edge water structure oilfield composed of multi-layer sandstone and ultra-deep rock. Long-term waterflood development is in the stage of ultra-high water cut. Individual contradictions in well area are prominent, and injection water wave efficiency is low. The process is as follows: the water content of oilfield injection reaches 99%, and the mixture system of 0.3PV cross-linking agent and polymer is injected to carry out chemical flooding. The polymer of weak gel system is high polymer. When flowing through high permeability core, water solution can enter the pore space of core. Coupled with the high viscosity of weak gel system, the water-oil fluidity ratio can be controlled. The larger the permeability range is, the lower the recovery rate of weak gel flooding is; on the contrary, the recovery rate increases, and the maximum recovery rate is achieved when the permeability range is 15.

4. Development effect

Since 2016, the oilfield has gradually used the fine water flooding technology to recombine the field well pattern, and the ASP flooding technology has also been used in different stages. By 2021, the accumulative utilized geological reserves have reached 50 million tons, producing 2.32 million tons of oil and increasing the recovery factor by 9%.

5. Conclusion

To sum up, the extra-high water cut stage of oilfield is faced with the problems of insufficient resources, increasing production decline pressure and serious imbalance between production and storage. It is necessary to lock the main direction, solve the bottleneck restricting oilfield development and deepen oilfield development through technological innovation. Therefore, in order to stabilize the output of oil field, fine water flooding technology, well pattern recombination technology and ASP flooding technology can be taken into consideration of the actual situation to rationally plan oil field production and improve oil field recovery.

References

1. Lin GUI, Haonan Dai, Qian Qian, Lin Deng. Current Chemical Industry, 201,50(09):2145-2148. (in Chinese)
2. WANG Fenglan, You Chunmei, Chen Baoyu. "Industry-university-research" joint solution to solve the eOR problem of super high water-cut old oilfields [J]. Petroleum Science and Technology Forum, 201,40(03):118-123.
3. Zhang Junting, Zhou Zheng, Xie Yue, Wang Meinan, Liu Bin. Petroleum Geology and Engineering, 201,35(02):56-61.
4. Wang Fenglan, SHA Zonglun, LUO Qing, Zhao Yunfei, Zhang Jifeng. Petroleum geology & oilfield development in daqing,2019,38(05):51-58.
5. QIAO Yong. Enhanced oil recovery technology in complex fault block reservoirs at high water cut stage: A case study of Mingyi West fault block in Wenzhzhai Oilfield [J]. Petrochemical technology,2019,26(07):52-53+19.

Research and application of the method based on the displacement gradient method to implement the fragmented breakpoint at the end of the fault

Jinyu Zhang*

No. 2 Oil Production Plant, Daqing Oilfield Co., Ltd., Daqing, Heilongjiang 163414, China

Abstract: The limitation of seismic resolution makes the interpretation of fault end locations based on seismic data unclear, and often can only be predicted qualitatively by human, which makes the combination relationship of breakpoints near or below seismic resolution unclear. The breakpoint combination rate is generally low, which affects the development injection and recovery relationship and remaining oil distribution.

Keywords: Displacement gradient; fault end; breakpoint combination rate; quantitative prediction.

1. Introduction

The study of fault ends is actually a kind of sub-earthquake study, and the so-called sub-earthquake faults are small displacement faults below seismic resolution, including isolated small faults and large fault ends, which control the subsurface fluid flow and affect the non-homogeneity of subsurface reservoirs in oil and gas fields as well as the effect of water injection and development [1-2], therefore, this study can not only improve the combination rate of block breaks and further clarify the tectonic features, but also Therefore, this study can not only improve the block breakpoint combination rate, further clarify the tectonic characteristics, but also improve the oil and gas field injection and extraction relationship, which is important to improve the development effect and optimize the well network deployment.

The study area is located in the northern part of Daqing Changyuan Sanan oil field, which is one of the most complex blocks on Daqing Changyuan. The fault combination rate of small faults in the study area is low, the overall fault combination rate of the block is 90.3%, but the fault combination rate of small faults below 3 meters is only 76.4%, which obviously lowers the overall accuracy of the block structure and affects the analysis of the development and injection relationship. How to improve the rate of fault combination is an important issue. Due to the limitation of seismic data resolution, fault ends are often predicted qualitatively by human, which makes most of the small fault points below 3 m uncertain in origin, and often treat them as scattered fault points within the neighboring fracture zone, and the cause is unclear. This paper quantitatively predicts the location of fault ends based on the displacement gradient method, and further attributes the fault end breakpoints correctly. The practical results show that the method can quantitatively predict the extension length of sub-seismic fault ends and effectively improve the combination rate of small breakpoints at fault ends.

2. Overview of the study area

The study area is located in the northern part of the Sanan oil field in Changquan, Daqing, with a development area of 41Km^2 and a stratigraphic dip angle of 5-23°. The study area is at the high point of the Changquan tectonics in Daqing, with complex fracture relationships (Figure 1). The 79 faults are developed on the top surface of the Sa II oil group, and the fault orientation is mainly north-east and south-west, with frequent mutual cutting. There are various combinations of fault planes and profiles, such as parallel, geese column and oblique, etc., and there are combinations of graben, basement, stacked tile and "Y" shape in the profile. The overall fault combination rate in the study area is 90.3%, of which 76.4% is below 3 meters, and there are many uncombined and scattered faults. It is important to understand the structural situation of this block, which is important for well network optimization, measures to explore potential and remaining oil analysis.

* Corresponding author: c2zhangjinyu@petrolchina.com.cn

 https://doi.org/10.1051/e3sconf/202235802024

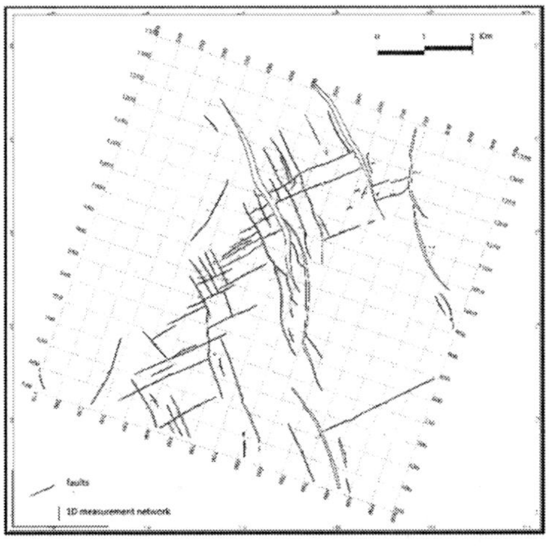

Figure 1 Polygon distribution of faults on the top surface of Sa II in the study area

3. 2 Based on the displacement gradient method to implement the end of the fault fragmentation breakpoint

The so-called displacement gradient method in this paper draws on numerical analysis methods [3] and the relative relationship between statistical fault extension length and fault distance size to establish a fault displacement gradient to estimate the length of the fault tail by plotting the relevant fault distance-distance curve, so as to quantitatively predict the length of the fault tail and its corresponding fault distance size (Figure 2).

Seismic fault end point is the point where the fault displacement is lower than the seismic resolution, and the fault displacement at this point can be considered as the seismic resolution, so the displacement gradient can be obtained from the fault displacement map, and practice shows that for the prediction of sub-seismic faults at the end of faults, the displacement gradient method can be used for quantitative prediction based on seismic data.

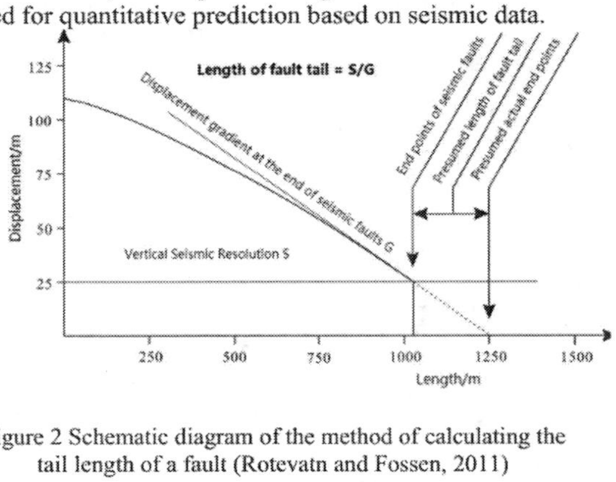

Figure 2 Schematic diagram of the method of calculating the tail length of a fault (Rotevatn and Fossen, 2011)

Using the seismic interpretation and logging data in the study area, the fault displacement gradient graph is established by counting the fault break distance, and then quantitatively predicting the extension length and break distance at the end of the fault based on the curve fitting formula, so that the suspected fault end break points can be combined.

The comparison method of "flat profile" is used to predict the tail of the fault quantitatively from the plane and profile respectively. In the plane, as shown in Figures 3 and 4, the location of the original interpreted end of fault No. 1 is shown as the red line, and the new extension location is predicted by making a displacement gradient graph of the fault and calculating the formula quantitatively based on the curve fitting formula as shown in the dashed line, but it is known that the fault is truncated by fault No. 2 based on the fault polygon, therefore, whether this prediction is reasonable, we further verified using logging data and found that N-A-2, the well, has a 4-meter uncombined scattered fault within the Sa II oil group, while a fault with a 3.1-meter break distance is predicted to exist at this location according to the displacement gradient map. It can be seen that the situation predicted by the displacement gradient method just happens to be very similar to the actual interpretation of this well N-A-2. The comprehensive comparison and verification of the above well seismic data leads to the conclusion that No. 1, after being truncated by No. 2 fault, does not end abruptly, but continues to develop for some distance through No. 2 fault, and also attributes the uncombined breakpoint with a break distance of 4 m in well N-A-2 to No. 1 fault.

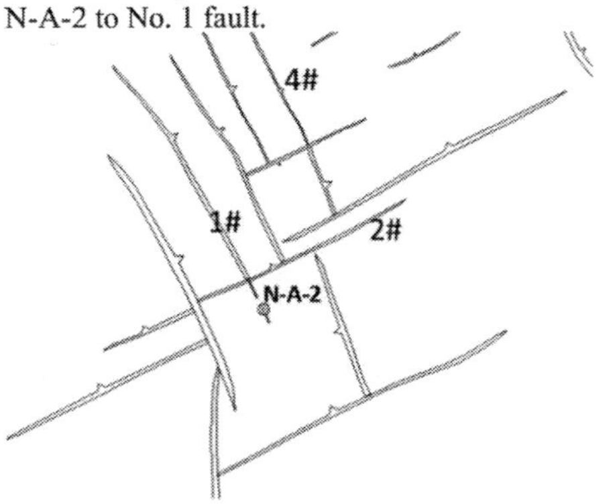

Figure 3. Fault polygons on the top surface of Sa II

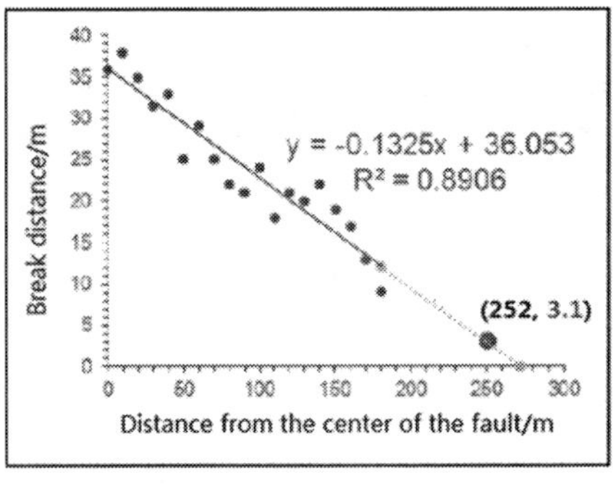

Figure 4. Quantitative prediction map of the end of fault No. 1

On the profile, as shown in Figures 5 and 26, the end of fault No. 4 is predicted to have developed a breakpoint with a break distance of 2 meters at a location with an extension length of 325 meters, as predicted by the

https://doi.org/10.1051/e3sconf/202235802024

quantitative prediction map of the fault end. By searching the log data, it was found that an uncombined scattered breakpoint with a break distance of 1.8 m was developed in the N-D-2 well within the Portuguese I formation, which further supports the accuracy of the prediction results, and therefore, the breakpoint was attributed to the end of the No. 4 fault.

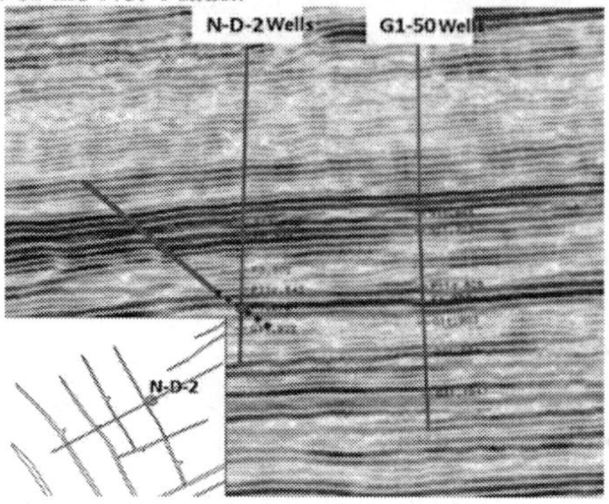

Figure 5 Seismic profile of the over N-D-2 well

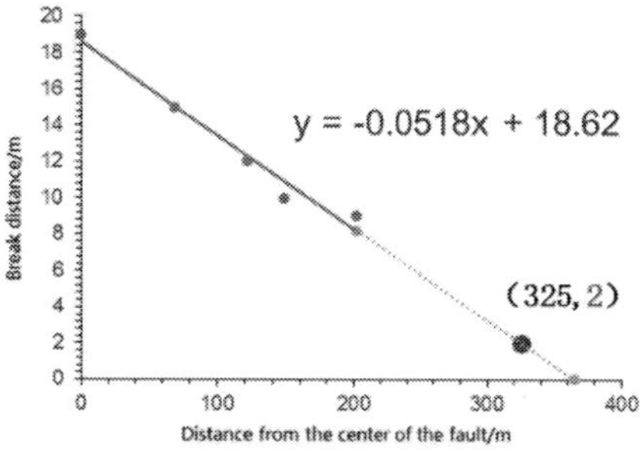

Figure 6 Quantitative prediction map of the end of fault No. 4

In the same way, by building the fault displacement gradient graphs one by one and quantitatively predicting the subseismic end extension length and break distance based on the curve fitting formula, the suspected fault end break points were compared with the results of this quantitative prediction, and a total of 16 suspected fault end break points were identified, of which the number of implemented combinations was 10 (Table 1), 7 combinations with break distance less than 3 m, and the other 6 were confirmed not to be fault end break points.

Table 1 Comparison of predicted results of fault ends and actual information of fault points

Number	Well No.	Depth /m	layer	Actual break distance /m	Predicted break distance /m	Relative Error %
1	G1-54	964.8	GI	1.8	1.4	22
2	G1-51	1004.2	GII	7.4	6	19
3	G1-33	1066.2	GIII	3	3.9	30
4	G1-03	1092.6	GIII	1.8	2.1	17
5	G1-544	1183	GIV	3.4	2.8	18
6	N-D-J4	1122.6	GIV	2.2	2.8	27
7	N-D-2	876.4	PI	1.8	2	11
8	N-A-2	772.4	SII	4	3.1	23
9	N-D-6	727.4	SII	2.8	2.2	21
10	N-5-4	913.4	SIII	2.4	1.7	29

4. Conclusion

(1) The seismic resolution controls the interpretation of the end of the fault, because the fault development is gradually pointed out, so when the fault end break distance is smaller than the seismic resolution, it is not explained by seismic data, and the artificial qualitative judgment of the fault location is actually the location where the seismic resolution can be distinguished, and the location is usually recognized as the end of the fault, while the actual fault end extension should be longer.

(2) The practical application of the displacement gradient method proves that the method can quantitatively predict the extension length of the end of the fault and its corresponding fault distance size, so as to determine the location of the end of the fault, and the method can break through the traditional practice of qualitatively predicting the end of the fault due to the constraints of the resolution of seismic data.

(3) The well interpretation breakpoint data can verify the reliability of the interpretation results of the fault end prediction by using this method, and the relative error is generally less than 30% by comparing the actual breakpoint distance and the predicted breakpoint distance at the breakpoint, which has a high degree of confidence.

(4) By this method, the attribution of fault end breakpoints was further implemented in the study area, and 10 new breakpoints were combined, including 7 with a break distance of less than 3 meters, which improved the combination rate of breakpoints in the study area.

References

1. Zhou Liye, Zhang Jianming, Li Wenzhong et al. Quantitative prediction technology for sub-seismic faults in BZ34 oil field. Fracture block oil and gas field [J].2018, 25(1): 48-51.
2. Fu Rongzhi, Quantitative prediction of sub-seismic faults and the impact on water injection development [D]. Daqing: Northeast Petroleum University, 2017.
3. Gong Lei, Zeng Lianbo, Miao Fengbin et al. Application of fractal geometry method in the description of complex fracture system [J]. Journal of Hunan University of Science and Technology (Natural Science Edition), 2012,27(4):6-10.

 https://doi.org/10.1051/e3sconf/202235802025

Application of new geosteering technology in thin reservoir

Lidong Yang

Management Office of Geological Research Institute of the tenth oil production plant of Daqing Oilfield Co., Ltd., Daqing, Heilongjiang 163001, China

Abstract: Because some oilfields in China belong to the thin bottom water reservoir model, the actual drilling process is technically difficult. In order to ensure the maximum drilling operation of the target layer and the basic development effect of the oilfield, technicians should introduce advanced steering tools and study new geosteering technology to fundamentally ensure the exploitation quality of thin oil reservoirs.After exploring the introduction of new geological guidance technology in detail, this paper further explores the principle of new geological guidance technology, and summarizes the application strategy of new geological guidance technology combined with the development of thin-layer reservoir.

Keywords: Thin reservoir; New geosteering technology; Energy depth; Operating principle

1. Introduction

The sandstone reservoir of an oil field in China is a typical bottom water structural reservoir, so the reservoir thickness is relatively low and the bottom water energy of oil and gas field is sufficient, so it is necessary to use horizontal well development technology. During the implementation of the new geological guidance technology, it mainly refers to the detailed exploration of the detection technology of wellhead trajectory based on the core conditions of basic geological research, aiming at the trajectory curve of drilling and exploration well, combined with the relevant data of downhole drilling, so as to give full play to the basic role of geological guidance technology, and combined with the drilling conditions, geological conditions The downhole curve predicts the structural change of the target layer, and timely adjusts the wellhead trajectory, so that the well body can cross the reservoir structure to the greatest extent.

2. Introduction to New Geosteering Technology

In the process of thin-layer reservoir exploitation, the new-type geosteering technology is developed on the basis of the guiding drilling technology, and with the continuous improvement of scientific and technological level, the application scope of the new-type geosteering technology is expanding.In the 1960s, enterprises have successfully developed natural gamma technology and resistivity detection technology, but remote sensing detection technology is still in the primary stage. This detection status leads to serious constraints and restrictions on the performance and function of downhole equipment.

Since the 1980s, the MWD/LWD technology has gradually matured, and the guided drilling technology has officially entered the exploitation of thin reservoirs, this situation has led to various types of geological information detection equipment, such as: electromagnetic induction resistivity detection equipment, neutron porosity detection equipment and formation density detection equipment, etc. , although the geological guidance equipment used in this period can meet the needs of a small part of geological structure exploration, it still can not carry out precise exploration for special geological conditions.

In recent years, with the continuous use of geological steering drilling technology, the basic functions of downhole steering equipment and instruments have been continuously improved and optimized, and bit geological drilling equipment have emerged one after another, such as EWR / ABI, DGR / ABI, etc. This exploration status also leads to the continuous research and production of various geological survey equipment with comprehensive functions and excellent operation performance. During this period, the new geological guidance technology and equipment have been growing rapidly and can evaluate the geology in real time and comprehensively. At this time, the geological guidance drilling technology has officially entered the peak development period.At present, more technicians are developing and exploring in an all-round way the more excellent Rotary Guidance Technology and closed-loop guidance technology on the basis of the already mature geological guidance technology, in the process of thin-layer reservoir exploitation, rotary steering and closed-loop steering will become the main

 https://doi.org/10.1051/e3sconf/202235802025

steering technology of exploration technology, which is widely used in various geological conditions. As shown in Table 1, the type of horizontal well and the Oriented Technical Model [2].

Table 1 Types of horizontal wells and oriented technical models

Serial number	Type of horizontal well	Main orientation mode
1	Conventional Horizontal well	three-dimensional
2	Thin layer horizontal well	three-dimensional
3	Horizontal well in tight reservoir	Two-dimensional geochemical/elemental logging
4	Regional horizontal well	3D + seismic
5	Horizontal well in Qianshan County Reservoir	3 d fracture analysis geochemical/elemental

3. Principle of new geological guidance technology

In the process of thin layer oil reservoir production, the new geosteering technology is mainly divided into two modes, such as SST and MWD/DWD, in which SST guiding equipment mainly provides technical support for downhole equipment through cable lines, in this way, the real-time collection of downhole reservoir data can be transmitted to the ground receiving equipment according to certain data and information patterns, and the ground structure instrument can decode and process the received signals in detail, finally, the accurate downhole data is displayed on the display equipment.As shown in Table 2, instrument comparison.

Table 2 comparison of instruments

Instrument type	SST	MWD/DWD
Signal transmission mode	Electric cable	Mud or formation medium
advantage	Low cost, fast data transmission The signal is less affected by drilling operation and has no special requirements for drilling fluid	Simple operation and effective occupation of drilling rig Less time, safe construction efficient
Faults	Construction Technology Complex tea, occupied the effective drilling time is too much, the safety can not be effectively guaranteed, the construction delivery rate is not high	The data transmission speed is slower than SST, the signal is affected by drilling operation and medium, and the cost is high
application	Small range use	It is in the mainstream position in China

When MWD/DWD technology is used for directional drilling and information measurement, it mainly uses mud or formation medium, and uses downhole detection equipment to carry out core information measurement or drilling operation, it is necessary to take out the downhole detection equipment and repeat the above construction and measurement process according to the project construction, but the equipment can not be effectively placed at the bottom of the well in the section with large displacement parameters or drilling angle, then need to use the opening pump to the bottom of the well, the opening pump base key and other construction technology.As shown in Figure 1, the implementation process of new geological guidance technology.

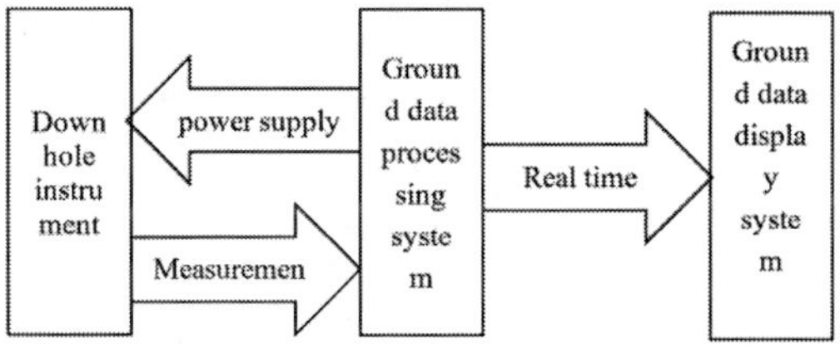

Fig. 1 implementation flow of new geosteering technology

4. Development of thin layer reservoir

In the actual energy development of thin layer reservoirs in a certain area of China, due to the special local natural environment and geological conditions, the reservoirs can not be effectively developed and used.

First, the reservoir foundation depth in this area is small, and the internal structure of the rock is very dispersed and loose. The shallowest landfill depth of the oil layer is only about 1700 meters underground, and the maximum energy depth is only about 1900 meters underground. Therefore, relatively little information and data are available.Under the background of incomplete data, it is impossible to get the drilling fluid structure system which matches the local formation better, and the Rock Structure Dispersion has put forward a higher technical requirement for bit equipment selection and operation technology.

Second, because the oil reservoir in this area belongs to thin geological structure, the oil layer is thin. The maximum oil layer thickness is only about 10m, while the minimum oil layer thickness is 3M, so the average oil layer in this area is no more than 5m. For this kind of thin-layer reservoir, when actually developing energy, each horizontal well should be controlled and managed by three location points, and the external shape of the horizontal development section needs to ensure bow shape, and the length of the construction section in the horizontal direction should be controlled at about 800m.Since the reservoirs in this area are bottom-water reservoir models, it is important to ensure that the entire horizontal section is kept away from the reservoir layer at the oil-water interface and that the excavation length is 1.2-1.4 m from the top of the soil layer, therefore, the orbit control of this class is relatively difficult.

Third, at present, the non-uniform characteristics of the vertical stratum of the thin-layer reservoir are relatively

 https://doi.org/10.1051/e3sconf/202235802025

high, so the geological structure of the reservoir fluctuates greatly, which is very easy to produce obvious structural fluctuations in local positions. Therefore, when using the geological guidance technology, it is required to be close to the top of the sand and gravel soil layer, so as to ensure a high level of drilling success rate.

5. Application strategy of new geological guidance technology

5.1 geological guidance technology for horizontal wells

1. Guiding technical operation principles

The base oil layer of an oil field in China is relatively thin, its average oil layer thickness is about 5 meters, so the oil field belongs to a typical thin layer oil reservoir, and the oil reservoir types in this area are mainly bottom water oil reservoirs, for this reason, during the actual development of an oil field, the project development needs to be carried out for a horizontal well, and the specific location of the horizontal section should be located at the top of the oil field about 1 meter, away from the interface between the oil and water resources at all times, ensure that the basic productivity of Horizontal Wells, the specific location of horizontal wells concentrated in the top of the oil.

When using the new geological guidance technology, due to the great impact of the oilfield hard interlayer on the landing, once the landing angle is too large, the rebound problem will occur at the interface when using the bit, resulting in the bit falling back into the soil and rock structure again. If the hard interlayer rock is relatively thick, even if the bit lands successfully, the angle will also be too large, resulting in the long-time drilling operation of the bit in the hard interlayer, resulting in serious wear of the bit equipment.If the hard interlayer is too thin, the vertical depth of the horizontal well will be lower because of the too small angle, which will affect the dynamic movement mode of the oil well [4] . Therefore, it is necessary to divide the landing mode of bit equipment carefully according to the actual situation of hard interlayer at the top of oil layer. If there is a hard interlayer at the top of the oil layer, the landing angle should be about 86-87 degrees, and the drilling speed should be controlled effectively during the actual drilling process, make sure that the deviation is gradually increased throughout the drilling process, so that the angle between the bit and the geological stratum can reach about 88-89 degrees after the bit penetrates the hard interlayer.

2. Operation cases

Due to the changes of geological structure and sedimentary environment in this area, the soil covered above the oil field shows a certain regularity. Therefore, during reservoir exploitation, it is necessary to use drill bits to distinguish in detail according to the changes of soil and rock distribution, and predict the specific landing depth in combination with the geological data of this area. Taking the A14 horizontal well as a practical case, it is necessary to distinguish the horizontal well from its own structure and the data of adjacent wells before drilling, and to divide at least 3 marked structural layers above the final exploration target structural layers, after drilling a marked layer, the drill bit needs to compare the actual drilling depth and angle with the expected measurement, which can be used as the basis to adjust the guidance parameter model and predict the landing depth.Because the sedimentary structure of the oil field in this area causes obvious changes in the local geological position, when the bit is drilled vertically to about 1700m, it is basically consistent with the scheme design. After drilling to the lower layer for re measurement and calculation, it is significantly different from the depth of the target landing layer. In addition, the changes and uncertainties caused by the local geological structure directly affect the production quality of the oil field, Therefore, if the drill bit wants to encounter the target layer, it needs to adjust the angle to drill at least 10m depth, and the drilling angle reaches more than 88 degrees, so the landing effect is ideal. As shown in Figure 2, downhole cross section.

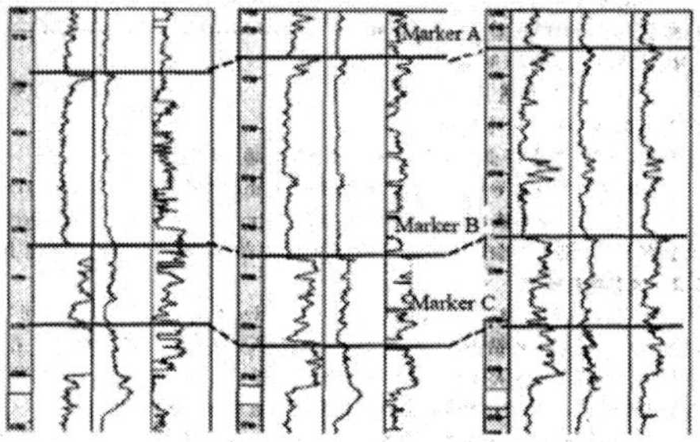

Fig. 2 underground cross section Fig. 2 underground cross section

5.2 Horizontal geosteering technology

1. Technology oriented operating principles

Since the oilfield is mainly a thin-layer bottom water reservoir, the geological stratum has sufficient water resources. In order to ensure that the drilling trajectory is always below the soil, rock and hard interlayer, it is necessary to stay away from the oil-water boundary in the horizontal section, and drill part of the oil layer structure above the vertical direction as far as possible, so as to ensure the basic productivity in the horizontal well, If there is a clay or calcareous interlayer below the oil layer, it cannot pass through the interlayer.

2. Case Study

In the process of drilling operation, on the one hand, the influence of seismic information on geological structure should be controlled as a whole, and on the other hand, the results of downhole detection curve should be analyzed and judged in detail, and on the basic conditions of comprehensive study, it can effectively judge the underground construction trajectory and the actual real-time position, and carry out the trajectory adjustment and control according to the situation of oil reservoir production, for this reason, well A11 should be taken as a practical case for the implementation of new geosteering technology. Because the base thickness of the oil layer in this horizontal well is about 4 meters, and the geology of this section belongs to thin sandstone reservoir, in general, the upper part of this geological oil layer will develop a mud interlayer structure with a thickness of about 1 meter, so horizontal well trajectory control, we need to choose the sudden intercalation and 1-1.5 meters between the top of the oil layer, and as close as possible to the top of the soil structure.When the drill bit is used to drill an oil well, the drilling angle should be controlled at about 87 degrees, and the maximum angle increases to about 90 degrees with the deepening of the drilling depth. After the drill bit is used to drill to 2400m, the gamma parameters generated by the drill bit will gradually increase. This situation directly shows that with the increasing depth of the drill bit, Soil quality and foundation content are increasing. At the same time, the resistivity and soil oil content are decreasing and decreasing.

According to the density curve of soil top structure, with the increase of drilling depth, the density of soil top structure shows an increasing trend, to this end, the technicians have combined the relevant data and information, such as Gamma Ray, resistivity and density curves, to further explain that the increase in drilling depth will result in the drilling location near the top of the soil and rock, the drilling angle should then be reduced to 89 degrees and the wellhead trajectory should be returned to the sandstone formation as far as possible.

Because the rock structure of the horizontal well is very scattered and loose, it is relatively difficult to control the well deviation. Therefore, it is necessary to continuously increase or reduce the well deviation angle to ensure that the bit can drill stably.The length of horizontal well is 860 meters, the effective length of horizontal section is 815 meters, and the vertical depth of drill bit should be controlled below 1 meter in order to ensure the quality of thin layer reservoir production. As shown in Fig. 3, downhole curve test chart [5].

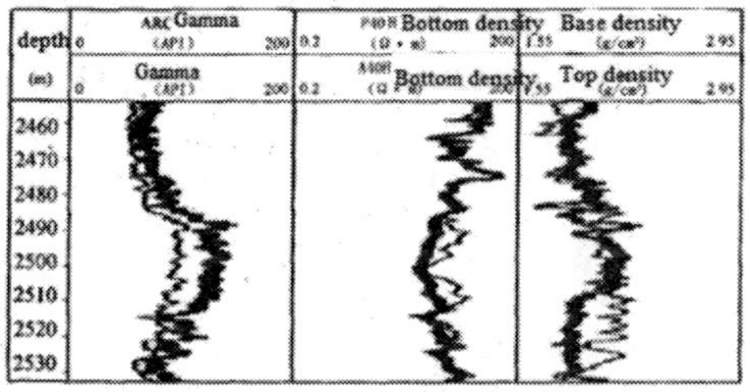

Fig. 3 downhole curve test diagram

6. Conclusion

It can be seen that in the process of thin-layer reservoir exploitation, the new geological guidance technology has become the key factor for the success of drilling, fundamentally ensuring the smooth landing of horizontal well drilling, better controlling the connecting track of horizontal section, and realizing the objectives of detecting geological structure.

References

1. Wang Wanqing, Yeung Kwong, Tian Fengjun. Application of Rotary geosteering technique in np-83 D horizontal well drilling [J]. Mine Engineering, 2020,8(2): 6-6.
2. Yang Yang. Application of horizontal well drilling technology in development of thin heavy oil reservoirs in Liaohe oil field [J]. Equipment maintenance technology, 2021(33): 1-1.
3. Zhang Bin. Application of reservoir prediction technology in reservoir evaluation of block j 31[j]. Heilongjiang Technology Information, 2020, 000(005): 17-18.
4. Golo. Application of horizontal well guidance and Fault Identification Technology in complex fault block reservoir [j]. Petrochemical Technology, 2021, 28(8): 2-2.
5. Wu Yue. Application of precise geosteering technology in horizontal well Development —— Taking Chang 8 reservoir in Zhenbei oilfield as an example [j]. Groundwater, 2021,43(4) : 2-2.

Practice and cognition of effective water-flooding development mode in Gaotaizi block of ultra-low permeability reservoir

Ping Cui

No.9 Oil Production Plant, Daqing Oilfield Co.Ltd., PetroChina 163000 China

Abstract. During nearly 25 years of water injection development in Gaotaizi reservoir of LHP oilfield, due to poor reservoir physical properties, poor adaptability of well pattern and difficulty in establishing effective displacement between oil and water Wells, linear water injection, infill adjustment and refracturing tests have been carried out successively. Although certain effects have been seen, the effects are short and unsatisfactory. In this paper, the microscopic pore structure, sensitivity and seepage characteristics of ultra-low permeability high platform reservoir are studied through core laboratory experiments. New technology of nonlinear test and low permeable physical simulation experiment, the research on the rules of the nonlinear characteristics of the reservoir fluid and the characteristics of movable oil through research has been clear about the dragon bubble oilfield high effective displacement distance platform layer, and the new fracturing technology and the development of the technology to adjust the organic combination, the displacement between well and well past into displacement between Wells and joint network, explore the ultra-low permeability reservoir to improve the single well production, new ways to improve the development effect, with the types of low permeability oilfield development effectively.

Key words: Ultra-low permeability;effective displacement; joint network; Development effect.

1. Block overview

1.1 Geological characteristics

The structure of LHP single production Gaotaizi oil layer block is located in the west of Qijia Gulong sag and the north of Longhupao Da'an terrace in the central depression of Songliao basin. It is the first ultra-low permeability oilfield in the western periphery of Changyuan that has been put into development in a large area. The developed oil-bearing area is 41km^2 and the produced geological reserves are 880 × 10^{4}t. The main reservoirs are G Ⅲ group and G Ⅳ group, which are characterized by poor reservoir physical properties and small single-layer thickness. The average porosity is 15.3% and the air permeability is 0.5mD, the thickness of perforated sandstone is 13.3m, and the effective thickness is 5.8m. From south to north, the thickness of single layer of oil layer becomes thinner, the number of layers increases, and there is no intensively developed interval. The reservoir lithology is mainly argillaceous siltstone, and the argillaceous content is as high as 23.0%, belonging to a tight reservoir with low porosity, low permeability.

1.2 Development profile

LHP Gaotaizi reservoir block was fractured and put into operation in 1998, and the oil well production decreased rapidly. In 2000, microbial injection, acidizing and plugging removal, repeated fracturing, hydraulic sand blasting perforation and other tests were carried out to reduce water injection pressure. Only 42.6% of the wells were effective, with an average effective period of 3.2 months. The cumulative oil increase of a single well was 78t; In 2001, the infill adjustment test area was opened up in L114-12 well block, with the well spacing reduced from 300m to 150m, and the daily liquid production of infill wells was only 0.32t; In L118-10 well block, in combination with the linear water injection well network from the original well pattern to the water injection flooded angle well, the oil well drainage is densified, the oil well drainage distance and well spacing are reduced to 212m, and the daily liquid production of the densified well is 0.5t; Since 2002, the water injection wells have been gradually inter injected, and the oil production wells have been changed to catch oil, so as to reduce the development cost of the oilfield.

2. Research on effective utilization mode

2.1 Core laboratory study

Through the indoor core test, the micro reservoir characteristics of Gaotaizi reservoir, such as pore structure, fluid and seepage characteristics, sensitivity, etc., are clarified.

The pore structure of LHP Gaotaizi reservoir is mainly composed of small throat, and the contribution of large throat to permeability is large; The permeability is less than 0.5mD, and the average throat radius is less than 1 μm. Permeability 0.5-1mD, throat radius 0.5-1.5 μm; The average permeability of Gaotaizi reservoir is 0.5mD, and the corresponding movable fluid percentage is 30.43%, belonging to low permeability class IV reservoir; The stress damage degree corresponding to permeability of 0.5mD is 21.74%. The core with permeability less than 0.5mD is mainly plastic deformation, and the reservoir as a whole is weakly pressure-sensitive; The two-phase co permeability ranges from 12.7% to 28.5%, and the average final recovery factor is 35.4%; Movable oil mainly comes from the pore space controlled by 0.1~1 μm throat, the two-phase co permeability zone is narrow and the oil displacement efficiency is low; Starting pressure gradient is 0.1685MPa/m.

According to the analysis of five parameter comprehensive evaluation method, LHP Gaotaizi reservoir belongs to low permeability class III reservoir. Compared with Putaohua and Fuyang oil layers, Gaotai sub layer has small throat radius, small percentage of movable fluid, high starting pressure gradient, high clay mineral content and low crude oil viscosity, which makes development difficult.

Table 1. Five parameter comprehensive evaluation results of low permeability reservoir

oil layer	block	permeability (mD)	Average throat half_path (μm)	Percentage of movable fluid (%)	Starting pressure gradient (MPa/m)	Clay mineral content (%)	Crude oil viscosity (mPa•s)
Gaotaizi	LHP	0.5	1.023(III)	30.43(IV)	0.1685(III)	23(IV)	1.40(I)
	QJN	1.18	1.839(III)	42.54(III)	0.0930(II)	10.33(III)	2.30(II)
Putaohua	AN	13.30	2.927(II)	51.70(II)	0.0067(I)	15.50(IV)	5.73(III)
	GLN	3.10	1.948(III)	51.73(II)	0.0240(II)	10.25(II)	2.12(II)
Fuyang	BYCG	2.30	1.987(III)	34.12(IV)	0.0887(II)	15.56(IV)	2.45(II)
	TLH	0.27	0.452(IV)	18.91(IV)	0.2209(III)	12.24(III)	1.74(I)

Calculation of effective driving well spacing by using the formula of effective driving distance in Daqing Peripheral Low Permeability oilfields:

$$L_c = \frac{P_w - P_f - \frac{cn\mu\eta}{kh}}{\lambda}$$

λ — Starting pressure gradient,MPa/m ; K — permeability, mD; P_w — flowing pressure of water injection well, MPa; P_f —Oil well flowing pressure, MPa; Lc —effective drive well spacing, m; c —unit conversion factor86.4; n —Ratio of row spacing to well spacing, dimensionless; μ —Formation viscosity of crude oil, mPa.s; η — Stable liquid production intensity, t/d.m; h —effective thickness, m。

It is calculated that the effective driving well spacing of Gaotai sub layer in LHP single high well area is 149m.

2.2 Network optimization

According to the idea of combining fracture network, due to the reservoir development status, combined with the existing injection production well network and stress direction, optimize the relationship between well network and fracture network, fracture network and sand body configuration, and transform the previous effective displacement between wells into effective displacement between wells and fracture network.

2.2.1 Optimization of large scale molded fracture network in vertical well

According to the direction of maximum principal stress, effective displacement distance and water drive front, the scale of single well fracture pattern is optimized to prevent premature flooding while realizing effective displacement; The designed fracturing half fracture length of the side well is 260m, the micro fracture extension width is 200m, and the displacement well spacing after fracturing is shortened to 85m; The designed fracturing half fracture length of angle well is 300m, the micro fracture extension width is 250m, and the displacement well spacing is shortened to 130m after fracturing.

2.2.2 Optimization of infill horizontal well spacing

According to the reservoir development, combined with the injection production well pattern and stress direction, the matching relationship between well pattern and fracture pattern is optimized, and a personalized perforation and fracturing scheme is prepared to establish effective displacement between fractures and well pattern. There is no perforation within 50m of the water well, the half fracture length of fracturing is 80-260m, and the displacement distance after fracturing is 50-139m, all within the range of effective displacement distance.

In order to ensure sufficient water injection energy at both sides of the horizontal well, the asymmetric diagonal injection mode is adopted to form a zigzag injection production well pattern. 8 wells are designed to be converted into injection. After the conversion, the water wells are reasonably matched with the water injection intensity. The water injection intensity is controlled in the vertical geostress direction, and the water injection intensity is controlled at about 1.0m^3/m.d. the water injection intensity is appropriately enhanced in the parallel geostress direction, and the water injection intensity is about 1.5 m^3/m.d.

According to different reservoir conditions, different methods are adopted for test exploration. Vertical well fracture pattern fracturing is adopted in the reservoir with a large number of layers and poor continuity (L118 test

 https://doi.org/10.1051/e3sconf/202235802026

area). Horizontal well infill and through layer fracturing tests are carried out in the stable sheet sand reservoir with prominent development in the main layer (L130 well area), so as to seek an economic and effective development method for different types of reservoirs.

3. Implementation effect and understanding

3.1 Vertical well fracture pattern fracturing

Five wells in L118 well cluster of longhupaogaotai sub layer are selected for vertical well fracture pattern fracturing test. The average fracturing fluid consumption of a single well is 3641m^3 and sand addition is 96m^3. After fracturing, the daily oil increase of a single well is 3.8t, and the cumulative oil increase of a single well is 2597t.
The following understandings have been obtained for vertical well fracture pattern fracturing:
First,a certain fracture network scale has been formed, and large-scale reservoir reconstruction has been realized. From the process of downhole microseismic real-time monitoring, during the fracturing process, the slippery water produced obvious main fractures in the first stage, and then the fractures were mainly generated in the fracture network in the clean water and sand adding stages. The internal volume of the early fracture grid was transformed. According to the underground microseismic monitoring results, the actual fracture network is 308-534m long, 63-93m wide and 66-83m high.
Second, reservoir production has been significantly improved, single well production has been increased, and oil production has been significantly increased. At the initial stage after fracturing, the thickness production ratio of sandstone in the whole well is increased from 62.9% to 97.0%, and the production degree of reservoirs with different thickness is improved; The output ratio of sandstone thickness of reservoir with effective thickness <0.5m has increased from 53.7% to 100%, and the daily fluid increase accounts for 39.8% of the whole well. The fluid increase effect is obvious. After 6 months, the output of thin and poor layers became worse, and the output ratio of sandstone thickness of reservoir with effective thickness <0.5m decreased from 100% to 18.5%; The liquid production of the main formation is stable, and the liquid production proportion increases. The liquid production proportion of GIII3 accounts for about 50% of the whole well. Compared with conventional fracturing, the daily oil increase of a single well at the initial stage of fracture pattern fracturing is 3.8t, which is 2.5 times that of conventional fracturing.
Third, effective displacement has not been established, and the liquid supply capacity is poor. From the water well, the injection pressure is stable at 22.5MPa, which is equal to the pump pressure. The apparent water absorption index is reduced from 1.07m^3/d.MPa before the measures to 0.90m^3/d.MPa; From the perspective of fractured oil wells, the work diagram shows that the liquid supply capacity gradually becomes worse, and the bottom hole flowing pressure drops from 3.4Mpa to 0.3MPa.

3.2 Horizontal well infill combined with layer penetrating fracturing

The length of single horizontal section of three horizontal wells in l130 well block is 803m, and the encountered sandstone and oil-bearing sandstone are 697m and 611m respectively, with the penetration rate of 86.8% and 76.0% respectively; The average perforation section of a single well is 7.7, and the fracturing section is 6. The average fracturing fluid consumption of a single well is 1033m^3, and the sand addition is 117m^3. At the initial stage of production, the daily oil production of a single well is 5.0t.
The following understandings have been obtained by infilling horizontal wells and fracturing through layers:
First, the output of infill horizontal wells is 3.3 times that of peripheral vertical wells in the same period, which has achieved good production effect.
Second, from the perspective of water injection, effective displacement can be achieved. Fracture monitoring was carried out for 14 sections of 2 horizontal wells. The fracture length was 108~238m, and the displacement distance with the surrounding water injection wells was 53~145m, both within the effective displacement distance of 149m. After the water injection of the surrounding wells is resumed, the obvious effect characteristics are seen within 13 days. After the effect is received, water is seen along the fracture rapidly. 3 days after the water injection of the wells is stopped, the water content is restored to about 40%, and the water injection adjustment effect is obvious. The formation pressure is 12.78MPa, 3.73MPa higher than the surrounding vertical wells, maintaining a high level.

4. Conclusion

(1) The microscopic characteristics and percolation law of Gaotaizi reservoir are clarified, which lays a foundation for scheme design and fracture network optimization.
(2) The development and design idea of fracture pattern matching has been formed, the three technologies of vertical well fracture pattern fracturing, horizontal well location deployment and completion optimization, and zigzag horizontal well injection production system adjustment have been improved, and effective displacement has been established.
(3) Although infill horizontal well combined with through layer fracturing technology can establish effective displacement, it is affected by small injection production well spacing. On the one hand, the number of fracturing sections and fracture length are limited, affecting the initial production; On the other hand, the fractured water breakthrough phenomenon seriously affects the development effect.

References

1. Lei Qun et al.Fracture network fracturing technology for improving the transformation effect of extra-low permeability reservoir[J].Acta Petrolei Sinica, 2009,03(30-2).

2. Weng Ding et al.Fracture network fracturing technology its field application[J]. Acta Petrolei Sinica, 2011,02.
3. Wang Haiqing et al.Application of volume fracturing in ultra-low permeability reservoir[J],China Petroleum and Chemical Standard and Quality[J],2012,02.
4. Tang Meirong et al.Research and test of new multi-fracture fracturing technology[J]. Oil drilling technology, 2010,03(32-2).
5. Cao Guoliang et al.Study on percolation law of long fracture fracturing in five-point pattern with ultra-low permeability[J].Western China science and technology, 2012,03(11-3).
6. Wang Guanhua.Research and application ong-seam fracturing completion technology for ertical Wells in ultra-low permeability reservoirs[J]. Neijiang science & technology,2012,08.
7. Wang Xianjun et al. Research and application of integrated fracturing technology in block 301, hailar oilfield [J]. Petroleum geology and engineering,2012.
8. Shi Qingxuan et al. Study on integrated fracturing scheme of low permeability oilfields -a case study of block tai1 [J]. Journal of Yangtze university (natural science edition),2010,7(2):62-65.
9. Liang Liping et al. Analysis on production decline law of fracturing Wells in low permeability reservoirs [J]. Natural gas geoscience,2012,08(23-4).
10. Yan Jianwen et al. Research and application of fracturing retrofit technology for horizontal Wells in low permeability oilfields [J]. Petroleum machinery,2009.
11. Hao Mingqiang etal. Production decline rule of fracturing horizontal Wells in ultra-low permeability reservoirs [J]. Acta petrolei sinica,2012,03(32-2).

E3S Web of Conferences 358, 02027 (2022)
GESD 2022
https://doi.org/10.1051/e3sconf/202235802027

Influence of abnormal water injection well on qualified rate and treatment countermeasures

Jianyu Zhang

The second operation area of the Third Oil Production Plant of Daqing Oilfield Co., LTD., Daqing 163000, Heilongjiang, China

Abstract: More heterogeneous oil reservoir in the process of waterflooding oil field, reasonable water injection is to improve the recovery efficiency, an important means to slow down the natural decline, strengthen the daily management of injection Wells, according to the influencing factors of injection Wells anomalies corresponding measures, improve the water time, improve the qualified rate of water injection, truly underground "good enough water and injection water" goal, It is the most economical and effective way to improve oil recovery.

Key words: Qualified rate Factors affecting injection well anomaly improve the qualified rate of water injection.

1. Question raising

At present, the test data of stratified injection well are in accordance with the data admission standards, and the normal service period is four months. In the four months, due to the complexity of underground geological conditions, the influence of the interlayer contradiction is very big, each section of the water quantity is very uneven, to make each interval water quantity can't meet the requirements of injection allocation, allocate water mouth or big, last time or congestion, or big or small, cannot under the existing injection pump pressure, injection oil pressure meet the requirements of injection allocation. The irrationality of underground water injection directly affects the development effect of oil well.

2. Analysis of abnormal cause of stratified injection well

Abnormal zonal water injection well refers to the zonal water injection well whose injection pressure or injection water volume changes suddenly under the condition of constant system pressure. The main causes of abnormal Wells are as follows.

2.1 The dynamic change of stratum results in abnormal water injection

According to the development of oil layer in reasonable injection interval and low layer of injection allocation, the time of leveling and underlining, historical water absorption and the change of surrounding oil Wells, the leveling and underlining can be divided into two categories: gradually worsening injection volume and suddenly decreasing injection volume. The abnormal causes of 9 test Wells were analyzed, and the abnormal changes of injection water after test were caused by the variation of formation water absorption in 4 Wells, which accounted for 44.4% of the total abnormal Wells. For example, after the test on February 27, 2013, the oil pressure and water injection volume of North 2-3-125 were 13.9mpa and 80m3. Since April 14, 2013, the water injection volume of this well has gradually decreased. After the test on April 24, 2013, the oil pressure and water injection volume were 13.9mpa and 48m3. On April 25, 2013, two pressure gauges were used to verify the injection pressure in the field. The pressure of the two pressure gauges was basically consistent with that of the original well. In addition, after testing, no spout spout was found to be large, and it was preliminarily determined that the reason for water volume rise was caused by the variation of water absorption in the formation.

Tab.1 Influence of stratum dynamic change on water absorption

NO	Injection allocation		Test			abnormal			After repair		
	pressure	The amount of water	Time	pressure	The amount of water	Time	pressure	The amount of water	Time	pressure	The amount of water
A	13.9	80	2013.2.27	13.9	80	2013.4.14	13.9	63	2013.4.24	13.9	48

2.2 Abnormal water injection due to test impact

At present, the average interval of injection Wells is 5.3. Especially in section 5-8, there is serious interzone interference during testing and deployment, which makes it difficult to form data. The analysis shows that the main reasons affecting the qualified rate of water injection are as follows: First, the efficiency of stratified testing and deployment is low. The average time of single well testing and deployment is 4.5 days due to the need for repeated quantitative testing, surface replay, and dropping and fishing nozzle. Second, because the interlayer interference has a great influence on the decline method test, the pass rate of test data is low, and the pass rate of reinspection is about 80%. Third, due to the small segment of some subdivided Wells, the flow rate of the fluid changes when it passes through the upper and lower tools, and the test instrument is difficult to locate in the stable flow rate section, resulting in the deviation of the test flow. In the testing process, restricted by factors such as the accuracy of the depth counter of the drawworks, the shrinkage degree of the steel wire and the wear of the centralizer of the flowmeter, the measured water volume deviation is large, which affects the allocation accuracy and shortens its stability period. After analyzing the abnormal causes of 9 test Wells, it was found that 1 well had abnormal injection water due to the large spout. For example, after the test of Well North 2-351-21 on May 16, 2013, the oil pressure was 14.1mpa and the water injection volume was 70m3. On August 25, the water injection volume reached 102m3 when the water injection pressure was 14.1mpa. The injection pressure is controlled at 12.5mpa and the injection volume is 73m3 due to the overflow scheme. On August 25, two pressure gauges were used to verify the injection pressure in the field. The pressure of the two pressure gauges was basically consistent with that of the original well gauge. It is preliminarily determined that the reason for the water increase is abnormal change of injection pressure caused by downhole pipe string or oil layer. On August 26, the water measured by the instrument under the test shift was basically consistent with the on-site water injection volume, indicating that the pressure was equal to the on-site pressure gauge. The test found that the partial II water injection volume exceeded the allocated water injection volume by a lot. After pulling out the water nozzle of this layer, it was found that the water nozzle was broken, and the water nozzle was restored to normal after replacement.

Tab.2 Influence of nozzle spout on water absorption

No	Injection allocation		Test			abnormal			After repair		
	pressure	The amount of water	Time	pressure	pressure	The amount of water	Time	pressure	pressure	The amount of water	Time
B	14.1	70	2013.5.16	14.1	70	2013.8.25	12.5	73	2013.8.26	14.0	77

2.3 The influence of the years of plugging nozzle and injection well string due to poor water quality results in abnormal testing

In recent years, the water quality of the outlet of the water injection station has been greatly improved, but the water injection quality is still poor due to the secondary pollution of oil layer caused by pipeline and pipe string corrosion. It has been shown that the worse the injection water quality is, the lower the qualified rate of injection water is. Injection water quality is poor, easy to cause the blockage of the plug. The whole well shows that the water injection pressure increases, the water injection volume decreases, and the water absorption of individual sections decreases or does not absorb water. Such abnormal water injection Wells are easy to be found and can be tested and verified in time. For abnormal Wells that cannot be found in time, individual intervals of injection Wells are blocked, but the amount of water injected into other intervals with strong water absorption capacity is increased, and the data application range of the whole well is not exceeded. Such abnormal water injection Wells can only be found in the normal test cycle, and the abnormal water injection takes a long time, causing great harm. In the process of water injection well, the water nozzle will be blocked due to poor water quality, so that the pressure of the well will rise and the water volume will drop. Although the proportion of Wells with poor water quality is not very large, it does affect the qualified rate of water injection, especially at the end of the water injection area. For example, in the test of Well North 2-361-23 on March 2, 2013, the partial v original water absorption was 15 m3. On April 22, 2013, the on-site water injection was abnormal. When tested on May 12, 2013, the water absorption of this section was only 5 m^3.

Tab.3 The effect of faucet plugging on water absorption

No	Injection allocation		Test			abnormal			After repair		
	pressure	The amount of water	Time	pressure	pressure	The amount of water	Time	pressure	pressure	The amount of water	Time
C	13.3	50	2013.3.2	13.3	52	12.4.22	13.3	19	2013.5.12	13.3	50

In some Wells, due to the long interval between operations, the pipe string is severely scaled and corroded underground, resulting in obstruction and blockage of the test instrument. Currently, the string is replaced only when the operation is required. In my opinion, a specific number of years should be specified for the string of water injection well. After a certain number of years, it is necessary to run the string again, so as to ensure the testing work. If you wait until the operation is blocked or stuck, it will affect the normal test and the qualified rate of water injection.

Tab.4 The test does not affect the water absorption

	NO	Time	Affect the number of days	The reason why water cannot be divided	Influence the number of layers	Affect the injection allocation (m^3/d)	Affect real note (m^3/ d)	The next step measures
1	X	2012.7.16	197	Partial 4 can not wait for homework	4	80	72	On the job
2	Y	2013.1.15	138	Partial 6 can not wait for homework	6	180	124	On the job
3	Q	2012.12.7	176	Partial 6 can not wait for homework	6	110	112	On the job
4	E	2013.4.28	33	Partial 3 can not wait for homework	4	40	37	On the job
5	R	2013.4.26	35	Partial 7 can not wait for homework	7	120	119	On the job
6	T	2013.4.18	43	Partial 5 can not wait for homework	5	110	75	On the job

2.4 The poor quality of the well washout caused the abnormal test

The quality of well washing directly affects the qualified rate of water injection. In the field management process, there are many reasons affecting the well washing quality. First, the current well washing mode is that the mine geological team organizes and arranges vehicles for well washing, and the oil production team supervises, and one tank car does not continuously wash well, so the tank car is used for well washing, and the well washing capacity is limited. Second, in the process of well washing, there is often a stop in the middle of the process, waiting for tank car sewage, which can not ensure the continuity of well washing, and the stolen goods in the well can not be cleaned. This is because the flushing fluid must be continuously circulated in order to carry the loot from the bottom hole and the string. As can be seen from the table below, well north 2-351-31 was washed twice by tank truck on March 24, 2013, but no effect was found after washing.

Tab.5 Influence of well washing quality on water absorption

No	Wash well before		After the wash well		effective
	Oil pressure (MPa)	The amount of water (m^3/d)	Oil pressure (MPa)	The amount of water (m^3/d)	
A	13.7	16	13.7	10	NO
B	13.4	69	13.4	80	YES

2.5 Low precision of pressure taking equipment and instrument in water injection well results in abnormal testing

The measurement deviation of the pressure gauge used in the field before and after checking is also the cause of the change of injection volume. There is a big gap between the accuracy of the pressure gauge and the on-site management requirements of the injection well. For example, the pressure gauge in the well group of North 2-351-19 was 12.9mpa during the test on January 23, 2013. After the test, the pressure gauge used in the well was checked, and the pressure value was 13.1mpa under the same injection volume. A total of four gauges were changed for verification. Under the same injection volume, only one of them showed injection pressure of 12.9mpa, while the other three gauges had the highest pressure of 13.2mpa and the lowest pressure of 12.8mpa, with the upper and lower error of 0.4mpa. But these tables are all in the prescribed range, a qualified table. At present, daily water injection volume comparison under the same pressure is required for on-site inspection of water injection well, but the existing pressure extraction equipment is difficult to meet the requirements. Abnormal Wells caused by unknown abnormal reasons account for 2.22% of the total number of abnormal Wells. Most of these Wells are Wells with reduced water absorption, and the accuracy of pressure taking equipment and pressure gauge has a great impact.

What is discussed above is that under the condition of constant system pressure, but in actual injection well management, we often encounter the pump pressure fluctuation has a certain influence (not all) on the injection water quantity, thus affecting the qualified rate of water injection. There are many production teams at the end of such well water injection network, about 7 Wells.

3. Countermeasures to improve the qualified rate of water injection

In view of the above factors affecting the qualified rate of water injection, we have taken corresponding measures:

3.1 Strengthen dynamic analysis, timely take fracturing, acidification and other stimulation and injection measures for oil and water Wells according to the water absorption status of the formation, and adjust the compliance rate of injection Wells; Whether a certain service life can be stipulated for the stratified pipe string of the injection well, after which new stratified pipe string shall be replaced; Strengthen the management of water injection station and

sewage station to ensure the quality of water injected qualified; Improve the accuracy of current gauges.
3.2 For Wells with poor water quality, wellhead filters should be installed, and a regular disassembly and washing system should be developed. At the same time, the drainage pipe network of certain Wells is flushed to ensure that the injected water quality is qualified.
3.3 In terms of well washing cycle, we changed the previous principle of well washing only when water absorption changes. For Wells with poor water absorption, we adjusted the well washing cycle to once every 2 months, and increased the total water volume of well washing to 50m3-80m3. In 2013, we tested 22 Wells. Using this cycle, the average water injection increased from 56.4m3 to 68.8m3, an increase of 12.48m3, and the time to water absorption deterioration increased from 17 to 31 days. The number of non-water-absorbing sections decreased from 13 to 8, and the number of water-absorbing sections recovered from 5.
3.4 At the same time, in order to effectively strengthen the management of abnormal water injection Wells and improve the qualified rate of water injection, the management of abnormal water injection Wells shall be "three determinations, three timely determinations, and four complete and accurate determinations". The three determinations are well determinations, table determinations and personnel determinations. Three timely is the position of workers in the timely detection of change; The reference staff focuses on timely implementation of information; Technicians focus on timely processing and resolution, reporting principles. At the same time, the goal of achieving accurate data is divided into four aspects, strict assessment, layer upon layer of checks. One is to implement the full and accurate admission of on-site data; The second is to implement the complete and accurate site data device; Three is the implementation of the site to take data instruments, instruments, all, accurate; The fourth is to implement the complete and accurate collation of on-site data. Check all on-site data of water injection Wells every month, do not estimate interest rate for inconsistent or false data, and strictly deal with them. Do a good job in testing and tracking, contact with the test team to jointly analyze the abnormal Wells and some key Wells, and deal with them in time, increase the number of days for water separation, and improve the qualified rate of layered water injection. The qualified rate of water injection was improved by taking comprehensive measures and strengthening the management of abnormal water injection Wells under the condition of increasing the number of stratified Wells and stratified Wells.

4. Conclusion

In the process of managing abnormal water injection well and improving the qualified rate of water injection, we have realized:
4.1 Reporting and tracking analysis of abnormal testing Wells will help solve and cure abnormal water injection Wells in the future.
4.2 The main means to ensure the days of water separation in stratified injection Wells and improve the qualified rate of stratified injection is to do a good job in the equipment management of water injection well data admission.
4.3 Pay attention to injection water quality and well washing quality, reduce the number of abnormal injection Wells, can effectively improve the qualified rate of stratified injection water.

References

1. Jiao Guiyan. The Method and Practice of Improving the Qualified Rate of Water Injection. Science and Technology Press, 2011.2.
2. Chen Taoping. Petroleum Engineering. Petroleum Industry Press, 2004.
3. Cheng Jiecheng, Liao Guangzhi, Yang Zhenyu, et al. Summary of ASP flooding test in Daqing Oilfield [J]. Daqing Petroleum Geology and Development, 2001,20(2):46-49.
4. Shao Zhenbo, Zhang Xiaoqin. Practice and understanding of polymer flooding in second-class oil layers in Daqing Oilfield [J]. Daqing Petroleum Geology and Development, 2009, 28(5):163-168.
5. Fu Xiaojun. Adjustment method and effect evaluation of injection and production parameters of weak alkali ternary flooding in second-class oil layers under small well spacing [J]. Chemical Engineering and Equipment, 2018(1): 143-146

https://doi.org/10.1051/e3sconf/202235802028

Analysis of reducing pressure and increasing injection in high pressure water injection block of low permeability reservoir

Qiong Wu

The 8th Operation Area of the 3rd Oil Production Plant of Daqing Oilfield Company Limited, Daqing 163000, China

Abstract: The plugging phenomenon near the well zone of oil injection well will seriously affect the water injection development of oil field, and then affect the exploitation of oil resources. The water injection development of low permeability reservoir is affected by various factors. Due to the large difference between layers and the serious interference between layers, the physical property of oil layer is relatively poor, and the water injection pressure shows a trend of increasing year by year, which has a great impact on the development of oil resources. As the oilfield development benefit is not ideal, it is necessary to take reasonable measures to reduce the pressure of water injection to ensure the effect of water injection to improve the recovery of the reservoir. At present, the technology has been mature, which can reduce the pressure of water injection and ensure the oil recovery rate. The application in several oil Wells has verified the correctness of this conclusion.

Key words: Oil pressure; Jam; Water injection.

1. Introduction

Permeability of low permeability reservoir development is affected by many factors, in the oilfield development part of the sensitive layer contains a large number of clay minerals, these components exist in water collision and migration, formation permeability, which to a large extent affect the specific development effect, so in production, after water injection geological contains clay layer formation permeability decline, It directly affects the actual production of oil Wells. However, many oil Wells still do not absorb water after water injection, and the formation operation cannot be carried out smoothly, and the pressure cannot produce, which brings great harm to the surface system and downhole tools. These unsafe factors directly affect the actual production. This paper analyzes how to achieve high efficiency production by reducing pressure and increasing injection technology.

2. Interpretation of low permeability reservoir production operations

In the current oilfield exploitation, the number of low permeability reservoirs is gradually increasing, because with the development of petroleum resources, people have a deeper understanding of reservoir development. Fracture-vuggy reservoirs are mostly low permeability reservoirs, showing the characteristics of diversification and deep burial. The underground oil reservoir space is large cave, reservoir heterogeneity is strong, and the relationship between oil and water and fluidity is complex. Under this trend, high-pressure water injection development becomes the main technology of oil exploitation. Current regional oil development shows the characteristics of the several problems: (1) reservoir physical property is poor, in the long existing in the development of "three highs" and "four low", the contradiction of the specific performance is high water injection pressure and injection-production well ratio is high, but the actual deficit is high also, in addition "four low" embodied in the low permeability and low productivity and other characteristics; (2) The continuity of the reservoir is poor, and the pressure relief of the oil layer is relatively fast in the development process. The water volume of water injection can not keep up, and the energy of the formation can not be supplemented in time, resulting in the phenomenon of insufficient energy, leading to the low productivity of the oil well. (3) low permeability reservoirs, the actual infiltration inefficient, pore is slightly fluid storage or stay region, but also the most fully contact with the formation water and the status of fouling in the most likely to occur, due to long-term under the influence of the reinjection sewage and homework workover job, cause of formation of suspended solids invasion, all sorts of bacteria, such as the stranded in the Wells. ④ The water quality of injection is poor. According to the water quality survey, the content of various substances in the water is relatively high, which affects the stability of water quality. The dissolved oxygen, oil content and bacteria content of water outside the station are relatively high. On the whole, all indicators are the main factors that restrict the water quality to reach the standard. (5) Secondary pollution of the reservoir is relatively serious, because impurities and related substances in the process of water injection production of low permeability reservoir are easy to form precipitation in the formation, resulting in blockage and secondary pollution to the formation [1].

3. Reducing pressure and increasing water injection process

3.1 Use surfactants

Because most of the domestic oilfield drilling for oil are taken injection, but according to the above content, and actual situation, the long-term serious influence to the actual production of high-pressure water injection technology, declining water injection after water injection pressure, lead to reservoir owe note, this phenomenon directly affect the actual production, so you need to take reasonable measures to reduce the injection pressure, increase injection The water injection effect was also significantly improved, which greatly improved the actual recovery [2]. Surfactants are generally used to reduce the starting pressure and injection pressure. In the mechanism of action, the relationship between pressure difference and flow rate is:

$$\Delta p = \frac{Q\mu \ln {}^{R_e}/_{R_w}}{2K_w h\Pi}$$

In the formula above, if Q、Re、h、μ、Rw is parameters are certain, In this case by increasing Kw Can reduce the Δp, And one of Kw=KxKrw, The following table describes the parameters and their meanings:

Table 1 Parameter meanings

KW	Effective penetration of injected water into the formation
K	The absolute permeability of the formation
KRw	The relative permeability of injected water in the formation

In the above formula, increasing the relative permeability and absolute permeability can make the injected water permeability. In terms of the tension and relative permeability of the oil-water interface, when the interface decreases, the relative permeability of oil and water will increase.

Surfactants reduce adhesion, in terms of technology, by reducing the tension of the oil-water interface and the adhesion of crude oil to the rock surface. Surfactants can dredge the formation, open up the network pores to increase the capillary number, enhance the flow performance of crude oil, thus reducing the water injection pressure, to achieve the purpose of increasing water injection. The use of surfactants to reduce the pressure of oil interfaces, increase the number of capillaries, and reduce or neutralize formation capillaries can reduce the amount of adhesion work required during oil recovery. On the other hand, the adhesion work of crude oil is reduced by improving the lubricity of rock surface. After adding surfactant, the interaction force between oil and water is reduced, and the lubrication contact Angle is reduced. The smaller the contact Angle is, the greater the interfacial tension is, which reduces the speed of rock to crude oil, increases the activity, and improves the exploitation efficiency of crude oil. The use of surfactants can reduce the adhesion of crude oil to rock during water injection. In addition, by improving the diffusion ability of crude oil in water, the emulsification of oil film can be increased after using surfactant. In principle, surfactants are very good at emulsifying and dispersing crude oil, emulsifying the oil dispersed on the rock surface and making it fall off quickly. In addition, the active agent is free on the oil-water interface to remove the active components in the oil phase, so that the formation deformation, similar to the increase in the number of capillaries. There are two infiltration modes of water flooding. In this way, water absorbs water and discharges oil in the same direction in the driving process, and the force between oil and water interface is small. Therefore, the main driving force of water flooding is the self-weight of oil and water. Reverse imbibition means that water is driven in opposite directions [3].

4. Reservoir geologic analysis

4.1 Project interpretation

This project is a low permeability reservoir. The geological details are shown in the following table:

Table 2 Geological parameters of reservoir

permeability	(1~10) x10^{-3}μm
Shale content	16~18%
Mean pressure coefficient	1.1~1.6
Injection Wells	24 □
Single well injection pressure	30.5MPA

The geology of this oil layer is complicated, the difference between layers is significant, the interference between geological operations is also very serious, the physical property of oil layer is relatively poor, so the water injection and water injection pressure are increasing with the development of geological exploitation, especially the water injection pressure shows a trend of increasing year by year. In oilfield development, some sensitive minerals expand and migrate when they meet groundwater, which leads to the decrease of formation permeability and affects the development and utilization of oil and gas. As water injection continues, bottom water coning and operating water entering the formation, the permeability of clay-containing formations decreases and affects well production. Mentioned in this project, the 24 hole injection Wells normal pressure is 30.5 MPa, but as the work goes on, there are still part of the water injection pressure more than the specification is still not bibulous, to produce a great impact on the actual exploitation, this particular disadvantage is poor reservoir property, poor permeability, poor connectivity, poor water quality, low permeability, existence of secondary pollution.

4.2 Reducing pressure and increasing injection process

In the process measures, the conventional measures are mud acid step-down augmented injection, multiple layer composite acid step-down augmented injection, litres of gas solution plugging technology, etc., these technologies in use will cause congestion phenomenon, and underground reservoir pollution hidden trouble, leading to the benefit of the step-down augmented injection is poor, the actual effect is not ideal, so in view of the reservoir blocking phenomenon, Adhering to the construction operation mode of "reservoir reconstruction first" and then "fine water injection", this project uses the technology of expanding, reducing pressure and increasing injection, polyhydroacid deep acidification and fine filtration to make up for the defects and deficiencies of conventional technology.

(2) **Technical analysis of reducing inflation, reducing pressure and increasing injection**

The mechanism of reducing inflation, lowering pressure and increasing injection is as follows: Contraction after inflation appears after the formation water absorption, water loss, the main reason is because some hydrophilic minerals in the water under the influence of the adsorption of water molecules, integrating with the inhalation and exhalation of expansion and contraction phenomenon, such as minerals such as montmorillonite and illite, between cell can absorb a large number of water molecules, water molecules into the silicon oxygen tetrahedron and alumina octahedral interlayer, When water increases crystal lattice will continue to expand and cause volume expansion, when water injection increases, so that clay expansion and migration phenomenon, the use of acid to resolve blocking measures, in this link using shrinkage expansion agent, which is a chemical synthesis of polymer material. In the process of using high molecular compound biological physical and chemical reaction and clay minerals content such as lattice modified, such clay under water won't expand, and have been passed by the expansion of the lattice modified release water molecules, the volume shrink back to the original state, change the characteristics of minerals by multi-point adsorption, make geological clay together with the original mineral formation, Avoid swelling and migration conditions that interfere with actual production. The oxidizing agent ionic group will destroy the clay lattice, and after absorbing water, the clay releases adsorbed water to shrink the volume and restore the dredging performance of the stratum. The analysis of the swelling agent interacts with the water molecule, and when water is added, the hydration group binds to the water molecule. Clay particles and water in compatible contact, hydrogen bonding, electrostatic adsorption clay, near the suction and expansion agent as have enough high molecular weight, and a larger linear properties and molecular structure, can adsorption on a na soil particles, and further connected to the adjacent clay, disperses more particles together is the organization of clay, The stability of clay is ensured so as to achieve the purpose of reducing pressure and increasing injection [4].

The principle of selecting well:

For the injection Wells with high clay quality and significant water sensitivity and speed sensitivity in the block; Or the oil and water well communication performance is good, the injection production process is more obvious; You can also choose Wells with no problems in initial exploitation, but secondary pollution in later period; After acidizing treatment, it has a good effect in the initial stage, but the actual pressure rises quickly and the injection-production period is short. In addition, there is a good well without sleeve change and sleeve leakage.

(2) Deep acidification of polyhydroacid

This technology is through the use of acid to remove the alternating injection reservoir blocking phenomenon, dissolve well layer to improve the physical properties of formation of minerals, on the mainland, in the subsequent acid of the expansion agent for formation of clay particles, forming new channels, in addition to reduce acid acidification efficiency filtration, can improve the permeability of the ground near wellbore, overall improve the water absorbing capacity of injection Wells in mainland layer. The use of this technology is characterized by the release of heat energy under the condition of removing pollution, remove the blockage phenomenon; Secondly, organic plugging agent can relieve the plugging phenomenon of polymer compounds to achieve cooling effect. The hydrochloric acid in the formation is removed when the pre-acid is injected to prevent the interaction between the hydrofluoric acid in the later stage and the calcium fluoride produced in the formation to precipitate. Organic soil acid is used to remove soil content and underground zone pollution near the well, which has the effect of deep acidification. Under the action of the construction equipment and the condition of not pressing open the stratum, the overall pump displacement should be improved as far as possible, and the water injection operation should be carried out without discharging residual acid. The low-damage, slow-moving acid in the acid has an impact as it enters the formation, extending the life of the acid and increasing the actual treatment radius. (3) Fine filtration technique

Principle of this technology is a deep filter together with the metal film surface layer separation mechanism, developed the oil field water injection terminal filtration equipment, in the concrete work to flow through the filter of fine fiber to the suspended particles in the water and oil and other substances to intercept, block large particles, after a two-stage filter to get clean water. The overall shape of the filter equipment is cylindrical, and the combined filter element can be operated. The filter element can be disassembled at any time, and the filter material can be used again after cleaning, with strong operability [5].

4.3 Process effect

The technology of expanding slurry injection is introduced into practical production, and the obvious effect of reducing pressure and increasing injection is achieved. Through the analysis of the target well, find out the cause of the well owe note, targeted to carry out the reduction of step-down augmented injection technology, with the use of the technology, the site construction injection pressure on the overall decline in 7 mpa, water injection increases 36 cubic meter a day, normal oil pressure normal stability after water injection, step-down augmented injection effect is obvious, and the failure of increase yield significantly. At the scene of the project, according to the actual circumstances of the oil Wells, selection pressure multistage acid acidification technology of construction, choose according to the actual circumstances of the oil well acidizing measure, when conventional acidification technology cannot change the character, depending on the production performance of oil Wells, demonstration and analysis, in the construction of plugging agent to reduce the overall water injection pressure, injection solution using deep hydrogen acid acidification technique, development efficiency significantly, It has the value of promotion. In addition, through the use of fine filtration technology, the overall efficiency of water injection is improved.

5. Conclusion

To sum up, in combination with the current oil exploitation situation, the application of pressure-reducing and injection increasing technology in oil Wells has achieved very significant effects. Specifically, the technology can be selected according to the actual situation of oil Wells to extend the service cycle of equipment and achieve breakthroughs in oil well exploitation technology. For the industry as a whole, the technology can be scaled up and applied to the entire industry as it achieves the desired benefits in target Wells.

References

1. LI Yan. Development and performance evaluation of compound surfactant system for reducing pressure and increasing injection in ultra-low permeability reservoirs in Ordos Basin [J]. Petroleum Geology and Oilfield Development in Daqing, 2022, 41(1):7-7.
2. LU Zhanguo. Experimental study on pressure reduction and injection increase of modified graphene oxide in low permeability reservoirs [J]. China Petroleum & Chemical Standards & Quality, 2021, 41(16):2-2.
3. WANG Zhengquan, Xie Tongping, ZHANG Huaijie, et al. Energy and Chemical Industry, 2021, 42(5):4-4. (in Chinese)
4. Wu Zhongbao, LI Li, Zhang Jialiang, et al. Research on transformation waterflooding development mode of low permeability reservoir: A case study of Kongnan GD6X1 block in Dagang Oilfield [J]. Petroleum Geology and Recovery Efficiency, 2020, 27(5):7-7.
5. WANG Yang, CHENG Shiqing, QIN Jiazheng, et al. Science China: Science and Technology, 2022, 52(4):14-14.

https://doi.org/10.1051/e3sconf/202235802029

Discussion on engineering geology of oil reservoir in oil and gas field development

Jing Yu

No. 4 Oil Production Plant of Daqing Oilfield Co., LTD., Daqing 163511, Heilongjiang, China

Abstract: As an essential resource in the process of national economic development, oil and gas has a very important strategic position. As a non-renewable resource, various countries and regions attach great importance to the development and use of oil and gas resources. Generally, oil and gas resources are buried deep underground and are the most critical resources in geological exploration. Reservoir development is a very important work, in the development process involves a lot of professional content, if the development is not reasonable, there will be some geological problems. Therefore, this paper chooses reservoir engineering geological problems in oil and gas field development as the research object, first expounds the impact of reservoir geological characteristics on oil and gas field development, then analyzes the existing problems, and finally proposes corresponding solutions to improve the quality and efficiency of oil and gas field development.

Key words: oil and gas field; Development; Reservoir engineering; Geological problems.

1. Introduction

In the process of social development, people's demand for petroleum resources gradually presents a rising trend, and petroleum, as a non-renewable energy, needs to be attached great importance in its development and use. There will be some geological problems in the development, resulting in the occurrence of land subsidence or earthquake, this problem has a long-term and difficult to reverse, which will have a certain impact on People's Daily production and life, relevant departments should pay attention to. Therefore, this paper chooses to discuss the reservoir engineering geological problems in the development of oil and gas fields, so as to ensure a higher safety in the development process and promote the development of China's oil and gas industry.

2. Distribution of oil and gas deposits

Different distribution of oil and gas will have different ways in the process of exploitation. Specific reservoir distribution is mainly reflected in the following aspects.

2.1 The reservoir distribution

The so-called reservoir mainly refers to some rock layers for oil and gas storage, in which a large number of pores are distributed, and these pores are important locations for oil and gas storage [1]. Therefore, in the process of oil and gas development, it is necessary to do a good job of reservoir distribution research. In the calculation of reservoir volume, it is necessary to use certain tools and technologies, on the basis of the study of reservoir distribution, and then according to the data of specific well points, the oil and gas distribution of underground reservoir is inferred. Before the development of oil and gas fields, if there is a lack of relevant data, it can not be easily exploited. Only with the support of a large number of favorable information and data can the exploitation work be carried out, so as to avoid the waste of manpower, material resources and financial resources in the exploitation process, and enhance the scientific and rational exploitation of oil and gas fields.

2.2 The pore distribution

Reservoir distribution is another important place in the pore distribution, pore as a storage medium, the cycle of oil and gas and has played an important role for oil and gas storage, underground in different regions will have different pore types, the role of these different pore in oil and gas storage also has a certain difference, in the oil and gas field development process, The permeability of different pores varies greatly, which is also an important factor affecting the development efficiency and quality of oil and gas fields. The oil and gas distributed in the pores need to be exploited with targeted measures to improve the quality and efficiency of production.

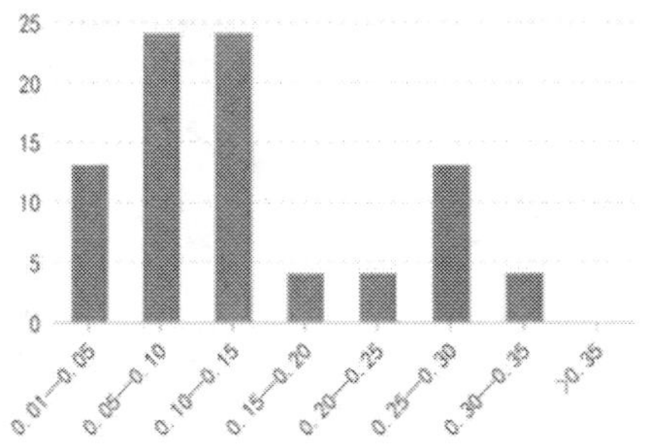

Figure 1 Frequency histogram of orifice throat radius

2.3 Fluid distribution

In the process of oil and gas exploitation, fluid distribution has a great influence, and the direction, quality and pressure of fluid are important factors affecting oil and gas development. There will be fluid distribution in oil and gas reservoirs, and the distribution of these fluids needs to be clear. In oil and gas reservoirs, if the fluid velocity is relatively fast, it will cause certain difficulties to the exploitation of oil and gas fields, which is not conducive to high-quality oilfield exploitation.

3. Influence of reservoir geological characteristics on oil and gas field development

3.1 Influence on oil displacement effect of oil and gas fields

As an important work, oil displacement has received great attention in the development of oil and gas fields. The development of oil and gas fields can directly affect the results of oil and gas development. Under normal circumstances, the reservoir geology has certain difference, but if it is a water reservoir, will have higher oil displacement effect, don't need too much in the process of the concrete oil displacement equipment and technologies involved in, at the time of water, water reservoir can be according to the surface, water oil washed out of the rock gap, can a good oil displacement effect with water. In the process of oil displacement of oil and gas fields, the hydrophilic zone will produce a huge oil displacement effect, but also have a certain impact on the oil layer, driving oil separation, and under the continuous influence, more oil and gas can be produced. However, in the process of oil displacement, it should be noted that it is not possible to continuously add water to the clean oil layer. If a large amount of clean water is injected without comprehensive consideration, the development of oil and gas fields will be affected to some extent and the oil displacement effect of oil and gas fields will be reduced [2]. Therefore, proper control of water injection and reduction of water content in oil are important contents to be considered. In the exploitation of oil and gas fields, if the time is too long, the reservoir pressure of oil and gas fields will be affected. In the case of low pressure, the development effect of oil and gas fields will be reduced.

3.2 Influence of oil reservoir on later development of oil and gas field

In the later stage of oil field development, some problems will gradually appear. For example, oil and gas reservoir pressure will gradually decrease in the later stage of development, which has an important impact on the overall development effect of oil and gas field. Relevant development enterprises will take certain measures to improve the reservoir pressure in the later stage of exploitation. According to the actual situation of oil and gas field development, appropriately increasing the reservoir pressure can assist the development of oil and gas fields and create favorable conditions for creating a good development environment. In the concrete solution measures, one of the most important is to inject water into the oil layer. In the process of water injection, we need to strictly control the amount of water injection and the speed of water injection, so that water can be accurately injected into the oil reservoir. The most important thing is not to affect the quality of oil and gas fields. In the process of water injection, the water injection process needs to be strictly controlled to improve the pressure of the oil field without affecting the overall quality of the oil and gas field. Only in this way can the effect of the later development of oil and gas fields be continuously improved [3].

4. Analysis of reservoir engineering geological problems in oil and gas field development

In the process of oil and gas field development, there will be some problems in reservoir engineering geology. These problems can have a huge impact on the surrounding environment, which is mainly reflected in the following two aspects.

4.1 Settlement problem

Before the oil and gas field is developed, the area in which the oil and gas field is located will present a relatively stable state, but in the process of gradual exploitation, this stable state is gradually destroyed [4]. And increase gradually with the oil and gas field development strength, its stability has also been gradually reduced, eventually led to the surrounding environment affected by huge, oil and gas fields in the process of development, the underground fluid will gradually reduce, volume in the underground was affected by the huge present a reduced state, as the volume of oil and gas layer, eventually can cause a wide range of strata subsidence, The serious settlement problem will not only affect the surrounding ecological environment, but also seriously affect people's life and health safety.

Along with the development of industrial production and people's life, oil and gas field exploitation activities are also increase gradually, gradually makes the strata subsidence problems tend to worse gradually, from the point of the settlement of the actual situation, mining has achieved the zenith of high oil and gas fields, both at home and abroad, due to excessive oil and gas field development strata subsidence problems, It has a very important impact on the surrounding ecological environment. The occurrence of the formation settlement problem will not only affect people's production and life, but also affect the urban construction and development. If the formation settlement is serious due to the over-exploitation of oil and gas fields, the urban development will be seriously affected. In the process of oil and gas production, there are certain characteristics of formation descent, which are mainly reflected in the following three aspects: first, the fluid injection volume is larger than the oil and gas production volume; Secondly, the subsidence phenomenon is caused by loose compaction of the overburden layer. Finally, the formation pressure drops, and the formation subsidence occurs during the exploitation of oil and gas fields.

4.2 Induced seismicity

An important influence in the development of oil and gas fields is to induce seismic activity. When not on oil and gas field development, the environment is stable, and in the oil and gas fields is developed gradually, after the local area underground situation must be changed, the earthquake problem gradually increased, especially in the oil and gas fields unrestrained development process, a large amount of fluid into the oil and gas fields, lead to the stress changes inside the rock, and will be gradually released, Over time, there will be the phenomenon of earthquakes. But in this case the tend to be lighter, the earthquake magnitude will be low, but is frequent, and if some oilfield is located in the special area, in a active area of the earth's crust, for example, is particularly vulnerable to earthquake and high frequency, magnitude in 2 above, will have an impact to people's life and property safety [5]. The increase in the frequency of small earthquakes, under the influence of a long time, will produce some large earthquakes, need to cause people's attention. The earthquake the main reason is that in some areas, deep when the stress reaches a certain limit, it will lead to a larger earthquake, so in the process of oilfield development, not only need to strictly grasp the flood situation, also need to strictly grasp of local earthquake situation before, need to work out some preventive measures against earthquakes, according to actual condition Ensure that earthquakes have a low impact when they occur and improve the ability to respond to earthquakes.

5. Measures to solve engineering geology problems in oil and gas field development

Oil and gas fields in the development process will produce some geological problems, the production of geological problems to the local ecological environment and people's living environment have a huge impact, therefore, need to take some measures to deal with reservoir engineering geological problems in oil and gas exploitation, make sure you are able to mining with high quality, specific solutions mainly reflects in the following respects.

5.1 Analyze the development of new and old oil fields

In the development of oil and gas fields, in order to minimize the relevant geological problems, it is necessary to analyze the development of new and old oil and gas fields in the process of exploitation. On the basis of comprehensive comparative analysis, it can ensure that it plays an important role in the development process of new oil and gas fields. By drawing lessons from the development of old oil and gas fields, a new development and improvement scheme is developed to ensure that the new development scheme can solve the disadvantages of the old scheme and improve the quality and efficiency of the development of new oil and gas fields. It is of great significance to make a comparative analysis with the old oil and gas fields, so as to judge the storage situation of the new oil and gas fields with the help of comparative analysis. According to the actual situation of the new oil and gas field, the actual plan of oil and gas field exploitation should be optimized or adjusted to promote the new oil and gas field to have higher quality in the development process. Therefore, in the process of developing new oil and gas fields, it is necessary to do a good job in the comparison and analysis of the development of old and new oil and gas fields, and constantly deal with the geological problems in the development process to improve the efficiency of the development of new oil and gas fields.

5.2 Strengthen the deep analysis of reservoir structure system

Generally, oil and gas fields are buried deep underground and their distribution is complicated. Therefore, in the process of oil and gas development, it is necessary to conduct in-depth analysis on the structural system of oil reservoirs according to the actual situation. The analysis of reservoir structure system needs to be carried out from many aspects. First of all, advanced technology and equipment can be adopted to analyze it. For example, big data technology can be used to integrate and analyze relevant data, and the original paper data can be converted into digital data to improve the utilization ability of data information. Secondly, it is necessary to constantly adopt new methods for in-depth analysis of reservoir structure system, which can be innovated on the basis of original methods to ensure that it can meet the requirements of the development of The Times and continuously improve the ability of reservoir structure system analysis. Therefore, in the process of oil and gas field development, it is necessary to conduct in-depth analysis of oil reservoir structure system according to the actual needs, confirm its distribution, and ensure that it can be highly standardized and scientific in the subsequent oil and gas field development process.

https://doi.org/10.1051/e3sconf/202235802029

5.3 Make full use of the development mode of large - section combined mining

The development mode of oil and gas fields can have the most direct impact on geological problems. In the traditional development mode of oil and gas fields, blind exploitation can often have a huge impact on the geological environment, resulting in serious geological problems. In the new era, the development of oil and gas fields needs to be combined with the actual needs to innovate the development mode, which can take the way of large stage development to ensure that oil and gas fields have higher development efficiency and quality. In the process of using this kind of exploitation method, it is necessary to make sufficient preparations, master the distribution structure system of oil reservoir becomes the primary link of exploitation, and build the comprehensive exploitation scheme on the basis of comprehensive analysis. At the same time, with the development of modern advanced technology, the exploitation of oil and gas fields can be carried out with the help of multiple technologies, such as information technology, big data technology and computer technology, to assist the development of oil and gas fields. Therefore, in the process of exploitation of oil and gas fields, it is necessary to combine the actual situation, adopt the way of large joint production, and actively combine advanced technology, so as to make oil and gas exploitation can have high quality and efficiency, and promote the sustainable and healthy development of oil and gas industry.

6. Conclusion

Oil and gas resources play a very important role in the process of social development, in its development process, will produce some geological problems and a threat to people's health and the local ecological environment, therefore, at the time of oil and gas field development need to according to the actual situation to solve the problems of reservoir geological engineering, through the reasonable process and specification, To ensure the security of oil and gas development.

References

1. Cao Guohua, Chen Peng, Li Caixia. Journal of Inner Mongolia normal university (natural science in Chinese edition),2019,48(05):465-470.
2. Yang Kai. Analysis of reservoir engineering geological problems in oil and gas field development [J]. Science & Technology Wind,2021(22):129-131.
3. SUN Tongtong. Discussion on reservoir engineering geology in development of Liaohe Oilfield [J]. Chemical Industry Management,2019(34):206.
4. LI Yixin. Analysis on some problems of reservoir engineering geology in oil and gas field development [J]. Yunnan chemical industry,2018,45(09):189-190.
5. Yang lidong. Analysis of reservoir engineering geological problems in oil and gas field development [J]. Petrochemical technology,2018,25(04):190+189.

https://doi.org/10.1051/e3sconf/202235802030

Analysis of low permeability reservoir development technology

Feng Su

Geological Research Institute of The Fourth Oil Production Plant of Daqing Oilfield Company Limited, Daqing 163511, Heilongjiang, China

Abstract: As an important energy source for social production and development, oil is non-renewable. In the process of development and utilization, attention should be paid to continuously improve the exploitation technology, enhance the ability of oil exploration, and strengthen the development of low permeability reservoir resources, so as to provide sufficient oil resources for social production and development. However, low permeability reservoir resources have their own characteristics and are limited by multiple factors in the process of exploitation. It is difficult to improve the exploitation efficiency of low permeability reservoir resources without adopting corresponding exploitation technology. This it is scientific and reasonable exploitation of low permeability oil reservoir resources technology are analyzed and discussed, in the comprehensive analysis based on the characteristics of low permeability oil reservoirs, mining characteristics, puts forward the well pattern optimization, waterflood recovery, fracturing technology and mining method, working for the effective implementation of the low permeability reservoir development provides the technical support, also need to continue well technology application in the construction process in the future.

Key words: oilfield; Low permeability; Reservoir development technology Exploration of low permeability reservoir development technology is helpful to solve all kinds of problems encountered in the process of oil exploitation and meet the needs of sustainable development of domestic oil and gas development. At present, low permeability reservoir resources are distributed in various oilfields, and there will be many difficulties in exploitation. In view of these problems, it is necessary to combine the characteristics of low permeability reservoir, realize scientific and reasonable well pattern layout, accurate water injection, and introduce various modern technological means to overcome the obstacles in the exploitation process. Therefore, it is necessary to select feasible technological means to solve the problem of water cut rise and improve the comprehensive development level of low permeability reservoir in combination with the characteristics of low permeability reservoir, so as to ensure the amount of crude oil recovery and improve the economic development benefit.

1. Low permeability oil field analysis

1.1 Low permeability oil field type

Low permeability oilfield is mainly refers to the reservoir permeability reservoir level and low level of abundance oil field of single well production capacity is relatively low, in carries on the classification, mainly according to the classification of low permeability reservoir in lower limit, it can usually be divided into the following three kinds: general low permeable oil fields, especially low permeability oilfield, and ultra low permeable oil fields. Each field has its own characteristics, which can be described as follows: The first is a general low-permeability field, where oil flow standards meet industrial production standards and fracturing measures can be used to increase productivity. The second is the ultra-low permeability oil field, the whole reservoir bound water saturation is relatively high, oil flow standards can not meet industrial standards, which requires large-scale fracturing methods to complete development. Finally, ultra-low permeability oil field, which has high oil layer density and comprehensive bound water saturation, is not of high development value due to its relatively low production capacity. However, due to its advantages of high oil quality and thick oil layer, it can be used as backup energy for development. The author sorted out the permeability of the three, and the specific values are shown in Table 1 below:

Table 1 Permeability values

Oil field type	Numerical range
Low permeability field	$0.01\mu m^2$-$0.05\mu m^2$
Extra-low permeability field	$0.001\mu m^2$-$0.01\mu m^2$
Ultra-low permeability field	$0.0001\mu m^2$-$0.001\mu m^2$

The production level of oil field will decrease gradually with the development time. For low permeability oil field, the comprehensive production may decrease, the water content of oil well gradually increases, the aging of mining equipment is obvious, and it is difficult to adapt to the more difficult mining work, and the comprehensive efficiency will decrease, which will lead to the comprehensive economic benefit of bottom hole oil field. Because low permeability oilfield is still an important object of petroleum development. It has the characteristics of rich reserves, diverse types and relatively wide distribution. In terms of the current proven oil reserves, the proportion of low permeability oil fields is more than two-thirds, which has good development potential. Therefore, in the future oil exploitation process, we should pay attention to the research and development of a variety of exploitation technologies, and give full play to the performance of low permeability oil fields.

1.2 Reservoir characteristics analysis of low permeability reservoir

Has significant characteristics of low permeability reservoir, first of all, the relatively poor reservoir physical property of low permeability reservoir, with more sandstone grain size, and distribution range is relatively wide, sandstone particles has obvious characteristic, such as poor sorting, clutter, and cement is more, it can cause internal structure presents obvious heterogeneity. Second, the pore channel radius is smaller than that of a normal reservoir, the pores are somewhat curved, and the entire outer surface is rough. Again, when the fluid in the reservoir is in contact with the demo, physical and chemical interactions occur, resulting in thinning and narrowing of the original pore channels within the rock structure. Over time, serious blockage occurs throughout the pores, and the reservoir permeability naturally decreases. Finally, the bound water saturation of oil layer is much higher than that of low permeability reservoir, that is, the maximum water saturation of low permeability reservoir can reach about 60%. In the process of exploiting low permeability reservoirs, the production staff should do a good job of characteristic analysis to improve the production efficiency [1].

1.3 Development characteristics of low permeability reservoir in oilfield

Low permeability reservoirs have the following characteristics: First, there is a direct proportional relationship between recovery and permeability, that is, when permeability increases, recovery increases, and vice versa. Second, there is an inverse relationship between start-up pressure and permeability, that is, the higher the permeability, the lower the start-up pressure, and vice versa. Thirdly, the comprehensive permeability level of reservoir is not high, and the abundance is relatively low, and the overall rate is relatively slow in the process of oil recovery. Fourth, the inner layer structure has natural cracks, and the cracks show irregular characteristics. When affected by external factors, the formation heterogeneity is significant, which affects the subsequent mining work. Fifth, the reservoir water shows poor hydrodynamic connectivity, which increases the difficulty of drainage work. Combined with the above characteristics, in the process of oil exploitation, we should adopt specific exploitation methods based on the characteristics of low permeability reservoir, so as to improve the production efficiency.

2. Current situation of low permeability reservoir development technology

2.1 Well pattern optimization technology

The use of well pattern can affect the development quality of low permeability reservoirs and requires accurate understanding of well location, which is determined by field conditions, oil fluid properties, and current development of the field. However, in the same layer, crude oil has a large drainage radius, which will exceed the normal radius range, making it relatively difficult for crude oil to return to the oil well. Therefore, it is necessary to pay attention to accurate measurement of the radius value and improve the efficiency of well pattern layout. Based on the analysis of previous experience, the current production capacity of oil Wells should be considered comprehensively when determining the location of well points. The following aspects can be considered: First, in the process of supplementing oil Wells, it is necessary to select oil Wells with high production and high potential to supplement formation capacity. Second, in the process of water injection, the advantage of geological conditions should be fully utilized to minimize the amount of loss at the production well point. Third, the production team needs to maximize the use of low-yield Wells to increase production at production points.

2.2 Advanced waterflooding technique

Water injection technology can effectively solve the problem of poor pressure conduction performance and low natural energy reservoir, improve the use value, so this technology is mainly used in the early development, so as to effectively ensure formation pressure, reduce the influence of reservoir permeability formation. When the reservoir has a certain distance between, you need to use the layered water injection form, increase the rate of crude oil in the reservoir, and in the process of water injection, water injection work along the oil analysis, so the choice of injection Wells location should be close to crack location, and injection Wells location to do a good job in uniform distribution to ensure the water flooding efficiency.

2.2.1 Water injection method

Through many years of experience in water injection, it is concluded that high pressure water injection, periodic water injection and cross pulse water injection have good water injection effects. However, the low pressure water injection is not ideal, and the comprehensive output of oil well does not reach the expected goal. Since fractures are not obvious in low permeability oil fields, it is more appropriate to use high pressure water injection, which can effectively remedy the problem of loss and form a water wall to ensure water flooding efficiency. Periodic water injection can be used when the well is multi-zone, but it can also be used when the combined water content of the well is not very high.

2.2.2 To ensure the water quality

Low permeable oil fields in the process of mining, should ensure that use water comprehensive water quality, which requires the staff before the formal water injection well water purification, disinfection, sterilization and anticorrosive processing, because if not for the corresponding processing, has been the problem of pollution of water quality of the oil would be caused by the bad influence. Therefore, attention should be paid to water treatment before water injection [2].

2.3 Acidizing technology

Acidizing technology is the most critical technical means to improve the productivity of low permeability reservoirs. It can effectively improve the production efficiency of oil Wells when it is applied to special well locations. Generally in the application of acidification process, the common ways include conventional acid acidification, acidification and fracturing and acidification in application technology, the main principle is to give play to the role of acid itself with the chemical, applied to the mining work, can have certain dissolution, blockage can be used for mining in the corrosion dissolution, Gradually restore the original permeability of the formation to ensure the effect of crude oil exploitation. Acidizing can reduce the combined pressure of the fluid, helping to increase the flow efficiency at the bottom of the hole.

2.4 Pressure system optimization technology

Pressure system optimization technology is the key technical means to ensure the production of low permeability oil fields. In the formal development of oil Wells, the production capacity of oil Wells should be comprehensively analyzed according to the production target starting pressure and oil production situation, and development plans should be made to monitor the net flow pressure of reservoir, so that the production efficiency of oil fields is gradually improved. In the application of this technology, the flow pressure is accurately calculated, the fluidity of the oil is reduced, and the overall viscosity of the oil is improved, which increases the sensitivity to pressure, improves the ease of extraction, and improves the utilization efficiency.

2.5 Microbial oil recovery technology

Microbial oil recovery technology mainly uses the degradation function of microorganisms to enhance the light components in crude oil. Because link of microorganisms in the degradation of crude oil, usually form some metabolic substances, which contains a large number of surface active agent, surfactant, interaction between crude oil and comprehensive annual can effectively control the oil, to improve the original crude oil flow state adverse situation, so as to gradually raise the overall production of crude oil Wells.

2.6 Fracturing and development technology

Generally speaking, the natural productivity of low permeability reservoirs is relatively low, which cannot meet the industrial oil production standards. In this case, fracturing can be adopted to improve the comprehensive development intensity, which is the main way to improve the development of low permeability reservoirs. Ongoing efforts to develop the technology have improved extraction efficiency. Fracturing technology includes many forms, including limited flow fracturing technology, carbon dioxide fracturing technology and composite fracturing technology. When applying the fracturing truck technology, the main principle is that the fracturing technology can effectively improve the existing fractures in the bottom hole reservoir, so that the irregular state of crude oil can be transformed into regular length fractures with sand filling fracture characteristics [3]. The reservoir also changes its permeability properties, allowing the direction of reservoir fluid flow and pressure changes to improve the overall well layout.

3. Low permeability reservoir development technology upgrade

3.1 Stimulation technique

In the whole process of crude oil exploitation, due to various factors, improper mining operation seriously polluted the shallow crude oil, so that it is difficult to use the shallow crude oil efficiently. Especially in ultra-low permeability fields, the combined permeability level near the well is lower, which makes it difficult for oil to flow into the wellbore. These factors can seriously affect the improvement of oil production efficiency of the well. If this problem is not solved, it is difficult to improve the efficiency of crude oil recovery. In this regard, the author believes that physical stimulation can be adopted to realize gravity separation between crude oil, water and rock strata, so as to improve the comprehensive flow level of crude oil, enhance the fluidity of crude oil and expand the comprehensive oil production efficiency of oil Wells [4].

3.2 Horizontal well development technology

The main form of traditional Wells is vertical well development mode, which is relatively low in the exploitation of low permeability oil fields. Therefore, the author believes that horizontal well development can be fully utilized in the future exploitation. Horizontal Wells can increase the fracturing gradient during oil production, so that the pressure around the well gradually decreases, and water injection operations can improve the efficiency of water injection by applying the necessary pressure. Compared with vertical Wells, the combined pressure level is relatively small and can reduce the energy consumption in the process of crude oil extraction [5].

4. Conclusion

In v, this article around the oil field of low permeability reservoir development technology is analyzed and discussed, it can be seen that the number of low permeability reservoirs is relatively abundant, but due to the complexity of this resource has certain, in the process of mining, using traditional mining technology, then nature will be affected by many factors, reduce overall production efficiency, such as congestion, pollution and other problems, It is not good for the performance of low permeability reservoir. In order to effectively solve such problems, combining with the existing types of low permeability reservoir development technology of oilfield, put forward the targeted the development of technology, into the exploitation in the future work, need to do a good job, well pattern arrangement to ensure scientific and reasonable distribution, guarantee the production efficiency, but also can take a variety of other forms of mining technology, multi-pronged, Only in this way can we deal with various technical problems that may occur in the mining process and improve the mining efficiency.

References

1. Wang Dongxue. Chemical Engineering and Equipment,2020(11):52-53. (in Chinese)
2. Yang Zhicheng, Yang Jian, Xu Zhixiong, Liu Huilin, Gao Shaofeng, Hu Jinbo, Chen Yuqin, Yao Yang. Petrochemical industry application,2020,39(10):58-61.
3. Wang Yudi, Jiang Houshun. Study on fine reservoir description of low permeability oilfield -- Review on Fine reservoir description and Development Evaluation Technology of low permeability sandstone reservoir [J]. Xinjiang geology, 2020, 38(01):131.
4. Pan Chen, Sun Junliang, Li Liang. China petroleum & chemical standards & quality,2019,39(06):183-184.
5. han hongxing. Research on development technology of low permeability reservoir in oilfield [J]. Chemical design communication,2018,44(11):47.

Research and application of water flooding injection and production technology after polymer flooding

Qingfeng Sun

Institute of Geology, No.6 Oil Production Plant, Daqing Oilfield Co., LTD Daqing 163000, Heilongjiang, China

Abstract. For reservoir water flooding after polymer flooding injection points existing in the mining process of high temperature resistant performance is poor and multistage unlock difficulty, examination, and the sealing problems such as in-depth study and analysis, innovation to a new set of return to spit points injection points production technology, to produce more suitable sealing high voltage can be found on a variety of well conditions return to spit points a column technology, is helpful for better improvement points injection tube service life, At the same time, it can effectively solve the problem of clogging downhole concentric test and adjustment unit. Oilfield exploration into the follow-up after water flooding stage, the well will show the water fast, fast production decline, the characteristics of a return so often need to take back to spit on mining technology, but the technology has high pressure resistance and multistage unlock difficulty problem, makes the life of the mining process is shorter, so according to the technical problems existing in the in-depth study and analysis, The effective outcome measures are put forward to provide necessary technical support for reservoir benefit after polymer flooding.

Key words: Reservoir water drive; Separate injection and production; Technology research.

1. Technical string composition and process principle

1.1 Technical string composition

In the process structure of reservoir water injection and production after polymer flooding, the development of the water distributor and the high pressure differential nozzle is helpful to improve the auxiliary tool technology such as the low-damage hydraulic anchor and the small diameter anti-reflux injection pipe. By this way can effectively adapt to high pressure difference and improvement points injection pressure, the purpose of the performance in the technical part of the tubing string, can pass the pressure check valve backflow prevention principle of ways, can effectively prevent the polymer to return to spit into the purpose of the tubing, and pressure from the oil lantern ring hole is able to be delivered to the tubing in the purpose of [1]. The pipe string of reservoir water flooding and fractional injection and production technology after polymer flooding is mainly composed of eccentric water distributor, packer, eccentric water distributor with pressure transmission and anti-reflux, pressure limiting single flow valve and slit screen.

1.2 Process principle

In the process of polymer reflux into the tubing, through the design of anti-backflow eccentric water distributor, in the eccentric water distributor outlet design of the corresponding spring return single-flow valve, can effectively achieve the compression of the spring during water injection, so as to achieve the purpose of opening the single-flow valve. In this way, it can not only effectively stop the spring reset and the purpose of closing the single flow valve when water injection, but also better prevent the formation spit polymer in the throttling plug into the water distributor caused by the exit and tubing design, and finally achieve the purpose of preventing the blockage of pipe and water distributor nozzle. In the design process of the bottom wash valve, a combined controllable wash valve must be designed because it needs to be considered not only to meet the opening of the bottom wash valve, but also to consider the way not to open when testing the step-down or injection stop [2]. For the components of the single flow valve and slit screen, according to the bottom water absorption condition, set a higher opening pressure way of the single flow valve, in order to realize the backflow phenomenon when the pressure difference between inside and outside the tubing is lower than the set pressure. The viscosity of the respitting polymer in the slotted screen section must be reduced during secondary shearing to prevent clogging of the string and common problems in the nozzle.

2. Research and analysis of the main supporting tools

2.1 Study on eccentric water dispenser for preventing vomiting

The anti-reflux eccentric water distributor is usually composed of an upper connector, a water distributor body, a rotating core, a fixed nozzle, an anti-clog mechanism and a lower connector. Its working principle is to ensure that the original fixed core plays an axial and radial role in limiting and positioning the electromagnetic concentric adjusting instrument through the anti-clog downhole adjustable unit and the special electromagnetic concentric measuring and adjusting instrument, and effectively adjust the fixed nozzle during the rotation of the rotating core [3]. For T seal, piston and piston accessories such as the use of, want to better achieve injection Wells water injection piston seat open, you will need to injection Wells in stops the piston can realize the function of automatic shutdown, can effectively realize the function of polymer Wells back to spit proof, thus preventing stop note when the underlying return sand spit out. In the study of anti-reflux eccentric water distributor, the corresponding technical parameter requirements include pressure resistance parameters of 35MPa, temperature resistance of 120°C, outer diameter of 100mm, inner diameter of 46mm, total length of 974mm, and maximum rotational torque of 82N·m.

2.2 Study on Pressure Limiting single flow valve

The structure of pressure limiting single flow valve is mainly composed of upper connector, baffle, reset spring, single flow valve ball and lower connector. In the working principle of pressure limiting single flow valve, since there is no external pressure action, the single flow valve ball of pressure limiting single flow valve sits on the seat on the lower joint, and the purpose of external hydraulic action can be realized by the way of positive closing and sealing, which can achieve high pressure design requirements. This is because the pressure limiting single-flow valve can realize internal and external communication by pushing the single-flow valve ball to compress the spring, and the function of the spring resetting the single-flow valve seal can be realized after the high-pressure well washing fluid enters the tubing interior from the outside for backwashing [4]. The single flow valve ball and the compression spring design of the centralizer can help to prevent the spring and the valve ball from misalignment. At the same time, an effective space is designed between the inner diameter of the centralizer and the working cylinder, which is conducive to the flow of liquid flow. The reasonable design of the structure, but also can effectively ensure the reliability of the single flow valve ball open, to avoid the single flow valve ball is blocked, the technical parameters of the pressure limit single flow valve, the pressure parameter is 35MPa, the opening pressure is between 2 and 3.5mpa, the outer diameter is 95mm, the total length is 450mm.

2.3 Study on Y341 step - by - step unseal waterflood packer

The structure of Y341 step by step unsealed water injection packer is mainly based on the requirements of the layered sand control process of the injection well after polymer flooding, and the water injection sealing is performed on the area of the well washing channel by improving the supporting outer diameter. Generally, the improved outer diameter is φ115mm. The structure of packer is composed of upper joint, center pipe, well washing sleeve, special rubber cylinder, well washing valve, piston, cylinder liner and lower joint. Therefore, the working principle of Y341 step-wise de-seal water-injection packer is as follows: the rubber cylinder can be compressed and set by internal and external pressure difference, and the locking claw and locking sleeve can be locked tightly. When de-sealing, the upper pipe string and the friction force of the rubber cylinder and casing will cut the de-seal pin, which will lead to the retaining ring being moved, and with the recovery of the rubber cylinder, The Y341 step-by-step unseal waterflood packer was unsealed and caused the string to be isolated [5]. In the technical parameters related to the Y341 step-wise unseal water-injection packer, the corresponding sealing and decompression parameters are 30MPa, the setting pressure difference is in the range of 7-8mpa, the operating temperature shall not exceed 125°C, and the backwash well displacement is in the range of 30-40m3 /h.

3. The establishment of numerical simulation model

3.1 An overview of oilfield engineering

In the study of water flooding and fractional injection and production technology after polymer flooding, a certain oilfield in northwest China is taken as an example to select the simulation model. The polymer area of the oilfield is 9.04km2, the geological reserve is 1367*104t, and the pore volume is 2477*104m3. The production layer belongs to the PI II-4 reservoir. The thickness of the drilled sandstone is 16.8m/ piece, the effective thickness is 12.8m, the effective permeability is 670*10-3μm2, and the original formation pressure is 11MPa. The saturation pressure is 10.06MPa, the average reservoir burial depth is 958.97m2, the formation temperature is 47°C, the viscosity of underground crude oil is 9.6mPs, the crude oil density is 0.803g/cm3, the original oil-gas ratio is 47.3m3/t, and the original formation water salinity is 7178mg/L. In the analysis of dynamic and static characteristics of the field, to take measures and test methods such as oil and water Wells, oil and water Wells and fully considering the heterogeneity, distribution of sedimentary facies, relative permeability, porosity, etc., and realize the fine simulation of field data, and with the aid of the Eclipse of the method of simulation software, also can realize the hermit, 3 d and 3-facies black oil simulator, So that the calculation results become more accurate.

3.2 Static data interpolation calculation

The formation parameters of oil field the actual thickness, effective thickness and porosity of related parameters, such as data interpolation calculation, can use phase interpolation approach, when top deep phase interpolation problem, need according to the library to extract data from layer related way, implement effective interpolation on the depth of the upper mouth and get the top data model USES. In the in-depth study and analysis of fine strata, the effective control of the precipitation phase zone diagram is helpful to realize the specific calculation and analysis of the interpolation of each formation parameter in the phase zone content, and the effective control of the heterogeneous parameters in the formation. For sand strata in the vertical permeability of the composite parameters was described, two contact network must be composite sand location is analyzed, according to 10% of the network block print permeability values at the same time, in the case of compound sand body anatomy, realize effective exchange of both material, in this case, the corresponding vertical permeability coefficient is zero, FIG. 1 shows the distribution of thickness, permeability, porosity and pore volume of each rock layer [6].

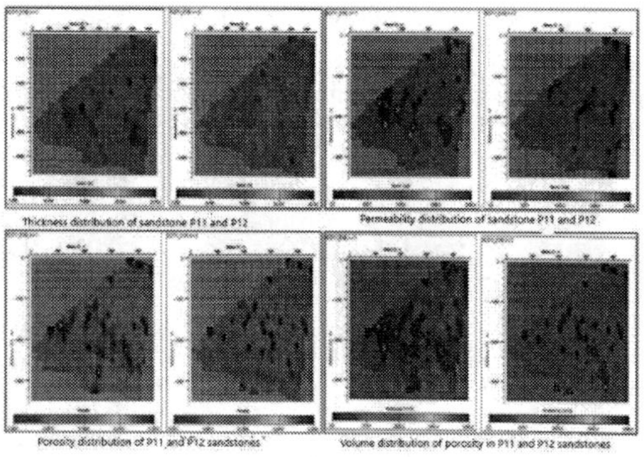

Figure 1: Distribution of parameters in interpolation calculation of static data

3.3 Fluid properties and their distribution

The distribution of fluid properties, aiming at the rate of the relative permeability curve of indoor result analysis comparison, usually need to larger data in the rate of penetration curve, to guarantee the precision of the data and curve smoothness requirements, to give time to calculate, then helps to realize the calculation precision of ascension. For different Wells between rock and fluid blocks through analyzing the relative permeability curve, generally take four different relative permeability curve, can realize the different between the geological conditions and rock fully consider regional quality analysis, finally can effectively achieve the liquid convection affect the thorough analysis, Corresponding permeability curves and adaptive formations are shown in Table 1.

Table 1: Relative permeability curves fit stratigraphic table

Relative permeability curve number	Adaptive formation
The first curve	Permeability above 800md
The second curve	Permeability between 400 and 800md
The third curve	Permeability between 150 and 400md
The fourth curve	Permeability less than 150md

3.4 Oil field production dynamic data analysis

After data collection and sorting according to the dynamic characteristics of the oilfield, the oil production and water mixing can be divided into each day according to the month, and then the data can be received by Eclipse software, so as to form the analysis of the corresponding dynamic data stream. At the same time, the relevant parameters such as oil production and water production can be corrected and analyzed. The corresponding time step division, treatment measures, isotope data and annulus water search data can be obtained [7]. And in the process of time step some effective division, want to better realize the oil Wells in the oil and water production status, the time step can be planning for a month's time, and within the minimum time step for the oil and water Wells, to ensure the production and shut down within the prescribed period of time, eventually be able to form the corresponding curve and data requirements. , and in the process of acidification and fracturing and fill hole processing, the problems of water shutoff must be carried out in accordance with the corresponding time input bit, in order to better realize the dynamic change of the parameters of effective judgment and isotopic parameters, and repeatedly in the process to modify the parameter simulation measures the formation parameters of control. In isotope data, the oil Wells in the reservoir pollution, assignments, and other measures of common water absorption layer change, isotope in the course of the simulation and the effective detection, finally combined with the actual situation to calculate the single well in oil Wells of interlayer contradictions, to reach the final change effectively in accordance with the results. For the annulus water search data, with the help of the current relevant data to carry out in-depth analysis of a single well layer, it can be found that the final simulation results are mainly based on the corresponding test data to achieve effective analysis, so the simulation experiment has a high credibility.

4. Conclusion

Reservoir water flooding after polymer flooding injection points mining technology can effectively solve the related problems of the original points injection process, such as resistance to high pressure difference, process defects such as short service life, through innovative perfect return to spit points a mining technology, to produce more effective resistance to high pressure sealing can be found on a variety of well conditions return to spit points process injection string, thus has excellent anti reflux vomiting and test functions, This has a great effect on improving the service life and test efficiency of water-flooding injection and production technology in reservoir after polymer flooding. After verification, the reliability of this technology makes the technology more perfect and developed.

References

1. ZHAO Na. Research and application of water flooding fractional injection and fractional production technology in reservoir after polymer flooding [J]. Inner Mongolia petrochemical industry,2020,46(08):105-107.
2. Liu Yigang, Meng Xianghai, Zhang Yunbao, Xia Huan, Cao Bao, Luo Yunlong. Numerical simulation of the effect of separate injection and production and its influencing factors in offshore oilfields: A case study of SZ36-1 Oilfield in Bohai Sea [J]. Petroleum geology and engineering,2020,34(04):95-101.
3. Luo Yunlong. Study on evaluation index and method of intelligent separate injection and production effect in Bohai Oilfield [D]. Northeast petroleum university, 2022. DOI: 10.26995 /, dc nki. GDQSC. 2022.000090.
4. Meng Xianghai, Xia Huan, Li Yanyue, Chen Zheng, Wang Nan, Luo Yunlong, Cao Bao. Intelligent points on mining technology application effect and its influencing factors research [J]. Journal of modern chemical industry, 2022 ploidy (01) : 156-159. The DOI: 10.13840 / j.carol carroll nki cn21-1457 / tq. 2022.01.034.
5. Li Baihong, Liu Wenlong, Cheng Zheng, Li Jun, Wang Yong, Shang Yangbing. Application and effect evaluation of water shutoff profile control technology [J]. Petrochemical application,2021,40(04):64-70.
6. Shu Shuang, Yao Lichao, Sun Haonan, Shu Kun. Application and effect evaluation of water shutoff profile control technology [J]. Chemical design communication, 2019,45(04):66+90.
7. Guo Zhihua, Yin Zhaojuan. Research and application of intelligent micro-production and injection distribution technology in low permeability reservoir [J]. Inner Mongolia petrochemical industry, 2021, 47(08):69-70+81.

 https://doi.org/10.1051/e3sconf/202235802032

Analysis of Favorable Injection-production Area Near the Fault Where the Directional Wells Were Deployed

Lili Meng

Fifth Oil Production Plant of Daqing Oilfield Co., Ltd., Daqing, China

Abstract. Through the large-scale application of directional well technology, X development zone has achieved good potential tapping effect near the fault. However, with the increase of production batches of directional wells, the proportion of high-yield wells decreases, and the difficulty of well location optimization increases gradually. In order to further improve the recovery near the fault, under the condition of large-scale development of sand body in the study area, this paper focuses on the analysis of the deployment area conducive to the high yield of directional wells from the perspective of injection-production relationship. Combined with the development practice of 48 directional wells that have been deployed and have achieved good development results, the well distribution effect is evaluated from three aspects of well pattern, well spacing and injection-production location, and the deployment principle of directional wells are determined. The favorable injection-production area for the deployment of directional wells near the fault is summarized, which indicates the direction for further improving oil recovery near the fault in the old oilfield.

Keywords: Fault; directional well; injection-production relationship; well location deployment.

1. Introduction

X Development Zone is a structural reservoir and structural-lithologic reservoir. At present, it has entered a period of ultra-high water cut development. The remaining oil is highly scattered, only enriched locally, and it is difficult to tap potential. In the past, when the dense well pattern was deployed near the fault, in order to avoid the loss of the target layer, the designed well position was far away from the fault as far as possible, resulting in low well pattern density near the fault and imperfect injection and production relationship, which was the main remaining oil enrichment area. Since 2013, based on the results of well-seismic combined with fine fault interpretation, 48 large displacement directional wells have been deployed near the fault of X development zone, and good development results have been achieved, which broadens the technical path of oilfield potential tapping in the late stage of ultra-high water cut. In practice, it is found that the structure, reservoir and injection-production relationship jointly affect the development effect of large displacement directional wells near the fault. This paper analyzes the three aspects of well pattern, well spacing and injection-production location, and summarizes the laws that they have a positive impact on the development effect of directional wells, providing guidance for the popularization and application of large displacement directional wells.

2. Analysis and evaluation of directional well production effect

The evaluation standard of directional well production effect is established. Because of the good correlation between the initial cumulative production, the initial daily oil production, the initial water content and the predicted ten-year cumulative production of the directional well with large displacement, the evaluation standard of directional well production effect is established by determining the parameters.

The wells that have been put into operation in X development zone are divided into four categories : the first type well has the best production effect, the second type well is relatively good, the third type well meets the design requirements, and the fourth type well has relatively poor production effect and low production, which does not meet the design requirements (Table 1).

 https://doi.org/10.1051/e3sconf/202235802032

Table 1. Statistical table of well distribution effect classification for large displacement wells

classification	First-year cumulative production (million tons)	Initial daily oil production (tons)	Water content in the first year (%)	Prediction of ten years cumulative production (million tons)	well numbers (pcs)	outcome evaluation
first kind	≥0.3	≥15	<80	≥1.5	7	very good
second kind	0.2~0.3	15~9	80~90	1~1.5	19	good
third kind	0.1~0.2	9 ~3	90~95	0.5~1	11	achieve design
fourth kind	<0.1	<3	≥95	<0.5	5	worse

3. Analysis of injection-production relationship in directional wells

The distance between different well patterns and the target layer of directional wells , and the position of directional wells in the injection-production relationship of different well patterns were analyzed to evaluate the well distribution effect.

Correlation analysis was carried out by using the evaluation type of directional wells and the initial oil production with the distance of adjacent wells and the injection-production location (Table 2).

Table 2. Statistical table of distance between target layer and adjacent wells, injection-production location

Serial design	well number	evaluation type	Oil production in early years	Basic oil wells	Basic wells	First Infill Well	first infill water well	Second infill oil wells	Second infill water wells	third infill oil wells	third infill water wells	Polymer flooding oil wells	Polymer flooding water wells
1	X1	2	0.8595	260	550 outside	262	199 edge	130	95 middle	222	313 outside		
2	X2	2	0.7251	335	662 outside	97	400 outside	132	345 edge	245	217 edge		
3	X3	1	0.6023	316	419 edge	75	334 outside	129	262 edge	208	362 edge		
	X4	2	0.5818	176	415 outside	75	272 edge	97	260 outside	195	239 edge		
4	X5	3	0.5719	220	480 edge	126	76 middle	81	315 outside	252	224 edge		
5	X6	2	0.8595	375	522 edge	208	130 middle	117	333 outside	222	686 outside		
…	…												
35	X7	4	0.1382				345 outside	98	316 outside			115	360 outside
36	X8	2	0.1288				344 outside	131	335 outside			141	349 outside
41	X9	3	0.0933			428	163 outside	102	200 outside			138	171 edge
42	X10	4	0.0908	658	183 outside	146	436 outside	343	180 outside			345	464 outside

https://doi.org/10.1051/e3sconf/202235802032

(1)Effect of well spacing on well distribution : basic well
The directional well and the basic well with relatively good effect are more than 200 meters, and the directional well is more than 250 meters away from the basic well.
(2) Effect of well spacing on well distribution : basic well
The distance between primary wells has little effect on the overall effect of directional well layout. When the primary well is more than 60 meters, and the primary well is more than 100 meters, it can meet the design requirements of directional wells.
(3)Effect of well spacing on well distribution:secondary well
The distance between the secondary wells has little effect on the overall effect of directional well layout. The secondary wells are more than 60 meters, and the secondary wells are more than 100 meters, which can meet the design requirements of directional wells.
(4) Effect of well spacing on well distribution : third infill well
Third infill well within 100 meters from the directional well will affect the well distribution effect.

4. The position of directional wells with different distribution effect in injection-production well pattern

One kind of directional well is more than 200 meters away from the foundation well and is located outside the main flow line of injection and production of the basic well.
If the directional well is more than 170 meters away from the first, second and third wells, it will not affect the well distribution effect at any injection-production location. The second type of directional wells are more than 100 meters away from the foundation wells, and are located outside the main streamline of injection and production of the foundation wells; directional well is more than 130 meters from first infill well and more than 100 meters from second well; when the distance from polymer flooding wells is more than 240 meters, the well distribution effect will be relatively good. The three types of directional wells are close to oil wells or wells, and most of them are located on the main flow line of water injection. If the four types of directional wells are within 200 meters from the basic well or are located on the injection-production relationship main streamline of the basic well , the well distribution effect will become worse .

Table 3. Influence of injection-production position on well location deployment

Well deployment evaluation	Directional well type	Basic oil wells	Basic water wells	The first、second、third oil wells	The first、second、third water well
good effect	Category I, II	≧180	≧250 and Outside main stream	≧60	≧150
best effect	Third type	≧100	≧200 and Outside main stream	≧50	≧100
poor effect	Fourth type	≧100	≦200 or inside main stream	poor effect	

5. Conclusions and recommendations

1. Basic well pattern has the greatest impact on directional well deployment, while first, second and third wells have relatively small impact on directional well deployment.
2. It is recommended that directional well locations be deployed over 180 metres from the basic well and over 200 metres from the basic well and outside the main stream line.

References

1. Author,WANGHui-yun.:Technical Difficulties in Directional Well Logging [J]. Science and Technology Information Development and Economy, 2009.
2. Author,GENG Qiang.:Directional well trajectory design principles and control strategies [J].China Science and Technology Expo, 2017.
3. Author,WANG Xue-jian.: Shallow directional well continuous control drilling technology [J].Petroleum drilling technology, 2004.

https://doi.org/10.1051/e3sconf/202235802033

Study of Fine Management Mode on Improving Potential Benefits of Measures

Yinfeng Shan

Fifth Oil Production Plant of Daqing Oilfield Co., Ltd., Daqing, China

Abstract. The oil field has entered the extra high water cut period, the oil well fracturing and oil increment have made great contributions to the fine tapping potential of water flooding. At the same time, the oil well fracturing and oil increment production has become an important part of the stable production of water flooding. Therefore, the fine oil well fracturing management mode and the improvement of oil well fracturing and oil increment effect are particularly important. Focus from the advance potential survey, the implementation of pre-pressure training; fine well selection and layer selection, optimize fracturing design ; strengthen process management, strengthen construction supervision ; do well after fracturing management, improve oil increment effect to fine oil well fracturing management mode. The comprehensive application of various technical means to continuously improve the accuracy of residual oil analysis is the basis to ensure the targeted fracturing measures; on the basis of the analysis of remaining oil, it is the prerequisite and necessary guarantee to improve the fracturing effect and prolong the effective period to do well in the pre-injection and post-injection recharge. The key to improve the fracturing effect of oil well is to combine the remaining oil potential and fracturing technology closely, design personalized scheme, refine the fracturing interval and expand the construction scale. Doing well in fracturing process management and refining each construction link is an effective guarantee to maximize the benefits of measures. On measures to tap potential, deepen potential research, optimize program design, give full play to the synergistic effect of measures, and strive to improve the pertinence of measures to tap potential in ultra-high water cut period, continuously improve the fracturing effect, improve the benefits of measures to tap potential, and achieve the goal of ' high efficiency, sustainable and guaranteed ' sustainable and stable production.

Keywords: Oil Well Fracture, Process Management, Potential Tapping Benefits.

1. Introduction

Fracturing, as an effective means of reservoir reconstruction and production increase, has been greatly developed both in supporting technology and application scale since the oilfield was put into development.

Especially in recent years, in the reservoir, the fracturing well selection criteria for different development blocks are studied and quantified. In the process, the matching fracturing technology suitable for different reservoirs and different residual oil types is developed, and the effective utilization of residual oil between layers and within layers is realized. When the oilfield enters the extra high water cut period, the oil production of oil well fracturing has become an important part of water drive stable production. Therefore, it is particularly important to refine the management mode of oil well fracturing and improve the oil production effect of oil well fracturing.

2. Main Practices

Facing these new challenges under the new situation, we should focus on the following aspects to fine oil well fracturing management mode:

2.1 Advance potential survey, pre-fracturing Culture

In the process of underground large-scale investigation, the layer by layer analysis was conducted to find out the well layer that may have fracturing potential. For the well layer that did not have fracturing conditions at that time, through the adjustment of injection-production structure, targeted pre-fracturing training was carried out, and the training situation was tracked and checked. The wells that did not have fracturing conditions were further cultivated, and the well scheme with fracturing conditions was prepared, so as to "mature batch, implement batch and cultivate batch ".

Establishment of a pool of potential measures. According to the potential size, well layer conditions and expected implementation effect, the measures potential classification standard is established, and the wells in the block are graded and evaluated according to the potential reservoir classification standard. The measure potential reservoir is maintained regularly to ensure that the number of measure potential wells remains at a certain scale.

The Culture of pre-fracturing. For the well group with imperfect injection-production system due to casing damage shut-in, the injection-production relationship of the well group is improved by means of water injection well overhaul and renewal ; for the wells with poor water absorption capacity of the target layer caused by interlayer interference of the surrounding wells, the water injection volume of the corresponding layer of the surrounding wells can be improved by adjusting the water injection scheme and increasing the injection volume of the corresponding layer. For thin and poor oil reservoirs and transition zone areas, due to poor reservoir physical properties, resulting in poor water injection efficiency of oil reservoirs, adjacent water injection wells can be fractured in advance to improve the permeability of the target layer ; for the oil wells with high water cut and high liquid volume, the interlayer contradiction is alleviated by controlling the main water breakthrough layer, water injection in the water direction, or blocking the liquid production in the high aquifer of the oil well.

2.2 Fine well and layer selection and optimization of fracturing design

With the continuous increase of comprehensive water cut in the block, the potential tapping objects continue to shift to thin and poor reservoirs. The original standard cannot meet the development needs and has poor applicability. It is necessary to explore new dynamic potential tapping standards.

The first is to break through the boundaries of water cut, change ideas, the whole well high water cut is not equal to the layers of high water cut, mining high water cut wells in low water cut reservoir remaining oil potential.

The second is to break through the thickness limit, do ' two combination ', weaken the thickness control, combined with the four types of monitoring data to optimize the vertical production difference horizon, combined with the four types of sand body morphology to optimize the plane low water content position.

Again is to break through the conventional fracturing design, according to the development of small layers in the fracturing layer, ' tailoring ', and ultimately determine the fracturing method, and finely quantify the amount of sand standard, increase the amount of sand per unit thickness, to ensure the fracturing effect laid a solid foundation.

Table 1. Optimization of vertical use difference layers based on four types of monitoring data.

Effective thickness	Three injection profiles		Three liquid production profiles		Logging Interpretation of New Well		Numerical simulation results		
	Water absorption times	Average relative water absorption	Number of liquid production	Average relative liquid production	flooding interpretation	water saturation	recovery percent of reserves	oil saturation	Layering Water
>1.0m	1time	<5%	1time	<15%	Unflooded	<40%	<40%	>50%	<90%
0.5~1.0m	≤2 times	<10%	≤2 times	<15%	Low flooding	<40%	<40%	>50%	<85%
0.2~0.4m	≤3 times	<15%	≤3 times	<10%	medium flooding	<40%	<40%	>50%	<80%
outside table layers	≥1 time	<15%	≥1 time	<10%	/	/	<40%	>50%	<80%

Table 2 Optimal Selection of Low Water Cut Area in Plane Combined with Four Types of Sand Body Morphology.

sandbody type	deposition diagram	Favorable distribution pattern	causes	Adjustment of well pattern
distributary channel sand		Narrow strip prominent parts or abandoned rivers	Well pattern cannot be controlled	basic well pattern
Main thin layer sand	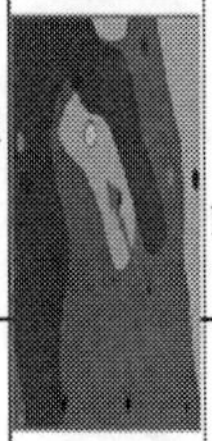	strip or large area of lump Saddle or finger protruding parts	Vertical upper level interference Thin Injection Thick Mining on Plane	one encryption
Non-main thin sand	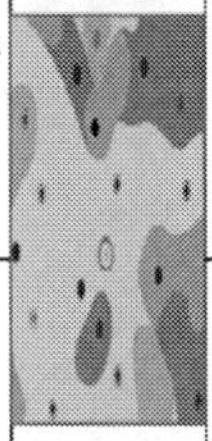	Banded Contact between lump distribution and main sand	Thick injection and thin mining area Interlayer interference	two encryption
outside table layers		Type II sandstone thickness greater than 2m Bridge with Interior Edge	Poor reservoir development There is a starting pressure gradient	three encryption

2.3 Strengthen process management and construction supervision

Strengthening the process management of measures is to strengthen the whole process management, that is, to pay attention to the whole process management of scheme design, team organization, operation construction, field supervision, tracking and evaluation. Firstly, it is necessary to clarify the working process and department responsibilities. Secondly, it is necessary to strengthen the supervision of the construction site. Finally, the responsibility is decomposed and the assessment objectives are formulated. In view of the actual situation of each block, the assessment objectives of measures to increase oil are formulated. In the quarterly or end-of-year performance appraisal, the effect of measures is taken as an assessment project, and the relevant responsible persons who fail to meet the standards are evaluated, and the corresponding rewards are given to the qualified personnel. The contribution rate of fracturing scheme and the compliance rate of oil increment effect of measures are mainly evaluated.

Well developed, can separate out of the layer, generally choose ordinary fracturing .

Large displacement fracturing is generally selected to increase fracture strength in thin and poor layers .

Multi - fracture fracturing is generally selected to increase the number of fractures by developing multiple thick layers in the surface .

Small Diameter Fracturing is preferred for casing wells with diameter more than 108 mm.

Table 3 Optimization of reservoir process parameters with different effective thickness

Effective thickness (m)	>1	0.5~1	0.2~0.5	0
Sand addition (m^3)	6~8	7~9	8~10	9~12
Displacement (m^3 / min)	2.4~2.8	2.6~3.0	2.8~3.2	3.0~3.5
Penetration ratio (%)	6~10	8~12	10~14	12~16

2.4 Do Well Post-fracturing Management and Improve Oil Increasing Effect

After oil well fracturing, it is not the termination of measures, but the start of production increase. The fracturing effect is finally reflected in the cumulative oil increase. Therefore, it is necessary to analyze the effect of production increase, the injection-production status and dynamic change of the well area, and take corresponding measures in time to extend the effective oil increase period.

For some wells with poor effect, we should carefully analyze the reasons, strengthen data acquisition and dynamic monitoring, and formulate targeted control measures

Table 4. Main means of tracking protection measures.

types	case analysis	governance ideas	main method
Low submergence depth, small liquid increase	insufficient for liquid	Strengthening fluid supply	measures to increase water injection
High submergence depth, small liquid increase	small Parameter	amplifying parameters	Adjusting parameters and changing pumps
High liquid increment and high water content	High aquifers	Water content control	water plugging of oil well

3. Some understandings

3.1 The quantitative standard of fracturing well selection can only play an auxiliary role, the real measures of well selection, or to rely on rich dynamic experience, which requires the design personnel to have a solid basic skills.
3.2 With the continuous increase of comprehensive water cut in oilfield, the remaining oil potential is continuously transferred to thin and poor reservoirs. The water cut limit and thickness limit should also be continuously broken through in the selection of fracturing wells and layers. The low aquifer should be found in high water-cut wells, the low water-bearing sand body should be found in high aquifer, and the low producing position should be found in high water-bearing sand body.
3.3 Based on the analysis of remaining oil, it is a prerequisite and necessary guarantee to improve the fracturing effect and extend the effective period to do well in the pre-injection training and post-injection supply.
3.4 Do a good job in the whole process of fracturing management, fine each construction link, is the effective guarantee to maximize the benefits of measures.

References

1. Christian oberwinkler,Michael J.Econmides.The definitive identification of candidate wells for refracturing[J].SPE84211,2003.
2. Yu Fenglin, Liu Min, Hou Jibo, etc. Discussion on well and layer selection method of repeated fracturing in oil wells at the late stage of high water cut [J]. Daqing Petroleum Geology and Development, 2005, 24 (4) : 47 – 49.
3. Song Yong, Liu Chunlin. Factors Affecting Fracturing Effect [J]. Daqing Petroleum Geology and Development, 1995,14 (2) : 43-46.
4. Author,WAN Xin-de., HU Guang-bin.,LV Xiang-jun.: Ways to maintain and improve the fracturing effect of oil wells in the later stage of high water cut. Daqing Petroleum Geology and Development, 1993.
5. Author,SUN Jian-ping., Author,Han Qing.: Principle and Analysis of Oil Well Fracturing Process [J]. China Petroleum and Chemical Standards and Quality, 2011,31 (12) : 94.
6. Author,YU Xiao-xia.: Study on the adaptability of water flooding technology for increasing production and injection in extra-high water cut period [D].Northeast Petroleum University, (2011).
7. Author,Liu Yongxi.:Research on Fracturing Technology of Thin and Poor Layer in Daqing Oilfield [D]. Daqing Petroleum University, (2006).
8. Characteristics and application of oil-based fracturing fluid performance. CNKI. 2009/12/15.

Geological and geochemical characteristics and genesis of carbonate rocks in a certain area

Tong Zhang

Geological Laboratory, Research Institute of Exploration and Development of Daqing Oilfield, Daqing, Heilongjiang 163712, China

Abstract: Carbonate rock is a general term for the rock composed of carbonate minerals, which is mainly composed of carbonate sediments, with calcite (limestone) and dolomite (dolomite) as the main forms. Carbonate rocks are important oil storage rocks, storing more than half of the world's natural gas and oil, and co-existing with various types of solid sedimentary deposits. They are themselves a kind of mineral that can be used in construction, chemical industry, metallurgy and other processes, so they have very important economic value. This paper analyzes the geological and geochemical characteristics and genesis of carbonate rocks in a certain area in order to provide theoretical support for related work.

Key words: Carbonate geology; Geological and geochemical characteristics.

1. Introduction

All kinds of geological are certain conditions and factors in the long-term effects after formed, and the process has many influence factors, often difficult to completely after carved from the details, but for all kinds of geological condition of study also has its unique significance, and can help people in some decisive conditions under the condition of accurate understanding of geologic types, This includes the geochemical characteristics of geological types. In terms of the crystal characteristics and crystal habits of carbonate minerals, the chemical characteristics of carbonate minerals are mixed with gypsum, barite, rock salt, potassium and magnesium salt minerals, and a small amount of opal, glauconite, phosphate minerals and various kinds of organic matter, which show certain differences. For the carbonate rocks mixed with terrigenous impurities, when the amount of terrigenous impurities and minerals is more than half, the carbonate rocks will transition to clay or clastic rocks. However, in order to ensure the accuracy of analysis, the specific geochemical characteristics of carbonate rocks should be comprehensively considered according to the geological environment where they are located.

2. Basic geological features of the region

2.1 Location and development history

It carbonate geological basin is located in a local place, found that after the access to previous research data, the geological early in a passive state there terrigenous extensional environment, in the condition of the terrigenous clastic rock and volcanic sedimentary, and form a bi-model volcanic construction, most of the iron ore in the area of the main forms in this period; After the beginning of the Late Carboniferous period, the geological environment in this area was in a state of convergence as a whole. At this time, there was obvious deposition of flaystone formation, including the deposition of intermediate acid volcanic rock formation, and the two kinds of formations were invaded by calc-alkaline granite in different degrees. Until the middle and late Carboniferous period, the process of molasse construction appeared. The last stable period occurred after the Permian period [1].

2.2 Distribution characteristics

After conducting a large number of geological sections (14) in the area and analyzing the results, the existence of carbonate rocks was found in all the section results, and the rock types were mainly marble, limestone, ribstone and so on. It includes microcrystalline limestone, fine crystalline limestone, fine crystalline marble, marbleized limestone, fine crystalline tuffy-bearing marble, calcareous mylonite, calcareous mylonite, calcite structural schist, etc. Except for calcite structural schist, other rock lithologies can be summarized as limestone. After investigation of its distribution location and analysis combined with existing relevant data, it is found that limestone is distributed in the vicinity of the ore body, but also in the periphery of the mining area [2]. After the investigation of their regional distribution patterns, it is found that most of their shapes present lenticular, layered and so on, and a large number of them exist in the volcanic

 https://doi.org/10.1051/e3sconf/202235802034

rock strata in the state of bedding distribution, and their thickness shows disorder, ranging from tens of centimeters to tens of meters.

3. Geological and geochemical characteristics of carbonate rocks in this area

3.1 Sampling and Testing

After sampling the central and peripheral carbonate rocks in the area, the samples were sent to the HeI Beijing Institute of Geology for analysis. The main elements and trace elements of the samples were analyzed by X-ray fluorescence spectrum and ICP-MS method, and the analysis results were shown in Table 1 and Table 2 below.

3.2 Principal element characteristics

Carbonate rocks in this area are mainly composed of calcite and contain a certain amount of volcanic material or silica on the basis of which the main component is silicate. Therefore, after combining CaO and SiO2 to draw a scatter plot and analyze, it is found that the main components in the samples are concentrated in a large number in the two areas. Among them, the content of CaO in microcrystalline limestone, fine crystalline limestone, marble, mylonite and calcalite structural schist ranges from 49.02% to 54.65%, which belongs to the high content of major elements (nearly half or more than half), while the content of SiO2 is between 0.72% and 8.27%, which belongs to the secondary content of major elements. Compared with CaO, the content of other major elements is generally low, not more than 1% and most of them not more than 0.5%. Considering the content characteristics of major elements, carbonate rocks in this area show high calcium content but low iron, magnesium and manganese content, and the variation of different element contents among different lithologies shows an irregular state [3]. For marble and mylonite containing tuff, the content of CaO ranges from 37.48% to 44.24%, which is relatively high. However, the content of SiO2 ranges from 18.29% to 22.98%, which is significantly higher than that of previous lithology, and the content of various compounds is also higher than that of carbonate rocks without tuff. However, the contents of TiO2, MgO, MnO and P2O5 compounds are not significantly different from those of carbonate rocks without tufts.

Table 1 Analysis results of major elements of carbonate rocks in a certain area

The sample name	ω (B) /%								
	SiO_2	Al_2O_3	TFeO	MgO	CaO	Na_2O	K_2O	MnO	Ignition loss
micrite	3.38	0.275	0.111	0.157	51.92	0.079	0.019	0.123	40.64
Fine grain limestone	1.77	0.123	0.510	0.312	53.46	0.037	0.060	0.049	41.79
marble	1.07	0.536	0.834	0.294	46.13	0.052	0.085	0.336	44.16
Tuffaceous marble	18.86	0.847	1.068	0.505	38.96	0.059	0.343	0.135	30.51
Calcite mylonite	7.18	0.505	0.252	0.248	50.61	0.066	0.037	0.149	43.93
Siliceous calcareous mylonite	18.38	0.352	0.219	0.145	44.24	0.035	0.038	0.158	38.52

Table 2 Analysis results of trace elements of carbonate rocks in a certain area

The sample name	ω (B) /10^{-6}								
	Sr	Ba	Rb	Li	Ca	Tl	Pb	Cu	Zn
micrite	765	13.46	0.68	2.31	0.06	0.13	15.65	3.86	5.35
Fine grain limestone	265	5.34	0.22	5.43	0.08	0.02	0.29	1.96	4.59
marble	199	17.89	1.13	2.52	0.06	0.00	6.38	63.17	8.38
Tuffaceous marble	608	49.13	12.3	13.46	2.93	0.16	0.05	6.38	5.07
Calcite mylonite	426	4.66	1.21	3.10	0.04	0.05	2.32	5.82	10.19
Siliceous calcareous mylonite	30.50	1.34	8.03	0.04	0.01	3.22	5.84	15.07	1.27

3.3 Characteristics of trace elements

It can be clearly seen from the data in Table 2 that the contents of trace elements (excluding rare earth elements) in the carbonate rocks in the central and peripheral areas of this area are significantly different, and show serious disorder, among which various ore-forming elements led by Cu and Zn have relatively obvious changes. In order to investigate the variation of trace elements in all carbonate rocks, the trace elements in microcrystalline limestone should be taken as the standard values, and then the trace elements in other samples should be processed. After this operation and data study, it is found that the contents of various trace elements in carbonate rocks containing tufts are significantly enriched compared with microcrystalline limestone. However, the content of trace elements in calcite schist showed obvious deficit. However, other types of carbonate rocks do not show the same characteristics, and there is no significant enrichment or deficit [4]. This situation indicates that during the tectonic-hydrothermal activity, impurities originally accumulated in the carbonate rocks are largely removed by the hydrothermal process, but the hydrothermal activity process basically does not bring new materials to the carbonate rocks.

4. Geological origin analysis of carbonate rocks in this area

4.1 Formation of the environment

After reviewing the previous relevant data, it is found that the carbonate geology in this area may be formed in the submersible environment above the wave base. The reference factors to provide this conclusion are mainly the constituent contents found in the carbonate rocks. The rocks composed of biodebris were found in the carbonate rocks in this area. After taking out the samples and disassembling them for analysis, it was found that the biodebris mainly consisted of coral, crinoid stem blocks and other Marine biological fossils, and the bamboo leaf-like gravel was also found in the samples. Considering that the distribution of carbonate rocks and volcanic rocks in this area shows obvious overlapping characteristics, and there are tuffite and carbonate tuffite rocks, which usually appear as overrock in the deposition process. The deposition of carbonate rocks and volcanic rocks on the surface of the above signs occurred successively in the same period, and it can be seen from the bioclasts that the sedimentary environment is shallow sea sedimentary environment, because the bioclasts do not contain deep-sea Marine biological fossils [5]. From the positive europium anomaly and negative cerium anomaly in carbonate rocks, it can be found that the seawater was relatively oxygen-rich during the sedimentary process, which coincides with the absence of deep-sea Marine fossils in the rocks.

4.2 Evolution process

The evolution of carbonate rocks in this area is mainly due to recrystallization process under the action of heat. The microcrystalline limestone is transformed into fine crystalline limestone or fine marble after recrystallization process. On the other hand, it also leads to the deformation of rock mass during the tectonic process, and the products are mainly calcite structural schist and calcite mylonite. After the analysis of its petrographic characteristics, it is found that the carbonate rocks in this area appear under the action of strong tectonic stress and heat at the same time, that is to say, the recrystallization is accompanied by the deformation of the rock strata. Combined with the analysis results of the main elements, the contents of the main elements in different lithologies show obvious disorder, which means that the later metamorphic process has not affected the main components in the lithology.

And characteristics of trace element analysis results, according to the area of fine limestone, calcite mylonite and marble on the trace elements were not significant change, that is to say, never happened in the process of carbonate in the subsequent deterioration of high strength of material exchange, and the only change of trace element calcite tectonic schist, its present as a result of the loss of all kinds of trace elements, This proves that some impurities may be removed from the calcite schist during the formation process, or that the calcite schist is formed because a part of the material is removed. In general, although the carbonate rocks in this area suffered a certain degree of material loss in the process of transformation, they did not receive other material inflow in the same period, and the content of ore-forming elements indicates that the ore-forming elements are not brought in.

4.3 The geological formation

Considering the formation environment and process of carbonate rocks in this area, its genesis may be related to skarn. This is because skarn is also accompanied by hydrothermal diffusion (or percolation metasomatism), and there is a large amount of material in and out of the process. Carbonate formation in the region, however, there is obvious difference with traditional silicon card lithification process, prospecting line profile skarn marble and between orebody and fault isolation, both almost no effective contact formation, and the field survey also found that contact with skarn, carbonate, basically did not show the characteristic of the silicon card lithification. Therefore, whether the carbonate rocks in this area are formed under the influence of skarn still needs to be further explored. But from a higher level, skarn type ore deposit can be divided into broad and narrow two kinds of concepts, domestic card lithification is usually determined by the silicon based on the special concept, namely must contact and have an impact to determine formation, skarn deposits, whereas in the broadest sense of the term in the skarn deposit, as a condition of determination not to contact or not. Therefore, from a broad perspective, the carbonate rocks in this area belong to the broad skarn category, which is similar to the continental volcanic lithologic iron ore and Hongyuntan iron ore in Ningwu area in the middle and lower reaches of the Yangtze River.

5. Conclusion

THE study on the formation process of carbonate rocks in this area shows that the carbonate rocks and volcanic rocks were deposited in shallow Marine environment at the same time. Combined with the results of trace element analysis, it is found that there is no obvious skarn mineralization in the carbonate rocks in this area, but it belongs to the generalized skarn mineralization due to the existence of skarn rocks. This analysis reasonably explains the stroke process of carbonate rocks in this area to a certain extent, so the final analysis results can provide reliable theoretical support for peers to a certain extent.

References

1. Ding Jianhua, Li Houmin, Li Lixing, et al. Geochemical Characteristics and Geochemical Characteristics of Carbonate Rocks in Yaman Cycas Deposit, Xinjiang, China [J]. Journal of Mineral Deposits, 2017 (1) : 219-236.
2. Sun Yuxiang, Song Changgui, Yang Jindong, et al. Geochemical characteristics and geological significance of carbonate rocks of the Middle Proterozoic Yangzhuang Formation in the northern margin of North China Craton [J]. Journal of Natural Gas Geoscience.2020 (2) : 268-281.
3. Huang Qinghua, Bai Xuefeng, Wang Hui, et al. Geochemical characteristics of trace elements and rare earth elements in the Middle and Upper Permian boundary carbonate rocks in Linxi Area, Inner Mongolia and their geological significance [J]. Global Geology.2019 (3) : 611-622.
4. Ma Haiou, Wang Lifa, Guo Song, et al. Geochemical characteristics and genesis of the overlying regolith of carbonate rocks in Eastern Yunnan Province [J]. Geological Journal of China Universities.2018 (2) : 222-232.
5. Lu Yong, Shan Keqiang, Pan Ming, et al. Geochemical characteristics and environmental significance of carbonate rocks in Qixia-Maokou Formation, Western Zhenxiong, Northeastern Yunnan Province [J]. Journal of Geomechanics. 2017 (3) : 348-357.

Exploration and Practice of CO2 Huff-N-Puff Combined CO2 Flooding in Shale Oil Well Cluster Group

Haifeng Wang [1], Xiaoyan Wang [1,*], Pengtao Wang [1], Yang Zhang [1], Wei Wang [1], Xi Yan [1], Nan Zhang [1], Jinzhu Li [2], Dehua Yang [1]

[1] Oil Production Technology Research Institute, Dagang Oilfield, PetroChina, Tianjin 300280, Chins
[2] Tianjin Dagang oilfield Bingang Group Bohong Petrochemical Co., Ltd, PetroChina, Tianjin 300280, China

Abstract. Shale oil production wells in Dagang Oilfield in China are put into production by means of horizontal well and segmented multi-cluster massive volume fracturing. The production characteristics are that the output is high in the initial phase of production, the output decreases rapidly with the continuous production, and there are no effective measures to increase production. There is an urgent need for stimulation technology to improve the recovery of shale reservoir. In the field of oilfield development, CO_2 has multiple advantages such as easy injection, viscosity reduction and energy increase. For the shale oil well cluster group with good inter well connectivity, the author designs an implementation mode of "CO_2 huff-n-puff combined CO_2 flooding" to improve shale oil production. This implementation mode can play the dual role of CO_2 huff-n-puff combined CO_2 flooding and CO_2 drive to realize the efficient production increase of shale oil well group. After two phases of implementation of the test well group, the cumulative oil increase of the well group has reached 3864 tons, and the comprehensive oil change rate has reached 1:1.22, which has achieved good stimulation effect. This implementation method can effectively improve the production of shale oil reservoirs and provide a new technical measure for improving production and efficiency of similar shale reservoirs.

Key words: Shale oil, CO_2 huff-N-puff, CO_2 Enhanced oil recovery, CO_2 flooding

1. Introduction

The thickness of shale oil reservoir in Dagang Oilfield is 150-200m and the buried depth is 3000-5000m; Dagang shale oil reservoir has poor producability and difficulty in development , which is characterized by higher proportion of nanometer pore throats and rather minute pore throats, and the flow capacity mainly comes from sub-micro pore throats with higher displacing pressure and lower maximum mercury-injection saturation. Shale oil production wells are put into production by means of horizontal well and segmented multi-cluster massive volume fracturing, and most of them have no artificial energy supplement. In the initial phase of self spraying production, the output is high, but the self spraying period is short and the decline is fast; Artificial lifting can rapidly increase production, but it is still in the dilemma of short high production period and rapid decline. There is an urgent need for the replacement technology of increasing production and efficiency to improve the recovery of shale reservoir.

2. Stimulation mechanism of CO_2 injection into shale oil

The multiple advantages of CO_2 in oilfield development, such as easy injection, viscosity reduction and energy increase, make use of the miscibility of CO_2 and shale oil to reduce the viscosity of shale oil, increase the fluidity of shale oil and improve the output of shale oil; At the same time, after entering the shale reservoir, CO_2 can diffuse, dissolve and extract shale oil in the matrix and supplement formation energy by volume expansion, which can further improve the recovery of shale reservoir.

Hawthorne divides carbon dioxide entering shale fracture matrix system into four steps: (1) Carbon dioxide passes through the crack quickly under the action of high pressure. (2) When the shale matrix is exposed to carbon dioxide, carbon dioxide penetrates into the matrix under the action of differential pressure. In this process, the carbon dioxide entering the matrix expands, forcing some shale oil to flow out of the matrix and enter the fracture, but at the same time, some carbon dioxide will also carry shale oil into the matrix. (3) Carbon dioxide entering the matrix is dissolved in shale oil while supplementing matrix energy, resulting in the expansion and viscosity

* Corresponding author: wxiaoy05@petrochina.com.cn

https://doi.org/10.1051/e3sconf/202235802035

reduction of some shale oil. (4) When the pressure system of matrix and fracture reaches equilibrium, shale oil enters the fracture from the matrix under the action of diffusion. Miscible flooding can also occur under appropriate conditions.

For shale matrix, Pierre et al. Pointed out that compared with methane, carbon dioxide is easier to be adsorbed in porous media with organic matter, which can replace the methane adsorbed on the surface of the media. Therefore, the mechanism of enhanced oil recovery of carbon dioxide in shale reservoir mainly includes pressurization, dissolution, extraction, expansion, adsorption and replacement, reducing capillary force and diffusion.

3. Well cluster group selection principle and implementation mode design

Combined with the stimulation mechanism of carbon dioxide injection in shale oil and the production status of Dagang shale oil wells, an implementation mode of "carbon dioxide swallowing drive" is designed for the shale oil well group that can establish displacement relationship, and the well selection principle is given. This embodiment can play the dual roles of huff and puff of injection wells and inter well displacement. CO_2 huff-n-puff combined CO_2 flooding can recover the remaining oil around the injection well from the wellhead, and CO_2 flooding can displace the remaining oil between wells near the production well and recover it.

3.1 Well selection principle

(1) There is a certain production in the early phase of the well group, and the production in the early phase of the well group is high, indicating that the shale reservoir has a certain reserve scale and has the potential of EOR;

(2) The production of the well group decreases rapidly, and the current low production shows that the well group meets the problem of low production and low efficiency, so it is necessary to take measures to increase production;

(3) The target formation of the well group has good connectivity and can establish displacement relationship. This is conducive to the realization of CO_2 flooding between wells and the production of remaining oil between wells.

3.2 Implementation mode design

According to the number of wells contained in the well cluster and the connectivity between wells, taking three well clusters as an example, the design and implementation methods are as follows:

(1) Implementation mode 1, Injection of intermediate wells and production of wells on both sides, as shown in Fig. 1. Shut in the wells on both sides and carry out pressure bearing treatment. Inject CO_2 into the intermediate well. When the three wells meet the well opening conditions at the same time, start the well for oil production at the same time. Until the oil production is lower than that before the measures, it is determined that the first round of measures has failed, and subsequent rounds can be implemented.

(2) Embodiment 2: Injection on both sides and intermediate production, as shown in Fig.2. Shut in the intermediate well and carry out pressure bearing treatment, and inject CO_2 into the wells on both sides. When the three wells meet the well opening conditions at the same time, start the well for oil production at the same time. Until the oil production is lower than that before the measures, it is determined that the first round of measures has failed, and subsequent rounds can be implemented.

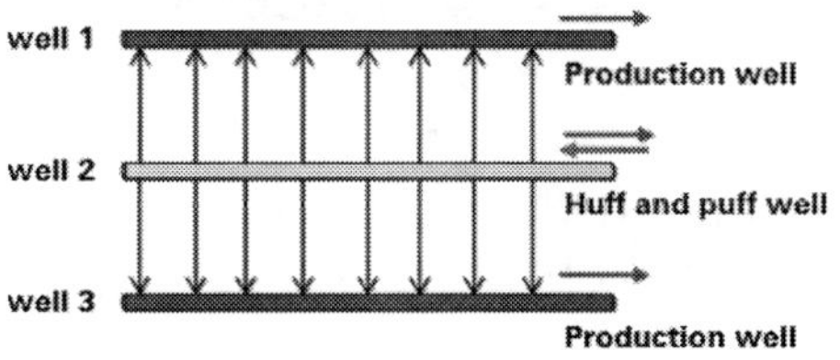

Fig. 1 Implementation mode 1

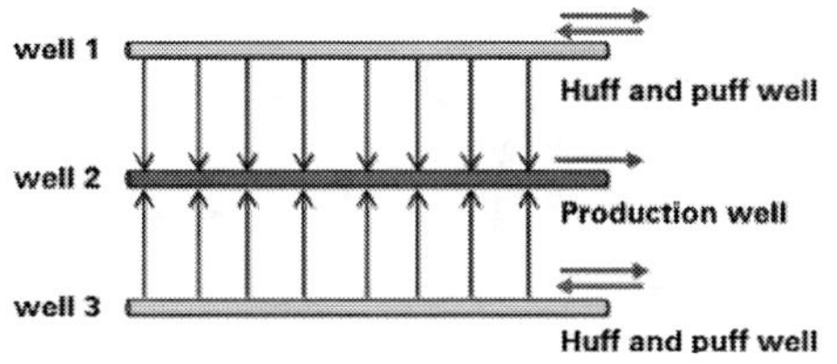

Fig. 2 Implementation mode 2

During the implementation process, in combination with the structural location and production requirements of each well, implementation mode 1 and implementation mode 2 can be carried out alternately to optimize and adjust the sequence. For multi well groups, the implementation methods are not limited to the above two, and the injection wells and production wells can be rotated according to the production needs. Considering the continuous reduction of shale oil near the well, it is necessary to increase the consumption of CO_2 round by round to expand the swept volume of CO_2,so that more remaining oil between wells and edges can be started and flow to the wellhead.

4. Field application effect and understanding

According to the principle of well selection, well group G1 of shale oil in Dagang Oilfield is selected to carry out field test of CO_2 huff-n-puff combined CO_2 flooding. The well cluster includes three horizontal wells arranged in parallel, all of which are put into production by segmented multi-cluster massive volume fracturing without artificial energy supplement. During fracturing, all adjacent wells react, and the well spacing is close, only 130m, with good connectivity between wells. The schematic diagram of test well group is shown in Fig. 3.

 https://doi.org/10.1051/e3sconf/202235802035

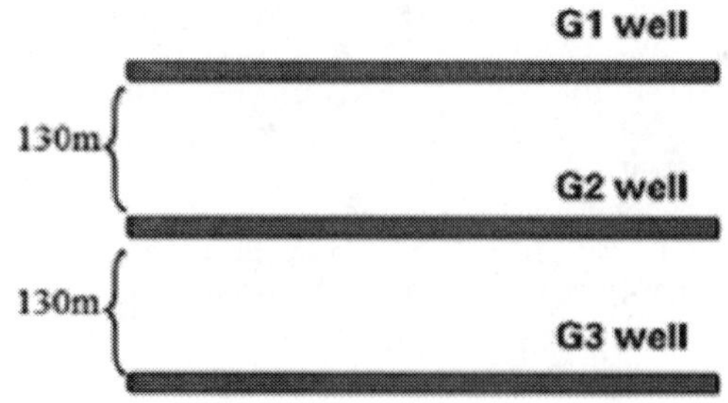

Fig. 3. Schematic diagram of test well group

In the early self-injection production of the well group, the maximum daily liquid production reached 172m^3 , and the maximum daily oil production reached 50t. However, the high-yield period was short, the decline was rapid, and the liquid production and oil production decreased significantly. In the later phase, artificial lifting was selected, and the liquid production and oil production recovered, but there was still a problem of rapid decline in development. The production dynamic curve of well group before test is shown in Fig. 4.

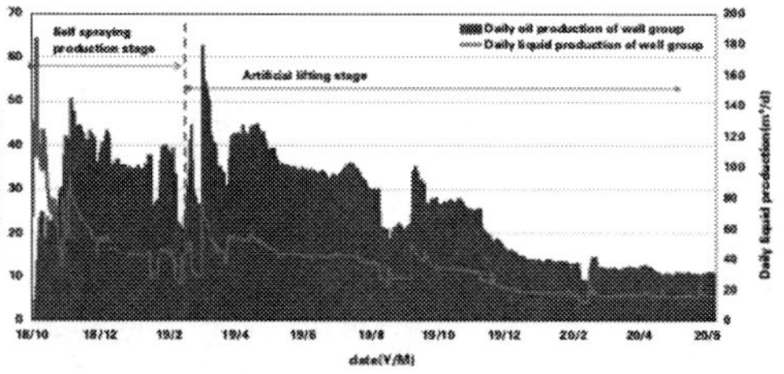

Fig. 4. Production dynamic curve of well group before test

4.1 Determination of implementation mode

The well group has good connectivity. The field test of CO_2 huff-n-puff combined CO_2 flooding is carried out in the well group by the combination of implementation mode 1 and implementation mode 2. Implemented in two phases:

(1) In the first phase, Well G1 and well G3 were subjected to pressure treatment, and CO_2 was injected into the well G2. After the three wells reached the blow off conditions at the same time, the self-injection production was carried out, and the oil recovery was resumed according to the original lifting process.

(2) In the second phase, the well G2l was subjected to pressure treatment, and CO2 was injected into well G1 and well G2. After the three wells reached the blow off conditions at the same time, the self-injection production was carried out, and the oil recovery was resumed according to the original lifting process.

4.2 Design of CO_2 injection amount

The CO_2 injection amount is determined according to the thickness of shale reservoir, fracturing volume of single well, horizontal section length, free hydrocarbon content, organic carbon content and other reservoir parameters of the test well group, as well as the process parameters such as CO_2 action radius and CO_2 solubility with shale oil under reservoir conditions. In the first phase of design, the CO_2 injection volume of G2 well is 1560 tons; In the second phase, 780 tons of CO_2 is injected into well G1 and 840 tons of CO_2 is injected into well G3.

4.3 Test effect

In the first phase, the total oil increment of the well group was 2087t, of which 922t were increased in well G2, 551t were increased in well G1, and 614t were increased in well G3, and the oil exchange rate in this phase was 1:1.34. In the second phase, the well group increased oil by 1777t, including 525t in Well G1, 645t in Well G3 and 607t in Well G2. The oil exchange rate in this phase was 1:1.10. Through the implementation of the two phases, the cumulative oil increase of the well group was 3864t, and the comprehensive oil exchange rate reached 1:1.22, which achieved good oil increase effect. The test results are shown in Table 1.

Table 1. Statistical table of field test results of CO_2 huff-n-puff combined CO2 flooding

Well number	First phase			Second phase		
	CO_2 injection amount(t)	Soaking time (d)	Oil increment(t)	CO_2 injection amount(t)	Soaking time (d)	Oil increment(t)
Well G1	0	28	551	740	25	525
Well G2	1560	20	922	0	29	607
Well G3	0	28	614	810	25	645
total	1560		2087	1550		1777

4.4 Understanding of the test

(1) The oil layer has good gas absorption performance. At the beginning of gas injection, the injection displacement is increased by stages, and the suction indicator curve is drawn according to the injection displacement and injection pressure. The test results show that the injection displacement is positively correlated with the injection pressure, indicating that the shale reservoir has a strong gas absorption capacity for CO_2.

(2) The mobility of shale oil increases. The viscosity and density of shale oil were tested before and after the test, as shown in Table 2. The test results show that the viscosity and density of shale oil decrease after the test, indicating that the injection of CO_2 can effectively reduce the viscosity and density of shale oil, improve the liquidity of shale oil, and facilitate the migration of shale oil in the reservoir and the lifting in the wellbore.

Table 2. Statistics table of shale oil viscosity and density test results before and after test

Well number	Test phase	Pre-test	Initial of the first stage	The middle of the first stage	Initial of the second stage	The middle of the second stage
Well G1	Density (g/cm³)	0.882	0.866	0.878	0.869	0.871
	Viscosity (mPa·s)	96.31	65.72	67.69	66.15	68.22
Well G2	Density (g/cm³)	0.873	0.859	0.862	0.865	0.864
	Viscosity (mPa·s)	67.92	60.94	49.54	52.47	50.31
Well G3	Density (g/cm³)	0.875	0.867	0.872	0.859	0.87
	Viscosity (mPa·s)	106.31	61.33	65.67	64.39	68.42

(3) Formation energy is supplemented. According to the monitoring data, the energy of all wells has been fully supplemented. In the first stage, the flowing production of G2 well will last for 35 days, and the flowing production of G1 well and G3 well will last for 25 and 28 days. In the second stage, the flowing production of G2 well will last for 30 days, and the flowing production of G1 well and G3 well will last for 26 days and 23 days;The dynamic liquid level of each well was tested. In the first stage, the dynamic liquid level on both sides of the well changed from no display before the test to 469M and 825m before blowout in the middle of the first stage; At the same time, the indicator diagram of each well shows that the liquid supply is sufficient.

5. Application prospect

The potential resources of shale oil in the world are considerable, with a total amount of 1.1×10^{13}~1.3×10^{13}t,mainly distributed in the United States, Russia, China, Australia and other countries. According to the statistics of the US energy information administration, the shale oil reserves of the United States, Russia and China account for 47.8% of the world's shale oil reserves.CO_2 huff-n-puff combined CO_2 flooding implementation mode can be implemented in shale reservoir well groups with good inter well connectivity or displacement relationship. It can significantly increase shale oil production and has broad application prospects.

References

1. Xianzheng Zhao,Lihong Zhou,Min Zhao,et al.Breakthrough and practice of industrial development on continental shale oil:a case study on Kong-2 Member in Cangdong sag, Bohai Bay Basin[J].China Petroleum Exploration, 2019, 24(05):589-600.
2. Lanlan Yao,Zhengming Yang,Haibo Li,et al.High pressure mercury injection and nitrogen adsorption tests for shale oil reservoirs in Lower sub-member of first member of Shahezi Formation of Dagang Oilfield[J].Petroleum Geology & Oilfield Development in Daqing,2021,40(04):162-168. DOI:10.19597/J.ISSN.1000- 3754.202005050.
3. Yin Shun li, Zhuang Tian lin,Yang Li yong,Jia Yun peng,Liu Xue wei,Liu Yu,Yan Yu. Exploration and Practice of Volume Fracturing Technology of Shale Oil in Dagang Oilfield[J]. E3S Web of Conferences,2021,252.
4. JIN L, SORENSEN J, HAWTHORNE S,et al.In inproving oil transportability using CO2 in the Bakken system:A laboratorial investigation[C].SPE 178948-MS ,2016.doi:10.2118/178948-MS
5. Haitao Wang,Zengmin Lun,Changyuan Lv,et al.Nuclear-magnetic-resongance study on oil mobilization in shale exposed to CO2[C].SPE 190185-PA,2019.doi:10.2118/190185-PA
6. SHEN Z, SHENG J J.Experimental and mumerical study of permeability reduction caused by asphaltene precipitation and deposition during CO2 huff-n-puff combined CO2 flooding injuection in Eagle Ford shale[J].Fuel,2018,211:432-445.doi:10.1016/j.fuel.2017.09.047
7. TODD H B, EVANS J G.In Improved oil recovery IOR pilot projects in the Bakken Formation[C].SPE 180270-MS,2016.doi:10.2118/180270-MS
8. Caineng Zou,Songqi Pan,Zhenhua Jing,et al.Shale oil and gas revolution an its impact[J].Acta Petrolei Sinica, 2020, 41 (1) : 1-12
9. Lu Ding,Hongjian Ni.Effects of Supercritical CO2 Immersion on Mechanical Properties of Shale[J].Science Technology and Engineering , 2019, 19 (30) : 122-127.
10. HAWTHORNE S B, GORECKI C D, SORENSEN J A,ety al.In hydrocarbon mobilization mechanisms form upper, middle,and lower Bakken reservoir rocks exposed to CO[C]. SPE 167200-MS, 2013.doi:10.2118/167200-MS
11. BILLEMONT P, COSNE B, WEIRELD G.Adsorption of carbon dio-xide,methane,and their mixtures in porous carbons:Effect of surface chemistry,water content,and pore disorder[J]. Aangmuir, 2013, 29(10): 3328-3338. doi:10.1021/la3048938
12. LAN Y, YANG Z, WANG P,et al.A review of microscopic seepage mechanism for shale gas extracted by supercritical CO2 flooding[J]. Fuel,2019,238:412-424.doi:10.1016/j.fuel.2018.10.130
13. ARGILLIER J F, COUSTET C,HENAUT I.Heavy oil rheology as a function of asphaltene and resin content and temperature[C].SPE 79496-MS, 2002.doi:10.2118/79496-MS
14. Rengui Zhang,Diren Liu,Cheng Peng,et al.Current status and prospects of China's continental shale oil exploration and development[J].Modern Chemical Industry,2022,42(03):6-10. DOI: 10. 16606/j. cnki. Issn 0253-4320.2022.03.002.

Suitability assessment of carbon dioxide storage sites in sedimentary basins

Pengtao Wang [1], Xi Wu [2], Gangke Ge [1], Xiaoyan Wang [1,*], Haifeng Wang [1], Yang Zhang [1], Wei Wang [1], Xi Yan [1], Yan Zheng [3]

[1] Oil Production Technology Institute, Dagang Oilfield Company, CNPC, Tianjin, 300280, China
[2] CNPC Dagang Oil Company, Tianjin 300280, China
[3] No.5 Oil Production Plant, Dagang Oilfeild Company, PetroChina, Tianjin, China

Abstract. The assessment of the suitability of CO_2 storage sites is the basis for large-scale deployment of Carbon Capture Utilization and Storage (CCUS) projects. However, the research on the suitability of CO_2 in sedimentary basins is relatively lagging behind in China. Previous studies have focused on CO_2 storage sites, while social, economic and environmental factors are ignored. This paper takes counties and districts as the research objects, and conducts a comprehensive evaluation of onshore sedimentary basins. The research results show that: (1) Onshore sedimentary basins cover 839 counties (districts) in China, of which 318 counties (districts) are prohibited from implementing CO_2 storage; 286 counties (districts) are Class II environmental risk areas; The suitable and suitable areas include 217 and 19 counties (districts) respectively, accounting for about 45% of the total onshore storage area. (2) The suitable and relatively suitable areas are mainly distributed in the Tarim Basin, the central and northern parts of the Songliao Basin, the northwestern part of the Ordos Basin, the northeastern part of the Erlian Basin, and the central and southern parts of the Bohai Bay Basin. Forbidden areas and general suitable areas are mainly distributed in Sichuan Basin, Jianghan-Dongting Basin, North Jiangsu Basin, northeastern Bohai Bay Basin, and southern Songliao Basin. This study establishes a method for assessing the suitability of CO_2 storage sites, and gives a suitable area for CO_2 storage in China, which provides a method and basis for the site selection of large-scale deployment of CCUS projects in China.

Key words: CCUS, CO_2 storage, Environmental risk, Suitable.

1. Introduction

Suitability assessment of CO_2 storage sites is a hot frontier issue at home and abroad. There are various evaluation methods, and they are developing more and more towards systematization and standardization. From a national perspective, CCUS geological storage technology plays an important role in achieving emission reduction targets, energy security, and improving environmental protection and economic benefits. Different countries and research institutions have carried out a number of studies in order to compete for the right to speak in international standard-setting. The National Energy Technology Laboratory (NETL) of the United States published a method for evaluating the location of CO_2 storage sites in the United States in 2016 and 2017[1, 2]. The European Union published the most advanced CO_2 storage site selection and characterization methods in 2013[3], as well as the latest progress in CO_2 storage site selection and characterization methods[4]. The British Geological Survey (BGS), through best practices, has also issued corresponding CO_2 storage specifications in saline aquifers[5]. In addition, some international organizations, such as the International Energy Agency (IEA), the World Resources Institute (WRI), and the US Environmental Protection Agency (EPA), have conducted preliminary research on the suitability of CO_2 storage sites[6, 7].
There is also a preliminary study on the suitability assessment of CO_2 storage sites through the China Geological Survey. In 2011, the geological assessment of reservoir and caprock, CO_2 geological storage safety and environmental risk assessment, and CO_2 geological storage site selection and assessment methods in deep saline aquifers were proposed. In 2012, the site selection stage of the deep saline CO_2 geological storage engineering site was divided, and combined with the main contents of the geological evaluation of the reservoir and caprock, the reservoir-caprock suitability evaluation index and its grading standard were preliminarily established. In 2017, the evaluation map of the suitability of CO_2 geological storage in major sedimentary basins in China and adjacent sea areas was drawn[8], which further advanced the theory of suitable evaluation of CO_2 storage sites. Although the assessment of the suitability of CO_2 in

* Corresponding author: wxiaoy05@petrochina.com.cn

sedimentary basins has made obvious progress in China. However, the CO_2 storage site not only involves complex geological problems, but also affects many aspects such as society, economy and environment. Finally, the standardization and standardization of CO_2 sequestration screening will be realized.

2. Data and Methods

2.1 Environmental assessment indicators of CO_2 storage sites

Considering the availability of evaluation data and referring to existing research[9], this paper selects 10 indicators that affect the suitability of CO_2 storage sites. These 10 indicators are divided into two categories: one is prohibited indicators, that is, the evaluation unit contains any prohibited indicators, which will be classified into the I-level control area. Such indicators mainly include national important groundwater resource areas and national important infrastructure construction areas (including dams, airports, etc.). The other category is the restrictive index, which is divided into 4 levels. The index value within the corresponding range will be divided into the corresponding level. The detailed index classification is shown in Table 1.

Table.1 Evaluation index system for environmental risks and environmental impacts of CO_2 geological storage

Indicator type	Social and Environmental Factors	Indicator description
Prohibit	water source	618 important drinking water sources in the country
	Airport	236 airports above 4D level nationwide
	Dam	922 important dams
Restrict	Population density	1km population grid data
	GDP	1km grid data
	Railway	Major railways in the country
	Highway	National main roads
	Surface water	Aboveground rivers and lakes
	Nature reserve	497 nature reserves nationwide
	Ecological reserve	4 types of ecological functional areas, 4 levels

2.2 Environmental assessment methods for CO_2 storage sites

According to the impact of social and environmental factors on CO2 storage sites, the suitability of CO2 storage sites is divided into four grades. Level I is unsuitable, level II is generally suitable, level III is more suitable, and level IV is suitable. In the existing CCUS source-sink matching and CO_2 transportation pipeline planning, cities [10, 11] or counties[12, 13] are often used as planning units.

For prohibited indicators, as long as such factors are included in the evaluation unit, the evaluation unit is defined as level I, that is, an unsuitable area. For the restricted index, generally according to the proportion of the unit area of the influencing factors, it is divided into four grades according to the size of the proportion, and the specific classification method is shown in Table 2. All evaluation units meet the "cask principle", that is, take the lowest level in the evaluation results of all influencing factors as the final evaluation result of the evaluation unit. For example, if a county contains important water sources, the environmental risk of CO_2 storage sites in the county is directly defined as level I (unsuitable).

Table.2 Evaluation imethod for environmental risks and environmental impacts of CO_2 geological storage

Indicator factors	Class I	Class II	Class III	Class IV
water source	√	/	/	/
Airport	√	/	/	/
Dam	√	/	/	/
Population density	>1000	500~1000	300~500	≤300
GDP	>1500	1000~1500	150~1000	≤150
Railway	>0.50	0.20~0.50	0.05~0.20	≤0.05
Highway	>1	0.5~1	0.3~0.5	≤0.3
Surface water	>20%	10%~20%	1%~10%	≤1%
Nature reserve	>50%	30%~50%	10%~30%	≤10%
Ecological windbreak and sand fixation	Extremely important area	Important areas	Medium area	General area
Ecological water conservation	Extremely important area	Important areas	Medium area	General area
Ecological soil conservation	Extremely important area	Important areas	Medium area	General area
Ecological biodiversity	Extremely important area	Important areas	Medium area	General area

2.3 Environmental suitability layer of CO2 storage site

The 1-kilometer grid data of China's GDP spatial distribution comes from the Resource and Environment Science and Data Center[14]. This data is based on the national county-level GDP statistics, correlates the spatial interaction law of residential density data, night light brightness, land use type and GDP, and establishes a relationship model between the three and GDP, and finally obtains . The 1×1 km grid data on the spatial distribution of China's population comes from the Resource and Environmental Science and Data Center[15]. During the processing, firstly, the type of land use, the brightness of night lights, and the population distribution weight under the density of residential sites are calculated, and then the Normalize the influence weights of these three aspects. On the basis of standardized processing, the total weight of each county-level administrative unit is calculated. Finally, on the basis of calculating the population proportion of the unit weight of each county-level administrative unit, using the grid space calculation, the population data on the unit weight and the total weight distribution map are combined to realize the spatialization of the population.

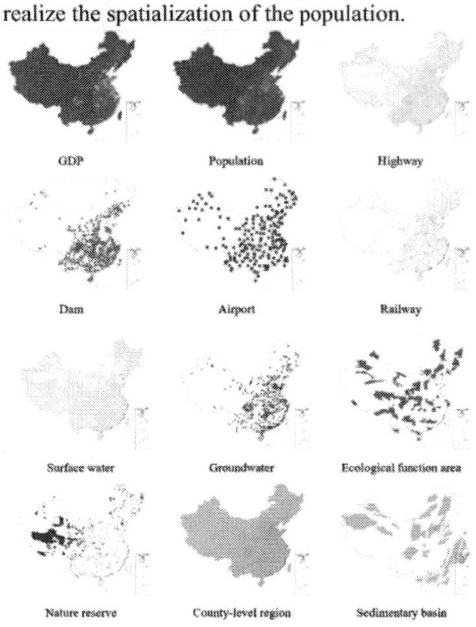

Fig. 1 Dataset for basic environmental elements

In addition, the airport data comes from the Civil Aviation Administration of China [16]. Using the name information of the airport, the specific latitude and longitude information is queried one by one in Google Maps, and the data linkage between Excel and ArcGIS is used to generate ArcGIS point data. The vector data of roads, railways and surface water systems come from the National Geographic Information Resource Catalog Service System[17]. The data of nature reserves comes from the Resource and Environmental Science and Data Center. The data includes 477 nature reserve boundaries, of which 167 nature reserves have core areas, buffer zones, and experimental area boundaries[14]. The ecological function is derived from the China Ecosystem Assessment and Ecological Security Database. The data is based on the characteristics of the ecosystem in western China, and factors such as biodiversity maintenance and protection, soil conservation, water conservation, windbreak and sand fixation, and natural and human landscape protection are selected to carry out ecological functions. Evaluate the importance of system service functions. Ecological functions are divided into general areas, moderately important areas, important areas and extremely important areas[18]. The dam data are derived from the Global Reservoir and Dams (GRanD) database[19]. The ArcGIS layer data distribution is shown in Figure 1.

3. Result

It is generally suitable for level II areas, including 286 counties (districts), accounting for 27.5% of the total onshore basin area. It mainly covers areas with a large area of ecological function zones and nature reserves in the county. For example, the Qaidam Basin, the Junggar Basin and the southern Ordos Basin contain a number of ecological function areas and nature reserves. However, some counties and districts in the Bohai Bay Basin are in the Class II area, mainly due to the high density of population, railways and highways. In addition, counties with more Grade II areas in the North Jiangsu Basin have higher general population and GDP unit values.

It is more suitable for level III areas, including 217 counties (districts), accounting for the largest proportion of the total onshore basin area, about 41%, with a total theoretical storage potential of 858.4 billion tons of CO_2. This shows that most of the major onshore storage basins in my country are within a suitable range. It can be found from Figure 2 that the more suitable areas for yellow are mainly distributed in the Tarim Basin, the northern Ordos Basin, the northeastern Erlian Basin and the northern Songliao Basin. These areas have less impact on society and the environment, and are ideal CO_2 storage sites.

It is suitable for level IV area, including 19 counties (districts), accounting for 4.4% of the onshore basin area, and its total theoretical storage potential of deep saline aquifers is about 90 billion tons of CO_2. It can be found from the map that these main counties (districts) are distributed in the Tarim Basin, Songliao Basin and Erlian Basin. For example, Aksu and Alar in the Tarim Basin, Tongyu County and Horqin Left Middle Banner in the Songliao Basin, etc. Although these 19 counties (districts) meet the CO_2 site selection conditions in terms of economy, society and environment, the geological storage of CO_2 is a complex systematic project, which also involves its own geological factors, whether water resources are sufficient, earthquake risks, and other factors. Factors such as distance from stable carbon emission sources. The social, economic and environmental suitability of the storage site is only a part of its preliminary site selection screening, and further field investigations are still required for practical engineering applications.

4. Conclusions and discussions

Based on selected social, economic and environmental indicators, a comprehensive evaluation of onshore sedimentary basins is carried out. in China. The suitability grades of CO_2 storage sites are divided into 4 grades. The area covered by the onshore basin includes 839 counties (districts). Among them, 317 counties (districts) are Class I areas, that is, areas where CO_2 storage is prohibited, accounting for 27% of the total onshore basin area. These 317 counties and districts are distributed in all onshore basins. The main reason is that, on the one hand, the population density of the main urban areas of these two cities is relatively high; or very important ecological areas. From the perspective of the basin, the main reasons for the prohibition of CO_2 geological storage areas in the Bohai Bay and the North Jiangsu Basin are social and economic factors, among which the high population density and dense infrastructure are the main reasons. The main reason for the occurrence of counties and districts in the Ordos Basin that is prohibited from sequestration is that there are many ecological functional areas. Several counties and districts in the Sichuan Basin are Class I areas, which are mainly due to the presence of multiple important water sources in this area. The reasons for the emergence of prohibited areas in the Tarim and Junggar basins include nature reserves and airports. Most of the Class I areas in other regions are densely populated areas and densely populated by railways and highways, mostly the core areas of provincial capital cities, or important water sources.

In order to better select suitable storage sites for CCUS source-sink matching. This chapter evaluates the social environment of CO_2 storage sites in onshore basins respectively. The evaluation results show:

(1) Onshore sedimentary basins cover 839 counties (districts) in China, of which 318 counties (districts) are prohibited from implementing CO_2 sequestration, that is, level I environmental risk areas; 286 counties (districts) are level II environmental risk areas The more suitable and suitable areas include 217 and 19 counties (districts) respectively, and these two areas account for about 45% of the total onshore storage area.

(2) From the perspective of geographical distribution, suitable and relatively suitable areas are mainly distributed in the Tarim Basin, the central and northern parts of the Songliao Basin, the northwestern part of the Ordos Basin, the northeastern part of the Erlian Basin, and the central and southern parts of the Bohai Bay Basin. Forbidden areas and general suitable areas are mainly distributed in Sichuan Basin, Jianghan-Dongting Basin, North Jiangsu Basin, northeastern Bohai Bay Basin, and southern Songliao Basin. Different basins have certain differences in the reasons that lead to forbidden areas in basins. The Sichuan Basin and the Jianghan-Dongting Basin are mainly due to the presence of important water sources and nature reserves. The Bohai Bay and the North Jiangsu Basin are mainly due to the high density of population, GDP and dense infrastructure.

References

1. Bachu S. Identification of oil reservoirs suitable for CO_2-EOR and CO_2 storage (CCUS) using reserves databases, with application to Alberta, Canada. International Journal of Greenhouse Gas Control. 2016;44:152-65.
2. Qin J, Han H, Liu X. Application and enlightenment of carbon dioxide flooding in the United States of America. Petroleum Exploration and Development. 2015;42:232-40.
3. Lv G, Li Q, Wang S, Li X. Key techniques of reservoir engineering and injection–production process for CO_2 flooding in China's SINOPEC Shengli Oilfield. Journal of CO_2 Utilization. 2015;11:31-40.
4. He L, Shen P, Liao X, Gao Q, Wang C, Li F. Study on CO_2 EOR and its geological sequestration potential in oil field around Yulin city. Journal of Petroleum Science and Engineering. 2015;134:199-204.
5. BGS. Best practice for the storage of CO_2 in saline aquifers. Nottingham, UK.2008.
6. IEA. CCS site characterisation criteria. 2009.
7. EPA. Class II oil and gas related injection wells. United States Environ Prot Agency2019.
8. Center for Hydrogeology and Environmental Geology Survey, CGS. Explanation of Suitability Evaluation Map of CO_2
9. Geological Storage in Main Sedimentary Basins in China and Its Adjacent Sea Area. Geological Publishing House BEIJING 2017.
10. Cai B, Li Q, Liu G, Liu L, Jin T, Shi H. Environmental concern-based site screening of carbon dioxide geological storage in China. Scientific Reports. 2017;7:7598.
11. Sun Liang. Research on China CCUS Source-Sink Matching Decision Support System: Tsinghua University; 2013.
12. Yong Li. Source-sink matching model for carbon capture and storage (CCS) and its application - a case study in Hebei Province: Tsinghua University; 2008.
13. Sanchez DL, Johnson N, McCoy ST, Turner PA, Mach KJ. Near-term deployment of carbon capture and sequestration from biorefineries in the United States. Proceedings of the National Academy of Sciences. 2018;115:4875.
14. Fan J-L, Shen S, Wei S-J, Xu M, Zhang X. Near-term CO_2 storage potential for coal-fired power plants in China: A county-level source-sink matching assessment. Applied Energy. 2020;279:115878.
15. Resource Environmental Science and Data Center. National Nature Reserve Boundary Data. 2018.
16. Xinliang Xu. A Kilometer Grid Dataset of Population Spatial Distribution in China. 2017 Data Registration and Publishing System of Resource and Environmental Science Data Center, Chinese Academy of Sciences.
17. Civil Aviation Administration of China. 2019 civil aviation airport production statistics bulletin. 2019.
18. National Geographic Information Resource Directory Service System. 1: 250000 national basic geographic database. 2019.
19. Ecological function zoning database. Ecological Function Zoning Database: Importance of Ecological Service Functions in China. 2020.
20. Lehner B, Liermann, C.R., Revenga, C., Vörösmarty, C., Fekete, B., Crouzet, P., Döll, P., Endejan, M., Frenken, K., Magome, J. Global reservoir and dam (grand) database. Technical Documentation. 1 ed2011.

New Progress in the Study of Siliceous Rock Oil and Gas in Nadanha Single Terrain

Guohui Qu*, Xuebin Tian, Bowen Li, Xiunan Li, Zilu Zhang, Weizhong He, Yikun Liu, Mingxing Bai, Kun Xie

Key Laboratory of Enhanced Oil Recovery (Northeast Petroleum University), Ministry of Education, Daqing163318, China

Abstract. The siliceous shale oil and gas in Nadanhada terrane has been studied through field geological survey, geochemistry and sedimentary petrology. The results show that the area belongs to a good shale reservoir, and the siliceous rock is brittle and suitable for exploitation. The abundance of organic matter is relatively high. From west to east and from south to north, the abundance of organic matter increases and the maturity of organic matter decreases. From west to east and from south to north, the dynamic metamorphism and thermal contact metamorphism of the siliceous rocks in the Nadanhada terrane are gradually weakened, and the folds and magmatic intrusion of the strata are alleviated.

Key words. Nadanhada terrane; Siliceous rock; Organic matter abundance; Geological section

1. Introduction

The Nadanhada terrane (Wanda mountain body) is located in the eastern part of Heilongjiang Province. The Nadanhada Yanshan fold belt located in the northeast of Heilongjiang Province is distributed to the east of Dahe Zhenshen fault. Its southern section is cut off by the Mishan Dunhua deep fault. The northern and eastern sections extend out of the country and into Russia and Japan. The region is part of the Sikhot Alin hyperplastic fold system at the edge of the northwest Pacific Ocean [1]. The area in China is about $4.2\times10^6 km^2$.

Siliceous rock is a new type of hydrocarbon source rock, and there are no special evaluation standards and specifications at home and abroad. On the basis of the research on hydrocarbon generation potential of siliceous rocks last year, this year, the hydrocarbon generation capacity of siliceous rocks in Sanjiang Basin and Nadanhada terrane was reevaluated by applying the evaluation standard of continental hydrocarbon source rocks in saline water environment in China and referring to the evaluation standard of marine hydrocarbon source rocks commonly used in the world, and the organic matter abundance was raised by a grade. At the same time, light oil and asphaltene asphalt are also found in siliceous rock fractures under microscope.

2. Petrological characteristics and thickness

2.1 Fuyuan Area of Sanjiang Basin

We carried out field petroleum geological survey on siliceous rocks in Fuyuan area of Sanjiang basin, Heilongjiang border, collected paleontological and organic geochemical samples, and carried out analysis and test. The siliceous rocks and siliceous shale in Sanjiang Basin are well developed without folds and metamorphism. Horizontal bedding and massive bedding are developed. The occurrence of the stratum is completely different from that of the siliceous rocks in Nadanhada terrane in Hulin area, and its genesis may be different [2]. According to the identification of Palaeontological Laboratory of China University of Geosciences (Wuhan), the radiolaria belongs to the Early Jurassic Middle Princeton Lower Torian.

The identification results of whole rock XRD show that there are both siliceous rocks (quartz content>50%) and siliceous shale (clay>50%) in the Dajiashan Formation of Sanjiang Basin [3]. The content of brittle minerals in siliceous rocks in Sanjiang Basin ranges from 59% to 95%, with an average of 75%; The content of brittle minerals in siliceous shale is 42%, meeting the conditions for shale gas reservoir [4].

Our measured profile results show that the thickness of the lower Jurassic siliceous rock in Fuyuan profile is more than 540m (Fig. 1), and the main lithology is siliceous rock, siliceous mudstone, siliceous siltstone, radiolarian siliceous rock, etc. [5].

* Corresponding author: quguohui1001@126.com

 https://doi.org/10.1051/e3sconf/202235802037

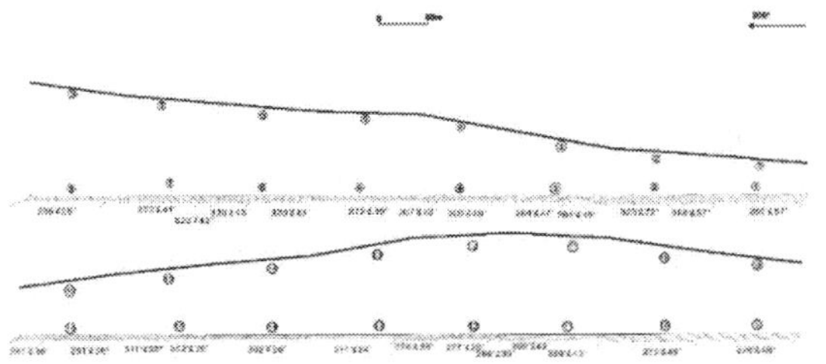

Figure 1. Measured Profile of Lower Jurassic in Fuyuan, Sanjiang Basin

2.2 Raohe region

The Dajiashan Formation (T3d -J1d) in Raohe area is developed with siliceous rocks (quartz content>50%), brittle minerals content between 75.6% and 82.5%, and the average value is 79.4%, which is very high, fully meeting the rock and mineral conditions of shale gas reservoirs. However, in the Nadanhada terrane of Raohe area, the siliceous rocks have been severely deformed and even formed sharp folds.

In conclusion, the main mineral composition of siliceous rocks and siliceous mudstones is quartz, followed by plagioclase. The content of other minerals is less than 10%, and the average content of brittle minerals is 75% and 79.4%. The brittle minerals of siliceous shale are up to 45%. Both can be used as shale gas reservoirs.

3. Organic geochemical characteristics of siliceous rocks

3.1 Organic matter abundance

3.1.1 Fuyuan District

The average content of organic carbon in siliceous rocks in Fuyuan area is 0.64%, mainly distributed in 0.07~1.89%; The hydrocarbon generation potential is 0.04-1.55 $mg\cdot g^{-1}$, with an average of 0.38 $mg\cdot g^{-1}$; The content of chloroform asphalt "A" ranges from 0.0051% to 0.0067%, with an average of 0.0059%; The total hydrocarbon value is 6~8ppm, with an average of 7ppm. According to the domestic organic matter abundance evaluation standard of continental saline ultra saline water source rock, it is a good source rock. However, due to the low content of clay minerals in siliceous rocks, which are similar to carbonate rocks and have poor adsorption capacity for organic matter, the standard of organic matter abundance should be lower than that of mudstone. According to the organic matter abundance standard of foreign marine carbonate [6], it also reaches the standard of good organic matter abundance of hydrocarbon source rocks.

3.1.2 Raohe region

The average content of organic carbon in the siliceous rocks in Raohe area is only 0.43%, mainly distributed between 0.09% and 0.81%; The hydrocarbon generation potential ranges from 0.01 to 0.07 $mg\cdot g^{-1}$, with an average of 0.04 $mg\cdot g^{-1}$; The content of chloroform asphalt "A" ranges from 0.0036% to 0.0127%, with an average of 0.083%; The total hydrocarbon value ranges from 16 to 54 ppm, with an average of 31 ppm. It belongs to the source rock with medium organic matter abundance.

3.1.3 Hongqiling region

The organic carbon content of siliceous mudstone in Hongqiling area ranges from 0.06% to 0.55%, with an average value of 0.28%; The hydrocarbon generation potential is mainly distributed in the range of 0.02~0.09 $mg\cdot g^{-1}$, with an average of 0.05 $mg\cdot g^{-1}$; The chloroform asphalt "A" is 0.007%; The total hydrocarbon is 25ppm, belonging to fair source rock.

The organic matter abundance of siliceous rocks in Nadanha terrane and Sanjiang basin is relatively high, and the hydrocarbon source rocks in Fuyuan, Raohe and Hongqiling regions reach the standard of good, medium and poor organic matter abundance respectively. The difference in the abundance of organic matter in siliceous rocks may be caused by weathering. During weathering and denudation of the source rocks, C and H elements are lost and O elements are enriched, leading to a decrease in the abundance of organic matter and a deterioration in the type [7]. In Raohe and Hongqiling regions of the Nadanhada terrane, siliceous rocks have suffered intense compression and deformation. In some places, there are even dynamic metamorphism and thermal contact metamorphism, and siliceous rocks have been lifted to the surface, It has suffered severe differentiation and denudation, so the abundance of organic matter is low. In addition, if the thermal evolution degree of the organic matter in the source rock is too high, the residual organic matter abundance can also be reduced if the organic matter in the source rock is migrated out after generating oil and gas. The organic matter maturity of the siliceous rocks in the study area is relatively high, which is currently in the high over mature stage. One reason for the low abundance of organic matter in rocks.

3.2 Types of organic matter

It can be seen from Figure 2 that most of the organic matters of siliceous rocks developed in Sanjiang Basin (Fuyuan area), Raohe area and Hongqiling area belong to Type III, and a few belong to Type II2 and Type I. However, the siliceous rocks belong to semi deep-sea facies sediments, in which the organic matter comes from radiolarians and other aquatic organisms, and the kerogen type should belong to Type I and Type II. The reason for this phenomenon is that weathering causes the loss of C and H elements in the organic matter of the source rock, the increase of O elements and the deterioration of the type, or the high maturity of the organic matter.

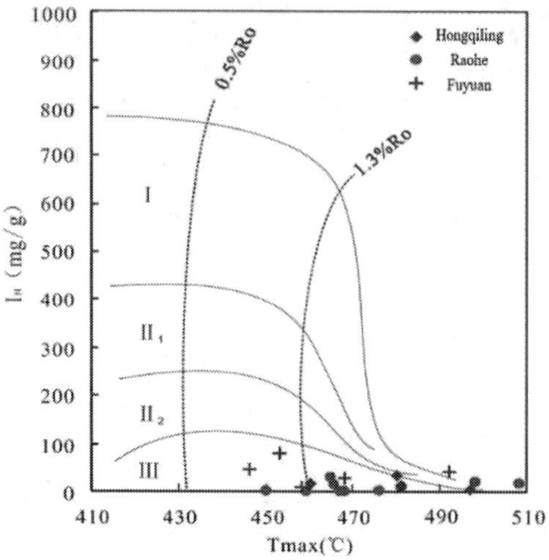

Figure 2. Organic Matter Types of Lower Jurassic Siliceous Rocks in Fuyuan, Raohe and Hongqiling Areas

3.3 Maturity of organic matter

3.3.1 Fuyuan District

In order to study the organic maturity and thermal metamorphism of siliceous rocks, Tmax, vitrinite reflectance Ro and illite crystallinity of source rocks were tested. The average Tmax of the siliceous rock in Fuyuan area is 463 °C, which shows that the siliceous rock is in a highly mature stage, while the organic matter maturity Ro is 6.26%, which shows that the siliceous rock has reached an over mature stage, with a crystallinity CIS of 0.28~0.43 and an average IS of 0.3525. According to the relationship between the illite crystallinity index and the diagenetic stage, the siliceous rock is in a near metamorphic to non metamorphic stage.

3.3.2 Raohe region

Due to the high maturity of organic matter in the source rock, the average Tmax of the siliceous rock in Raohe area is 484 °C, and the siliceous rock is still in the high maturity stage, with the maturity Ro of 5.7%. According to the oil and gas industry standard (SY/T 5735-1995), the siliceous rock has reached the over mature stage, with the stone crystallinity of CIS=0.39. The relationship between the crystallinity index and the diagenetic stage shows that this set of siliceous rock is in the near metamorphic to non-metamorphic stage.

3.3.3 Hongqiling region

The maturity of organic matter Ro in Hongqiling area is 2.06%, which is in the over mature stage, both at 493 ℃, and in the high mature stage. The crystallinity of lith is CIS=0.28~0.43, and the average CIS is 0.39. According to the relationship between illite crystallinity index and diagenetic stage, this set of siliceous rocks is close to the metamorphic—unmetamorphic stage

According to the three maturity indicators of each region, the siliceous rocks found in the three regions of Yuanhe, Raohe and Hongqiling are obviously in the high over mature stage, but the thermal metamorphism is relatively low, and they still have a certain hydrocarbon generation potential

4. Conclusion

It can be seen from the above that the organic matter abundance of siliceous rocks in Fuyuan area of Sanjiang basin and Raohe and Hongqiling areas of Nadanhada terrane is relatively high, reaching the standard of good, medium and poor organic matter abundance respectively; Type III kerogen is the main type of organic matter, which is currently in the high over mature stage with low metamorphic degree. However, siliceous rocks belong to deep-sea sediments, and their organic matter types should mainly be Type I and Type II, and their organic matter abundance should also be high. The test results may be caused by weathering and high thermal evolution.

In the process of field petroleum geological survey, it is found that from west to east and from south to north, the dynamic metamorphism and thermal contact metamorphism of the siliceous rocks in the Nadanhada terrane are gradually weakened, and the folds and magma intrusion of the stratum are alleviated. The organic matter and maturity of siliceous rocks are affected by dynamic metamorphism and thermal contact metamorphism. From west to east and from south to north, the abundance of organic matter increases and the maturity of organic matter decreases. The organic matter abundance of the siliceous rocks in Hongqiling area (near the Dahezhen fault, the westernmost), Raohe area and Fuyuan area (the northernmost) reaches the level of poor, medium and good respectively; Tmax is 493 °C, 483 °C and 463 °C respectively.

Acknowledgments

This work was financially supported by Heilongjiang Provincial Undergraduate University Central Support Local University Reform and Development Fund(2020YQ17).

References

1. Huan Xu, Yongqing Liu, Hongwei Kuang, et al. Characteristics and tectonic significance of Wandashan accretive complex in Northeast China [C]// 2015 National Sedimentology Congress. 2015:318-319.
2. G.H. Qu,N. Liu,Y.K. Liu Y.L. Meng,Z.L. Zhang,W.Z. He(2021). Simulation Study of Hydrocarbon Generation Characteristics in the Siliceous Rock. CHEMISTRY AND TECHNOLOGY OF FUELS AND OILS, vol.56, no.6, p. 985-993.

3. G.H. Qu,N. Liu, Y.L. Meng and X.N. Li(2019). Research progress on migration and accretion of Nadanhada terrane. International Conference on Oil & Gas Engineering and Geological Sciences, vol.384,no.1,p.12120.
4. G.H. Qu,N. Liu, Y.L. Meng and X.N. Li(2019). Advances in tectonic genesis of Nadanhada terrane. International Conference on Oil & Gas Engineering and Geological Sciences, vol.384,no.1,p.12121.
5. G.H. Qu,N. Liu, Y.L. Meng and X.N. Li(2019). Evaluation of Source Rocks in Sanjiang Basin of Nadanhada Terrane, vol.6,no.5,p.276-279.
6. Jianping Chen, Digang Liang, Shuichang Zhang, Chunping Deng, Dijia Zhang. Evaluation Criterion and Methods of the Hydrocarbon Generation Potential for China' s Paleozoic Marine Source Rocks [J]. ACTA GEOLOGICA SINICA, 2012, 86(07):1132-1142.
7. Yuanlin Meng, Lihua Xiao, Junsheng Yang, Jianguo Wang, Yuanlin Sun. Influences of weathering on organic matter of outcrop and correcting methods in Xining Basin [J].GEOCHIMICA,1999(01):42-50.DOI:10.19700/j.0379-1726.1999.01.006.

https://doi.org/10.1051/e3sconf/202235802038

Construction of demand response model of integrated energy system based on machine learning algorithm

Yunyi Hou

School of Electrical and Electronic Engineering, North China Electric Power University, Baoding, Hebei, 071003, China

Abstract: The "multi-energy era", which is complementary to new energy and fossil energy, has arrived. Demand-side management (DSM) has gradually gained worldwide attention because of its advantages of high flexibility and great response potential. However, the appearance of integrated energy system (IES) has broken the existing mode of independent operation of traditional energy systems, and made different forms of multi-energy flow more and more closely coupled. Demand response is the key measure to stimulate demand-side resources to participate in scheduling. IES can integrate various forms of energy, which brings new development to demand response. This paper studies the demand response project of IES, introduces the basic concept and popularization value of integrated demand response, and builds the demand response model of IES based on machine learning algorithm.

Keywords: Machine learning algorithm, Demand response model of integrated energy system

1. Introduction

Machine learning as a major component of modern intelligent technology, has been continuously innovated, improved and developed until now, and it still maintains a vigorous vitality today, setting off a wave of research frenzy of machine learning. It is widely used in almost all popular fields, including unmanned driving, robotics, medical treatment, finance, machinery, energy, network security, etc. It can be said that it is the most revolutionary science and technology in this era [1]. Learning, as its name implies, is to let the machine realize people's learning behavior, thus improving the performance of the system through continuous self-learning, and finally giving a relatively correct judgment according to the new conditions [2]. It integrates knowledge of many disciplines, including cross probability theory, statistics, cybernetics, algorithm complexity, computational theory in complex environment and other interdisciplinary fields. The core of machine learning is algorithm. In the learning process, the data is iteratively analyzed and calculated by some algorithm, and finally a reasonable judgment and prediction is made on the matter [3].

Building a clean and low-carbon energy system and promoting the transformation of the energy industry has become the focus of the competitive development of the energy industry all over the world [4]. The report of the 19th National Congress of the Communist Party of China has positioned the future development direction of China's energy, and it is necessary to establish an energy supply system that meets diversified consumption needs and is energy efficient. The IES couples various energy forms such as electric energy, thermal energy, natural gas and hydrogen energy. Different from the traditional energy system, it can not only meet the diversified energy supply demand, but also greatly improve the capacity of renewable energy consumption, with stronger reliability and flexibility [5]. The IES can realize multi-energy coupling transformation and utilization, which provides the possibility to improve the utilization rate of renewable energy, enhance the comprehensive utilization rate of energy and solve the energy crisis and environmental pollution problems.

The primary goal of the IES is to realize a large amount of renewable energy consumption, improve the comprehensive utilization of energy and ensure the safe, stable, flexible and efficient operation of the system [6]. However, there are a lot of uncertainties in the IES, such as the fluctuation of renewable energy output and the fluctuation of different load demand [7]. These uncertain factors will bring challenges to the safe and reliable operation of the system. The uncertainty of renewable energy output means that it is influenced by various factors such as meteorology and environment during power generation, which makes it difficult for us to predict the output sequence and amplitude of renewable energy. With the increasing penetration of renewable energy into the grid, this challenge is becoming more and more severe. Therefore, how to solve the uncertainty of supply and demand side in the IES and bring more stable optimization scheme to the system optimization scheduling is still a problem to be solved at present.

2. Research status of IES and demand response

2.1 Research status of IES

The IES effectively integrates various energy sources such as cold, heat, electricity and natural gas, and can fully tap the potential of each energy system, which has become an important direction of the current energy revolution and a new strategic competition focus of various countries [8]. The U.S. Department of Energy put forward a comprehensive energy system development plan, aiming at improving the utilization rate of renewable energy, promoting the popularization and application of combined cooling, heating and power technology, and improving the reliability of social energy supply [9]. According to the Energy Independence and Security Act of the United States, comprehensive energy planning must be placed in an important position in energy development, and a large amount of funds will be invested in the special research and implementation of comprehensive energy system planning in the later period. Countries all over the world have also launched related research and system construction projects. In the seventh R&D framework of the European Union, the requirements for the security and reliability of high-quality energy networks are put forward, so as to achieve further energy conservation in various industries [10]. The pilot project is based on heating technologies such as cogeneration, and a multi-energy energy system with high permeability and renewable energy is built. The ELECTRA demonstration project of the European Union is based on modern power communication technology and market mechanism, aiming to achieve the local consumption of large-scale renewable energy and the stable operation of large power grid. Germany has always listed energy transformation as one of the major national projects. Aiming at the typical energy problems in China, the German demonstration project in langenfeld, based on modern communication, automatic control and measurement technologies, is committed to building wireless city and minimizing the energy utilization of users [11]. Manchester demonstration project in England has set up an interactive platform between IES and users, which can solve the problems of cascade utilization of energy, comprehensive coordination among energy sources and demand side management of various energy sources.

2.2 Present situation of demand response research

Traditional power demand response is simply the load transfer or energy consumption reduction in time scale, which greatly changes users' energy consumption habits, with low enthusiasm of users and limited response potential. In order to cope with large-scale distributed renewable energy access, people also need a long-term sustainable solution. With the continuous enrichment of research results and the gradual landing of key technologies, the researcher put forward the concept of demand response for the first time in 2015, and defined demand response as the derivative and expansion of traditional electric power on the demand response side in the integrated energy network. The multi-energy intelligent management system, which combines energy conversion with time transfer, enriches the response mode of demand side, strengthens the flexibility of demand side response, and is an important embodiment of the convergence of energy flow, information flow and value flow on the user side. At present, there are many research achievements on traditional demand response strategies at home and abroad. Due to the early start of foreign research, a relatively formed market system has been put into operation. At present, many countries have carried out research on demand response, Europe and North America are at the forefront of this field, and small-scale experiments have been carried out at the community level, and certain research results have been achieved in response types, basic devices, interactive ways and user participation forms. " To sum up, at present, the research on demand response is still in the initial stage, lacking perfect interaction mechanism, and the transfer relationship between information flow and energy flow is not clear; In the process of interaction, the end user's adjustable ability is difficult to describe, and the superior network can't accurately issue the control instruction; In addition, the complementary characteristics of cold, heat, electricity and other energy sources in time and space are less considered, which leads to the failure to further tap the response potential of users. Therefore, it is an important exploration direction of demand response in the future to clarify the interaction mode and regulation strategy of demand response, to model demand response resources in a refined way and to quantify the regulation ability, which is of great significance for demand response to really participate in the optimal scheduling of IES.

3. Build the demand response model of IES based on machine learning algorithm

3.1 Model Construction Based on BP Neural Network

Before constructing the demand response model of IES based on machine learning algorithm, before choosing machine learning methods, the first thing to do is to choose the machine learning mode according to the characteristics of the subject. In this paper, the BP neural network is chosen to build and optimize the model, because of the following reasons: the neural network can use a simple single hidden layer network to update the weights and thresholds by iteration to fit arbitrary functions; Neural network can deal with high-dimensional feature vectors and extract features. Neural network has learning ability similar to human brain, and the trained network evaluation model can be generalized to untrained samples, which is suitable for evaluation research. The neural network is easy to expand. When the amount of information to be stored exceeds the capacity, the storage capacity of the network can be expanded by increasing the network scale. It is self-adaptive and can dynamically

adapt to all kinds of samples as the training samples change.
First, the determination of network layers. On this issue, a famous theorem gives a judgment, that is, a three-layer neural network with only input layer, output layer and hidden layer can approach any nonlinear system to the maximum extent, as long as the number of nodes in the hidden layer is not limited. Therefore, in necessity, the three-layer network model is enough. In addition, if the number of layers of the network is increased, the accuracy of the training results will be improved to some extent, but at the same time, the complexity of the system will be greatly increased, and the training efficiency of the neural network will be reduced. The result is probably not worth the candle. Therefore, the basic three-layer network structure is selected for the number of BP network layers in this paper. In the selection of the number of neurons in each layer, the number of nodes in the input layer is determined to be 25 by the above research on the demand response of the IES; According to the purpose and demand of this topic, that is, the performance of demand response model of IES, the number of nodes in the output layer is determined to be 4. The flow chart of BP neural network training is shown in Figure 1.

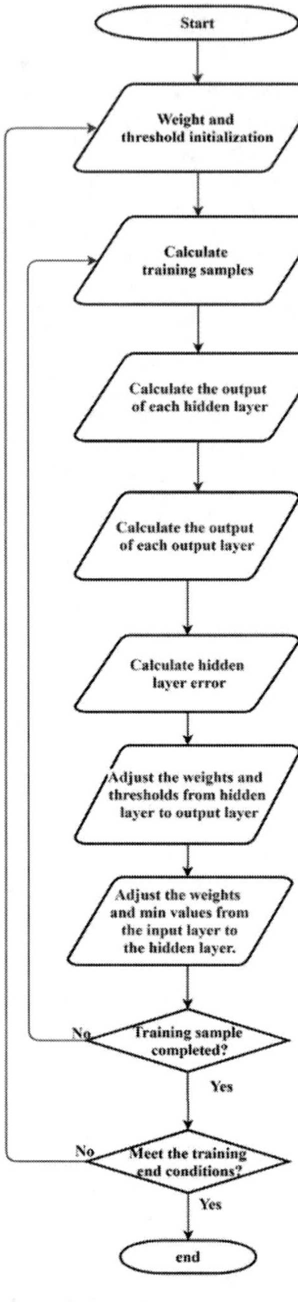

Figure 1 BP neural network training flow chart

3.2 System structure design of demand response model of IES

Firstly, the purpose and core of the demand response model of IES are defined, that is, the evaluation of the demand response model of IES is realized. In order to achieve this goal, the required structural modules mainly include: data management module, improved BP neural network configuration module and evaluation result and analysis module. The following is a description of the specific contents and functions of each main module. The first block is the data management module, which mainly manages information data, including the input of collected data, the basic operation and processing of the input data, the viewing and deletion of the input data, etc. This module is the premise and foundation for the follow-up module or the whole evaluation model to run smoothly. The second block is the improved BP neural network configuration module, which is the key to realize the system goal. This module embeds the improved BP neural network model which has been constructed before, and users can use this module to obtain the evaluation results. This module is where the evaluation prototype system of this system is continuously maintained and updated. There will definitely be a better evaluation model in the future, and there will definitely be more optimized research results in the current evaluation index system, weights and evaluation methods. At that time, based on the latest research, the evaluation model and related indexes and coefficients can be optimized to update this system. The third block is the evaluation result and analysis module, which is the purpose module of the evaluation prototype system. It is used to display the evaluation results, thus providing analysis materials and foundation for future researchers to improve and optimize the energy system demand response model, and promoting the development of the research field of energy system demand response. The schematic diagram of the functional modules of the IES demand response model is shown in Figure 2.

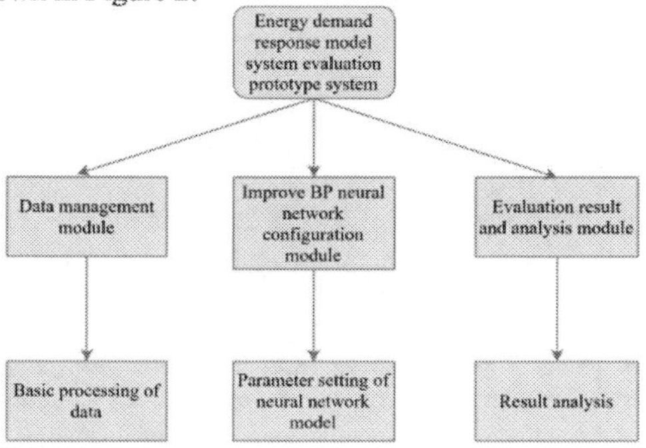

Figure 2 Schematic diagram of functional modules

4. Conclusions

In order to improve the consumption of renewable energy, maximize the utilization of comprehensive energy, solve the uncertainty of demand response, and realize the economic and stable operation of comprehensive energy system. Based on the machine learning method, this paper focuses on the demand response model of IES, and studies the evaluation system of demand response model of IES. On the basis of previous studies, from understanding the definition and connotation of IES demand response model, an evaluation index system of IES demand response model is obtained. By studying the IES demand response model and its evaluation methods in related fields, the IES demand response model is combined with the development status and bottleneck of IES, and the shortcomings and shortcomings of traditional evaluation methods are realized. For the first time, the demand response model of IES is introduced into several mature and widely used machine learning evaluation methods, and then the improved optimization strategy is added to construct the evaluation model of demand response model of IES, which is relative to the objective subject of demand response model of IES. In a sense, it adds a research attempt of machine learning method evaluation to the development of demand response model of IES.

References

1. Zhong J, Cao Y, Li Y, et al. Distributed Modeling Considering Uncertainties for Robust Operation of Integrated Energy System[J]. Energy, 2021(1):120179.
2. Zhuang W, Zhou S, Gu W, et al. Optimized dispatching of city-scale integrated energy system considering the flexibilities of city gas gate station and line packing[J]. Applied Energy, 2021, 290(3):116689.
3. Liu Z, Yu H, Liu R. A novel energy supply and demand matching model in park integrated energy system[J]. Energy, 2019, 176(JUN.1):1007-1019.
4. Li J, Yu T, Zhang X, et al. Efficient experience replay based deep deterministic policy gradient for AGC dispatch in integrated energy system[J]. Applied Energy, 2021, 285:116386.
5. Li Z, Guo W. Optimal operation of integrated energy system considering user interaction[J]. Electric Power Systems Research, 2020, 187:106505.
6. Zeng, Jiang, Liu, et al. Optimal Dispatch of Integrated Energy System Considering Energy Hub Technology and Multi-Agent Interest Balance[J]. Energies, 2019, 12(16):3112.
7. Li G, Bie Z, Yu K, et al. Reliability evaluation of integrated energy systems based on smart agent communication[J]. Applied Energy, 2015, 167:397-406.
8. Wang Y, Song F, Ma Y, et al. Research on Capacity Planning and Optimization of Regional Integrated Energy System Based on Hybrid Energy Storage System[J]. Applied Thermal Engineering, 2020, 180(6390):115834.
9. Ali R, Lee S, Chung T C. Accurate multi-criteria decision making methodology for recommending machine learning algorithm[J]. Expert Systems with Applications, 2017, 71:257-278.
10. Challagulla V, Bastani F B, Yen I L, et al. Empirical assessment of machine learning based software defect prediction techniques[J]. International Journal of Artificial Intelligence Tools, 2015, 17(02):389-400.
11. MAIRAL, JULIEN. INCREMENTAL MAJORIZATION-MINIMIZATION OPTIMIZATION WITH APPLICATION TO LARGE-SCALE MACHINE LEARNING. [J]. SIAM Journal on Optimization, 2015, 25(2): 829-855.

 https://doi.org/10.1051/e3sconf/202235802039

Design of core thermocouple adaptive plug-and-pull device

Jianhua Zhou [1], Yongqiang Guo [1], Guangzheng Tian [2], Songdi Ji [2], Jifu Zhang [3,*]

[1] China Nuclear Power Operations Co.,Ltd .,Guangdong ,China
[2] Daya Bay Nuclear Power Operations and Management Co.,Ltd. ,Guangdong ,China
[3] China Nuclear Power Technology Research Institute Beijing Division, Beijing, China

Abstract: In the process of change and maintenance material of nuclear power plants, due to the uncontrollable driving force of the existing core thermocouple insertion and removal tools, it is easy to cause the thermocouple to be installed in place, the thermocouple to be easily damaged, or even to have poor sealing, and the pressure boundary of the primary circuit is incomplete and other issues. Aiming at the above problems, this paper takes FPGA as the core controller, and adopts a closed-loop control method with multiple modes of force, speed and displacement, and develops a set of automatic insertion and removal device for nuclear power plant core thermocouple. The device has the advantages of safety and reliability, high control precision, and online observation of multiple parameters.

Keywords: Core thermocouple; adaptive plug and pull; automatic control

1. Introduction

At present, during the change and maintenance material of domestic nuclear power plants, the torque of the existing plug and pull tools used for the core thermocouple is uncontrollable, resulting in easy damage to the thermocouple and incomplete sealing of the primary circuit. The wall is rough or the burr is generated when the thermocouple is pulled out, which causes the resistance of the thermocouple to pass here to become larger; there are impurities in the end plug at the end of the thermocouple conduit or the hole at the entrance of the end plug becomes smaller due to deposits, which leads to the insertion of the thermocouple difficulty. And RIC thermocouple replacement is a delicate operation, requiring the use of a special tool to insert and remove thermocouples: a thermocouple insertion and removal tool. The power station uses a manual plug and pull tool, which is characterized by: the operator is manually driven, which is easy to fatigue; the stroke is less than 30mm adjustable, and intermittent movement; the installation and fixing of the tool and the switching of the driving direction are complicated; the driving force is manually controlled by the operator, The amount of force varies from person to person; a large operating space is required; the thermocouple at the catheter port is easily bent and deformed, causing the joint seal to fail. Need to develop new plug and pull device to meet on-site needs. According to this background, a set of torque controllable and automatic control automatic plugging and plugging device is developed.

2. Analysis

If the core thermocouple is found to be faulty during the operation of the unit, the thermocouple of this group of signals needs to be replaced during the unit change materal repair. Replacing a thermocouple involves pulling out the failed thermocouple and replacing it with a new one.The thermocouple spare is reinserted into the guide tube to the design depth. Since the core thermocouple does not run in a straight line in the guide tube, taking the M310 reactor as an example, the core thermocouple will have one or two 90° corners in the guide tube, and other places will also be bent. So that when the thermocouple is inserted or pulled out, there are more contact surfaces between the thermocouple and the inner wall of the guide tube, and as the contact surface increases, the friction between the thermocouple and the inner wall of the guide tube becomes larger; Due to irradiation, the material of the thermocouple and the guide tube will also have a hardening effect and ion dislocation, which increases the static friction between the surface of the thermocouple and the guide tube, and the dynamic friction during the insertion and removal of the thermocouple also increases accordingly. Therefore, it is necessary to control the force during the insertion and extraction of the thermocouple, so as to protect the thermocouple from being broken in the guide tube when it is pulled out, and to prevent the thermocouple from being damaged due to inaccurate control of the entry and exit force during insertion. Therefore, a control system is needed to control the strength of thermocouple insertion and removal.

* Corresponding author: zhangjifu@cgnpc.com.cn

3. Design Guidelines

3.1 Performance

•RIC Thermocouple self-adaptive insertion and extraction device has the continuity of thermocouple insertion and extraction process

• Adaptive regulation and maximum driving force protection function

• The maximum driving force is not less than 100kg, Driving force accuracy 1%;

• The device has the function of real-time monitoring of driving force

• The device has a thermocouple insertion depth display function, and the displacement monitoring device accuracy 1mm;

• can store at least 100 sets of data, each set of data contains the position and driving force of the thermocouple during the whole process from 0 to insertion. The data can be exported for post-processing

3.2 Actuator

The actuator is fixed by clamping the thermocouple conduit, and is designed with a handle for the operator to support with one hand. The shape design should conform to ergonomics. The specific design guidelines are as follows

• The actuator clamps the thermocouple through two sets of synchronous belts arranged symmetrically to achieve continuous insertion/pull out

• The power of the actuator is provided by the drive unit, which is connected to the actuator through the 10mm standard drive square head;

• The driving force input by the actuator is not less than 100kg;

• The actuator can be manually plugged in and unconnected without the drive and equipped with the corresponding drive handle;

• Synchronous wheel timing belt: The power of the drive device needs to be transmitted to two sets of synchronous belts, keeping the two sets of synchronous belts rotating in opposite directions, so that the driving force direction of the thermocouple is the same; the contact surface of the timing belt and the thermocouple is made of irradiation-resistant rubber material to improve its service life; the clamping force between the timing belt and the thermocouple can be adjusted by the knob; the timing belt is supported by 4 pairs of synchronous wheels to provide reliable support pressure; the synchronous wheel is arranged in front of each other to ensure that the friction force is continuously reliable and prevents bending thermocouple; the adhesive layer is thickened. And the use of increased friction surface design; the gap between the two sets of timing belts can be

The adjustment range is 0mm~3.2mm, and the gap between the thermocouple catheter clamping parts can be adjusted from 7.5mm~20mm;

• DC motor:DC12V Brushless DC motor, with typical mechatronics design, higher output power, working efficiency, lower failure rate;

• Reducer: Planetary reducer, compact structure, large output torque, combined with DC motor, maximum output 35NM;

• Torque sensor: high detection accuracy, good stability, strong anti-interference, small size, light weight, can be measured in any direction, and can measure rotational torque (normal work) , can also be measured static torque(under locked-rotor) ;

• Displacement sensor: Realized by the rotation encoder integral of the torque transducer, the number of turns of the output shaft rotation is calculated by the controller, and the circumference of the actuator is calculated as the product calculation.

3.3 Controller

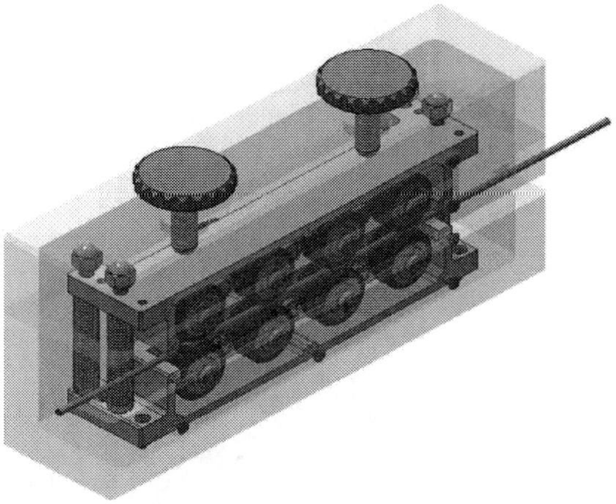

Fig. 1 Schematic diagram of the actuator

https://doi.org/10.1051/e3sconf/202235802039

• Control the forward, reverse, start and stop of the motor to realize the function of plug and pull;
• Real-time acquisition of encoder signals, calculation of velocity and displacement;
•
Real-time acquisition of torque sensor signals, calculation of torque, and speed control based on torque;
• With the driving force adaptive adjustment function and the maximum driving force protection function, when resistance torque is less than or equal to 6Nm, the drive speed of the driving device is 60rpm;
• when the resistance moment is greater than the 6Nm and less than or equal to 10Nm, the speed 30rpm;
• when the resistancemoment is greater than the 10Nm, the drive is shut down protectively
•Real-time output working status;
• Displacement-Torque correspondence data conversion

3.4 Control System

The entire control system controls the power output of each torque sensor, driving motor and other components, but also has the sensor feedback signal reception function, the system controls the operation of the drive motor to achieve the continuous operation of the thermocouple plugging and unplugging action, in the control system can be set according to the actual work needs of the motor speed to meet the time requirements of thermocouple insertion and removal operation. At the same time, according to the feedback of the real-time torque sensor of the thermocouple in the insertion and removal operation to judge the resistance of the thermocouple in the insertion and removal process, the size of the insertion force can be set in the system to achieve the protection of the thermocouple during the insertion and removal process, in addition, the control system has its own storage function, and each thermocouple insertion and removal data will be stored in a fixed position and stored. The controller uses the internal architecture of Compact RIO as shown in Figure 2 below, and the LabVIEW FPGA code can run directly on the hardware.

High reliability: LabVIEW FPGA code running on an FPGA is very reliable because the logic is compiled into the physical hardware design. In this way the physical hardware becomes a very reliable hardware chip;

High certainty: the FPGA does not require an operating system, which greatly improves the reliability of parallel execution and deterministic hardware, using NI FPGA-based hardware, you can make key components obtain up to 25ns of accurate timing, this advantage allows the thermocouple to control synchronization during the docking process to ensure the accuracy of subsequent solutions;

True parallel operation: Every piece of parallel code throughout the FPGA can be executed simultaneously. Even through in executes programs in a pipelined manner on the FPGA;

Reconfigurability: Because the FPGA chip can be reconfigured, it can be configured according to different types of probes, and the algorithm can also be modified according to different measurement objects and needs. As a product or mature system, its performance can be improved without spending time redesigning hardware or modifying panel layouts;

Instant Power On: Because LabVIEW FPGA code runs directly on the FPGA without relying on any operating system, code downloaded to the FPGA's flash memory is ready to run within milliseconds of powering up the CompactRIO chassis. This code can execute a control loop or set a start, calculating the output value monitoring and alarm signals based on the frequency signal output by the sampled sensor;

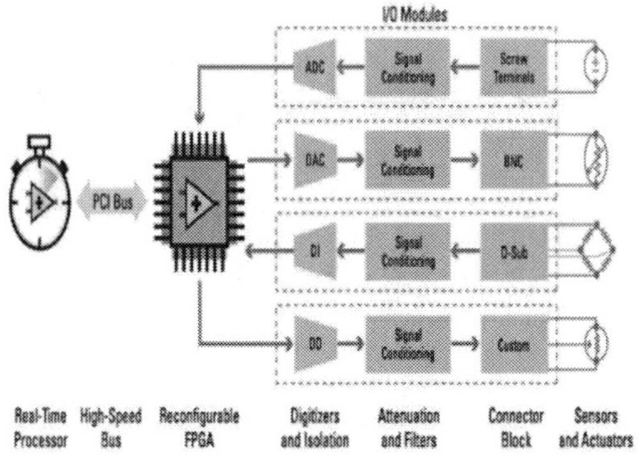

Fig. 2 Internal Architecture of the Controller Using Compact RIO

4. Summarize

In terms of function, the RIC thermocouple adaptive insertion and removal device can realize the continuous insertion and removal operation of the thermocouple in the conduit, and dynamically control the driving torque and speed by monitoring the resistance moment and displacement of the thermocouple during the insertion and removal process in real time, so as to quickly complete the insertion and removal operation without damaging the thermocouple body.

In terms of control, the use of FPGA to drive the motor can not only be freely set in a certain range of stepper motor speed, round-trip rotation state and number of rotations within a certain range through programming methods, but also can easily and flexibly control the running state of the motor to meet the requirements of different operating states. This simplifies the circuitry and reduces the cost. The use of FPGA control, no need to connect external D / A converter, so that the peripheral control circuit is greatly simplified, optimize the equipment structure, reduce the post-maintenance cost and improve the work efficiency. It also reduces the risk of foreign bodies in the pool.

References

1. karen. Design of intelligent motor drive control system based on FPGA [J]2018
2. Wang Junjie. Research on a new type of brushless DC motor drive control system[D] Nanjing University of Aeronautics and Astronautics2008
3. Zhao Zhipu. Neural Network Control Design Based on FPGA Brushless DC Motor [D] Gangnam University,2006
4. Yang Wang. FPGA-based DC Motor Control Design [D] Nanjing University of Science and Technology,2012

Determination of reasonable test time of periodic injection well in Punan Oilfield

Wanxing Gu

Geological Brigade, No.7 oil production plant, Daqing Oilfield Co., Ltd, Daqing, China

Abstract: There were 232 stratified injection Wells in Bunan Oilfield in 2020, including 69 year-round injection Wells and 163 periodic injection Wells. There are many periodic injection Wells and the pressure recovery is slow after injection. After the test, the pressure fluctuates greatly, and the data cannot be reported outside in time, which affects the water distribution. In 2020, according to our actual statistics, the water injection pressure of 38 Wells had a difference of more than 2MPa from the qualified pressure point of the test data before the test of stratified injection well. The reason of pressure discrepancy is mainly related to periodic water injection. In view of this situation, we conducted an investigation on the cycle injection Wells of the whole mine in 2020. The pressure recovery curve was plotted according to the daily injection volume and the length of periodic injection stop time. At the same time, the method of determining the reasonable testing time of periodic injection well is put forward, which provides scientific basis for determining the testing time of periodic injection well.

Key words: Cycle water injection; pressure recovery; reasonable test time.

1. Pressure recovery of periodic water injection well after water injection

There are three reasons for periodic water injection in punan oilfield. One is to take periodic water injection directly at the initial stage of transfer; Second, intermittent water injection can supplement energy in order to control the water cut rising rate of oil Wells. Third, because the water absorption can not pass the winter, the winter cycle stop injection. Therefore, we made statistics on the recovery of water injection pressure in 38 Wells after water injection was resumed (see Table 1).

Table 1. Recovery statistics of periodic injection well pressure in Punan Oilfield

Well	Injection allocation (m^3)	Stop injection cycle (m)	Pressure recovery days (d)	Well	Injection allocation (m^3)	Stop injection cycle (m)	Pressure recovery days (d)
7P2 64-86	30	3	31	7P2 14-98	50	2	38
7P2 74-80	30	3	30	7P1 72-130	50	2	35
7P1 86-138	30	1	15	7P2 36-88	40	3	38
7P2 78-78	20	2	40	7P1 78-44	55	3	45
7P1 64-66	30	3	50	7P1 94-12	50	3	48
7P2 30-96	20	5	35	7P1 80-28	40	2	40
7P1 72-20	20	5	36	7P1 68-14	40	3	25
7P1 78-28	30	5	38	7P1 43	40	3	35
7P2 26-90	30	5	36	7P1 74-134	50	2	25
7P2 68-78	20	5	38	7P2 62-80	40	2	40
7P2 40-92	30	4	30	7P1 60-36	40	2	35
7P2 74-63	25	2	30	7P1 92-140	55	2	35
7P2 94-76	30	3	25	7P2 48-78	60	1	35
7P2 44-90	25	2	40	7P1 70-24	60	2	40
7P2 78-63	20	2	27	7P1 64-28	65	3	55
7P1 66-16	35	3	38	7P2 78-74	65	2	45
7P1 70-52	50	5	48	7P1 48-50	75	2	60

7P1 80-56	30	3	36	7P2 20-98	70	2	60
7P2 14-100	50	2	35	7P2 02-86	75	2	45

The time required for pressure to return to normal for different types of Wells was calculated based on different injection allocation and shutdown cycles (see Table 2).

Table 2. Summary of pressure recovery of periodic injection Wells after water injection

Daily water injection	Stop injection cycle (m)	Pressure recovery days (d)
<30 m^3	≤3 months	24
	5 months	33
30 m^3～60 m^3	≤3 months	36
	5 months	45
>60 m^3	≤3 months	50

According to the data in the table, if the daily injection is less than 30m3 and the injection stop period is less than 3 months, the average days of single well pressure recovery is 24 days; if the injection stop period is 5 months, the average days of single well pressure recovery is 33 days. With daily injection of 30m3 to 60m3 and injection stop period <3 months, the average days of single well pressure recovery was 36 days, and the average days of single well pressure recovery was 45 days for Wells with injection stop period of 5 months.

The daily injection allocation was greater than 60m3, the injection stop period was less than 3 months, and the average single well pressure recovered stable days was 50 days. The data in Table 2 can be used to make curves of the average pressure recovery days of Wells with different injection volumes and stoppage periods (see Figure 1). It can be seen that the longer the stoppage time, the longer the pressure recovery time.

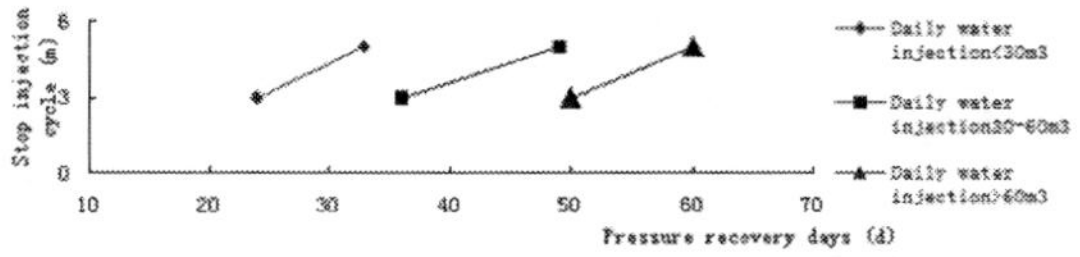

Figure 1. Curve of average well pressure recovery days with different daily injection rates and shut-down periods

2. Determination of reasonable test time of periodic injection well

2.1 Determine a reasonable test time on the pressure recovery curve

According to the above investigation, we drew a single well pressure recovery curve for Wells with different daily injection volume and injection shutdown period, and determined a reasonable test time through the pressure recovery curve.

For example, well 7P166-16, with daily injection of 35 m3, will be shut down in August and September 2020. In fact, the well was washed and transferred to injection on September 27. The initial injection pressure was 4.7mpa, and the pressure gradually recovered with the extension of injection time. The pressure recovered to 7.5MPa and stabilized on October 25. The pressure recovery time was 30 days. The stratification test was carried out on November 1st and finished on November 3rd. The qualified pressure point is 7.7mpa (see Figure 2). Water injection was basically stable under this pressure when tested in the first half of 2003.

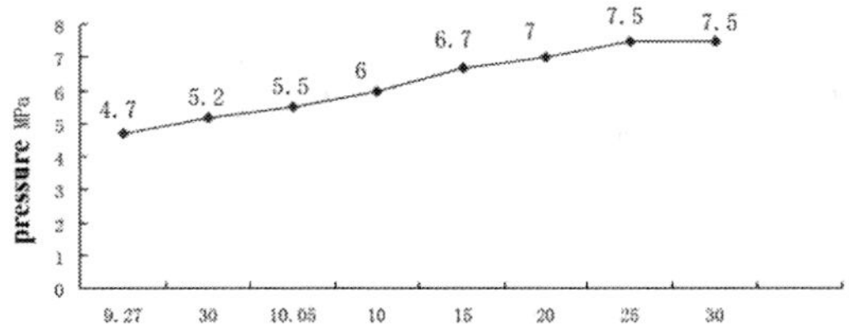

Figure 2. 7P166-16 Water injection pressure recovery curve

The injection was 35 m3 per day, the injection stop period was 2 months, and the pressure recovered and stabilized for 30 days.

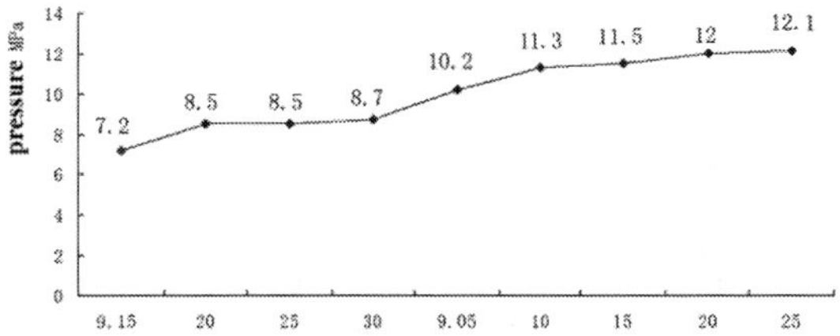

Figure 3. 7P172-130 Water injection pressure recovery

The injection was 50m3 per day, the injection stop period was 2 months, and the pressure recovered steadily for 35 days

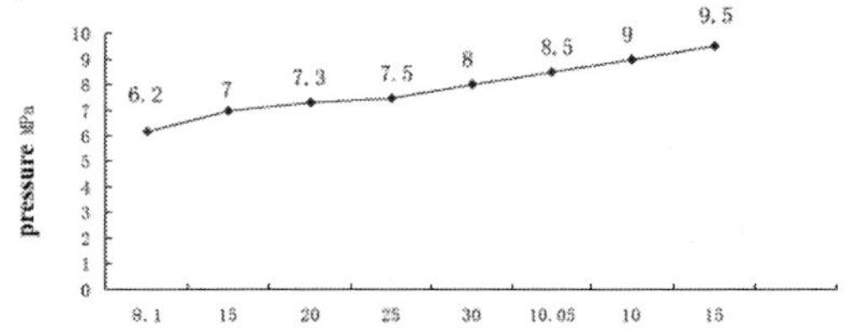

Figure 4 7P143 Water injection pressure recovery

The injection was 40m3 per day, the injection was stopped for 3 months, and the pressure recovered steadily for 35 days

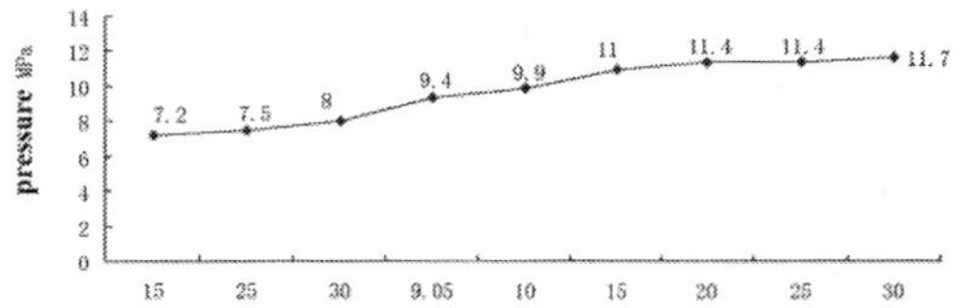

Figure 5. 7P278-74 Water injection pressure recovery

The injection was 65m3 per day, the injection was stopped for 2 months, and the pressure recovered steadily for 45 days

2.2 Determine the reasonable test time with the meter algorithm

It can be seen from the above Wells with different injection volume and stop-injection cycle that there is a good linear relationship between the recovery time of injection well pressure and injection volume and stop-injection cycle. The longer the injection stoppage period, the longer the recovery time of injection pressure; The larger the injection amount is, the longer the recovery time of water injection pressure is. In the same stop-injection period, the longer the recovery time of injection pressure is for Wells with large injection volume. The quantitative relationship of reasonable test time of periodic injection well can be obtained by analyzing and processing the regular data in the production process by using multiple regression method.

It is analyzed and processed by multiple regression, and its normal equation is:

$$L_{11}b_1+L_{12}b_2+\ldots\ldots+L_{1m}b_m=L_{10}$$
$$L_{21}b_1+L_{22}b_2+\ldots\ldots+L_{2m}b_m=L_{20} \quad (1)$$
$$L_{m1}b_1+L_{m2}b_2+\ldots\ldots+L_{mm}b_m=L_{m0}$$

$$b_0=\overline{Y}-b_1\overline{X}_1-\ldots\ldots-b_m\overline{X}_m \quad (2)$$

The regression equation is $Y=b_0+b_1X_1+b_2X_2+b_mX_m$ (3)

L_{ij}--- The value to be calculated can be obtained by the method of extremum of multivariate function.

Bi -- regression coefficient, obtained from equations (1);

B0 -- the constant term in the sum of regression squares is obtained from Equation (2);

-- Mean value of observed value of dependent variable;

I -- Mean value of observed value of independent variable;

Xi-- Statistical variables.

Through the statistical arrangement of relevant data, the coefficients of B1 and B2, L10 and L20, were respectively solved by using the method of extremum of multivariate function (the calculation data was omitted), and the following binary linear equations were obtained:

$$2672.5b_1+240b_2=2181.5 \quad (4)$$
$$240b_1+3960b_2=906$$

Solving equations: B1 =0.70005 B2 =-0.1803 B0 = 14.2600

When b0, B1 and B2 are substituted into Equation (3) and relevant parameters are substituted for Y, X1, X2 and X3, the following regression equation can be obtained:

$$Th = 14.2600 + 0.70005\ Pz\ 0.1803\ St \quad (5)$$

Where Th --, days for water injection pressure to recover and stabilize, d;

Pz -- Daily injection distribution of water injection well, t;

St - Injection stop period, d.

Equation (5) is the mathematical model of quantitative calculation of the time of periodic injection well pressure recovery and stability, i.e. reasonable test time.

2.3 Comparison of the two methods

In order to verify the error between the calculation method and the curve method (determining the reasonable test time on the curve) in determining the reasonable test time, we selected 12 Wells and used the two methods to determine the reasonable test time respectively. The comparison results are shown in Table 3.

Table 3 Calculation method and curve method to determine the reasonable test time comparison table

WELL	In 2020 Stop injection cycle (m)	Injection allocation (m^3)	Calculation of pressure Restore the number of days	The actual pressure Restore the number of days	poor
7P164-34	2	60	45	40	-5
7P170-16	2	30	25	27	2
7P172-130	1	50	43	40	-3
7P174-132	3	70	47	43	-4
7P182-128	2	40	31	35	4
7P224-92	4	50	27	30	3
7P230-90	2	60	45	43	-2
7P206-90	3	60	40	41	1
7P174-66	2	40	31	34	3
7P166-80	4	35	17	20	3
7P214	2	55	41	39	-2

It can be seen from the table that compared with the curve method, the error of single well is within 5 days, which is basically consistent with the actual situation and can be used to guide production.

3. Conclusion

3.1 after the application of this method to determining cycle injection Wells open hole pressure stable, without special funds input, the method is simple and easy, simply enter stop water injection cycle and injected water volume of data, to calculate the pressure after the well open well stability returns the number of days, when the actual validation, error in about five days, can be used to guide cycle injection well test of time.

3.2 With the extension of well opening time, the pressure of periodic injection water well gradually increases. After a period of time, the pressure gradually recovers and stabilizes at this pressure point, which is the best time for testing.

3.3 For the same well, the pressure recovery time is different at different time periods or with different injection amounts. It is not fixed, but dynamic and constantly changing. According to the pressure recovery time of periodic injection well after well opening, selecting appropriate test time is an effective way to reduce retesting Wells and ensure the normal water distribution of injection well.

https://doi.org/10.1051/e3sconf/202235802040

3.4 The pressure recovery law of under-injection Wells is inconsistent with the above formula. All the Wells involved in calculation can complete injection allocation, even if injection allocation is consistent with actual injection.

References

1. Zhenqi Jia. Methods of Oilfield Development design and Analysis, Harbin Institute of Technology Press, 1994.
2. Guo Weifeng, Fang Yujin, Yang Yongxia, et al. Research and application of periodic water injection in low permeability oil fields. Tukha Oil and Gas, 2004, 9(3):4.
3. Wang Fengqin, Liao Hongwei, Jiang Fenghua, et al. Analysis of the causes of declining water injection capacity in low-permeability oil fields and its countermeasures. Journal of Xi'an University of Petroleum: Natural Science Edition, 2009, 24(1):5.

The reform direction of property right system of natural resources assets in China format

Xiaoyong Zhong, Dongyan Guo *

Chinese Academy of Natural Resources Economics, 101149 Beijing, China

Abstract. The reform of property right system of natural resources is the only way to promote the construction of ecological civilization in China. By systematically combing the construction process of China's natural resources assets property rights system and drawing lessons from foreign property rights system experience, this paper can systematically analyze the main problems faced by China's natural resources assets system at present, and put forward targeted policy suggestions for reform.

1 The construction process of China's natural resources asset property rights management system

1.1 The state fully and directly exercise the property rights of natural resources assets

China's 1954 Constitution stipulates that "mineral deposits, water flows, forests, wasteland and other resources that are state-owned by law are all owned by the whole people." From 1960s to the end of 1970s, China's property right system of natural resources assets was replaced and managed by the natural resources administrative management system for a long time. Natural resources were prohibited from circulation, and resource products also became restricted circulation. The government was responsible for the supply, distribution, operation and management of natural resources, and all the production and operation activities of state-owned natural resources enterprises were arranged and undertaken by the government. At this stage, China has no relevant provisions on the property right system of natural resources assets in law.

1.2 The stage of separation of ownership and use right of natural resources assets and free use

The 1982 Constitution not only emphasizes the ownership of state-owned natural resources, but also stipulates the collective ownership of natural resources assets and clarifies the dual structure of property rights of natural resources assets. But at the same time, it stipulates that "state-owned mineral deposits and water flows, forest land, mountains, grasslands, wasteland and tidal flats owned by the state and collectively owned by law shall not be bought, sold, leased, mortgaged or illegally transferred in other forms." Under the premise that the ownership of natural resources assets prohibits trading, state-owned and collective-owned natural resources can be developed and utilized by individuals and units according to law, including the use, income, exploration, exploitation and logging of natural resources. Through the creation of various natural resources use rights, such as exploration rights, mining rights, forestry rights, contracting rights and so on. Grant natural resource assets to developers and users free of charge.

1.3 The stage of separation of ownership and use right of natural resources assets and paid use of use right

During this period, China revised several major natural resources management laws, including the Urban Real Estate Management Law of 1994, the Mineral Resources Law of 1996, the Forestry Law of 1998, the Water Law of 2002, the Land Management Law of 2004 and the Fisheries Law of 2004. At the same time, a series of laws, regulations and rules on various natural resources were formulated and promulgated. At this stage, the law stipulated the state ownership of natural resources, and made it clear that the State Council exercised the ownership on behalf of the state, and established the system of separation of ownership and use right of natural resources assets and paid use of use right, which laid the foundation of China's current property right system of natural resources assets.

2 The property rights system of natural resources assets in major countries in the world

2.1. The registration system

In most countries, the registration of natural resources started from land at the earliest, and most of them were under the responsibility of one department, mainly rights-

* Corresponding author: 360248692@qq.com

https://doi.org/10.1051/e3sconf/202235802041

related registration. Not only in many common law countries but also in civil law countries, the concept of "land" includes the surface and various mulches such as forests, grasslands, minerals and water attached to the surface. Other countries implement comprehensive and unified registration of natural resources, represented by British Columbia, Canada, which implements a unified registration system of natural resources. Natural resources such as forests and land are managed by the Forest, Land and Natural Resources Operation Department of British Columbia. Among the natural resources rights and interests that need to be registered in Canada, the transfer of government administration and control rights is included. Government agencies can transfer the administrative management rights, control rights and incidental income rights of resources to other agencies or hand them over to the federal government.

2.2 Property rights system

It has become a common trend of all countries in the world that natural resources are owned by the state or classified as public property. The nationalization of natural resources makes the attribution of natural resources in the world converge, and makes the system of the right to use natural resources have a common ownership foundation. The laws of all countries (regions) in the world basically stipulate that the right to use natural resources can be obtained through franchise, license and agreement, but the most important form is through administrative license, which aims to strengthen the management of natural resources utilization. Because natural resources are also the source of human basic means of subsistence and the characteristics of some natural resources shared, the laws of various countries also stipulate that it is not necessary to apply for a license to meet the basic needs of survival and non-commercial utilization. For example, citizens may not apply for water intake permits in order to meet their basic living needs and raise livestock. Citizens can stay freely and free of charge in the forest and pick wild fruits according to law; Aborigines can cut trees and hunt wild animals without permission to make a living. Residents do not need permission to use minerals for their own use on their own land.

2.3 Investigation and evaluation system

With the needs of economic and social development, countries have deepened their understanding of the role of comprehensive investigation and evaluation, and its role field has been continuously expanded, and its position in economic and social development and resource management has also been rising day by day. It has not only become the basis for solving many major comprehensive problems, but also an indispensable part of resource strategy, resource planning and resource decision-making. The Japanese government promulgated the Law on Land Survey in 1951 and enacted the Law on Special Measures for Promoting Land Survey in 1962, aiming at "promoting land development, land protection and high utilization of land, seeking clear cadastre and obtaining land facts scientifically and comprehensively". Australia's national land survey began in 1946, which carried out a comprehensive survey and investigation of land resources in the whole country, serving for land use planning and management decision-making. Founded in 1879, the purpose of the U.S. Geological Survey changed from "investigation and data collection" to "investigation and data comprehensive analysis", which plays an extremely important role in improving economic development, environmental health and basic human living conditions.

3 The current problems faced by China's property rights system of natural resources assets

3.1 The legal system of the right to use natural resources needs to be improved

First, “China's Property Law” and “Single-category Natural Resources Law” leave gaps in the provisions on the right to use. Although the Property Law of China promulgated in 2007 has detailed regulations on land use rights and further clarifies the system of paid use of natural resources in Article 119, it only selectively confirms several kinds of natural resource use rights, but has no regulations on other natural resource use rights. Second, there is a lack of coordination between the right to use natural resources. Due to the classified management of various natural resources in China, the current legal system lacks overall consideration of the right to use various natural resources, which leads to the lack of coordination between the right to use natural resources, the vague boundary of rights and a large number of conflicts of rights. Such as the relationship between prospecting right, mining right and land use right, sea area use right, fishery right and sea area use right, water right and so on.

3.2 The overlapping contradiction between the administrative management of natural resources and the supervision and management rights

Natural resources are owned by the state, which makes the state have the dual identities of civil subject and public power subject. For a long time, due to cognitive reasons and institutional problems, the administrative power of the state as the manager of natural resources is mixed with the property rights of the state as the owner of natural resources. In practice, natural resources owned by the whole people are often represented by the administrative departments of governments at all levels. On the one hand, as the owner of natural resources, the government has various powers of ownership, on the other hand, as the manager of natural resources, it has administrative functions. In some cases, in order to obtain asset income, the protection of resources and ecological environment is deliberately ignored, and administrative power is also used to intervene in the normal process of asset transfer and transfer.

3.3 The degree of market allocation of natural resources assets is not high

First, the system of paid use of natural resources assets has not been fully established. At present, with the exception of state-owned construction land and mineral resources, the system of paid use of other natural resources assets such as water and forest is relatively slow, and the real market-oriented operation has not been realized. Administrative means are still the main management and control mode, and the proportion of market-oriented transfer such as bidding, auction and hanging is low. Second, the construction of China's natural resources factor market is backward. At present, the entry rules, competition rules, transaction rules and exit mechanism of natural resources property rights transaction in China are imperfect, which leads to the confusion of natural resources property rights transfer. For example, the primary market of land is highly monopolized by the government, the secondary market is imperfect, and the transaction price of property rights does not truly reflect the value, scarcity, intergenerational relationship and ecological environment damage of natural resources, resulting in long-term inefficient market allocation of natural resources and extensive utilization of resources. Third, the administrative allocation of natural resources assets is high. At present, China basically replaces property right management with administrative management of natural resources. Many state-owned natural resources continue to use the traditional management mode of "who develops, who benefits", resulting in low-cost or even free, excessive and disorderly development and utilization, and the owner's rights and interests of the country cannot be fully guaranteed. Fourth, the income distribution of natural resources assets is unreasonable. The development of natural resources involves the interests of the central government, local governments, development enterprises and the public. Due to the lack of a reasonable benefit distribution mechanism, natural resources are nominally owned by the state, but most of the development benefits are taken away by the development subjects, and the governments and the public where the resources are located do not share the due benefits. There is a phenomenon that "resources are taken away and pollution is left behind; Wealth is taken away, poverty is left behind".

4 The future reform trend of China's natural resources asset property rights system

4.1 Further improve the system of using natural resources owned by the whole people

Clarify the main rights of property rights of various natural resources, innovate the realization forms of national ownership and collective ownership of natural resources, and promote the further separation of ownership and use right. For the natural resources that can be developed and utilized, the ownership relationship and rights and responsibilities of rights such as possession, use, income and disposal should be clearly defined, and the rights and responsibilities of transferring, leasing, mortgaging, guaranteeing and taking shares of the right to use should be moderately expanded. Accelerate the formulation of measures for the administration of the right to use state-owned agricultural land, clarify the rights of owners and users of state-owned rural areas, forest farms and pastures, and improve the rights of state-owned agricultural land transfer, lease, mortgage, and investment (shareholding). Improve the possession, transfer, lease, mortgage, investment (shareholding) and other powers of prospecting and mining rights. Improve the right to take water, moderately expand the power of transfer and lease, and actively and steadily explore various forms of water rights trading. Improve the powers of transfer, mortgage, lease, investment (shareholding) of sea area use rights, and explore the three-dimensional establishment of sea area use rights. Accelerate the construction of the right system of the right to use uninhabited islands, and give the power of transfer, lease, mortgage, investment at a fixed price (shareholding), etc., so as to coordinate and seamlessly connect with the land use right of inhabited islands.

4.2 Deepening the market-oriented reform of the development and utilization of natural resources

Further deepen the reform of paid use of water resources, state-owned forest resources and state-owned grassland resources, and explore the establishment of multi-level property rights system arrangements for nature reserves with important ecological functions through the reform of property rights system, which is conducive to protecting resources and reducing the transaction costs of resource management and development. According to the actual needs, we can deal with the relationship with the main body of property rights of natural resources in nature reserves by means of expropriation, lease, replacement, agreement management and protection, easement contract and franchise, so as to realize the effective protection and rational utilization of natural resources. At the same time, it is necessary to adhere to the unity of rights and obligations, and clarify the obligations of natural resources property owners such as use control, resource protection, rational utilization, ecological restoration, pollution prevention and control by perfecting relevant laws and regulations, so as to strengthen the ecological protection responsibility of property owners.

4.3 Establish a centralized and efficient natural resources property rights supervision mechanism

Improve the classification standards of natural resources such as land, rationally divide ecological land, and clarify the boundaries between forest land and grassland, rivers and lakes and wetlands, so as to lay the foundation for unified registration of ownership. On the basis of real estate registration, we will accelerate the pilot project of unified registration of natural resources and gradually cover all natural resources. Strengthen the unified registration of independent natural resources registration units such as national parks, nature reserves, wetlands and water flows, draw a clear line between national ownership and collective ownership, draw a clear line between national ownership and ownership exercised by governments at different levels, and draw a clear line between different collective owners. Improve and broaden the channels of public supervision, establish a timely feedback and processing mechanism, and realize the organic combination of social supervision and government supervision. Innovate supervision means, give full play to the role of modern technical means such as mobile internet, big data, cloud computing and Internet of Things, and build a supervision system that highlights prevention and early warning beforehand, strengthens monitoring and restriction in the event, and focuses on accountability and feedback afterwards, so as to realize dynamic supervision of the whole process of natural resource ownership and use rights.

References

1. Liu, Z.Q., Zhang, J.H. Thoughts on Perfecting the Legal System of Natural Resources [J]. China Land and Resources Economy, 2019, 32(3):23-26.
2. Gao, B., Zhong, X.Y., Xu, B. Reform process of natural resources management system [J]. China Land, 2018, (10):14-17.
3. Zhang, F.G. How to break the reform of property rights system of natural resources [J]. China Land, 2017, (12):14-15.
4. Xie, G.D., Cao, S.Y., Wang, H., etc. Development Trend of Property Rights System of Natural Resources Assets [J]. Journal of Shaanxi Normal University (Philosophy and Social Sciences Edition), 2015 (05):162-166.

https://doi.org/10.1051/e3sconf/202235802042

Mix-ratio optimization for air-tight tunnel concrete

Xiaoliang Huang [1,*], Yong Zhang [1], Dingge Hou [1], Cheng Xue [1], Guang Li [1], Bo Li [1], Zhendi Wang [2] and Bin Li [2]

[1]CCCC-SHEC Meng Xing Engineering Co., Ltd., Xi'an 710076, China
[2] State Key Laboratory of Green Building Materials, China Building Materials Academy, Beijing 100024, China

Abstract. In this paper, we studied the effect of fly ash and mix ratio on the compressive strength of air-tight concrete, the introduced fly ash would decrease the early compressive strength of concrete. With curing age increasing, the gape in the compressive strength between concrete without fly ash and concrete containing fly ash is decreasing. Moreover, concrete containing fly ash showed smaller air permeability, the introduced fly ash improved the airtightness of concrete.

1 Introduction

With the quick development of transportation network construction, the tunnel construction has been an important part. During the tunnel construction process, especially in the gas zone, gas leaking is a common problem. For avoiding gas leakage, it is necessary to enhance the air-tightness of the tunnel concrete, air-tight concrete is a kind of concrete with excellent air tightness performance. However, the performance of air-tight concrete is influenced by materials [1-2] and mix ratios [3-4]. In order to determine the influence of different materials and mix ratios on the performance of air-tight concrete. In this paper we studied the influence of concrete's mix ratios on the performance of air-tight property of concrete to provide theoretical reference for the subsequent application of air-tight concrete.

2 Materials and experiment

2.1 Materials

2.1.1 Cement

The type of cement used in this study is P.O. 42.5. The performance test is followed by Chinese Standard Testing Method GB175-2007 and test results is shown in Table 1.

Table 1 The properties of cement

Density g/cm^3	Specific surface area m^2/kg	Setting time min		Flexural strength/ MPa		Compressive strength/ MPa	
		Initial setting time	Final setting time	3d	28d	3 d	28 d
3.09	310	308	341	5	8.3	23.0	49.0

2.1.2 Sand

The sand used in this study is coarse sand, which is tested by Chinese Standard Testing Method GB/T14684-2011 and test results are shown in Table 2.

Table 2 The properties of sand

Fineness modulus	Stone powder content/ %	Particle gradation and cumulative sieve residue / %					
		4.75 mm	2.36 mm	1.18 mm	600 μm	300 μm	150 μm
3.5	9.6	9.6	47.9	68	77.6	84.6	88.7

2.1.3 Gravel

The gravel used in this study is crushed stone. The properties of crushed stone are tested by Chinese Standard Testing Method GB/T14685-2011 and the test results are shown in Table 3.

Table 3 The properties of graded crushed stone

Clay content/ %	Particle gradation and cumulative sieve residue / %				
	26.5 mm	19.0 mm	9.5 mm	4.75 mm	2.36 mm
0.2	0	5	72	99	100

* Corresponding author: 244692606@qq.com

2.1.4 Admixture

Air-tight agent is produced by Shanxi Kent fine chemical Co. Ltd.; Water reducing agent are two kinds of polycarboxylate superplasticizer (PCE1 water reduction ratio = 32% and PCE2 water reduction ratio = 35%), which is produced by Jiangsu Sobute New Materials Co. Ltd.; Accelerator agent is produced by Jiangsu Sobute New Materials Co. Ltd. Concrete preservative is produced by Jiangsu Sobute New Materials Co. Ltd.

2.1.5 Water

The water used in this study is tap water, which is in line with JGJ 63-2006.

2.2 Specimen preparation and test method

2.2.1 Concrete specimen preparation

The mix proportion of concrete specimen in this study is shown in Table 4 and Table 5. The slump of concrete is fixed (200 mm) via adjusting the water-cement ratio of concrete.

Table 4 The mix ratio of C25 concrete

Sample number	cement	fly ash	sand	Gravel (5~10) mm	water	Airtight agent	PCE1	Accelerator agent
C25-1	382	68	965	755	178	25	2	0
C25-2	382	68	965	755	178	25	2	27
C25-3	450	0	965	755	176	25	2	27
C25-4	460	0	965	755	178	25	2	27
C25-5	470	0	965	755	182	25	2	27

Table 5 The mix ratio of C40 concrete

Sample number	Cement	Fly ash	Sand	Crushed stone (5~10) mm	Crushed stone (10~20) mm	water	Air-tight agent	Concrete preservative	PCE1	PCE2
C40-1	328	132	775	280	700	148	25	38	1.91	
C40-2	328	132	775	280	700	142	25	38		1.91

2.2.2 Experiment method

The test method of concrete's performance

Test method for the performance of concrete is followed by Chinese Standard Testing Method GB 8076-2008.

Air permeability test

The air permeability of concrete was determined by the Autoclam test, it was designed by Basheer et.al to determine the air permeability of concrete. There is a linear relationship between the natural logarithm of pressure and time, and the slope of the fitting line is autoclam permeability coefficient (API) in Ln (pressure)/ min. Then the air permeability coefficient k_{Au} (m^2) of concrete was calculated according to Eq. (1).

$$k_{Au} = 8.395 \times 10^{-16}(\mathrm{API})^{0.8754} \quad (1)$$

3 Results and discussion

The compressive strength of concrete (C25) with different mix ratio at different curing ages is shown in Figure 1.

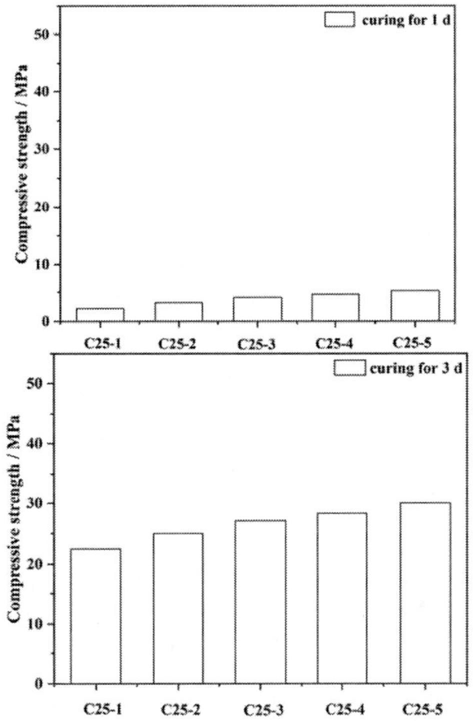

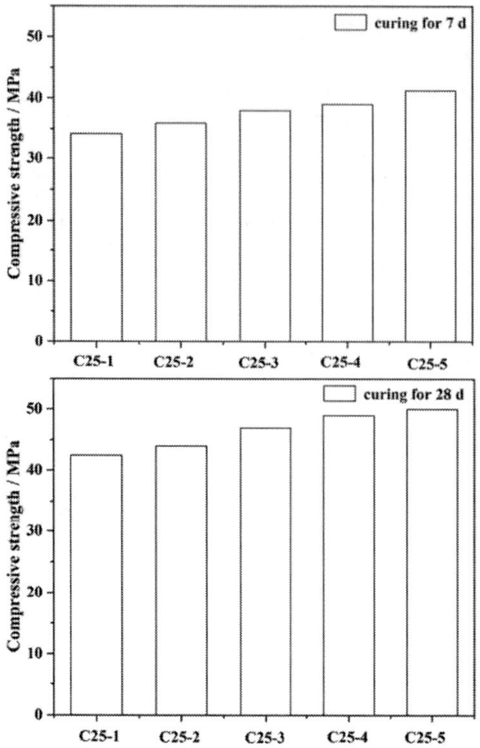

Figure 1 The compressive strength of C25-1~C25-5 specimen at different curing age

It can be seen that with curing age increasing, the concrete' compressive strength is increasing. When the curing age was 1 d, the C25-5 showed the maximum strength, due to the incorporation of fly ash in C25-1 and C25-2. At the early hydration, the degree of hydration is lower, so their compressive strength is smaller.

Meanwhile, C25-2 shows larger compressive strength than C25-1, which can be attributed to the introduction of accelerator agent. The introduced accelerator agent increases the hydration degree of C25-2, at the same curing age, C25-2 produces more hydration products. For C25-3, C25-4 and C25-5, with the content of cement increasing, their 1 d compressive strengths are larger, the greater the cement content, the greater the hydration products produced.

When the curing age at 3 d and 7 d, the C25-5 still shows the largest compressive strength, and other concretes show the similar tendency like curing for 1d. However, with the curing age increasing, the gap of compressive strength of different concrete samples is getting smaller, which indicates that the amount of produced hydration product is similar. When the curing age is 28 d, C25-3~C25-5 shows larger compressive strength than C25-1~C25-2, the introduced fly ash decreases the early compressive strength of concrete, due to the lower hydration speed of fly ash.

The compressive strength of concrete (C40) with different mix ratio at different curing ages is shown in Figure 2.

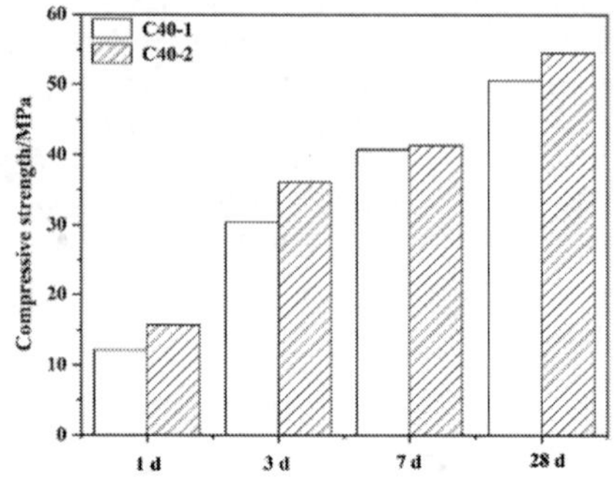

Figure 2 The compressive strength of C40-1 and C40-2 specimen at different curing age

As shown in Figure 2, it can be seen that C40-2 shows larger compressive strength at the same curing age. Comparing the mix ratio of two concrete samples, the only difference is the type of water reducing agent, These two PCEs has a few differences in their molecular parameters. PCE2 has larger water reduction ratio, which is beneficial to the compressive strength of concrete.

The air permeability of concrete (C25-1~C25-5) and C40-1~C40-2 cured for 28 d is shown in Table 6.

Table 6 The air permeability of C25-1~C25-5 after curing for 28 d

Sample number	Air permeability/$\times 10^{-17}$ m^2
C25-1	14.2
C25-2	13.4
C25-3	17.2
C25-4	16.8
C25-5	15.2
C40-1	11.3
C40-2	10.9

For C25-1~C25-5 five concrete samples, it can be seen that comparing with C25-3~C25-5, C25-1 and C25-2 show smaller air permeability, the introduced fly ash has good micro-aggregate effect. With the continuous advancement of the hydration reaction, the active micro-aggregate effect of fly ash shows good filling and compacting effect on the pores of the cement paste, which decreases the air permeability of concrete. For C25-3~C25-5, C25-5 shows smaller air permeability, which may be attributed to the higher content of cement and larger amount of hydration product. For C40-1 and C40-2, C40-2 shows smaller air permeability, which can be attributed to the higher density of C40-2.

https://doi.org/10.1051/e3sconf/202235802042

4. Conclusion

In this research, we prepared air-tight concrete with different mix ratios and studied their air tightness and compressive strength. It can be found that the introduced fly ash can improve the airtightness of concrete.

References

1. H. D. Liu, R. H.Zhang, X. Y. Jiang, G. S. Gui, X. Wu, China Concrete and Cement Products. **6**, 46-49 (2021)
2. J. P. Kou, Cement and concrete production. **1**, 9-10 (2021)
3. Y. F. Wang, Highways & Transportation in inner Mongolia. **4**, 29-33 (2020)
4. C. Aldea, S. Shah, A. Karr, Mater. Struct. **32**, 370–376 (1999)

Research progress of tunnel air-tight concrete

Junjie An [1,*], Guang Li [1], Gang Liu [1], Cheng Xue [1], Dingge Hou [1], Shaolong Wang [1], Zhendi Wang [2], Bin Li [2]

[1]CCCC-SHEC Meng Xing Engineering Co., Ltd., Xi' an 710076, China
[2]State Key Laboratory of Green Building Materials, China Building Materials Academy, Beijing 100024, China

Abstract. With the quick development of transportation network construction, the tunnel construction has been an important part. During the tunnel construction process, especially in the gas zone, gas leaking has been a common problem. For avoiding gas leakage, it is necessary to enhance the air-tightness of the tunnel concrete, air-tight concrete is a kind of concrete with excellent air tightness performance. In this paper, we introduced the research progress of tunnel air-tight concrete, including the effect of mix ratio, composition on the air-tight performance of tunnel concrete and corresponding characterization methods.

1 Background

In recent years, the rapid development of economic construction and the western region of large construction further implement in China make traffic infrastructure construction develop rapidly, especially in the construction of roads and railways. Concrete as one of the most important materials in engineering, the usage amount of concrete is extremely huge. However, due to the difference of the regional geological terrain and environment, the demanding of concrete, such as strength, durability and other requirements are increasing, especially in its durability.

Due to the rapid development of construction in the west of China, tunnel under construction and crossing the gas zone have become an inevitable problem. If the gas in the rock is leaking through the tunnel lining, and accumulates to a certain concentration, it may induce to a serious gas explosion. The main component of gas is methane, when the volume concentration of methane in the air reaches 5%~6%, the methane will cause the explosion. Therefore, various protective measurements should be taken to prevent the gas leakage in tunnels. Like the corrosion resistance of concrete in various erosion environments are proposed, the air tightness of concrete is also one of the important indicators to investigate the durability of concrete, especially for tunnel concrete.

2 The influencing factors of tunnel concrete's airtightness

For improving the air tightness of concrete, it is necessary to consider the type of cement, the clay content, maximum particle size of aggregate, water to cement ratio, curing age and so on.

Impact of water cement ratio on concrete confidentiality. The water binder ratio is an important index in evaluating the air tightness of concrete. In order to reduce the porosity of concrete, the concrete with low water binder ratio should be considered first to reduce the capillary pore of concrete. In addition, the water binder ratio can also affect the workability of concrete, thereby affecting the air tightness of concrete. Excessive water binder ratio will cause the concrete slump to drop, while too small water binder ratio will affect the workability of concrete. With the increase of water binder ratio, the air permeability coefficient of concrete will increase significantly, especially when the water binder ratio reaches 0.45, so the effect of admixture is more obvious when the water binder ratio is small. Liu [1] has studied the effect of water/cement ratio and mineral admixture' dosage on the air tightness of tunnel concrete. They found that when the water/cement ratio was higher than 0.42, the air permeability of concrete increased obviously, which would result in a poor air tightness of concrete. Moreover, the mineral admixture also influenced the air tightness of concrete, when the dosage of mineral admixture was over 25%, the air permeability would decrease markedly.

Kou [2] has studied the effect of air-tight agent on the permeability of tunnel concrete, it can be found that for C30 air-tight tunnel concrete, with the dosage of air-tight agent dosage increasing, the air-tightness performance of concrete would increase significantly. Moreover, the sand ratio of tunnel concrete should be lower than 45.7%.

Wang [3] studied the dosage of air-tight agent dosage on the air permeability of tunnel concrete, he found that when the dosage of air-tight agent was 6.05%, the concrete had the lowest air permeability, and the dosage of air-tight agent did not influence the concrete's air content and workability.

Porosity is an important component of concrete, which has an important impact on the performance of concrete. The air tightness of concrete is directly related to its porosity. However, there are few studies on the relationship between pore structure and concrete permeability. From some research results, the air tightness

* Corresponding author: 244692606@qq.com

of concrete is related to porosity, but it is not a simple function

The air tightness of concrete depends on the connectivity of the internal porosity. Based on this, the pore structure of concrete can be densificated to improve the air tightness of concrete. Adding water reducing agent to concrete can improve the structure of cement mortar to a certain extent, and the use of cement with silica fume can also significantly improve the air tightness of concrete.

According to the influence of admixture on the confidentiality of concrete, the air permeability coefficient of concrete will increase with the addition of fly ash, which is mainly due to the contact between fly ash and aggregate. Although the fly ash will increase the volume of the hole, the low early strength of the fly ash causes the air tightness of the concrete.

The concrete with fly ash shows better air tightness can be attributed to the introduced fly ash reduces the bleeding of concrete, which can be evenly distributed inside macrostructure of the concrete and improve the structural composition of concrete holes. Silica fume also has positive effect on the improvement of the air tightness and impermeability of cement concrete. When the water binder ratio is unchanged, adding silica fume can improve the compactness at the interface of concrete aggregate, reduce the gap in the mixture, and improve the air tightness of concrete.

The sand ratio also make effects on the air tightness of concrete. The sand ratio has a direct impact on the fluidity of concrete. Increasing the sand ratio can improve the workability of concrete in the pumping process. However, the increase of sand ratio will directly affect the drying shrinkage of concrete. With the increase of sand ratio in concrete, the air tightness of cement concrete will be increased. When the total sand ratio of concrete reaches a certain value, the air tightness of concrete will be significantly increased. Generally, the sand ratio of air tight concrete is about 38%~42%.

The influence of admixture on the air tightness of cement concrete. Water reducing agent is an additive commonly used in cement concrete, which can reduce the water binder ratio of concrete on the premise of ensuring the compactness of concrete. Thus, the gas permeability of concrete is reduced. Concrete surfactant can also increase the bonding property between concrete aggregate and binder, thus reducing the air permeability of concrete.

Effect of air permeability pressure on air permeability coefficient of concrete. Through relevant indoor tests, it can be found that when the air tight pressure value of concrete is increased from 0.3 MPa to 0.6 MPa, the air permeability coefficient of concrete will decrease, which indicates that the air permeability coefficient of concrete measured with large seepage pressure is smaller at the same time. Therefore, in the process of studying the air tight concrete of gas tunnel, appropriate test pressure should be selected in combination with the gas concentration, pressure and other conditions at the construction site.

Water binder ratio is the main factor to control the compactness of concrete. The research results of the original gas tight concrete of Dongjiashan Gas Tunnel show that when the water binder ratio is reduced from 0.55 to 0.49, the compressive strength of concrete is increased from 33.1 MPa to 37.4 MPa, and the air permeability coefficient of concrete is increased from 0.88×10^{-11} cm/s.

Gradually reduce to 0.60×10^{-11} cm/s. TB 10424-2010 stipulates that the water binder ratio of concrete should not exceed 0.45, otherwise the air permeability coefficient of concrete cannot meet the design requirements.

The amount of air sealing agent has a great influence on the air permeability of concrete. When the content of air sealing agent increases from 4% to 8%, the air permeability coefficient of concrete changes from 1.12×10^{-11} cm/s gradually decreases to 0.50×10^{-11} cm /s. The optimum dosage shall be determined through multiple tests.

The sand ratio has a great influence on the properties of concrete mixture and hardened concrete. In a certain range, the air permeability coefficient of concrete decreases gradually with the increase of sand ratio. The sand ratio of air tight concrete shall not be less than 36%. The sand ratio shall be as large as possible under the condition of meeting other performance indexes of concrete to improve the air tightness of concrete.

3 Testing methods

At the same time the airtightness measurements have also became an important part of modern building evaluation process. The measurement can help to evaluate the quality of the air tightness of concrete, which helps to reduce the energy consumption of the building. A few studies have reported the test method of permeability of concrete and rocks in labs [4–6] as well as in situ [7–9], and they reported that concrete permeability could change, depending on the details of the test conditions, such as water–cement ratio [10, 11], moisture content [12, 13], temperature [14], cement preparation methods [15], measuring scale [16] and so on. Thus, the concrete permeability, which is the most critical parameter for air tightness assessment. The air tightness of concrete can be measured through indoor tests. The measuring devices include air compressor, air tank, impermeability meter and measuring instrument. The measuring principle uses the relevant contents of Darcy's seepage law. Place the test piece in the sealed mold, apply 0.3 MPa and 0.6 MPa air pressure to the concrete, keep the pressure stable for 8 hours, and record the measured value.

There are some other common methods to determine the air-tight performance of concrete:

3.1 Constant pressure method

In this test method, the working pressure is unchanged, the air tightness index is the time needed for a certain amount of gas to flow through the specimen or the volume of gas to flow through the specimen in a certain time. This testing method is suitable for the common specimens.

3.2 Variable pressure method

In this test method, the pressure gradient is stable and the amount of air flowing through the specimen is monitored within a specified time interval. This method is suitable for box-shape specimen.

Previous researchers have studied the void ratio of ordinary concrete by constant pressure method. They found that the ordinary concrete's air tightness index K≥ 10^{-3} (cm^3•cm). Other researchers studied the effect of water/cement ratio on the air tightness of concrete and found that the air tightness index was increased with the water/cement ratio increasing (when the W/C is 0.4, the $K=2\times10^{-10}\sim7.5\times10^{-11}$ cm/sec. However, when the W/C is 0.7, the $K=5.8\times10^{-3}\sim1.0\times10^{-9}$ cm/sec). Li et. al found that the diffusion coefficient of methane in concrete is similar with the air. They showed the same magnitude, so it is reasonable to use the air tightness index of air to show the methane's air tightness index.

3.3 Research on construction technology of air tight concrete

Influence of mixing time on the performance of air tight concrete. The sufficient mixing process of concrete is the key link to ensure the air tightness of concrete. Through the research on the mixing time in the actual construction process, it can be found that when the concrete mixing time reaches about 3 minutes, the air tightness of the concrete is the best. If the mixing time is too long, the water in the concrete will evaporate, which will reduce the air tightness of the concrete.

Effect of standing time on the performance of air tight concrete. Pumping concrete is generally used in the construction of gas tunnels, which requires consideration of the influence of the static time of air tight concrete on its performance. Air tight concrete will produce certain segregation during transportation, and cement concrete will also start to try partial hydration heat. These concrete quality problems will have a certain impact on its air tightness. In general, the longer the standing time of concrete is, the lower the air tightness of concrete will be. Generally, the air tightness of concrete can be effectively guaranteed if the standing time of air tight concrete is kept within 45 minutes, so the concrete pouring shall be completed within 45 minutes as far as possible during the construction of air tight concrete.

Effect of vibration time on the performance of air tight concrete. Vibration is an important means to ensure the compactness of concrete during concrete construction. At present, the vibration methods for concrete are mainly for ordinary concrete, while ordinary vibration operations may not be suitable for air tight concrete. From the actual situation, the air permeability of air tight concrete is still relatively high after 10 s of vibration time, and the air tightness of concrete will be significantly improved when the vibration time reaches about 20~30 s. At the same time, it can also be predicted that the longer the vibration time of air tight concrete is, the better its air tightness will be. However, long-term vibration will cause the aggregate to sink and cause segregation. Therefore, in the actual construction process, the mixing time of air tight concrete should be determined comprehensively according to the actual situation.

Influence of concrete strength on the performance of air tight concrete. On the whole, there is a good linear relationship between the strength of concrete and the air tightness of concrete. This is mainly because the strength of concrete is directly affected by the internal porosity. With the occurrence of cement hydration heat, the internal porosity of concrete can be filled, which improves the strength of concrete and reduces the permeability of concrete. From the strength of concrete, there is a natural logarithmic curve between the strength curve and the permeability coefficient of concrete after 28 days of curing. However, the strength of air tight concrete will be improved when adding additives or admixtures, but the change of its air permeability coefficient may show different trends. Specific analysis should be made on the relationship between the strength and air tightness of air tight cement concrete with admixtures

4 Summary

Summary of previous researches, the type of cement, aggregate gradation and silt content, the maximum particle size of aggregate, water/cement ratio, curing age, admixture, etc., all of these factors have different influence degrees on the air tightness of concrete. However, in this research field, only a few scholars have done some research on the basic theory of concrete air tightness and the influencing factors of air tightness. In their researches, they proposed some detection methods. However, their results have not formed the standard, and the relating research data and results are limited.

References

1. H. D. Liu, R. H.Zhang, X. Y. Jiang, G. S. Gui, X. Wu, China Concrete and Cement Products. 6, 46-49 (2021)
2. J. P. Kou, Cement and concrete production. 1, 9-10 (2021)
3. Y. F. Wang, Highways & Transportation in inner Mongolia. 4, 29-33 (2020)
4. C. Aldea, S. Shah, A. Karr, Mater. Struct. 32, 370–376 (1999)
5. C. Desmettre, J. Charron, Mater. Struct. 44, 1713–1723 (2011)
6. J. Figg, Mag. Concr. Res., 25, 213–219 (1973)
7. Y. Billard, G. Debicki, L. Coudert, Nucl. Eng. Des., 235, 2109–2123 (2005)
8. D. White, P. Vennapusa, M. Suleiman, C. Dahren, Geotech. Test. J., 30, 282–291 (2007)
9. H. M. Kim, Y. Lettry, D. Park, D. W. Ryu, B. H. Choi, W. K. Song, Eng. Geol., 137–138, 97–106 (2012)

10. A. Alshamsi, H. Imran, Cem. Concr. Res., 32, 923–929 (2002)
11. M. Sanjuan, R. Munoz-Martialay, J. Mater. Sci. 31, 2829–2832 (1996)
12. A. Abbas, M. Carcasses, J. Ollivier, Mater. Struct. 32, 3–8 (1999)
13. M. Romer, Mater. Struct., 38, 541–547 (2005)
14. M. Choinska, A. Khelidj, G. Chatzigeorgiou, G. Pijadier-Cabot, Cem. Concr. Res., 37, 79–88 (2007)
15. L. Schwarz, R. Krizek, Geotech. Test. J., 17, 434–443 (1994)
16. L. Alarcon-Ruiz, M. Brocato, S. Dal Pont, A. Feraille, Transp. Porous. Media. 85, 541–564 (2010)

Research on the technology of increasing the output of thermal power units in Shandong Province under the demand of ensuring supply

Ziwei Zhong*, Lingkai Zhu, Wei Zheng, Zhiqiang Gong, Panfeng Shang, JunshanGuo, Jun Liu, Yanpeng Zhang

Electric Power Research Institute of State Grid Shandong electric power company, Jinan, China

Abstract: Under the background of "carbon peaking and carbon neutralization", China proposes a new power system based on new energy. Due to the strong randomness, volatility and anti peak regulation characteristics of renewable energy, in order to promote the consumption of new energy and ensure the power supply, coal-fired power generation units will undertake the dual tasks of deep peak regulation and power supply guarantee in the future, and gradually transform from electricity type power supply to regulation type power supply.

key word: Deep peak shaving, thermal power contradiction

1. Introduction

Under the background of "carbon peaking and carbon neutralization", China proposes a new power system based on new energy. Due to the strong randomness, volatility and anti peak regulation characteristics of renewable energy, in order to promote the consumption of new energy and ensure the power supply, coal-fired power generation units will undertake the dual tasks of deep peak regulation and power supply guarantee in the future, and gradually transform from electricity type power supply to regulation type power supply [1].

In order to further improve the clean, efficient and flexible level of coal-fired power units, we will comprehensively consider the energy-saving and consumption reduction transformation, heating transformation and flexibility transformation of coal-fired power units, and promote the "three transformation linkage" of coal-fired power units [2]. After the heat supply transformation of the thermal power unit, the peak output of the unit will decrease due to heating and industrial steam supply in winter. It is predicted that the peak output of the thermal power unit in Shandong Province will decrease after the implementation of the heat supply transformation, as shown in the following table.

In the second half of 2021, it is difficult to balance power supply and demand due to factors such as the decline of coal-fired power generation willingness, the obvious increase of external power uncertainty, and the insufficient peak capacity of new energy. It is estimated that the maximum power gap in 2022 will reach 7 million kilowatts, especially in the heating period. The thermal power units are faced with the double pressure of peak power consumption and peak heating, and are forced to further reduce output due to meeting the heating demand. Most units in Shandong Province have reduced their output due to heat supply, and the contradiction between heat supply and peak regulation is particularly prominent in the evening peak [3]. From November 1, 2021 to December 26, 2021, a total of 107 units applied for output reduction, including winter heating and industrial steam extraction, with a total of 5.87 million KW; Among them, the output reduction due to heating in winter is 5.445 million kilowatts, and the output reduction due to industrial steam supply is 425000 kilowatts.

Table 1 Peak output drop after implementation of heat supply transformation

year	2022	2023	2024	2025
Shandong power grid maximum power load forecast (100 million KW)	1.23	1.31	1.38	1.45
Provincial installed capacity forecast (100 million KW)	1.80	2.07	2.34	2.62
Limited output value of heating transformation unit (10000 kW, heating period)	28.35	24.23	41.26	30.79
Limited output value of heating transformation unit (10000 kW, non heating period)	14.11	7.53	12.15	11.67

* Corresponding author: dkyzhongziwei@163.com

https://doi.org/10.1051/e3sconf/202235802044

2. Thermal inertia characteristics of heat supply network

The district heating pipe network connects the heating source and the user's heat load, and realizes heat energy through water circulation in the pipeline Transmission to ensure the balance of heat supply and demand. Relevant thermodynamic studies show that the heating network has great potential in the coordination and optimization of electric heating. The continuous and stable operation mode of the heating system is essentially changed by using the thermal inertia of the heating network [4]. When the heat generation of the thermal power unit changes, the heating medium and indoor temperature fluctuate, forming the thermal dynamic change process of the heating system, and achieving the purpose of the cogeneration system participating in the peak load regulation without affecting the heating quality.

In order to adapt to the thermal inertia of the heat supply network, the upper limit of heat storage of the heat supply network is the upper limit of the design temperature of each pipe, and the lower limit of heat storage is the minimum temperature requirement of the return water pipe [5]. It can be seen from Fig. 1 that the decreasing trend of the return water temperature of the heat supply network lags behind the inlet water temperature of the heat supply network, and the cooling range is smaller. Therefore, the heat storage and release of the heat supply network can be optimized before the time when the power consumption of the power grid needs to be increased; There is a lot of heat energy in the heating pipe, so the heat network can be regarded as an energy storage device. Make use of the thermal storage capacity of the thermal system to "shift the peak and fill the valley" of the thermal load, improve the peak output capacity of the unit, and ensure the stability of the power supply.

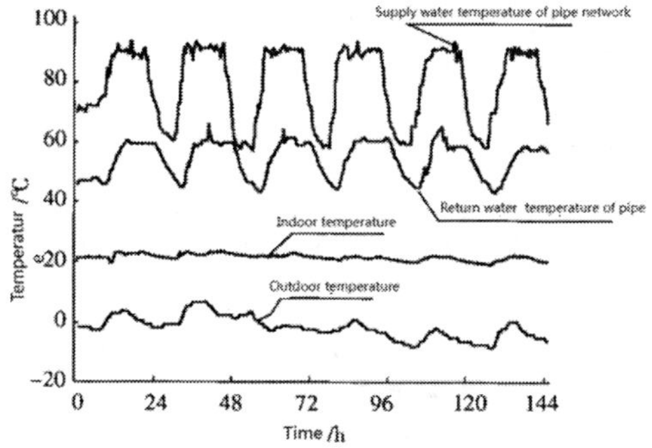

Figure 1. Curve of supply and return water temperature and indoor and outdoor temperature of heat supply network

2.1 Shandong power supply situation and Countermeasures

With the rapid development of ultra-high voltage external power and new energy in Shandong power grid, the difficulty of system peak regulation is increasing. At present, the peak regulation is mainly based on direct regulating thermal power units. The capacity of pumped storage power stations is limited and there are no gas-fired generating units, which also leads to serious shortage of other flexible regulation resources that Shandong power grid can call. In 2022, Shandong Province plans to implement the renewable energy multiplication plan, and it is expected that the installed capacity of new energy will account for 38%. With the large-scale access of new energy, the "double high" characteristics of the power system are highlighted, and the technical characteristics and stability mechanism of the power grid will undergo profound changes. In the new power system with new energy as the main body, the role of thermal power generation has gradually changed from the electric power source to the regulating power source.

In the second half of 2021, due to the decline of coal-fired power generation intention, the obvious increase of external power uncertainty, the insufficient peak capacity of new energy and other factors, it is difficult to balance power supply and demand. It is estimated that the maximum power gap in 2022 will reach 7 million kilowatts. Especially in the heating period, the proportion of thermal power units in Shandong Province's direct dispatching units has exceeded 86%, and the power load adjustment capacity of thermal power units is greatly restricted by the heat supply. Under the "three public dispatching" mode, The thermal power unit guarantees the income in the form of "guaranteed quantity and guaranteed price". In the spot mode, the thermal power unit is no longer "guaranteed quantity and guaranteed price". Due to its poor adjustment ability, the thermal power unit accepts the settlement of the spot market deviation, and the reduced output part has no capacity compensation fee. The income in the spot mode is far lower than that of the pure condensing unit. At the same time, the thermal power unit is also subject to the assessment of "two detailed rules". Under the condition of continuous high coal price, This has greatly increased the difficulty of power generation enterprises. The thermal power unit faces the double pressure of power consumption peak and heat supply peak, and is forced to further increase and decrease the output due to meeting the demand of heat supply. Most units in the province have reduced their output due to heat supply, and the contradiction between heat supply and peak regulation is particularly prominent in the evening peak. From November 1, 2021 to December 26, 2021, a total of 107 units applied for output reduction, including winter heating and industrial steam extraction, with a total of 5.87 million KW; Among them, the output reduction due to heating in winter is 5.445 million kilowatts, and the output reduction due to industrial steam supply is 425000 kilowatts.

At the present stage, the external heating regulation mode of the thermal power plant is relatively extensive, and the influence of factors such as the peak valley difference of the power grid and the inertia of the heat grid is not fully considered, which not only affects the heating experience of the heat users, but also further aggravates the difficulties of the power grid regulation. Generally speaking, the temperature at noon in the heating season is significantly higher than that in the morning and evening. The heat users use less heat at noon, and the heat consumption increases significantly in the morning and evening. If the heating network is not adjusted, it will inevitably lead to overheating at noon and insufficient heating in the morning and evening. At the same time, with the continuous increase of the photovoltaic power supply capacity of Shandong power grid, the photovoltaic power generation load in the evening has dropped significantly, forming an obvious peak output demand of thermal power units at night. At this time, due to the lack of research on coordination and optimization of electric heating, the continuous high-intensity heating of thermal power units will inevitably lead to insufficient peak output capacity and aggravate the difficulty of grid regulation. If the peak valley difference of the power grid and the new energy consumption demand are combined in the power grid dispatching, the thermal inertia of the heat grid can be used to reduce the heat supply in a short time, so as to realize the "heat reduction and power increase" of the thermal power unit. On the premise of no additional technical transformation and cost increase, the peak output demand of the power grid in the evening peak can be met, the power supply capacity in winter can be improved, and the construction of new power systems can be assisted.

2.2 Case analysis of using thermal inertia to improve unit output capacity

The current installed capacity of the power plant is 2 × 330MW subcritical, single shaft, double cylinder, double exhaust condensing steam turbine, equipped with 2 × 1036t / h Subcritical Pressure primary intermediate reheat control circulating drum boiler# Unit 1 was put into commercial operation on October 25, 2006 and unit # 2 was put into commercial operation on May 22, 2007, with a total installed capacity of 660MW.. The heating of the power plant is mainly divided into residential heating and industrial steam supply:

1、 Heating:

1) Unit extraction.

Each unit adds steam extraction on the connecting pipe, with the extraction pressure of 0.8MPa and 330 °C, and the maximum extraction capacity of a single unit is 300t / h. During the reconstruction, the structure of medium and low pressure connecting pipe is modified, and a DN800mm extraction pipe is led upward from the middle horizontal section to serve as the heating steam source, and three-way, safety valve, extraction regulating valve and check valve are installed in the connecting pipe. Each unit has only one heating steam extraction circuit and is equipped with pressure, temperature and flow measuring points. The steam extraction pipeline of each unit is connected to the dn1000mm steam main pipe to supply steam to the initial heating station.

2) The first heating station.

The 0.8MPa and 330 °C extraction steam of the unit is transmitted to the sub cylinder of the initial station through the steam header and divided into two paths. One path of superheated steam is heated into 110 °C high-temperature hot water through the high-pressure steam water heat exchanger, and the 90 °C condensate enters the condensate recovery device and is sent back to the inlet of No. 6 low-pressure heater of the unit by the condensate pump; The other way of superheated steam is used as the driving steam source of the small steam turbine. After the exhaust steam is heat exchanged with the circulating water of the heat supply network through the low-pressure steam water heat exchanger, the condensate returns to the condensate recovery device. The return water at 60 °C in the heat supply network is pressurized by the decontaminator and circulating water pump and then enters the steam water heat exchanger to be heated into 110 °C high-temperature hot water, which is sent to the secondary heat exchange station in the urban area through the hot water pipe network. The process flow of the initial heat exchange station is as follows:

a.Steam heat exchange system:

steam extraction of the unit in the turbine room → steam pipe in the turbine room → cylinder separation in the initial station → steam branch pipe → high-pressure steam water heat exchanger → condensed water branch pipe → condensed water main pipe → condensed water collector → condensed water pump → inlet of No. 6 low-pressure heater of the unit.

b.Steam drive system:

steam extraction of the unit in the turbine room → steam pipe in the turbine room → sub cylinder of the initial station → steam branch pipe → small turbine → exhaust steam pipe → low-pressure steam water heat exchanger → condensate pipe → condensate water collector → condensate pump → inlet of No. 6 low-pressure heater of the unit.

c.Hot water system:

return water main pipe of heat supply network → decontaminator → circulating water pump of heat supply network → steam water heat exchanger → high temperature hot water branch pipe → water supply main pipe.

Explore the feasibility of several technical measures to realize "heat reduction and power increase" at night peak by using the characteristics of heat supply network: first, directly reduce the steam extraction capacity of the unit without changing the parameters of heat supply network; Second, adjust the heating temperature of the primary heat supply network; Third, regulate the circulating water volume of the primary heat supply network. The influence of the above three regulation modes on the unit output and the heat supply of heat users is studied. Explore the effect of reducing or interrupting the heat source of the central heating system on the heating of heating users and the improvement of the supply and regulation capacity of the power system, study the continuous full output power generation time of the heating unit after the suspension of heating, study the ability of the heating unit to reduce the heat supply and guarantee the power peak, and demonstrate the heating unit to improve the supply capacity in winter to adapt to the development trend of large load in winter.

Figure 2. Electric heating coordination curve of a power plant

3. Conclusion

During the period from November 19 to December 31, 2021, a power plant applied for output reduction due to heating in winter, and declared that the adjustable upper limit of the whole plant was 500000 kW, and the blocked capacity was 160000 kW. At the end of the heating season in 2022, the relevant data of the heating season will be obtained through the thermoelectric online monitoring system. During 16:00-18:00 of the evening peak on December 27, 2021, the maximum output of the whole plant can reach 520000 kilowatts without affecting the people's livelihood. The sustainable and stable operation of the evening peak is above the maximum adjustable limit (500000 kilowatts) applied to the control center.

Acknowledgements

This work is supported by the science and technology project of Electric Power Research Institute of State Grid Shandong electric power company (ZY-2022-15), which studies Research on peak capacity improvement of thermal power unit based on electric heating coordination optimization under guaranteed supply demand.

References

1. Ding X L, Han X S, Zhang H, et al. Analysis on Electrothermal Coordination Power Flow and Transmission Line Temperature Variation Process[J]. Zhongguo Dianji Gongcheng Xuebao/Proceedings of the Chinese Society of Electrical Engineering, 2008, 28(19):138-144.
2. Banakar H, Alguacil N, Galiana F D. Electrothermal Coordination Part I: Theory and Implementation Schemes[J]. IEEE Transactions on Power Systems, 2005, 20(2):798-805.Haiying Dong, Lei Fang, Kun Ding, etc. Peak shaving strategy of solar thermal power generation based on Cogeneration operation mode [J]. Journal of solar energy,2019,40(10):2763-2772.
3. Alguacil N, Banakar M H, Galiana F D. Electrothermal coordination part II: case studies[J]. IEEE Transactions on Power Systems, 2005, 20(4):1738-1745.
4. Silva A, Dias L F, Saint'Pierre T D, et al. Determination of Hg and Tl in environmental reference materials using slurry sampling electrothermal vaporization inductively coupled plasma mass spectrometry with permanganate as modifier and calibration against aqueous standards[M]. Pergamon Press, 2003.
5. Mengxia, Wang, Ming, et al. Optimal power flow considering transient thermal behavior of overhead transmission lines - ScienceDirect[J]. International Journal of Electrical Power & Energy Systems, 114:105396-105396.

A Comparative Study of AutoML Approaches for Short-Term Electric Load Forecasting

Zhaorui Meng*, Xiaozhu Xie, Yanqi Xie, Jinhua Sun

School of computer and information engineering, Xiamen University of Technology, Xiamen, Fujian, 361024, P.R. China

Abstract. Deep learning is increasingly used in short-term load forecasting. However, deep learning models are difficult to train, and adjusting training hyper-parameters takes time and effort. Automated machine learning (AutoML) can reduce human participation in machine learning process and improve the efficiency of modelling while ensuring the accuracy of prediction. In this paper, we compare the usage of three AutoML approaches in short-term load forecasting. The experiments on a real-world dataset show that the predictive performance of AutoGluon outperforms that of AutoPytorch and Auto-Keras, according to three performance metrics: MAE, RMSE and MAPE. AutoPytorch and Auto-Keras have similar performance and are not easy to compare.

Keywords: Automated machine learning, load forecasting, deep learning.

1. Introduction

Accurate load forecasting is of great significance to power network planning and management decision of economic development department. Power consumption prediction based on the study of different users' power consumption characteristics can help power enterprises better understand users' personalized service demand, and provide data support for future power grid development [1].

However, due to the characteristics of intelligent power distribution big data, such as multiple types, large volume, high dimension and fast generation speed, the traditional load forecasting method has certain limitations in mining massive data information, and it is difficult to accurately grasp the relevant factors and change rules of users' electricity consumption. How to study users' electricity consumption characteristics and related factors of electricity consumption in the big data environment and predict their electricity consumption is a challenge for researchers [2].

At present, machine learning approaches have been widely applied to load forecasting, including: XGBoost [3], recurrent neural network (RNN) [4], and long short-term memory (LSTM) [5] etc. To implement the best prediction model in each data set, there are several procedures required in traditional machine learning methods, for instance: data pre-processing and cleaning, selecting and constructing appropriate features, selecting appropriate models, optimizing model super parameters, post process processing of machine learning models, etc. The processing of these tasks is usually very complex, which cannot be handled by non-machine learning experts. With more and more models being developed, how to choose an appropriate model becomes more and more difficult, and parameter tuning needs to traverse all possible values, which requires a lot of manual operations, resulting in low process efficiency and difficult management.

Auto Machine Learning (AutoML) technology is emerging in the past two years. Although it is still in its infancy, it has already achieved good results. AutoML technology is an algorithm that enables machine learning to learn how to learn. In traditional machine learning, it's mainly achieved by automatic parameter tuning, while in deep learning, it's mainly achieved by Neural Architecture Search (NAS) so far. Because the most important parameter in deep learning is the number of layers of network structure and the number of neurons in each layer.

* Corresponding author: mengzhaorui@xmut.edu.cn

In recent years, the application of AutoML in various industries has been extensively studied. Tsiakmaki et al. [6] proposed an approach using autoML to predict student academic performance. Johnson et al. [7] explored the AutoML usage in digital forensic of crime analysis and prediction. Gerassis et al. [8] introduced an AutoML-based approach for the construction of environmental impact assessment. In general, AutoML has lowered the threshold for artificial intelligence applications. With its low threshold and automated nature, AutoML is expected to revolutionize traditional machine learning methods in the coming years and make artificial intelligence truly universal.

In this paper, three AutoML methods are used for short-term load forecasting, including AutoGluon, Auto-Keras and AutoPytorch. For load forecasting data of different regions, the network structure suitable for the data set are designed automatically, without the need to manually select the network structure and parameters. On the premise of ensuring accuracy, manpower is saved. A real-world load forecasting data set is used to compare the performance of the three AutoML methods.

2. Automatic machine learning methods

2.1 Auto-Keras

AutoKeras uses a repetitious trained RNN controller to sample candidate sub-models and then train them to measure their performance in the desired task [9]. Then, the controller uses performance as internal metrics to find a more promising architecture. However, neural architecture search is computationally expensive and time-consuming. To solve this problem, AutoKeras uses efficient neural architecture search (NAS). NAS adopts a concept similar to transfer learning. The parameters learned for a specific model on a specific task can be used for models on other tasks. Therefore, NAS forces all generated sub-models to share weights, to deliberately prevent training each sub-model from scratch. ENAS can not only share parameters among sub-models, but also obtain outstanding performance.

2.2 Auto-Pytorch Tabular

Early AutoML methods are implemented by optimizing machine learning pipelines and hyperparameters. More recent autoML models use neural architecture search. To better utilize positive interaction effects between the best learning hyperparameter and the best network architecture, Zimmer et. al. proposed Auto-Pytorch Tabular framework [10], which combines meta-learning, ensemble learning and multi-fidelity optimization. In addition, Auto-Pytorch Tabular combines traditional machine learning baselines with tuned deep neural networks to achieve state-of-the-art performance on tabular data.

2.1 AutoGluon-Tabular

Unlike other frameworks that focus on hyperparametric search optimization, AutoGluon-Tabular [11] tends to use multi-model ensemble, multi-layer stacking and K-fold bagging to achieve better and more stable model results. Furthermore, in the selection of the basic model, AutoGluon-Tabular includes the gradient ascending tree model with excellent performance in various tabular data competitions to achieve better performance for tabular data, instead of blindly pursuing the lofty deep learning model. Finally, AutoGluon-Tabular optimized the neural network model particularly for tabular data. For example, per-variable embedding for class variables is added, and skip-connection is used to further improve model training.

3. Experiment

3.1 Dataset description

To evaluate the performance of AutoML in short-term load forecasting, a real-world dataset from GETCom2012 [12] was used in this paper. Originally GETCom2012 dataset was used for load forecast competition 2012. The dataset includes hourly load from 20 geographical zones and hourly temperature from 11 weather stations in the United States. As the purpose of this paper is short-term load forecasting, instead of using full GETCom2012 dataset, data from 5 of the 20 zones are selected as experimental data. And the training dataset is from January 1, 2007 to June 30,2007, and the testing dataset is from July 1, 2007 to July 7, 2007.

3.2 Performance metrics

To evaluate the performance of short-term load forecasting models, three performance metrics were used in this paper, including mean absolute error (MAE), root mean square error (RMSE) and mean absolute percentage error (MAPE). The definitions of these three metrics are shown as below:

$$MAE = \frac{\sum_{i=1}^{n}|\widehat{y_i}-y_i|}{n} \quad (1)$$

$$RMSE = \sqrt{\frac{\sum_{i=1}^{n}(\widehat{y_i}-y_i)^2}{n}} \quad (2)$$

$$\text{MAPE}=\frac{100\%}{n}\sum_{i=1}^{n}\left|\frac{\widehat{y_i}-y_i}{y_i}\right| \quad (3)$$

where n is the total number of testing data, y_i is the actual value and $\hat{y}_i$ is the predicted value.
Above all three metrics are widely used metrics. Each metric has its own advantages and disadvantages. In a general way, MAE is less sensitive to outliers. RMSE has greater penalties for large error samples. MAPE is the average error percentage. If the actual value at one point is very low and the error is large, it can have a big impact on the value of MAPE. The selection of metrics should be based on data distribution and experimental results. Generally speaking, the lower value of the above metrics, the closer are the predicted values to the actual values.

3.3 Experimental results

In this section, we present the experiment results for comparing three proposed AutoML short-term load forecasting approaches. Table 1 summarizes the prediction results of each model compared with MAE, RMSE and MAPE. In Table 1, the minimum values of the evaluation metrics for each zone are identified in bold. From Table 1, we observe that AutoGluon model has the lowest value for all three metrics in all five zones. As for AutoKeras and AutoPytorch, it is not easy to distinguish between the performance of the two models. In some zones, models generate by AutoKeras have lowest MAPE value, while models generate By AutoPytorch have lowest RMSE value in some zones.

Table 1. Comparison of MAE, RMSE and MAPE for three AutoML approaches.

	AutoGluon			Auto-Keras			AutoPytorch		
	MAE	RMSE	MAPE	MAE	RMSE	MAPE	MAE	RMSE	MAPE
Zone 1	**1127.61**	**1434.87**	**6.53**	2314.07	2843.26	24.32	1668.86	2243.59	22.36
Zone 11	**6872.97**	**8583.76**	**7.01**	10098.78	12205.02	20.21	7640.68	9295.38	22.25
Zone 12	**7383.35**	**9366.14**	**6.08**	15808.85	19269.01	19.34	10819.81	13340.86	21.78
Zone 15	**5057.28**	**6148.74**	**10.53**	6818.92	8863.6	16.39	5168.67	6087.19	19.01
Zone 20	**5116.38**	**6152.78**	**6.41**	7263.26	9661.15	16.4	5127.29	6186.55	17.14

4. Conclusion

In this paper, three AutoML methods are applied to short-term load forecasting, which solves the current problem of retraining and network selection for different data sets. Through the analysis of the experimental results, it can be seen that the method based on AutoML can automatically construct the network structure, and at the same time ensure the same accuracy as the manual selection of the classical network. In addition, different AutoML methods show different forecasting performance. The experiment on a real-world data set shows that the forecasting accuracy of AutoGluon method is significantly better than the other two methods: Auto-Keras and AutoPytorch. The AutoML approaches will be tested in more load forecasting data sets for further comparison.

References

1. Mamun, A. A., et al. "A Comprehensive Review of the Load Forecasting Techniques Using Single and Hybrid Predictive Models." IEEE Access PP.99(2020):1-1.
2. Mujeeb, S., et al. "Big Data Analytics for Load Forecasting in Smart Grids: A Survey." International Conference on Cyber Security and Computer Science (ICONCS), 2018 2019.

3. Bae, D. J., B. S. Kwon, and K. B. Song. "XGBoost-Based Day-Ahead Load Forecasting Algorithm Considering Behind-the-Meter Solar PV Generation." Energies 15(2021).
4. Shi, H., M. Xu, and R. Li. "Deep Learning for Household Load Forecasting – A Novel Pooling Deep RNN." IEEE Transactions on Smart Grid (2017):1-1.
5. Zhang, Y., et al. "Short-Term Residential Load Forecasting Based on LSTM Recurrent Neural Network." IEEE Transactions on Smart Grid (2019).
6. Tsiakmaki, M., et al. "Fuzzy-based active learning for predicting student academic performance using autoML: a step-wise approach." Journal of Computing in Higher Education (2021):1-33.
7. Johnson, S. A., and S. Ananthakumaran. "Smart Digital Forensic Framework for Crime Analysis and Prediction using AutoML." International Journal of Advanced Computer Science and Applications 12.3(2021).
8. Gerassis, S., et al. "AI Approaches to Environmental Impact Assessments (EIAs) in the Mining and Metals Sector Using AutoML and Bayesian Modeling." Applied Sciences 11.17(2021):7914.
9. Jin, H., Song, Q., & Hu, X.. (2019). Auto-Keras: An Efficient Neural Architecture Search System.
10. Zimmer, L., M. Lindauer, and F. Hutter. "Auto-PyTorch Tabular: Multi-Fidelity MetaLearning for Efficient and Robust AutoDL." (2020).
11. Erickson, N., et al. "AutoGluon-Tabular: Robust and Accurate AutoML for Structured Data." (2020).
12. https://www.kaggle.com/c/global-energy-forecasting-competition-2012-load-forecasting (Accessed on April 15th, 2022)

Calculation of Energy Storage Cost and Benefit Based on Units-of-production Method

Shuo Yin, Xing Chen, Zhe Chai, Yao Lu, Xingwu Guo, Man Jin

State Grid Henan Electric Power Company Economic and Technical Research Institute, Zhengzhou, Henan, 450052 China

Abstract: Since 2021, the state has successively issued a series of energy storage development policies. Among them, the guiding opinions on accelerating the development of new energy storage, and the notice on further promoting the participation of new energy storage in the power market and dispatching and utilization have defined the boundaries of the energy storage industry in terms of planning and construction, allocation scale, and income mode. The Henan provincial government issued relevant policies in combination with the actual situation, clarifying the direction for the development of energy storage in the province. In order to analyze the economy of electrochemical energy storage, we use units-of-production method to calculate energy storage cost and benefit.

Keywords: Electrochemical energy storage; cost and benefit analysis; units-of-production method.

1. Introduction

The national development and Reform Commission and the National Energy Administration jointly issued the guiding opinions on accelerating the development of new energy storage, put forward the target of installed capacity, and made it clear that by 2025, the installed capacity of new energy storage will reach more than 30million kW, close to 10 times the current installed capacity of new energy storage. In terms of energy storage allocation, the national development and Reform Commission and the National Energy Administration issued the notice on encouraging renewable energy power generation enterprises to build or purchase peak shaving capacity to increase the scale of grid connection. In terms of energy storage ratio requirements, the peak shaving capacity should be allocated and built according to the linkage ratio of 15% of power (more than 4 hours) at the initial stage of the scale beyond the guaranteed grid connection of power grid enterprises, Priority grid connection shall be given to those built according to the linkage ratio of more than 20%.

Henan Province issued the 14th five year plan for modern energy system and carbon peak and carbon neutralization in Henan Province, which proposed to strive to achieve the installed scale of new energy storage of 2.2 million KW. Second, in terms of energy storage allocation, on June 11, 2021, Henan Provincial Development and Reform Commission and Henan Energy Bureau jointly issued the guiding opinions on accelerating the construction of energy storage facilities in Henan Province, which made it clear that the proportion of energy storage allocation for supporting construction of new energy projects is not less than 10%, and the continuous energy storage duration is more than 2 hours.

2. Energy storage construction cost

lithium iron phosphate batteries are used to calculate the construction cost of energy storage, because lithium batteries have comprehensive advantages in energy conversion efficiency, continuous discharge time, self-discharge rate, service life, cycle times, response speed, heat resistance and energy cost. Among them, lithium iron phosphate batteries are safer than ternary lithium batteries. The National Energy Administration issued the twenty five key requirements for preventing power production accidents (2022 Edition) (Exposure Draft), The ternary lithium battery and sodium sulfur battery are excluded from the medium and large-scale electrochemical energy storage power stations, so the lithium iron phosphate battery is more practical and worth popularizing. Lithium iron phosphate batteries account for 95.5% of the battery shipments of energy storage projects in China's power system.

https://doi.org/10.1051/e3sconf/202235802046

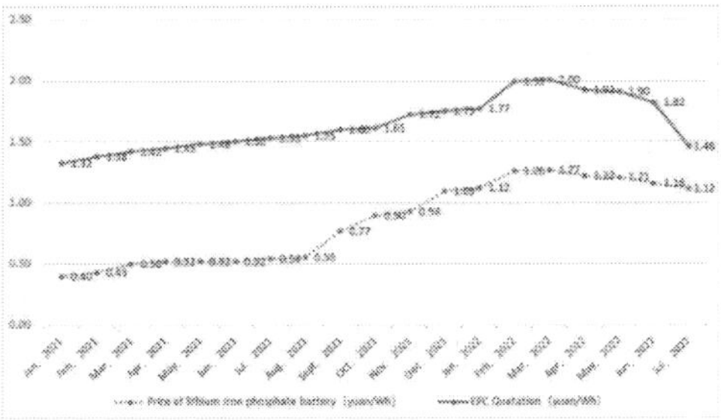

Figure 1 Trends of battery cost

The change of battery cost is the main reason for the change of EPC project contracting cost of energy storage. According to the analysis conclusion of Bloomberg new energy finance, the cost of lithium iron phosphate battery accounts for 55% of the construction cost of energy storage system on average. The cost of lithium iron phosphate battery has increased first and then decreased since 2021. As shown in Figure 1 above, it was the lowest in January 2021, at 0.4 yuan / watt hour, and reached the highest in March 2022, at 1.27 yuan / watt hour, which is highly correlated with the average contracting cost of energy storage EPC project.

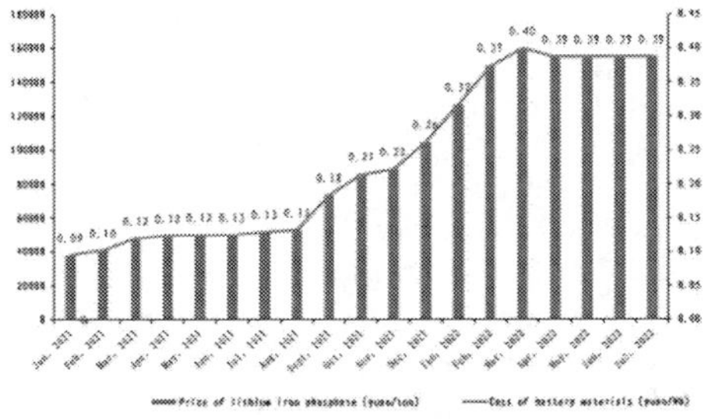

Figure 1 Trends of battery raw material cost

The change of raw materials is the main reason for the price change of lithium iron phosphate energy storage battery. According to the cost analysis of each link in the production of lithium iron phosphate battery, the cost of cathode lithium iron phosphate raw materials accounts for about 20%. From 2021 to now, lithium iron phosphate raw materials have shown an upward and downward trend. As shown in Figure 2 below, the lowest in January 2021 was 37983 yuan / ton, the highest in March 2022 was 159866 yuan / ton, and fell to 155000 yuan / ton in July 2022. The production of 1GWh battery generally requires 2200-2500 tons of lithium iron phosphate. According to this conversion ratio, the raw material cost of lithium iron phosphate in the battery can be obtained. As shown in Figure 2 above, the lowest in January 2021 is 0.09 yuan / watt hour, and the highest in March 2022 is 0.4 yuan / ton, an increase of about 4 times, and decreased to 0.39 yuan / ton in July 2022.

3. Energy storage cost analysis

The operation and maintenance management of the energy storage power station is carried out according to the decentralized layout, centralized operation and maintenance and unattended mode. According to the calculation of material cost of 7.5 yuan / kW and repair cost of 27 yuan / kW, the average annual operation and maintenance cost of 10000 kW / 20000 kwh is 345000 yuan.

Based on the above analysis, we can get the energy storage investment cost boundary of 10000 kW /20000 kwh, as shown in Table 3 below. For the 10000 kW /20000 kwh energy storage project, the self owned capital investment at the beginning of the period is 5.84 million yuan, the loan is 23.36 million yuan, and the annual loan interest is 1.03952 million yuan. The loan principal is repaid at the end of the project operation period. Considering the annual operation and maintenance cost of 345000 yuan, the total cost is 40.27616 million yuan.

Table 1 Calculation results of energy storage investment cost

	EPC capital investment	Loan interest expense	Repayment of loans at maturity	Operation and maintenance cost
Year 1	584	103.952	-	34.5
Year 2	-	103.952	-	34.5
Year 3	-	103.952	-	34.5
Year 4	-	103.952	-	34.5
Year 5	-	103.952	-	34.5
Year 6	-	103.952	-	34.5
Year 7	-	103.952	-	34.5
Year 8	-	103.952	2336	34.5
Total	584	831.616	2336	276

4. Analysis of investment benefit of energy storage

4.1 Reduce peak shaving cost sharing

After the new energy is configured to store energy, it can be charged during the opening period of auxiliary services and discharged during the closing period of auxiliary services, which can reduce the sharing cost of deep peak shaving. The specific calculation method is shown in formula (1):

$$\Delta F_i = \sum_{d=1}^{365} \left\{ Max[\int_{t=dk}^{t=dj} C_t, \overline{M}] \bullet \frac{Y_{di}}{\int_{t=dk}^{t=dj} P_{dit}} \right\} \quad (1)$$

In the formula, it represents the reduction of the deep peak shaving sharing cost within one year after the new energy type I (wind power or photovoltaic) is configured to store energy, D represents the date, T represents the time, DK represents the opening time of the auxiliary service on day D, DJ represents the end time of the auxiliary service on day D, represents the charging power of the stored energy at time t, represents the charging amount of the stored energy during the opening period of the auxiliary service, and represents the maximum capacity of the stored energy, Represents the maximum charging amount of stored energy during the opening period of auxiliary service. Represents the apportionment amount of new energy type I in the auxiliary service market on day D, represents the output of new energy type I at time t on day D, and represents the power generation of new energy type I in the auxiliary service opening period on day D, that is, the integral of new energy output power in the auxiliary service opening period. Divide the penalty of the day by the wind power and photovoltaic power generation when the auxiliary service is turned on to obtain the unit power penalty corresponding to the wind power and photovoltaic power generation when the auxiliary service is turned on. By multiplying the penalty per unit of electricity of the day by the amount of wind power and photovoltaic power generated when the auxiliary service of energy storage reduction is turned on, the peak shaving sharing fee can be reduced on the day after the new energy is configured and stored. By adding up the daily reduced peak shaving sharing costs throughout the year, we can get the annual reduced peak shaving sharing costs after the new energy is allocated and stored.

4.2 Energy storage income from user side investment

Taking the users with industrial and commercial transformer capacity of more than 315 KVA and voltage class of 110 kV as an example, according to the electricity price table of electrical commercial users purchased by State Grid Henan electric power company in July (see attached table 5), the benefits of 10000 kW /20000 kwh energy storage in one day to save power purchase costs are:one point four four × one point zero four four two two four × 2-1.6 × 0.675599-1.6 × 352244 = 136281632 yuan. Taking 300 days a year as the basis for charging and discharging, the income from saving power purchase costs is about 408844896 yuan, and the income in the 8-year investment cycle is 32.7759168 million yuan. Compared with the total investment of 40.27616 million yuan, the cost cannot be recovered. The specific results are shown in Table 2.

Table 2. Energy storage income of industrial and commercial users of various capacity and voltage levels by 1 kW /2 kwh

Classification of electricity consumption		Voltage level	Electricity price			Annual income	Total revenue
			peak	flat	trough		
Industrial, commercial and other power consumption	Power consumption with transformer capacity below 315 KVA	<1	1.109222	0.716999	0.372944	435.20	3481.56
		1-10	1.066047	0.689499	0.359194	417.69	3341.54
		35-110	1.023971	0.662699	0.345794	400.63	3205.07
		>110	0.982052	0.635999	0.332444	383.64	3069.12
	Power consumption with transformer capacity above 315 KVA	<1	1.097604	0.709599	0.369244	430.49	3443.88
		1-10	1.072484	0.693599	0.361244	420.30	3362.41
		35-110	1.044224	0.675599	0.352244	408.84	3270.76
		>110	1.028524	0.665599	0.347244	402.48	3219.84

4.3 Income from residential users' investment in energy storage analysis

According to the current peak valley time of use tariff policy, only one charge and one discharge can be realized. Take the direct supply and wholesale sale of household electricity for one meter residents, less than 1 kV users as an example, 1 kW /2 kwh energy storage can save the cost of purchasing electricity in one day, and the income of the first level of electricity (2160 degrees and below) is:0.1456 yuan， In the second gear, the power (2160 degrees -3120 degrees) is:0.1376 yuan， In the third gear, the power (3120 degrees and above) is:0.0976 yuan.Based on 300 days a year as the basis of charging and discharging, it is charged once a day, with a single charge of 1.6 degrees, and a total charge of 480 degrees. Assuming that the first level of electricity price is adopted, the cost of power purchase can be saved, and the income is 300 × 0.1456=43.68 yuan, with a return of 349.44 yuan in the 8-year investment cycle. Considering that the cost of installing energy storage batteries for residential users is basically the cost of lithium iron phosphate batteries, it was 1.12 yuan / watt hour in July, and the investment cost of 1 kW / 2 kwh storage was 2240 yuan. The income of resident users' investment in energy storage fails to exceed the cost..

5. Conclusions and recommendations

The notice on further improving the time of use tariff mechanism (Exposure Draft) clearly proposes to encourage industrial and commercial power users to reduce the peak power load, increase the low power consumption, and reduce the power consumption cost by changing the power consumption period by allocating energy storage and carrying out comprehensive energy utilization. Taking the users with industrial and commercial transformer capacity of more than 315 KVA and voltage class of 110 kV as an example, according to the electricity price table of electrical commercial users purchased by State Grid Henan electric power company in July (see attached table 5), the benefits of 10000 kW /20000 kwh energy storage in one day to save power purchase costs are:136281632 yuan. Taking 300 days a year as the basis for charging and discharging, the income from saving power purchase costs is about 408844896 yuan, and the income in the 8-year investment cycle is 32.7759168 million yuan. Compared with the total investment of 40.27616 million yuan, the cost cannot be recovered.

In terms of rental income, according to the implementation plan of new energy storage in Henan Province, it is proposed that the reference rental price of centralized shared electrochemical energy storage capacity in 2022 is 260 yuan / kwh per year, the annual income of 10000 kW / 20000 kWh energy storage is 5.2 million yuan, and the income within the eight-year investment cycle is 41.6 million yuan. Relative to the total investment of 40.27616 million yuan, the cost can be recovered.

If it is considered that the independent energy storage entity can obtain an internal rate of return of 6.5% on the basis of recovering the investment principal through leasing services, the 8-year investment cycle needs to obtain income: 4027.616+584 × [1.065^8-1] =44.10133472 yuan. According to the annual average cost of 5.51266684 yuan, the price of leasing services provided by energy storage power stations for new energy enterprises should not be less than 275.63 yuan / kWh • year.

References

1. Thang Ngoc Cong.Progress in electrical energy storage system:A critical review[J].Progress in Natural Science,2009,19(03):291-312.
2. Xuli Chen, Rajib Paul,Liming Dai.Carbon-based supercapacitors for efficient energy storage[J]. National Science Review,2017,4(03):453-489.
3. Giray Gozgor, Chi Keung Marco Lau, Zhou Lu, 2018. Energy consumption and economic growth: New evidence from the OECD countries. Energy, 153(15), 27-34. Doi:https://doi.org/10.1016/j.energy.2018.03.158
4. Kraft, J. and Kraft, A., 1978. On the Relationship between Energy and GNP. Journal of Energy Development, 3, 401-403. Doi: http://www.jstor.org/stable/24806805
5. Krausmann, F., H. Haberl, 2002. The process of industrialization from the perspective of energetic metabolish: Socioeconomic energy flows in Austria 1830-1995. Ecological Economics, 41(2), 177-201. Doi:https://doi.org/10.1016/S0921-8009(02)00032-0

Research on self-balancing transaction scheduling strategy of new energy power in County considering section load rate

Yao Lu [1,*], Meng Yang [1], Shuo Yin [1], Xing Chen [1], Xiaoxi Gu [2], Man Jin [1], Zhe Chai [1], Xingwu Guo [1]

[1] State Grid Henan Economic Research Institute, Zhengzhou, China
[2] Marketing Service Center of State Grid Henan Electric Power Company (Measurement Center), Zhengzhou, China

Abstract: Under the dual carbon target, with the gradual increase of the scale of new energy grid connection, the randomness and volatility of new energy output will further affect the safe and stable operation of the system. In view of the frequent fluctuation of the section load rate of the county external liaison line, this paper proposes a self balancing transaction scheduling strategy of new energy power within the county considering the section load rate, and establishes a day ahead real-time two-tier transaction joint optimization model considering the section load rate. The simulation calculation shows that this strategy can significantly improve the local consumption rate of new energy compared with the simple physical method using only energy storage technology, Reduce the power flow fluctuation of tie line, optimize the section load rate and maintain the stability of power grid.

Keywords: new energy; Section load rate; County; Self balance; scheduling strategy.

1. Introduction

In recent years, in view of global warming, energy conservation and carbon reduction has become the subject of continuous attention and research by experts and scholars in various countries. With the acceleration of carbon peak and carbon neutralization, in 2021, China proposed to build a new power system with new energy as the main body for the first time, and the dominant position of new energy in the future power system was clarified for the first time. With the increasing installed capacity of new energy, large-scale grid connection will bring more security risks to the stability of the system. First, the mismatch between regional new energy output and demand will lead to the abandonment of wind and light; Second, the load rate of the tie line section caused by the residual power passing through the tie line outside the area increases, causing network congestion and increasing the system backup cost and operation risk.This paper focuses on the interior of the county, takes the interior self balancing of the county as the starting point, proposes a new energy power self balancing transaction scheduling strategy within the county considering the cross-section load rate, considers the means of market trading to achieve the optimal balance of the interior new energy power resources within the county, and establishes a joint optimization model of day ahead and real-time two-tier transaction considering the cross-section load rate. The example verification shows that, Compared with the simple physical method using only energy storage technology, this trading strategy can effectively reduce the cross-sectional load rate of county and external connecting lines and improve the internal consumption level of new energy.

2. County self balancing transaction scheduling framework

This paper considers the establishment of a county self balancing transaction scheduling framework including County grid operators, new energy power producers (wind power generation, photovoltaic power generation, battery, biomass power generation), load aggregators (energy storage system, demand response), as shown in Figure 1. According to the design idea of "centralized optimization and decentralized control", promote the smart IOT and multi-source information perception of wind power, photovoltaic, power grid, energy storage, energy consumption equipment, etc., build a county self balancing transaction scheduling framework based on big data cloud computing platform and energy management system, form a coordinated and optimized operation strategy of energy resources within the county, and promote the local and nearby consumption of new energy.

* Corresponding author: hnjyyly@126.com

https://doi.org/10.1051/e3sconf/202235802047

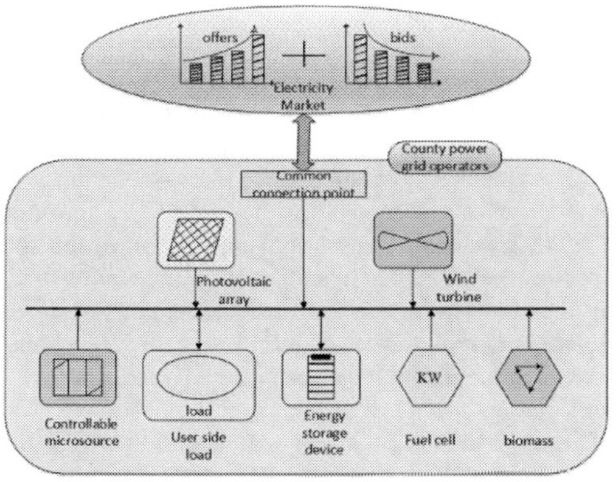

Figure 1. County self balancing transaction scheduling framework

Since the production and use of electric energy have time continuity, the goods sold or purchased are "energy blocks" with a certain duration, as shown in Figure 1. Therefore, only by adding the time dimension of electric energy commodity (power curve) and considering the time dynamic characteristics of power curve can the quality of electric energy commodity be accurately measured (note that there is no contradiction between the homogenization of electric energy in physics and the heterogeneity of electric energy commodity in economics, and the latter is mainly used for commodity pricing). In the trading process of power market, different (horizontal or vertical) "energy blocks" are used to fill the area under the load curve to achieve power and electricity balance. In this process, prices with different connotations will also be formed.

The schematic diagram of the county self balancing transaction scheduling level is shown in Figure 2. The whole county self balancing transaction scheduling system realizes the data access and data processing of various controllable resources such as source, load and storage within the county through the big data cloud computing simulation platform , carries out the modeling of the characteristics of the coordinated operation of source, network, load and storage, visual timing simulation, and provides evaluation services, production services, the evaluation of the adjustable capacity and trading capacity of power grid resources Auxiliary decision-making such as adjustable load application service; By building an energy management system, formulate optimized operation strategies, carry out energy control, and realize the monitoring and regulation of new energy resources within the county.

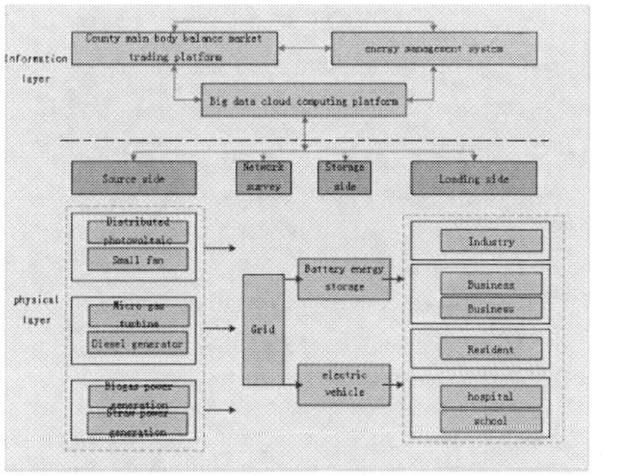

Figure 2. Schematic diagram of county self balanced dispatching level

3. Multi market agents self balancing power transaction scheduling strategy considering section load rate

Based on the cross-section load rate, this paper constructs a two-level joint optimization model including day ahead and real-time transactions. First, based on the cross-section load rate, the day ahead market transaction clearing optimization model is established, and then according to the system operation in the day ahead market to minimize the real-time adjustment, the day ahead real-time two-level transaction joint optimization model considering the cross-section load rate is constructed.

(1)According to the output prediction curve of the scenery, the first level trading strategy takes the economic benefits of the county market entities and the optimal section load rate as the goal, coordinates and schedules the output of the internal market entities, constructs a double objective day ahead trading joint optimization benefit model considering the section load rate, and takes the optimized results as the basis for the second level trading optimization. For convenience, this paper uses the absolute value of the section net flow to represent the section load rate, Its objective function is:

$$\max Y = \max \sum_{t=1}^{T} (R_t^{\mathrm{LD}} + R_t^{\mathrm{MA}} - C_t^{\mathrm{ES}} - M_t^{\mathrm{DG}} - R_t^{\mathrm{DR}}) - \min \left| P_{\mathrm{G},t}^{\mathrm{DA}} - P_{\mathrm{L},t}^{\mathrm{DA}} \right| \quad (1)$$

Where $P_{G,t}^{DA}$ 、 $P_{L,t}^{DA}$ respectively represents the generation power and total load power of new energy units in the whole county at t time before the day, C_t^{ES} 、 M_t^{DG} 、 R_t^{DR} respectively represents the depreciation cost of battery energy storage, the operation and maintenance cost of wind power and photovoltaic units, and the compensation cost of user demand side response, R_t^{MA} 、 R_t^{LD} represents the trading income with the electricity market and the electricity sales income to users respectively.

(2) In order to minimize the error costs such as cost deviation caused by the connection of the two-tier trading strategy model, according to the optimization results of the first tier trading strategy, combined with the day ahead market system to minimize the real-time adjustment of cost deviation, the corresponding objective function is:

$$\min P_{fee,t} = \sum_{j=1}^{4}\left(\Delta M_{t,j}^{DG} + \Delta C_{t,j}^{CG} + \Delta C_{t,j}^{ES} - \Delta R_{t,j}^{LD}\right) - \Delta R_t^{MA} + \Delta P_t^{DR} \quad (2)$$

$$\Delta R_{t,j}^{LD} = \pi_t^{ret}\left(\Delta P_{t,j}^{RT\text{-}LD\sim} + P_t^{RT\text{-}LD}\right)\Delta T \quad (3)$$

$$\Delta R_t^{MA} = 4\pi_t^{RT} P_t^{RT} \Delta T \quad (4)$$

Where:For t period, π_t^{RT} means the real-time market forecast electricity price; π_t^{RT} represents the predicted load deviation at time J (15min); $P_t^{RT\text{-}LD}$ indicates the power reduction of user load in the real-time market. See the notes above for other repeated parameters.

(3) According to the upper level transaction optimization results and the actual deviation data generated by the day ahead real-time market operation, a county entity maximization benefit model considering the real-time market section load rate is constructed, and its objective function is:

$$\max G^{RT} = \max\left[\frac{1}{N}\sum_{\omega=1}^{N}\left(R^{RT}(\omega) - C^{RT}(\omega)\right) - R_{clear}^{DA}\right] - \min\left|P_{G,t}^{RT} - P_{L,t}^{RT}\right| \quad (5)$$

Where: G^{RT} is the maximization utility evaluation function considering the section load rate in the real-time market, $P_{G,t}^{RT}$ 、 $P_{L,t}^{RT}$ are the unit generating power and total load power of the whole county real-time market at time t, $R^{RT}(\omega)$ is the real-time market revenue function, $C^{RT}(\omega)$ is the real-time market cost function, and R_{clear}^{DA} is the minimum adjustment cost of the day ahead market.

4. Conclusion

In this paper, through market means, a new energy power self balancing transaction scheduling strategy within the county considering the section load rate is proposed. Firstly, the self balancing market mechanism and composition structure within the county are introduced, and the day ahead transaction joint optimization model based on system security and stability constraints and the day ahead real-time two-tier transaction joint optimization model considering cross-section load rate are established respectively. Through simulation analysis, it is verified that the scheduling strategy can significantly improve the high proportion of new energy consumption level, achieve high-level local balance of new energy, and reduce grid congestion and fluctuation, Reducing the section load rate of tie line, maintaining the stability of large power grid and reducing power grid investment are of great significance to clarify the transmission and distribution price pricing transmission mechanism and improve the power grid investment capacity.

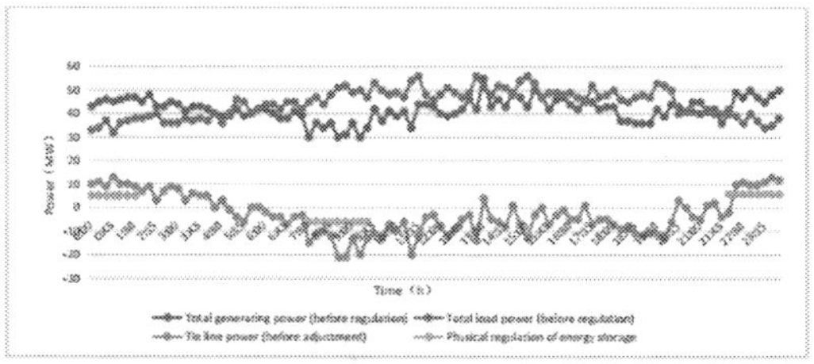

Figure 3. Scheme 1: power curve of tie line before and after optimization

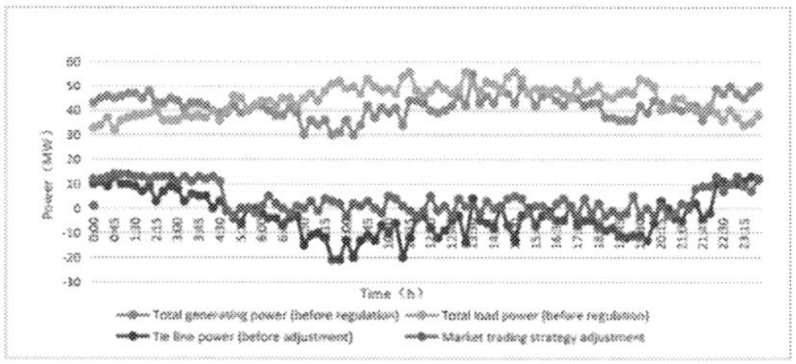

Figure 4. Scheme II: power curve of tie line before and after optimization

Acknowledgments

This work was supported by the Science and Technology Project of SGCC .

References

1. Comprehensive evaluation of shared energy storage towards new energy consumption scenarios under the "double carbon" goal [J]. Power automation equipment, 2021, 41 (10): 244-255.
2. Distribution network optimization and reconfiguration based on BSA under the background

https://doi.org/10.1051/e3sconf/202235802047

of non-cooperative game [J]. Journal of Southwest Jiaotong University: 1-9 [2021-11-29].

3. C Microgrid hybrid energy storage capacity allocation method based on non cooperative game [J]. Global energy Internet, 2021,4 (05): 454-463.
4. Research on substation wireless network resources based on non cooperative game [J]. Computing technology and automation, 2021,40 (03): 173-176.
5. Electricity segmentation model and electricity price scheme design based on transaction curve [J]. China Southern Power Grid technology, 2020,14 (03): 71-78.

Analysis and Prospect of Key Technologies for Crop Straw Recycling

Jianzhi Tuo, Qingsong Zhao and Wei Zhang

School of Mechanical and Electrical Engineering, Weifang Engineering Vocational College, Weifang 261000, China.

Abstract. For the whole process of straw recycling and processing of various general crop harvesters, summarize and analyze the current key technology research in the fields of crop straw recycling and processing technology, foreign matter detection and positioning technology, crop cleaning technology, robot technology, etc., aiming at the efficient recovery of crop straw, the efficient and accurate detection and positioning methods of foreign matters such as stones, soil blocks and metals, the rapid removal of foreign matters. Prospects are made in several key technologies, corresponding technical innovation suggestions are put forward for designers in related fields. It will help to obtain the straw recovery and precise foreign matter detection system that is universal for general agricultural machinery and help the development of efficient and precise agriculture all over the world.

Keywords: Straw recycling; foreign matter detection; crop cleaning; removal of foreign matter.

1. Introduction

With the continuous improvement of residents' living standards and quality, the demand for dairy products and meat such as cattle and sheep is also increasing year by year. High-quality products cannot be separated from efficient and high-quality feeding. The green fodder is the main feed of the animals, the mechanization and high-quality preparation of its harvesting has become a key link in the efficient development of animal husbandry [1]. Silage or yellow silage is one of the most important ways to reserve feed, which can not only save the cost of feed, but also adjust the supply period of forage, solve the problem of seasonal shortage of feed, reduce straw burning, and promote a virtuous circle of agriculture. It is a sustainable development technology [2].

2. Research Status and Problem Analysist

Most straw harvesting is to install straw picking and harvesting device on the original harvester. Metals, stones, clods and other foreign objects in the field enter the header with the straw, which will damage the mechanical transmission mechanism and the fixed cutter, reduce production efficiency, and seriously reduce the production efficiency. The quality of silage has a serious impact on the health of animals [1].

Germany CLAAS Agricultural Machinery Company successfully developed a metal detection device in 2006, which has many functions such as stone, metal and other foreign body detection, automatic knife sharpening, self-adjustment of cutting length and grain crushing, representing the highest technical level and future development of silage harvesters' trend [3]. The metal detection system developed by John Deere Company in the United States, combined with the special roller mechanism design, can quickly disengage the feed roller within 1s to prevent metal foreign objects from entering. The Metal III system of New Holland Company detects and displays the position of foreign objects in real time, the drum reverses, and the metal is taken out [1]. The Menoble uses metal detection sensors from WTK Electronic, Chinese Academy of Agricultural Mechanization Sciences has developed a metal foreign body detection system for green feeders, which must be stopped manually to remove foreign bodies [1]. Most of the above detection methods can only detect metal, even iron foreign objects, have strict requirements on vibration and magnetic fields, and have poor anti-interference ability, requiring shutdown and manual removal, low efficiency, and great potential safety hazard [4].

3. Prospect of Key Technologies for Crop Straw Recycling

Based on the above status and problem analysis, the whole process of straw recycling and processing of general crop harvesters is taken as the core, combined with the key technologies in the fields of crop straw recycling and processing, detection and positioning of adulterated foreign objects, crop cleaning, robots and other fields. The

https://doi.org/10.1051/e3sconf/202235802048

efficient and accurate detection and positioning methods, efficient cleaning and rapid autonomous removal of foreign objects such as stones, clods and metals are prospected in detail.

3.1 Technical Prospect of Crop Straw Recycling System

Aiming at the harvesting principle and steps of the straw recovery mechanism of the crop harvester, it focuses on the technical prospect of the mechanical structure and characteristics, dynamic characteristics and control process related to the processing of silage raw materials such as straw and ear bark.

Taking corn harvesters as an example, most corn harvesters have the function of collecting and crushing straw after picking ears. Research and improve the design of the transmission mechanism after straw collection and before crushing. After the ear is peeled, the ear bark and straw are collected together. For the harvester with the stubble height adjustment function, reduce the stubble height within the allowable range. For the harvester without this function, research and design the stubble left separate harvesting and collection mechanism to improve straw recovery rate. The collected silage raw materials such as straw, ear skin and field stubble after ear picking are concentrated on the foreign body detection and transmission workbench before the straw is crushed. The transfer table must have the functions of coarse vibration filtering, accurate detection and positioning of residual foreign objects, and robot grasping and elimination.

The design of the straw recycling system must be deeply combined with the mechanical structure characteristics of the harvester, and strive to achieve the recycling of valuable straw and ear skins simply and efficiently, so as to achieve seamless connection with the original mechanical parts and highly match the power system of the original machinery, to avoid excessive increase in mechanical structure and power configuration.

3.2 Technical Prospect of Foreign Matter Frequency Conversion Vibration Filtration System

By studying the cleaning principles and cleaning methods of crops such as grains commonly used in general agricultural machinery, we can grasp the obvious difference characteristics of foreign objects such as stones, clods and metals in the straw, and estimate the possible foreign objects and sizes of foreign objects. Combined with the cleaning style of soybean and wheat, design a vibrating filter screen with variable frequency for foreign matter, it is necessary to design appropriate screen holes and structures, and determine key parameters such as the vibration frequency and inclination angle of the screen through experiments to efficiently filter out a certain volume of stones and soil blocks or foreign objects such as metals.

During the working process of the filtration system, the vibration frequency and inclination angle of the sieve can be adjusted in real time according to the characteristics of the foreign body volume and density detected by the detection system, and strive to filter out most foreign bodies through frequency conversion vibration. Through the foreign body vibration filtering system, a certain volume of foreign bodies such as stones, clods or metals are filtered out, and the crop straw enters the next stage of the foreign body detection and removal process.

3.3 Technology Prospect of Residual Foreign Object Accurate Detection and Positioning System

Considering that the detection objects are foreign objects such as stones, clods and metals mixed in crop straw, the location of foreign matter is uncertain, and a single detection method such as metal detection cannot detect foreign objects such as stones and clods. Ordinary image processing technology is not suitable for the detection of foreign matter hidden under the straw or inside the straw. Considering the characteristics of the straw containing foreign matter laid on the transmission table, the methods including laser triangulation, eddy current method, ultrasonic measurement method, machine vision measurement, infrared detection, X-ray detection and other non-contact detection methods are used for detection and positioning. Through experiments to determine the detection parameters of various common foreign objects and straw, and establish a detection feature library, the following three methods can be adopted to complete the accurate measurement of parameters such as the volume and position of residual foreign objects in straw.

3.3.1 Comprehensive detection and positioning method of the transmission workbench

When the detection method is achievable and the cost allows, a comprehensive detection and positioning system covering the entire transfer table is adopted. The detection system will perform full-coverage real-time detection on the straw containing foreign matter laid on the transmission workbench, and feed back the detected value of the volume and position of the foreign matter to the control system in the form of absolute coordinates.

3.3.2 Feed port detection and positioning method

Limited by the detection method or cost control, the detection and positioning method of the feed inlet is adopted to simplify the detection mechanism and production cost. The sensor of the detection system adopts a linear design and installation method, and is installed at a reasonable position near the feed port of the transmission workbench or the feed port. The volume and position detection values of foreign objects are fed back to the control system by relative coordinates. The parameters related to the running displacement of the transmission table must be fed back to the control system in real time, and the control system can obtain the precise parameters of the foreign object through the conversion of coordinates and volume.

For the displacement detection of the workbench, full-closed loop or semi-closed loop feedback can be used. If full closed-loop feedback is used, a displacement detection device such as a grating ruler should be installed in the direction of the parallel transmission table for detection, and the detection accuracy is high. If the semi-closed-loop feedback method is adopted, the uniform running speed and running time of the transmission table can be fed back together in real time, and obtained through conversion by the control system, or a detection device such as an encoder to measure the rotation angle can be installed on the transmission table drive shaft or motor, which is obtained by conversion in combination with the transmission parameters of the worktable.

3.3.3 Mobile scanning detection and positioning method

Install the non-contact detection device at the appropriate position of the robot grasping platform, reasonably plan the running trajectory of the robot, conduct real-time dynamic and comprehensive detection of foreign objects, and cooperate with the robot to achieve efficient and accurate foreign object positioning, grasping and elimination through relative coordinate information.

3.4 Technology Prospect of Foreign Object Removal System

Through the processing of the first few processes such as primary screening and foreign object detection and positioning, combined with the characteristics of the straw containing foreign objects laid on the transfer table and the volume and position information of foreign objects obtained by the detection unit, a robot grasping and elimination system must be designed. It consists of X-Y motion subsystem, robot motion subsystem and grasping subsystem, which are installed directly above the transmission table. The motion of the robot is driven by a three-dimensional motion platform, and its working range effectively covers the entire working area.

3.4.1 X-Y working subsystem

The X-Y working subsystem is installed just above the transmission table, and its mechanical system is driven by a synchronous belt or a rack-and-pinion mechanism, and the drive system is controlled by a high-precision stepper motor or AC servo motor with its own encoder. The secondary position sensor performs safety detection, alarm and emergency stop to ensure the safe and efficient operation of the system. The X-Y working subsystem combines the precise position and volume of the foreign body obtained by the control system to accurately locate the foreign body.

3.4.2 Robot motion subsystem

Combined with the harvester's own power situation and cost control requirements, the mechanical mechanism of the robot motion subsystem can realize the telescopic motion control of the mechanical mechanism through pneumatic, hydraulic or AC servo motors. According to the positioning data of foreign objects, the XY working subsystem efficiently and accurately operates the mechanical gripper of the robot motion subsystem to the effective grasping working range. Limited by the characteristics and space of the mechanical part of the harvester, a two-axis or three-axis robot motion system can be designed according to specific user needs, which is convenient for accurately and efficiently grasping and discharging foreign objects.

3.4.3 Robotic grasping subsystem

Considering the mixing of foreign objects and crop straws, the foreign objects may be exposed on the surface of the straw, or mixed in the middle of the straw or covered by the straw at the bottom of the transmission table. Combined with the volume data of the foreign objects, a special robot grasping mechanism is designed. The mechanism must have practical functions such as crop straw cutting and foreign matter grasping. After the X-Y working subsystem moves the robot to the effective grasping working range, the straw cutting and grasping mechanism of the robot cuts off part of the straw around the foreign object radially under the control of the robot motion subsystem, and grabs it together with the foreign object, and then passes through a special discharge channel discharges the foreign matter and discards it.

4. Summary

Taking the general crop harvester as the research object, he has consulted many Chinese and foreign papers and works on the recycling technology of valuable straw such as crop straw, ear bark and the mainstream detection technology, positioning technology and removal technology of foreign objects such as stones, clods and metals. And research results, in-depth investigation and research on the domestic and foreign crop harvesting machinery market and the mainstream products and technical characteristics of medium and large agricultural machinery enterprises, and accurately grasp the most cutting-edge research directions in the world and the current problems in the field of harvesting machinery. Facing the general crop harvester, the key technical fields such as efficient recovery of crop straw, detection, positioning and cleaning of straw-doped foreign matter are prospected, and some suggestions for technological innovation are put forward.

References

1. Wang Jizhong: Development of Metal Foreign Body Detection System in Harvesting Machine (Master, Chinese Academy of Agricultural Mechanization Sciences, China, 2016). p.1-3.
2. Wang Zequn, Teng Shaomin, Zhao Ming.Development Status of Silage Harvesting Machinery in my country. Chinese Dairy Cow. Vol. 2 (2007) No. 02, p. 43-44.

3. Shao Heyong. Development of Silage Harvester in Domestic and Abroad. Agricultural Engineering. Vol. 7 (2017) No. 04, p. 19-21.
4. Hong Chao, An Guixiang, Pan Mengliang. X-ray machine automatic detection and removal equipment. Plastics Technology and Equipment. Vol. 46 (2020) No. 03, p. 35-39.

Geological recognition and potential analysis of D block

Hongmei Man

Hulunbeier subsidiary of daqing oilfield Co., Ltd, Inner Mongolia, Hulunbeer, 021000, China

Abstract. Under the guidance of petroleum geology, on the basis of in-depth understanding of the geological characteristics of D block such as structure and sedimentation and the law of oil and gas enrichment, this paper discusses the reservoir potential of D block and selects the target block. The study shows that D block is a fault nose structure controlled by reverse normal fault, and the fault has a controlling effect on its provenance, reservoir formation and oil and gas distribution.

Keywords: Structural characteristics of D block; strata sedimentary; fault control.

1. Overview

D block of Huhe noren oilfield is located in beirsumu, new balhu Right Banner, Hulunbuir League, Inner Mongolia Autonomous Region. The structural location belongs to huhe-9 structure of Beixi sub sag and noren structure in Beier Sag of beierhu depression, Hailar Basin, and it is the southward extension of B fault block. The main target horizons are N and X Formation.

2. Geological recognition of D block

2.1 Structural characteristics

The Hohhot noren structural belt is located in the southwest of Beier Sag and distributed in NE direction. It is a fault nose structure developed under the control of the main control normal fault F1 extending in nee direction and falling to the West. The Nantun Formation was a complete structural belt during the sedimentary period, and the main ontrol normal fault F2 extending NE in the later period cut the structure into two blocks D and B.The fault offsets of both F1 and F2 faults increase from southwest to northeast.

All faults in D fault block are normal faults, and most of them are reverse faults. Among them, the main control fault F1 is a reverse growth fault extending to the basement, the main body is shovel type, gradually changing to a seat type with steep upper and lower parts and gentle middle in the East, with large scale and long development time. The extension length in the study area is 8.9km, the fault displacement gradually decreases from west to East, with a maximum of 400m, strike nee, NW tendency and an inclination of 51°; The extension length of fault F2 in the study area is 4.4km, the maximum fault displacement is 60m, the strike is ne, the dip angle is 55°, and the dip is NW.

Secondary small faults are developed in the area, mainly in NE direction, which are normal faults, most of which are reverse faults, with poor inheritance; On the plane of Nantun Formation, it mainly intersects with the main control fault in an oblique way, and on the section, it presents a ladder like, Y-shaped or anti-y-shaped combination; The zhixing'anling formation is mainly parallel to the main control fault in plane.

2.2 Stratum sedimentation

The basement of Hailar Basin is Paleozoic and pre Paleozoic strata, and its sedimentary cover is Mesozoic Cretaceous and Cenozoic Neogene and Quaternary strata. The maximum thickness of sedimentary rock is about 6000m, and the main oil bearing series is Lower Cretaceous strata. At present, the sedimentary strata of Hailar Basin are divided into 6 groups and 10 lithologic sections.The main target horizons are N and X Formation.

N Formation in D3 block of Huhe Noren oil field is mainly fan delta facies deposit in nearshore shallow water environment. Deposits formed by the traction hydrodynamic mechanism are more common, and the proportion of deposits formed by the gravity hydrodynamic mechanism is relatively small, which indicates that the hydrodynamic conditions in this area have both traction flow and gravity flow, and the occurrence of deposits is the result of the combined action of the two hydrodynamic forces, of which the traction hydrodynamic force occupies a dominant position.

There are two sedimentary cycles developed from top to bottom in N Formation of D block. According to the characteristics of sedimentary cycles, two oil-bearing groups I and II are divided. Oil layer group I is mainly composed of thick sandy conglomerate, mudstone and transitional lithology, and the bottom is divided by low resistivity and high gamma thin sand mudstone interbedding and the middle oil layer group; The resistivity curves of II oil layer group are mostly wide, slow and sharp peaks, and the natural gamma is slightly higher than that of the upper oil layer group. The horizontal distribution of strata in D block is characterized by gradual thinning from east to west and annihilation. Reservoir group II is annihilated in the west of D3 well block, and reservoir group I is missing near D8 well block. The N Formation oil layer of D block is developed with glutenite, with a thickness of 39.2 ~ 61.5m and an average thickness of 52.3m. The plane distribution of glutenite is controlled by the distribution of sedimentary fans and topography. The thickness of sandy conglomerate is large in the main body of fan delta and the thick part of stratum sedimentation. On the whole, the thickness of sandy conglomerate gradually thins from north to south. Sandstone in the East and west well areas is mainly developed, and sandstone in well bay D7 of oil formation II. The sandstone of oil formation II is more developed than that of oil formation I, with good continuity and large single-layer thickness.

Table 1 Sandstone Data Statistics of N Formation in D Block

Well	Sandstone (m)	
	Groups I	Groups II
D3	18.4	20.4
D2	5.4	23.3
D4	20.6	36.8
D6	8.8	27.3
D8	26.4	33.8
D5	16.8	31
D6	24.2	12.2
D65	27	21.2
D67	23	32.2
D73	19.2	38.4
D7	15	58.2
D62	\	16.6
D64	8.4	19
D66	9.6	51.6
D68	3.4	32
D72	55.6	62
D74	33.4	19
D65	7.8	38
Average	19	31.8

Table 2 Statistics of Effective Sandstone Data of N Formation in D Block

Well	Effective sandstone (m)	
	Groups I	Groups II
D3	4.2	15.4
D2	1.1	9.4
D4	6.5	16.5
D6	2	18.3
D8	8.6	9.4
D5	1.8	0
D6	\	8.4
D65	6.2	15.6
D67	9	9.5
D73	2.7	5.6
D7	8.5	3.4
D62	\	11.8
D64	6.8	16.3
D66	2.1	40.5
D68	\	21.6
D72	6.9	0
D74	9.4	0
D65	\	24
Average	5.4	12.5

The reservoir lithology of X formation in D block is mainly basalt and pyroclastic rock, which is thick layered, and the thickness of basalt stratum is relatively stable. According to the research on the volcanic rock reservoir in X, Hohhot Norren area, basalt is of volcanic fissure eruption origin. The magma overflows the surface along the fault and covers the unconformity surface of the lower stratum, forming a flood basalt sheet with a wide distribution, relatively gentle occurrence, relatively stable thickness and composition. Volcanic clastic rocks are developed on the flanks and local top surfaces of the lava sheet. Lithofacies are vertically divided into three parts: overflow facies dense basalt, rock flow self clastic rock and pyroclastic sedimentary rock.A total of 24 wells in D block are drilled into the X formation, and sandstone is developed in all the wells. Due to the influence of drilling depth, the thickness of sandstone varies greatly, with the maximum thickness of 124m, the thinnest only 4m, and the average thickness of 41.6m. According to the seismic attribute analysis, sandstone is generally developed in D block, and the thickest is developed in D8 well block.

2.3 Reservoir type

During the sedimentation period of the N Formation of the Beixi sub sag, fan delta deposits of east and west provenances are developed. The well area is generally located in the fan delta front facies belt, and the underwater distributary channel sand bodies of multi-phase front are developed. The sand bodies are distributed in a wide range and are staggered, providing a good reservoir space for oil and gas accumulation.

In the early stage of the Bexi sub sag, NEE trending faults controlled the formation and distribution of structural belts, while NE trending faults controlled the formation of

https://doi.org/10.1051/e3sconf/202235802049

local structures. The distribution of oil and gas is controlled by NEE and NE trending faults, and it is most abundant at the structural highs. The lithology in the area near the trough also controls the reservoir to a certain extent.

There are three types of oil-water distribution in the vertical direction of N Formation in well block D3: oil layer dry layer water layer, oil layer water layer and oil layer dry layer. The above oil is mainly distributed in the lower dry layer. The oil reservoirs of N Formation are mainly glutenite reservoirs controlled by reverse normal faults, which are lithologic structural reservoirs mainly controlled by structures. The logging curve of high-yield wells in N Formation is box shaped, with a resistance of more than 15 Ω. m and a density of less than 2.4 g/cm^3, which is a typical underwater distributary channel feature. High yield wells are distributed at the structural high point of the pinch out line of the 15 and 111 layers in the II oil formation of the South Second Member. For example, the effective thickness of D66 drilling is 42.6m, and only 5.9m is effective for shooting. At the initial stage, the daily oil salvage is 3.3t, and the cumulative oil production is 4660t.

X oil layer in D block is mainly distributed in upper oil and lower dry, with local upper oil and lower water. Oil and gas are mainly concentrated in the structural highs near the two overlapping lines that are not developed in the strata of N Formation. Because of long-term exposure to the surface, weathering, and local holes and fractures, it is an unconformity reservoir controlled by unconformity. Oil testing and production of high-yield wells are all in Well Block D8. Well Blocks D12 and D3 have overlying formation of the second section of the south, and the fractures and caves at the top of the X Mountains are poorly developed. Oil and gas are enriched and high-yield only near local high point faults. For example, the fault on the west side of D68 controls the formation of local structural high point oil and gas enrichment, which is put into production by conventional shooting. At the initial stage, the daily oil production is 6.3t, and the cumulative oil production is 7325t. As the formation in Well Block D8 has been exposed to the surface for a long time, fractures and caves are developed with high production, and fractures and caves are mainly distributed in the structural highs controlled by the overlap line of the upper II oil formation.

3. Fault control

According to the sand ground ratio and large fault superposition diagram of the South second section of the exploration and evaluation well in Hohhot noren oilfield, the original entrance of block B is the intersection of large faults, and the growing large fault from the north to the West controls the raw material supply of Hohhot noren oilfield.

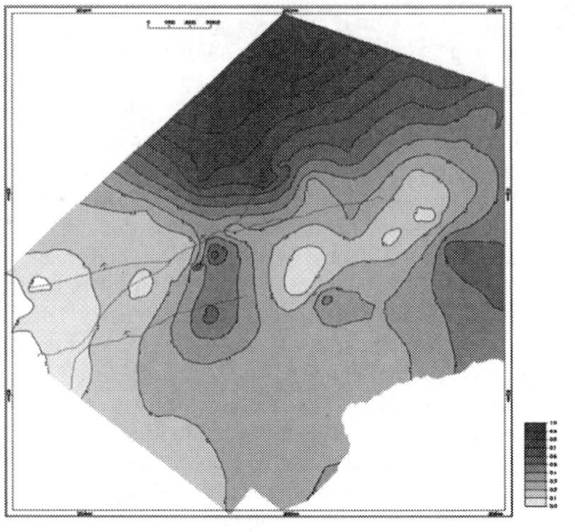

Fig. 1 Sand ground raio and fault superposition diagram of Hohhot noren Oilfield

B block and D block in Hohhot noren oilfield are lithologic structural reservoirs controlled by the same boundary fault. The oil layer of N Formation in D block is vertically divided into two sets of oil-water systems. Oil group I is oil layer dry layer, and oil group II is oil layer water layer dry layer. The oil bottom elevation is -567.8~-785.53m, and the oil bottom elevation of well D7 is the lowest.The high-yield well in well D3 area is located at the high part of the fault structure,Well D6 is located at the rising wall of graben.

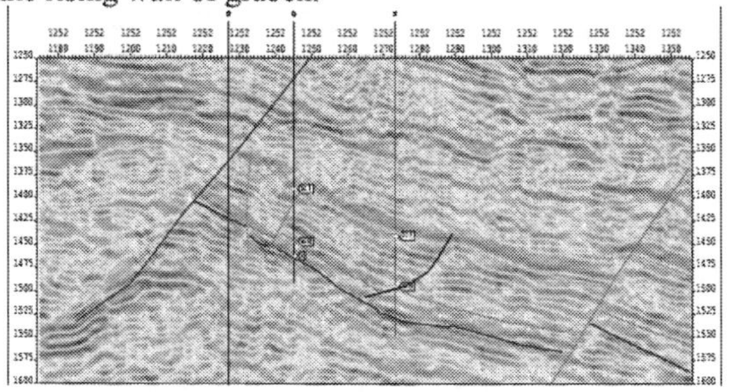

Fig. 2. Seismic profile of well D6

4. Potential analysis

By analyzing the oil and gas distribution, structural location and sedimentary facies development in the high-yield well area of D block, it is found that the high-yield wells are distributed in the high part of the structure and the sandstone is developed; The well logging of X formation in well D3 area shows no perforation with high-yield oil flow. According to the above characteristics, it is found that the sand body of X formation is developed according to the logging interpretation of well D5 in the old well, and it is recommended to shoot the sandstone of X formation in well D5; A potential block is selected in the north of well D3, which is located on the Horst composed of boundary faults and internal small faults.

5. Conclusion and understanding

Block D and block B are separated by faults, which are fault nose structures under the common control of two boundary faults; The main control fault F1 is a reverse growth normal fault with large scale and long activity time. Its main body extends to the basement in a shovel type and turns into a seat fault with steep upper and lower parts and gentle middle in the East; Small faults are developed in the block, all of which are normal faults, most of which are reverse faults, with small fault displacement and poor inheritance.

Two oil formations I and II are developed in N Formation of D block, and the bottom stratum is missing; From east to west, the sedimentary thickness of the strata gradually thinned to extinction. Sandstone of X formation is widely developed, among which well bay D is the most developed.

Faults have strong control over Hohhot noren Oilfield: regional large faults control the material supply of Hohhot noren oilfield; Boundary faults control the formation of lithologic structural reservoirs in D and B blocks; The local microstructure controls the distribution of oil and gas in D3 well area, and the high-yield well is located in the rising wall of the fault.

The area with good seismic attributes and the high tectonic position of the fault rising wall is the lower potential tapping area.

References

1. Ying Danlin, Pan Mao, Lizhongquan,et al.Reverse faults in Wuerxun depression and their relationship with oil and gas.Journal of Chengdu University of technology.Vol.39 (2012) No.3, p.244-250.
2. Wang Peng, Wang Weifeng, Huang Longwei,et al.The relationship between the formation mechanism of reverse normal fault and hydrocarbon accumulation in Huqing oilfield.Fault block oil and gas field.Vol. 11 (2004) No.3, p.7-9.
3. Wu Yinye, Zhangtianshu, Zhangzhijie, et al.Types and characteristics of sedimentary system tracts and their petroleum geological significance.Journal of paleogeography.Vol. 12 (2010) No. 1, p.69-81.
4. Xu Xiansheng,Wang Jianxin,Yang Ting,et al.Characteristics of sublacustrine fan and favorable reservoir facies in Nantun Formation, Beier Depression.Xinjiang Petroleum Geology.Vol.31(2010),No.4,p.361-364.
5. Pang Lin'an,Wang Zhixin,Zhu Guanghui,et al.Analysis of Petroleum Geological Characteristics and Exploration Potential in Bonaparte Basin.Xinjiang Petroleum Geology.Vol.31(2010),No.4,p.444-447.
6. Li Songmao, Xiao Ping, Tang Dazhen,et al.Evaluation of coal formed gas resource potential in Huhehu sag, Hailar basin.Geology of China.Vol.36(2009),No.6,p.1350-1358.
7. Liu Jiying, Sun Yonghong, Huang Chunyan,et al.Analytical technique of nitrogen compounds in crude oil and its application in the study of oil and gas migration.China's high-tech enterprises. (2009), No.6,p.34-36.
8. Wang Jialiang,Wang Yunzeng,Cao Zhengwei,et al.Laboratory Feasibility Study on Water Injection Development of Nantun Formation Strong Water Sensitive Reservoir in Bei301 Block, Hailar Basin.Inner Mongolia Petrochemical Industry. Vol.29(2002),p.89-91.
9. Liu Jinlong.Study on Sedimentary Microfacies of Nantun Formation in Bei301 Block, Beier Sag, Hailar Basin.Inner Mongolia Petrochemical Industry. (2010),No.5,p.124-125.
10. Lv Ming.Study on Main Controlling Factors of Hydrocarbon Accumulation in Hailar Basin, Luming.Journal of shijiazhuang vocational and technical college.Vol.22 (2010), No.2, p.40-42.
11. Wu Ying,Chen Junliang,Zhang Ying.Hailar_ Relationship between Tectonic Belt and Oil and Gas in Tarim Chage Basin.Journal of Daqing Petroleum Institute.Vol.33 (2009), No.3, p.31-35.
12. Wu Kaisong, Luo Jinjun, Zhang Qingsheng,et al.Study on the sleeve deformation law of rock salt layer under earthquake. Geotechnical Mechanics.Vol.30 (2009), Supp. ,p.231-234.
13. Meng Lixin, Ren Baosheng, Ju Binshan,et al.Study on water drive law of fault block reservoir in high water cut stage.Journal of Southwest Petroleum University (Natural Science Edition).Vol.32 (2010), No.6 ,p.139-142.

Optimization of Solar Cells Based on Perovskite Materials

Liufeige Lu

Suzhou Foreign Language School, Suzhou, China

Abstract. With the rapid advancement of economy, human's demand for energy continues to grow, and energy shortage has become the primary problem hindering further economic development. Therefore, changing the existing energy structure and developing and utilizing sustainable clean energy are the research directions of all countries in the world. Perovskite solar cells have rapidly become a research hotspot in the field of solar cells worldwide in recent years due to their remarkable advantages such as low manufacturing cost and high efficiency. However, PSC (Perovskite solar cell) have many problems in the stability, reproducibility and performance evaluation of high-efficiency battery devices. By introducing the structure and performance of perovskite, this paper summarizes the research progress of solar cells based on this kind of materials, analyzes its working mechanism, summarizes the key issues affecting the performance of PSC, points out the direction of efforts to further improve the performance of PSC, and looks forward to the optimization development of PSC.

Keywords: Perovskite, Material, Solar cell, Optimization

1. Introduction

With the growth of the world's population, the development of industry and the acceleration of urbanization, energy consumption in the world is increasing, and environmental pollution is becoming increasingly serious [1]. It is urgent to replace traditional energy with clean renewable energy. However, how to use these energy sources efficiently and environmentally is still a challenge in theory and technology. Solar energy has attracted much attention because of its extensive distribution and inexhaustible use [2]. Over the years, various researchers have devoted themselves to the research and development of various types of solar cells, trying to develop solar cells with low cost, high efficiency, no pollution and easy industrialization [3]. PSC represent the third generation of solar cells. Because of their high photoelectric conversion efficiency and low production cost, PSC have been comprehensively studied as cheaper alternatives to silicon solar cells [4]. Moreover, it combines the advantages of high power factor, cost-effectiveness and processability of solutions, providing power for exploring large-scale applications [5]. Compared with traditional batteries, perovskite batteries have many advantages, such as high open circuit voltage, high efficiency, simple preparation process and low material cost [6]. However, the poor environmental stability of perovskite materials has become one of the main obstacles to the practical application of perovskite batteries [7]. When the battery is placed under high temperature, high humidity, oxygen and ultraviolet light for a long time, it will cause irreversible damage to the perovskite layer, resulting in serious performance degradation [8].

2. Development status

As a light absorption layer, perovskite layer is a main component of PSC, where photogenerated electrons and holes are generated, and plays a vital role in the photovoltaic performance of cells [9]. Perovskite light absorbing layers widely used at present contain lead, and sometimes toxic organic solvents are used in the preparation process [10]. The stability of PSC is greatly affected by ultraviolet light, water, heat and organic molecules. Perovskite materials with different structures and properties can be obtained by element replacement of perovskite materials [11]. The stability of the materials also changes when the optical properties change. Perovskite materials have many advantages, such as bipolar charge transfer, long electron / hole diffusion length and adjustable band gap, which show very good photoelectric performance after being applied in the field of solar photovoltaic. PSC have been developed rapidly. The power conversion efficiency of this battery has increased from 1.5% to 18.4% in the past 10 years, as shown in Figure 1, surpassing the dye-sensitized solar cells and organic solar cells that have been developed for many years.

Fig. 1 Development history of conversion rate of PSC

With the advent of PSC, articles on improving the photoelectric conversion efficiency of PSC continue to appear, while the research on the stability of PSC is relatively lagging, and the stability of PSC has become a bottleneck restricting its further development. PSC still face many problems if they want to truly realize industrialization and market product production:

(1) Perovskite materials currently used have problems of decomposition in air and dissolution in water and organic solvents, resulting in short device life. Therefore, it is necessary to develop battery materials that are stable to air and water or solve this problem from packaging technology.

(2) Commercial device development. Because large-area thin films are difficult to maintain uniformity, the working area of the reported high-efficiency PSC is only about 0.15 cm, which is still a long way from practicality. Therefore, it is necessary to develop perovskite solar cell device preparation technology with stable performance from the scale of laboratory square centimeter to the scale of application square meter.

3. Working principle and structure of perovskite solar cells

The research on PSC mainly focuses on two kinds of structures. The first is a mesoscopic solar cell based on porous structure, which is evolved from dye-sensitized solar cells. The second is a thin film solar cell based on a planar heterojunction structure, in which perovskite is both a light absorption layer and a transport layer of electrons and holes. Perovskite materials will undergo the following basic chemical reactions:

$$PbI_2(s)+CH_3NH_2(\mathrm{a}q)\leftrightarrow CH_3NH_2PbI_3(s) \quad (1)$$

The forward reaction is the synthesis reaction of perovskite material, and the reverse reaction is the decomposition reaction of the solar material. Environmental factors also have certain influence on the stability of perovskite materials. The environmental conditions that affect the stability of PSC include: water and oxygen environment, temperature change, device processing conditions, illumination conditions, etc. At present, the research on the stability of PSC is still in the primary stage of development. In order to further expand the optical absorption spectrum of perovskite materials and increase the utilization of near infrared light to generate higher photocurrent, recently developed new titanium-type solar materials have smaller band gap, good thermal stability and photoelectric conversion performance.

The basic structure of PSC mainly includes a light absorption layer composed of perovskite materials, an electron transport layer / hole barrier layer, a hole transport layer / electron barrier layer, and a working electrode. The basic photoelectric conversion process is that under illumination, photons with energy greater than the band gap of the light absorption layer excite valence band electrons in the light absorption layer to the conduction band, and leave holes in the valence band. Holes are transported to the cathode and the external circuit. In addition to the above photoelectric energy conversion process, there are also some energy loss processes. There are charge recombination centers at the interfaces on both sides of the light absorption layer, resulting in unnecessary charge and energy loss. Improving these energy loss problems can effectively improve the efficiency of the device.

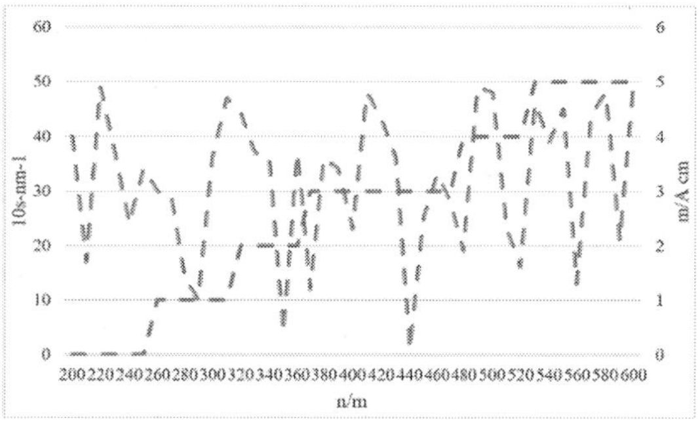

Fig. 2 The relationship between the theoretical photocurrent of the solar cell and the cut-off wavelength of the absorption spectrum of the light absorbing layer

The scheme used for the electron transport layer in the perovskite solar cell in the initial study is based on the research results of the dye-sensitized solar cell, that is, the porous layer composed of TiO2 nanocrystals, whose role is to accept the conduction band electrons in the photogenerated excited state in the perovskite light absorption layer and transport them to the transparent conductive electrode. The hole blocking layer uses a dense layer of TiO2, because its valence band energy level is far lower than the valence band top of the perovskite light absorption layer, which can effectively prevent hole injection (see Fig. 2). Changing the concentration of perovskite precursor additive and the preheating temperature of perovskite film during the preparation process will only cause changes in the surface morphology of the film, but will not change the perovskite phase structure. Changes in the surface morphology will affect the photoelectric performance of the perovskite film and improve the light absorption, charge transport and charge collection process of the light absorption layer.

4. Optimization development trend of solar cells

With the continuous deepening of the research on PSC, people are increasingly eager to find out the internal reasons behind the excellent photoelectric performance of PSC, so as to achieve further optimization and even industrial trials. The key to prepare excellent performance solar cells is to improve the effective light trapping ability of perovskite materials to obtain long-life excitons with small thermodynamic loss, which is the fundamental source of high voltage. In terms of improving the optical performance of perovskite, many research groups have established optical models of perovskite solar cell devices to calculate the EQE of different devices, so as to analyze the optical loss in the devices and optimize the devices. Among these models, the simulation of external quantum efficiency based on optical admittance can provide an important basis for the interpretation of many external quantum efficiency spectra, short-circuit current density JSC and the optimization of device performance. So far, new materials and technologies such as new structures, new perovskite materials, new hole transport layers, new electron transport layers, solution engineering and energy band engineering have achieved lead-free, flexible and large-area PSC. Obviously, this is not the end point. By further increasing the electron hole transport rate, matching the energy levels of each layer, reducing the resistance of the transport layer and other measures, perovskite cells will certainly achieve more efficient photoelectric conversion efficiency. Its development trend is as follows:

(1) In order to fundamentally improve the stability of perovskite materials and realize large-scale commercial production, it is necessary to find more stable new materials (perovskite absorption layer materials and hole transport layer materials), which are not only efficient and stable, but also low-cost and non-toxic.

(2) By changing the additive concentration of perovskite precursor and the preheating temperature of perovskite in the preparation process, combined with the volt ampere characteristic curve, ultraviolet visible absorption and steady-state fluorescence, it was found that the perovskite phase of Perovskite Thin Films did not change when the additive concentration and preheating temperature were changed, but the thin films with the additive concentration of 0.69 M and the preheating temperature of 100 °C were compact and flat without holes, and the perovskite had the most photogenerated electrons.

According to the current research results, PSC with high efficiency are promising photovoltaic converter devices. After comprehensive optimization, PSC are expected to achieve large-scale applications. At present, researchers have carried out research on all solid-state PSC from different perspectives such as the preparation method of materials, the development of new perovskite materials and hole transport materials, the selection of substrate materials, and the study of working mechanism, which is expected to achieve efficient and clean utilization of solar energy.

5. Conclusions

The rapid development of PSC has brought a ray of light to solve the future energy problem. However, PSC in the reproducibility, stability and performance evaluation of high-efficiency battery devices. The stability of solar cell materials and devices directly limits the application and development of PSC. Therefore, improving the stability of PSC is an urgent problem for the application and development of subsequent battery devices. At the same time, the battery packaging process needs to be further improved. So far, the main components of all high-performance PSC contain toxic lead, so toxicity problems will occur in the process of device manufacturing and disposal, which is a very unfavorable factor for the commercialization of PSC in the future. Therefore, the future research focus in the field of PSC also needs to explore the preparation of high-efficiency lead-free PSC. In short, in order to achieve large-scale solar power generation as soon as possible to solve the energy and environmental problems worldwide, it is a very urgent scientific and engineering problem to develop low-cost and high stability solar cells.

References

1. Liu Cheng, Shen Luying, Xu Zhengyu, et al. Research progress of perovskite solar cells [J]. Progress in Chemical Industry, 2017, 33(12):7.
2. Yao Xin, Ding Yanli, Zhang Xiaodan, et al. Review of perovskite solar cells [J]. Acta Physica Sinica, 2018(3):8.
3. Cui Xinghua, Xu Qiaojing, Shi Biao, Hou Fuhua, Zhao Ying, Zhang Xiaodan. Research progress of wide-bandgap perovskite materials and solar cells [J]. Acta Physica Sinica, 2020, 69(20):21.
4. Ke Xiaolong, Wang Xiaoping, Wang Lijun, et al. Key issues in the development of perovskite solar cells [J]. Functional Materials, 2017, 47(12):6.
5. Wang Yanxiang, Luo Jun, Guo Pingchun, et al. Application and development of hybrid perovskite materials in solar cells [J]. Journal of Inorganic Materials, 2019, 30(7):10.
6. Wei Jing, Zhao Qing, Li Heng, et al. Perovskite Solar Cells: New Hope in Photovoltaic Field [J]. Science in China: Technological Science, 2019, 44(8):801-821.
7. Zhang Bowen. Preparation and optimization of carbon-based perovskite solar cells with high stability [J]. Jilin University, 2019(7):039.
8. Wang Nan, Liang Zhurong, Wang Junxia, et al. Optimization of the fabrication process of perovskite sensitized solar cells [J]. Progress in New Energy, 2018, 3(6):6.
9. Dong Qiao. Structural design and optimization of perovskite solar cell materials [J]. North China Electric Power University (Beijing) North China Electric Power University, 2017(9):141.
10. Qian Liu, Ding Liming. The working mechanism of perovskite solar cells and the main factors affecting

 https://doi.org/10.1051/e3sconf/202235802050

the performance [J]. Journal of Chemistry in Universities, 2019, 36(4):13.

11. Li Xinli, Li Lihua, Huang Jinliang, et al. Research progress of novel perovskite solar cells [J]. Modern Chemical Industry, 2018(9):5.

 https://doi.org/10.1051/e3sconf/202235802051

Electrochemical Preparation of Novel Perovskite Solar Cells

Yiping Sun*

Guangdong University of Technology, Guangzhou, China

Abstract. With the development of human society, the shortage of traditional fossil fuels has intensified, and due to sustainable development and ecological environment, new green and pollution-free energy has entered the public's field of vision. Thin-film solar cells have high photoelectric conversion efficiency, stable performance and low consumption of raw materials. However, some elements in raw materials are toxic and their reserves are small, which limits their large-scale popularization and use. Perovskite solar cells have attracted public attention because of their low cost and simple production process. The traditional energy content is limited, and the demand for energy is increasing with the development of the world economy, so the world is bound to face the dilemma that single fossil energy cannot meet the demand for energy for economic maintenance and development. As a clean energy, extensive use of solar energy is also an effective solution to the pollution caused by traditional energy utilization. This paper mainly analyzes the electrochemical preparation methods of perovskite solar cells, and studies its performance optimization.

Key words: Perovskite solar cell; Electrochemical preparation; Manufacturing technique

1. Introduction

Energy and environment are two major issues of great concern to the whole society today, which are of great significance to the sustainable development of mankind. In the 21st century, the disadvantages of fossil fuels are gradually revealed, the problems of energy crisis have surfaced, and environmental problems have followed. It is extremely important to find renewable and environmentally friendly new energy, among which solar energy is widely used because of its advantages of cleanness and large energy release [1]. Photovoltaic, as a forward-looking and strategic new energy technology to support the sustainable development of China's national economy, is the key support and priority development direction in the national long-term scientific and technological development plan [2]. Low material cost, simple preparation process and high photoelectric conversion efficiency are the key factors for solar cells to market and realize industrialization. Solar energy is one of the important new energy sources, and the medium of converting solar energy into electric energy is solar cell [3]. In recent years, solar cells made of perovskite have been widely welcomed. Thin-film solar cells have high photoelectric conversion efficiency, stable performance and low consumption of raw materials. However, some elements in raw materials are toxic, and their reserves are small, which limits their large-scale popularization and use [4]. Perovskite solar cells have attracted public attention because of their low cost and simple production process.

Solar energy is a kind of renewable energy with abundant reserves. Inside the sun, hydrogen is converted into helium by nuclear fusion. During the reaction, energy passes through the universe and the earth's atmosphere in the form of electromagnetic waves, reaching the earth's surface [5]. Collecting and using this energy will create a huge wealth for mankind. This paper mainly analyzes the electrochemical preparation methods of perovskite solar cells, and studies its performance optimization.

2. Preparation of perovskite solar cells

2.1 Electronic transport layer

Solar energy is an important new type of energy. In this paper, a new perovskite solar cell is studied. When preparing perovskite solar cells, the emphasis is on the fabrication of electron transport layer, dense layer, skeleton layer and light absorption layer. The electron transport layer is one of the important structures in perovskite solar cells. The electron transport layer extracts the electrons from the photoelectric conversion layer for the next further photoelectric reaction. At the same time, it can prevent the hole injection reaction in the photoelectric conversion layer, which will reduce the density of photoelectrons and affect the conversion efficiency of solar cells [6]. The energy level structure and working principle of the battery are shown in Figure 1.

* Corresponding author: sun_yiping@foxmail.com

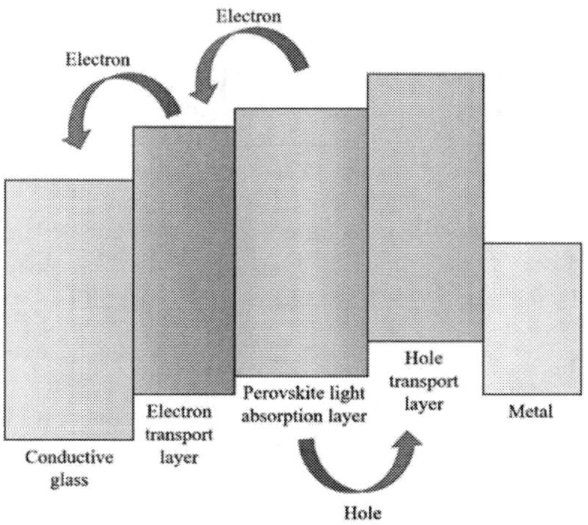

Figure 1 Battery energy level structure and working principle

Under illumination, excitons generated by photons in the goethite body can be separated at the interface of electron transport layer/perovskite layer/hole transport layer. Electrons are collected by FTO through electron transport layer, and holes are collected by metal electrode through hole transport layer. The barrier layer can effectively prevent the recombination of electrons and holes collected by FTO. The material of the electron transport layer is required to be high. First of all, it is necessary to ensure that the material can match the energy level of perovskite, so as to ensure that the electron extraction reaction from the photoelectric conversion layer can proceed smoothly. Each layer of the calcium iron ore solar cell has different energy band structures. The holes are transported from the goethite layer to the hole transport layer, and finally collected by the metal electrode, and FTO is connected with the metal electrode to form a loop.

2.2 Compact layer

Low material cost, simple preparation process and high photoelectric conversion efficiency are the key factors for solar cells to market and realize industrialization. Solar energy is one of the important new energy sources, and the medium of converting solar energy into electric energy is solar cell. In perovskite solar cells, the dense layer can conduct photogenerated electrons, which can ensure the smooth progress of photoelectric reaction by transmitting photogenerated electrons to the conductive substrate of the solar cell. The transparent FTO conductive glass was etched with zinc powder and hydrochloric acid to form a 1mm wide slit, resulting in FTO open circuit on both sides of the slit. After that, in order to prevent the adhesion on the surface of conductive glass from affecting the performance of solar cells, the etched FTO was cleaned [7]. The process of cutting and cleaning conductive glass is shown in Figure 2.

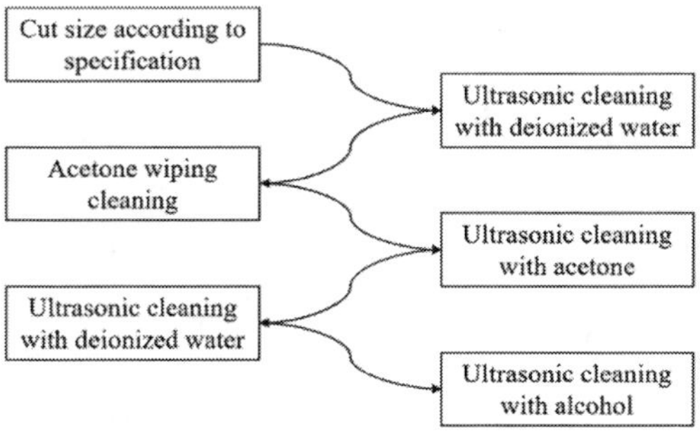

Figure 2 Conductive glass cutting and cleaning process

The dense layer can isolate the photoelectric conversion layer from the conductive substrate, and prevent photo-generated electrons from reacting with holes in the photoelectric conversion layer. Among several main materials for preparing dense layer, ZnO is a well-balanced material. The preparation process of ZnO is simple, the requirement for temperature is low, and it has a good conversion rate. Therefore, ZnO dense layer is also regarded as a good substitute for TiO2. When the dense layer is too thick, it is not conducive to the effective conduction of electrons while preventing holes from contacting FTO glass. However, when the dense layer is too thin, it can't play a good role in blocking holes.

2.3 Skeleton layer

At present, perovskite solar cells have two branches. The first one has a skeleton layer structure. The skeleton layer is divided into those that can transmit electrons, such as TiO2 and ZnO, and those that cannot transmit electrons, such as Al2O3. The initial research of calcium iron ore solar cell is based on the skeleton structure, so it has made great progress and is relatively mature. The other is a skeleton-free layer, that is, a flat plate structure. At present, TiO2 as the barrier layer is the mainstream. The role of skeleton layer in perovskite solar cells is mainly to support absorption layer and conversion layer, and also to conduct electrons [8]. Generally speaking, the main materials for preparing nanorods are ZnO and TiO2, in which ZnO nanorods have lower temperature requirements and can be applied to flexible solar cells. Besides semiconductor materials, nanorods can also be used as excellent materials for preparing the skeleton layer. Compared with traditional semiconductor materials, the new nanorod materials can improve the performance of perovskite-type solar cells.

2.4 Light absorbing layer

The light-absorbing layer is the most important structure in solar cells. The light-absorbing layer absorbs sunlight and converts it into photo-generated electrons, and then carries out the next reaction [9]. In the application field of perovskite solar cells, TiO2 can be used in porous materials to collect electrons, that is, the electrons generated by light excitation of perovskite light absorption layer are transported to FTO conductive glass. The function of perovskite light-absorbing layer is to absorb sunlight, generate photo-generated carriers, and realize the separation of electron-hole pairs at the interface. This requires that perovskite layer should have proper thickness, proper density and good interface connectivity, and poor perovskite layer will cause recombination of electrons and holes [10]. In perovskite solar cells, the main raw material of the light-absorbing layer is perovskite thin film. The performance of the light-absorbing layer can be improved by improving and upgrading the preparation process of perovskite thin film.

3. Performance optimization of perovskite solar cells

3.1 Optimization of ZnO electron transport layer

Under the condition of general light intensity, in the whole material, the influence of photo-generated carrier concentration on multi-carrier concentration is relatively small, but it has a great influence on minority carrier concentration, so the movement of photo-generated minority carrier is dominant [11]. In order to obtain the stability and reliability of low transmission resistance, high parallel connection and low reverse saturation, a large number of waste materials caused by human errors can be reduced, and pollution can be reduced. Compared with TiO_2, ZnO has many advantages. For example, the superior electronic/optical characteristics make the morphology easy to control and dope, the electron mobility is high, which promotes the faster transfer of charge, and high-quality thin films can be prepared at low temperature. Therefore, ZnO is considered as an ideal ETL material for PSCs.

3.2 Stability optimization of perovskite solar cells

The factors affecting the stability of perovskite solar cells are as follows: perovskite decomposition caused by annealing process; Perovskite decomposition caused by natural humidity; The reaction of electron transport layer affects the transmission of photo-generated electrons; Ultraviolet rays affect the work of solar cells, etc. The main idea to improve the stability of perovskite solar cells is to use the materials with stable chemical properties and difficult reaction, which are the preparation materials of solar cells. For example, using Al_2O_3 as the decorative layer of perovskite film can prevent the decomposition of perovskite layer caused by natural humidity. Graphene nanosheets are used as the transport layer to improve the performance of transporting photogenerated electrons, thus improving the conversion rate of solar cells. The doping of these hybrid perovskite can not only make the perovskite phase structure more stable, but also effectively improve the conversion efficiency and device stability of the corresponding perovskite solar cells.

3.3 Hole transport layer optimization

PEDOT is the most common hole transport layer at present, but due to the existence of many limiting factors, it is urgent to find a new material with strong substitutability. The porous support layer in the solar cell with porous support structure can be regarded as an extension of the electron transport layer. Due to the thick layered structure of the scaffold layer, it is beneficial to the combination of perovskite materials and electron transport materials, and provides a larger contact surface for perovskite materials. In addition, the sufficient contact between semiconductor metal oxide and perovskite material can improve the efficiency of photo-generated carrier separation and photo-generated electron injection into porous layer to some extent. Compared with PEDOT, the application of NiOx improves the filling factor of the device. In the two-step PbI_2-DMSO composite device, the performance of the device using NiOx is better than that using PEDOT. The use of NiOx improves the efficiency of the device to some extent.

4. Conclusions

Solar energy is an important new type of energy. In this paper, a new perovskite solar cell is studied. When preparing perovskite solar cells, the emphasis is on the fabrication of electron transport layer, dense layer, skeleton layer and light absorption layer. In order to obtain the stability and reliability of low transmission resistance, high parallel connection and low reverse saturation, a large number of waste materials caused by human errors can be reduced, and pollution can be reduced. The main idea to improve the stability of perovskite solar cells is to use materials with good chemical properties as the preparation materials of solar cells, so as to reduce the interference of natural environment on the panels. The sufficient contact between semiconductor oxide and perovskite material can improve the efficiency of photo-generated carrier separation and photo-generated electron injection into porous layer to some extent. It is an important task for today's social development how to develop solar energy and how to make more efficient use of it. In the future, perovskite solar cells should continue their current development advantages, continue to focus on research quality, and constantly optimize and upgrade the structure of each carrier transport layer.

References

1. Correa-Baena J P, Abate A, Saliba M, et al. The rapid evolution of highly efficient perovskite solar cells[J]. Energy & Environmental Science, 2017, 10(3):710-727.
2. Domanski K, Roose B, Matsui T, et al. Migration of cations induces reversible performance losses over day/night cycling in perovskite solar cells[J]. Energy & Environmental Science, 2017, 10(2):604-613.
3. Nam J K, Chai S U, Cha W, et al. Potassium Incorporation for Enhanced Performance and Stability of Fully Inorganic Cesium Lead Halide Perovskite Solar Cells[J]. Nano Letters, 2017, 17(3):2028.
4. Cho K T, Paek S, Grancini G, et al. Highly efficient perovskite solar cells with a compositionally

engineered perovskite/hole transporting material interface[J]. Energy & Environmental Science, 2017, 10(2):621-627.

5. Bishop J E, Mohamad D K, Wong-Stringer M, et al. Spray-cast multilayer perovskite solar cells with an active-area of 1.5 cm2[J]. Scientific Reports, 2017, 7(1):7962.
6. Cao X, Li H, Li G, et al. Electrocatalytically active MoSe2 counter electrode prepared in situ by magnetron sputtering for a dye-sensitized solar cell[J]. Chinese Journal of Catalysis, 2019, 40(9):1360-1365.
7. Paek S, Schouwink P, Athanasopoulou E N, et al. From Nano- to Micrometer Scale: The Role of Antisolvent Treatment on High Performance Perovskite Solar Cells[J]. Chemistry of Materials, 2017, 29(8):3490-3498.
8. Luo Q, Deng X, Zhang C, et al. Enhancing photovoltaic performance of perovskite solar cells with silica nanosphere antireflection coatings[J]. Solar Energy, 2018, 169(10):128-135.
9. Zhang X, Ren X, Liu B, et al. Stable high efficiency two-dimensional perovskite solar cells via cesium doping[J]. Energy & Environmental Science, 2017, 10(10):2095-2102.
10. Yang Guangwu, Cheng Yuanzhuang, Tian Zhongcheng, et al. Comprehensive experimental design of perovskite solar cell material preparation, device assembly and performance test [J]. Experimental Technology and Management, 2022, 39(3):7.
11. Lu Yue, Ge Yang, Sui Manling. Difference of degradation mechanism of lead methylaminoiodide-based perovskite solar cells irradiated by white light and ultraviolet light [J]. Acta Chemica Sinica, 2021, 79(3):9.

Research and demonstration of 20kW wind and solar complementary coupling power supply in Xiangmao Township, Tibet

Junle Wang*, Li Wan, Xiaoyu Zhang

Tibet Autonomous Region Energy Research Demonstration Center, Lasa 850000, China

Abstract: In this paper, through technological research and development, we will improve the existing technology of domestic wind / light complementary power stations, apply wind power generation equipment with a single unit capacity of 5kW and 10kW in Xiangmao Township, Naqu City, Tibet, and optimize the system design. The completed power station will supply power to 42 herdsmen in Xiangmao Township through centralized power supply, and use the form of wind and light complementary coupling power generation to reduce the construction cost of the power station and improve the reliability of the system, It will provide a scientific basis for the next promotion and application of wind and solar projects.

Keywords: Pv/ Wind hybrid; Coupling power supply; Research and demonstration; Tibet region

1. Introduction

Xiangmao township of Naqu County is located in the area with abundant solar energy resources from the central and eastern part of Naqu to Changdu on the surface of Himalayas. It has long sunshine time, large annual total radiation and stable resources. Due to geographical reasons, the remote villages and towns in Naqu city are not covered by the power grid, and these areas are remote and poor areas. Due to the lack of conventional energy, less coal and oil, the electricity and other domestic energy of towns and herdsmen need to be imported from other regions, and the cost is relatively expensive, As shown in Figure 1.

Xiangmao Township, Naqu County, Naqu City, 4683 meters above sea level, is located 72 kilometers south of Naqu County, Tibet, with 272 villagers from 42 households in the village. According to our investigation, the village has used the solar household system with a capacity of 150W as early as 1999, but many household systems have been unable to use due to the long history. Zongrege village, Xiangmao Township, is dominated by animal husbandry, of which 42 households live relatively concentrated. According to the field measurement, these herdsmen live within a radius of 800 meters, and the other 20 households live very scattered, with a distance between households of more than 300 meters. Some villagers do not use electricity, but still use candles or butter for lighting, and some even use firewood for heating and lighting, The lack of electricity has restricted the local economic development and the improvement of people's production and living conditions. This village urgently needs to solve the problem of electricity. According to the local solar energy resources and wind resources, solar power stations with complementary wind and solar energy coupling are used to solve the problem of power consumption.

Fig.1 Local scattered farmers' and herdsmen's houses

2. Design of 20kW wind solar complementary coupling power station

2.1 Basic principles

The wind / light complementary power generation system consists of PV array and fan array of battery modules, charging controller, inverter, battery, energy consuming load, AC distribution cabinet and other components, As shown in Figure 2[1-3].

* Corresponding author: Junle Wang 853969982@qq.com

Working principle: in the daytime, under the irradiation of sunlight, solar cells generate direct current; Driven by the wind, the alternating current generated by the wind turbine is rectified and turned into direct current; Part of the direct current generated by solar energy and fan is transmitted to the inverter through their respective controllers to convert into current and supply it to users, and the other part is stored after charging the battery; When sunlight or wind energy is insufficient, the battery sends power to the inverter through the DC control system, which is converted into AC for AC load[4].

Fig.2 20kW wind solar complementary coupling power station after completion (left: 5kW fan, middle: 5kW photovoltaic, right: 10kW fan)

2.2 Solar module design

According to the provisions of GB50797-2012 code for design of photovoltaic power stations, in the photovoltaic array, the electrical performance parameters of each photovoltaic module in the same photovoltaic module string should be consistent. In this paper, a total of 32 165Wp solar panels are used, which are connected by 8 series as a group, 4 groups and a charging circuit. The output charging circuit is 4 channels, and the actual power output of the photovoltaic array is 5.28kW, as shown in Figure 3[5-6].

Fig.3 Site photo of solar photovoltaic array

2.3 Fan selection

According to the site conditions, the annual average wind speed in Xiangmao township is 5.1m/s, of which the wind speed is the largest in winter and spring, and the wind speed is slightly weaker in August and September every year. The fan is installed in the open ground in the south of the power station control room, with an interval of 55m. There is high wind speed and no shelter. Because the fan is installed in the open ground, it is convenient to maintain in the later stage. During installation, the cable tower is used for installation. The controller of the fan rectifies the three-phase unstable AC into 220V DC, and has the functions of emergency braking, short-circuit protection, reverse connection protection, etc, as shown in Figure 4[7].

Fig.4 Fan installation site

2.4 Control and inverter design

The controller is a control unit connecting the solar photovoltaic module array and the energy storage battery pack. It charges the battery by controlling the current and voltage, and has the function of protection and monitoring. Photovoltaic inverter is mainly used to convert DC power supply into AC power supply. The inverter configured in the system of this power station adopts dc220v power supply. The function of the inverter is to convert DC power from solar cells into AC power for AC loads. The controller is matched with the photovoltaic module, and the capacity of the inverter should be determined according to the actual load, with additional consideration of capacity reduction at high altitude[8].

The inverter capacity in this paper is 30KVA. Inverter is the core equipment of photovoltaic power station, and the load is mainly motor inductive load. The impulse current of motor load is large when starting. Considering the guarantee of actual power supply, it is generally 1.2-5 times. A total of 3 sets of 10KVA off grid inverter equipment are selected for this project, and the power supply mode of two for use and one for standby is adopted, as shown in Figure 5[9].

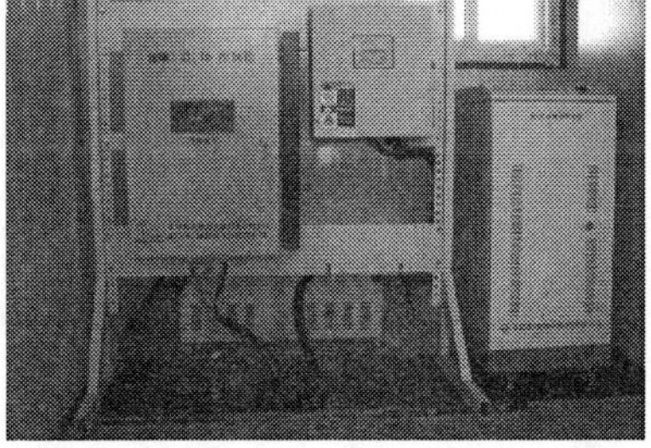

Fig.5 Site photos of controller and inverter

https://doi.org/10.1051/e3sconf/202235802052

2.5 Battery pack design

In this paper, the DC voltage of the power station system is designed according to dc220v. In this project, the discharge depth coefficient of the battery is 0.7, the efficiency is taken as 90%, and the continuous rainy days are considered as 2 days. The 20kW off grid photovoltaic power station generates an average of 4.6 hours of electricity per day. Based on the calculation of 2-day daily electricity consumption, the batteries in this paper are installed in two groups, with 108 batteries in each group, a total of 216 batteries, single batteries of 800ah/2v, and the net weight of the whole battery group is about 14.47 tons, as shown in Figure 6[10-11].

Fig.6 Site photo of battery pack

2.6 Transmission line design

Tab.1 Transmission line design

Order Number	Name	Design content
1	Voltage level	AC220
2	Frequency	50Hz
3	Power supply radius	800 meters
4	Trunk line	$16mm^2$ armored cable
5	Incoming line	$4mm^2$ power cable
6	DC incoming line	$2.5mm^2$ special cable, laid with PVC sleeve for entry
7	Power installation of each household	One set of three 18W energy-saving lamps, lamp caps and hangers, one drag line board and one set of lines
8	Trunk line to branch line, branch line to user, use trunk line box and branch line box for connection and protection.	

(a) Armored cable buried laying

(b) PVC casing laying

Fig.7 Photos of on-site implementation of transmission lines

3. Benefit analysis

The 20kW wind solar complementary coupling power station demonstrated in this project is located in Xiangmao Township, Naqu City, Tibet. The power station is expected to generate 34310 kWh per year. According to the actual data, the average annual power generation in recent years is about 36100 kWh. According to the average on grid electricity price of Naqu City, Tibet is 0.7 yuan / kWh (regardless of dry season, wet season and peak valley level), and the annual income is 36100 kWh × 0.7 yuan/degree =25270 yuan ≈ 25000 yuan. After the completion of the project, the daily electricity consumption of 272 people in 42 households has been solved, and their living conditions have been improved. This is of great significance to solve the problem of electricity consumption of farmers and herdsmen in Naqu power free area of Tibet, improve the quality of life of farmers and herdsmen, build a harmonious and prosperous Tibet, promote the stability and prosperity of Tibet, and promote the construction of a new socialist countryside in Tibet. The social benefits of the project are very significant.

According to the calculation of 4 tons of standard coal consumed per 10000 kwh of electricity, 0.5 tons of dust and 10 tons of carbon dioxide emitted into the atmosphere, it is equivalent to saving 14.4 tons of standard coal per year, and reducing 1.8 tons of polluted dust and 36 tons of carbon dioxide emissions at the same time.

4. Conclusion

Photovoltaic and wind energy are clean and renewable energy. The development and wide application of photovoltaic and wind energy technology plays an important role in alleviating the shortage of conventional energy in Tibet, solving the problem of electricity consumption by the population without electricity, and reducing environmental pollution. It is very important both from the perspective of energy conservation and environmental protection. The ecological status of Tibet Plateau is very important, and the ecological environment is fragile. It is an important ecological security barrier in China. Once damaged, it cannot be restored. After the construction of the photovoltaic power station, it will make a certain contribution to the protection of the ecological environment in Tibet. Vigorously developing and utilizing solar energy is of great significance to solve the daily electricity consumption problem of people in border areas without electricity, improve the living conditions of local people, and improve the quality of life.

References

1. Amy H I L, Chen H H, Kan H Y . Multi-criteria decisi on making on strategic selection of wind farms[J]. Renewable Energy, 2009, 34(1): 120-126 .
2. Rob V H, Vasilis F . GIS-based wind farm site selection using spatial multi-criteria analysis (SMCA): evaluating the case for New York State[J]. Renewable and Sustainable Energy Reviews, 2011, 15(7): 3332-3340 .
3. Ramirez-Rosado I, Garcia-Garrido E, Fernandez-Jimenez L, et al . Promotion of new wind farms based on a decision support system [J]. Renewable Energy, 2008(33): 558-566 .
4. Shimy M E . Optimal site matching of wind turbine generator: case study of the gulf of suez region in Egypt[J]. Renewable Energy, 2010(35), 1870-1878 .
5. Yoreley C S, Antonio J G, Jorge X B . Analytical methods for wind persistence: their application in assessing the best site for a wind farm in the State of Veracruz, Mexico[J]. Renewable Energy, 2010(35): 2844-2852 .
6. Liang Guilan , Xu Weiya , Tan Xiaolong . Application of extension theory based on entropy weight to rock quality evaluation[J]. Rock and Soil Mechanics, 2010(2):535-540 (in Chinese) .
7. Xia Shu, Zhou Ming, Li Gengyin. Multi-objective optimization algorithm for distributed generation locating and sizing[J]. Power System Technology, 2011, 35(9): 115-121(in Chinese) .
8. Boonbumroong U, Pratinthong N. Model-based optimization of stand alone hybrid power system[C]//World Renewable Energy Congress. Sweden: Linkoping University, 2009: 1-10 .
9. Shen Hai, Jie Jiancang, Li Jianxun . Multiple attribute to hydrological forecasting based on fuzzy matter-element method[J]. Northwest A&F University Press: Natural Science, 2011(11): 209-215(in Chinese) .
10. Kong Xiangxing, Xia Caichu, Qiu Yuliang, et al. Health diagnosing method for shield tunnel based on extension theory[J] . Journal of Tongji University: Natural Science, 2011(11): 1610-1615(in Chinese) .
11. Hajibashi M, Ebrahimi A . Selecting the wind site location and wind turbine rated power based on reliability indices of power system[C]// 10th International Conference on Environment and Electrical Engineering (EEEIC) . Rome: Wroclaw University of Technology, 2011: 1-4 .

Study on current characteristic detection method of hydraulic solenoid valve in nuclear power plant

Haibao Wang [1], Xudong Tan [1], Qunxiong Xiao [1] and Xincai Chang [2,*]

[1] Yangjiang Nuclear Power Co., ltd. China
[2] Shanghai Engineering Science &Technology Co., Ltd Cgnpc, Beijing, 100086, China

Abstract. In this paper, a set of electromagnetic valve current characteristic test system is designed to improve the reliability of the trip solenoid valve in the field steam turbine system of nuclear power plant. The test system starts from the field requirements, working principle of solenoid valve, software control and other aspects to complete the pressure suppression function of the hydraulic test system and the functions of data acquisition, storage, display, playback and so on. The test system is composed of a hydraulic system and an electrical control system. The system verifies the reliability of the tested object by detecting the important parameters of the solenoid valve, calibrating the current characteristic points, calculating the characteristic parameters, and judging whether the test is qualified. The test results show that the system has been successfully applied in nuclear power field, meets the needs of the site, and works well.

1. Introduction

The solenoid structure is shown in Fig. 1. Solenoid valve P port is the oil inlet, T port is the oil outlet, under normal conditions, A port, T port is usually closed, when the solenoid excitation, A port, P port is conductive.

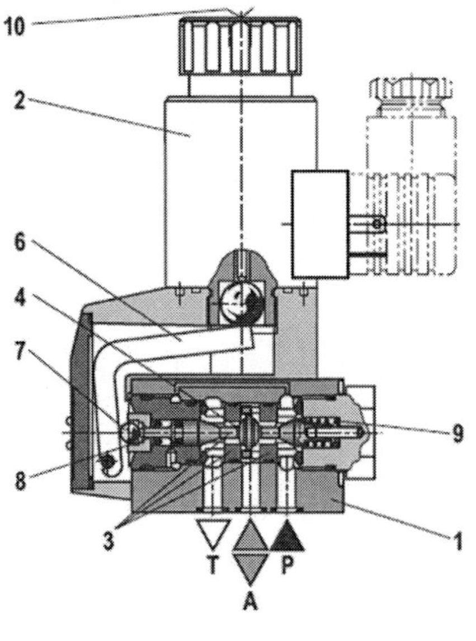

Figure 1. Solenoid Valve Diagram.

According to the working principle and technical standard of the jumper solenoid valve in the field steam turbine system [1-2], and in combination with the field conditions, the off-line test system of the solenoid valve is designed. Characteristic parameters such as current ratio and operating time are calculated. Firstly, the pressure of outlet A and outlet P of the solenoid valve in Fig. 1 is checked. Secondly, the current curve of the solenoid valve in on-off state is tested, and the characteristic parameters are recorded again [3]. Finally, the current curve fluctuation graph of the coil during the switching process of the hydraulic solenoid valve is calculated by drawing [4].

* Corresponding author: changxincai@cgnpc.com.cn

https://doi.org/10.1051/e3sconf/202235802053

2. Composition and function of test system

2.1 Hydraulic system

The system is composed of hydraulic oil pump, pressure transmitter, pressure gauge, ball valve, stainless steel pipeline, test solenoid valve, oil tank, etc. The main function of this system is to complete the pressing function of the hydraulic test platform. The system works under certain pressure by hydraulic oil pump, which makes the stainless steel pipeline form a high pressure working environment. The pressure transmitter completes the measurement of pressure throughout the pipeline.

2.2 Electric control system

The system is composed of upper computer, high-speed acquisition card, etc.

1) High-speed acquisition card: to achieve the system's data collection and control functions.

2) Develop Hose Computer through programming, including display module, control module, data processing module, collection module, storage module and alarm module.

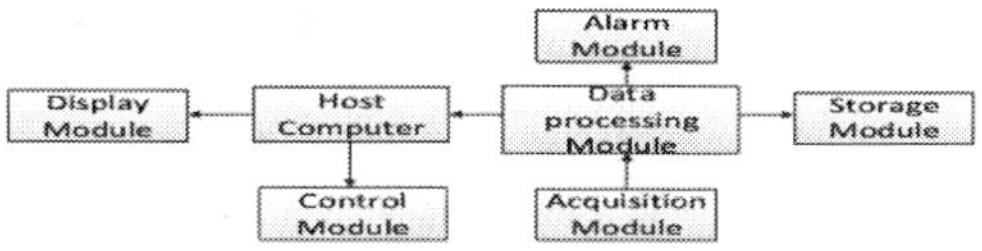

Figure 2. Diagram of electrical control system

The upper computer controls the pressure of the pipeline system by the oil pump through the control card. The current characteristic curve of the test solenoid valve is collected by the high speed acquisition card and uploaded to the upper computer. The upper computer calculates, processes and stores the collected information through the test software. At the same time, the upper computer can automatically judge whether the test solenoid valve is normal based on the measured data. The upper computer can also output relay control signal through the control card to control the start and stop of the measured solenoid valve. In addition, the system set up interlock protection, when the pipeline pressure is higher than the maximum working pressure, the program will automatically disconnect the hydraulic oil pump power supply, so that the pipeline pressure does not continue to rise, in addition, set the overflow valve higher than the set value for pressure relief.

After the start of the program, real-time test the level signal of the tank, judge if the filter is blocked or needs to stop in an emergency, etc. In case of an emergency, the program control automatically disconnects the power supply of the hydraulic oil pump and stops the pressure.

3. Software Control Design

3.1 Real-time testing process

Test the coil current curve fluctuation during the solenoid valve switching process. Solenoid valve P port is pressure medium inlet, this system is oil inlet, and A port is solenoid outlet, connecting downstream equipment. T is the oil return port. The pressure of the P and A ports of the solenoid is tested by installing a pressure sensor, and the measured solenoid is rated at 24VDC by programmed power supply. Current characteristic test interface. As shown in Figure 3.

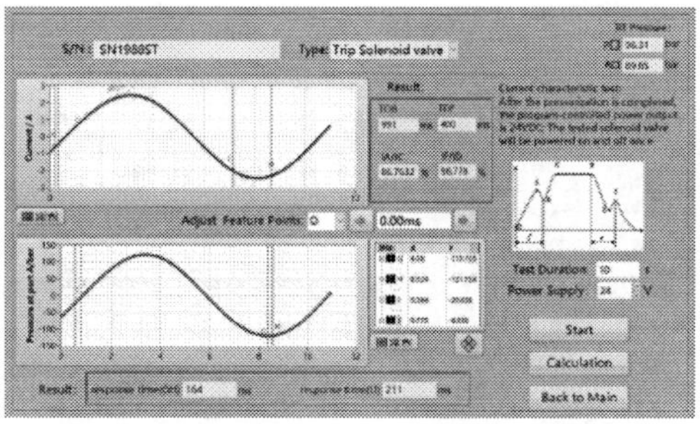

Figure 3. Current characteristic test interface

https://doi.org/10.1051/e3sconf/202235802053

Operation steps:

1) Install the test solenoid valve, start the program to start the test, run the software to start the hydraulic test system, set up the pressure of the hydraulic system first, and keep at 150 bar (factory set 150bar, can increase or decrease the test pressure as required by the test)
2) Set alarm parameters, set IA/IC and action time TOB limits respectively. When the test results are outside this range, the program sets automatic alarm.
3) Set the supply voltage of the solenoid valve to 24VDC;
4) Manually open the test solenoid P-port ball valve and keep the test solenoid A-port ball valve closed;
5) Real-time test the pressure and current of the P and A ports, and observe the real-time change of the curve through the interface of Figure 3.
6) Set the test time, such as 10s, with a duty cycle of 50%, to control the solenoid valve to be tested to power on for 5S and power off for 5s, to view the current curve and the pressure change curve of port A in real time;
7) After one test, manually adjust the characteristic points of the current curve, and mark the position of each point (A~F) according to the current characteristic curve table of the solenoid valve in Fig. 4. The computer automatically calculates the current ratio IA/IC and the action time TOB and other characteristic parameters, and judges whether the test is qualified according to the set alarm value. Measure the current characteristic parameters of the solenoid valve 5 times in succession. The average value of 5 measurements is the current characteristic parameters of the test solenoid. Save the test results to the database.

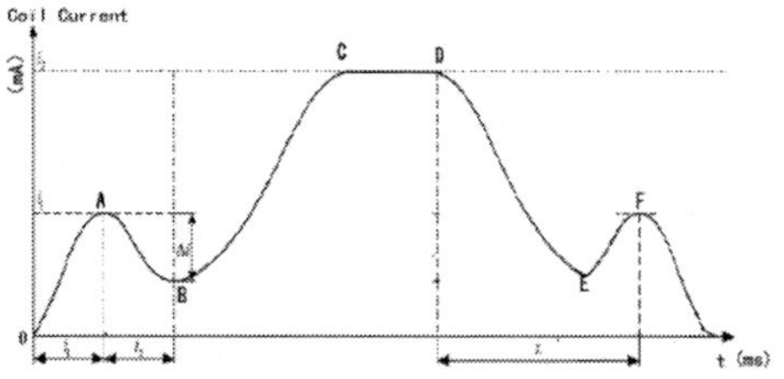

Figure 4. Chart of solenoid current characteristics

3.2 History Playback

The program has a history playback function to export the test data from the database to the display interface for experts to view the test data of solenoids. The playback interface is shown in Figure 5, where you can continue to adjust the feature points and calculate the relevant parameters. The response time can also be adjusted according to the pressure curve at port A and the report can be saved and exported.

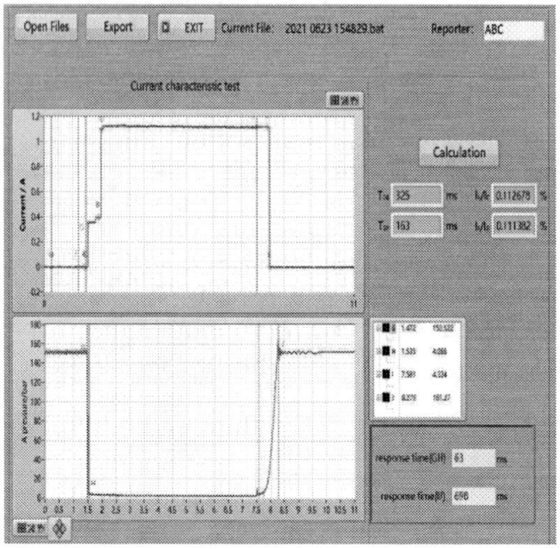

Figure 4. History playback interface

4. Conclusion

In this paper, the test platform of the hydraulic solenoid is set up, and the real-time data of the current of the hydraulic solenoid and other parameters related to the current are detected, so that the abnormal situation of the hydraulic solenoid can be accurately analyzed. The performance of the solenoid valve is determined in real time by the test results, which provides data support for the overhaul of the steam turbine system in nuclear power plant, and provides guidance for the determination of the availability of the solenoid valve in nuclear power site.

References

1. Jian-Wei Yang; Jin-Hai Wang; Qiang Huang; Ming Zhou; Reliability Assessment for the Solenoid Valve of a High-Speed Train Braking System under Small Sample Size [J]. Chinese Journal of Mechanical Engineering. 2018 (3)
2. Angadi S V, Jackson R L, Choe S Y, et al. Reliability and life study of hydraulic solenoid valve. Part 2: Experimental study[J]. Engineering Failure Analysis, 2009, 16(3):944-963.
3. Zai-Qing, Ren , and W. Jian-Hui . "Test-system design for hydraulic solenoid valve of a certain aircraft." IOP Conference Series: Materials Science and Engineering 1043.4(2021):042023 (6pp).

4. KalinowskiMarcinmarcin@o.plKamińskiZbigniewz.kaminski@pb.edu.plFaculty of Mechanical Engineering,Bialystok University of Technology,ul. Wiejska C, -Biaystok,Poland. (2021). Measurement and evaluation of functional and operational coefficients of hydraulic solenoid valve prototypes used for variable valve timing control in combustion engines. Acta Mechanica et Automatica, 15(2), 74-81.

https://doi.org/10.1051/e3sconf/202235802054

Research on security protection strategy of flexible interactive platform between captive power plant and power grid

Lingkai Zhu [1,*], Jiyan Liu [2], Weishuai Wang [2], Ziwei Zhong [1], Wei Zheng [1], Yue Han [1], Zhiqiang Gong [1], Junshan Guo [1], Panfeng Shang [1]

[1] State Grid Shandong Electric Power Research Institute, Jinan, China
[2] State Grid Shandong Electric Power Company, Jinan, China

Abstract. In order to strengthen the security protection of the flexible interactive platform between the self owned power plant and the power grid, ensure the security of business data, resist the malicious attacks of hackers, improve the robustness of the platform operation, and ensure the safety and reliability of users' power consumption, it is necessary to take a number of security protection technical means to build a comprehensive security defense system of the power grid platform.

1. Introduction

The application of computer, network and information technology in power system has effectively improved the quality of power production and power utilization efficiency. The information communication system of power system has made good progress in the direction of systematization, intelligence and automation. Modern power system has developed into a complex coupling network system composed of physical power system and information communication system[1][2].
The captive power plant has jumped out of the traditional self-sufficiency mode and participated in the power market competition, which has been strongly supported by national policies, and effectively improved the power balance of the grid from the power supply side[3]. The "flexible interactive platform between captive power plant and power grid" is connected with the system of State Grid Corporation of China. Its network security protection is very important, which will directly affect the normal operation of the power system. The influencing factors of power system network security in the big data environment include: lack of citizen security awareness, data storage security, system security vulnerabilities, virus intrusion and imperfection of network security mechanism[4].
In order to strengthen the security protection of the flexible interactive platform between self owned power plants and power grids, ensure the security of business data, resist malicious attacks by hackers, and improve the robustness of the platform, it is necessary to take a number of security protection technical means to build a set of in-depth comprehensive security defense system.

2. Research on safety protection methods

Security protection methods mainly include border security protection, terminal security protection, network security protection, and comprehensive security protection.

2.1 Border security protection

The goal of border security protection is to protect the interior of the border from external attacks. At the same time, it is also used to prevent malicious internal personnel from carrying out attacks across the border, or external personnel from entering the internal network through open interfaces and hidden channels[5].The captive power plant realizes boundary safety from three levels: terminal credibility, transmission reliability and boundary controllability, including horizontal boundary protection, vertical boundary protection and third-party boundary safety protection.
Horizontal security protection is a security protection measure formulated for the transmission protection of communication data flow between security domains, including network access control and information intrusion protection. Network access control generates data access matrix based on the analysis of application data access requirements, and then sets access control policies. The log information generated when the access control policies conflict due to inter domain data access should be analyzed and processed in time. Information intrusion protection adopts intrusion protection system to detect and protect the business data flow transmitted between security domains.
The vertical security boundary protection is the vertical defense line of the system. Certain technical measures are

* Corresponding author: zhulingkai@woyoxin.com

adopted to realize the remote safe transmission of data and the security protection of the vertical boundary. Security measures include border network access control, information intrusion prevention and defense of hidden borders. Border network access control deploy border network access control measures at the vertical network boundary; Information intrusion prevention adopts the intrusion prevention system to detect and protect the information flow flowing through the boundary. When an attack is detected, the intrusion prevention system should record the attack source IP, attack type, attack purpose, attack time and block the attack behavior. In case of serious intrusion events, it should be able to provide timely alarm information; Hidden boundary refers to the hidden non-public network interface caused by illegal outreach. Management means and special technical measures are adopted to prevent the information intranet host from illegally connecting to the Internet, and its behavior is located and blocked.

The third-party boundary security protection mainly refers to the network boundary between the interactive platform and the Internet, and the network boundary between the interactive platform and other stakeholders. It is divided into the third-party boundary of the information extranet and the third-party boundary of the information intranet to analyze the security protection strategy. The third-party boundary of information extranet refers to the security boundary between information extranet and Internet and the network boundary formed by dial-up connection with other units. The security protection measures for such boundary mainly include: boundary network access control, remote security access, external release service security, information intrusion protection, content filtering and defense of hidden boundary. The third-party boundary of the information intranet refers to the network boundary formed by the connection between the information intranet and other third-party networks. The relevant security protection measures mainly include: boundary network access control, information intrusion protection and defense of hidden boundaries. Firewall, VPN and leased line shall be adopted between the platform and external network to ensure the security of boundary and data transmission.

2.2 Terminal safety protection

Terminal security protection refers to the security protection of all terminals in the platform system [6]. Due to the characteristics of large mobility, scattered location, large number and difficult management of terminal equipment, in addition to installing anti-virus software for computer virus protection, terminal level firewall control, intrusion protection and patch management are also required to better ensure the security of all kinds of terminals, and also reduce the security risks faced by the entire platform system to a great extent.

The terminals related to the platform are divided into intranet terminals and extranet terminals. The intranet terminals are used for the business operation and information processing of the information intranet, and the extranet terminals are used for the information access of the extranet. For the internal and external network terminals, their security protection should be considered. The terminal security should deploy host based security measures, and the relevant security protection design should be carried out from the following aspects: terminal virus protection, malicious code protection, patch management Terminal security management.

2.3 Network security protection

Network security is the ability to prevent attacks, intrusion, interference, destruction, illegal use and accidents on the network, keep the network in a stable and reliable state, and ensure the integrity, confidentiality and availability of network data. The network security protection research of the platform includes network equipment security protection, network information security protection and network channel protection. The data communication used by the system should be protected by encryption, identity authentication and other technical measures.

Network equipment security protection includes the security protection of routers, switches, wireless devices, firewalls, security gateways and other security devices that provide network operation proppant security defense in the basic network and security domain. The main protection strategies include security access control, equipment security management, equipment security reinforcement, equipment link redundancy, etc.

Network information security protection refers to taking security measures at the network level to ensure the security of information transmitted through the network. When the business information is transmitted across the security domain through the network, encryption and decryption measures such as encryption or inserting digital signatures into the information flow should be adopted based on the confidentiality requirements of the information flow, so as to ensure that sensitive information will not be illegally intercepted when it is transmitted through the network. If the application does not support the above functions and it is difficult to transform, it can also be implemented by means of IPSec VPN and other third-party technical means.

Network channel security protection deploy intrusion protection system in the network segment where the main servers are located for network monitoring, map the traffic of important business servers to the listening port of network intrusion protection system, and customize the intrusion protection rule base according to the specific application data.

In addition, the construction of network security protection system should adopt the strategy of defense in depth of multi-layer defense and multi-point defense. According to the requirements of level protection, the comprehensive protection of physical security, computing environment security, regional boundary security and active defense can effectively improve the network security protection ability of the platform.

2.4 Comprehensive safety protection

Integrated protection is a process of information security protection for the power monitoring system from the host, network equipment, malicious code prevention,

 https://doi.org/10.1051/e3sconf/202235802054

application security inspection and audit analysis, backup and disaster recovery and other levels in combination with the relevant requirements of the national information security level protection work. It mainly formulates protection strategies from eight aspects: intrusion detection, host and network equipment reinforcement, malicious code prevention, security inspection and analysis, network security management, backup and disaster recovery, user interface security protection, and data interface security protection.

3. Formulation of platform security protection strategy

According to the principle of hierarchical and sub domain security protection, the protection objects are mapped to each security domain, and the protection measures required for each domain are classified into boundary security protection, network security protection, terminal security protection, and finally the security protection measures are mapped to the corresponding security types. The specific security protection measures include: network access control, system security reinforcement, system vulnerability scanning, intrusion protection measures, wireless security measures Remote access control, content security measures, virus detection measures, log audit measures, backup and recovery measures, identity authentication and access management related control measures and physical security measures.

Combined with the above-mentioned protection methods and measures, the platform system is divided into security zones, and the overall security protection strategy system is shown in Figure 1.

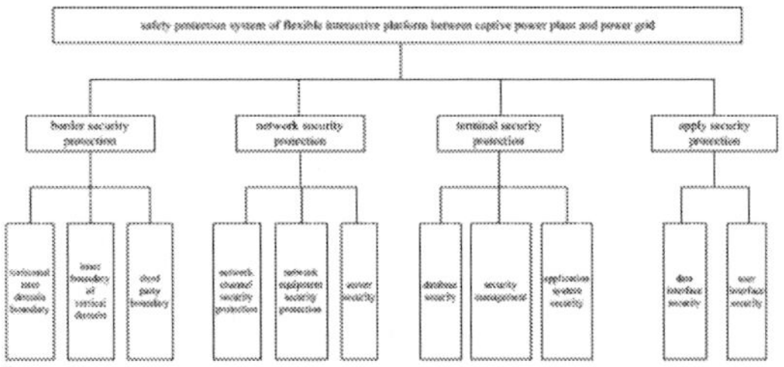

Figure 1. Overall strategy system of security protection.

According to the overall security protection strategy mentioned above, the security protection analysis is carried out for the flexible interactive platform between the captive power plant and the power grid, and a one-way access security protection scheme is formulated. The one-way access scheme has less information interaction and is relatively safe. The specific security protection deployment strategy is shown in figure 2.

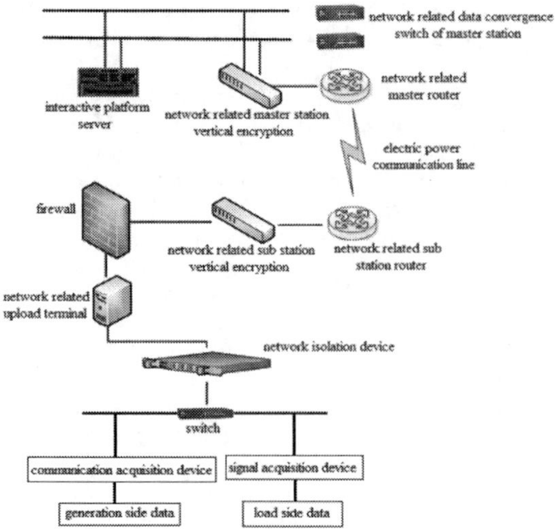

Figure 2. Overall strategy system of security protection.

4. Summary

The era of big data has put forward higher requirements for the security of network systems. Formulating effective network security protection strategies is crucial to the robust operation of the system. Aiming at the self-developed flexible interactive platform between self owned power plant and power grid, this paper formulates network security protection strategies from four aspects: boundary security protection, terminal security protection, network security protection and comprehensive security protection, so as to ensure the safe operation of the system. This work was supported by The Science and Technology Project of State Grid Corporation of China - Research on low-carbon regulation oriented coal-fired captive power plants' participation in power grid supply and demand regulation and source grid load coordinated operation technology（5400-202236173A-1-1-ZN）.

References

1. Sisi Qiu. Security and protection analysis of power system information communication network[J]. China New Communications, 2022,24(01):18-19.
2. Zi Wang, Zhihua Wang, Yong Han, etc. Research on multi-layer cooperative defense model for power system network security [J]. Computer Engineering, 2021,47(12):131-140.
3. Yifan Huang, Haijing Zhang, Lei Wang, etc. Evaluation method for flexibility regulation of captive power plants based on production characteristics of enterprises [J]. Power demand side management, 2021,23(1):61-66.
4. Ke Duan. Research on network information security protection strategy in big data environment[J]. Digital Communication World, 2022,(03):137-139.

5. Zheng Chen. Construction of network security protection system for power monitoring system [J]. Electrotechnics, 2019,(03):106-107.
6. Jian Chen, Haoming Wu. Research on information security protection measures of power network terminal system [J]. Digital technology and Application,2019,37(10):201-202.

Study on an on-line monitoring method for peak shaving capacity of thermal power unit

Lingkai Zhu*, Qian Wang, Ziwei Zhong, Yue Han, Zhiqiang Gong, Wei Zheng, Panfeng Shang, Junshan Guo

Shandong Electric Power Research Institute, Jinan, China

Abstract. Relying solely on pure condensing units can not meet the demand for power grid peak shaving, and thermoelectric units must also participate in peak shaving. In view of the dynamic change of peak shaving capacity of thermoelectric units with steam extraction, an on-line monitoring method of thermoelectric units is proposed in this paper. The upper and lower limits of peak shaving capacity of thermoelectric units under different steam extraction are calculated in real time, and the peak shaving capacity of thermoelectric units is evaluated on-line, so as to provide data support for scientific dispatching of power grid.

1. Introduction

With the economic and social development and industrial structure adjustment, the proportion of industrial power consumption decreases, the proportion of municipal and commercial power consumption increases, the peak valley difference of power grid increases, and the demand for peak regulation increases. In view of the increasing pressure of environmental protection, the state has increased its efforts to develop new energy power generation. It is expected that China's energy cleaning rate will reach 50% in 2050. Large scale new energy units occupy the space of conventional power units, resulting in serious shortage of peak shaving capacity of receiving power grid [1]. The rapid development of new energy units in some regions and the particularity of new energy unit power generation, especially wind power generation has the "inverse peak regulation characteristics" of continuous fluctuation within a certain range during the day, gradual increase of wind power output at night, the highest average output in winter and the lowest average force in summer. This inverse peak regulation characteristic will aggravate the peak regulation pressure of the system with wind power grid connection, and the demand for peak shaving capacity of the system is greater, which brings greater challenges to power grid peak shaving [2, 3].

At present, the peak shaving of power grid mainly depends on thermal power units. In northern China, the proportion of thermal power units accounts for more than 50%. In the heating season, the units should give priority to meeting the heat load demand. There is a working condition constraint of "determining electricity by heat", and the peak shaving capacity is greatly reduced [4]. In the past, thermoelectric units rarely participated in peak shaving, but with the large number of new energy connected to the grid, the demand for peak shaving of power grid is increasing. Relying solely on pure condensing units can not meet the demand, and thermoelectric units must also participate in peak shaving [5].

Different from pure condensing unit, the peak shaving capacity of thermoelectric unit is not a fixed value, but changes dynamically with steam extraction. The heating and steam extraction capacity of the unit changes with the weather conditions, and the upper and lower limits of peak shaving also change [6]. In view of the shortcomings of the above existing technologies, this paper proposes an on-line monitoring method for the peak shaving capacity of thermoelectric units, which can realize the on-line monitoring of thermoelectric units, calculate the peak shaving upper and lower limits of thermoelectric units under different air extraction, evaluate the peak shaving capacity of thermoelectric units on-line, and provide reliable data support for the scientific dispatching of power grid.

2. Innovative on-line monitoring method of peak shaving capacity of thermoelectric units

The on-line monitoring method of peak shaving capacity of thermoelectric units proposed in this paper includes four steps: one is to upload the real-time heating extraction capacity of the monitored units; the other is to find the maximum power and minimum power corresponding to the heating extraction capacity according to the heating working condition diagram; the third is to calculate the maximum power or minimum power that can be generated by the monitored thermal power unit under the real-time heating extraction capacity according to the formula, that is, the upper or lower limit of peak shaving. Fourth, data such as real-time heating extraction volume and calculated power value are uploaded to the system and displayed.

* Corresponding author: zhulingkai@woyoxin.com

 https://doi.org/10.1051/e3sconf/202235802055

2.1 Upload real-time heating and steam extraction capacity of thermal power unit

Through the dedicated optical fiber channel of the power system, the real-time heat supply steam extraction and active power data in the DCS system of the monitored thermoelectric unit are uploaded to the database of the main monitoring equipment, and the real-time data is collected from the server of the main monitoring device and calculated for two times. Table 1 shows the real-time heating and steam extraction capacity transmitted to the main monitoring equipment server in a certain period of time.

Table 1. Real time heating steam extraction capacity.

Real time heating steam extraction t/h				
68.4	90.7	105.7	105.7	140.6

2.2 Find the power corresponding to the heating extraction volume according to the heating working condition diagram

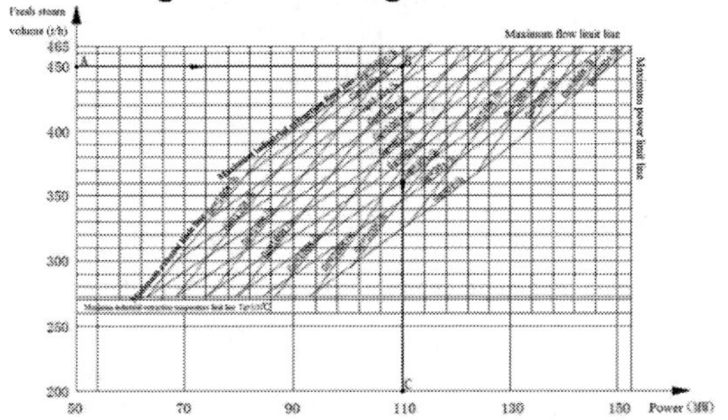

Figure 1. Structural diagram of steam feed pump system

Taking a 135MW Steam Extraction Unit of a thermal power Co., Ltd. as an example, the unit meets the operating conditions of the calculation conditions, and the design heating condition diagram of the steam turbine of the monitored thermal power unit is shown in Figure 1.

In Figure 1, G_Z is the exhaust steam volume of LP cylinder and G_g is the extraction steam volume. Main steam pressure $P_0 = 13.24$ MPa, main steam temperature $T_0 = 535$ °C, extraction pressure $P_g = 0.981$ MPa, reheat steam temperature $T_r = 535$ ℃. In Figure 1, another parameter can be found according to any two of the three parameters: fresh steam volume, industrial extraction volume and power. For example, it is known that the fresh steam volume is 450 t/h (point A) and the industrial extraction volume is 160 t/h. in Figure 1, the horizontal line passing through point A intersects with the industrial extraction volume line of 160 t/h, and the intersection point is point B. Point C is obtained by crossing point B vertically downward and intersecting the power axis. The value of point C is 110.1mw, which is the checked power.

From the steam turbine design heating working condition diagram of the monitored thermal power unit in Figure 1, there are 10 equal extraction steam lines of 0 to 180 t/h respectively. The maximum power or minimum power corresponding to 10 groups of heating steam extraction of the monitored thermal power unit can be checked from 10 equal steam extraction lines, as shown in Table 2. Input 10 groups of the heating steam extraction volume and the corresponding maximum power or minimum power into the real-time calculation database.

Table 2. The maximum power and minimum power corresponding to the heating steam extraction volume obtained from Figure 1.

Heating extraction capacity t/h	0	20	40	60	80	100	120	140	160	180
Maximum power MW	152.7	147.4	143.1	138.6	133.9	129.1	124.7	119.7	114	111.6
Minimum power MW	92.8	84.4	79.1	73.9	68.4	59.4	68.6	75	81.6	85.7

2.3 Calculate the power (upper and lower limits of peak shaving) of the monitored thermal power unit under real-time heating steam extraction according to the formula

Adjust the real-time heating extraction volume, judge that the real-time heating extraction volume is between the heating extraction volume of group i and group i+1, and take the maximum power or minimum power corresponding to the heating extraction volume of group i and group i+1 as the calculation interval, where i= 1, 2,... 9. The linear interpolation method is used to calculate the maximum power or minimum power that can be generated by the monitored thermal power unit under the real-time heating extraction capacity, that is, the upper limit or lower limit of peak shaving under the real-time heating extraction capacity of the monitored thermal power unit.

The calculation formula of maximum power or minimum power that can be generated by the monitored thermal power unit under real-time heating and steam extraction is shown in (2-1).

$$P = P_i + \frac{P_{i+1} - P_i}{Q_{i+1} - Q_i}(Q - Q_i) \qquad (1)$$

Wheres: Q -real time heating and steam extraction capacity of monitored thermal power unit, t/h ; $Q_i(i = 1,2,...9)$ -the steam extraction capacity of group i in the design heating working condition diagram of the monitored thermal power unit, t/h ; P -maximum power or minimum power corresponding to real-time heating and steam extraction of monitored thermal power unit, MW ; $P_i(i = 1,2,...9)$ -the maximum power or minimum power corresponding to the heating extraction capacity of group I in the design heating working condition diagram of the monitored thermal power unit, MW .

According to formula (2-1), the upper or lower limit of peak shaving under the real-time heating and steam extraction capacity of the monitored thermal power unit is obtained from the real-time heating and steam extraction capacity uploaded in Table 1, as shown in Table 3.

Table 3. Upper or lower limit of peak shaving under real-time heating extraction capacity.

Real time steam extraction t/h	68.4	90.7	105.7	105.7	140.6
Maximum power(Peak-shaving upper limit) MW	136.7	131.3	127.8	127.8	119.5
Minimum power (Peak-shaving lower limit) MW	71.6	64.0	60.2	60.2	75.2

2.4 Upload data for system display

The above real-time heating steam extraction volume, active power and the calculated peak shaving upper or lower limit are output to the man-machine interface for display, so as to complete the on-line monitoring process of peak shaving capacity of thermal power units to be monitored. The dispatcher can monitor the upper limit or lower limit of peak shaving of the monitored thermal power unit under the real-time heating extraction capacity in real time from the man-machine interface.

Compared with the traditional calculation methods, the on-line monitoring method proposed in this paper has obvious advantages: for example, it can monitor multiple groups of thermal power units in various regions at the same time, with large monitoring intensity and wide range, which provides important data support for multiple groups of thermal power units in various regions at the same time; Transmitting real-time data to the server of the main monitoring equipment through special optical fiber channel can not only make full use of the resources of the power system, but also increase the reliability of data transmission; The linear interpolation method is simple, convenient and fast.

3. Conclusion

This paper presents an on-line monitoring method of thermoelectric units. By uploading the real-time data in the DCS system of the monitored thermal power unit to the main monitoring equipment, find multiple groups of heating extraction capacity and the corresponding maximum power or minimum power from the design heating working condition diagram of the monitored thermal power unit, and use the actual calculation database in the main monitoring equipment to calculate the maximum power or minimum power of the monitored thermal power unit under the real-time heating extraction capacity in the background, then output it to the human-machine interface for the reference of the power supply network dispatcher, so as to provide reliable data support for the scientific dispatching of the power grid. This method can monitor the peak shaving upper and lower limits of thermoelectric units under different real-time heating steam extraction in the heating season, and provide important data support for the power grid dispatching department to scientifically and reasonably formulate the load dispatching curve of thermoelectric units.

Acknowledgements

This work is supported by the science and technology project of Shandong Electric Power Research Institute (ZY 2020-05), which studies the calculation method and platform application of minimum operation mode of thermal power plant based on power grid peak shaving dispatching.

References

1. Rong Sum, Chengzhi Zhang, Bing Chen, etc. Research on comprehensive evaluation of peak shaving mode of receiving power grid under large-scale new energy access [J]. Power automation equipment, 2021, 41(6): 149-155,163.
2. Yiqi Jin, Wu Zhu, YunHao Hua, etc. Energy storage peak shaving optimization strategy considering wind power reverse peak shaving characteristics [J]. Science, technology and Engineering, 2021, 21(20): 8498-8503.
3. Shengqiong Tan, Cheng Lang, Zhanqi He, etc. Summary of research and application prospect of wind power generation system [J]. Machine design, 2021, 38(8): 1-8.
4. Haiying Dong, Lei Fang, Kun Ding, etc. Peak shaving strategy of solar thermal power generation

based on Cogeneration operation mode [J]. Journal of solar energy, 2019, 40(10): 2763-2772.

5. Qinglin Gao, Jiaqi Gao, Qingzhuan Huang, etc. Optimization strategy of steam turbine operation regulation mode [J]. Hunan Electric Power, 2020, 40(3): 56-61.
6. Lingkai Zhu, Qian Wang, Wei Zheng, etc. Peak shaving capacity test of a 330MW thermal power unit under steam extraction condition [J]. Power station system engineering, 2020, 36(6): 37-40.

Study on Calculation Method of Inlet Steam Flow of External Steam Cooler in Feedwater Heating System

Lingkai Zhu*, Qian Wang, Wei Zheng, Ziwei Zhong, Yue Han, Panfeng Shang, Junshan Guo, Zhiqiang Gong

Shandong Electric Power Research Institute, Jinan, China

Abstract. In order to improve the thermal economic performance of coal-fired units in thermal power industry, supercritical units above 600MW generally adopt the method of installing external steam cooler system to improve the utilization of reheat. In view of the difficulties and problems in calculating the three extraction flow of the external steam cooler high pressure heater system by the traditional method, In this paper, a new calculation method of medium flow at the inlet of external steam cooler is proposed based on medium fluid dynamics. This method does not need to install a measuring device, the treatment method is simple and clear, and improves the test efficiency while saving the cost.

1. Introduction

In recent years, with the development of thermal power units to large capacity and high parameters, how to further improve the thermal economy of coal-fired units has become an urgent problem to be solved in the thermal power generation industry, especially the utilization of reheat [1]. For the ultra supercritical reheat unit, the temperature of the exhaust steam of the high-pressure cylinder of the steam turbine after reheating is high, resulting in the increase of the superheat of the extraction steam of the intermediate pressure cylinder of the steam turbine regenerative system. The extraction steam of the intermediate pressure cylinder with high superheat used to preheat the low-temperature boiler feed water will bring irreversible losses and affect the thermal economy of the unit [2, 3]. In order to reduce the superheat of extraction steam and improve the regenerative effect, high-pressure heater (high-pressure heating) system and some low-pressure heater (low-pressure heating) systems are equipped with steam coolers. The installation of external steam cooler system with flexible layout is an effective scheme to reduce the superheat of regenerative extraction steam [4, 5]. The external steam cooler has a large heat exchange area, which can not only reduce the end difference of the heater at this level, but also directly heat the feed water. With higher thermal economy, it is gradually used in supercritical units above 600MW [6].

It is difficult to calculate the three extraction flow of the external steam cooler high-pressure heater system with the traditional method. The three extraction flow can also be calculated directly from the heat balance equation of the external steam cooler. However, this method needs to install a flow orifice to measure the feed water flow of the external steam cooler, which increases the test cost, changes many measuring points and increases the test time. In view of the above problems, a calculation method of inlet steam flow of external steam cooler in feedwater heating system is proposed in this paper. This method does not need to install measuring device, the treatment method is simple and clear, and the test cost and time are saved.

* Corresponding author: zhulingkai@woyoxin.com

2. Calculation principle of inlet steam flow of external steam cooler system

The structure of the external steam cooler system studied in this paper is shown in Figure 1.

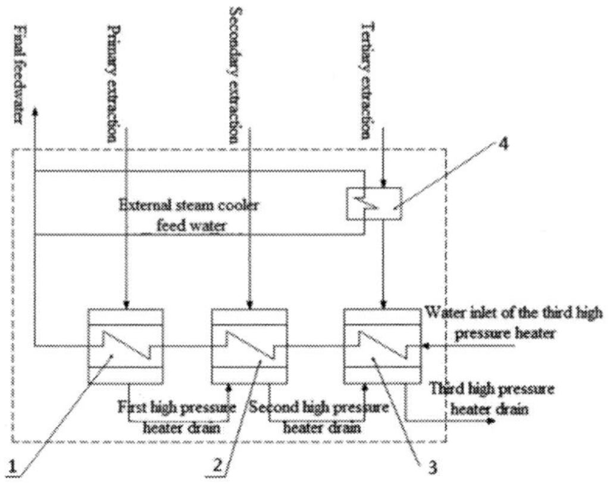

Figure 1. Design heating condition of steam turbine of monitored thermal power unit

Figure 1 shows a single-stage series external steam cooler. The three-stage extraction steam enters the external steam cooler 4 for cooling and heat release, and then enters the third stage high-pressure heater 3 for heat release, which can effectively reduce the steam superheat, reduce the heat exchange temperature difference and heat loss of the third stage high-pressure heater 3, and increase the outlet water temperature of the third stage high-pressure heater 3, The regenerative steam extraction capacity of higher pressure stage is reduced and the heat consumption is reduced. On the other hand, the feed water flow at the outlet of the first stage high-pressure heater 1 is partially through the external steam cooler 4, which increases the final feed water temperature and further reduces the heat consumption. For the thermal test, the three-stage extraction steam flow (three extraction flow), that is, the inlet steam flow of the external steam cooler 4, is used whether the condensate flow is used to calculate the feedwater flow or a kind of correction.

It is difficult to calculate the three extraction flow in the external steam cooler high-pressure heater system by traditional methods. Firstly, the feed water flow into the external steam cooler 4 is uncertain, and the three extraction flow cannot be calculated by the direct heat balance formula of the external steam cooler 4.In addition, since the external steam cooler 4 is provided with a drain pipe to the third stage high-pressure heater 3 for startup and shutdown drainage, if the valve is not tight, the steam inlet volume of the third stage high-pressure heater 3 may not be completely equal to the three extraction flow. Therefore, only the heat balance calculation of the third row of the third stage high-pressure heater cannot obtain the accurate three extraction flow.

In addition, the direct four row heat balance equation of the external steam cooler can also calculate the three extraction flow, but this method has two problems: first, the flow orifice needs to be installed to measure the feed water flow of the external steam cooler 4, which increases the test cost; Second, it is necessary to measure the feed water pressure and temperature at the inlet and outlet of the external steam cooler 4 and the inlet pressure and temperature of the third stage high-pressure heater 3. There are many replacement measuring points, which increases the test time.

3. Innovative calculation method of inlet steam flow of external steam cooler

In order to solve the above problems, this paper proposes a technical scheme, which regards all the heaters and external steam coolers in the feedwater heating system as a whole module, lists the heat balance equation for the heat in and out of the whole module, and obtains the inlet steam flow of the external steam cooler. The specific steps are as follows:

(1) Measure the following parameters: pressure and temperature of primary extraction, pressure and temperature of secondary extraction, pressure and temperature of tertiary extraction, inlet water temperature of primary high-pressure heater 1, secondary high-pressure heater 2 and tertiary high-pressure heater 3, outlet water temperature of primary high-pressure heater 1, primary high-pressure heater 1 Drain temperature, final feedwater temperature, final feedwater pressure and feedwater flow of the second stage high pressure heater 2 and the third stage high pressure heater 3. Among them, replace the pressure transmitter at the pressure measuring point to measure the pressure, replace the temperature sensor at the temperature measuring point to measure the temperature, and use the orifice flowmeter to measure the differential pressure at the flow measuring point.

(2) According to the pressure and temperature measured in step (1), the first stage extraction enthalpy, second stage extraction enthalpy, third stage extraction enthalpy, inlet enthalpy and drain enthalpy of the first stage high pressure heater 1, the second stage high pressure heater 2 and the third stage high pressure heater 3, outlet enthalpy and final feed enthalpy of the first stage high pressure heater 1 are calculated by using IFC-97 industrial water and steam thermal property model loaded in Excel. According to the feedwater flow differential pressure measured in step (1), the feedwater flow is calculated by using the orifice flow calculation model loaded in Excel.

IFC-97 thermodynamic property model of industrial water and steam and orifice flow calculation model are the existing calculation software in the industry. Through the calculation model to calculate and generate results, it can greatly save manpower, improve efficiency and accuracy. After loading the calculation model into excel and inputting corresponding parameter values, it can quickly generate results, which is simple and clear.

(3) Calculate the first stage extraction steam flow for the first column of heat balance equation of the first stage high-pressure heater. For the first stage high-pressure heater, column 1 heat balance equation is as follows:

$$G_1h_1 + G_{gs}h_{1j} = G_1h_{1s} + G_{gs}h_{1c} \quad (1)$$

Wheres: G_1-primary extraction flow, G_{gs}-feedwater flow, h_1-primary extraction enthalpy, h_{1j}-inlet enthalpy of the first stage high pressure heater 1, h_{1s}-first stage high pressure heater 1 drain enthalpy, h_{1c}-Outlet enthalpy of first stage high pressure heater 1.

According to the heat balance equation of the first stage high pressure heater 1, the first stage extraction steam flow is obtained as follows:

$$G_1 = \frac{G_{gs}(h_{1c} - h_{1j})}{h_1 - h_{1s}} \quad (2)$$

(4) The second stage extraction steam flow is calculated for the two rows of heat balance equations of the second stage high-pressure heater. For the second stage high-pressure heater, the heat balance equation of column 2 is as follows:

$$G_2h_2 + G_1h_{1s} + G_{gs}h_{2j} = (G_1 + G_2)h_{2s} + G_{gs}h_{1j} \quad (3)$$

Wheres: G_2 -secondary extraction flow, h_2 -secondary extraction enthalpy, h_{2j}-inlet enthalpy of the second stage high pressure heater 2, h_{2s}-second stage high pressure heater 2 drain enthalpy.

According to the heat balance equation of the second stage high pressure heater 2, the second stage extraction steam flow is obtained as follows:

$$G_2 = \frac{G_{gs}(h_{1j} - h_{2j}) + G_1(h_{1s} - h_{2s})}{h_2 - h_{2s}} \quad (4)$$

(5) The first stage high-pressure heater 1, the second stage high-pressure heater 2, the third stage high-pressure heater 3 and the external steam cooler 4 are regarded as a closed integral module, and the heat entering the integral module is the first stage steam extraction heat, the second stage steam extraction heat, the third stage steam extraction heat and the water inlet heat of the third stage high-pressure heater 3, The heat flowing out of the overall module is the drainage heat and final feed water heat of the third stage high-pressure heater 3. For the overall module, formulate a heat balance equation and calculate the third stage extraction steam flow, that is, obtain the inlet steam flow of the external steam cooler.

For the overall module train, the heat balance equation is as follows:

$$(G_1 + G_2 + G_3)h_{3s} + G_{gs}h_{gs} = G_1h_1 + G_2h_2 + G_3h_3 + G_{gs}h_{3j} \quad (5)$$

Wheres: G_3 -tertiary extraction flow, h_3 -tertiary extraction enthalpy, h_{3j}-inlet enthalpy of the third stage high pressure heater 3, h_{3s} -third stage high pressure heater 3 drain enthalpy, h_{gs} -final feedwater enthalpy.

According to the heat balance equation of the whole module, the third stage extraction steam flow, i.e. the extraction steam flow of external steam cooler 4, is obtained as follows:

$$G_3 = \frac{G_{gs}(h_{gs} - h_{3j}) + G_1(h_{3s} - h_1) + G_2(h_{3s} - h_2)}{h_3 - h_{3s}} \quad (6)$$

4. Conclusion

In this paper, a new calculation method of steam inlet flow of external steam cooler in feedwater heating system is proposed. When used in high-pressure heating system with only two-stage high-pressure heaters, the steam inlet flow of external steam cooler can be calculated after calculating the first stage extraction flow. In the low-pressure heating system with external steam cooler, the low-pressure heaters at all levels and external steam cooler are regarded as an integral module, and the heat balance equation is listed for the integral module. The parameters to be measured in this method are the measuring points to be replaced in the conventional steam turbine thermal test, and there is no need to add a new feed water flow orifice of the external steam cooler and measure the feed water pressure and temperature at the inlet and outlet of the external steam cooler and the inlet pressure and temperature of the high-pressure heater connected with the external steam cooler, which saves the test cost and time.

Acknowledgements

This work is supported by the science and technology project of Shandong Electric Power Research Institute (ZY 2020-05), which studies the calculation method and platform application of minimum operation mode of thermal power plant based on power grid peak shaving dispatching

References

1. Lin Yue. Dynamics of clean coal-fired power generation development in China[J]. Energy Policy, 2012,15(1):136-142.
2. Luyao Zhou, Mengjiao Wang. Analysis of external steam cooler system (exergy) of 1000 MW Unit [J]. Thermal power engineering, 2021,36(2):24-30.
3. Weiping Yan. Review on development of supercritical steam parameter power generation

technology [J]. Power science and Engineering, 2014,30(1):1-7.

4. Zhongmin Niu, Yiyu Ding. Thermal economic analysis of external steam cooler for ultra supercritical 1000MW unit [J]. Thermal power generation, 2011,40(12):67-69.
5. Junfeng He. Parallel application of No. 0 high pressure heater and external steam cooler [J]. Auxiliary equipment of power station, 2021,42(2):10-13.
6. Lei Qian. Thermal economic analysis and optimization of 1000MW unit [D]. Shanghai: Shanghai Jiaotong University, 2015.

Ture three-dimensional fracture simulation of geology-engineering integration in low permeability and tight reservoirs in southern of Ordos Basin

Zhigang Li, Jiahao Wang*, Jinyuan Yang, When He

School of Energy, Chengdu University of Technology, Chengdu 610059, China

Abstract: In view of the characteristics of the Mesozoic oil reservoirs in the southern Ordos block such as large buried depth, strong heterogeneity, unclear understanding of oil and gas distribution, and immature prediction technology for engineering construction mechanism and effects, the fracture propagation morphology is studied. The 3D geomechanical parameters are truly 3D distributed by the integrated fracturing simulation technology of geological engineering and the research results of reservoir evaluation and rock mechanics. The spatial distribution of rock mechanics parameters of tight sandstone reservoir and interlayer is accurate. Considering the heterogeneity between layers, the distribution law of true 3D fractures in the real formation environment of high temperature and high pressure is simulated. The simulation results show that the fractures simulated by the 3d geomechanical model updated in real time are irregular and asymmetrical, and have a high degree of matching with the seismic data. This fracture model can guide the field construction and optimize the fracturing parameters of subsequent Wells to ensure that fractures occur in the most favorable position, which has high practical value and economic benefits.

Keywords: hydraulic fracturing; true three-dimensional fracture; fracture simulation; engineering geology integration

1. Introduction

The hydraulic fracturing fracture propagation model has gradually developed from the two-dimensional model before the 1980s to the true three-dimensional and pseudo three-dimensional models after the 1990s: In 2008, Chen et al. proposed a pseudo-three-dimensional model of hydraulic fracturing fracture morphology in layered formations, and established a full three-dimensional hydraulic fracture propagation theoretical model under non-uniform geological conditions [1]. In 2015, Salimzadeh et al. used the extended finite element method to establish a true three-dimensional hydraulic fracturing mechanics model, and studied the fracture propagation criterion of the cohesive fracture model [2]. In the same year, Liu and others used the cohesion zone method to establish a three-dimensional finite element model to predict the initiation and propagation of hydraulic fractures, and studied the influence of stress shadows [3]. In 2016, Li et al. based on the hydraulic fracturing numerical calculation method of the dispersive fracture model, and simulated the characteristics of the hydraulic fracturing propagation of rocks containing different inclusions [4]. In 2018, Sobhaniaragh et al. took into account that the coupling of elastic deformation and fluid flow and stress shadow in hydraulic fracturing, simulated fracture propagation with a cohesive fracture model [5]. The in-situ stress of artificial fracture simulation reservoir is mainly generated by interpolation of logging curve, and the in-situ stress is irregular vector. The in-situ stress generated by interpolation lacks a true representation of the reservoir condition and cannot show its heterogeneity [6]. Considering longitudinal lithological changes and lateral reservoir heterogeneity, the fracture propagation form is simulated in the three-dimensional geomechanical model to make the fracture simulation results more efficient and accurate, so as to provide theoretical guidance and technical support for subsequent fractured wells .

The traditional homogeneous reservoir hydraulic fracturing fracture extension simulation method will produce large errors in the simulation of fracture extension [7]. One is because some reservoirs have strong heterogeneity, complex geological conditions, and large differences in the mechanical properties of multiple layers of rock. The formation profile interpreted by traditional fracturing software has a fixed value for each layer. Moreover, the formation profile discretized by the logging curve lacks stress research based on the diagenetic process, which will cause certain errors, and the in-situ stress in the formation cannot be generated by interpolation; Secondly,

* Corresponding author: 821688985@qq.com

the fracture explained by traditional fracturing software is a regular oval symmetrical fracture [8]. However, in real formation conditions, due to the difference in the elastic modulus and Poisson's ratio of the formation, artificial fractures will extend and expand vertically and longitudeto the formation at different rates, with irregular morphology. With the increasing difficulty of reservoir development, in order to cope with complex geological conditions, there is an urgent need for a method to accurately guide the field hydraulic fracturing construction and simulate the expansion and extension of hydraulic fractures in the reservoir. The geology-engineering integration true three-dimensional fracture simulation technology can effectively solve this problem.

2. Three-dimensional geomechanical parameter model and stress simulation

2.1 Establishment of geological model

The southern part of Ordos is located in the western part of the North China Platform, and its main pay zone A8 is a distributary channel sedimentary microfacies. The structure of the top surface of the pay zone is simple, and it is a west-dipping monocline, with slightly nasal uplift structure locally. The reservoir has medium-strong heterogeneity. In order to accurately estimate the reservoir changes and various reservoir parameter values in the undrilled area between the wells, and to minimize the uncertainty of the stochastic simulation, the algorithm must be constrained. Based on sedimentary microfacies, seismic body attribute range and single well constraints, combined with well-seismic, the geological model of southern Ordos is established.

While establishing the geomechanical model, it is necessary to control the geologic body shape to ensure modeling accuracy and calculation speed, so as to meet the needs of later reservoir prediction accuracy and fracturing simulation. Through discretization processing and data analysis of single well data, while using sedimentary microfacies, seismic bodies and single well logging data for comprehensive constraints, the sequential Gaussian simulation algorithm is used to establish a geomechanical model in southern Ordos. The elastic modulus is mainly distributed in the range of 10~35Gpa, and the plane heterogeneity of the Chang A layer is the strongest (Fig. 1). The Poisson's ratio of the entire stratum is mainly distributed between 0.16 and 0.28. From the Chang A layer to the Chang B layer, the Poisson's ratio generally shows a decreasing trend.

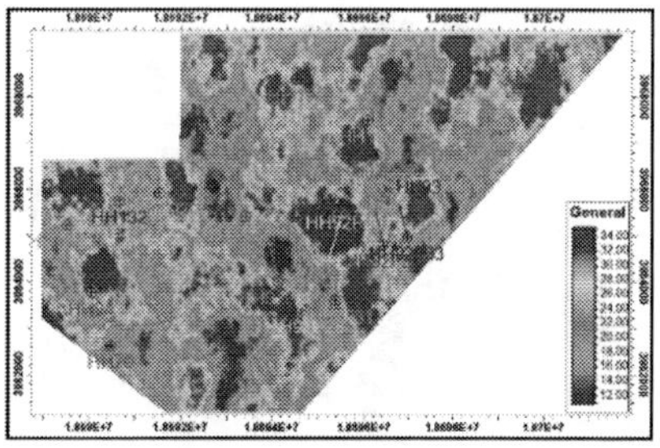

Fig 1. Simulation results of elastic modulus of Chang A layer

2.2 2.2 Stress simulation technology

Under the premise that the lithology data is not distorted, the established 3D data volume of rock mechanics parameters is coarsened to meet the calculation requirements of the finite difference software as much as possible. Assuming that the reservoir is elastic medium, the constitutive equation of linear elastic-stress-strain relationship is used to establish the geomechanical parameter model of the study area. The equivalent medium is used to add surrounding rock and bedrock around the reservoir and the top and bottom of the reservoir, while steel plates are added to the outside of the reservoir surrounding rock to eliminate the effects of stress concentration. The process of rock diagenesis is simulated by spatial differential finite difference approximation and convergence of maximum equilibrium force. For a calculation unit, if the internal and particle velocity is a continuous vector field, according to the principle of Gauss integral

$$\int_{v}\frac{\partial v_{i}}{\partial x_{i}}dv=\int_{s} v_{i}\, n_{j}ds \qquad (1)$$

In the formula, v_i is velocity vector, m/s; x_i is node location, m; n_j is external normal vector unit vector field, dimensionless; s is unit distance, m.

Then the difference formula of node and displacement is used

$$x_i^{<i>}(t+\Delta t)=x_i^{<i>}(t)+\Delta t v_i^{<i>}(t+\frac{\Delta t}{2}) \qquad (2)$$

$$u_i^{<i>}(t+\Delta t)=u_i^{<i>}(t)+\Delta t v_i^{<i>}(t+\frac{\Delta t}{2}) \qquad (3)$$

In the formulas, t is node time, s; Δt is time variation, t; u_i is displacement distance, m.

The generated rock mechanics parameter model is coarsened, and the output data file is transferred to the finite element simulator. After the structural model and the attribute parameter model are established, the stress is continuously loaded to simulate the diagenesis process of the formation. When the maximum unbalanced force converges, the work area tends to be stable. By continuously adjusting the rock mechanics parameters, when the attributes of the target well are consistent with the single-well logging curve data, the stress simulation of the work area is completed (Fig.2).

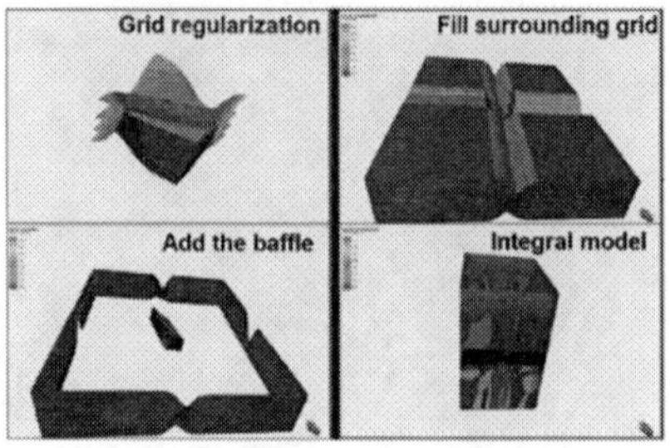

Fig 2. Stress simulation process

After stress simulation, the non-destructive conversion of the three-dimensional geological model is realized, which provides a geological basis for the study of rock mechanics parameters and in-situ stress plane, profile distribution characteristics, natural fracture distribution and direction of the block, and revises the rock mechanics parameters and in-situ stress. Fig.3 is the modified section of stress simulation (take the minimum horizontal principal stress as an example).

Fig 3. Minimum horizontal principal stress simulation results

3. True 3D fracture simulation of conventional fracturing

Based on the three-dimensional grid structure of the reservoir, full consideration of factors such as formation anisotropy, multi-phase and multi-dimensional flow, proppant transportation, fracturing fluid rheology and dynamic fluid loss, acid rock reaction (including fracturing fluid concentration, gel aging, gel breaking, shearing and other factors), using discrete element simulator to dynamically calculate and simulate, so that the expansion of three-dimensional fractures would be more accurate.

3.1 Establishment of rock mechanics profile

Based on the current regional principal stress direction, the maximum principal stress direction reflected by fracturing dynamic data and the imaging logging, it is considered that the maximum horizontal principal stress direction is 245° and the minimum horizontal principal stress direction is 155°. In the three-dimensional geomechanical model established before, combined with the direction of the minimum horizontal principal stress in the region, the geomechanical parameter profile of well A is sliced to form the basis for the true three-dimensional simulation of artificial fractures. The properties of slices (taking Poisson's ratio as an example) are shown in Fig.4.

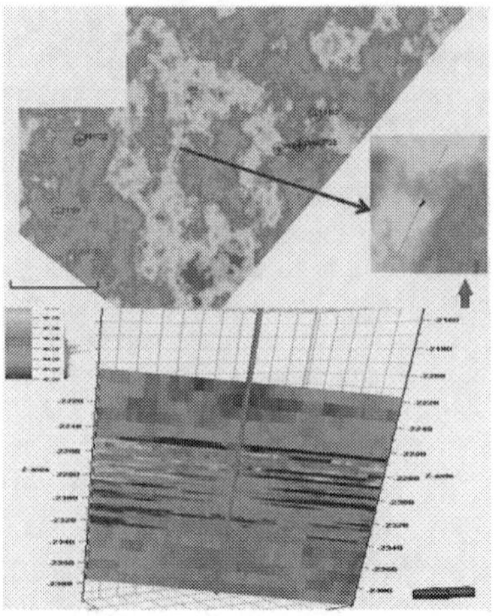

Fig 4. Slice of Poisson's ratio of well A

3.2 True 3D simulation of fractures

Based on the actual geological characteristics of the southern of Ordos Basin, perpendicular to the direction of the minimum horizontal principal stress in the region, slices of the rock mechanics characteristic parameters of the structural model are generated by combining the layered data of the well points, converted by the corner grid formula, and imported into the rock mechanics grid of discrete element simulator, the revised Young's modulus, Poisson's ratio and formation stress profile (taking the minimum horizontal principal stress as an example) are obtained. Compared with the geological attributes before the correction, and the properties of each layer after the correction will be different according to the planar heterogeneity of the reservoir, which is more consistent with the real situation of the formation.

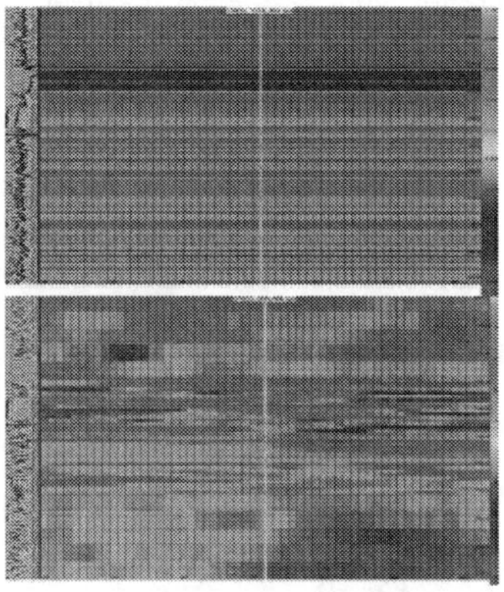

Fig 5. Minimum horizontal principal stress of well A (before and after correction)

The fracture propagation mechanism under in-situ stress constraint.

$$p_c = \frac{v}{(1-v)}[D_{tv}\Upsilon_{ob} - \alpha_v(D_{tv}\Upsilon_p + p_{off})] + \alpha_h(D_{tv}\Upsilon_p + p_{off}) + \varepsilon_x E + \sigma_t \quad (4)$$

In the formula, p_c is the closure pressure, MPa; v is Poisson's ratio, dimensionless; D_{tv} is the vertical depth, m; Υ_{ob} is the pressure gradient of the overlying rock, MPa/m; Υ_p is the pore fluid gradient, MPa/m; α_v is the vertical boit porous elastic constant, dimensionless; α_h is the horizontal boit porous elastic constant, dimensionless; p_{off} is the added value of pore pressure, MPa; ε_x is the initial horizontal tensile strength, MPa; E is the elastic modulus, GPa; σ_t is the horizontal stress, MPa.

Import the revised rock mechanics section into the discrete element simulator, fully consider the heterogeneity of the reservoir in the spatial distribution, and improve the inadequacie that only considered the two-dimensional planar heterogeneity in the past. Used the three-dimensional grid structure for true three-dimensional fracture simulation to further simulated the artificial asymmetric fracture distribution pattern under the real reservoir environment.

The length of simulated fracture in Fig.6 is 180m, the height is 32m, and the average width is 6.521mm, which is in good agreement with the microseismic results. When designing the pump program, the longitudinal lithology of the reservoir is taken into account, and specific proppant combinations are used to control fracture height to avoid unnecessary fracturing of the reservoir, resulting in more accurate and efficient fracture modeling. Figure 6 shows the comparison of symmetrical fractures before the uncorrected geological section and after the modified geological section.

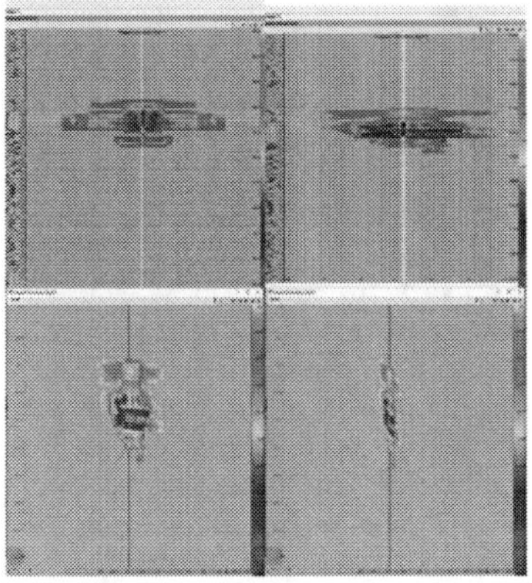

Fig 6. The conventional 2D model of hydraulic fracture (before and after correction) and 3D model(front and side view)

True three dimensional simulation technology of fractures based on the geolog-engineering integration can show the real distribution of fractures underground. According to its real morphology, the communication with natural fractures and the connection of reservoir can be determined, providing a reliable fracture foundation for subsequent productivity simulation, and a direct theoretical basis for subsequent well fracturing design in this area, and filling the gap in fracturing technology.

=

4. Conclusion

Combined geomechanical modeling, stress simulation, rock mechanics profile, true three-dimensional fracture simulation and other technologies to build the true three-dimensional fracture simulation technology based on geology-engineering integration, which accurately simulates the initiation and propagation of fractures under real conditions. While revising geomechanical parameters, the attribute model is improved, an accurate geomechanical section is established, so that the complete and reliable asymmetric fractures model can be obtained, which matched the microseismic data to a high degree. It realizes the integrated process from geomechanical modeling, stress simulation to the true three-dimensional fractures simulation, which is convenient for guiding operations, ensuring fractures are generated in the most favorable position, and determining the formulation of construction plans and optimized design of fracturing parameters.

Acknowledgments

This work was financially supported by the National Natural Science Foundation of China (No. 51504042); Subitem of National Science and Technology Major Project (2016ZX05048-001-04-LH).

References

1. M. Chen, J. Zhou, Y. Jin, et al. Experimental study on fracture characteristics of random fractured reservoirs[J]. Editorial office of ACTA PETROLEI SINICA, 2008, 29(3).
2. S. Salimzadeh, N. Khalili. A Three-Phase XFEM Model for Hydraulic Fracturing with Cohesive Crack Propagation[J]. Computers and Geotechnics, 2015, 69:82-92.
3. L. Liu, K.M. Wang, W.H. Wang, et al. Numerical simulation of quasi-brittle rock hydraulic fracturing based on cohesive zone mechanics[C]//2015 China Mechanics Conference.
4. M. Li, P.J. Guo, X. Li, et al. Modeling of heterogeneous rock based on level set method and propagation characteristics of hydraulic fracturing[J]. Rock and Soil Mechanics, 2016(12).
5. B. Sobhaniaragh, W.J. Mansur, F.C. Peters. The role of stress interference in hydraulic fracturing of horizontal wells[J]. International Journal of Rock Mechanics and Mining Sciences, 2018, 106:153-164.
6. J.J. Zhou. A review of the development of three-dimensional model of hydraulic fracturing [J]. Science and Technology Innovation Guide, 2017, 14 (08): 71+73.
7. M. Haddad, K. Sepehrnoori, M. Haddad, et al. XFEM-Based CZM for the Simulation of 3D Multiple-Stage Hydraulic Fracturing in Quasi-brittle Shale Formations[C]//Us Rock Mechanics/Geomechanics Symposium. 2015.
8. A. Settari, and H.S. Price. Simulation of Hydraulic Fracturing in Low-Permeability Reservoirs. Society of Petroleum Engineers. 1984. doi:10.2118/8939-PA.

Application of inorganic chemicals in industrial wastewater treatment

Zhien Zhang

Zhenhai High School, Ningbo 315000, China

Abstract. With the development of social economy, environmental protection projects have attracted wide attention from all walks of life, especially in industrial development, which can pay attention to environmental problems. Inorganic chemicals have important value in industry and can be well used for scientific treatment of industrial wastewater. It mainly includes the specific application of inorganic chemicals in industrial treatment of printing and dyeing wastewater, heavy metal wastewater, papermaking wastewater, fur, slaughter and other organic wastewater, and electroplating wastewater.Understand the contents of chemical agents themselves, discuss the specific classification of chemical agents, and think about the specific application ways and directions of inorganic chemical agents in the process of industrial wastewater treatment, hoping to have a deeper understanding of inorganic chemical agents, provide a good foundation for the smooth and stable development of industrial wastewater treatment, and show the corresponding technologies to improve the future industrial level.

Keywords: Inorganic chemicals; Industrial wastewater; deal with

1. Introduction

The development of society has promoted the progress of industry, and at the same time, it has caused serious environmental pollution problems, which is unfavorable to the overall development of our country, especially for the atmosphere, soil and water quality. In the process of industrial wastewater treatment, it is necessary to adopt a more scientific and reasonable way to reduce wastewater problems and prevent and reduce the pollution phenomenon caused by industrial development. Actually, there are various ways to treat industrial wastewater. Treating industrial wastewater with inorganic chemicals is a new way of treatment, which shows great difference compared with the traditional physical way, and is also a key part of the former industrial wastewater treatment process. It is hoped that through the analysis of this paper, we can understand the application of inorganic chemicals in industrial wastewater treatment, discuss related problems, and promote the smooth and stable development of environmental protection.

2. Overview of chemical agents

2.1 The actual meaning of chemical agents

Chemical agents are agents that can inhibit bacteria. Chemical agents have a wide range of applications, and their uses are relatively wide. Chemical agents can not only be used for the effective treatment of industrial wastewater, but also be related to people's daily life. In the process of scientific use of chemical agents, first of all, we should pay attention to the performance of chemical agents themselves, so that they can show specific characteristics in the process of application, and select more suitable chemical agents in combination with different application environments, so as to make them play their corresponding values, so as to prevent the use effect of chemical agents from being not obvious because the same agents are used in different environments or in environments that do not meet their application conditions, which may also cause further damage and impact on the environment. In addition, in the process of using chemical agents, it is necessary to scientifically limit the amount of chemical agents, which must not be too much or too little.

2.2 Scientific and reasonable classification of chemical agents

In the process of treating industrial wastewater, the chemicals used include organic chemicals and inorganic chemicals. There are obvious differences between organic chemicals and inorganic chemicals, mainly because the molecular structures of these two chemicals are not exactly the same. For some inorganic chemicals, their applications in the treatment of industrial wastewater will be more extensive. At the same time, the results of industrial wastewater treatment will be relatively better, so it is a kind of chemical agent which is widely used in

https://doi.org/10.1051/e3sconf/202235802058

the process of industrial development at present, and it is also a more scientific and effective method to treat industrial wastewater. At the same time, it should be understood that inorganic chemical treatment agents mainly include iron-based treatment agents and aluminum-based treatment agents in the process of treating industrial wastewater, and there are differences between high and low molecules. At present, some scholars have comprehensively used these two kinds of treatment agents to treat industrial wastewater, worked out corresponding treatment schemes, and achieved obvious wastewater treatment effects. However, this treatment method is not very common and applicable in the application process, mainly because this treatment method will consume relatively high cost and low cost performance in its specific application process. In the process of treating different industrial wastewater, different inorganic chemical treatment agents need to be adopted, and a reasonable industrial wastewater treatment scheme should be selected in combination with the actual treatment situation and requirements of industrial wastewater to ensure the actual treatment effect of industrial wastewater.

2.3 Classification of chemical agents

Chemical agents refer to agents that can inhibit bacteria (agents beneficial to bacterial growth are classified as nutrients). Chemical agent is a relatively broad concept, and its use is also relatively extensive. It can not only be used in the treatment of industrial wastewater. At present, the use of chemical agent seems to have little to do with people's daily life, but actually has a lot to do with it. The following points should be paid attention to during the use of chemical agents. First of all, the use of chemical agents must be targeted and reasonable for different use environments. In addition, we must pay attention to the principle of accurate quantification in the use of chemical agents, and neither too much nor too little. Water treatment chemicals include organic and inorganic chemicals. The difference between the two is mainly due to the difference in molecular composition. Relatively speaking, inorganic chemicals are more widely used in industrial wastewater treatment, and the treatment effect is also better. At present, inorganic chemicals are widely used in industrial wastewater treatment. Inorganic chemical water treatment agents mainly include iron series and aluminum series, and have low molecular weight and high molecular weight. At present, there are some industrial wastewater treatment schemes using inorganic chemical agents combined with the two. The effect of wastewater treatment is relatively good, but the cost is relatively high and the cost performance is relatively low. In different industrial wastewater treatment processes, the selection of inorganic chemical agents will be different, and specific plans need to be formulated according to specific conditions.

3. Specific application of inorganic chemicals in industrial wastewater treatment

3.1 Application of inorganic chemicals in the treatment of heavy metal wastewater

In the current social situation, heavy industry has become a powerful driving force for the development of the whole society and an important foundation for economic development. During the development of heavy industry, a large amount of heavy industry wastewater is produced, and at the same time, the environmental pollution phenomenon is extremely serious. There are corresponding difficulties in the specific treatment of heavy metal wastewater, and the treatment process is more complicated and cumbersome. The radioactive substances contained in this heavy industry wastewater are extremely rich, so it can't be easily discharged to the outside world, otherwise it may cause a corresponding impact on the overall environment, the living conditions of human beings and the overall air quality. In addition, after the heavy metal wastewater is produced, it also needs to be treated in time to ensure the treatment effect of heavy metal wastewater. In the treatment of heavy metal wastewater, the chemical substances that need to be adopted also show diversity. In application, it is necessary to choose an appropriate treatment method according to the differences of industrial wastewater itself, which shows the effect of needle alignment. Specifically, first of all, it is necessary to analyze the components of particles in industrial wastewater, understand the pH and pH of industrial wastewater, and take appropriate materials in combination with the basis of industrial chemistry to form oxide precipitation through comprehensive reaction. Generally speaking, the presence of iron oxidants in inorganic oxidants can precipitate some heavy metal wastewater, which can reduce and control the pollutants. Iron inorganic chemicals can save the treatment time of industrial wastewater more effectively in the specific application process, and relieve the pressure of industrial development, making the treatment of industrial wastewater simpler and more convenient.

3.2 Application of inorganic chemicals in printing and dyeing wastewater treatment

With the development of economy, people's living standard is getting higher and higher, people have higher requirements for material life, and they also pay more attention to their own image. On this basis, clothing manufacturers have gradually developed and produced more obvious effects in the fierce market competition. The development of garment manufacturing industry can produce all kinds of garments and garments of all kinds of materials and quality for people, but at the same time, it will also produce printing and dyeing wastewater. The existence of these printing and dyeing wastewater will cause serious pollution to articles, but the treatment process will be more complicated. In the process of treating printing and dyeing wastewater, it is often necessary to use more technologies and more professional tools. This kind of wastewater treatment is extremely difficult. If the printing and dyeing wastewater is not treated scientifically and reasonably, it will inevitably lead to a more serious impact on the overall environment and threaten people's lives. Therefore, it is necessary to decontaminate the main components existing in printing and dyeing wastewater, including sulfate ions and acid ions, etc. For different printing and dyeing ion components, it is necessary to finish the treatment regularly and make corresponding plans. In the process of treatment, it is necessary to first condense the printing and dyeing wastewater, then stir it properly, and after a period of precipitation, add inorganic chemicals properly to precipitate it again, and then filter to obtain more pure water resources, so as to achieve the basic requirements and objectives of inorganic chemicals in the treatment of printing and dyeing wastewater, and make the wastewater treatment effect better.

3.3 Application of inorganic chemicals in organic wastewater treatment process

In the development of society, people will have a higher pursuit of material life, hoping to use some more advanced materials in their lives, reflecting their own characteristics and unique pursuit and taste, which also enables a number of fur manufacturers to achieve the corresponding development and make the fur market more advanced. The industrial wastewater produced by fur manufacturers in the process of development is mainly produced in the process of fur processing. The fur processing often needs to be cleaned, divided, assembled and dyed, and the discharged wastewater will include rich organic matter, fat and protein. At the same time, organic wastewater will also come from slaughter plants, and people will slaughter animals and then clean them, and the blood discharged during the cleaning process will be rich in organic matters. The existence of organic matter will have a serious impact on water bodies. Although some organic wastewater can be treated and controlled by simple inorganic chemicals, some organic wastewater produced by fur processing will include many suspended solids, which are often difficult to achieve specific goals and requirements only by inorganic chemicals, so it is necessary to use iron inorganic chemicals for corresponding treatment to achieve the ultimate goal.

3.4 Application of inorganic chemicals in industrial wastewater treatment after electroplating

In the development of people's daily life, many articles are finished by electroplating, and the difference of electroplating ingredients makes the quality of electroplated materials different. At the same time, the pollutants in industrial wastewater produced after electroplating also show the difference. In view of the current electroplating wastewater, there are some toxic elements in many industrial wastewater. If these toxic substances are discharged to the outside environment, it will inevitably lead to the corresponding impact on the environment. Therefore, it is necessary to pay attention to the treatment of industrial wastewater after electroplating. At the same time, it should also be clear that if electroplating industrial wastewater really meets the treatment objectives and requirements, it must be treated with inorganic chemicals to realize the function and effect of adsorption reaction. On this basis, some toxic substances can be separated from the chaotic structure of water molecules, and after that, these toxic substances can be precipitated and adsorbed, and finally the filtered pure water resources can be obtained, thus achieving the specific goal of industrial wastewater treatment. After the first treatment, it will be found that these electroplating wastewater molecules still exist in the wastewater, and they need to be adsorbed again to achieve complete treatment, which reflects the specific application value of inorganic chemicals in industrial wastewater treatment.

3.5 Application of inorganic chemicals in papermaking wastewater treatment

With the development of modern science and technology, internet plus's model has become more and more advanced and modern, and people's dependence on electronic equipment is getting higher and higher, far exceeding that on paper. At the same time, there is less and less waste water produced by the paper industry in the society, but there are many areas where the knowledge of waste water is not comprehensive enough. They still pay attention to and use paper books, and many people are highly dependent on paper reading, so there is still the problem of paper waste water. In the process of treating papermaking wastewater, the operation process is complicated, and some inorganic chemicals will be used. Actually, compared with some other wastewaters, papermaking wastewater mainly includes bleaching agents and pulp, etc. In the process of treatment, inorganic chemicals that can be used are more common and common. The actual process of papermaking wastewater treatment is complicated, including sedimentation, flocculation, etc. Inorganic chemicals need to be added in sedimentation and other aspects, so that it can be fully combined with wastewater, and then the whole wastewater treatment process can be completed by filtration.

4. Conclusion

To sum up, inorganic chemicals are the main application method of industrial wastewater treatment. This wastewater treatment method has a good treatment effect, and the cost of some inorganic chemicals is relatively low. Therefore, it is a cost-effective industrial wastewater treatment method on the whole. In the specific treatment process, different treatment schemes are required for different types of industrial wastewater, so as to effectively improve the effectiveness of treatment. Especially in the treatment process of heavy metal wastewater, electroplating wastewater and other wastewater containing toxic substances, it is more necessary to improve the wastewater treatment process, so that the wastewater can meet the discharge standard, To achieve industrial development without causing excessive damage to the environment.In the process of modern industrial development, the wastewater problem is a common phenomenon. Through the application of inorganic chemicals, it can better assist in the treatment of industrial wastewater, solve the impact of pollutants caused by industrial wastewater problems on the development of the whole society, and comprehensively treat industrial wastewater after it is produced, so as to reduce environmental pollution.

References

1. Guo Jie, Li Caixia, Xiao Jianqiang, Bai Yuling, Li Ruijuan. Application methods of inorganic chemicals in industrial wastewater treatment [J]. cleaning world, 2022,38(06):46-47.
2. Wang Xiaoli. Application of inorganic chemicals in industrial wastewater treatment [J]. Heilongjiang Environmental Bulletin, 2022,35(02):144-145.
3. Wang Xiaoli. Application of inorganic chemicals in industrial wastewater treatment [J]. Chemical Design Communication, 2021,47(10):190-191.
4. Yang Xiaojuan. Application of inorganic chemicals in industrial wastewater treatment [J]. Chemical Design Communication, 2020,46(03):243+271.

Influence of the Heating Temperature on the Electrochemical Performance of Coal-Based Needle Coke Anode for Lithium Ion Batteries

Jing Xia [1,*], Yuan Gao [1], Shuxiang Li [1], Qiongguang Li [2] and Yanhong Wang [2]

[1] Ansteel Beijing Research Institute CO., LTD., Beijing, China
[2] Institute of Process Engineering, Chinese Academy of Sciences, Beijing, China

Abstract. In this study, the effect of the heating temperature on the initial coulombic efficiency, capacity and cycle stability of the coal-based needle coke was examined. The needle cokes were heated to different temperatures, and their microstructures were characterized by the scanning electron microscope, the X-ray diffraction and Raman spectroscope. The results indicated that the defect and the degree of disorder were reduced with increasing the heating temperature. Moreover, the electrochemical properties of these needle cokes were tested. A decrease of the capacity from 368.3 to 257.3 mAh/g, while an increase of the initial coulombic efficiency was observed, when the heating temperature was improved to 1100 oC. This is because fewer defects and more ordered crystal structures reduce active sites for the storage of Li ions and decrease the irreversible consumption of Li ions, as the heating temperature was higher.

1. Introduction

To solve the problems of environmental pollution and energy crisis, flexible technological solutions including energy supply and storage have been developed rapidly [1,2]. The lithium ion battery, an environmentally friendly secondary battery with high capacity, could contribute strongly to the energy storage picture [3-5]. The anode material, one of the most critical factors to improve the electrochemical performances, has always been a hot topic in lithium ion batteries [6-8]. Carbon materials, such as mesocarbon microbead, graphite and needle coke, are the main anode materials for lithium ion batteries [9,10]. Coal-based needle cokes with a variety of sources and low price, have attracted increasing attention of researchers, recently. However, due to their complex components, the method to improve their electrochemical performances has been a problem. In this research, the influence of heating temperatures on the electrochemical performances of coal-based needle cokes were revealed, through analyzing the microstructures, capacities, initial coulombic efficiency and cycle stability of coal-based needle cokes fabricated at different temperatures. It affords clues to improve electrochemical properties of coal-based needle cokes.

2. Experimental

2.1 Materilas

The coal-based needle coke, fabricated at 600 °C, was purchased from Ansteel Group CO., LTD. Its volatile, ash content, density and sulfur residue value were shown in table 1.

Tabel 1. The properties of the coal-based needle coke.

	Volatile (%)	Ash (%)	Density (g/cm^3)	S (%)
Needle Coke	5.2	0.01	1.41	0.45

2.2 Heat Treatment of the Needle Coke

The pristine needle coke (Nc) was heated to 850 °C at a rate of 10 °C/min under the protection of nitrogen, and kept at this temperature for 6 h. The product was named as Nc850. The sample Nc1100 was produced at the same condition but the heating temperature was changed to 1100 °C.

* Corresponding author: xiaj0001@126.com

https://doi.org/10.1051/e3sconf/202235802059

2.3 Characterization

The microstructure of the sample was observed by the field emission scanning electron microscope (FESEM, Gemini 560), and the X-ray diffraction (XRD, Bruker D8 Advance) spectroscope with a Cu Kα radiation source (1.5405 Å) was employed to analyze the crystalline structures of samples. The Raman spectra were recorded by a Horiva (JY-HR800 micro-Raman) spectrometer. All the electrochemical tests were conducted using coin cells (CR2025). The working electrodes were prepared by spreading the mixed slurry of active material, Super P and sodium alginate in water with a weight ratio of 7:2:1 onto copper foil. Then they were dried at 120 °C in vacuum for 12 h. The electrolyte was a solution of 1 M $LiPF_6$ in ethylene (EC) and dimethyl carbonate (DMC) (v:v=1:1). A lithium foil was used as the counter electrode. All the operations were performed in a glove box under argon atmosphere. The discharge/charge tests were carried out on a Land BT2000 battery test system at room temperature. The voltage ranged from 0.01 to 2 V with a current density of 100 mA/g.

3. Results and Discussion

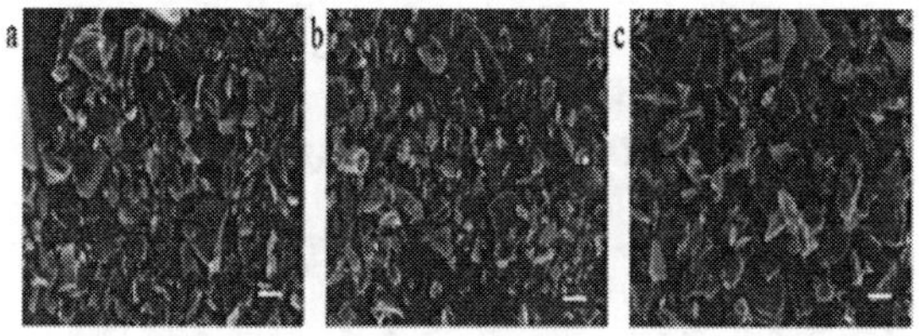

Figure 1. Typical SEM images of needle cokes heated at different temperatures. a) Pristine needle coke, b) 850 °C and c) 1100 °C (scale bar: 10 μm).

The morphologies of the pristine needle coke, needle cokes heated at 850 °C and 1100 °C were presented in Fig. 1. The sizes of these needle cokes were 2-20 μm. The XRD patterns of needle cokes prepared at different temperatures were shown in Fig. 2. The needle cokes Nc, Nc850 and Nc1100 have broad peaks of (002) at approximately 25.2°, 25.6° and 25.8°, respectively. It confirms that these samples have amorphous structure. However, as the increase of heating temperature, the peaks of (002) are sharper, the intensities are stronger, and the 2θ is closer to the graphite. These results indicate a more ordered structure of the needle coke with the improvement of heating temperature.

The structure of needle cokes fabricated at different temperatures was further investigated by Raman spectroscopy, as shown in Fig. 3. The samples Nc, Nc850 and Nc1100 all exhibited a defect-induced band at about 1343 cm^{-1} (D-band) and a crystalline graphite band at approximately 1589 cm^{-1} (G band) (Fig. 3a-c). The intensity ratios of G-band and D-band (I_G/I_D) for Nc, Nc850 and Nc1100 were 1.06, 1.12 and 1.21, respectively (Fig. 3d). It demonstrated that the disorder degree decreased as the increasing of heating temperature, which was consistent with the XRD patterns.

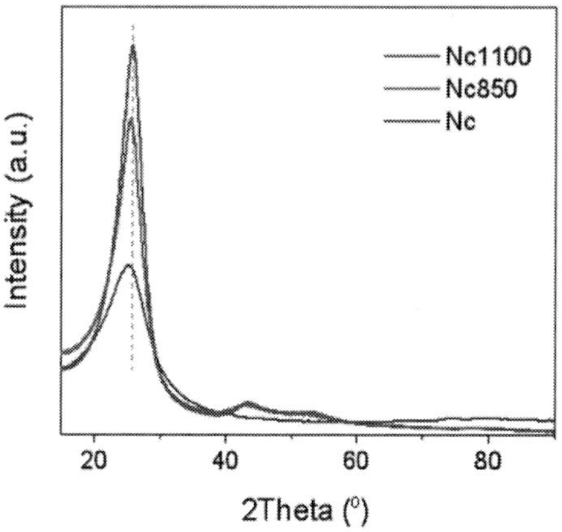

Figure 2. XRD patterns of needle cokes heated at different temperatures.

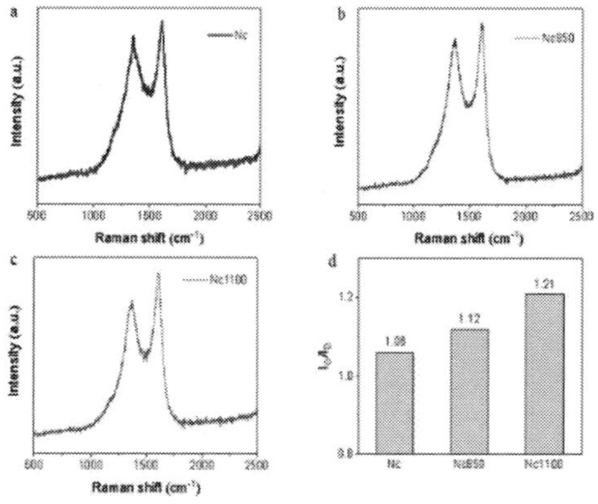

Figure 3. Raman spectra of a) Nc, b) Nc850 and c) Nc1100. d) The ratios of the I_G/I_D.

The electrochemical performances of needle cokes heated at different temperatures were analyzed in half cells using Li foil as the counter electrode. As presented in the Fig. 4a, the pristine needle coke Nc showed a reversible capacity of 368.3 mAh/g and a coulombic efficiency of 44.9% in the first cycle at a current density of 100 mA/g. After heating at 850 °C and 1100 °C, the needle cokes Nc850 and Nc1100 delivered a reversible capacity of 306.7 mAh/g and 257.3 mAh/g in the first cycle, respectively (Fig. 4b,c). However, the initial coulombic efficiency increased to 53.7% and 65.0% as the improvement of the heating temperature to 850 °C and 1100 °C, respectively (Fig. 4b,c). Table 2 summarized the crystal structure data and electrochemical performances of needle cokes heated at different temperatures. As discussed above, increasing heating temperature led to less defects and more ordered crystal structure, which reduced active sites for the storage of Li ions and

 https://doi.org/10.1051/e3sconf/202235802059

decreased the irreversible consumption of Li ions. Thus, a higher heating temperature caused less capacity and higher initial coulombic efficiency [11].

After 50 cycles, the Nc, Nc850 and Nc1100 showed a good cycling stability, and the reversible capacities were 352.3 mAh/g, 341.4 mAh/g and 267.4 mAh/g, respectively (Fig. 4a-c). Fig. 4d showed the first discharge-charge profiles of the Nc, Nc850 and Nc1100. All the samples presented sloping charging curves, also demonstrating their amorphous structures (consistent with the XRD patterns).

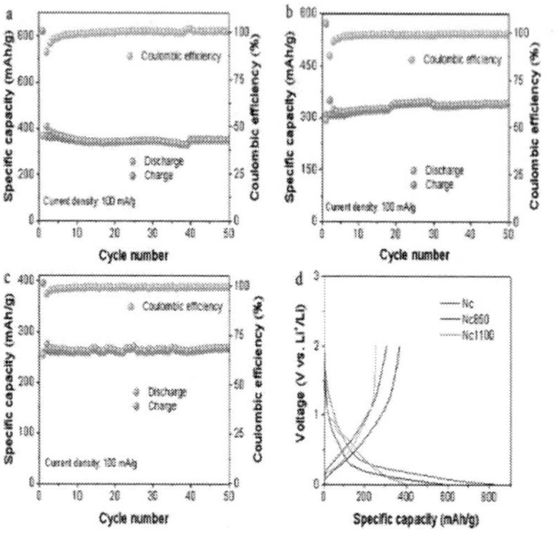

Figure 4. Cycle performance and the corresponding coulombic efficiency of a) Nc, b) Nc850 and c) Nc1100 at the current density of 100 mA/g. d) Voltage profiles of Nc, Nc850 and Nc1100 at the current density of 100 mA/g in the first cycle.

Table 2. The crystal structure data and electrochemical performances of needle cokes heated at different temperatures.

Samples	$2\theta_{(002)}$	I_G/I_D	Capacity (mAh/g)	Initial Coulombic Efficiency (%)
Nc	25.2°	1.06	368.3	44.9%
Nc850	25.6°	1.12	306.7	53.7%
Nc1100	25.8°	1.21	257.3	65.0%

4. Conclusion

In this research, the coal-based needle cokes were heated at different temperatures to prepare anode materials of the lithium ion battery. SEM, XRD and Raman spectroscopy were utilized to characterize the microstructures of these needle cokes. Also, their capacity, initial coulombic efficiency and cycling stability were analyzed in coin cells. The results demonstrated that the capacity decreased from 368.3 to 257.3 mAh/g, but the initial coulombic efficiency increased from 44.9% to 65.0%, as the improvement of heating temperature to 1100 °C. Since less defects and more ordered crystal structures were formed at a higher temperature, which reduced active sites for the storage of Li ions and decreased the irreversible consumption of Li ions. This work revealed the influence of the heating temperature on the electrochemical performances of coal-based needle cokes, providing clues to adjust coal-based needle cokes for the application in the lithium ion battery.

Acknowledgments

The authors are very grateful for the financial support from the Ansteel Beijing Research Institute CO., LTD..

References

1. S. Chu, and A. Majumdar, Opportunities and challenges for a sustainable energy future, Nature, vol. 488, pp. 294-303, 2012.
2. X. Gao, Y. Gao, Q. Li, and Y. Wang, Scalable synthesis of high-performance anode material SiOx/C for lithium-ion batteries by employing the Rochow reaction process, Journal of Alloys and Compounds, vol. 902, pp. 163668- 163677, 2022.
3. N. Du, H. Zhang, and B. D. Chen, Porous Co3O4 nanotubes derived from Co4(CO)12 clusters on carbon nanotube template: a highly efficient material for Li battery applications, Advanced Materials, vol. 19, pp. 4505-4509, 2007.
4. P. J. Zuo, G.P. Yin, and Z. L. Yang, Improvement of cycle performance for silicon/carbon composite used as anode for lithium ion batteries, Materials Chemistry and Physics, vol. 115, pp. 757-760, 2009.
5. C. K. Chan, R. Ruffo, and S. S. Hong, Surface chemistry and morphology of the solid electrolyte interphase on silicon nanowire lithium-ion battery anodes, Journal of Power Sources, vol. 189, pp. 1132-1140, 2009.
6. A. Concheso, R. Santamaria, and R. Menendez, Iron-carbon composites as electrode materials in lithium batteries, Carbon, vol. 44, pp. 1762-1772, 2006.
7. S. Lee, S. Yoon, and C. M. Park C M, Reaction mechanism and electrochemical characterization of a Sn-Co-C composite anode for Li-ion batteries, Electrochimica Acta, vol. 54, pp. 364-369, 2008.
8. S. S. Zhang, K. Xu, and T. R. Jow, EIS study on the formation of solid electrolyte interface in Li-ion battery, Electrochimica Acta, vol. 51, pp. 1636–1640, 2008.
9. G. X. Wang, J. Yao, and H. K. Liu. Characterization of nanocrystalline Si-MCMB composite anode materials, Electrochem Solid-state Lett., vol. 7, pp. 250-253, 2004.
10. N. Ohta, K. Nagaoka, and K. Hoshi, Carbon-coated graphite for anode of lithium ion rechargeable batteries:Graphite substrates for carbon coating,

https://doi.org/10.1051/e3sconf/202235802059

Journal of Power Sources, vol. 194, pp. 985-990, 2009.

11. D. J. Wang, Y. L. Wang, L. Zhan, and X. Y. Zhang, Graphitization mechanisms and electrochemical performance of needle coke anode for Li-ion battery, Journal of Inorganic Materials, vol. 26, pp. 619-624, 2011.

Analysis on the Formation of Water/Oil Emulsion System of High Water-cut Crude Oil and Its Profile and Flooding Effect

Hongjiang Ge, Jing Cheng*, Dangke Ge, Weihua Yang, Zhiqiang Guo, Lihua Ren, Jianjun Bao, Qian Li, Xiaoyan Ma and Xiaoliang Liang

Oil Production Technology Institute, Dagang Oil field Company, Tianjin Binhai New Area

Abstract: In view of the limitations of modulated flooding technology in low permeability high temperature reservoirs, the water/oil emulsion with the effect of expanding sweep volume and its influence on the effect of modulated flooding were studied. Under the condition of high water content (80% water content), the dispersion system formed by emulsifier 55# and crude oil is mainly water/oil emulsion. At the same time, the influence of surfactant type, water content, temperature and other factors on the formation of water/oil emulsion system was investigated, and the influence of emulsion properties on the effect of displacement was analyzed. The low permeability core experiment shows that the emulsifier with the effect of expanding sweep volume can improve the oil recovery by 17.93% on the basis of water flooding, which is obviously higher than that of polymer flooding. At the same time, emulsifier has good temperature resistance, which is of great significance for enhancing oil recovery in high temperature and low permeability water drive reservoirs.

1. Overview

At present, most of my country's water flooding development reservoirs have entered a period of high water cut. As a technical means of enhancing oil recovery, with the change of the properties of underground reservoirs, the effective period and effect of control flooding are also decreasing. The existing modulate flooding (profile) systems [1-2] mainly include polymers, gels, bulk swellable particles and precipitation agents, which cannot be effectively expanded the effect of expanding sweep volume due to the adsorption and retention of formation porous media and the dilution of formation water, low permeability reservoirs and high temperature conditions also limit the application of organic modulate flooding systems [3]. Therefore, there is an urgent need to find new oil displacement methods to greatly enhance oil recovery.

Emulsion flooding has become a hot research topics of chemical flooding in recent years [4-9]. Under different water-oil ratios, emulsifiers can form water-in-oil (W/O) or oil-in-water (O/W) emulsions with crude oil. When the O/W type emulsion is formed, the crude oil molecules are solubilized in the active agent micelles; When the W/O type emulsion is formed, the emulsion has a higher viscosity, which reduces the mobility ratio of water-flooding oil, increases the swept volume of the subsequent injected water, and improves the heterogeneous formation profile. At the same time, there is a low interfacial tension between oil and water, which plays the role of oil displacement. Wanfen Pu, et al [10] developed a W/O emulsifier OB-2 when the volume ratio of oil to water is 7:3, and the viscosity increase rate of OB-2 emulsion with a concentration of 0.3 wt% reaches 370%. Studies have shown that viscous W/O emulsions are used as a very promising oil displacement system [11-12].

We have developed a water-in-oil emulsion system with good water solubility that can easily form a high water content after mixing with crude oil, Taking the S reservoir of Dagang Oilfield as the research object, the causes and influencing factors of the water-in-oil system of high water-cut crude oil were investigated, and the emulsification, viscosity increase and oil displacement were realized, which is expected to be applied to the high-water-cut crude oil reservoir for enhanced oil recovery [13-14].

2. Experiment part

2.1 Materials and Instruments

Cocamidopropyl betaine (effective content 35%), fatty alcohol polyvinyl ether sulfate (effective content 70%), industrial products, provided by Linyi Yiqun Chemical Co., Ltd.; polyoxyethylene sorbitan monostearate, polyoxyethylene fatty alcohol ether, alkanolamide, oleic acid diethanolamide, industrial products, provided by Nantong Chenrun Chemical Co., Ltd.Emulsifier 55# is self-made in the laboratory, and the main component is non-ionic surfactant mixed in proportion;The oil used in the experiment is the crude oil from the S reservoir of Dagang Oilfield, and the viscosity of the crude oil is 80 mPa·S (laboratory test at 90°C); The experimental water is the formation water of S reservoir in Dagang Oilfield, the salinity is 28630 mg/L, and the ion content is as follows: Cl-, 15267 mg/L; HCO3-, 251 mg/L, SO42-, 368

* Corresponding author: dg_chengjing@petrochina.com.cn

mg/L; Ca2+, 340 mg/L, Mg2+, 48 mg/L; Na+, K+, 9540 mg/L.
The experimental instruments mainly include TX500C full rotation drop meter/interfacial tension meter, Brookfield DV-III+Pro viscometer, magnetic stirrer, DGM-III core displacement device, etc.

2.2 Experimental Method

2.2.1 Preparation of Water-in-oil Emulsion System.

Formation water was used to prepare 55# aqueous solution of emulsifier with different concentrations and crude oil of S reservoir according to different water-oil ratios. The total volume was 50 mL, and the mixture was stirred with a magnetic stirrer at 90°C. The prepared emulsion was placed in a 90°C incubator to observe the emulsification.

2.2.2 Emulsion viscosity.

Emulsion viscosity refers to the viscosity value of oil-water emulsion at a certain temperature. The unit is mPa.S. The viscosity of dehydrated crude oil at 90°C is 80 mPa.S.

2.2.3 Evaluation of plugging performance.

A sand-packed pipe model with a length of 30 cm and a diameter of 3.12 cm was used to investigate the variation law of the displacement pressure before and after the injection of the emulsifier 55# solution. The specific experimental steps are as follows: First, the formation water is saturated in vacuum, the porosity and PV volume of the sand-packing pipe are calculated, and the water phase permeability is measured. At the displacement rate of 0.23 mL/min, the water content of the oil from water flooding to the produced fluid was 98%, and the displacement pressure was stable, and the displacement pressure in the water flooding stage was recorded; After injecting 0.5 times the pore volume of the emulsifier 55# at the same displacement rate, the water flooding was continued until the water content of the produced fluid was 98%, and the displacement pressure in the subsequent water flooding stage was recorded.

2.2.4 Evaluation of oil displacement performance.

A sand-packing tube model with a length of 30 cm and a diameter of 3.12 cm was used to investigate the oil displacement performance of the injected emulsifier 55#. The specific experimental steps are: a) vacuum saturated pure water, weigh the core mass before and after saturated water, and calculate the porosity and PV volume; b) Load the core into the core holder, and measure the water permeability K; c) Saturate crude oil at a temperature of 104°C; d) Water flooding to 98% water content, and measure its water flooding efficiency; e) Inject 0.5PV 0.3% emulsifier 55# solution, continue water flooding to 98% water content, calculate the oil displacement efficiency of the emulsifier and improve the oil recovery factor.

3. Experimental results and analysis

3.1 Component Selection for Forming High Water-cut Crude Oil Water-in-oil Emulsions

The formation of water-in-oil emulsions in high water-cut reservoirs requires that the surfactant solution injected on the ground has a certain solubility. At the same time, the water-in-oil emulsions formed also have emulsification and viscosity enhancement, control the mobility of the displacement fluid, and play a role in profile control and oil increase effect.

3.2 Conditions and Influencing Factors for the Formation of High Water-cut Crude Oil Water-in-oil Emulsion Systems

3.2.1 Formation conditions.

Einstein viscosity formula:

$$\eta_r = \frac{1}{1-\sqrt[3]{h\varphi}}\eta_0 \qquad (1)$$

Among them, ηo: viscosity of external phase; φ: volume fraction of internal phase; h: volume factor related to the type of emulsion.
According to the Einstein viscosity formula, the viscosity increase of the emulsion is mainly related to the volume of the internal phase, that is, we need to choose a suitable emulsifier to wrap as much internal phase water as possible. To this end, under the premise of taking into account the interfacial tension and lipophilicity, it was determined experimentally that nonionic surfactants with HLB values between 6 and 8 can form water-in-oil emulsions that can drive oil. Further, amide-based surfactants containing different fatty acids and different alkanols were targeted as thickening emulsifiers.

3.2.2 Factors influencing the viscosity-enhancing performance of water-in-oil emulsions of high water-cut crude oil

3.2.2.1 Surfactant Type.

Reducing the oil-water interfacial tension is the most basic requirement of surfactants. In the experiment, the low interfacial tension system LAB with betaine as the main agent was selected, and on this basis, its emulsifying and viscosity-increasing properties were further optimized. Use different fatty acid R and alcohol amine NH (CH2CH2OH) x (X value is an integer of 1 to 3) composed of alkanolamide surfactants, the structure is:

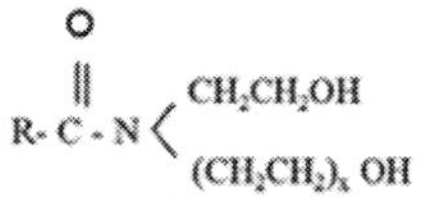

The compound experiment of alkanolamide surfactant and low interfacial tension system LAB was carried out. The results are shown in Table 1.

Table 1. Emulsifying and thickening properties of different surfactants and crude oil

R	X	Solubility	Emulsifying & thickening	Lower interfacial tension
Stearic acid	1	√	-	√
	2	√	√	√
Oleic acid	1	√	-	√
	2	√	√	√
Lauric acid	1	√	-	√
	2	√	√	√
coconut oil	1	√	-	√
	2	√	√	√
Palmitic acid	1	√	-	√
	2	√	√	√

√ stands for Yes; - stands for No.

It can be seen from the table that ethanolamide surfactants have good water solubility, and can form a uniform emulsion after mixing with crude oil. The viscosity test results show that the emulsification and viscosity-increasing effect of diethanolamide surfactant is better than that of monoethanolamide surfactant. The viscosity enhancement of diethanolamide surfactants is largely due to the pair of polar head groups of fatty acid-ethanolamine. The viscosity increase rate has a lot to do with the matching between the two.

3.2.2.2 Moisture content.

In order to analyze the viscosity increase of water-in-oil emulsion, the effect of different water content on the viscosity of 3% emulsifier 55# emulsion was investigated under the condition of experimental temperature of 90℃. As can be seen from Figure 1, within 75% of the water content, the viscosity of the emulsion increases greatly with the increase of water content. At this time, the emulsion is mainly water-in-oil type, and the emulsification and viscosity increase rate can reach more than 400%. When the water content is greater than 80%, the oil-water two phases are obviously delaminated, and the viscosity of the emulsion decreases sharply, and the water-in-oil emulsion undergoes phase inversion, mainly the oil-in-water type. When the water content reaches more than 85%, the viscosity of the emulsion is lower than that of the crude oil itself.

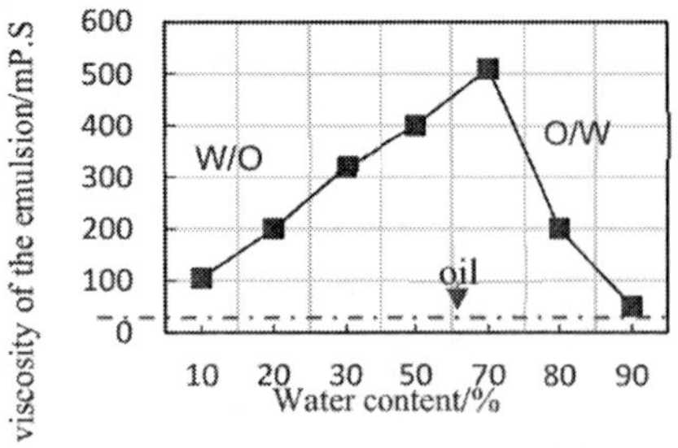

Fig. 1. Viscosity curve of emulsion with different water content

3.2.2.3 Temperature.

Under the condition of 50% moisture content and 3% mass fraction, and the experimental temperature is 90, 100, 110, 120, 130℃, the emulsion viscosity of emulsifier 55# is 380, 370, 362, 355, 347 mPa.S respectively. The experimental results show that when the experimental temperature is 90-130℃, the viscosity of the emulsion changes within 10%. It shows that the emulsifier 55# has better temperature resistance and can be used in high temperature oil reservoirs.

3.3 Analysis of the control and flooding effect of the water-in-oil emulsion

Three groups of cores with different permeability were designed to investigate the plugging performance and oil displacement performance of emulsifier 55# with a mass fraction of 3% on the cores. In this way, the influence of the formation of the viscosifying water-in-oil emulsion on the improvement of the control and flooding effect in the later stage of water flooding reservoir development was analyzed.

3.3.1 Evaluation of plugging performance.

In order to eliminate the fluctuation effect of different cores on the displacement pressure during the displacement process, the displacement pressure is normalized. It can be seen from Figure 2 that the displacement pressure of the core with a permeability of 45 mD has been rising during the process of surfactant injection, indicating that a strong flow resistance is established in the core pores and a high-viscosity water-in-oil emulsion is formed in the core. After switching to waterflooding, with the migration of the oil-water emulsion, under the influence of the accumulation effect, the water injection pressure also continued to increase, with a maximum increase of 2.38 times. The internal displacement pressure of the core with permeability of 80 mD and permeability of 150 mD did not increase, and no strong flow resistance was established. The reason for the analysis may be that the size of the formed water-in-oil emulsion droplet does not match the size of the core pore throat, which is insufficient to establish the flow resistance of its subsequent injection; It is also possible that due to the strong heterogeneity of the core itself, a water flow channel is formed and an oil-in-water emulsion is formed, resulting in no increase in the displacement pressure.

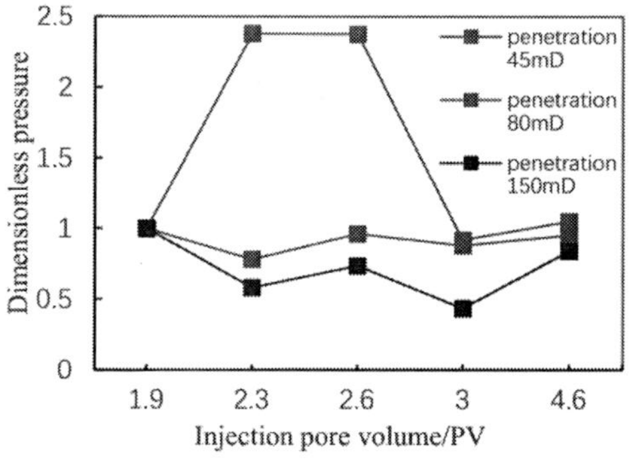

Fig. 2. Relationship between injection pore volume and dimensionless pressure

3.3.2 Experiment of oil displacement performance.

From the experimental results, the enhanced oil recovery values of cores with a permeability of 45 mD and a permeability of 80.2 mD are similar, which are 17.93 and 15.85 percentage points, respectively, indicating that the water-in-oil emulsion formed by the emulsifier is compatible with the core pore throat and other parameters better. Among them, the core with a permeability of 150.4 mD has a low enhanced oil recovery ratio of 10.91 percentage points. The reason for the analysis should be that during the water flooding process of this group of cores, it may be that the texture of the core itself is highly heterogeneous, resulting in the formation of water flow aisle. When the viscosified emulsion enters the high water-cut water channeling channel, since the water content is 80%, the water-in-oil emulsion undergoes phase inversion at this time, mainly oil-in-water emulsion, so the final enhanced oil recovery value is low. Therefore, we also get a little understanding that it is not suitable to use the water-in-oil emulsion system to enhance oil recovery in the formation with obvious water channeling channels.

Table 2. The experimental results of emulsifier 55# on cores with different water contents

core number	3-1	3-2	3-3
Porosity (%)	13.5	14.8	17.2
Oil saturation(%)	45.4	52.3	47.8
Penetration(mD)	45.0	80.2	150.4
Waterflood recovery(%)	38.79	37.12	20.55
System concentration(%)	0.3	0.3	0.3
Increased recovery(%)	17.93	15.85	10.91

4. Conclusion

Under the condition of high water cut and low permeability reservoir (water cut 80%), emulsifier 55# and crude oil can form a water-in-oil emulsion-based dispersion system. Low-permeability core experiments show that the water-in-oil emulsion with the effect of expanding the swept volume can further enhance oil recovery on the basis of water flooding. At the same time, the emulsifier has good temperature resistance, which overcomes the disadvantage of polymer flooding that is easily degraded under high temperature conditions, and points out a technical approach for further improving oil recovery in low-permeability and high-temperature water flooding reservoirs.

References

1. H.J.Ge, H.F.Zhang, J.Cheng and W.H.Yang, Oilfield Chemistry, 36,1 (2019).
2. H.J.Ge, X.J.Liu and Q.L.Lei, Chinese Journal of Colloid& polymer, 21, 4 (2003).
3. B.Q.Zeng, L.S.Chen, C.L.Li, et al. Journal of Petroleum Science and Engineering, 80, 1 (2011).
4. M.S.Karambeigi, R.Abbassi, E.Roayaei, et al. Journal of Industrial and Engineering Chemistry, 29, 2 (2015).
5. M.K.Zhang, Y.G.Wu, X.Y.Chang, et al. Journal of liaoning shihua university, 39, 2 (2019).
6. L.Wu, Advances in fine petrochemicals, 19, 6 (2018).
7. C.H.Ren, Y.Lu,L.S.Wang, et al. Advances in fine petrochemicals, 16, 4 (2015).
8. Y.K.Li, F.L.Zhao and P.Yang, Oilfield Chemistry, 1 (2013).
9. X.L.Cao, H.N.Zhao, C.Ma, et al. Acta Phys. Chim. Sin. 30, 7 (2014).
10. W.F.Pu, Z.L.Mei, Y.Yang, et al. Reservoir evaluation and development , 9, 1 (2019).
11. Y.Z.Zhou, D.Y.Yin, R.Cao Rui, Oilfield Chemistry, 33, 2 (2016).
12. M.Z.Tian, X.Zhao, X.M.Song, et al. Reservoir evaluation and development,44, 8 (2015).
13. X.L.Cao, B.D.Ma, J.C. Petroleum Geology and Recovery Efficiency, 23,1(2016).
14. G.S.Cao, X.Y.Yang, Z.Q.Zhang, Contemporary Chemical Industry, 46, 7 (2017).

AUTHOR INDEX